Fundamental Statistics for the Social and Behavioral Sciences

Second Edition

To my kids, Meagan and Will, and my parents,
Katsumi and Grayce Tokunaga.

Fundamental Statistics for the Social and Behavioral Sciences

Second Edition

Howard T. Tokunaga
San Jose State University

SAGE

Los Angeles | London | New Delhi
Singapore | Washington DC | Melbourne

SAGE

FOR INFORMATION:

SAGE Publications, Inc.
2455 Teller Road
Thousand Oaks, California 91320
E-mail: order@sagepub.com

SAGE Publications Ltd.
1 Oliver's Yard
55 City Road
London EC1Y 1SP
United Kingdom

SAGE Publications India Pvt. Ltd.
B 1/I 1 Mohan Cooperative Industrial Area
Mathura Road, New Delhi 110 044
India

SAGE Publications Asia-Pacific Pte. Ltd.
3 Church Street
#10-04 Samsung Hub
Singapore 049483

Acquisitions Editor: Leah Fargotstein
Content Development Editor: Chelsea Neve
Editorial Assistant: Claire Laminen
Production Editor: David C. Felts
Copy Editor: Gillian Dickens
Typesetter: C&M Digitals (P) Ltd.
Proofreader: Jennifer Grubba
Indexer: Marilyn Augst
Cover Designer: Candice Harman
Marketing Manager: Susannah Goldes

Printed in the United States of America

Library of Congress Cataloging-in-Publication Data

Names: Tokunaga, Howard, author.

Title: Fundamental statistics for the social and behavioral sciences / Howard T. Tokunaga, San Jose State University.

Description: Second Edition. | Thousand Oaks : SAGE Publications, [2018] | Revised edition of the author's Fundamental statistics for the social and behavioral sciences, [2016] | Includes bibliographical references and index.

Identifiers: LCCN 2018006464 | ISBN 978-1-5063-7748-3 (pbk. : alk. paper)

Subjects: LCSH: Social sciences—Statistical methods. | Psychology—Statistical methods.

Classification: LCC HA29 .T634 2018 | DDC 519.5—dc23
LC record available at https://lccn.loc.gov/2018006464

This book is printed on acid-free paper.

SUSTAINABLE FORESTRY INITIATIVE
Certified Chain of Custody
Promoting Sustainable Forestry
www.sfiprogram.org
SFI-01268
SFI label applies to text stock

18 19 20 21 22 10 9 8 7 6 5 4 3 2 1

BRIEF CONTENTS

DETAILED CONTENTS

ABOUT THE AUTHOR

Howard T. Tokunaga is Professor of Psychology at San Jose State University, where he serves as Coordinator of the MS Program in Industrial/Organizational (I/O) Psychology and teaches undergraduate and graduate courses in statistics, research methods, and I/O psychology. He received his bachelor's degree in psychology at UC Santa Cruz and his PhD in psychology at UC Berkeley. In addition to his teaching, he has consulted with a number of public-sector and private-sector organizations on a wide variety of management and human resource issues. He is coauthor (with G. Keppel) of *Introduction to Design and Analysis: A Student's Handbook.*

PREFACE TO THE SECOND EDITION

Perhaps the most appropriate way to introduce the second edition of this book is to thank the instructors who adopted the first edition and were kind enough to send me their thoughts and suggestions, as well as the reviewers of the first edition who were amazingly insightful and thoughtful in their comments and observations. It is my sincere hope that they see the incorporation of their input in this edition.

The primary changes made to the book involve both formatting and content. In terms of formatting, the readability of the book was hopefully enhanced through such things as reorganizing the chapter outlines at the start of each chapter, reducing the size of figures, and more clearly distinguishing the learning checks and end-of-chapter summaries. In terms of content, in addition to making minor editing and grammatical changes throughout the book, the following changes were made to selected chapters:

Chapter 1 (Introduction to Statistics) and Chapter 2 (Examining Data: Tables and Figures): The published research studies used to illustrate the concepts discussed in these two chapters (research hypotheses, independent and dependent variables, sampling, levels of measurement, experimental vs. nonexperimental research methods, different types of distributions [unimodal, bimodal, multimodal, skewed]) have been updated to highlight current research trends on topics relevant to students' lives.

Chapter 4 (Measures of Variability): "'Eyeball Estimating' the Mean and Standard Deviation" has been added to emphasize the value in having students visually estimate the mean and standard deviation for a set of data in order to confirm the correctness of their mathematical calculations. Also, the "Measures of Variability for Nominal Variables" section discusses a statistic (the index of qualitative variation) that may be used to measure the variability of categorical variables.

Chapter 6 (Probability and Introduction to Hypothesis Testing): This chapter provides a greater elaboration of the logic used to make the decision whether to reject the null hypothesis, as well as the implication of this decision (reject vs. not reject) for a stated research hypothesis.

Chapter 8 (Estimating the Mean of a Population): The main research example, which involves data from an actual national salary survey, has been updated with 2017 salary data in order to provide students information about current occupational information.

Chapter 13 (Correlation and Linear Regression) from the first edition has been divided into two chapters. The new Chapter 13 (Correlation) covers the concept of correlation and correlational statistics (Pearson and Spearman rank-order correlation). Chapter 14 (Linear Regression and Multiple Correlation) discusses linear regression and also provides an in-depth introduction to multiple correlation and regression, using an example with two predictor variables to demonstrate the calculation, interpretation, and application of the multiple correlation coefficient and the multiple regression equation. Also, a new research example is used to illustrate correlation and linear regression; this example, which focuses on children's eating behaviors, addresses a topic students can relate to in their own lives.

Ultimately, the main goals of this second edition were to update the research examples used to illustrate statistical procedures and respond to users' and reviewers' reactions to the first edition. As with the first edition, it is our hope that this book will help students understand the use of statistics to answer questions and test ideas and to appreciate how research studies are conceived, conducted, and communicated. Any and all questions, concerns, suggestions, and complaints you have about the book are most welcome, and can be sent to me at howard.tokunaga@sjsu.edu.

PREFACE TO THE FIRST EDITION

It may surprise students to learn they have something in common with writers of books such as this one: When you get close to finishing a writing assignment, you get a bit tired and a bit lazy. The first attempt at this preface was written shortly after final drafts of chapters were sent to my editor at SAGE, Vicki Knight. After reading it, she said, "It's not bad, but it reads like the 'typical' Preface. I think it would be useful for the reader to have a sense of *why* you wrote this book and *why* you wrote it the way you did."

In responding to my editor's plea for self-analysis, I found that this book's journey began in college. When I entered college, I thought my path would take me to law school; however, taking an Intro to Psych class my freshman year made me realize I enjoy the challenge of trying to understand the human mind. The school I attended, UC Santa Cruz, was a fairly unconventional university, but somehow in the midst of a sea of humanistic psychologists, I became attracted to the empirical and methodological aspects of psychology. This was a result of taking classes with instructors such as David Harrington and Dane Archer, who showed me that statistics could appeal to students if taught using a gentle, guiding approach that addresses questions relevant to students' lives. After graduating from college, I was able to get a job as a research assistant for a human resource consulting firm. Despite my lack of work experience, I was hired primarily as a result of having taken statistics and research methods courses, which taught me that learning statistics has benefits both inside and outside of the classroom.

Several years later, I started grad school at UC Berkeley, where two events critical to this book took place. First, serving as a teaching assistant, I found I really enjoyed helping students, particularly in statistics and research methods classes that were often viewed with fear and suspicion. Second, I took graduate classes from Geoff Keppel, who had developed his own unique method and system for analyzing experimental research designs. His lectures and books were instrumental in showing me that statistics can be taught in a systematic way that highlights similarities rather than differences between different research situations. Geoff managed to transform something as daunting sounding as a "3 × 2 × 4 research design" into the mathematical equivalent of playing with wooden toy alphabet blocks labeled "A," "B," and "C." For a long time, I thought my gratitude to Geoff was an isolated occurrence. However, the appreciation others have for his approach to teaching statistics was made apparent to me several years later when I watched him receive an American Psychological Association (APA) Lifetime Achievement award.

After leaving Cal, I took on a teaching position at San Jose State, where my teaching responsibilities included an introductory statistics course aimed at students with a wide range of background, ability, and motivation. As I needed to select a textbook to use in this course, for the first time I looked carefully at the wide range of offerings. What I found striking (and still find striking) was that the majority of books focused on providing formulas and very small sets of data designed to demonstrate how to correctly calculate the correct numbers from these formulas. Little emphasis, however, was given to what these numbers meant or implied. Given my own experiences learning statistics, I thought a book was needed that discusses statistics in a thematic manner, focusing on how they are used to answer questions and test ideas within the larger research process.

The primary purpose of this book is to not just teach students how to calculate statistics but how to interpret the results of statistical analyses in light of a study's research hypothesis and to communicate one's results and interpretations to a broader audience. Hopefully, this book will not only help students understand the purpose and use of statistics but also give them a greater understanding of how research studies are conceived, conducted, and communicated.

The 14 chapters of this book may be placed into three general categories. The first four chapters are designed to introduce students to the research process and how data that have been collected may be organized, presented, and summarized. Chapters 5 through 10 discuss the process of conducting statistical analyses to test research questions and hypotheses, as well as issues and controversies regarding this process. The final four chapters of this book, Chapters 11 to 14, discuss different statistical procedures used in research situations that vary in the number of independent variables in the study as well as how the independent and dependent variables have been measured.

A FEW TIPS FOR STUDENTS

To you, the college student about to read this book as part of taking a statistics course: "Welcome!" and "Great job!" I welcome you because you're about to embark on a semester-long journey that I hope will enhance your skills and widen your perspective; I congratulate you because it's a journey not everyone is willing to take.

At the present moment, I know my encouragement and appreciation may be of little comfort to you as you might be somewhat anxious about having to learn statistics. Some of you might be anxious about having to learn statistical concepts and formulas; some of you might be anxious because the research process seems pretty complicated. Talking with students who have taken my courses over the years has helped me assemble the following advice:

Master the material presented in the early chapters

Chapters 1 through 6 discuss how research is often conducted, how data are summarized and described, and the process by which researchers conduct statistical analyses to test their ideas. It is absolutely critical for you to have a firm understanding of these chapters as they lay the groundwork for later chapters that discuss a variety of statistical procedures used by researchers to test hypotheses.

What does it mean to master this material? First, *read the chapters both before and after they're discussed in class.* By reading the chapters beforehand, you'll be able to identify anything that's unclear to you and have your questions addressed by your instructor. Rereading the chapters after they're discussed in class will help confirm your understanding of the material. Next, *be able to define and explain key concepts presented in these chapters.* These concepts are highlighted in bold-faced type, and it is important to learn them when they're first introduced because they'll appear throughout the remainder of the book. Next, *do the learning checks within the chapters and the exercises at the end of the chapters.* The learning checks include both exercises and review questions you can ask yourself to assess your understanding of the material. Most important, *do not miss class during the early part of the semester!* It's been my experience that students who miss critical lectures at the beginning of the term often have difficulty keeping up as the semester continues.

Review high school algebra

If you read newspapers or watch television, you might conclude that the most complicated data analysis people can comprehend is a pie chart. As frustrating as this is to researchers and statistics instructors, they understand that statistics can be confusing. If you happen to move beyond the introductory statistics course for which you are reading this book, you'll find that statistical procedures are often conducted using computers and statistical software rather than hand calculations. Consequently, some statistics textbooks no longer include mathematical formulas but instead have students conduct analyses using statistical software or Microsoft Excel. However, this book has chosen a different approach for two main reasons. First, I believe that to comprehend the results of data analysis, one must understand the underlying foundation of statistical procedures. Learning statistics via computer software can lead to a "brain-dead" approach in which students are at the mercy (rather than control) of their computers. I refer to this as the "I'm only as smart as my printout" method of learning statistics, also known as "Because this is what my computer told me."

Second, my steadfast and somewhat stubborn adherence to an approach emphasizing mathematical formulas and calculations is based on a simple reason: The statistics in this book are not hard to calculate! To assess whether you're adequately prepared to read this book, take the following test:

(1) Do you know how to add, subtract, multiply, and divide?

(2) Do you know how to calculate the "square" or "square root" of a number?

(3) Do you understand the "order of operations" and how to use parentheses within mathematical calculations?

If your answers to these questions are yes, congratulations! You possess the ability to conduct every mathematical calculation in this book. None of the formulas in this book requires knowledge of geometry, trigonometry, or calculus. However, if you're unsure of your mathematical ability, you may want to review your high school algebra. At the end of this book is an appendix that includes the mathematical concepts and operations needed to conduct the statistical analyses in this book. I highly recommend you read this appendix to review and assess your mathematical skills.

Learn to use a statistical calculator

Although learning the statistics in this book doesn't require the use of computers or software, you will need a hand calculator. Calculators with statistical capabilities are typically identified as statistical or scientific calculators. Although there's a wide selection of brands and models, one simple criterion to use in selecting a calculator is price. The statistical calculators most appropriate for this book cost somewhere in the $10 to $15 range at the time of this writing. I do *not* recommend you use or purchase a more expensive calculator! I prefer these simpler calculators because I've found students are able to conduct the calculations in this book more quickly and with fewer errors. I also recommend you purchase a calculator similar to the one your instructor uses as he or she will be more familiar with its features and idiosyncrasies.

In selecting a calculator, look closely at the keyboard; you'll need a calculator whose keys have labels such as "$\bar{X}$," "σ," and "ΣX" (Chapters 3 and 4 discuss what these symbols represent). When you go calculator shopping, you'll find calculators that have keys with labels such as "ΣY" and "XY." Although calculators with these "Y" symbols are only slightly more expensive than calculators with just the "X" symbols, I do *not* recommend their purchase because these features are not needed to perform the calculations in this book. In fact, I've found students with "Y" calculators have a more difficult time performing simpler calculations.

Ask questions

Conducting research is the process of asking and answering questions. Accordingly, throughout this book, issues are often framed in the form of questions, which I hope encourages you to ask questions as well. The first person you should direct any questions to is yourself. As you work your way through the chapters, get into the habit of asking yourself, "What does this mean?" A good test of your level of comprehension is whether you can explain the material to yourself in a meaningful way.

Also, be sure to work with your instructor to confirm your level of understanding and clarify unanswered questions. Instructors often say to students, "The only 'bad' question is the one you don't ask." Don't be hesitant about asking questions in class. Instructors will tell you they use one student's question to assess an entire class' level of understanding. This is what I call the "pencil test"—when one student asks a question, I look to see how many other pencils get raised, which reflects the number of students who had the same question and are ready to write down the response. Asking questions enhances the learning experience for yourself, your classmates, and your instructor.

Form a study group

No matter how much I encourage students to ask questions in class, it seems they hesitate for fear of drawing attention to themselves. As a result, I turn to another well-traveled saying: "Misery loves company." Forming a study group to meet with other students on a regular basis will help you keep up with reading assignments, confirm your comprehension of the material, clarify any confusion you have, prepare for exams, and receive an invaluable source of social support. I do suggest that your group contain at least one person who will raise the group's questions and concerns to the instructor in class or office hours.

Practice, practice, practice

The best way to learn the topics covered by this book is to obtain as much practice as possible. Throughout the chapters are examples in which the calculations have been worked out for you; in the middle and end of each chapter are a number of exercises. The answers to some of the exercises are provided at the back of this book, and your instructor has access to the answers to the other exercises. Do the end-of-chapter exercises even if your instructor does not assign them as homework in order to identify any recurring mistakes you make. We find that the vast majority of errors students make are simple computational errors, committed when students do the calculations too quickly.

WELCOME!

GREAT JOB!

ACKNOWLEDGMENTS

This book has traveled a long journey, with many people playing both transient and persevering roles in its development. First and foremost, I'd like to thank my kids, Meagan and Will, for providing me perspective and inspiration amid all of the twists and turns we faced while this book was being written. I'm so very blessed to have you as my "burden." Away from home, I'm grateful to my graduate students at San Jose State—working on this book has given me much empathy for their thesis-related travails. However, they should be aware that the completion of this book ends the unspoken agreement that I not ask, "So how's your thesis?" if they don't ask, "So how's your book?" A number of friends have provided most welcome relief after long days staring at a computer screen, with particular thanks going to Ellie, Eileen, Susan, Bonnie, Ann, Lisa, and Terry. I would like to thank Charles Linsmeier and David Shirley for their many helpful comments and advice on earlier versions of this book, and Julie West for her help in developing chapter exercises.

I would like to express my appreciation to everyone at SAGE who has played a role in shepherding this book to completion. First and foremost, I must express my gratitude to my editor, Vicki Knight. It has been a joy and an honor to work with someone with such a truly exceptional combination of experience, knowledge, perspective, and humor. In production I was fortunate enough to work with others on the SAGE team who made the process a pleasure: Yvonne McDuffee, Editorial Assistant; Katie Guarino, Assistant Editor; Nicole Elliott, Executive Marketing Manager; Candice Harman, Cover Designer; and the copyeditor, Gillian Dickens. Furthermore, I am grateful to the reviewers, who took the time to read the chapters, raise questions, and provide constructive comments and suggestions to assist me in clarifying and strengthening the manuscript. Reviewers for second edition:

Thomas S. Alexander, University of Maryland College Park

Steve Bounds, Arkansas State University

Jeff W. Bruns, Southeast Missouri State University

Derrick M. Bryan, Morehouse College

Robert M. Carini, University of Louisville

Janine Everett, Franklin & Marshall College

Rebecca Ewing, Western New Mexico University

Keith Feigenson, Albright College

Chad M. Galuska, College of Charleston

Theresa Garfield, Texas A&M University–San Antonio

Gary Hackbarth, Valdosta State University

Casey Jakubowski, SUNY Albany

Seth Jenny, Winthrop University

Leslie Anne Keown, Carleton University

Athena King, Jackson State University

Laura L. King, Boise State University

Jennifer Kwak, University of Hawaii

Robert G. LaChausse, California Baptist University

Kyle C. Longest, Furman University

Spencer MacAdams, University of South Carolina

Sheila MacEachron, Ferris State University

Margaret Maghan, Rowan University

Favian Martin, Arcadia University

Daniel S. McConnell, University of Central Florida

Arturo Olivarez, University of Texas at El Paso

JoEllen Pederson, Longwood University

Cristine Rego, Fleming College

Jonathan Ring, University of Tennessee

Evan Roberts, University of Minnesota

Karen Jeong Robinson, California State University San Bernardino

Hannah Rothstein, Baruch College

Pamela Schuetze, SUNY Buffalo State

Jeff Seger, Cameron University

Maurice Sheppard, Madison Area Technical College

Kimberly Twist, San Diego State University

Timothy Victor, University of Pennsylvania

R. Shane Westfall, University of Nevada, Las Vegas

Pi-Ming Yeh, Missouri Western State University

Egbert Zavala, University of Texas at El Paso

Haley Zettler, University of Memphis

Cathleen Zucco-Teveloff, Rider University

COMPANION SITE

The SAGE edge companion site for *Fundamental Statistics for the Social and Behavioral Science*, Second Edition, is available at **edge.sagepub.com/tokunaga2e**.

SAGE edge for Students provides a personalized approach to help students accomplish their coursework goals.

- Mobile-friendly **eFlashcards** strengthen understanding of key terms and concepts.
- Mobile-friendly **self-quizzes** allow for independent practice and assessment.
- **Multimedia content** includes video and audio links, plus relevant websites for practice and research.
- **Data sets** to accompany exercises in the book are available for download.

SAGE edge for Instructors supports teaching by making it easy to integrate quality content and create a rich learning environment.

- Chapter-specific **test banks** provide a diverse range of prewritten questions, available in **Word and electronic format**.
- **Sample syllabi** provide suggested course models.
- Editable, chapter-specific **PowerPoint slides and lecture notes** assist in class preparation.
- **Multimedia content** includes video and audio links, plus relevant websites for practice and research.
- **Data sets** to accompany exercises in the book are available for download.
- **Answers to even-numbered questions** from the book help assess student progress.
- **Tables and figures** from the book are available to download for use in your course.

1

INTRODUCTION TO STATISTICS

CHAPTER OUTLINE

1.1 What Is Statistics?
1.2 Why Learn Statistics?
1.3 Introduction to the Stages of the Research Process
- Developing a research hypothesis to be tested
  - o Identifying a question or issue to be examined
  - o Reviewing and evaluating relevant theories and research
  - o Stating a research hypothesis: Independent and dependent variables
- Collecting data
  - o Drawing a sample from a population
  - o Determining how variables will be measured: Levels of measurement
  - o Selecting a method to collect the data: Experimental and nonexperimental research methods
- Analyzing the data
  - o Calculating descriptive statistics
  - o Calculating inferential statistics
- Drawing a conclusion regarding the research hypothesis
- Communicating the findings of the study

1.4 Plan of the Book
1.5 Looking Ahead
1.6 Summary
1.7 Important Terms
1.8 Exercises

In introducing this book to you, we assume you're a college student taking what is perhaps your first course in statistics to fulfill a requirement for your major or a general education requirement. If so, you may be asking yourself two questions:

- What is statistics?
- Why learn statistics?

The ultimate goal of this book is to help you begin to answer these two questions.

1.1 WHAT IS STATISTICS?

Whether or not you are aware of it, you encounter a variety of "statistics" in your day-to-day activities: the typical cost of going to college, the yearly income of the average college graduate, the average price of a home, and so on. So what exactly is "statistics"? The Merriam-Webster dictionary defines **statistics** as a branch of mathematics dealing with the collection, analysis, interpretation, and presentation of masses of numerical data. When people think about statistics, they often focus on only the "analysis" aspect of the above definition—that is to say, they focus on numbers that result from analyzing data. However, statistics not only is concerned about how data are analyzed but also recognizes the importance of understanding how data are collected and how the results of statistical analyses are interpreted and communicated. The main purpose of this book is to introduce, describe, and illustrate the role of statistics within the larger research process.

1.2 WHY LEARN STATISTICS?

We believe there are a variety of reasons why you should learn statistics. First, not only do you currently encounter statistics in your daily activities, but throughout your life you have been and will continue to be affected by the results of research and statistical analyses. Which college or graduate school you attend is based in part on test scores developed by researchers. You may have to take a personality or intelligence test to get a job. The choices of drugs and medicines available to you are based on medical research and statistical analyses. If you have children, their education may be affected by their scores on standardized tests. Learning about statistics will help you become a more informed and aware consumer of research and statistical analyses that affect many aspects of your life.

A second reason for learning statistics is that you may be required to interpret the results of statistical analyses. Many college courses require students to read academic research journal articles. Evaluating published research is complicated by the fact that different people studying the same topic may come up with diverse or even opposing conclusions. Understanding statistics and their role in the research process will help you decide whether conclusions drawn in research articles are appropriate and justified.

Another reason for learning statistics is that it will be of use to you in your own research. College courses sometimes have students design and conduct mini-research studies; undergraduate majors might require or encourage students to do senior honors theses; graduate programs often require masters' theses and doctoral dissertations. Learning to collect and analyze data will help you address your own questions in an objective, systematic manner.

A final reason for learning statistics is that it may help you in your future career. The website Careercast.com conducts an annual survey in which they evaluate jobs on four dimensions: environmental factors, income, employment outlook, and stress. In 2017, the number one rated job was "statistician." They noted, "A statistician's skill set can be used to break down and analyze large quantities of data. The demand for these skills spans a variety of industries, including marketing, banking, government, sports, retail, and even healthcare."

It is generally a good idea for students to maintain a healthy level of curiosity or even skepticism in regards to their education. However, we find that when it comes to learning statistics, students' frame of mind may be characterized as one of fear or anxiety. Although we understand these feelings, we hope the benefits of learning statistics will become clear to you and help overcome any concerns you have.

1.3 INTRODUCTION TO THE STAGES OF THE RESEARCH PROCESS

Much of scientific research involves asking questions. Throughout this book, we'll examine a broad range of questions regarding human attitudes and behavior that contemporary researchers have asked and attempted to answer. Below are research questions we'll address in this chapter to introduce the stages of the research process:

- Is college students' performance on tests more influenced by their learning strategies (*how* they learn) or their motivation (*why* they learn)?
- How do parents' feeding practices affect whether their children will eat new foods?

- Does coercive interrogation of eyewitnesses affect their accusations of crime suspects?
- Are there gender differences in terms of who contributes to Wikipedia?

How might you try to answer questions such as these? You could base your answers on your personal beliefs, or you could adopt the answers given to you by others. But rather than relying on subjective beliefs and feelings, researchers test their ideas using science and the scientific method. The **scientific method** is a method of investigation that uses the objective and systematic collection and analysis of empirical data to test theories and hypotheses.

At its simplest, this book will portray the scientific method as consisting of five main steps or phases:

- developing a research hypothesis to be tested,
- collecting data,
- analyzing the data,
- drawing a conclusion regarding the research hypothesis, and
- communicating the findings of the study.

Accomplishing each of these five steps requires completing a number of tasks, as shown in Figure 1.1. Because this sequence of steps will serve as the model for the wide assortment of research studies we'll review and discuss throughout this book, each step is briefly introduced below. It's important to understand that the research process depicted in Figure 1.1 represents an ideal way of doing research. The "real" way, as you may discover in your own efforts or from speaking with researchers, is often anything but a smooth ride but rather is filled with false starts, dead ends, and wrong turns.

Developing a Research Hypothesis to Be Tested

The initial stage—and the first step—of the research process is to develop a research hypothesis to be tested. A **research hypothesis** may be defined as a statement regarding an expected or predicted relationship between variables. A **variable** is a property or characteristic of an object, event, or person that can take on different values. One example of a variable is "U.S. state," a variable with 50 possible values (Alabama, Arkansas, etc.).

Research hypotheses are usually developed through the completion of several tasks:

- identifying a question or issue to be examined,
- reviewing and evaluating relevant theories and research, and
- stating a research hypothesis.

Each of these three tasks is described below.

Identifying a Question or Issue to Be Examined

Most research starts with a question posed by the researcher that often comes from the researcher's own ideas and daily observations. Although this may not seem terribly scientific, there is an advantage in using one's own experience as a starting point: People are generally much more motivated to explore issues that concern them personally. In teaching statistics, we advise students developing their own research projects to study something of interest to them. Conducting research can be tedious, difficult, and frustrating. At various points, you may

FIGURE 1.1 ● STEPS IN THE RESEARCH PROCESS WITHIN THE SCIENTIFIC METHOD

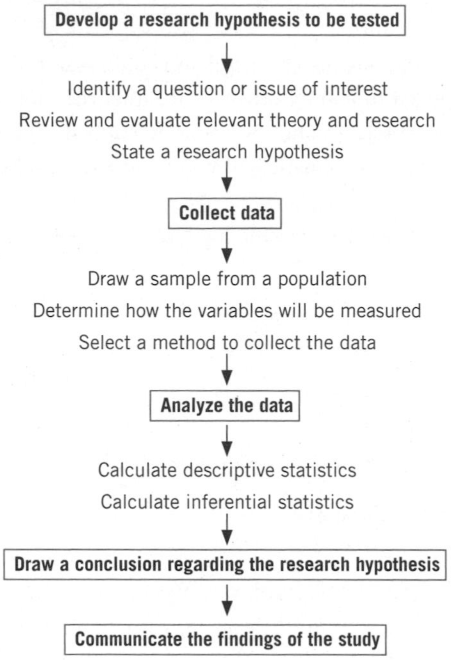

ask yourself, "Why am I doing this?" Being able to provide a satisfactory answer to this question will help you overcome obstacles you encounter along the way.

Reviewing and Evaluating Relevant Theories and Research

Beyond the researcher's own curiosity, research questions often arise from an examination of the theories, ideas, and research of others. A **theory** is a set of propositions used to describe or explain a phenomenon. The purpose of a theory is to summarize and explain specific facts using a logically consistent framework. Placing a question within a theoretical framework provides guidance and structure to research.

Reviewing and evaluating existing theories and research helps the researcher decide whether it is worth the time and energy required to conduct the study. By seeing what others have done, the researcher may decide that a particular idea has already been investigated and there is no reason to duplicate earlier efforts. However, the researcher may conclude that the current way of thinking is incomplete or mistaken. By doing this review, researchers are able to ensure that the studies they undertake add to and improve upon an existing body of knowledge.

Stating a Research Hypothesis: Independent and Dependent Variables

Understanding and evaluating an existing literature not only helps articulate a question of interest but also may lead to a predicted answer to that question. Within the scientific method, this answer is stated as a research hypothesis, defined earlier as a statement regarding an expected or predicted relationship between variables. Table 1.1 lists the research questions and research hypotheses for the studies listed at the beginning of this section. For example, the first research hypothesis states that "students who are taught effective learning skills will perform better on tests than students offered incentives to do well."

One characteristic of research hypotheses such as those listed in Table 1.1 is that they identify the variables that are the focus of their research studies. As mentioned earlier, a variable is a property or characteristic with different values. Variables can be classified in several ways. In specifying a research hypothesis, researchers often speak in terms of "independent" and "dependent" variables. An **independent variable** may be defined as a variable manipulated or controlled by the researcher. A **dependent variable**, on the other hand, is a variable measured by the researcher that is expected to change or vary as a function of the independent variable. The independent variable is seen as the "cause" or "treatment," whereas the dependent variable is the "effect" or "outcome." Researchers are interested in examining the effect of the independent variable on the dependent variable.

Consider the first research hypothesis provided in Table 1.1: "Students who are taught effective learning skills will perform better on tests than students offered incentives to do well." Here the independent variable is the instructional method by which students are taught, which consists of two values: learning skills and incentives. The dependent variable is the test performance that will be measured during the research. In this study, the effect of the independent variable on the dependent variable is that differences in students' test performance (the dependent variable) may "depend" upon which instructional method (learning skills or incentives) a student receives. Table 1.2 lists the independent and dependent variables for each of the research hypotheses in Table 1.1.

In addition to identifying variables, a second characteristic of research hypotheses is that they specify the nature and direction of the relationship between variables. For example, the first research hypothesis in Table 1.1 states that "students who are taught effective learning skills will perform *better* on tests." The word *better* indicates the nature and direction of the relationship between the independent variable (instructional method) and the dependent variable (test performance). The direction of the relationship would not have been stated if the hypothesis had included the less specific phrase "will perform *differently* on tests," which simply indicates that the two groups are not expected to be the same. Table 1.3 provides directional and nondirectional research hypotheses for the research studies from Table 1.1.

The ability to form a directional research hypothesis is dependent on the state of the existing literature on the question of interest. If little or perhaps conflicting research has been conducted, researchers may not be able to form a directional hypothesis before they begin their research. One study, for example, examined the relationship between exercise deprivation (not allowing people to exercise) and tension, depression, and anger

TABLE 1.1 ● EXAMPLES OF RESEARCH QUESTIONS AND HYPOTHESES

Research Question	Research Hypothesis
Is college students' performance on tests more influenced by their motivation (*why* they learn) or their learning strategies (*how* they learn)?	Students who are taught effective learning skills will perform better on tests than students offered incentives to do well.
How do parents' feeding practices affect whether their children will eat new foods?	Greater use of forceful feeding practices leads to greater food refusals.
Does coercive interrogation of eyewitnesses affect their accusations of crime suspects?	Coercive interrogation increases eyewitnesses' false accusations.
Are there gender differences in terms of who contributes to Wikipedia?	Women are less confident in contributing to Wikipedia than are men.

TABLE 1.2 ● RESEARCH HYPOTHESES AND THEIR INDEPENDENT AND DEPENDENT VARIABLES

Research Hypothesis	Independent Variable	Dependent Variable
Students who are taught effective learning skills will perform better on tests than students offered incentives to do well.	Instructional method	Test performance
Greater use of forceful feeding practices leads to greater food refusals.	Use of forceful feeding practices	Frequency of food refusals
Coercive interrogation increases eyewitnesses' false accusations.	Type of interrogation	Making a false accusation
Women are less confident in contributing to Wikipedia than are men.	Biological sex	Level of confidence

(Mondin et al., 1996). The researchers for the study reported, "We did not have a directional hypothesis, and when participants asked what we expected to find in this study, we replied: 'We really are not sure since the results of earlier work on exercise deprivation are mixed'" (p. 1200).

Collecting Data

Once a research hypothesis has been formulated, researchers are ready to proceed to the second stage in the research process: collecting data relevant to this hypothesis. This step is seen as being composed of three tasks:

- drawing a sample from a population,
- determining how the variables will be measured, and
- selecting a method by which to collect the data.

Drawing a Sample From a Population

The first step in collecting data is to identify the group of participants to which the research hypothesis applies. The group to which the results of a study may be applied or generalized is called a population. A **population** is the total number of possible units or elements that could potentially be included in a study. For

TABLE 1.3 ● DIRECTIONAL AND NONDIRECTIONAL RESEARCH HYPOTHESES

Directional Research Hypothesis	Nondirectional Research Hypothesis
Students who are taught effective learning skills will perform *better* on tests than students offered incentives to do well.	Students who are taught effective learning skills will perform *differently* on tests than students offered incentives to do well.
Greater use of forceful feeding practices leads to *greater* food refusals.	Use of forceful feeding practices *leads to* food refusals.
Coercive interrogation increases eyewitnesses' false accusations.	Coercive interrogation *affects* eyewitnesses' false accusations.
Women are *less* confident in contributing to Wikipedia than are men.	Women are *different* in confidence in contributing to Wikipedia than men.

LEARNING CHECK 1:

Reviewing What You've Learned So Far

1. Review questions
   a. What are the main steps involved in the research process?
   b. Why is it useful to review and evaluate theories and research before conducting a study?
   c. What are the two main characteristics of research hypotheses?
   d. What is the difference between an independent variable and a dependent variable?
2. Listed below are several research hypotheses from published studies. For each research hypothesis, identify the independent and dependent variable.
   a. "College students will rate instructors who dress formally (i.e., business suit and tie) as having more expertise than instructors who dress casually (i.e., slacks and shirt)" (Sebastian & Bristow, 2008, p. 197).
   b. "It was expected that . . . greater amounts of television viewing . . . would predict greater . . . posttraumatic stress symptoms" (McLeish & Del Ben, 2008, p. 417).
   c. "We . . . hypothesized persons who estimated the HSAS level to be red (severe) or orange (high) . . . when the HSAS level was [in fact] yellow (elevated), would report greater worry about terrorism" (Eisenman et al., 2009, p. 169).
   d. "We hypothesized that prekindergarten children who participated in the 6-week intervention would perform better [on a test of literacy skills] than their peers in a control group who did not participate in the program" (Edmonds, O'Donoghue, Spano, & Algozzine, 2009, p. 214).

example, researchers could define the population of interest for the first research hypothesis in Table 1.1 as "college students," "college students in the United States," "college students in Georgia," or "college students at the University of Georgia." Researchers typically try to define their populations as broadly as possibly (e.g., "college students in the United States" rather than "college students in Georgia") to maximize the applications or implications of their research.

It is typically difficult to collect data from all members of a population. Imagine, for example, the time and money needed to collect information from every college student in the United States! For this reason, researchers typically draw conclusions about populations based on information collected from a sample of the population. A **sample** is a subset or portion of a population. Table 1.4 describes the samples used in the four studies introduced in Table 1.1. As you can see from Table 1.4, samples greatly vary in terms of their targeted population and size (the number of participants). The extent to which the results of a study can be generalized to the target population depends on the extent to which the sample is representative of the population.

Determining How Variables Will Be Measured: Levels of Measurement

The research hypotheses described in Table 1.1 involve variables such as instructional method, frequency of food refusals, type of interrogation, and level of confidence. To conduct a research study, the researcher must determine an appropriate way to measure the variables stated in the research hypothesis. **Measurement** is the assignment of numbers or categories to objects or events according to rules.

For example, to measure the variable "height," a researcher might use "number of inches from the ground in bare feet" as a form of measurement. To measure "success in college," a student's grade point average (GPA) may be obtained from school transcripts. "Self-esteem" might be measured by having people complete a questionnaire and, on the basis of their responses, be categorized as having either "low" or "high" self-esteem. As these examples demonstrate, the result of measurement is an assignment of a number or category to each participant in the research study. Different types of variables require different forms of measurement. Consequently, researchers have identified four distinct levels for measuring variables: nominal, ordinal, interval, and ratio.

The values of variables measured at the **nominal level of measurement** differ in category or type. The word *nominal* implies having to do with "names," such that we use first and last names to distinguish between people. Gender is an example of a nominal variable, in that it consists of categories or types (male and female) rather than numeric values. In the first research hypothesis in Table 1.1, the independent variable of instructional method is measured at the nominal level, consisting of two categories: learning skills and incentives. It is important to understand that although numbers can be assigned to nominal variables, this does not allow for comparisons based on these numbers. For example, saying "1 = Male and 2 = Female" does not mean that females have "more" gender than males.

Variables measured at the **ordinal level of measurement** have values that can be placed in an order relative to the other values. Rankings (such as finishing first, second, or third in a race), size (small, medium, large, or extra large), and ratings (below average, average, or above average) are familiar examples of an ordinal scale. Ordinal scales allow researchers to indicate that one value represents more or less of a variable than do other values; however, it is not possible to specify the precise difference between values. For example, although you can say that a runner who finishes "first" in a race is faster than the runner who finishes "second," you cannot specify the exact difference between the two runners' times. Also, similar to nominal variables, assigning numbers to ordinal variables does not provide the ability to make numeric comparisons between the values of these variables.

The values of variables measured at the **interval level of measurement** are equally spaced along a numeric continuum. One example of an interval variable is the Fahrenheit scale of temperature. Here, a difference of 5 degrees has the same meaning anywhere along the scale; for example, the difference between 45°F and 50°F is the same as the difference between 65°F and 70°F. Many variables studied in the behavioral sciences (e.g., personality characteristics or attitudes) are considered to be measured at the interval level of measurement. Interval variables not only provide more precise and specific information than do ordinal variables but also fulfill the requirements of the most commonly used statistical procedures.

Variables at the **ratio level of measurement** are identical to interval variables, with one exception: Ratio scales possess what is known as a true zero point, for which the value of zero (0) represents the complete absence of the variable. Variables that describe a physical dimension (such as height, weight, distance, and time duration) typically have a true zero point. In the first research hypothesis in Table 1.1, the researchers in this study measured the variable "test performance" by recording the number of correct answers in a test of reading comprehension. Test performance is a ratio variable because it has a true zero point, where zero would indicate the complete absence of correct answers.

TABLE 1.4 ● RESEARCH HYPOTHESES AND STUDY SAMPLES

Research Hypothesis	Study Sample
Students who are taught effective learning skills will perform better on tests than students offered incentives to do well.	"109 juniors and seniors . . . enrolled in three sections of an educational psychology course" (Tuckman, 1996, p. 200).
Greater use of forceful feeding practices leads to greater food refusals.	"60 . . . families who . . . had at least one toddler between the ages of 12–36 months" (Fries, Martin, & van der Horst, 2017, p. 94).
Coercive interrogation increases eyewitnesses' false accusations.	"Sixty university undergraduates studying introductory psychology at a provincial university in the Greater Toronto Area" (Loney & Cutler, 2016, p. 31).
Women are less confident in contributing to Wikipedia than are men.	"1,598 participants (17.15% women) . . . classified as being occasional contributors to Wikipedia" (Bear & Collier, 2016, p. 258).

One advantage of ratio measurement versus interval measurement is that ratio variables allow for a greater number of comparisons among values. Consider, for example, the value 6 for the ratio variable "inches." Not only is the difference between 4 and 6 inches the same as the difference between 6 and 8 inches (thereby involving addition and subtraction, which also can be used with interval variables), 6 inches is also twice as long as 3 inches and half as long as 12 inches (involving multiplication and division). Multiplication and division comparisons cannot be made with interval variables. You cannot, for example, say that a temperature of 90°F is three times as much temperature as 30°F.

The boldfaced and italicized text in Table 1.5 illustrates how the authors of the four studies described in the earlier tables chose to measure their independent and dependent variables. As you can see, the variables in these studies have been measured in a variety of ways at different levels of measurement.

What difference does the level of measurement make? A variable's level of measurement has important implications for researchers in that it influences how research hypotheses are stated and how data are analyzed. For example, imagine you were interested in studying the variable "success in college." You could measure this variable at the *nominal* level of measurement by having faculty members classify students into one of two groups: successful or unsuccessful. Researchers typically employ nominal variables when they are interested in questions involving differences between groups, such as, "Can differences in study habits predict whether a student will be successful or unsuccessful?"

You could instead measure success in college as an *ordinal* variable by having faculty members rank their students from top to bottom. Ranking enables a researcher to study relative differences with less concern for the precise magnitude of these differences. For example, using ranked data, you could ask, "How similar are younger and older faculty members' rankings of their students?"

To study success in college using the *interval* level measurement, you could have faculty members rate each student on a scale from 1 (low) to 100 (high). You could then use these ratings to address a research question such as, "Is there a relationship between faculty members' ratings of their students and students' ratings of themselves?"

Finally, to measure success in college using the *ratio* level of measurement, you could use students' salary after graduation (in dollars) as an indicator of success. Salary is a ratio variable because it contains a true zero point. In practice, ratio variables can be used to ask many of the same questions as those involving interval variables. For example, you could ask, "Is there a relationship between faculty members' ratings of students and students' salaries after graduation?"

TABLE 1.5 ● RESEARCH HYPOTHESES AND MEASUREMENT OF VARIABLES

Research Hypothesis	Independent Variable	Dependent Variable
Students who are taught effective learning skills will perform better on tests than students offered incentives to do well.	Instructional method **Learning strategy, incentive motivation *(nominal)***	Test performance **Number of items correct *(ratio)***
Greater use of forceful feeding practices leads to greater food refusals.	Use of forceful feeding practices **Number of forceful practices *(ratio)***	Frequency of food refusals **Number of refusals *(ratio)***
Coercive interrogation increases eyewitnesses' false accusations.	Type of interrogation **Coercive, not coercive *(nominal)***	Accuracy of accusation **Accurate, inaccurate *(nominal)***
Women are less confident in contributing to Wikipedia than are men.	Biological sex **Man, woman *(nominal)***	Level of confidence **1–5 scale (1 = disagree fully, 5 = agree fully) *(ordinal)***

As you see, how researchers choose to measure their variables influences how they state their research questions and hypotheses. The level of measurement for a variable has another important implication: It helps determine the statistical procedures researchers use to analyze the data. Certain statistical procedures can be applied only to variables measured at the interval or ratio levels of measurement, whereas other procedures are appropriate for variables measured at the nominal or ordinal level. As we move from chapter to chapter of this book in order to discuss different statistics, the level of measurement of the variables being analyzed will be noted.

Selecting a Method to Collect the Data: Experimental and Nonexperimental Research Methods

In addition to drawing a sample from a population and determining how the variables in a study may be measured, the third step in collecting data is to determine the type of research method to use to collect the data on these variables. Research methods may be classified into two main types:

- experimental research methods, and
- nonexperimental research methods.

Experimental research methods. **Experimental research methods** are methods designed to test causal relationships between variables—more specifically, whether changes in independent variables produce or cause changes in dependent variables. To make inferences about cause-effect relationships, researchers conducting an experiment must first eliminate other possible causes or explanations for changes in the dependent variable besides the independent variable. If it can be claimed that a variable other than the independent variable created the observed changes in the dependent variable, the research results are said to be confounded. A **confounding variable** is a variable related to an independent variable that provides an alternative explanation for the relationship between the independent and dependent variables.

To illustrate the concept of confounding variables, consider the following question: Is there a relationship between a mother's ethnicity and the birth weight of her baby? Although research has shown differences in the birth weights of babies of different ethnicities, one study noted that some of this research did not eliminate the possibility that these differences may be partly due to differences between ethnic groups on other factors that are also associated with low birth weights, such as the mother's average age at the time of pregnancy and behaviors such as smoking and drinking (Ma, 2008). These other factors are considered confounding variables in that they provide alternative explanations for any causal relationship between ethnicity and babies' birth weights. Confounding variables make it difficult for researchers to draw firm conclusions regarding the relationship between the independent and dependent variable.

LEARNING CHECK 2:

Reviewing What You've Learned So Far

1. Review questions
   a. What is the difference between a population and a sample?
   b. Within the research process, what is the relationship between populations and samples?
   c. What are the four levels of measurement? How do they differ?
2. For each of the following variables, name the scale of measurement (nominal, ordinal, interval, or ratio).
   a. Type of school attended (public, private)
   b. Probability of graduating college in 4 years (0% to 100%)
   c. Rating of television (good, better, best)
   d. Number of computers in home

Researchers minimize the influence of confounding variables in two main ways. First, researchers exercise experimental control by making the research setting (i.e., characteristics of the research participants, the instruments or measures administered, the methods used to collect data) the same for all participants. In the birth weight study, for example, the research design might control for the effect of parental smoking on birth weight by only including nonsmokers in the study sample, thereby excluding those who smoke.

Another way researchers exert control over the research setting is to include a condition known as a **control group**, which is a group of participants in an experiment not exposed to the independent variable the research is designed to test. For example, if you conducted an experiment designed to examine the effects of caffeine on one's health, one group (the experimental group) might be instructed to drink coffee while a second group (the control group) drinks decaffeinated coffee. By contrasting the results of the two groups, researchers can then assess the impact of caffeinated coffee on the health of those who drink it.

Researchers cannot possibly identify and control for the effects of all potential confounding variables. Consequently, a second strategy used to minimize the influence of confounding variables is **random assignment**, which is assigning participants to each category of an independent variable in such a way that each participant has an equal chance of being assigned to each category. For example, a researcher could randomly assign a participant to the experimental or control condition of a study by flipping a coin: heads for experimental, tails for control. The purpose of random assignment is to equalize or neutralize the effects of confounding variables by distributing them equally over the levels of the independent variable.

Nonexperimental research methods. Experimental research designs are one of the best tools researchers have for making causal inferences. However, the ability to make causal inferences requires a great deal of control over the situation, control that may not always be possible or desirable. For this reason, researchers often employ nonexperimental research methods (sometimes referred to as correlational research methods). **Nonexperimental research methods** are research methods designed to measure naturally occurring relationships between variables without having the ability to infer cause-effect relationships. Some of the most common types of nonexperimental research designs include quasi-experiments, survey research, observational research, and archival research.

Quasi-experimental research compares naturally formed or preexisting groups rather than employing random assignment to conditions. For example, suppose a researcher wanted to study the effects of different methods of teaching reading on children's verbal skills. Ideally, the researcher would randomly assign a sample of schoolchildren to receive the different teaching methods. However, because children are taught together in classes, it would be difficult to have children in the same classroom receive different methods. Implementing multiple methods in the same classroom would not only place an unreasonable burden on the teacher, but the children would also see classmates being treated differently, which might influence their behavior. To address these concerns, a researcher might assign entire classrooms of children to receive a particular method. Comparing the classrooms is an example of quasi-experimental research.

Survey research methods obtain information directly from a group of people regarding their opinions, beliefs, or behavior. The goal of survey research, which can involve the use of questionnaires and interviews, is to obtain information from a sample that can then be used to represent or estimate the views of a larger population. As one example of survey research, political pollsters attempt to predict how people will vote in an election by asking a sample of voters about their preferences. Because the researcher does not directly manipulate any variables, survey research is not conducted to make causal inferences but is instead used to describe a phenomenon or predict future behavior.

Observational research is the systematic and objective observation of naturally occurring behavior or events. The purpose of observational research is to study behavior or events, with little, if any, intervention on the part of the researcher. Observational research is often used to study phenomena that the researcher cannot or should not deliberately manipulate. For example, researchers studying interpersonal conflict among children would never force children to argue with each other. One study resolved this issue by videotaping groups of adolescents discussing topics such as competition and peer pressure and then counting the number of times the adolescents teased or mocked each other (Connolly et al., 2015).

Rather than observing or measuring behavior directly, **archival research** is the use of archives (records or documents of the activities of individuals, groups, or organizations) to examine research questions or hypotheses. One example of an archival research study studied the question of whether men are perceived to be more creative than women (Proudfoot, Kay, & Koval, 2015). In this study, the 100 most viewed talks on TED.com were identified, and the researchers compared viewers' posted ratings of the talks given by men versus those given by women.

Labeling variables in nonexperimental research. One issue that often confuses students when learning about nonexperimental research is related to the practice of referring to the variables as the "independent variable" and the "dependent variable." For example, in the study described above that involved biological sex and TED.com ratings, biological sex may be called the "independent variable" and TED.com ratings the "dependent variable." However, because biological sex isn't manipulated by the researcher, it would be inappropriate to say that biological sex "causes" differences in TED.com ratings. To avoid confusion, researchers sometimes refer to the independent variable and the dependent variable in a nonexperimental study as the **predictor variable** and **criterion variable**, respectively.

An example of combined experimental and nonexperimental research. Both experimental and nonexperimental research methods have strengths and weaknesses. The strength of experimental research is the ability to demonstrate cause-effect relationships between independent and dependent variables; however, the control that experiments require creates situations that may not resemble the real world. Nonexperimental research methods don't allow the researcher to make causal inferences; however, they have the advantage of allowing researchers to study variables as they naturally occur. Given the strengths and limitations of both research methods, one solution is to compare the findings from both methods in examining the same question.

Does playing violent video games lead to aggressive behavior? Two researchers addressed this important question by using both experimental and nonexperimental research methods, saying, "We chose two different methodologies that have strengths that complement each other and surmount each others' weaknesses" (Anderson & Dill, 2000, p. 776).

For their experiment, participants were randomly assigned to play either a violent video game or a nonviolent game; "type of video game" was their independent variable. Next, they played another type of game in which they could punish their opponent (who was actually a computer) by delivering a loud blast of noise. The loudness and duration of the noise delivered by participants represented the dependent variable of "aggressive behavior."

For their nonexperimental method, the researchers used a survey methodology, asking students to fill out a questionnaire about the number of hours they played violent video games each week (the predictor variable). For the criterion variable of aggressive behavior, students reported the number of times in the previous year that they had performed different aggressive acts, such as hitting or stealing from other students.

What did the researchers find? In reporting their findings, they wrote, "In both a correlational investigation using self-reports of real-world aggressive behaviors and an experimental investigation using a standard, objective laboratory measure of aggression, violent video game play was positively related to increases in aggressive behavior" (Anderson & Dill, 2000, p. 787). In evaluating their study, they noted the advantages of using both types of research methods, explaining that the nonexperimental method "measured video game experience, aggressive personality, and delinquent behavior in real life . . . [whereas] an experimental methodology was also used to more clearly address the causality issue" (p. 782).

Analyzing the Data

Once data collection has been completed, the next step in the research process involves data analysis. This part of the research process addresses the primary topic of this book: statistics. Because this is the first chapter of this book, we will not describe analyzing data in detail. Instead, we introduce the notion that there are two main purposes of analyzing data: (1) to organize, summarize, and describe data that has been collected and (2) to test and draw conclusions about ideas and hypotheses. These two purposes are met by calculating two main types of statistics: descriptive statistics and inferential statistics.

Calculating Descriptive Statistics

Descriptive statistics are statistics used to summarize and describe a set of data for a variable. For example, you have probably heard of crime statistics and unemployment statistics, statistics used to describe or summarize certain aspects of our society. Using a research-related example, Caitlin Abar, a researcher at Brown University, conducted a study examining increases in students' alcohol-related behaviors after entering college (Abar, 2012). To measure students' level of alcohol use, she created a variable called "typical weekend drinking," which "was measured as the sum of drinks consumed on a typical Friday and Saturday within the past 30 days" (p. 22). One way to summarize students' responses to the typical weekend drinking variable would be to calculate the mean, which is the mathematical average of a set of scores. The mean, described in Chapter 3, is an example of a descriptive statistic. The first part of this book describes a variety of descriptive statistics used by researchers to summarize data they have collected.

Calculating Inferential Statistics

Besides summarizing and describing data, a second purpose of statistics is to test hypotheses and draw conclusions. **Inferential statistics** are statistical procedures used to test hypotheses and draw conclusions from data collected during research studies. Using inferential statistics, researchers make inferences about the relationships between variables in populations based on information gathered from samples of the population.

As an example of an inferential statistic, let's return to the example of students' alcohol-related behaviors introduced above (Abar, 2012). This study was interested in seeing whether there was a relationship between these behaviors and aspects of these students' relationships with their parents. One aspect was "alcohol communications," defined as "the extent that they discussed alcohol related topics with their parents at some point during the past several months" (p. 22). To test a hypothesis regarding the relationship between alcohol communications and students' alcohol use, we could use an inferential statistic known as the Pearson correlation coefficient, which will be discussed in Chapter 13. The chapters in the last half of this book discuss a broad variety of inferential statistics, along with detailed instructions for analyzing and drawing conclusions from statistical data and presenting research findings.

Drawing a Conclusion Regarding the Research Hypothesis

Once statistical analyses have been completed, the next step is to interpret the results of the analyses as they relate to the study's research hypothesis. More specifically, do the findings of the analyses support or not support the research hypothesis? The word *support* in the previous sentence is very important. Students and researchers are sometimes tempted to conclude that their findings "prove" their hypotheses are either "true" or "false." However, as we'll discuss in Chapter 5, because researchers typically don't collect data from the entire population, they can't know with 100% certainty whether their hypothesis is in fact proven to be true or false. Also, given the complexity of the phenomena studied by researchers, it is extremely difficult for any one research study to prove a hypothesis is completely true or false. Drawing appropriate conclusions from statistical analyses is critical for building and testing theories that are based on the findings of research studies.

Communicating the Findings of the Study

Conducting research requires a variety of different skills: conceptual skills to develop research hypotheses, methodological skills to collect data, and mathematical skills to analyze these data. Another integral part of the research process is the communication skills needed to inform others about a study. Researchers must be not only skilled scientists but also effective writers.

LEARNING CHECK 3:
Reviewing What You've Learned So Far

1. Review questions
   a. What are the main differences between experimental and nonexperimental research methods?
   b. What are confounding variables? What do researchers do to minimize the effects of confounding variables?
   c. What are the main types of nonexperimental research methods?
   d. What are the relative strengths and weaknesses of experimental and nonexperimental research?
   e. What are the main purposes of descriptive and inferential statistics?

For many years, the American Psychological Association (APA), the professional association for psychologists, has provided researchers guidance on how to communicate the results of their research. In 1929, the APA published a seven-page article in the journal *Psychological Bulletin* entitled, "Instructions in Regard to Preparation of Manuscript." In contrast, the sixth edition of the *Publication Manual of the American Psychological Association,* published in 2010, consists of 272 pages. This increase in length highlights the challenges faced by writers to communicate their research in an effective and efficient manner.

1.4 PLAN OF THE BOOK

The primary purpose of this chapter was to introduce statistics and place it within the larger research process. The remainder of the book will discuss a number of different statistical procedures used by researchers to examine various questions of interest. Although it is critical for you to be able to correctly calculate statistics, it is equally important to understand the role of statistics within the research process and appreciate the conceptual and pragmatic issues related to the use (and sometimes misuse) of statistics.

The first half of the book (Chapters 2 through 6) introduces conceptual and mathematical issues that are the foundation of statistical analyses. Chapters 2, 3, and 4 discuss how data may be examined, described, summarized, and presented in numeric, visual, and graphic form. Descriptive statistics are introduced and discussed in these chapters. Chapters 5 and 6 introduce critical assumptions about and characteristics of the inferential statistical procedures used to test research hypotheses. The chapters in the second half of the book cover a number of different inferential statistical procedures, along with key issues surrounding the use of these procedures.

In most cases, the chapters in this book share a uniform format and structure and are centered on the research process. In making our presentation, we will provide various examples of actual published research studies that address questions with which you may be familiar. For each example, we will clearly state the research hypothesis, briefly describe how the data in the study were collected, introduce and describe the calculation of both descriptive and inferential statistics, and discuss the extent to which the results of the statistical analyses support the research hypothesis. Finally, we will illustrate how to communicate one's research to a broader audience.

1.5 LOOKING AHEAD

We began our presentation by defining statistics and providing reasons why learning about them may be of value to you. Next, we placed statistics within the stages of the larger research process, a process that will be the foundation of this textbook. As we have explained, the research process is centered on a research hypothesis, a predicted relationship between variables. Once a hypothesis has been stated, the next step is to collect data about the variables included in the research, after which the process of statistical analysis begins. Analyzing data involves the calculation of two main types of statistics, one designed to describe and summarize the data for a variable (descriptive statistics) and one designed to test research hypotheses (inferential statistics). However, before conducting analyses on a set of data, there is preliminary work that must be completed. The next chapter will focus on methods used by researchers to examine data.

1.6 Summary

Statistics may be defined as a branch of mathematics dealing with the collection, analysis, interpretation, and presentation of masses of numerical data. As such, statistics not only is concerned about how data are analyzed but also recognizes the importance of understanding how data are collected and how the results of analyses are interpreted and communicated. There are a variety of reasons why students should learn statistics: Statistics are encountered in a wide variety of daily activities, students may be asked or required to read and interpret the results of statistical analyses in their courses, students may use statistics in conducting their own research, and statistics may help one's career.

Researchers conduct research using the ***scientific method*** of inquiry, a method of investigation that uses the objective and systematic collection and analysis of empirical data to test theories and hypotheses. The research process used within the scientific method of inquiry consists of five main steps: developing a research hypothesis to be tested, collecting data, analyzing the data, drawing a conclusion regarding the research hypothesis, and communicating the findings to the study.

A ***research hypothesis*** is a statement regarding an expected or predicted relationship between variables. Developing a research hypothesis involves identifying a question or issue to be examined, reviewing and evaluating relevant theories and research, and stating the research hypothesis to be tested in the study. A ***theory*** is a set of propositions used to describe or explain a phenomenon. A research hypothesis contains ***variables***, which are properties or characteristics of some object, event, or person that can take on different values. More specifically, a research hypothesis states the nature and direction of a proposed relationship between an ***independent variable*** (a variable manipulated or controlled by the researcher) and a ***dependent variable*** (a variable measured by the researcher that is expected to change or vary as a function of the independent variable).

Collecting data involves drawing a sample from a population, determining how the variables will be measured, and selecting a method to collect the data. A ***population*** is the total number of possible units or elements that could potentially be included in a study; a ***sample*** is a subset or portion of a population. Variables can be measured at one of four levels of measurement: ***nominal*** (values differing in category or type), ***ordinal*** (values placed in an order relative to the other values), ***interval*** (values equally spaced along a numeric continuum), or ***ratio*** (values equally spaced along a numeric continuum with a true zero point).

There are two main types of methods used to collect data: ***experimental research methods***, designed to test whether changes in independent variables produce or cause changes in dependent variables, and ***nonexperimental research methods***, designed to examine the relationship between variables without the ability to infer cause-effect relationships.

In experimental research, researchers are concerned about possible ***confounding variables***, which are variables related to independent variables that provide an alternative explanation for the relationship between independent and dependent variables. To minimize the impact of confounding variables, a research study may exercise experimental control over various aspects of the situation, including the use of a ***control group***, which is a group of participants not exposed to the independent variable the research is designed to test, or ***random assignment***, which involves assigning participants to each category of an independent variable in such a way that each participant has an equal chance of being assigned to each category.

Examples of nonexperimental research methods are ***survey research***, in which information is directly obtained from a group of people regarding their opinions, beliefs, or behavior; ***observational research***, which involves the systematic and objective observation of naturally occurring events; and ***archival research***, which uses archives (records or documents of the activities of individuals, groups, or organizations) to examine research questions or hypotheses. In nonexperimental research, the independent variable and dependent variable may be referred to as the ***predictor variable*** and ***criterion variable***, respectively.

Once data have been collected, the next step is to analyze them using two main types of statistics: ***descriptive statistics***, which summarize and describe a set of data, and ***inferential statistics***, which are statistical techniques used to test hypotheses and draw conclusions from data collected during research studies.

Once the statistical analyses have been completed, the results of the analyses are interpreted regarding whether they support or not support the study's research hypothesis. The final step in the research process is to communicate the study to a broader audience.

1.7 Important Terms

statistics (p. 1)
scientific method (p. 3)
research hypothesis (p. 3)
variable (p. 3)
theory (p. 4)
independent variable (p. 5)
dependent variable (p. 5)
population (p. 6)
sample (p. 7)
measurement (p. 7)
level of measurement (nominal, ordinal, interval, ratio) (p. 8)
experimental research methods (p. 10)
confounding variable (p. 10)
control group (p. 11)
random assignment (p. 11)
nonexperimental research methods (p. 11)
quasi-experimental research (p. 11)
survey research (p. 11)
observational research (p. 11)
archival research (p. 12)
predictor variable (p. 12)
criterion variable (p. 12)
descriptive statistics (p. 13)
inferential statistics (p. 13)

1.8 Exercises

1. Listed below are a number of hypothetical research hypotheses. For each hypothesis, identify the independent and dependent variable.
   a. Male drivers are more likely to exhibit "road rage" behaviors such as aggressive driving and yelling at other drivers than are female drivers.
   b. The more time a student takes to finish a midterm examination, the higher his or her score on the examination.
   c. Men are more likely to be members of the Republican political party than are women; women are more likely to belong to the Democratic political party than are men.
   d. Students who receive a newly designed method of teaching reading will display higher scores on a test of comprehension than the method currently used.
   e. The more time a child spends in daycare outside of the home, the less he or she will be afraid of strangers.
2. Listed below are a number of research questions and hypotheses from actual published articles. For each hypothesis, identify the independent and dependent variable.
   a. "The use of color in a Yellow Pages advertisement will increase the perception of quality of the products for a particular business when compared with noncolor advertisements" (Lohse & Rosen, 2001, p. 75).
   b. "We hypothesized that parents who use more frequent corporal and verbal punishment . . . will report more problem behaviors in their children" (Brenner & Fox, 1998, p. 252).
   c. "It was hypothesized that adolescents with anorexia nervosa would . . . be more respectful when compared with peers with bulimia nervosa" (Pryor & Wiederman, 1998, p. 292).
   d. "The purpose of the present research was to assess brand name recognition as a function of humor in advertisements. . . . It was predicted that participants would recognize product brand names that had been presented with humorous advertisements more often than brand names presented with nonhumorous advertisements" (Berg & Lippman, 2001, p. 197).

e. "The purpose of this study was to explore the relation between time spent in daycare and the quality of exploratory behaviors in 9-month-old infants . . . it was hypothesized that . . . infants who spent greater amounts of time in center-based care would demonstrate more advanced exploratory behaviors than infants who did not spend as much time in center-based care" (Schuetze, Lewis, & DiMartino, 1999, p. 269).

f. "It was hypothesized that students would score higher on test items in which the narrative contains topics and elements that resonate with their daily experiences versus test items comprised of material that is unfamiliar" (Erdodi, 2012, p. 172).

3. Listed below are additional research questions and hypotheses from actual published articles. For each hypothesis, identify the independent and dependent variable.

   a. "It is expected that achievement motivation will be a positive predictor of academic success" (Busato, Prins, Elshout, & Hamaker, 2000, p. 1060).

   b. "Men are expected to employ physical characteristics (particularly those that are directly related to sex) more often than women in selecting dating candidates" (Hetsroni, 2000, p. 91).

   c. "The purpose of our study was to gain a better understanding of the relationship between social functioning and problem drinking. . . . We predicted that problem drinkers would endorse more social deficits than nonproblem drinkers" (Lewis & O'Neill, 2000, pp. 295–296).

   d. "We predicted that people in gain-framed conditions would show greater intention to use sunscreen . . . than those people in loss-framed conditions" (Detweiler, Bedell, Salovey, Pronin, & Rothman, 1999, p. 190).

   e. "Students with learning disabilities who received self-regulation training would obtain higher reading comprehension scores than students (with learning disabilities) in the control group" (Miranda, Villaescusa, & Vidal-Abarca, 1997, p. 504).

   f. "We hypothesized that consumers would think that a 'sale price' presentation would generate a greater monetary savings than an 'everyday low price' presentation" (Tom & Ruiz, 1997, p. 403).

   g. "We hypothesize that physical coldness (vs. warmth) would activate a need for psychological warmth, which in turn increases consumers' liking of romance movies" (Hong & Sun, 2012, p. 295).

4. Name the scale of measurement (nominal, ordinal, interval, ratio) for each of the following variables:

   a. The amount of time needed to react to a sound
   b. Gender
   c. Score on the Scholastic Aptitude Test (SAT)
   d. Political orientation (not at all conservative, conservative, very conservative)
   e. Political affiliation (Democrat, Republican, Independent)

5. Name the scale of measurement (nominal, ordinal, interval, ratio) for each of the following variables:

   a. One's age (in years)
   b. Size of soft drink (small, medium, large, extra large)
   c. Voting behavior (in favor vs. against)
   d. IQ score
   e. Parent (mother vs. father)

6. A faculty member wishes to assess the relationship between students' scores on the Scholastic Aptitude test (SAT) and their performance in college.

   a. What is a possible research hypothesis in this situation?
   b. What are the independent and dependent variables?
   c. How could you measure the variable "performance in college" at each of the four levels of measurement?

7. A researcher hypothesizes that drivers who use cellular phones will get into a greater number of traffic accidents than drivers who do not use these phones. For this example,
   a. What are the independent and dependent variables?
   b. How could you measure the dependent variable?
   c. How could you conduct this study using an experimental research method? How could you conduct this study using a nonexperimental method?

Answers to Learning Checks

Learning Check 1

1. a. IV: Instructor's dress; DV: Expertise
   b. IV: Amount of television viewing; DV: Posttraumatic stress symptoms
   c. IV: Estimation of HSAS level; DV: Worry about terrorism
   d. IV: Participation in intervention program; DV: Literacy skills

Learning Check 2

2. a. Nominal
   b. Ratio
   c. Ordinal
   d. Ratio

Answers to Odd-Numbered Exercises

1. a. Independent variable (IV): Gender; dependent variable (DV): "Road rage" behaviors
   b. IV: Time; DV: Exam score
   c. IV: Gender; DV: Political affiliation
   d. IV: Method of teaching; DV: Comprehension test scores
   e. IV: Time in daycare; DV: Fear of strangers
3. a. IV: Achievement motivation; DV: Academic success
   b. IV: Gender; DV: Emphasis on physical characteristics
   c. IV: Alcohol drinking; DV: Social deficits
   d. IV: Framing; DV: Intention to use sunscreen
   e. IV: Training program; DV: Reading comprehension scores
   f. IV: Price presentation; DV: Perception of savings
   g. IV: Physical temperature; DV: Liking of romance movies
5. a. Ratio
   b. Ordinal
   c. Nominal
   d. Interval
   e. Nominal

7. a. IV: Cellular phone use; DV: Number of traffic accidents
   b. Example: Determine the number of traffic accidents each person has experienced in the past year.
   c. Example of experimental: Use a driving simulation with the "driver" talking on a phone and measure the number of potential driving mistakes or accidents.

      Example of nonexperimental: Take a survey of the number of accidents people have been in and whether or not they were using a phone during the accident.

2

EXAMINING DATA: TABLES AND FIGURES

CHAPTER OUTLINE

Chapter 1 described a process by which scientific research may be conducted. This process begins by stating a research hypothesis regarding an expected relationship between variables. The next step involves collecting data, data that will ultimately be used in statistical analyses designed to test the research hypothesis. However, researchers do not conduct statistical analyses until they have first examined the data to ensure appropriate conclusions can be drawn from the analyses. This chapter describes why researchers examine data and how data may be examined using tables, charts, and graphs. As with many of the chapters in this book, this discussion will be guided by the findings from a published research study.

2.1 AN EXAMPLE FROM THE RESEARCH: WINNING THE LOTTERY

A friend of yours flips a coin and asks you to guess whether the coin has landed "heads" or "tails." You guess "tails," but the answer is "heads." Your friend flips the coin a second time, and again you predict "tails." But again the answer is "heads." On the third coin flip, you guess "tails" once more—but once again you are wrong. As your friend prepares to flip the coin a fourth time, you tell yourself there is little chance the coin could land on "heads" again. But the reality is that the fourth coin flip is equally likely to be "heads" or "tails." This is because the outcome of the first three coin flips has absolutely no impact on the outcome of the fourth flip. The outcome of a coin flip is a random event that cannot be controlled or predetermined. But do you really believe this?

The impact of random events can take on greater significance than the outcome of a coin flip. For example, some government-run lotteries provide the opportunity to win huge amounts of money through the random selection of numbers. In 2001, Dr. Karen Hardoon, a researcher at McGill University in Canada, studied the thought processes people might use when purchasing lottery tickets (Hardoon, Baboushkin, Derevensky, & Gupta, 2001). As previous research had found some gamblers believe they can control random events such as the rolling of dice or the spinning of a roulette wheel, Dr. Hardoon and her colleagues were interested in determining "whether individuals with gambling problems perceive the purchase of lottery tickets in a similar manner as non-problem gamblers" (p. 752). On the basis of the earlier findings, Dr. Hardoon hypothesized that problem gamblers are less likely to believe lottery outcomes are random than are non–problem gamblers.

To test their research hypothesis, Dr. Hardoon and her associates located a group of gamblers and asked whether they had ever experienced problems as a result of their gambling. The research team then classified the participants into two groups: problem gamblers and non–problem gamblers.

Each participant in the study, which we will refer to as the lottery study, was shown four lottery tickets and was asked, "If you were to buy a ticket to play in the lottery, which one would you select?" Each ticket contained six numbers selected from among the numbers 1 through 49; the four tickets were given a particular label. The *sequence* ticket had numbers in consecutive order (30-31-32-33-34-35); numbers on the *pattern* ticket increased by 5s (5-10-15-20-25-30); the *nonequilibrated* numbers (35-37-40-43-44-49) were clustered at the upper end of the 49 numbers, and the *random* ticket (7-8-23-34-36-42) shared none of the characteristics of the other three tickets. We will call the variable in this study "lottery ticket," a variable with four values or categories: sequence, pattern, nonequilibrated, and random.

The researchers recorded which of the four tickets was chosen by each participant. Table 2.1 lists the data for the lottery ticket variable for the non–problem gamblers (the data for the problem gamblers in the study are provided in the exercises at the end of this chapter). Looking at this table, we see that the non–problem gamblers differed in their lottery ticket choices; however, it is difficult to describe the extent or nature of these differences given the unorganized nature of the collected data. Consequently, before conducting any statistical analyses on their data, researchers typically organize and examine them. In the following sections, we will discuss why and how data are organized and examined.

2.2 WHY EXAMINE DATA?

There are a variety of reasons why researchers examine data they have collected before conducting statistical analyses on the data. These reasons include the following:

- to gain an initial sense of the data,
- to detect data coding or data entry errors,
- to identify outliers,
- to evaluate research methodology, and
- to determine whether data meet statistical criteria and assumptions.

Each of these reasons is discussed below.

TABLE 2.1 ● THE LOTTERY TICKET CHOICES OF 22 NON-PROBLEM GAMBLERS

Participant	Lottery Ticket	Participant	Lottery Ticket	Participant	Lottery Ticket
1	Random	9	Pattern	17	Nonequilibrated
2	Random	10	Pattern	18	Random
3	Nonequilibrated	11	Random	19	Random
4	Pattern	12	Pattern	20	Pattern
5	Random	13	Random	21	Random
6	Sequence	14	Nonequilibrated	22	Random
7	Random	15	Random		
8	Random	16	Pattern		

Gaining an Initial Sense of the Data

Researchers spend a great amount of time and energy designing their research studies; consequently, they are eager to analyze the data they collect so that they may test their research hypotheses. Students taking courses in statistics, on the other hand, face a different challenge: They are given sets of data with which they have no prior experience and have only a short amount of time to learn and apply statistical formulas to the data. Although both researchers and students may be motivated to analyze their data as quickly as possible, it is important to first examine the data in order to gain an initial sense of them.

As an example, imagine you conduct a study regarding the fuel efficiency of automobiles. In your study, you collect data on the miles per gallon (MPG) of different types of cars. You might expect the scores in your data set to range from 10 to 40 MPG, with the majority of scores between 15 and 25 MPG. Examining data helps researchers gain an initial sense of the data they have collected and whether the data are in line with their expectations. Examining data also provides a way of confirming the results of later statistical analyses. We've found that students' errors in their calculations could have been avoided had they first looked at the set of data. For example, examining data will help you avoid making statements such as, "The MPG for the cars in my sample ranged from 10 to 324 MPG," "The average GPA in my sample of college students was 6.36," or "The students in my sample owned an average of 13.24 cellphones."

Detecting Data Coding or Data Entry Errors

Another reason for examining data before conducting statistical analyses is to detect any errors made in the coding of data that have been collected. Imagine, for example, that a group of students is asked to complete a 20-item test of mathematical calculations; each student's score on the test is the number of items answered correctly. Once the test scores are calculated, they are entered into a computer file. In this situation, it's crucial to ensure that no mistakes are made in calculating the test scores or in data entry. Data coding and data entry errors are not at all uncommon. Left unattended, seemingly minor errors can lead to a great loss of time and energy—including the need, in some cases, to repeat the statistical analysis of the data.

Identifying Outliers

Researchers also examine their data to identify **outliers**, defined as rare, extreme scores that lie outside of the range of the majority of scores in a set of data. Returning to our example of the 20-item test of mathematical calculations, suppose the test is given to 15 students. In scoring the tests, a researcher finds that 14 students correctly answered between 8 and 17 questions. However, one student provided the wrong answer to every question on the test, for a score of 0. In this case, the score of 0 would be considered an outlier.

If not identified, outliers may distort conclusions researchers draw about their sample and their data, particularly when the sample is relatively small. In the above example, the single score of 0 might lead an observer to conclude the class as a whole performed more poorly on the test than they actually did.

Evaluating Research Methodology

Examining data is also useful in assessing the effectiveness of the methods researchers use to collect their data. For the 20-item math test, for example, the possible scores range from 0 to 20 questions answered correctly. Having administered the test to 15 students, it might be reasonable to expect the majority of scores to fall in the middle of the 0 to 20 range. But what if every student answered at least 16 of the 20 questions correctly? Based on such a finding, we might conclude the test was too easy and that more questions are needed to accurately assess students' level of knowledge. By comparing the collected data for a variable with the range of possible values, researchers are able to evaluate and modify their measurement tools.

Determining Whether Data Meet Statistical Criteria and Assumptions

Researchers also examine their data to see whether they meet certain statistical criteria and assumptions. Many of the statistical procedures discussed in this book are based on specific assumptions regarding the shape and nature of a set of data that has been collected. These procedures generally assume, for example, that in most data sets, the majority of the data points will fall in the middle of the range of possible values, with a relatively small amount of data at the highest and lowest possible values. In the example of the 20-item math test, it might be assumed that most scores will be between 7 and 13, with smaller number of scores either less than 7 or greater than 13. When assumptions such as this are not met, it is more likely a researcher may draw inappropriate conclusions from any statistical analyses that are conducted.

2.3 EXAMINING DATA USING TABLES

One part of examining data that have been collected for a variable involves answering a series of questions:

- What are the values of the variable?
- How many participants in the sample have each value of the variable?
- What percentage of the sample has each value of the variable?

As it would be difficult to answer questions such as these just by looking at the data in Table 2.1, researchers typically organize their data by creating a table known as a **frequency distribution table**, a table that summarizes the number and percentage of participants for the different values of a variable. This section discusses the steps involved in creating frequency distribution tables using the questions listed above.

What Are the Values of the Variable?

The first step in organizing the data for a variable is to identify all of the possible values for the variable. Table 2.2 begins the construction of a frequency distribution table by listing the four values for the lottery ticket variable.

How Many Participants in the Sample Have Each Value of the Variable?

Once the values of the variable have been identified, the next step in organizing data that have been collected is to determine the **frequency** of each value, which is the number of participants in a sample corresponding to a

TABLE 2.2 ● IDENTIFYING THE VALUES OF THE LOTTERY TICKET VARIABLE

Lottery Ticket
Sequence
Pattern
Nonequilibrated
Random

TABLE 2.3 ● USING THE TICK MARK METHOD TO DETERMINE THE FREQUENCIES OF THE LOTTERY TICKET VARIABLE

Lottery Ticket	Tick Marks	*f*
Sequence	I	1
Pattern	~~IIII~~ I	6
Nonequilibrated	III	3
Random	~~IIII~~ ~~IIII~~ II	12
Total		22

value of a variable. These frequencies provide an indication of the nature and shape of the distribution of values for a variable in a sample.

One way to determine the frequencies for a variable is to sort the data into the different values and then count the number of participants having each value using a "tick mark method." Using this method, a tick mark is placed next to a value of the variable each time that value appears in the sample. As shown in Table 2.3, the tick marks are then counted to determine the frequency for each value. For example, the number of tick marks and therefore the frequency (f) for the sequence ticket is 1 because it was picked by one participant (Participant 6 in Table 2.1). Similarly, the tick marks for each of the other three types of tickets can be recorded and counted until all of the data in the sample have been accounted for.

What Percentage of the Sample Has Each Value of the Variable?

Although it is important to determine the frequency for each value of a variable, these frequencies by themselves are not always particularly informative. For example, the statement "There are 35 Democrats in my sample" may be interpreted differently depending on whether the total sample consists of 40 voters versus 400 voters.

Because the interpretation of frequencies depends on the size of the sample, it is useful to calculate the percentage of the sample having each value of a variable. These percentages can be calculated using the following formula:

$$\% = \frac{f}{total\ number\ of\ scores} * 100 \tag{2-1}$$

where f is the frequency for a value of a variable.

From Table 2.3, we found that 1 of the 22 non–problem gamblers chose the sequence ticket. Therefore, the percentage of the sample choosing the sequence ticket is

$$\begin{aligned}\% &= \frac{f}{total\ number\ of\ scores} * 100 \\ &= \frac{1}{22} * 100 = .05 * 100 \\ &= 5\%\end{aligned}$$

After calculating the percentage for each of the four tickets, Table 2.4 provides the frequency distribution table for the lottery ticket variable.

Once a frequency distribution table has been constructed for a variable, researchers can begin to draw conclusions about the data in their sample. For the lottery ticket variable, the random ticket had the highest frequency and percentage (f = 12, 54%) compared to the pattern (f = 6, 27%), nonequilibrated (f = 3, 14%), and sequence (f = 1, 5%) tickets. As a result, the researchers in the lottery study made the following observation:

> The results of the present study indicated that, for the entire sample, the most commonly cited reason for selecting a lottery ticket was perceived randomness. Furthermore, with respect to actual ticket selections, irrespective of explanations, the greatest percentage of tickets selected by the entire sample . . . were random tickets. (Hardoon et al., 2001, p. 760)

TABLE 2.4 ● FREQUENCY DISTRIBUTION TABLE FOR THE LOTTERY TICKET VARIABLE

Lottery Ticket	*f*	%
Sequence	1	**5%**
Pattern	6	**27%**
Nonequilibrated	3	**14%**
Random	12	**54%**
Total	22	**100%**

LEARNING CHECK 1:

Reviewing What You've Learned So Far

1. Review questions
   a. What are some reasons for examining data before conducting statistical analyses?
   b. Why are outliers of concern to researchers?
   c. What questions do you address about a set of data for a variable in constructing frequency distribution tables?
   d. Why is it useful to calculate percentages in addition to frequencies for each value of a variable?
2. For each of the following situations, create a frequency distribution table.
   a. A college counselor asks a group of 116 seniors what their plans are after graduating from college; 71 say they are going to "work," 34 are planning on going to "graduate school," and 11 say they are "not sure" what their plans are.

(Continued)

(Continued)

b. A high school instructor teaching a course finds that of the students she teaches, 69 are Freshmen, 18 are Sophomores, 12 are Juniors, and 9 are Seniors.

c. A pollster stops 12 people and asks if they are either "in favor" or "against" a local proposition:

Person	Position	Person	Position	Person	Position
1	**In favor**	5	**Against**	9	**In favor**
2	**In favor**	6	**Against**	10	**Against**
3	**Against**	7	**In favor**	11	**In favor**
4	**In favor**	8	**Against**	12	**In favor**

d. The company that makes M&Ms candy (www.mms.com) conducted a survey asking people which new color they would like to have added: purple, aqua, or pink. Below are the votes of a hypothetical sample:

Person	Color	Person	Color	Person	Color	Person	Color
1	**Pink**	7	**Purple**	13	**Pink**	19	**Aqua**
2	**Purple**	8	**Aqua**	14	**Purple**	20	**Purple**
3	**Pink**	9	**Purple**	15	**Aqua**	21	**Purple**
4	**Aqua**	10	**Pink**	16	**Pink**	22	**Pink**
5	**Purple**	11	**Aqua**	17	**Purple**	23	**Pink**
6	**Purple**	12	**Pink**	18	**Purple**	24	**Purple**

2.4 GROUPED FREQUENCY DISTRIBUTION TABLES

The frequency distribution table for the lottery ticket variable in Table 2.4 was sufficient for presenting the frequencies for each of the four lottery ticket choices. However, other situations may involve numeric variables with a large number of possible values. Consider, for example, the variable grade point average (GPA). The values for GPA typically range from 0.00 to 4.00, with hundreds of possible values in between. Creating a frequency distribution table for a variable such as GPA would result in an extremely large table, with many values having frequencies of zero.

To examine a variable with a large number of values, it is often useful to construct a **grouped frequency distribution table**, a table that groups the values of a variable measured at the interval or ratio level of measurement into a small number of intervals and then provides the frequency and percentage for each interval. To illustrate how grouped frequency distribution tables are constructed, Table 2.5(a) provides a sample of 42 hypothetical GPAs. In Table 2.5(b), each of these 42 GPAs has been placed into one of eight intervals that comprise a grouped frequency distribution table.

To illustrate how to read and interpret a grouped frequency distribution table, the lowest interval in Table 2.5(b) consists of students with GPAs less than (<) 2.00. The frequency in this interval ($f = 1$) consists of one student: Student 29 (GPA = 1.96). The next interval consists of students with GPAs ranging from 2.00 to 2.29 and contains three students (4, 22, and 38). This grouping continues until all of the GPAs in the sample have been accounted for.

To create a grouped frequency distribution table, the real limits of each interval need to be identified. The **real limits** of an interval are the values of the variable that fall halfway between the top of one interval and the bottom of the next interval. For example, because of how values of GPA are rounded, the lowest value of GPA that would lead someone to be placed in the interval 2.30–2.59 is not 2.30 but rather 2.295, which is halfway between the top of the 2.00–2.29 interval and the bottom of the 2.30–2.59 interval. The **real lower limit** is the smallest value of a variable that would be grouped into a particular interval. In this case, 2.295 represents the real lower limit for the 2.30–2.59 interval. On the other hand, the **real upper limit** is the largest value of a variable that would be grouped into a particular interval. For the 2.30–2.59 interval, the real upper limit is 2.594. This is the real upper limit because it falls halfway between the top of this interval (2.59) and the bottom of the next interval (2.60).

Once the grouped frequency table for the GPA example has been created (Table 2.5(b)), what preliminary conclusions might be drawn about these students? First, because the 2.90–3.19 interval has the greatest frequency ($f = 10$), you might conclude that the typical student in this sample has a GPA close to 3.00. Second, moving in both directions away from the 2.90–3.19 interval, the frequencies of the different intervals grow progressively smaller, with fewer and fewer students receiving either very high or very low GPAs.

Table 2.6 provides a list of guidelines for creating grouped frequency distribution tables. Grouped frequency distribution tables are useful in summarizing variables that have a large number of values. However, by combining values of a variable, these tables provide little detail and specificity about individual values. For example, although Table 2.5(b) reveals that 10 students have GPAs in the 2.90–3.19 interval, it does not provide the exact GPA of any particular student. For example, there is no way of determining whether any student had a GPA of exactly 3.00.

Cumulative Percentages

In addition to determining the percentage of the sample that has each individual value or interval for a variable, it is sometimes useful to combine these percentages. For the GPA example, you may be interested in knowing, "What percentage of the sample had GPAs less than 2.60?" To answer this question, the percentages in the "%" column of the frequency distribution table must be combined. This type of percentage is referred to as a **cumulative percentage**, defined as the percentage of a sample at or below a particular value of a variable.

TABLE 2.5 ● GRADE POINT AVERAGE (GPA), 42 STUDENTS

a. Individual GPAs

Student	GPA	Student	GPA	Student	GPA	Student	GPA
1	**3.05**	12	**3.10**	23	**2.65**	34	**2.71**
2	**2.83**	13	**3.06**	24	**3.40**	35	**3.43**
3	**3.26**	14	**3.83**	25	**2.87**	36	**3.95**
4	**2.19**	15	**2.39**	26	**2.92**	37	**3.32**
5	**3.52**	16	**3.16**	27	**2.89**	38	**2.28**
6	**3.34**	17	**3.92**	28	**3.41**	39	**3.08**
7	**3.02**	18	**2.37**	29	**1.96**	40	**3.25**
8	**2.97**	19	**3.65**	30	**3.77**	41	**2.40**
9	**3.71**	20	**3.70**	31	**3.20**	42	**3.22**
10	**2.50**	21	**3.00**	32	**2.86**		
11	**2.77**	22	**2.26**	33	**2.94**		

(Continued)

TABLE 2.5 ● (CONTINUED)

b. Grouped Frequency Distribution Table

GPA	*f*	%
3.80+	3	7%
3.50–3.79	5	12%
3.20–3.49	9	21%
2.90–3.19	10	24%
2.60–2.89	7	17%
2.30–2.59	4	10%
2.00–2.29	3	7%
< 2.00	1	2%
Total	42	100%

TABLE 2.6 ● GUIDELINES FOR CREATING GROUPED FREQUENCY DISTRIBUTION TABLES

1. *Variables are grouped into approximately 10 intervals.*

   The exact number of intervals depends on the data in each sample. For example, data that fall within a small range of values may be accurately represented with a relatively small number of intervals, whereas data extending across a wide range may require a larger number of intervals.

2. *The number of intervals should accurately represent the data.*

   The intervals should represent the nature of the data as accurately as possible. For example, there should not be so large a number of intervals that many intervals have frequencies of zero (e.g., 3.00–3.05) or so few intervals that a large majority of the sample falls into only one or two intervals (e.g., 3.00–3.99).

3. *Intervals should be of equal size.*

   To make the intervals comparable, they should be of equal size or width. For example, you would not want one interval of 2.01–2.50 (a width of .50) and another of 2.51–2.70 (a width of .20). One exception to this guideline is the lowest or highest interval (e.g., less than 2.00 or greater than 3.70), which is typically wider than the others because it contains a relatively small frequency.

4. *Intervals should not overlap.*

   Each interval should be distinct from the other intervals, such that each score is included in only one interval. For example, you would not want one interval to be 2.30–*2.60* and the next to be *2.60*–2.90 because the GPA 2.60 appears in both intervals. This would lead to confusion regarding which interval to place someone with a GPA of 2.60.

Table 2.7 includes the cumulative percentages for the GPA example. As an example of a cumulative percentage, the cumulative percentage for the 2.30–2.59 interval (19%) has been calculated by combining the percentages for the < 2.00, 2.00–2.29, and 2.30–2.59 intervals (2% + 7% + 10% = 19%). As a result, we could conclude that 19% of this sample had GPAs less than 2.60. The accumulating of percentages continues until all (100%) of the scores in the sample are combined. As shown in Table 2.7, cumulative percentages are typically located in the final column of a frequency distribution table and labeled "Cum %." Although we used a grouped frequency distribution table to illustrate cumulative percentages, cumulative percentages may also be calculated for frequency distribution tables in which the values for the variable are not grouped.

TABLE 2.7 ● FREQUENCY DISTRIBUTION TABLE WITH CUMULATIVE PERCENTAGES FOR THE GPA VARIABLE

GPA	*f*	%	Cum %
3.80+	3	7%	**100%**
3.50–3.79	5	12%	**93%**
3.20–3.49	9	21%	**81%**
2.90–3.19	10	24%	**60%**
2.60–2.89	7	17%	**36%**
2.30–2.59	4	10%	**19%**
2.00–2.29	3	7%	**9%**
< 2.00	1	2%	**2%**
Total	42	100%	

2.5 EXAMINING DATA USING FIGURES

In addition to organizing data into frequency distribution tables, researchers also examine data visually using figures such as charts or graphs. There are four types of figures often used to visually portray data for variables:

- bar charts,
- pie charts,
- histograms, and
- frequency polygons.

Which type of figure a researcher may use depends on the variable's level of measurement (nominal, ordinal, interval, ratio). Data for variables measured at the nominal or ordinal level of measurement are typically displayed using bar charts or pie charts. The values of interval and ratio variables are graphed using histograms and frequency polygons. Each of these types of visual illustration is discussed below.

Displaying Nominal and Ordinal Variables: Bar Charts and Pie Charts

The values of variables measured at the nominal level of measurement differ in category or type; one example of a nominal variable is the lottery ticket variable discussed earlier, which consisted of four categories: sequence, pattern, nonequilibrated, and random. The values of variables measured at the ordinal level of measurement can be placed in an order relative to the other values; age (young, middle age, elderly) is one example of an ordinal variable.

One way to display nominal and ordinal variables is to use a bar chart. A **bar chart** is a figure that uses bars to represent the frequency or percentage of a sample corresponding to each value of a variable, with the different bars not touching each other. Figure 2.1 shows a bar chart for the data from the lottery ticket variable. In this figure, each bar corresponds to one of the four types of lottery tickets; the height of each bar corresponds to the frequency (f) of non–problem gamblers who selected that type of ticket.

FIGURE 2.1 ● BAR CHART SHOWING THE FREQUENCIES OF THE LOTTERY TICKET VARIABLE FOR THE NON–PROBLEM GAMBLERS

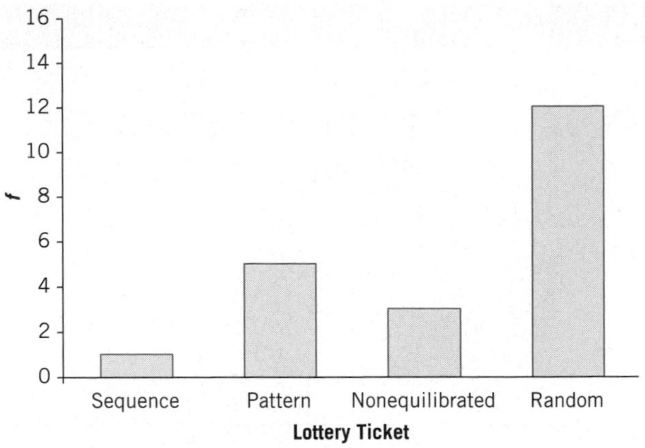

When creating a bar chart, the values of the variable are placed along the horizontal (X) axis and the frequency (f) or percentage (%) for each value is placed along the vertical (Y) axis. Note that the different bars in the bar chart in Figure 2.1 do not touch or intersect. The gap between bars indicates that the values of the variable represent distinct groups or categories that cannot be connected together along a numeric continuum.

A second type of figure used to display nominal or ordinal variables is a pie chart. A **pie chart** is a figure that uses a circle divided into proportions to represent the percentage of the sample for each value of a variable. Pie charts are created by dividing the 360 degrees of a circle into pieces (or proportions) corresponding to the percentage of each value.

Figure 2.2 shows a pie chart for the data for the lottery ticket variable. To illustrate how pie charts are constructed, the frequency distribution table in Table 2.4 shows that 5% of non–problem gamblers chose the sequence ticket. Given that a pie chart is a circle consisting of 360 degrees, 5% of the 360 degrees are needed to represent the sequence ticket. As 5% * 360 = 18, 18 degrees of the pie is assigned to the sequence ticket. Similarly, 97 degrees of the pie chart (27% of 360) are used to represent the pattern ticket. This process continues until all 360 degrees of the pie chart are accounted for.

Displaying Interval and Ratio Variables: Histograms and Frequency Polygons

Interval and ratio variables have values equally spaced along a numeric continuum and are identical to each other with one exception: Ratio scales possess a true zero point, for which the value of zero represents the complete absence of the variable. Examples of ratio variables include weight (measured in pounds) and speed (measured in miles per hour). The values of interval and ratio variables may be illustrated by using histograms and frequency polygons, the choice of which depends on the number of values for the variable.

A **histogram** is a figure in which bars are used to represent the frequency of each value of a variable, with the different bars touching each other. The values of the variable are located on the X-axis, and the frequency or percentage for each value is on the Y-axis. Histograms are similar to bar charts, with one important difference: The bars in a histogram touch each other, indicating that the values of the variable are connected to each other along a numeric continuum.

Imagine a teacher is interested in studying aggressive behavior in children. She decides to measure this variable by counting the number of fights each of her students is involved in during a 3-day period. "Number of fights" is a ratio variable with a small number of values; it is a ratio rather than interval variable because the

FIGURE 2.2 ● PIE CHART SHOWING THE PERCENTAGES OF THE LOTTERY TICKET VARIABLE FOR THE NON-PROBLEM GAMBLERS

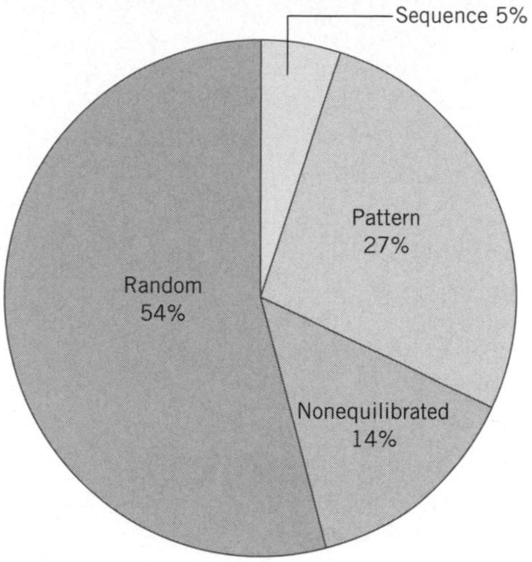

value of 0 represents the complete absence of fights. Figure 2.3 provides a frequency distribution table and histogram for this variable.

A histogram is used when an interval or ratio variable consists of a relatively small number of values. But what if, for example, the number of fights could be anywhere from 0 to 20 rather than 0 to 5? In order to have the histogram fit on one page, the 21 bars would need to be very thin and crowded together, making the histogram confusing and difficult to read. In situations such as this, the values of the variable may be more clearly illustrated using a figure known as a frequency polygon. A **frequency polygon** is a line graph that uses data points to represent the frequency of each value of a variable, with lines connecting the data points. In a frequency polygon, the X-axis represents the values of the variable, while the Y-axis represents the frequency for each value.

FIGURE 2.3 ● FREQUENCY DISTRIBUTION TABLE AND HISTOGRAM FOR THE NUMBER OF FIGHTS VARIABLE

# Fights	*f*
4	1
3	2
2	5
1	1
0	3
Total	12

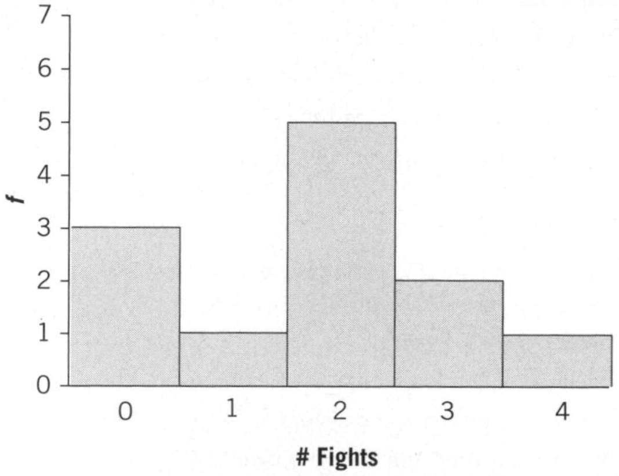

FIGURE 2.4 ● FREQUENCY POLYGON SHOWING THE FREQUENCY OF STUDENTS IN EACH INTERVAL OF THE GRADE POINT AVERAGE (GPA) VARIABLE

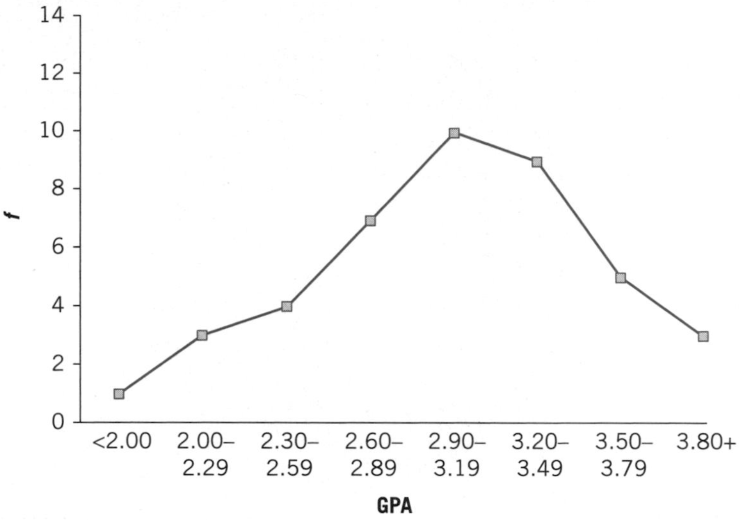

Figure 2.4 shows a frequency polygon based on the grouped frequency distribution table of the earlier example of students' GPA (Table 2.5(b)). In this frequency polygon, the data points representing the frequencies of the eight GPA intervals are connected using straight, sharp lines. These lines indicate that the values of the variable can be connected together along a numeric continuum.

Although a frequency polygon like the one in Figure 2.4 may look jagged as it moves from one data point to the next, it provides a general sense of the shape of the distribution of scores for a variable. For example, looking at Figure 2.4, you might conclude that the majority of GPAs in this sample are in the middle of the 1.00 to 4.00 range, with relatively low frequencies of students with either low or high GPAs.

LEARNING CHECK 2:

Reviewing What You've Learned So Far

1. Review questions
   a. What is the difference between a frequency distribution table and a grouped frequency distribution table?
   b. For what levels of measurement would you create a pie chart, bar chart, histogram, or a frequency polygon?
   c. What is the difference between a bar chart and a histogram?
   d. Why would it be inappropriate to create a histogram or frequency polygon for a nominal or ordinal variable?
   e. Under what circumstances might you use a frequency polygon rather than a histogram to graph an interval or ratio variable?
2. For each of the following variables, identify whether you could use a bar chart, pie chart, histogram, and/or a frequency polygon to visually display the data.
   a. Miles per gallon (MPG)
   b. Living situation (on-campus, off-campus)
   c. Number of brothers and sisters
   d. Marital status (single, married, divorced)

Drawing Inappropriate Conclusions From Figures

A well-constructed pie chart, bar chart, histogram, or frequency polygon can greatly facilitate the understanding of complex information. Conversely, a poorly designed figure may not only confuse the reader, it can lead to inappropriate or misleading conclusions. Table 2.8 provides guidelines for constructing figures based on recommendations from the American Psychological Association (APA).

A simple example may demonstrate how altering the features of a figure can lead to different interpretations of the same set of data. Imagine a pollster assesses how voters with different political affiliations plan to vote on a ballot measure. She asks a sample of Democrats, Republicans, and Independents the question, "Do you plan on voting in favor of Measure A?" In examining her data, the pollster finds 56% of Democrats, 47% of Republicans, and 52% of Independents intend to vote "yes" on the measure.

Figure 2.5 graphs these three percentages using two bar charts; bar charts are used to display the data because the variable of interest (political affiliation) is a nominal variable (Democrat, Republican, or Independent). The bars in each chart do not touch each other because the three political affiliations are distinct from each other and cannot be placed along a numeric continuum.

Although the basic structure of the two bar charts is similar, there is one important difference in how they are constructed. Following the guidelines in Table 2.8, the bar chart in Figure 2.5(a) starts the Y-axis with the value 0% and ends with 100%. The bar chart in Figure 2.5(b), however, starts its Y-axis with the value of 45% and ends with 65%. By limiting the range of values for the percentages, the bar chart in Figure 2.5(b) exaggerates differences in the heights of the three bars. As a result, someone looking at Figure 2.5(b) may interpret the differences between the three political affiliations as being larger than someone looking at Figure 2.5(a).

The saying "a picture is worth a thousand words" is particularly relevant for the presentation of statistical data. Figures provide a vivid representation of data. It is the responsibility of researchers to portray their data in an accurate manner and to avoid creating distorting images that may mislead or confuse the reader.

SUMMARY BOX 2.1 SOME APA GUIDELINES FOR CREATING FIGURES

1. The figure should be formatted in a portrait rather than landscape page orientation (the reader should not turn the page sideways to read the table).
2. The figure is numbered using Arabic numbers (Figure 1, not Figure *I* or Figure *A*).
3. The figure includes a title that describes the contents of the figure. For example, it is not enough to name a figure simply "Figure 2.1" or "Figure 2.1. Bar chart."
4. Figures have a rectangular shape, with the height of the vertical (Y) axis two thirds to three fourths the length of the horizontal (X) axis.
5. The horizontal (X) axis contains the values of the variable; the vertical (Y) axis contains the frequency (*f*) or percentage (%) of the sample having each value of the variable.
6. The X- and Y-axes include descriptive labels written in both upper- and lowercase letters (as opposed to all capital letters).
7. The labels for the X- and Y-axes are short enough to fit on one line along the axes.
8. The label of the Y-axis is placed parallel to the Y-axis (the letters face sideways rather than vertically).
9. The Y-axis starts with the value zero (0) and is divided into equally spaced values or intervals.
10. The Y-axis is divided into 6 to 10 values (just enough values to accommodate bar length). The bars should neither be flattened along the bottom of the Y-axis nor bumped up against the top of the Y-axis.

FIGURE 2.5 ● DIFFERENT DISPLAYS OF THE SAME DATA

Political Affiliation	% "Yes"
Democrat	56%
Republican	47%
Independent	52%

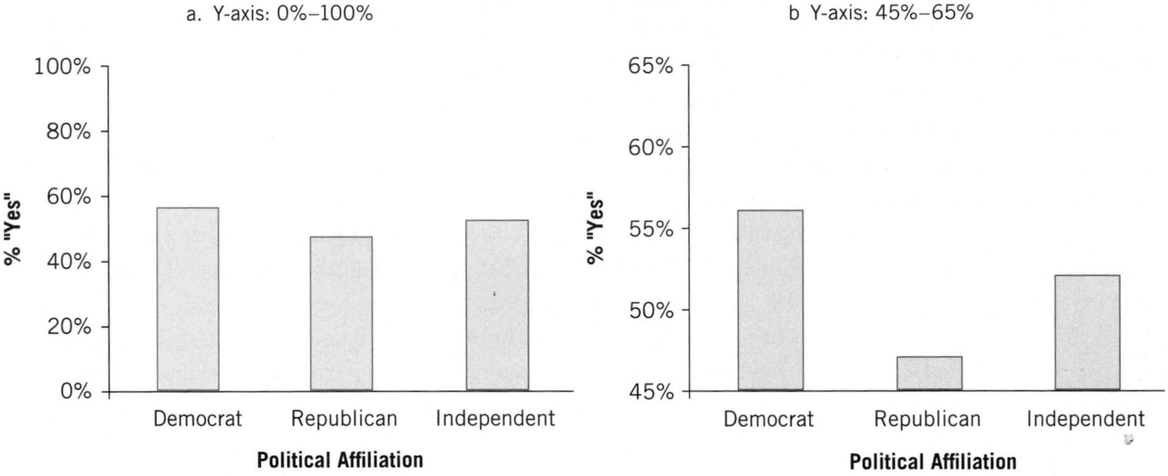

2.6 EXAMINING DATA: DESCRIBING DISTRIBUTIONS

We have discussed how researchers construct tables and figures to organize and examine data they have collected. Looking at a table or figure enables researchers to describe the distribution of values or scores for a variable. Describing a distribution involves addressing questions such as, "Which values of the variable are the most common or occur most often?" "How do the frequencies of the values of the variable change in relation to the most common values?", and "What is the amount or nature of differences among the different values?" These questions pertain to three aspects of distributions:

- modality,
- symmetry, and
- variability.

These three aspects of distributions are introduced below; they will also be discussed in greater detail in the next two chapters of this book.

Modality

The first feature of a distribution of data for a variable is its modality. The **modality** of a distribution refers to the value or values of the variable that have the highest frequency or occur most often in a set of data. Using the lottery study as an example, identifying the modality of the lottery ticket variable involves answering the question, "Which of the four lottery tickets was chosen the most often?"

The modality of a distribution can be identified by looking at a frequency distribution table; the modality is the value(s) of the variable with the highest frequency (*f*). The modality may also be determined by examining a figure and locating the tallest bar in a bar chart or histogram, the highest point in a frequency polygon, or the biggest "slice" in a pie chart.

In identifying the modality of a distribution, a researcher may discover that one value or a limited number of values adjacent to one another clearly have the highest frequency. In such cases, the distribution is referred to as a unimodal distribution. A **unimodal distribution** is a distribution where one value occurs with the greatest frequency. Looking at the frequency distribution table (Table 2.4) or figure (Figure 2.1) for the lottery ticket variable, this distribution would be considered unimodal as one value, the random ticket, clearly occurred more often than the other three ticket choices.

In contrast to a unimodal distribution, it is possible that two values may both have relatively high frequencies; if so, the distribution is referred to as a bimodal distribution. A **bimodal distribution** is a distribution where two values clearly distinct from each other have the greatest frequency. A bimodal distribution sometimes resembles the two humps on a camel's back. As an example of a bimodal distribution, in February 2017, the Pew Research Center (www.pewresearch.org) asked a sample of 1,500 voters the extent to which they approved of newly elected Donald Trump, with five responses ranging from "strongly disapprove" to "strongly approve." Most voters in this sample either strongly approved (38%) or strongly disapproved (34%), with only 7% choosing the middle response of "don't know."

And, in addition to unimodal and bimodal distributions, it's also possible for a distribution to have more than two values with relatively equally high frequencies; this type of distribution is known as a **multimodal distribution**—a distribution where more than two values have the greatest frequency. In examining people's attitudes toward gun control, in January 2016, CBS News and the *New York Times* asked 1,276 people the question, "How much do you think stricter gun laws would do to help prevent gun violence?" They found people differed greatly in their answers to this question ("a lot" [25%], "some" [28%], "not much" [16%], "not at all" [27%]), with none of the choices being selected more than the others. Figure 2.6 presents an example of a unimodal, bimodal, and multimodal distribution.

Symmetry

The second feature of a distribution of data is its symmetry. The **symmetry** of a distribution refers to how the frequencies of values of a variable change in relation to the most common or frequently occurring values. Describing the symmetry of a distribution means answering the question, "What is the shape of the distribution?"

The symmetry of a distribution can be determined by looking at the frequencies in a frequency distribution table or the bars, points, or pieces of a figure. By doing so, a distribution may be described as being either symmetric or asymmetric. A **symmetric distribution** is one in which the frequencies change in a similar manner moving away in both directions from the most frequently occurring values. An example of a symmetric

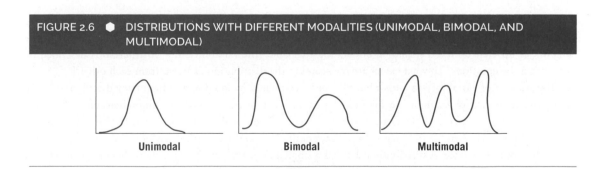

FIGURE 2.6 ● DISTRIBUTIONS WITH DIFFERENT MODALITIES (UNIMODAL, BIMODAL, AND MULTIMODAL)

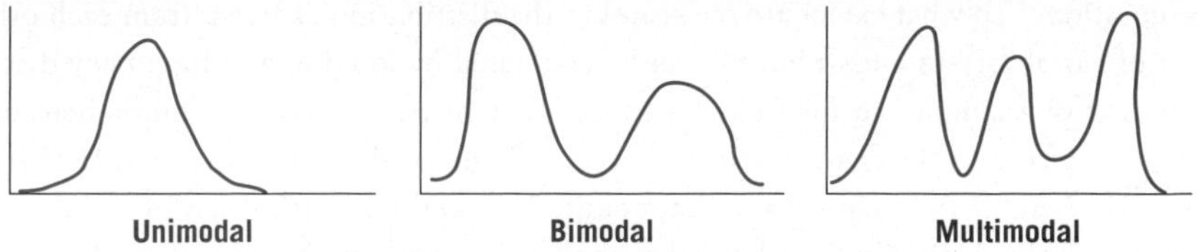

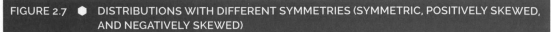

FIGURE 2.7 ● DISTRIBUTIONS WITH DIFFERENT SYMMETRIES (SYMMETRIC, POSITIVELY SKEWED, AND NEGATIVELY SKEWED)

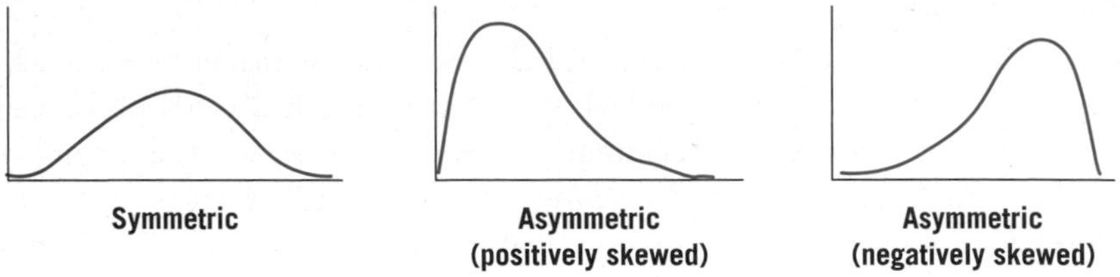

distribution is illustrated in Figure 2.7. One way to determine whether a distribution is symmetric is to draw the distribution on a piece of paper and then fold the distribution vertically down the middle so that the two halves are on top of each other. A distribution is symmetric when the two halves are similar in shape such that they resemble mirror images of each other.

However, a distribution may also be asymmetric. An **asymmetric distribution** (also known as a skewed distribution) is one in which the frequencies change in a different manner moving away in both directions from the most frequently occurring values. In asymmetric distributions, the highest frequencies are located at one end of the distribution rather than in the middle, such that if you fold an asymmetric distribution down the middle, the shapes of the two halves are not the same and do not resemble each other.

In an asymmetric distribution, the frequencies take on the shape of a long tail as the values move from the values with the highest frequencies toward the other end of the distribution. Given that the majority of scores are at one end of the distribution, scores in the tail of the distribution are often considered to be outliers. For example, in the 2016 Current Population Survey, the U.S. Census Bureau (www.census.gov) estimated that most Americans (75%) had annual incomes less than $50,000. However, a very small percentage (less than 2%) had annual incomes greater than $200,000; those with these higher incomes would be considered outliers.

Depending on the location of this tail, asymmetric distributions are referred to either as positively skewed or negatively skewed. A **positively skewed distribution** is one in which the higher frequencies are at the lower end of the distribution, with the tail on the upper (right) end of the distribution; a **negatively skewed distribution** is one in which the higher frequencies are at the upper end of the distribution, with the tail on the lower (left) end of the distribution. Figure 2.7 provides an illustration of a symmetric distribution and two asymmetric distributions (one positively skewed and one negatively skewed). Annual income, illustrated by the U.S. Census Bureau example, is an example of a positively skewed distribution.

Variability

The third way in which distributions may be described is in terms of the amount or nature of differences among the different values for a variable—what is known as variability. The **variability** of a distribution refers to the amount of differences in a distribution of data for a variable. The variability in a distribution is described by answering the question, "To what extent are the scores in the distribution different from each other?"

The amount of variability in a distribution can be estimated by looking at a frequency distribution table or figure. Imagine, for example, an instructor gives an 11-item test of reading comprehension to her class and finds almost all of the students answered six, seven, or eight of the items correctly. In this situation, the distribution would resemble the shape of a tall mountain: thin and tall, with many of the scores residing in a few values of the variable. A distribution that has a small amount of variability such as this one is referred to

FIGURE 2.8 ● DISTRIBUTIONS WITH DIFFERENT VARIABILITY (PEAKED, FLAT, AND NORMALLY DISTRIBUTED)

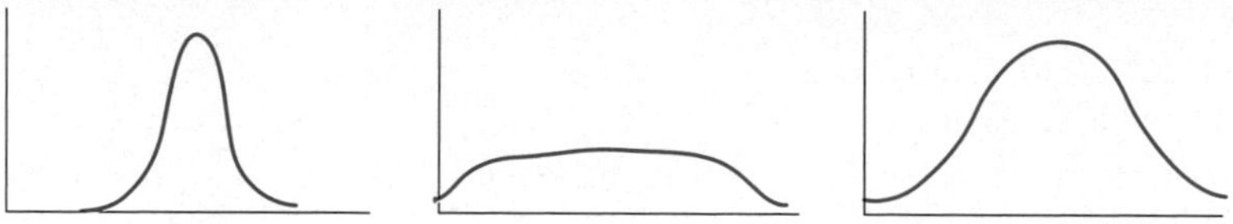

as a **peaked distribution**, defined as a distribution where much of the data are in a small number of values of a variable.

On the other hand, what would the distribution have looked like if she had found her students' scores were distributed relatively evenly across the entire range of possible scores? In this situation, the distribution would not be tall and thin like a mountain but rather would be short and wide like a plateau. This type of distribution is referred to as a **flat distribution**, which is a distribution where the data are spread evenly across the values of a variable. The CBS News/*New York Times* gun control example described earlier is an example of a flat distribution.

In discussing modality, symmetry, and variability, distributions may have more than one mode, may be asymmetric, and may be peaked or flat. But what if a distribution has none of these qualities? A distribution that is in essence unimodal, symmetric, and neither peaked nor flat may be described as "bell shaped." If you look at a bell, you will notice that it is the highest in the middle, and as you move away from the top of a bell in both the left and right directions, the shape of the bell slopes downward in the same manner. A bell-shaped distribution for a variable is given a special name: a normally distributed variable.

A **normally distributed variable** is a distribution for a variable that is considered unimodal, symmetric, and neither peaked nor flat. Note that the word *normal* is a statistical concept and not a value judgment. In other words, asymmetric, peaked, or flat distributions are not considered "abnormal." The concept of normal distributions will be discussed in greater detail in Chapter 5. Figure 2.8 provides an illustration of three distributions with different amounts of variability: peaked, flat, and normally distributed.

LEARNING CHECK 3:

Reviewing What You've Learned So Far

1. Review questions
   a. What are the three aspects used to describe distributions of variables?
   b. What is the difference between unimodal, bimodal, and multimodal distributions?
   c. What is the difference between symmetric and asymmetric (skewed) distributions?
   d. What is the difference between positively and negatively skewed distributions?
   e. What is the difference between peaked and flat distributions?
   f. What are the characteristics of a normally distributed variable?

(Continued)

(Continued)

2. For each of the following sets of data, determine whether the distribution of scores is unimodal, bimodal, or multimodal.
   a. 4, 2, 11, 6, 10, 2, 6, 13, 2, 10, 6, 10, 1
   b. 15, 17, 13, 18, 17, 14, 20, 18, 17, 14, 17
   c. 20, 10, 50, 10, 80, 30, 70, 10, 60, 10, 80, 0, 30, 80, 10, 100, 80
3. For each of the following sets of data, determine whether the distribution of scores is symmetric or asymmetric (skewed).
   a. 4, 5, 1, 4, 2, 4, 3, 5, 3, 4, 4, 3
   b. 220, 230, 250, 230, 200, 240, 210, 230, 270, 220, 240, 230, 240, 230
   c. 3, 1, 9, 2, 3, 6, 2, 7, 3, 5, 2, 6, 3, 2, 2
4. For each of the following sets of data, determine whether the distribution of scores is peaked or flat.
   a. 16, 10, 19, 12, 13, 15, 14, 17, 15, 20, 11, 13, 19, 16, 15
   b. 12, 15, 12, 11, 13, 11, 12, 12, 14, 10, 12, 11, 16, 12, 13, 12, 11
   c. 14, 12, 16, 14, 13, 15, 12, 14, 14, 15, 14, 14, 13, 14

2.7 LOOKING AHEAD

This chapter has discussed why and how researchers examine data. Examining data by creating tables and figures requires a number of different skills, including organizational skills (the ability to take a set of data and prepare it for statistical analysis), communication skills (the ability to present data in tabular or graphic form), and conceptual skills (the ability to describe a distribution). The next two chapters begin to focus on mathematical skills related to the statistical analysis of data, describing how researchers examine and describe data and distributions numerically rather than visually. Given that this is a statistics textbook, we will focus on calculating statistics to describe and analyze data. However, and we cannot say this strongly enough, it is *extremely* important to examine a set of data before conducting statistical analyses as this will prevent errors both in your calculations of statistics as well as your interpretation of the results of these calculations.

2.8 Summary

Before conducting statistical analyses, researchers typically examine their data to gain an initial sense of the data, detect data coding or data entry errors, identify ***outliers*** (rare, extreme scores that lie outside of the range of the majority of scores in a set of data), evaluate research methodology, and determine whether the data meet statistical criteria and assumptions.

One common way of examining variables involves creating tables known as ***frequency distribution tables*** (tables that summarize the number and percentage of participants that express the different values for a variable) and ***grouped frequency distribution tables*** (tables that group the values of a variable measured at the interval or ratio level of measurement into a small number of intervals and then provide the frequency and percentage within each interval). To create a grouped frequency table, the ***real limits*** of each interval (the values of the variable that fall halfway between the top of one interval and the bottom of the next interval) need to be identified.

Data for a variable may also be examined by creating figures such as bar charts, pie charts, histograms, and frequency polygons. ***Bar charts*** and ***pie charts*** are used for variables measured at the nominal or ordinal level of measurement; a bar chart uses bars to represent the frequency or percentage of a sample corresponding to each value of a variable, with the different bars not touching each other; a pie chart is a circle divided into proportions that represent the percentage of the sample corresponding to each value of a variable. ***Histograms*** and ***frequency polygons*** are used for variables measured at the interval or ratio level of measurement; a histogram is a figure in which bars are used to represent the frequency of each value of a variable, with the different bars touching each other; a frequency polygon is a line graph that uses data points to represent the frequency of each value of a variable, with lines connecting the data points.

From examining their data, researchers can describe the distribution of values or scores for a variable. There are three main aspects of distributions: modality, symmetry, and variability.

The ***modality*** of a distribution is the value(s) of the variable that have the highest frequencies or occur most often in a set of data. ***Unimodal distributions*** are distributions where one value occurs with the greatest frequency, ***bimodal distributions*** have two values with the greatest frequency, and ***multimodal distributions*** have more than two values with the greatest frequencies.

The ***symmetry*** of a distribution refers to how the frequencies of values of a variable change in relation to the most common or frequently occurring values of the variable. In ***symmetric distributions***, the frequencies change in a similar manner, moving away in both directions from the most common values of the variable. In ***asymmetric (skewed) distributions***, the frequencies change in a different manner, moving away in both directions from the most common values of the variable.

The ***variability*** of a distribution refers to the amount of differences in a distribution of data. In ***peaked distributions***, much of the data are in a few values of a variable; in ***flat distributions***, the data are spread evenly across the values of a variable. A ***normally distributed variable*** has a symmetric distribution that is neither peaked nor flat.

2.9 Important Terms

outlier (p. 22)
frequency distribution table (p. 23)
frequency (f) (p. 23)
grouped frequency distribution table (p. 26)
real limits (p. 27)
real lower limit (p. 27)
real upper limit (p. 27)
cumulative percentage (p. 27)
bar chart (p. 29)
pie chart (p. 30)
histogram (p. 30)
frequency polygon (p. 31)
modality (p. 34)
unimodal distribution (p. 35)
bimodal distribution (p. 35)
multimodal distribution (p. 35)
symmetry (p. 35)
symmetric distribution (p. 35)
asymmetric (skewed) distribution (p. 36)
positively skewed distribution (p. 36)
negatively skewed distribution (p. 36)
variability (p. 36)
peaked distribution (p. 37)
flat distribution (p. 37)
normally distributed variable (p. 37)

2.10 Formula Introduced in This Chapter

Percentage (%)

$$\% = \frac{f}{total\ number\ of\ scores} * 100 \tag{2-1}$$

2.11 Using IBM SPSS Software*

Creating Data Files: Defining Variables and Entering Data

1. Open a new data file.

*IBM SPSS software. SPSS is a registered trademark of International Business Machines Corporation.

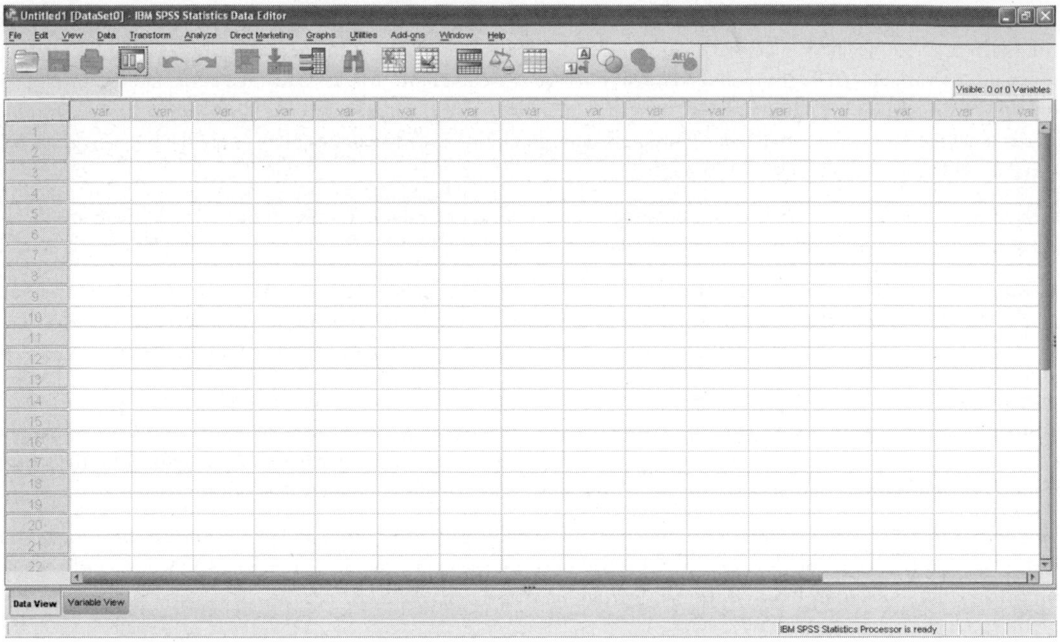

2. Define and name variable(s).

   How? (1) Click **Variable View** tab, (2) click below **Name**, and (3) type name of variable in the cell.

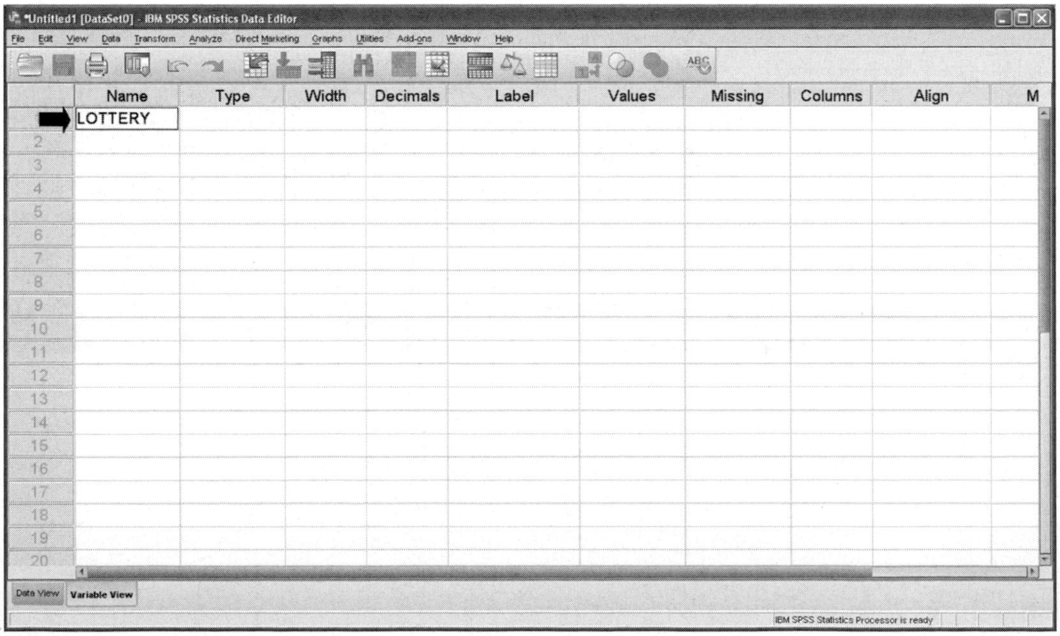

3. Determine number of decimals for your variable(s).

How? Click desired number of decimal places in the **Decimals** box.

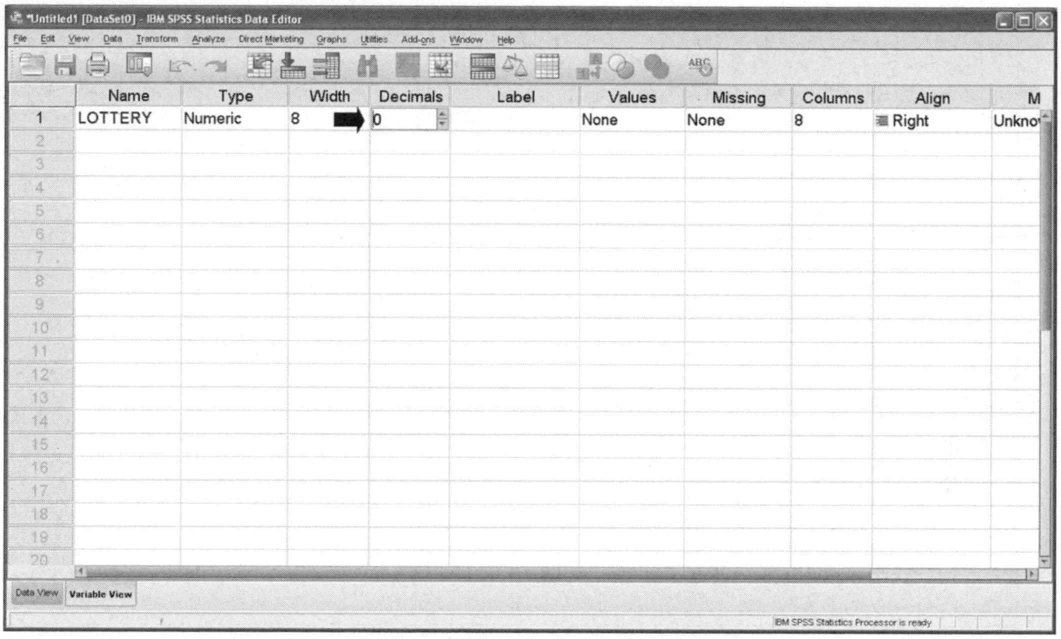

4. Provide labels for your variable(s).

How? (1) Click below **Label**, and (2) type the label for the variable in the cell.

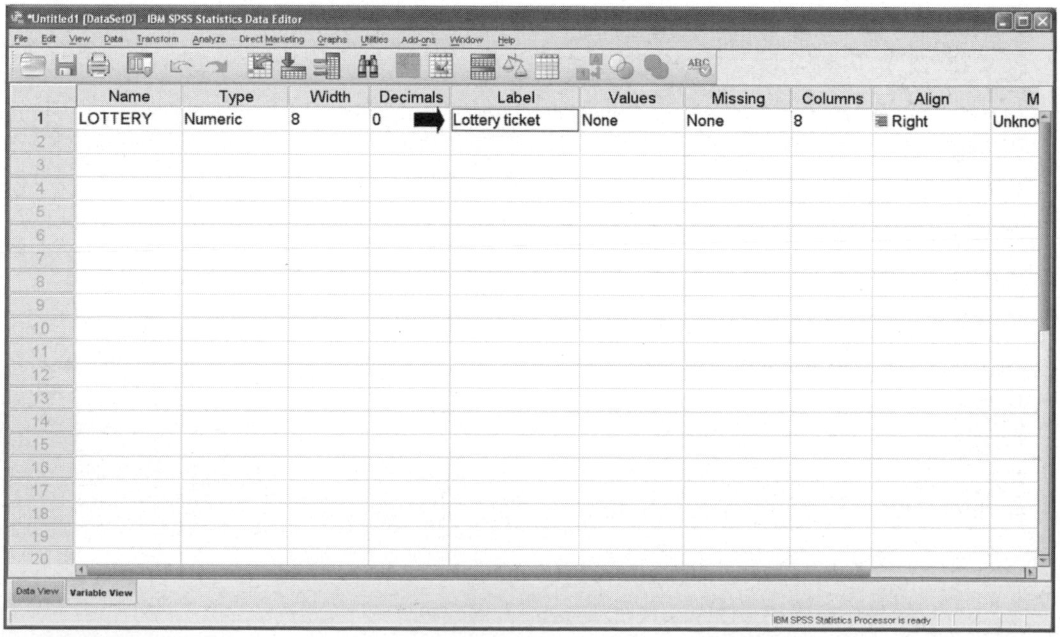

5. Provide labels for values of nominal (categorical) variable(s).

   How? (1) Click below **Values**, and (2) type labels for coded values of the variable.

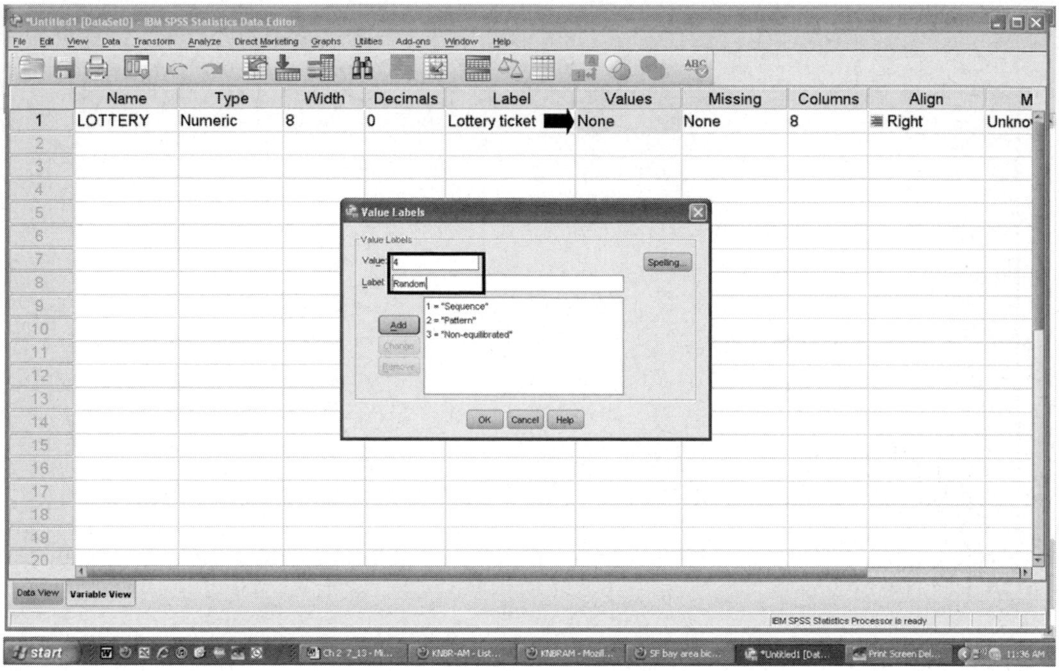

6. Enter data.

   How? (1) Click **Data View** tab and (2) enter data for the variable in the cells.

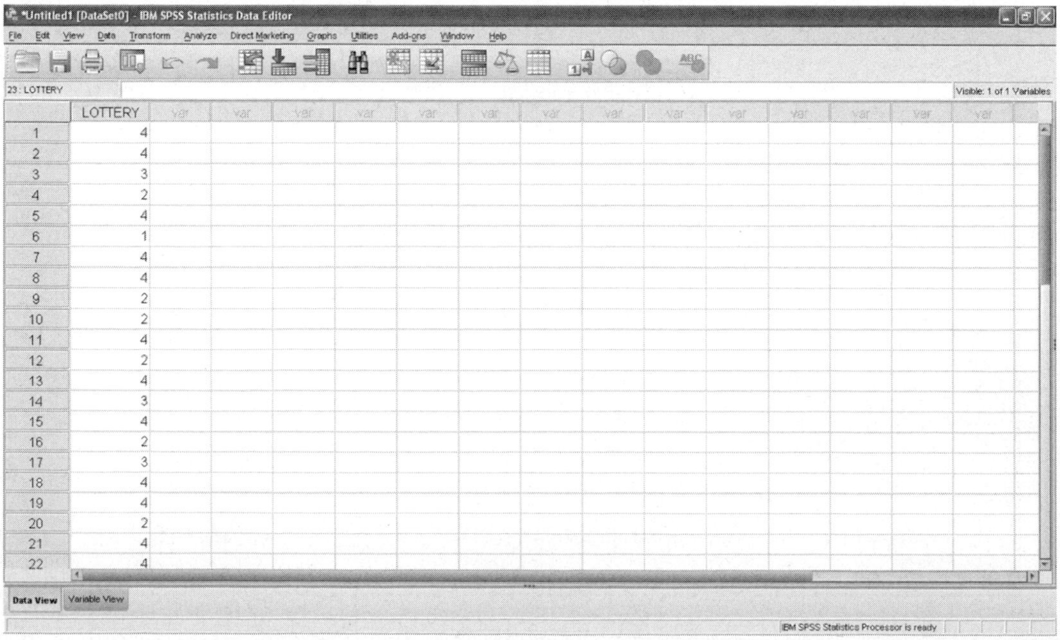

Examining Data: Frequency Distribution Tables and Figures: The Lottery Study (2.1)

1. Select the frequency distribution procedure within SPSS.

   How? (1) Click **Analyze** menu, (2) click **Descriptive Statistics**, and (3) click **Frequencies**.

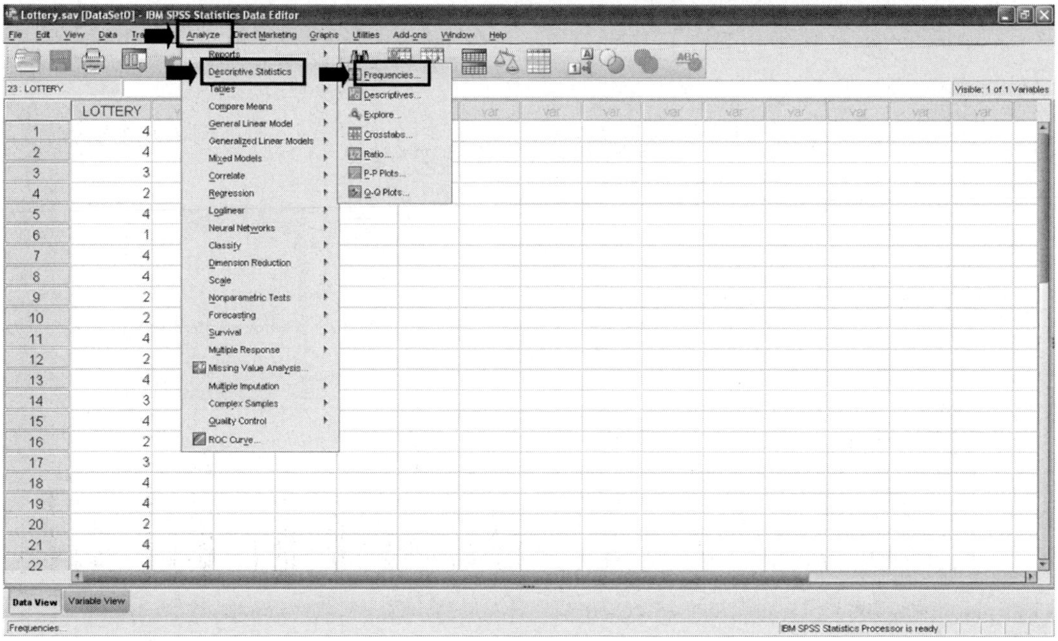

2. Select the variable to be analyzed.

   How? (1) Click variable and→.

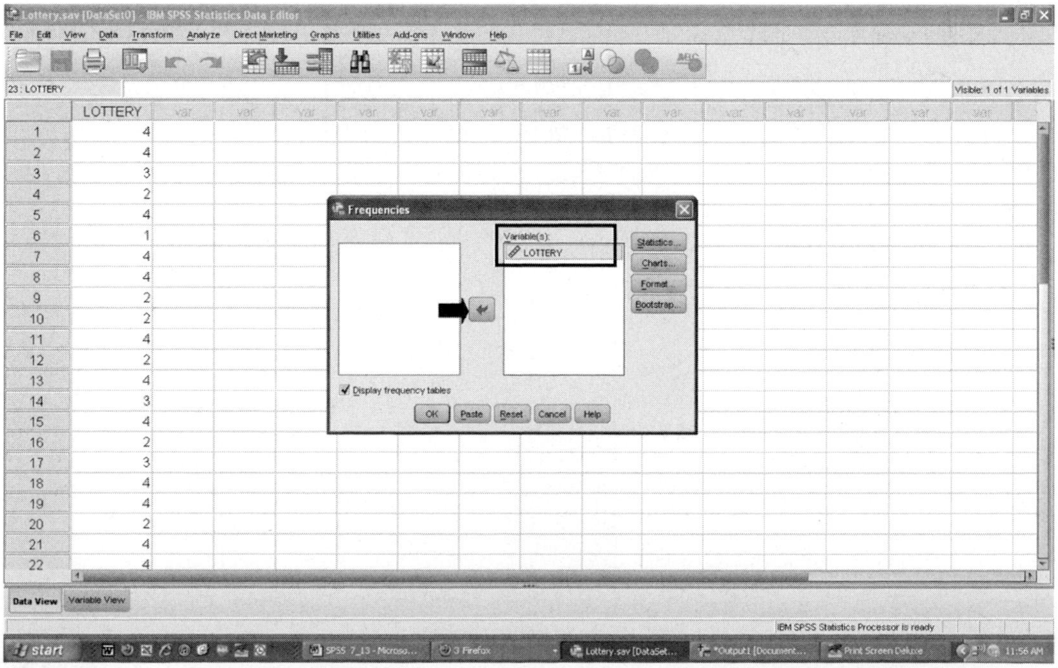

3. Select the appropriate type of figure for the variable.

How? (1) Click **Charts**, (2) select the desired type of figure, (3) click **Continue**, and (4) click **OK**.

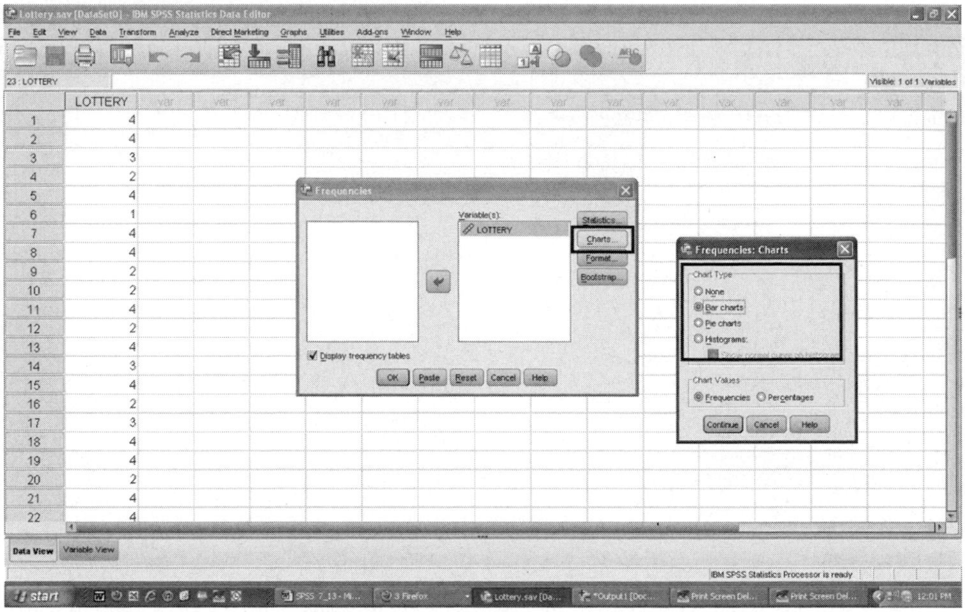

4. Examine output.

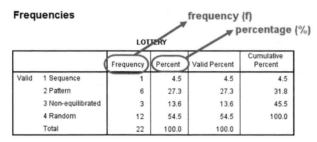

Frequencies

frequency (f)

percentage (%)

LOTTERY

		Frequency	Percent	Valid Percent	Cumulative Percent
Valid	1 Sequence	1	4.5	4.5	4.5
	2 Pattern	6	27.3	27.3	31.8
	3 Non-equilibrated	3	13.6	13.6	45.5
	4 Random	12	54.5	54.5	100.0
	Total	22	100.0	100.0	

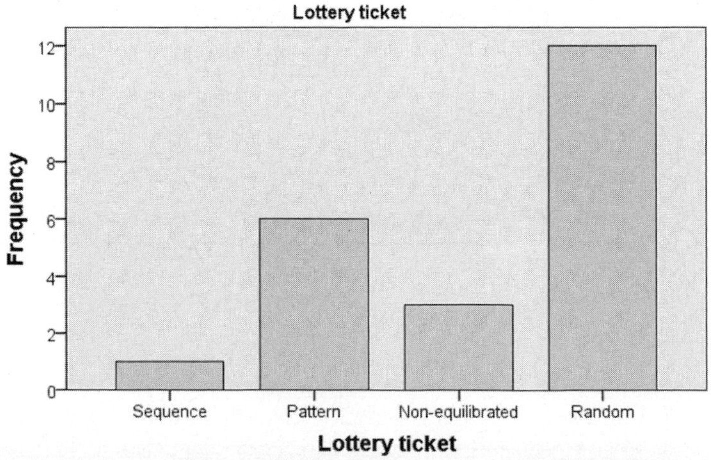

2.12 Exercises

1. For each of the following variables, identify whether you could use a pie chart, bar chart, histogram, or frequency polygon to visually display the data:
   a. Gender (male, female)
   b. Weight (in pounds)
   c. Number of children in families
   d. Baseball batting averages
   e. Eye color
   f. Temperature

2. For each of the following variables, identify whether you could use a pie chart, bar chart, histogram, or frequency polygon to visually display the data:
   a. Shoe size
   b. College major
   c. Favorite radio station
   d. Midterm exam score
   e. Zip code
   f. Marital status

3. The owner of an ice cream store asks 75 people which flavor of ice cream they prefer. Thirteen of them say strawberry, 11 say chocolate, 24 say vanilla, and 27 provide a flavor other than strawberry, chocolate, or vanilla. Create a frequency distribution table for these data.

4. A political pollster approaches people on the street and asks them to describe their political affiliation. Twenty-eight people describe themselves as Democrats and 25 as Republicans, 8 people provide a political party other than Democrat or Republican, 13 label themselves as Independent, and 10 people say they do not belong to a political party. Create a frequency distribution table for these data.

5. Below is a frequency distribution table for a hypothetical variable:

Value	*f*	%
100	3	10%
90	8	26%
80	5	17%
70	6	20%
60	3	10%
50	2	7%
40	0	0%
30	2	7%
20	1	3%
10	0	0%
Total	30	100%

   a. How many of the scores for this variable have the value 70?
   b. What percentage of the scores has the value of 30?

6. Below is a frequency distribution table for a hypothetical variable:
   a. How many of the scores for this variable have the value 2?
   b. What percentage of the scores has the value of 7?

Value	*f*	%
7	5	15%
6	8	24%
5	10	29%
4	6	18%
3	3	9%
2	2	5%
1	0	0%
Total	34	100%

(Exercises 7–10 pertain to the following situation.) You're standing in line to see a movie with two of your friends (who also happen to be taking a statistics class). You are not sure whether you want to see this particular movie, so you decide to stop people coming out of the theater and ask them for their opinion about the movie.

7. You stop 10 people and ask them whether they would or would not recommend the movie to others. They give you the following answers:

Person	Recommend?	Person	Recommend?
1	**No**	6	**No**
2	**Yes**	7	**Yes**
3	**No**	8	**Yes**
4	**No**	9	**No**
5	**Yes**	10	**No**

   a. Construct a frequency distribution table of these data (be sure to include columns representing the frequency and percentages).
   b. What level of measurement is this variable?
   c. Construct an appropriate figure for these data.

8. One of your friends feels the responses of your 10 people did not result in a clear recommendation or nonrecommendation of the movie. She decides to ask 15 people to give the movie one of three ratings: above average, average, or below average. Their ratings are listed below:

Person	Rating	Person	Rating	Person	Rating
1	**Average**	6	**Above average**	11	**Average**
2	**Below average**	7	**Average**	12	**Above average**
3	**Average**	8	**Average**	13	**Average**
4	**Above average**	9	**Average**	14	**Average**
5	**Average**	10	**Above average**	15	**Below average**

a. Construct a frequency distribution table of these data.

b. What level of measurement is this variable?

c. Construct an appropriate figure for these data.

9. Your other friend asks 20 people to rate the movie using a 1- to 5-star rating: the higher the number of stars, the higher the recommendation. Their ratings are listed below:

Person	# Stars	Person	# Stars	Person	# Stars
1	★★★	8	★★	15	★★★
2	★★★★★	9	★★★	16	★★
3	★★	10	★★★★★	17	★★★
4	★★★★	11	★	18	★★
5	★★★	12	★★★★	19	★★★★
6	★★★	13	★★★★	20	★★★
7	★★★★	14	★★★		

a. Construct a frequency distribution table of these data.

b. What level of measurement is this variable?

c. Construct an appropriate figure for these data.

10. Another friend asks 30 people to rate the movie by their likelihood of seeing the movie again, ranging from 0% to 100%. Their ratings are listed below:

Person	Likelihood	Person	Likelihood	Person	Likelihood
1	**50%**	11	**99%**	21	**60%**
2	**10%**	12	**50%**	22	**50%**
3	**66%**	13	**35%**	23	**80%**
4	**60%**	14	**75%**	24	**70%**
5	**50%**	15	**60%**	25	**50%**
6	**90%**	16	**25%**	26	**65%**
7	**33%**	17	**75%**	27	**20%**
8	**95%**	18	**90%**	28	**30%**
9	**75%**	19	**30%**	29	**70%**
10	**5%**	20	**33%**	30	**33%**

a. Construct a grouped frequency distribution table of these data, with intervals of 10 (0%–9%, 10%–19%, 20%–29%, etc.).

b. What level of measurement is this variable?

c. Construct an appropriate figure for these data.

d. Construct a grouped distribution of these data with intervals of 33 (0%–33%, 34%–66%, and 67%–100%). How does this table compare with one having intervals of 10?

11. Given the growing popularity of video games among young people, one study examined how women are portrayed in these games (Dietz, 1998). Dr. Dietz was particularly interested in whether video games maintain gender role stereotypes or contain violence against women. She viewed 25 video games with human characters popular at that time and defined the primary role of women in each of them. Each video game was placed into one of five groups: no female characters, female characters portrayed as sex objects, females as the victim, females as the hero, or females in traditional feminine roles.

Video Game	Role of Women	Video Game	Role of Women	Video Game	Role of Women
1	**Victim**	10	**Hero**	19	**Hero**
2	**No females**	11	**Sex object**	20	**Feminine**
3	**No females**	12	**No females**	21	**No females**
4	**Victim**	13	**No females**	22	**Victim**
5	**No females**	14	**No females**	23	**Sex object**
6	**Sex object**	15	**No females**	24	**Hero**
7	**Sex object**	16	**Sex object**	25	**Victim**
8	**Sex object**	17	**No females**		
9	**No females**	18	**Hero**		

a. Create a frequency distribution table for these data.

b. What level of measurement is this variable?

c. Construct an appropriate figure for these data.

12. An instructor administers a 27-item quiz to her class of 25 students. Each student's score on the quiz is the number of items answered correctly. These scores are listed below:

Person	Quiz Score	Person	Quiz Score	Person	Quiz Score
1	**22**	10	**17**	19	**20**
2	**15**	11	**18**	20	**21**
3	**11**	12	**14**	21	**17**
4	**19**	13	**10**	22	**18**
5	**12**	14	**6**	23	**15**
6	**21**	15	**16**	24	**24**
7	**22**	16	**20**	25	**22**
8	**19**	17	**21**		
9	**23**	18	**19**		

a. Construct a grouped frequency distribution table of these data, following the guidelines presented in this chapter.

b. What level of measurement is this variable?

c. Construct an appropriate figure for these data.

d. How would you describe the shape of the distribution in terms of its symmetry?

e. How would you characterize the students' performance? Did they do well or poorly? Would you say the quiz was easy or hard?

13. Imagine you are in your local department store to buy a sweater. Which of the following signs would make you think you're getting a good deal: "Everyday Low Price $15.00" or "Regularly $20.00/Sale $15.00"? Because the actual price in both signs is the same, you should be equally likely to pick either sign. One study investigated whether the presentation of a price influences customers' perceptions of savings (Tom & Ruiz, 1997). They asked 49 college students to select one of the two signs. Twelve students selected the everyday low price sign, and 37 selected the sale price sign.

a. Create a frequency distribution table for these data.

b. The researchers "hypothesized that consumers would think that a sale price presentation would generate a greater monetary savings than an everyday low price presentation" (p. 403). Do the percentages in your frequency distribution table provide preliminary support for this hypothesis?

14. The lottery ticket variable discussed in this chapter (Hardoon et al., 2001) examined the ticket preferences of non–problem gamblers. One of the purposes of this study was to compare the preferences of non–problem gamblers and problem gamblers (people experiencing personal, professional, financial, or legal problems as a result of their gambling). The lottery ticket choices for the 24 problem gamblers in this study are listed below:

Problem Gambler	Ticket Choice	Problem Gambler	Ticket Choice	Problem Gambler	Ticket Choice
1	**Pattern**	9	**Sequence**	17	**Pattern**
2	**Random**	10	**Random**	18	**Nonequilibrated**
3	**Nonequilibrated**	11	**Random**	19	**Random**
4	**Pattern**	12	**Pattern**	20	**Random**
5	**Random**	13	**Nonequilibrated**	21	**Nonequilibrated**
6	**Random**	14	**Random**	22	**Pattern**
7	**Nonequilibrated**	15	**Random**	23	**Random**
8	**Random**	16	**Sequence**	24	**Pattern**

a. Construct a frequency distribution table of these lottery ticket choices.

b. Construct an appropriate figure for these data.

c. Looking at your table and figure, how would you summarize the lottery ticket choices of the problem gamblers?

d. Compare the percentages in this table with the table for the non–problem gamblers (Table 2.4). In what ways are the lottery ticket choices of the two groups similar and in what ways are they different?

15. Another published study of government-sponsored lotteries examined reasons why people purchase lottery tickets (Miyazaki, Langenderfer, & Sprott, 1999). The researchers approached a number of people in different parts of one city and asked them, "At the times that you decide to go ahead and purchase tickets, what are the reasons why you personally do buy them?" (p. 6). The researchers grouped these people's responses into the following categories:

Reason	*f*
Desire to win money	70
Impulse purchase	28

(Continued)

(Continued)

Reason	*f*
Feeling lucky	18
Enjoyment and fun	15
To help schools	7
Miscellaneous/other reasons	14
Total	152

a. What level of measurement is this variable?

b. Construct an appropriate figure for these data.

c. Looking at the table and figure, how would you summarize the reasons why people purchase lottery tickets?

16. How information is presented may influence people's decision-making processes; this was examined in a study of the doctor-patient relationship (Gurm & Litaker, 2000). This study examined how the manner in which the potential risks and benefits of a medical procedure are presented affect patients' willingness to undergo the procedure. In one condition of this study, 63 participants were told that "99% of patients undergoing the procedure do not have any of these complications." When asked to describe the likelihood they would undergo the procedure, 16 said "definitely," 36 said "probably," 9 said "probably not," and 2 said "definitely not."

    a. Create a frequency distribution table for these data.

    b. How would you summarize the likelihood these participants would undergo the procedure?

17. In a second condition of the study described in Exercise 16, 53 participants were told, "These complications are seen in 1 out of 100 who undergo the procedure." Note that both conditions contain the same degree of risk (99% chance of no complications is equal to 1% chance of complications). When participants in the second condition were asked whether they would undergo the procedure, 4 said "definitely," 23 said "probably," 24 said "probably not," and 2 said "definitely not."

    a. Create a frequency distribution table for these data.

    b. How would you summarize the responses of the participants in this second condition?

    c. How would you describe the difference between the responses of the participants in the two conditions? What implications might these findings have for physicians?

18. For each of the following sets of data, determine whether the distribution of scores is unimodal or bimodal.

    a. 17, 10, 8, 7, 14, 7, 6, 10, 9, 10, 7, 8

    b. 12, 10, 9, 11, 9, 20, 9, 4, 9, 16, 9, 10

    c. 9, 6, 8, 5, 8, 7, 9, 10, 8, 7, 8

    d. 160, 197, 170, 115, 158, 170, 175, 115, 112, 115, 104, 170, 136, 115, 102, 181, 170

19. For each of the following sets of data, determine whether the distribution of scores is symmetric or asymmetric (skewed).

    a. 9, 7, 6, 8, 6, 13, 7, 6, 5, 6

    b. 10, 17, 16, 14, 16, 21, 19, 15, 18, 15, 16, 17, 16

    c. 22, 8, 17, 3, 19, 14, 19, 6, 19, 23, 16, 19, 22, 17, 19, 20

    d. 101, 109, 99, 112, 109, 110, 109, 107, 109, 108, 117, 109, 108, 110, 109

20. For each of the following sets of data, determine whether the distribution of scores is peaked or flat.
    a. 18, 15, 19, 18, 23, 18, 17, 19, 18, 17, 18
    b. 15, 8, 12, 6, 9, 11, 7, 17, 9, 10, 8, 5, 4, 8, 9, 7
    c. 11, 13, 15, 14, 17, 16, 13, 9, 13, 12, 7, 13, 12, 14, 13
    d. 72, 56, 63, 66, 49, 63, 60, 56, 60, 63, 53, 63, 70, 60, 44, 60, 56

Answers to Learning Checks

Learning Check 1

2. a.

Plans After Graduation	*f*	%
Work	71	61%
Graduate school	34	29%
Not sure	11	10%
Total	116	100%

b.

Year in College	*f*	%
Freshman	69	64%
Sophomore	18	17%
Junior	12	11%
Senior	9	8%
Total	108	100%

c.

Voting Intention	*f*	%
In favor	7	58%
Against	5	42%
Total	12	100%

d.

New M&M Color	*f*	%
Aqua	5	21%
Pink	8	33%
Purple	11	46%
Total	24	100%

Learning Check 2

2. a. Frequency polygon or histogram
   b. Bar chart or pie chart
   c. Histogram or frequency polygon
   d. Bar chart or pie chart

Learning Check 3

2. a. Multimodal
   b. Unimodal
   c. Bimodal
3. a. Asymmetric (negatively skewed)
   b. Symmetric
   c. Asymmetric (positively skewed)

Answers to Odd-Numbered Exercises

1. a. Pie chart or bar chart
   b. Frequency polygon or histogram
   c. Histogram or frequency polygon
   d. Frequency polygon or histogram
   e. Bar chart or pie chart
   f. Frequency polygon or histogram

3.

Flavor	*f*	%
Strawberry	13	17%
Chocolate	11	15%
Vanilla	24	32%
Other	27	36%
Total	75	100%

5. a. 6
   b. 7%

7. a.

Recommend?	*f*	%
Yes	4	40%
No	6	60%
Total	10	100%

b. Nominal

c. Bar chart

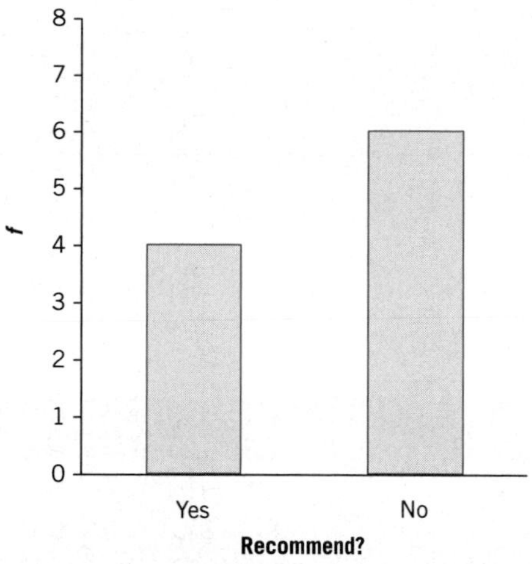

9. a.

# Stars	f	%
5	2	10%
4	5	25%
3	8	40%
2	4	20%
1	1	5%
Total	20	100%

b. Interval

c. Histogram

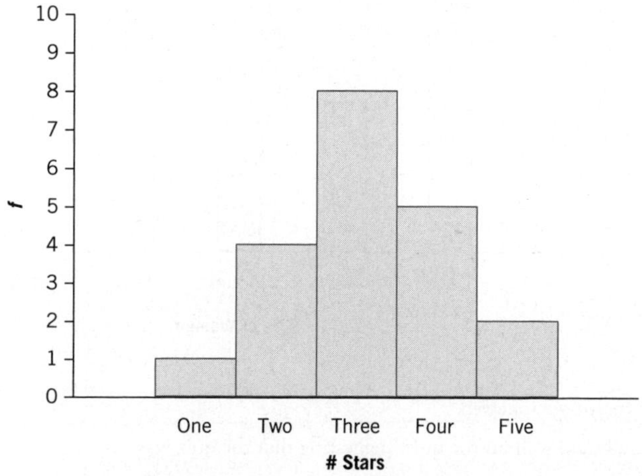

d.

Likelihood	*f*	%
67%–100%	10	33%
34%–66%	10	33%
0%–33%	10	33%
Total	30	100%

e. Grouping the ratings into only three categories gives the impression people's ratings are equally distributed when in fact there is a fair amount of variability.

11. a.

Role of Women	*f*	%
No females	10	40%
Sex object	6	24%
Victim	4	16%
Hero	4	16%
Traditional	1	4%
Total	25	100%

b. Nominal

c. Bar graph

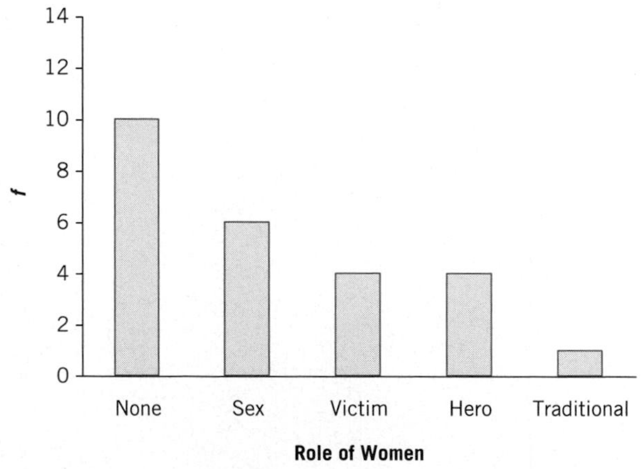

d. The distribution is negatively skewed.

e. The students generally did well on the quiz, suggesting that the quiz was easy.

13. a.

Sign	*f*	%
"Everyday Low Price $15.00"	12	24%
"Regularly $20.00/Sale $15.00"	37	76%
Total	49	100%

b. Yes, the table provides preliminary support for this hypothesis in that a majority of the sample (76%) picked the sale price presentation ("Regularly $20.00/Sale $15.00").

15. a. Nominal

b. Bar chart

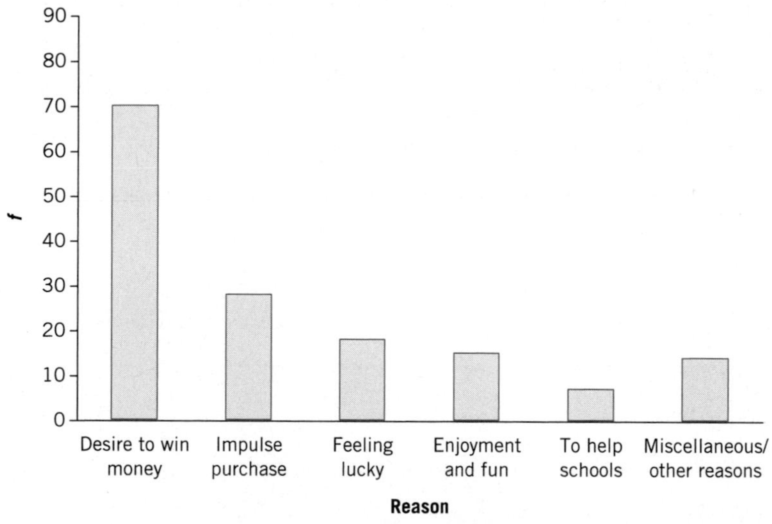

c. In this sample, the desire to win money is clearly the most frequently given reason for buying lottery tickets; helping schools is the least often given reason.

17. a.

Likelihood	*f*	%
Definitely	4	8%
Probably	23	43%
Probably not	24	45%
Definitely not	2	4%
Total	53	100%

b. When told that 1% of patients experience complications, the participants are almost equally likely to say they would "definitely" or "probably" (51%) undergo the medical procedure as they would "probably not" or "definitely not" (49%).

c. The wording doctors use in presenting the risk factors influences the likelihood the participant will undergo medical treatments such that emphasizing the possibility of complications may lower the likelihood a patient will undergo a medical procedure.

19. a. Positively skewed

b. Symmetric

c. Negatively skewed

d. Symmetric

3

MEASURES OF CENTRAL TENDENCY

CHAPTER OUTLINE

Chapter 2 discussed the importance of examining data using tables and figures. Researchers examine frequency distribution tables and figures to draw initial conclusions regarding three aspects of distributions of data: modality, symmetry, and variability. In addition to visually examining data, researchers typically describe data numerically by calculating statistics. Recall from Chapter 1 that there are two main types of statistics: descriptive and inferential. The purpose of descriptive statistics is to numerically describe or summarize data for a variable. This chapter discusses one of the main types of descriptive statistics, known as measures of central tendency. The first section of this chapter introduces a published research study we'll use to introduce and illustrate our discussion of these measures.

3.1 AN EXAMPLE FROM THE RESEARCH: THE 10% MYTH

One reason for conducting research is to test ideas systematically and objectively rather than relying on personal perceptions or beliefs. One example of a popular belief related to the field of psychology is the statement, "Most people use only 10% of their brains." This implies that people are not living up to their full potential and that we could all have phenomenal mental abilities if only we were able to use more of our brains. This belief has received widespread attention through the popular media, appearing in movies, advertisements, and magazines. For all its influence and popularity, however, there is one problem with the statement: It is not true. In his study, "Whence Cometh the Myth That We Only Use Ten Percent of Our Brains?" (Beyerstein, 1999), psychologist Barry Beyerstein demonstrated there is no scientific evidence to support this popular belief, referring to it as the "ten-percent myth."

The origin of the 10% myth is not precisely known. It may be a misrepresentation of a statement made by William James, one of the pioneers of psychology. In his 1911 book, *The Energies of Men*, James stated that "as a rule men habitually use only a small part of the powers which they actually possess. . . . We are making use of only a small part of our possible mental and physical resources" (pp. 11–12).

Researchers Kenneth Higbee and Samuel Clay decided to investigate the potential impact of higher education on dispelling misconceptions such as the 10% myth. "We expected that the kind of academic training psychology students receive . . . would give them a healthy skepticism" (Higbee & Clay, 1998, p. 471). More specifically, they hypothesized that "psychology majors who had a significant amount of training in psychology would be less likely to believe the ten-percent myth than would non-majors who had no training in psychology" (p. 471).

To test their research hypothesis, Higbee and Clay located two groups of college students: 38 psychology majors and 39 non–psychology majors. The students in the study, which we will refer to as the 10% myth study, were asked the following question: "About what percentage of their potential brain power do you think most people use?" The students were asked to choose from among 21 scores, ranging from 0% to 100%, with scores increasing in 5% increments (0%, 5%, 10%, etc.); we will refer to this variable as "estimated brain power."

The example from the 10% myth study included in this chapter will focus on the data provided by the psychology majors. (The data for the non–psychology majors are included in the exercises at the end of this chapter.) The estimated brain power chosen by the 38 psychology majors is listed in Table 3.1(a); to organize these data, Table 3.1(b) arranges the data into a frequency distribution table.

Figure 3.1 illustrates the data from the frequency distribution table using a frequency polygon; a frequency polygon was used because estimated brain power is measured at the ratio level of measurement, with the values of the variable equally spaced along a numeric continuum with a true zero point. As we can see from the table and figure, the majority of the data are at the low end of the 0% to 100% scale, with many psychology majors reporting the belief that most people use 5% or 10% of their potential brain power. This distribution is best described as asymmetric; more specifically, it is positively skewed because the higher frequencies are at the lower end of the distribution, and the tail is at the upper end of the distribution.

Examining the frequency distribution table and frequency polygon for the estimated brain power variable provides an initial indication of the extent to which the sample of psychology majors believed in the 10% myth. However, to test the study's research hypothesis, the researchers needed to statistically analyze the data. A common first step in statistical analyses is to summarize a set of data by calculating descriptive statistics. The next section introduces one type of descriptive statistic, measures of central tendency, which summarize a set of data by identifying the most common or frequently occurring values of variables.

3.2 UNDERSTANDING CENTRAL TENDENCY

For many variables studied by researchers, the majority of data cluster or center on the middle of the distribution. For this reason, the first statistic used to describe data numerically is a **measure of central tendency**, defined as a statistic that identifies the center of a distribution; a measure of central tendency is a single score that is the most typical, common, or frequently occurring value for a variable.

TABLE 3.1 ● THE ESTIMATED BRAIN POWER BY 38 PSYCHOLOGY MAJORS

(a) Raw Data

Psychology Major	Estimated Brain Power	Psychology Major	Estimated Brain Power	Psychology Major	Estimated Brain Power
1	**45**	14	**5**	27	**30**
2	**40**	15	**5**	28	**40**
3	**10**	16	**10**	29	**5**
4	**30**	17	**10**	30	**30**
5	**10**	18	**15**	31	**35**
6	**40**	19	**20**	32	**75**
7	**10**	20	**10**	33	**10**
8	**10**	21	**5**	34	**5**
9	**5**	22	**35**	35	**10**
10	**15**	23	**5**	36	**90**
11	**75**	24	**5**	37	**20**
12	**10**	25	**10**	38	**55**
13	**10**	26	**20**		

(b) Frequency Distribution Table

Estimated Brain Power	f	%
100%	0	0%
95%	0	0%
90%	1	3%
85%	0	0%
80%	0	0%
75%	2	5%
70%	0	0%
65%	0	0%
60%	0	0%
55%	1	3%
50%	0	0%

Estimated Brain Power	f	%
45%	1	3%
40%	3	8%
35%	2	5%
30%	3	8%
25%	0	0%
20%	3	8%
15%	2	5%
10%	12	31%
5%	8	21%
0%	0	0%
Total	38	100%

The following sections describe three measures of central tendency:

- the mode,
- the median, and
- the mean.

FIGURE 3.1 ● FREQUENCY POLYGON FOR THE ESTIMATED BRAIN POWER VARIABLE

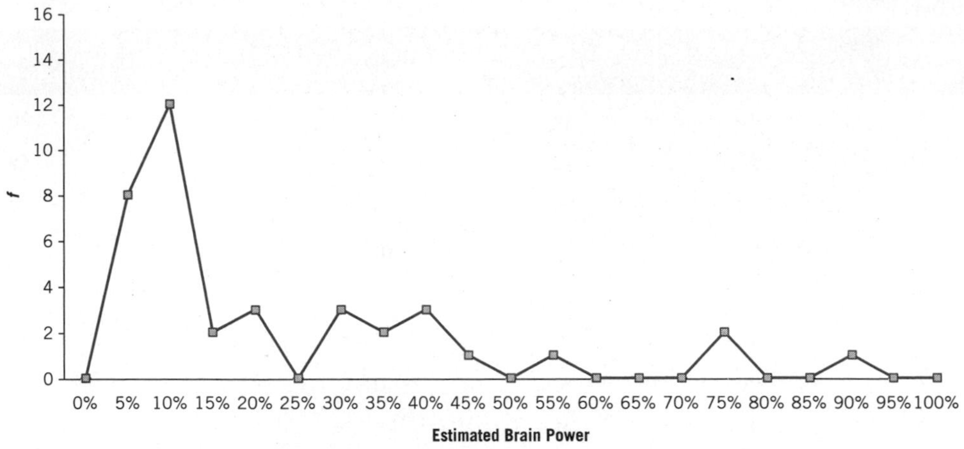

After describing and illustrating the different measures, we will compare them in terms of their relative strengths and weaknesses and the conditions under which it may be preferable to use a particular measure.

3.3 THE MODE

One way to describe or summarize a set of data for a variable is to ask, "What values of the variable are the most common or occur the most often?" The first measure of central tendency that we will consider, the **mode**, sometimes referred to as the *modal score*, is the score or value of a variable that appears most frequently in a set of data.

Determining the Mode

The simplest way to determine the mode is to examine a frequency distribution table or figure for a variable. For the estimated brain power variable, the mode may be identified by looking at the frequency distribution table presented earlier in Table 3.1(b). By scanning down the frequency (f) column in this table, we see that 12 of the 38 psychology majors believed that most people use 10% of their potential brain power. Because the value of 10% has the highest frequency, 10% is the mode or modal score for this set of data. Keep in mind that the mode is the value of the variable with the greatest frequency, not the frequency itself (in this example, the mode is equal to 10%, not 12). Following American Psychological Association (APA) style, the abbreviation Mo is used in presenting the mode, such as "*Mo* = 10%." (Note that APA style requires the word *Mo* be italicized.)

To locate the mode in a figure such as a bar chart, histogram, or frequency polygon, we identify the tallest bar in the bar chart or histogram or the highest data point in the polygon. Looking at the frequency polygon for the estimated brain power variable in Figure 3.2, because the highest data point corresponds to the value of 10%, it is the mode for this data set.

FIGURE 3.2 ● IDENTIFYING THE MODE FOR THE ESTIMATED BRAIN POWER VARIABLE

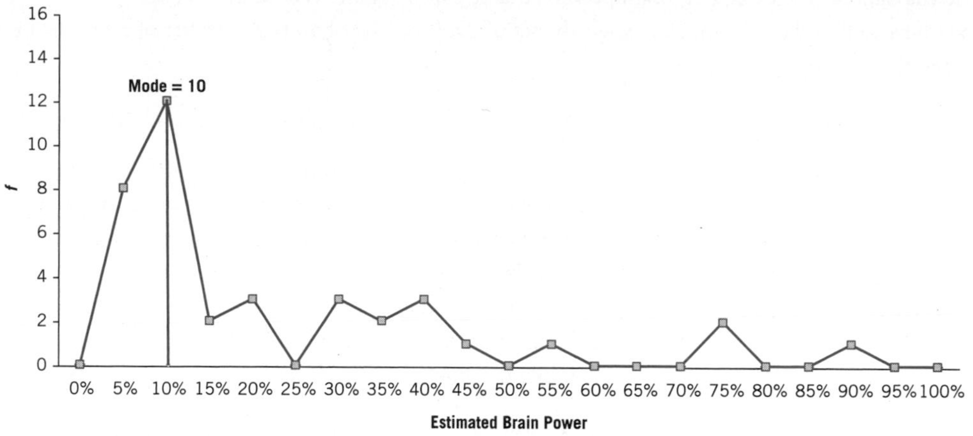

LEARNING CHECK 1:

Reviewing What You've Learned So Far

1. Review questions
   a. What is the main purpose of a measure of central tendency?
   b. How do you identify the mode from a frequency distribution table or figure?
2. Determine the mode for each of the following sets of data.
   a. 2, 1, 4, 2, 7, 2, 3
   b. 6, 7, 6, 4, 5, 4, 2, 6
   c. 12, 5, 9, 12, 10, 11, 12, 8, 15, 12, 7
   d. 4, 2, 2, 1, 5, 4, 5, 2, 3, 2, 4, 2
   e. 3.0, 2.8, 3.2, 3.6, 3.7, 3.2, 3.5, 3.6, 3.1, 3.2, 3.0, 3.2, 3.8, 3.3

3.4 THE MEDIAN

In addition to the mode, another way of describing a distribution of data for a variable involves looking at the entire distribution. One way to do so is by asking the question, "What value of the variable is located in the center of the distribution?" A measure of central tendency designed to answer this question is the **median**, defined as the value of a variable that splits a distribution of scores in half, with the same number of scores above the median as below.

Determining the median for a set of data involves some simple calculations. However, because determining the median involves locating the precise center of the distribution, the calculation of the median varies depending on whether the set of data consists of an odd number of scores or an even number of scores. These two situations are discussed in turn below.

Calculating the Median: Odd Number of Scores

Because the data for the estimated brain power variable consist of an even number of scores ($N = 38$), we will illustrate how to calculate the median when the set of data consists of an odd number of scores using the following set of nine scores:

2 3 1 4 3 1 4 6 3

There are three steps involved in calculating the median for a set of data: (1) sort the scores from lowest to highest, (2) calculate the median score in the set of data, and (3) determine the value of the median score.

The first step involves sorting the score from lowest to highest. Sorting the scores is necessary to ensure that the median represents the middle of the distribution. Imagine, for example, you wanted to determine the median height of a group of men standing in a line. If you did not sort the men by height, you might incorrectly end up saying the tallest person represented the "middle" height simply because he happened to be standing in the middle of the line. The set of nine scores presented above is sorted from lowest to highest below:

1	1	2	3	3	3	4	4	6
Lowest								*Highest*

The second step is to calculate the median score in the set of data. The median score is the score that has the same number of scores above it as below it. Formula 3-1 presents the formula for calculating the median with an odd number of scores:

$$\text{Median} = \frac{N+1}{2}\text{th score} \tag{3-1}$$

where N is the total number of scores. Because there are nine scores ($N = 9$) in this set of data, the median is calculated as follows:

$$\begin{aligned}\text{Median} &= \frac{N+1}{2}\text{th score} \\ &= \frac{9+1}{2}\text{th score} = \frac{10}{2}\text{th score} \\ &= 5\text{th score}\end{aligned}$$

Therefore, when $N = 9$, the median is located at the fifth score in the set of data. At this point, it is important for you to keep in mind that the median for this variable is not 5 but is rather the value of the fifth score.

The third step is to determine the value of the median score. Looking at the nine sorted scores, we find the value of the fifth score is equal to 3:

1	1	2	3	**3**	3	4	4	6
1st	2nd	3rd	4th	**5th**	6th	7th	8th	9th

The median in the set of nine scores is equal to 3; it is the median because the same number of scores (four) is above this score as below it. When reporting the median following APA format, use the abbreviation *Mdn* and write "*Mdn* = 3."

Calculating the Median: Even Number of Scores

If a data set has an odd number of scores, the median is the value of the one score located in the center of the sorted scores. However, if a set of data consists of an even number of scores, the formula for the median changes because no single score has the same number of scores above and below it. In this situation, the median is determined by identifying the two scores in the middle of the distribution and then calculating the number halfway between the values of the two scores; this is represented in Formula 3-2:

$$\text{Median} = \frac{\frac{N}{2}\text{th score} + (\frac{N}{2}+1)\text{th score}}{2} \quad (3\text{-}2)$$

where N is the total number of scores.

As a simple example, let's say we had the following eight scores: 39, 15, 11, 24, 45, 31, 42, and 19. To calculate the median, the first step is to sort the scores from lowest to highest:

11	15	19	24	31	39	42	45
Lowest							*Highest*

The second step is to calculate the median scores in the set of data using Formula 3-2:

$$\begin{aligned} \text{Median} &= \frac{\frac{N}{2}\text{th score} + (\frac{N}{2}+1)\text{th score}}{2} \\ &= \frac{\frac{8}{2}\text{th score} + (\frac{8}{2}+1)\text{th score}}{2} \\ &= \frac{\text{4th score} + \text{5th score}}{2} \end{aligned}$$

Therefore, for $N = 8$, the value of the median falls halfway between the values of the fourth and fifth scores. Similar to the situation with an even number of scores, the median in this example is not equal to (4 + 5)/2 or 4.50; the median is the number halfway between the values of the fourth and fifth scores.

The third step is to calculate the value of the median. In this example, we need to identify the values of the fourth and fifth scores:

11	15	19	**24**	**31**	39	42	45
1st	*2nd*	*3rd*	***4th***	***5th***	*6th*	*7th*	*8th*

Because the fourth score in this distribution is 24 and the fifth score is 31, the value for the median is the number halfway between these two scores:

$$\begin{aligned} \text{Median} &= \frac{\text{4th score} + \text{5th score}}{2} \\ &= \frac{24+31}{2} = \frac{55}{2} \\ &= 27.50 \end{aligned}$$

In this set of eight scores, the median is 27.50, or "$Mdn = 27.50$." There are two things to note about this value of the median. First, the same number of scores falls below it (11, 15, 19, 24) as above it (31, 39, 42, 59). Second, because 27.50 is halfway between the scores below it and above it (24 and 31), it lies precisely in the center of the set of scores.

11	15	19	24	31	39	42	45
1st	*2nd*	*3rd*	*4th*	*5th*	*6th*	*7th*	*8th*

27.50

As a second example, let's calculate the median for the 38 scores for the estimated brain power variable. Once the 38 scores have been ranked from lowest to highest (see below), the two median scores are calculated using Formula 3-2:

$$\text{Median} = \frac{\frac{N}{2}\text{th score} + (\frac{N}{2}+1)\text{th score}}{2}$$
$$= \frac{\frac{38}{2}\text{th score} + (\frac{38}{2}+1)\text{th score}}{2}$$
$$= \frac{19\text{th score} + 20\text{th score}}{2}$$

For the estimated brain power variable, the median falls between the 19th and 20th scores:

5	5	5	5	5	5	5	5	10	10	10	10	10
1st	2nd	3rd	4th	5th	6th	7th	8th	9th	10th	11th	12th	13th

10	10	10	10	10	10 ---|---10		15	15	20	20	20	30
14th	15th	16th	17th	18th	19th	20th	21st	22nd	23rd	24th	25th	26th

30	30	35	35	40	40	40	45	55	75	75	90
27th	28th	29th	30th	31st	32nd	33rd	34th	35th	36th	37th	38th

As both the 19th and 20th scores are equal to 10, the value for the median is the following:

$$\text{Median} = \frac{19\text{th score} + 20\text{th score}}{2}$$
$$= \frac{10+10}{2} = \frac{20}{2}$$
$$= 10.00$$

So, for this sample of 38 psychology majors, the median for the estimated brain power variable is 10.00 ($Mdn = 10.00$). Figure 3.3 identifies the median on the frequency polygon for the estimated brain power variable.

Calculating the Median When Multiple Scores Have the Median Value

The median of 10.00 for the estimated brain power variable suggests that, in the sample of 38 psychology majors, an equal number of the scores fall below and above the value of 10.00. This is not precisely true, however, as more than two scores have the median value of 10.00. Returning to Table 3.1(b), we can see that, because 12 of the 38 psychology majors gave the response of 10%, only 8 of the 38 scores actually fall below the median value of 10.00.

When there are multiple scores with the median value, the method for calculating the median presented in this book does not provide the exact value for the median but rather an approximation. This is because the method does not consider all of the scores with the median value but only one (in samples with an odd number of scores) or two (in even-numbered samples) of the scores. Calculating the exact median when multiple scores have the median value requires using a mathematical operation known as interpolation. This book will not cover interpolation, however, because the differences between the approximate and exact values of the median are typically small.

FIGURE 3.3 ● IDENTIFYING THE MEDIAN FOR THE ESTIMATED BRAIN POWER VARIABLE

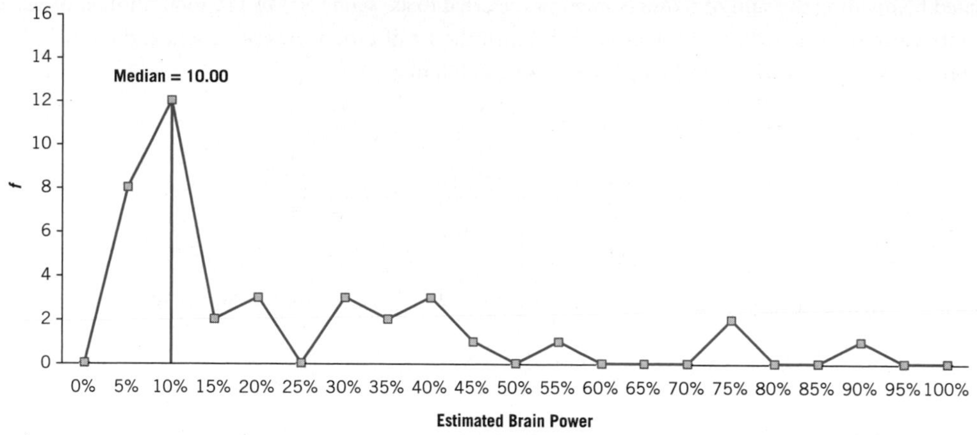

LEARNING CHECK 2:

Reviewing What You've Learned So Far

1. Review questions
   a. Why are there different formulas for calculating the median depending on whether there are an odd or even number of scores for a variable?
2. Calculate the median for each of the following sets of data.
   a. 2, 3, 3, 4, 5, 6, 8
   b. 21, 14, 13, 17, 30, 17, 14, 11, 13, 14, 19
   c. 5, 6, 7, 9, 12, 15, 16, 17
   d. 73, 66, 91, 84, 69, 87, 62, 79, 82, 90
   e. 11, 9, 12, 9, 10, 7, 9, 8, 9, 10, 9, 8

3.5 THE MEAN

A third measure of central tendency used to represent a distribution concerns the question, "What is the average value of the variable in this set of data?" The descriptive statistic that answers this question is the **mean**, which is defined as the arithmetic average of a set of scores. The mean is the measure of central tendency with which you are most likely familiar. For example, you may have heard or read about the "average life expectancy," "average miles per gallon," or "earned run average (ERA)." Because this book will refer to scores for a variable using the symbol X, the mean of scores in a set of data is represented by the letter X with a bar above it ($\overline{X}$). This symbol is referred to as the "mean of X" or "X bar." In reporting the mean in a paper or journal article, many academic disciplines use the italicized letter M rather than $\overline{X}$.

Calculating the Mean

Mathematically, the formula for the mean (Formula 3-3) is the following:

$$\overline{X} = \frac{\Sigma X}{N} \tag{3-3}$$

where X is a score for a variable and N is the total number of scores. To calculate the mean, we first calculate the sum of the scores, which is represented by ΣX (the Σ [sigma] symbol represents "the sum of"). The mean is then calculated by dividing the sum of scores (sometimes referred to as "sigma X") by the total number of scores (N).

To introduce the calculation of the mean, we return the set of nine scores presented earlier: 2, 3, 1, 4, 3, 1, 4, 6, and 3. The mean for this set of data is calculated as follows:

$$\begin{aligned}\bar{X} &= \frac{\Sigma X}{N} \\ &= \frac{2+3+1+4+3+1+4+6+3}{9} = \frac{27}{9} \\ &= 3.00\end{aligned}$$

For the 10% myth study, let's calculate the mean for the 38 scores for the estimated brain power variable in Table 3.1(a):

$$\begin{aligned}\bar{X} &= \frac{\Sigma X}{N} \\ &= \frac{45+40+10+\ldots+90+20+55}{9} = \frac{870}{38} \\ &= 22.89\end{aligned}$$

The ellipses (. . .) included in the middle of this calculation (45 + 40 + 10 + . . . + 90 + 20 + 55) represent mathematical operations conducted to correctly carry out the formula but have not been printed to save space.

On the basis of the above calculations, we could report that the sample of 38 psychology majors believe people use an average amount of brain power of 22.89% ($M = 22.89$). The mean for the estimated brain power variable is illustrated in Figure 3.4.

Reporting Calculated Numbers

As calculating statistics involves a large number of mathematical formulas, this section discusses several issues concerning the reporting and rounding of numbers resulting from these calculations.

In reporting calculated numbers, the general guideline is to round calculated numbers to *two* decimal places.

Including two decimal places indicates that the reported number is the result of a mathematical calculation. For the sake of consistency, calculated numbers are rounded to two decimal places even when the result does not require two decimal places (i.e., $\frac{21}{3} = 7.00$ [rather than 7], $\frac{38}{10} = 3.80$ [rather than 3.8]).

In some situations, more than two decimal places may be needed to accurately describe a variable. For example, a study looking at how long it takes people to respond to a stimulus presented on a computer screen may produce scores such as .007 and .014 seconds. Here you would want to report the results of any calculations using these data with three or perhaps four decimal places (e.g., "the sum of scores is equal to .021").

Rounding Calculated Numbers

When rounding numbers to two decimal places, we will use the method employed by most computers and calculators:

If the number in the thousandths column is <u>0, 1, 2, 3, or 4,</u> the number in the hundredths column remains *unchanged*.

If the number in the thousandths column is <u>5, 6, 7, 8, or 9,</u> the number in the hundredths column is *increased by 1*.

FIGURE 3.4 ● IDENTIFYING THE MEAN FOR THE ESTIMATED BRAIN POWER VARIABLE

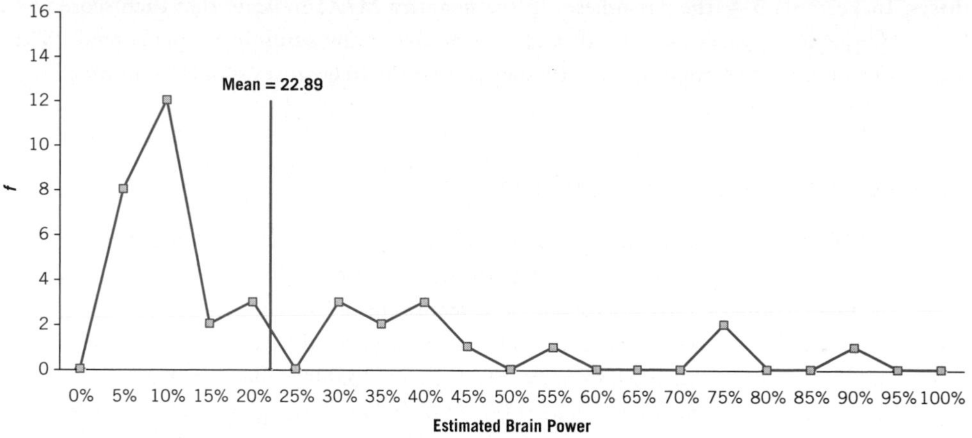

Using these rounding rules, the number 5.26<u>3</u> would be rounded *down* to 5.26 because the number in the thousandths column is 3. The number 3.41<u>7</u> would be rounded *up* to 3.42 because the number in the thousandths column is 7.

Students sometimes become concerned about a possible loss of accuracy due to rounding, choosing instead to report numbers using many decimal places. For example, a student may wish to report a number as 4.2316 rather than 4.23. However, for most situations, rounding has very little impact on the conclusions drawn from statistical analyses.

Calculating the Mean From Frequency Distribution Tables

When a data set is large, an alternative way to calculate the mean involves first organizing the data into a frequency distribution table. The mean can then be calculated using the following formula:

$$\frac{\Sigma(fX)}{N} = \quad (3\text{-}4)$$

where X is a score for the variable, f is the frequency of the score in the set of data, and N is the total number of scores. To calculate the mean using Formula 3-4, each score (X) is multiplied by its frequency (f) within the set of data; the results of these multiplications (fX) are then summed, and this sum ($\Sigma(fX)$) is divided by the total number of scores (N).

Using the frequency distribution table for the estimated brain power variable in Table 3.1(b), the mean for this data set may be calculated using Formula 3-4 as follows:

$$\begin{aligned}\bar{X} &= \frac{\Sigma(fX)}{N} \\ &= \frac{(0*0)+(8*5)+(12*10)+\ldots+(1*90)+(0*95)+(0*100)}{38} \\ &= \frac{0+40+120+\ldots+90+0+0}{38} = \frac{870}{38} \\ &= 22.89\end{aligned}$$

The value of the mean ($M = 22.89\%$) is the same as the one earlier obtained using Formula 3-3.

In the above calculations, parentheses are used to indicate the order in which calculations are made. Mathematical operations placed within parentheses are calculated *before* operations that are outside of parentheses. In Formula 3-4, the parentheses in the quantity $\Sigma(fX)$ indicate that each score (X) is multiplied by its frequency (f) (such as (0 * 0) and (8 * 5)) *before* any summing is performed. Without the parentheses, the quantity ΣfX might incorrectly suggest that the frequencies should be summed (Σf) before performing any multiplication.

The Mean as a Balancing Point in a Distribution

One way to think of the mean as a measure of central tendency is as a balancing point in a distribution of scores for a variable. To illustrate this concept, let's return to the simple example of nine scores (2, 3, 1, 4, 3, 1, 4, 6, 3) used earlier. The first column of Table 3.2 lists the nine scores, the second column provides the mean of these scores (M = 3.00), and the numbers in the third column represent the difference between each score and the mean. For example, subtracting the mean from the first score (2 – 3.00) results in a value of –1.00. It is noteworthy that some of these differences are positive (1.00, 3.00) and some are negative (–1.00, –2.00).

The mean is a balancing point in a distribution because the sum of the negative differences of scores from the mean is *always* equal to the sum of the positive differences from the mean (ignoring the minus sign). Using the third column in Table 3.2:

$$\text{Negative differences from mean} = \text{Positive differences from mean}$$
$$-1.00 + -2.00 + -2.00 = 1.00 + 1.00 + 3.00$$
$$-5.00 = 5.00$$

Figure 3.5 plots the three negative differences (–1.00, –2.00, –2.00) and the three positive differences (1.00, 1.00, 3.00) across the length of a balance beam. Notice that the balance beam does not tilt to the left or the right; this is because the sums of the weights on each side of the balance beam are equal to each other. This simple example illustrates the mean as a measure of central tendency: Mathematically, the mean is in the center of the distribution of any set of data. For any set of data, the sum of the negative differences from the mean is equal to the sum of the positive differences from the mean.

TABLE 3.2 ● EXAMPLE OF THE MEAN AS A BALANCING POINT IN A DISTRIBUTION

Score	Mean	Score – Mean
2	3.00	–1.00
3	3.00	.00
1	3.00	–2.00
4	3.00	1.00
3	3.00	0.00
1	3.00	–2.00
4	3.00	1.00
6	3.00	3.00
3	3.00	.00

FIGURE 3.5 ● ILLUSTRATION OF THE MEAN AS A BALANCING POINT

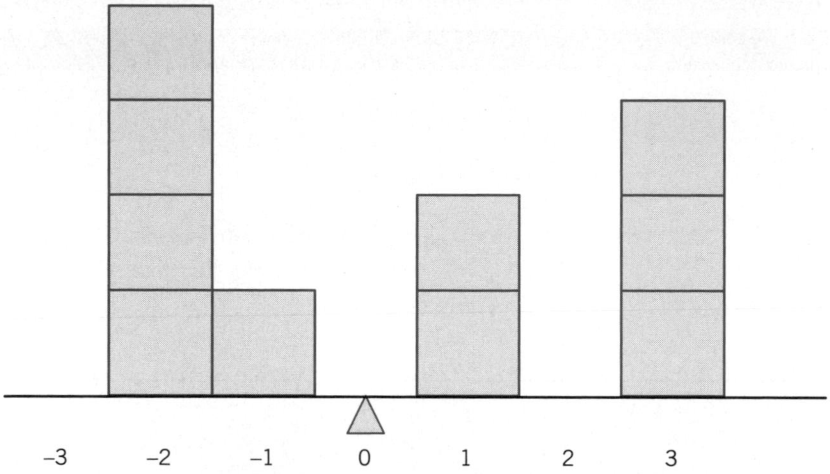

LEARNING CHECK 3:

Reviewing What You've Learned So Far

1. Review questions
   a. Why do we say that the mean is the balancing point in a distribution of scores?
2. Calculate the mean using Formula 3-3 for each of the following sets of data.
   a. 2, 3, 3, 5, 7
   b. 8, 6, 4, 8, 3, 7
   c. 71, 84, 65, 78, 89, 72, 60, 85
   d. 2.5, 1.0, 3.5, 3.0, 2.0, 2.5, 4.0, 1.0, 2.0
   e. 10, 8, 12, 8, 9, 7, 23, 6, 8, 11, 10, 6
3. Calculate the mean using Formula 3-4 for the data in each of the following frequency distribution tables.

a.

Score	*f*	%
5	1	10%
4	1	10%
3	2	20%
2	4	40%
1	2	20%
Total	10	100%

(Continued)

(Continued)

b.

Score	*f*	%
6	3	11%
5	7	25%
4	5	18%
3	8	29%
2	0	0%
1	4	14%
0	1	4%
Total	28	100%

4. For the sets of data in 2(a) and 2(b), calculate the difference between each score and the mean (use Table 3.2 as an example). Next, compare the sum of the positive and negative differences to see if they balance each other.

The Sample Mean vs. the Population Mean: Statistics vs. Parameters

For the estimated brain power variable, a mean of 22.89% was calculated for the sample of 38 psychology majors. A mean calculated from data for a sample may be referred to as a **sample mean**, which we have chosen to represent using the symbol $\overline{X}$. But what if, instead of a sample drawn from the population, the researchers had collected data from the entire population of psychology majors? A mean associated with an entire population is called the **population mean**, represented by the Greek symbol μ (pronounced "mew"). Because researchers typically collect data from samples rather than entire populations, when we refer to "the mean" in this book, we will be referring to the sample mean.

A numeric characteristic of a sample (such as $\overline{X}$) is called a **statistic**; a numeric characteristic of a population (such as μ) is called a **parameter**. This relationship between statistics and parameters is summarized below:

Target	Numeric Characteristic	Mean
Sample	Statistic	$\overline{X}$
Population	Parameter	μ

As we discussed in Chapter 1, one goal of research is to test hypotheses regarding what is believed to exist in the population. To test their hypotheses, researchers collect data from samples of the population; the statistics calculated from these samples may be used to estimate population parameters. For example, in the 10% myth study, the sample mean ($\overline{X}$) of 22.89% could be used to estimate the mean for the entire population of psychology majors (μ).

The relationship between samples and populations, as well as the use of statistics to estimate parameters, is critical to testing research hypotheses. The process of hypothesis testing and statistics used to test research hypotheses will be illustrated in greater detail starting in Chapter 6 of this book.

3.6 COMPARISON OF THE MODE, MEDIAN, AND MEAN

In this chapter, we introduced three measures of central tendency: the mode, median, and mean. Given that these three measures represent different things and are calculated differently, it is perhaps not surprising that each of them has strengths and weaknesses. This section evaluates and compares the mean, median, and mode, identifying situations in which the use of one may be preferable to the others.

Strengths and Weaknesses of the Mode

The first comparative advantage of the mode is that it can be determined relatively easily and quickly because it does not require any mathematical calculations. Unlike the median and mean, the mode is identified simply by looking at a frequency distribution table or figure. Another advantage of the mode is that it may be used to summarize categorical variables, which are variables measured at the nominal level of measurement. Imagine, for example, 50 people are asked to indicate their favorite flavor of ice cream and their data are organized into the following frequency distribution table:

Flavor	*f*	%
Vanilla	18	36%
Chocolate	23	46%
Strawberry	9	18%
Total	50	100%

At a single glance at this table, we could state that "the modal flavor was chocolate."

Another strength of the mode is that it is always an actual score for a variable, something that may not be true for the mean or median. You may have heard, for example, that "the average American family has two and a half children." If these data had been summarized using the mode rather than the mean, you would not lose sleep trying to figure out what "half" a child looks like.

An additional advantage of the mode is that it accurately describes distributions that have more than one mode. Below is a small set of data that has been sorted from lowest to highest:

3 4 4 4 6 8 10 10 10

A histogram has been created for this set of data in Figure 3.6. Looking at the histogram, we see that this distribution has two modes: 4 and 10; in Chapter 2, we referred to this type of distribution as a bimodal distribution. As we will demonstrate shortly, the mode is a more appropriate measure of central tendency for bimodal distributions than the median or the mean.

Finally, because the mode is the single most common value for a variable, determining the mode is not affected by outliers in the distribution. In Chapter 2, outliers were defined as rare, extreme scores for a variable lying outside the range of the majority of the scores.

Despite these strengths, the mode is of limited use to researchers. First, because the mode is based on just one value of a variable, it may not adequately represent the entire set of data. Consider the following two sorted sets of data, each of which consists of eight scores:

Variable 1: 5 7 10 **11** **11** 13 15 18

Variable 2: **11** **11** 16 18 19 20 22 25

FIGURE 3.6 ● IDENTIFYING THE MODES IN A BIMODAL DISTRIBUTION

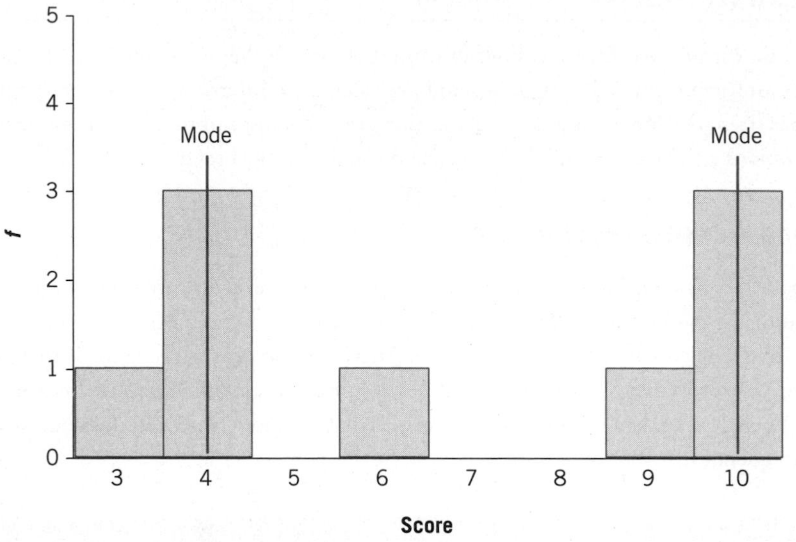

Although the mode for the two variables is the same (Mo = 11), the two sets of data are clearly different from each other. For Variable 1, the modal value of 11 lies near the center of the distribution of scores and provides a useful representation of the distribution. For Variable 2, however, the modal value is at one end of the distribution and does not accurately represent the set of data. Because the mode does not take into account all of the scores in the distribution, it cannot be used in statistical analyses that are designed to test hypotheses about distributions.

Strengths and Weaknesses of the Median

The primary strength of the median is that it provides a more accurate description of skewed (asymmetric) distributions than does the mean. Recall from Chapter 2 that a skewed distribution is one in which the majority of data for a variable are located at one end of the range of values. Skewed distributions are different from symmetric distributions, where the highest frequencies are in the center of the distribution, with the frequencies decreasing in a similar manner as we move away from the center in both the left and right directions.

A good example of a variable with a skewed distribution is income. As a simple yet representative example of income, Table 3.3 lists the annual salaries for the players on the starting offensive unit of the San Francisco 49ers professional football team, sorted from lowest to highest. The distribution of salaries in this table is positively skewed because most are at the lower end of the distribution, with the tail at the upper end consisting of the particularly high salaries of \$4,675,000 and \$11,800,000.

Suppose we decided to describe the distribution of salaries in Table 3.3 using the mean as the measure of central tendency. The mean of the 11 salaries is calculated below:

$$\begin{aligned}\bar{X} &= \frac{\Sigma X}{N} \\ &= \frac{449{,}800 + 791{,}720 + \ldots + 4{,}675{,}000 + 11{,}800{,}000}{11} = \frac{34{,}354{,}800}{11} \\ &= 3{,}123{,}163.64\end{aligned}$$

TABLE 3.3 ● SALARIES OF THE STARTING OFFENSIVE UNIT OF THE SAN FRANCISCO 49ERS

Player	Salary ($)
1	449,800
2	791,720
3	1,250,000
4	1,575,520
5	2,000,000
6	2,330,000
7	2,662,000
8	2,820,760
9	4,000,000
10	4,675,000
11	11,800,000

Although this mean of $3,123,163.64 may be a balancing point of the distribution, it is not located in the center of the distribution because the salaries of 8 of the 11 players (73%) are below the mean. The mean is affected by outliers, such as the one player's salary of $11,800,000, and hence does not accurately portray a skewed distribution such as this one.

An alternative way to describe the salaries of these players would be to use the median. Because there are an odd number of scores in the variable ($N = 11$), Formula 3-1 is used to calculate the median as follows:

$$\begin{aligned}\text{Median} &= \frac{N+1}{2}\text{th score}\\ &= \frac{11+1}{2}\text{th score} = \frac{12}{2}\text{th score}\\ &= 6\text{th score}\end{aligned}$$

Starting from the lowest salary, the sixth player's salary is $2,330,000. Table 3.4 illustrates the mean and median of the salaries of the 11 players. As we can see, the median (Mdn = $2,330,000) is located closer to the center of the distribution than is the mean (M = $3,123,163.64).

Although the median is useful for describing skewed distributions, it does have several limitations. First, like the mode, the median is not based on all of the scores in a distribution of data but rather is based only on the one or two scores in the center of the distribution. Consequently, the median cannot be used in statistical analyses that are designed to test hypotheses about distributions in the population.

A second limitation of the median is that it does not accurately describe bimodal distributions. Let's return to the set of nine numbers that comprise the bimodal distribution in Figure 3.6. What if we were to calculate the median for this set of data? Because there are an odd number of scores ($N = 9$), we apply Formula 3-1 to the data as follows:

$$\begin{aligned}\text{Median} &= \frac{N+1}{2}\text{th score}\\ &= \frac{9+1}{2}\text{th score} = \frac{10}{2}\text{th score}\\ &= 5\text{th score}\end{aligned}$$

TABLE 3.4 ● SALARIES OF THE STARTING OFFENSIVE UNIT OF THE SAN FRANCISCO 49ERS FOOTBALL TEAM, WITH THE MEAN AND MEDIAN INDICATED

Player	Salary ($)	
1	449,800	
2	791,720	
3	1,250,000	
4	1,575,520	
5	2,000,000	
6	2,330,000	← **Median (2,330,000)**
7	2,662,000	
8	2,820,760	
9	4,000,000	← **Mean (3,123,163.64)**
10	4,675,000	
11	11,800,000	

Finding the fifth score, the median of the set of nine scores is equal to 6. But as we can see from the histogram in Figure 3.7, the two modes of 4 and 10 much more accurately represent the most common values of this bimodal distribution than does the median value of 6.

Strengths and Weaknesses of the Mean

Compared with the mode and median, the mean has one critical strength: It is calculated from *all* of the scores in a distribution of data (ΣX) rather than just the most frequently appearing scores or the one or two scores in the center of the distribution. As such, the mean possesses important mathematical properties that allow it to be used in statistical analyses to test hypotheses regarding populations. Because testing hypotheses is a critical aspect of research, the mean will be used as the primary measure of central tendency for the remainder of this book.

However, the mean has several limitations. As illustrated in the football salary example in Table 3.4, the mean does not accurately represent skewed distributions because it is affected by outliers. In the 10% myth study, the sample mean for the estimated brain power variable was calculated to be 22.89%; however, the mode and the median were both 10%, which was in fact the most frequently occurring value in the distribution. Why was the mean so much higher than the mode and median? As the frequency distribution table in Table 3.1(b) reveals, the value of the mean was influenced by several outliers, particularly one student's score of 90%.

The mean is also limited in its ability to provide an accurate representation of bimodal distributions. Returning once more to the bimodal distribution in Figure 3.6, below we calculate the mean for the set of nine scores:

$$\begin{aligned}\bar{X} &= \frac{\Sigma X}{N} \\ &= \frac{3+4+4+4+6+9+10+10+10}{9} = \frac{60}{9} \\ &= 6.67\end{aligned}$$

As was the case with the median, the mean of 6.67 represents the distribution less accurately than the two modes of 4 and 10. These comparisons illustrate an important principle for statistical analysis: It is helpful to calculate more than one measure of central tendency to understand the shape and nature of a distribution.

FIGURE 3.7 ● IDENTIFYING THE MODES AND THE MEDIAN IN A BIMODAL DISTRIBUTION

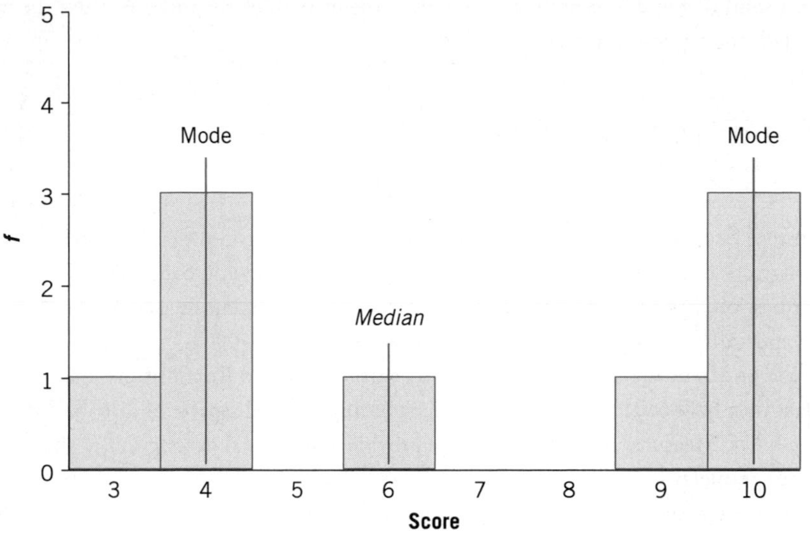

Choosing a Particular Measure of Central Tendency

Table 3.5 summarizes the relative strengths and weaknesses of the mode, median, and mean. Under what conditions might one use a particular measure of central tendency? As the previous examples have demonstrated, the choice will be largely based on the shape of the distribution.

Figure 3.8 displays the mode, median, and mean for three distributions: symmetric, asymmetric (skewed), and bimodal. The distribution in Figure 3.8(a) is symmetric and unimodal. In this situation, the mode, median, and mean are all in the center of the distribution. When the distribution is symmetric and unimodal, the mean is used as the measure of central tendency because it can be used to conduct statistical analyses designed to test hypotheses. The two distributions in Figure 3.8(b) are both skewed. In portraying skewed distributions,

TABLE 3.5 ● STRENGTHS AND WEAKNESSES OF THE MODE, MEDIAN, AND MEAN

Measure of Central Tendency	Strength	Weakness
Mode	• Quick and easy to identify (no calculations necessary) • Can be used with categorical variables • An actual score for the variable • Describes bimodal or multimodal distributions • Not affected by outliers	• Not based on all of the scores in a set of data • Cannot be used in statistical analyses to test hypotheses
Median	• Describes skewed distributions • Not affected by outliers	• Not based on all of the scores in a set of data • Cannot be used in statistical analyses to test hypotheses • May not describe bimodal distributions
Mean	• Based on all of the scores in a set of data • Can be used in statistical analyses to test hypotheses	• Affected by outliers • May not describe skewed distributions • May not describe bimodal distributions

the median is the preferred measure of central tendency because it lies closer to the center of the distribution of scores than does the mean or the mode. Finally, for the bimodal distribution illustrated in Figure 3.8(c), the mode is a more useful measure of central tendency than the median or the mean because the mode highlights the two scores with the highest frequencies.

3.7 MEASURES OF CENTRAL TENDENCY: DRAWING CONCLUSIONS

One of the last steps in conducting statistical analyses is to draw conclusions that interpret the results of statistical analyses in light of a study's research questions or hypotheses. Drawing conclusions from analyses involves answering the question, "So what does it mean?" Answering this question accurately and appropriately is perhaps one of the most difficult aspects of the research process.

What conclusions can be drawn from a measure of central tendency? Recall from Chapter 2 that researchers examine the data they have collected to gain an understanding of three aspects of distributions: modality, symmetry, and variability. Measures of central tendency provide a numerical description of the modality and, to a lesser degree, the symmetry of a variable.

In terms of modality, the most common value for a variable, the information provided by a measure of central tendency can be used to characterize the sample from whom the data were collected. For example, the finding that "the mean GPA of a sample of students was 3.80" may lead to the conclusion that this is a very capable group of students, given that the highest possible value of GPA is 4.00.

In addition to modality, conclusions about the symmetry of the distribution can be drawn from a measure of central tendency. For example, the real estate sections of newspapers often report the "median price of a home." Using the median (rather than the mean) to describe housing prices helps us understand that the distribution of home prices is skewed rather than symmetric, consisting mostly of lower-priced homes with a small percentage of extremely expensive homes.

The study introduced in the beginning of this chapter hypothesized that studying psychology would lead to a lowered belief in the 10% myth. However, the mode and the median of the data collected from the sample of 38 psychology majors indicated that they were in fact more likely to say that people use 10% of their brain power than any other amount. On the basis of their calculation and examination of measures of central tendency, the researchers for the study proposed the following implication:

> Students trained in psychology may not be more critical of the ten-percent claim than are other students. . . . The results of this study suggest that for this popular misconception, additional psychology courses taken by psychology majors may not have much effect. (Higbee & Clay, 1998, p. 472)

LEARNING CHECK 4:

Reviewing What You've Learned So Far

1. Review questions
   a. What is the difference between the sample mean $(\bar{X})$ and the population mean (μ)?
   b. What is the difference between statistics and parameters?
   c. What are the relative strengths and weaknesses of the mode, median, and mean as measures of central tendency? For what types of situations might you use one rather than another?
   d. For which of the three aspects of distributions (modality, symmetry, and variability) do measures of central tendency provide information?

FIGURE 3.8 ● THE MODE, MEDIAN, AND MEAN IN DIFFERENT DISTRIBUTIONS

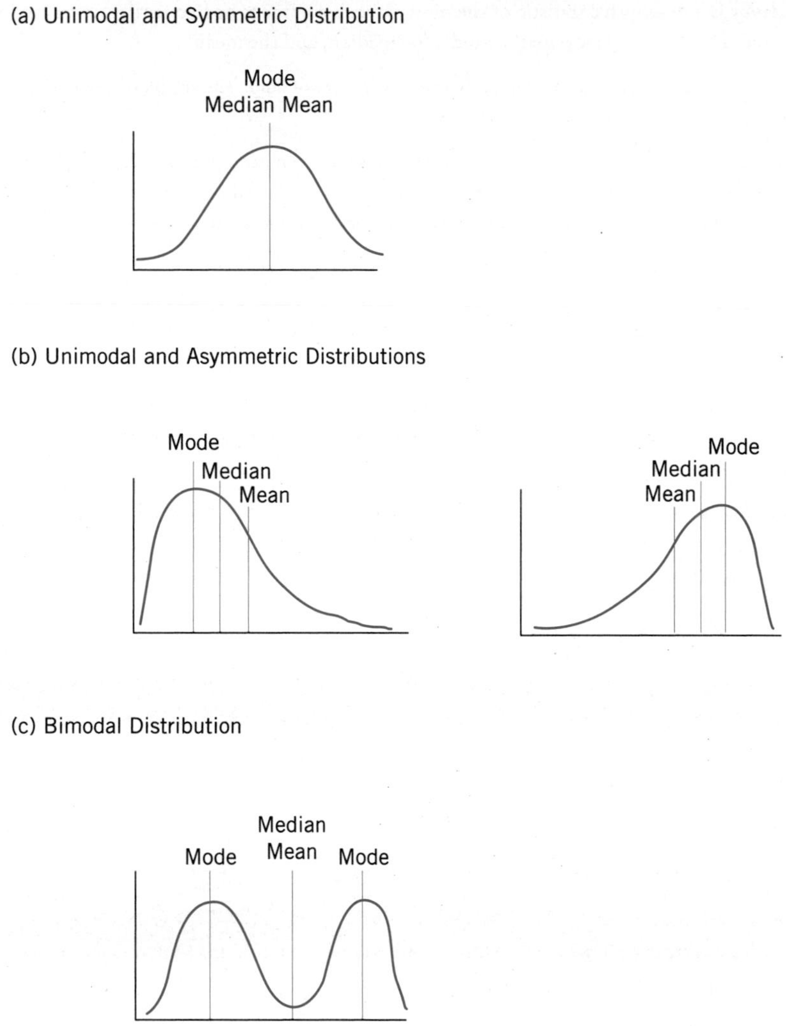

3.8 LOOKING AHEAD

Measures of central tendency provide one way to summarize data by identifying the most common or central score in a set of data. In essence, a measure of central tendency is the score that best represents what the scores have in common with each other. However, to more comprehensively describe a set of data, we must also represent the extent to which the scores are different from each other. For example, imagine you take a midterm and get a score of 39. Your instructor tells the class, "The average score on the midterm was 34.75." How can you evaluate your midterm score? To answer this question, you must take into account the third aspect of distributions: variability, which refers to the amount of differences in a distribution of data. How you evaluate your midterm score depends on the amount of variability in the midterm scores of your class. In Chapter 4, we will discuss how to numerically describe variability.

3.9 Summary

A ***measure of central tendency*** is a descriptive statistic of the most typical, common, or frequently occurring value for a variable. Three common measures of central tendency are the mode, the median, and the mean.

The ***mode*** is the score or value of a variable that appears most frequently in a set of data. The simplest way to determine the mode is to examine a frequency distribution table or figure.

The ***median*** is the number, located at the center of a set of numbers that has been arranged in ascending or descending order, that splits a distribution of scores in half, with the same number of scores above the median as below. The median is determined by rank ordering the scores and uses the appropriate formula based on whether the total number of scores is odd or even.

The ***mean*** is the arithmetic average of a set of scores. The mean is determined by calculating the sum of a set of scores and dividing this sum by the total number of scores. The mean may also be calculated from data that have been organized into a frequency distribution table. The mean is a balancing point because the sum of the negative differences of scores from the mean is always equal to the sum of the positive differences of scores from the mean.

A mean calculated from data for a sample may be referred to as a ***sample mean*** ($\overline{X}$); a mean associated with a population is called the ***population mean*** (μ). A numeric characteristic of a sample (such as $\overline{X}$) is called a ***statistic***; a numeric characteristic of a population (such as μ) is called a ***parameter***.

The mode, median, and mean each have relative strengths and weaknesses. As a result, the mode is the preferred measure of central tendency when a distribution of scores for a variable is bimodal, the median is preferred when the distribution is skewed, and the mean is preferred when the distribution is symmetric and unimodal.

3.10 Important Terms

measure of central tendency (p. 58)
mode (modal score) (p. 60)
median (p. 61)
mean (p. 65)
sample mean (p. 70)
population mean (p. 70)
statistic (p. 70)
parameter (p. 70)

3.11 Formulas Introduced in This Chapter

Median (Odd Number of Scores)

$$\text{Median} = \frac{N+1}{2}\text{th score} \quad (3\text{-}1)$$

Median (Even Number of Scores)

$$\text{Median} = \frac{\frac{N}{2}\text{th score} + (\frac{N}{2}+1)\text{th score}}{2} \quad (3\text{-}2)$$

Mean

$$\overline{X} = \frac{\Sigma X}{N} \quad (3\text{-}3)$$

Mean (Frequency Distribution Table)

$$\overline{X} = \frac{\Sigma(fX)}{N} \quad (3\text{-}4)$$

3.12 Exercises

1. Identify the mode for each of the following sets of data:
   a. 7, 3, 7, 1, 7
   b. 3, 1, 1, 2, 3, 3
   c. 8, 22, 8, 10, 9, 8, 12, 7, 8
   d. 5, 3, 6, 4, 5, 3, 1, 5, 2, 5, 6, 5
   e. 11, 9, 13, 6, 14, 12, 7, 12, 10, 8, 15, 5, 12
   f. 7, 3, 14, 9, 7, 6, 4, 7, 4, 10, 4, 2, 8, 5, 11, 12
2. Identify the mode for each of the following sets of data:
   a. 6, 2, 7, 6, 6, 4
   b. 13, 19, 12, 13, 7, 13, 20, 13, 15
   c. 4, 2, 5, 1, 2, 2, 5, 5, 3, 1, 5, 2, 6, 3
   d. 40, 10, 35, 30, 10, 25, 5, 10, 15, 10, 5, 35, 10, 40, 20
   e. 1, 4, 2, 4, 4, 1, 5, 1, 4, 2, 3, 1, 4, 2, 3, 5, 1
3. Calculate the median for each of the following sets of data:
   a. 3, 4, 6, 7, 10
   b. 14, 16, 19, 21, 22, 27, 36
   c. 6, 6, 9, 9, 9, 10, 13, 16, 16, 21, 24
   d. 4, 2, 6, 3, 8, 3, 8
   e. 27, 24, 35, 30, 41, 32, 36, 35, 27
   f. 3, 5, 15, 6, 9, 10
   g. 11, 13, 13, 16, 17, 22, 24, 27, 28, 30
   h. 12, 10, 9, 11, 9, 15, 10, 9, 8, 13, 9, 10
4. Calculate the median for each of the following sets of data:
   a. 2, 2, 3, 4, 6, 9, 9
   b. 7, 7, 8, 12, 13, 14, 17, 21, 23
   c. 11, 6, 3, 14, 12, 15, 8, 10, 7
   d. 9, 6, 7, 5, 8, 7, 8, 9, 10, 8, 7
   e. 10, 13, 13, 14, 15, 16, 17, 24
   f. 6.0, 8.5, 7.0, 3.0, 4.5, 2.5, 3.5, 7.5, 3.5, 8.0
5. Calculate the mean for each of the following sets of data:
   a. 2, 2, 3, 5
   b. 3, 5, 5, 7, 7, 8, 8, 11
   c. 9, 6, 7, 5, 8, 7, 8, 9, 10, 8, 7
   d. 11, 13, 14, 15, 16, 16, 16, 17, 18, 20, 22, 25
   e. 2.57, 3.62, 2.84, 1.90, 3.15, 3.41, 3.03
   f. 5, 2, 3, 6, 3, 4, 7, 2, 2, 8
   g. 12.50, 10.00, 9.75, 11.50, 9.00, 15.00, 10.25, 9.50, 8.75, 13.75, 9.50, 10.25
6. Calculate the mean for each of the following sets of data:
   a. 1, 1, 2, 3, 3
   b. 1, 4, 4, 5, 6, 7, 8

c. 47, 56, 62, 69, 70, 73

d. .75, .23, .48, .60, .98, .65, .08, .12, .39

e. 11.02, 13.67, 17.39, 14.74, 16.10, 9.28, 12.31, 8.66

f. 12, 10, 9, 11, 9, 15, 10, 9, 8, 13, 9, 10

7. Identify the mode in each of the following frequency distribution tables.

a.

Score	f	%
4	5	31%
3	3	19%
2	6	38%
1	2	12%
Total	16	100%

b.

Score	f	%
5	4	16%
4	8	32%
3	6	24%
2	3	12%
1	2	8%
0	2	8%
Total	25	100%

c.

Score	f	%
8	0	0%
7	2	6%
6	3	8%
5	6	17%
4	10	28%
3	8	21%
2	5	14%
1	2	6%
Total	36	100%

d.

Score	f	%
100	0	0%
90	1	3%
80	2	7%
70	0	0%
60	2	7%
50	3	10%
40	6	20%
30	5	17%
20	8	26%
10	3	10%
Total	30	100%

8. Using Formula 3-4, calculate the mean for the variables in the frequency distribution tables in Exercise 7.

9. If a variable with $N = 25$ has a mean of 3.25, what is the value of ΣX?

10. For the sets of data in Exercise 5(a) and (b), calculate the difference between each score and the mean (use Table 3.2 as an example). Next, compare the sum of the positive and negative differences to see if they balance each other.

11. What makes you happy? Participants in a research experiment reported their levels of happiness and the amount of time spent alone versus with friends or family. The researchers found that people who were reportedly "very happy" spent more time with friends and family and concluded that sociability may be related to happiness (Diener &

Seligman, 2002). The following data represent the number of hours eight "very happy people" spend with friends and family per day (the data below are representative of the study's findings):

Participant	# Hours
1	5
2	3
3	7
4	5

Participant	# Hours
5	4
6	5
7	5
8	7

a. Calculate the number of scores (N) and the sum of scores (ΣX).

b. Calculate the mode, median, and the mean.

12. (This example was introduced in Chapter 2.) A friend of yours asks 20 people to rate a movie using a 1- to 5-star rating: the higher the number of stars, the higher the recommendation. Their ratings are listed below:

Person	# Stars	Person	# Stars	Person	# Stars
1	★★★	8	★★	15	★★★
2	★★★★★	9	★★★	16	★★
3	★★	10	★★★★★	17	★★★
4	★★★★	11	★	18	★★
5	★★★	12	★★★★	19	★★★★
6	★★★	13	★★★★	20	★★★
7	★★★★	14	★★★		

a. Calculate the number of scores (N) and the sum of scores (ΣX).

b. Calculate the mode, median, and the mean.

c. Looking at the mode, median, and mean, how would you describe the shape of the distribution of star ratings?

13. One study stated the research hypothesis, "Violence behavior in children may be reduced by teaching them conflict resolution skills" (DuRant et al., 1996). The variable "violence behavior" was measured by the number of fights in which each student was involved. Below is a frequency distribution table for the number of fights for 13 students in this study.

# Fights	*f*	%
4	1	8%
3	2	15%
2	3	23%

# Fights	*f*	%
1	3	23%
0	4	31%
Total	13	100%

a. Calculate the mode, median, and the mean for these data.

b. Comparing the mode, median, and the mean, how would you describe the shape of the distribution?

14. The 10% myth study discussed in this chapter measured the beliefs of both psychology majors and non–psychology majors. The 39 non–psychology majors in this study provided the following values for the estimated brain power variable:

Non–Psychology Major	Estimated Brain Power	Non–Psychology Major	Estimated Brain Power	Non–Psychology Major	Estimated Brain Power
1	15	14	10	27	15
2	45	15	20	28	25
3	10	16	50	29	10
4	40	17	25	30	20
5	5	18	45	31	10
6	10	19	40	32	5
7	50	20	25	33	15
8	45	21	10	34	10
9	10	22	5	35	5
10	10	23	15	36	60
11	35	24	5	37	10
12	5	25	40	38	5
13	15	26	30	39	15

a. Calculate the number of scores (N) and the sum of scores (ΣX).

b. Calculate the mode, median, and the mean.

c. Looking at the mode, median, and mean, how would you summarize the beliefs of the non–psychology majors? How would you compare their beliefs to those of the psychology majors discussed earlier in this chapter?

15. At an ice skating competition, the score for each skater is the mean of the different judges' scores. However, before this mean is calculated, the lowest and the highest scores among the judges are discarded.

a. Why are the highest and lowest scores not included in the calculation of the mean?

b. Which measure(s) of central tendency would eliminate the need to discard scores?

16. Given the following values for the mode, median, and mean, determine whether you believe the distribution is symmetrical, positively skewed, or negatively skewed.

a. Mode = 4, median = 5, mean = 8

b. Mode = 9, median = 8, mean = 4

c. Mode = 6, median = 6, mean = 6

17. On your own, generate a set of data where the mean, median, and mode are the same, and then create a figure for these data. How would you describe the shape of this distribution?

18. On your own, generate a set of data that is positively skewed, and then create a figure for these data. If you calculate the mean and median of these data, which is larger?

Answers to Learning Checks

Learning Check 1

2. a. $Mo = 2$
   b. $Mo = 6$
   c. $Mo = 12$
   d. $Mo = 2$
   e. $Mo = 3.2$

Learning Check 2

2. a. $Mdn = 4$
   b. $Mdn = 14$
   c. $Mdn = 10.50$
   d. $Mdn = 80.50$
   e. $Mdn = 9$

Learning Check 3

2. a. $M = 4.00$
   b. $M = 6.00$
   c. $M = 75.50$
   d. $M = 2.39$
   e. $M = 9.83$

3. a. $M = 2.50$
   b. $M = 3.61$

4. a.

Score	Mean	Score – Mean
2	4.00	–2.00
3	4.00	–1.00
3	4.00	–1.00
5	4.00	1.00
7	4.00	3.00

Negative differences from mean = Positive differences from mean

$$-2.00 + -1.00 + -1.00 = 1.00 + 3.00$$

$$-4.00 = 4.00$$

b.

Score	Mean	Score – Mean
8	6.00	2.00
6	6.00	.00

(Continued)

(Continued)

Score	Mean	Score – Mean
4	6.00	–2.00
8	6.00	2.00
3	6.00	–3.00
7	6.00	1.00

$$\text{Negative differences from mean} = \text{Positive differences from mean}$$
$$-2.00 + -3.00 = 2.00 + 2.00 + 1.00$$
$$-5.00 = 5.00$$

Answers to Odd-Numbered Exercises

1. a. $Mo = 7$
   b. $Mo = 3$
   c. $Mo = 8$
   d. $Mo = 5$
   e. $Mo = 12$
   f. $Mo = 4$ and 7

3. a. $Mdn = 6$
   b. $Mdn = 21$
   c. $Mdn = 10$
   d. $Mdn = 4$
   e. $Mdn = 32$
   f. $Mdn = 5.50$
   g. $Mdn = 19.50$
   h. $Mdn = 10.00$

5. a. $M = 3.00$
   b. $M = 6.75$
   c. $M = 7.64$
   d. $M = 16.92$
   f. $M = 2.93$
   g. $M = 4.20$
   h. $M = 10.81$

7. a. $Mo = 2$
   b. $Mo = 4$
   c. $Mo = 4$
   d. $Mo = 20$

9. $\Sigma X = (3.25 * 25) = 81.25$

11. a. $N = 8$, $\Sigma X = 41$
    b. Mode = 5, median = 5, mean = 5.13

13. a. Mode = 0, median = 1, mean = 1.46
    b. The fact that the mode ($Mo = 0$) is less than the median ($Mdn = 1$), which in turn is less than the mean ($M = 1.46$), suggests that the distribution of the number of fights is positively skewed, with the majority of the sample having relatively few fights.

15. a. The highest and lowest scores are not included to prevent the mean from being affected by possible outliers.
    b. If they used the median or the mode rather than the mean, they would not need to discard any scores because the median and mode are not affected by outliers.

17. Note: A distribution with equal values for the mode, median, and mean should be symmetrical.

4

MEASURES OF VARIABILITY

CHAPTER OUTLINE

Chapter 3 introduced descriptive statistics, the purpose of which is to numerically describe or summarize data for a variable. As we learned in Chapter 3, a set of data is often described in terms of the most common or frequently occurring score in the set of data by using a measure of central tendency such as the mode, median, or mean. Measures of central tendency numerically describe two aspects of a distribution of scores, modality and symmetry, based on what the scores have in common with each other. However, researchers also need to describe the amount of *differences* among the scores of a variable. This chapter introduces measures of variability, which are designed to represent the amount of differences in a distribution of data. In this chapter, we will present and evaluate different measures of variability in terms of their relative strengths and weaknesses. As in the previous chapters, our presentation will rely on the findings and analysis from a published research study.

4.1 AN EXAMPLE FROM THE RESEARCH: HOW MANY "SOMETIMES" IN AN "ALWAYS"?

Throughout our daily lives, people are constantly asked to fill questionnaires and surveys. Restaurants ask customers to evaluate the quality of the food and service they provide; websites want to know how satisfied shoppers are with their online purchases; political organizations are interested in learning what voters think about their elected officials. In responding to these types of scenarios, people are often asked to describe their feelings or beliefs using words such as *excellent, good,* or *poor.* But do words such as these have the same meaning for different people? If research participants disagree about the meaning of key terms and phrases, it's difficult to combine or compare their responses.

Two researchers, Suzanne Skevington and Christine Tucker, conducted a study assessing differences in interpretation of labels for questionnaire items measuring the frequency of health-related behaviors, such as, "How often do you suffer pain?" (Skevington & Tucker, 1999). As a part of their study, they examined how British people define frequency-related terms such as *seldom, usually,* and *rarely.* More specifically, they assigned numeric values to participants' evaluation of these words in order to measure the degree to which people differ in their interpretation of these words.

In their study, Skevington and Tucker (1999) presented 20 British adults a variety of frequency-related words. For each word, they included a line 100 mm (about 4 inches) in length. At the two ends of the line were the labels *Never* and *Always,* representing the lowest (0%) and highest (100%) possible values for frequency. After reading each word, participants were asked to place an X on the point in the line that they felt best represented the frequency of the word, relative to the end points of Never and Always. An example of their methodology is provided in Figure 4.1.

After the participants completed their responses, the researchers used a ruler to measure the distance from the word *Never* to the "X" provided by the participant. This distance, which was measured in millimeters, represented the perceived "frequency" of the word. For example, in Figure 4.1, the X for the word *sometimes* is in the exact middle of the line; consequently, this word is given a frequency rating of 50% because it is 50 mm from the left end. In this study, the variable, which we will call "frequency rating," had possible values ranging from 0 to 100.

In this example, we will focus on the study's findings regarding the word *sometimes.* The frequency ratings for the word *sometimes* for the 20 participants are listed in Table 4.1(a). Table 4.1(b) organizes the raw data into a grouped frequency distribution table, and Figure 4.2 illustrates the distribution using a frequency polygon. A frequency polygon is used rather than a histogram or pie chart because the variable "frequency rating" is measured at the ratio level of measurement, with the value of 0 representing the complete absence of distance from the left end of the 100-mm line.

As discussed in Chapter 3, the data for a variable can be summarized using a measure of central tendency. Using Formula 3-3 from Chapter 3, the mean frequency rating of the word *sometimes* for the 20 participants is calculated as follows:

$$\begin{aligned}\bar{X} &= \frac{\Sigma X}{N} \\ &= \frac{46+15+21+\ldots+22+66+26}{20} = \frac{666}{20} \\ &= 33.30\end{aligned}$$

From this calculation, we conclude that the participants in the study believed, on average, the word *sometimes* represents a 33.30% frequency. However, the grouped frequency distribution table in Table 4.1(b) reveals that

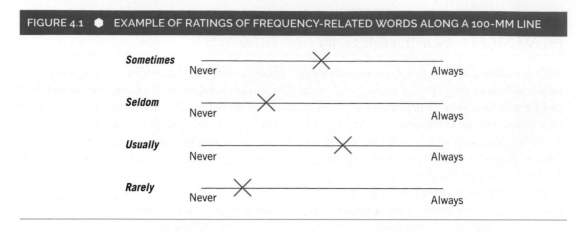

FIGURE 4.1 ● EXAMPLE OF RATINGS OF FREQUENCY-RELATED WORDS ALONG A 100-MM LINE

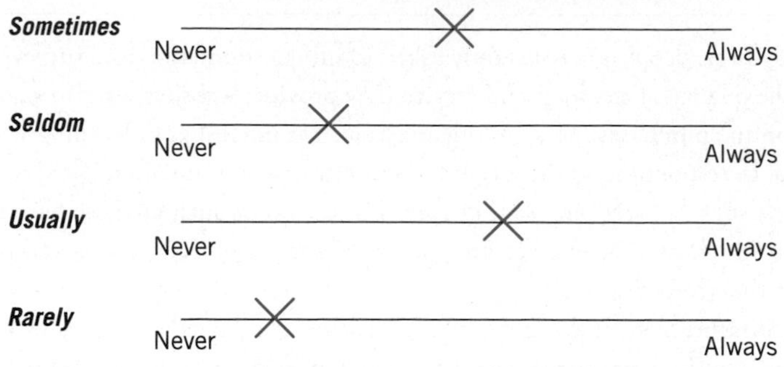

participants provided a wide range of ratings, some of which were far from the average rating of 33.30%. For example, the ratings provided by 11 of the 20 participants were lower than 30%, while 6 of the 20 participants provided ratings of 40% or higher. In other words, the people in this sample greatly varied in their interpretation of the word *sometimes.*

Given the variety of ratings these participants gave the word *sometimes,* in order to accurately describe the distribution of responses for this variable, one must not only include a measure of central tendency (which focuses on what the scores have in common with each other) but also represent the degree to which the scores differ, or vary. The next section introduces the concept of variability.

4.2 UNDERSTANDING VARIABILITY

We begin our discussion of variability by examining the three distributions presented in Figure 4.3. Chapter 2 discussed three aspects of distributions: modality, symmetry, and variability. The modality and symmetry of a distribution can be described using measures of central tendency such as those introduced in Chapter 3. However, because all three distributions in Figure 4.3 are unimodal and symmetric, they would be described as having the same mode, median, and mean. It is obvious, however, that the three distributions are very different. As the remainder of this chapter will illustrate, these differences can be portrayed using a statistic that describes the third aspect of distributions: variability.

The word *variability* typically evokes words such as *differences* or *changes.* In a statistical sense, variability refers to the amount of spread or scatter of scores in a distribution. The concept of variability is a critical issue in the behavioral sciences, where research frequently examines differences in such things as characteristics, attitudes, and cognitive abilities. Different people, for example, express different levels of extroversion, different attitudes toward capital punishment, and different learning styles. Ultimately, the primary goal of a science such as psychology is to describe, understand, explain, and predict variability.

Researchers have developed statistics designed to measure variability. A **measure of variability** is a descriptive statistic of the amount of differences in a set of data for a variable. The purpose of measures of variability is to numerically represent a set of data based on how the scores differ or vary from each other. Similar to measures of central tendency, there are multiple measures of variability. The next part of this chapter presents and discusses four measures of variability:

- the range,
- the interquartile range,

TABLE 4.1 ● FREQUENCY RATING OF THE WORD SOMETIMES BY 20 PARTICIPANTS

(a) Raw Data

Participant	Frequency Rating	Participant	Frequency Rating	Participant	Frequency Rating
1	**46**	8	**27**	15	**49**
2	**15**	9	**20**	16	**32**
3	**21**	10	**23**	17	**45**
4	**49**	11	**58**	18	**22**
5	**23**	12	**24**	19	**66**
6	**39**	13	**36**	20	**26**
7	**25**	14	**20**		

(b) Grouped Frequency Distribution Table

Frequency Rating	*f*	%
70–100	0	0%
60–69	1	5%
50–59	1	5%
40–49	4	20%
30–39	3	15%
20–29	10	50%
10–19	1	5%
0–9	0	0%
Total	20	100%

- the variance, and
- the standard deviation.

Each of these measures of variability will be defined, illustrated in terms of their necessary calculations, and evaluated based on their relative strengths and weaknesses.

4.3 THE RANGE

One way to describe the amount of variability in a distribution of data for a variable is to focus on the two ends of the distribution. Therefore, the first measure of variability we will discuss is the **range**, defined as the mathematical difference between the lowest and highest scores in a set of data:

$$\text{Range} = \text{highest score} - \text{lowest score} \tag{4-1}$$

The range is computed by identifying the lowest and highest scores in a set of data and then subtracting the lowest score from the highest score to compute the difference between the two scores.

FIGURE 4.2 ● FREQUENCY POLYGON, FREQUENCY RATINGS OF *SOMETIMES*

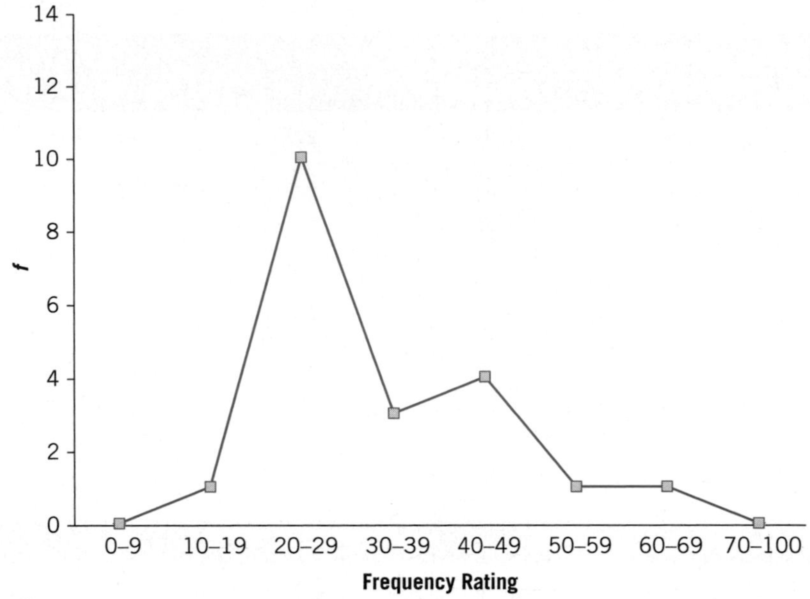

FIGURE 4.3 ● THREE DISTRIBUTIONS WITH THE SAME MODALITY AND SYMMETRY BUT DIFFERENT VARIABILITY

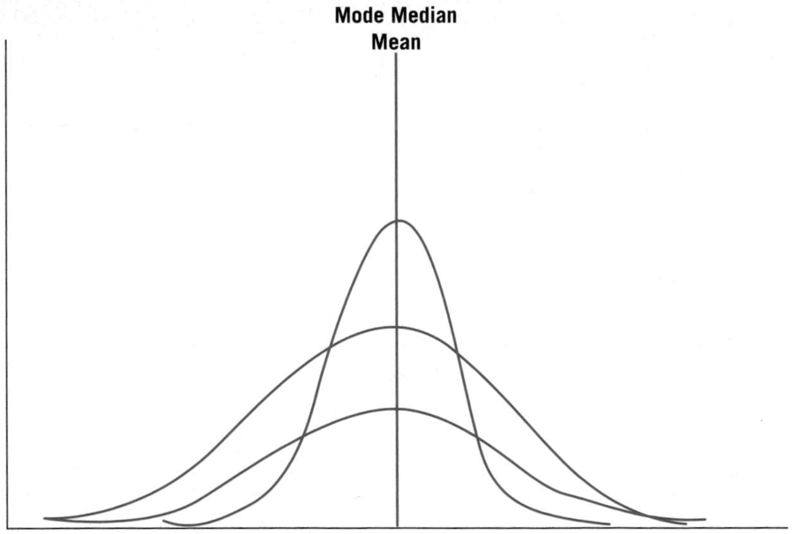

To illustrate how to calculate the range, let's return to the small set of data introduced in Chapter 3 to discuss measures of central tendency: 2, 3, 1, 4, 3, 1, 4, 6, and 3. Among the nine scores in this set of data, the lowest score is 1 and the highest score is 6. Using Formula 4-1, the range for this set of data is the following:

$$\begin{aligned} \text{Range} &= \text{highest score} - \text{lowest score} \\ &= 6 - 1 \\ &= 5 \end{aligned}$$

As a second example, to calculate the range for the frequency rating variable in Table 4.1(a), the ratings of the 20 participants are sorted from lowest to highest:

Lowest	*2nd*	*3rd*	*4th*	*5th*	*6th*	*7th*	*8th*	*9th*	*10th*	*11th*	*12th*	*13th*	*14th*	*15th*	*16th*	*17th*	*18th*	*19th*	*Highest*
15	20	20	21	22	23	23	24	25	26	27	32	36	39	45	46	49	49	58	66

Among the 20 participants, the lowest and highest ratings are 15 and 66, respectively. Therefore, the range for the frequency rating variable is

$$\begin{aligned} \text{Range} &= \text{highest score} - \text{lowest score} \\ &= 66 - 15 \\ &= 51 \end{aligned}$$

When researchers report the range for a variable, they generally provide the actual values of the lowest and highest scores. For the frequency rating variable, for example, the range might be reported as the following: "In this sample of 20 participants, frequency ratings for the word *sometimes* had a range of 51% (low = 15%, high = 66%)." Including the lowest and highest scores may provide information about the sample from which the data were collected. For example, even though the range of incomes ($35,000) may be the same in two samples, a sample in which the lowest and highest incomes are $5,000 and $40,000 would be considered differently from a sample in which the lowest and highest incomes are $175,000 and $210,000.

Strengths and Weaknesses of the Range

As a measure of variability, the range has comparative strengths and weaknesses. The primary strength of the range is that it is easy and quick to compute, particularly if the sample is small or if computer software is used to sort the scores from lowest to highest. A second strength of the range, as mentioned earlier, is that indicating the lowest and highest scores for a variable provides information about the sample from which the data were collected.

However, because the range is calculated from only two scores in the distribution (the lowest and highest), it may not accurately reflect the amount of variability in the entire distribution of scores. Consider, for example, the following two sets of scores sorted from lowest to highest:

Set 1: ***1*** 2 2 3 3 3 4 4 ***5***

Set 2: ***1*** 3 4 5 5 5 5 5 ***5***

Although the range for both sets of data is the same (5 − 1 = 4), there is much more variability in the first set than in the second. The data in Set 2 illustrate another weakness of the range: It is affected by outliers (in this case, the one score of 1). Because the range does not take into account all of the scores in the distribution, a fundamental weakness of the range is that it cannot be used in statistical analyses designed to test hypotheses about distributions.

4.4 THE INTERQUARTILE RANGE

One of the limitations of the range as a measure of variability is that it is affected by extreme scores known as outliers. One way to overcome the influence of outliers on the range is to calculate the **interquartile range,** which is the range of the middle 50% of the scores in a set of data. The interquartile range is calculated using Formula 4-2:

$$\text{Interquartile range} = (N - \frac{N}{4})\text{th score} - (\frac{N}{4} + 1)\text{th score} \qquad (4\text{-}2)$$

where N is the total number of scores. The interquartile range is calculated by removing the highest and lowest 25% of the distribution and then calculating the range of the remaining scores. The primary purpose of the interquartile range is to decrease the influence of outliers in representing the variability in a set of data.

As a simple example, consider the following eight scores ($N = 8$), sorted from lowest to highest:

Lowest	*2nd*	*3rd*	*4th*	*5th*	*6th*	*7th*	*Highest*
11	**15**	**19**	**24**	**31**	**39**	**42**	**89**

Using Formula 4-2, the first step is to identify the $(N - \frac{N}{4})$ and the $(\frac{N}{4} + 1)$ th scores. Because $N = 8$ in this example,

$$\begin{aligned}\text{Interquartile range} &= (N - \frac{N}{4})\text{th score} - (\frac{N}{4} + 1)\text{th score}\\ &= (8 - \frac{8}{4})\text{th score} - (\frac{8}{4} + 1)\text{th score}\\ &= (8 - 2)\text{th score} - (2 + 1)\text{th score}\\ &= 6\text{th score} - 3\text{rd score} = 39 - 19\\ &= 20\end{aligned}$$

So, for this small set of data, the interquartile range is the difference between the values of the sixth and the third scores. Among the sorted scores, the sixth score is 39 and the third score is 19; therefore, the interquartile range is equal to (39 – 19), or 20. Note that the interquartile range of 20 is much smaller than the range of this set of data, which is (89 – 11), or 78.

To calculate the interquartile range for the frequency rating variable, because $N = 20$ in this example, we start by entering the value 20 into Formula 4-2:

$$\begin{aligned}\text{Interquartile range} &= (N - \frac{N}{4})\text{th score} - (\frac{N}{4} + 1)\text{th score}\\ &= (20 - \frac{20}{4})\text{th score} - (\frac{20}{4} + 1)\text{th score}\\ &= (20 - 5)\text{th score} - (5 + 1)\text{th score}\\ &= 15\text{th score} - 6\text{th score} = 45 - 23\\ &= 22\end{aligned}$$

To calculate the interquartile range for the frequency rating variable, the 15th and 6th scores must be identified. Earlier, the 20 ratings were sorted from lowest to highest to calculate the range; looking at the sorted ratings, we find the 15th score is equal to 45 and the 6th score is equal to 23. Therefore, the interquartile range for this set of data is equal to (45 – 23), or 22.

Strengths and Weaknesses of the Interquartile Range

Compared with the range, the primary purpose of the interquartile range is to represent the variability in a set of data while lessening the influence of outliers. However, using the interquartile range has the potential to misrepresent a set of data by ignoring half (the top and bottom 25%) of the scores. It is somewhat counterintuitive to measure the variability in a set of data with only half of the data. Also, because the interquartile range, like the range, does not take into account all of the scores in the distribution, it cannot be used in statistical analyses designed to test hypotheses about distributions.

Given its limitations, under what conditions is the interquartile range most appropriately used? Concern about the impact of outliers was discussed in Chapter 3, where one advantage of the median as a measure of central tendency is that it is not affected by outliers. Therefore, the interquartile range is typically reported along with the median to represent the variability and central tendency in distributions that are skewed or have outliers.

LEARNING CHECK 1:

Reviewing What You've Learned So Far

1. Review questions
   a. Why is it important to calculate measures of variability for a variable in addition to measures of central tendency?
   b. What are the relative advantages and disadvantages of the range as a measure of variability?
   c. What is the difference between the range and the interquartile range?
   d. For what types of distributions might you report the interquartile range?
2. Calculate the range for each of the following sets of data:
   a. 2, 1, 4, 2, 7, 3
   b. 15, 6, 17, 9, 12, 5, 16, 7
   c. 21, 14, 13, 17, 30, 17, 14, 11, 13, 14, 19
   d. 3.0, 2.8, 3.2, 3.6, 3.7, 3.2, 3.5, 3.6, 3.1, 3.2, 3.0, 3.2, 3.8, 3.3
3. Calculate the interquartile range for each of the following sets of data:
   a. 9, 3, 2, 7, 15, 10, 14, 8
   b. 15, 6, 17, 9, 12, 5, 16, 7
   c. 13, 9, 9, 16, 10, 9, 5, 7, 9, 8, 10, 9
   d. 3.0, 2.8, 3.2, 3.6, 3.7, 3.2, 3.5, 3.2, 3.0, 3.2, 3.8, 3.3

4.5 THE VARIANCE (s^2)

The range and interquartile range describe the variability of a distribution of data based on two scores at or near the ends of the distribution. However, other measures of variability are based on all of the scores in a set of data. These other measures are an essential component of statistics designed to test research hypotheses.

To illustrate measures of variability based on all of the scores in a set of data, let's return to the simple data set of nine scores used earlier to illustrate the range: 2, 3, 1, 4, 3, 1, 4, 6, and 3. The mean $(\overline{X})$ of this sample of nine scores is calculated below:

$$\begin{aligned}\overline{X} &= \frac{\Sigma X}{N}\\ &= \frac{2+3+1+4+3+1+4+6+3}{9} = \frac{27}{9}\\ &= 3.00\end{aligned}$$

One way to measure the variability in a set of data is based on the extent to which each score differs from the mean. You could, for example, ask the question, "On average, how different is each score from the mean?" In the set of nine scores, variability could be represented by the average difference between each score (X) and the mean of 3.00 ($\bar{X} = 3.00$). Referring to this difference as a "deviation," the third column in Table 4.2 calculates the deviation of each of the nine scores from the mean, symbolized by $(X - \bar{X})$.

Calculating the average of the deviations consists of dividing the summed deviations by the number of deviations, which is equal to N. The formula for the average deviation from the mean is provided below:

$$\text{Average deviation from the mean} = \frac{\Sigma(X - \bar{X})}{N}$$

The numerator of this formula is "the sum of deviations from the mean." Note that parentheses are placed around $X - \bar{X}$ to indicate that this deviation should be calculated for each score *before* adding the deviations together; if we did not include the parentheses, the notation $\Sigma X - \bar{X}$ would imply that the first step is to calculate the sum of the scores (ΣX) and then subtract the mean ($\bar{X}$) from this sum. As in all mathematical calculations, the placement of parentheses indicates the order in which mathematical operations are performed.

Using the deviations calculated in Table 4.2, the average deviation from the mean for the set of nine scores is calculated below:

$$\begin{aligned}\text{Average deviation from the mean} &= \frac{\Sigma(X - \bar{X})}{N} \\ &= \frac{-1.00 + .00 + -2.00 + \ldots + 1.00 + 3.00 + .00}{9} = \frac{0}{9} \\ &= .00\end{aligned}$$

Here, the average deviation from the mean is equal to zero (0). From this, we could conclude, "There is zero variability among the nine scores." However, concluding there is "zero" variability implies that all nine scores are exactly the same, which we know is not true. How did we reach the erroneous conclusion that all the scores are identical?

In *any* set of data, the average deviation from the mean is *always* equal to zero. This is because, as we explained in Chapter 3, the sum of the positive deviations from the mean is always equal to the sum of the negative deviations, which results in the sum of the deviations being equal to zero. Because the sum of the deviations is always equal to zero, the average deviation from the mean is also always equal to zero, regardless of the actual amount of variability among the scores.

Definitional Formula for the Variance

As we've just discussed, calculating a measure of variability that is based on the deviations from the mean is complicated by the fact that the sum of the negative and positive deviations will always be equal to each other. How can we overcome this issue? One way to resolve the "balancing act" of negative and positive deviations is to eliminate the negative deviations by squaring each deviation: $(X - \bar{X})^2$. The square of any number, negative or positive, is always a positive number.

So, rather than calculating the average deviation from the mean, the amount of variability in a sample of data may be measured by calculating the **variance** (**s^2**), defined as the average squared deviation from the mean. Formula 4-3 provides what is known as the definitional formula for the variance:

$$s^2 = \frac{\Sigma(X - \bar{X})^2}{N - 1} \tag{4-3}$$

where X is a score for the variable, $\bar{X}$ is the mean of the sample, and N is the total number of scores.

Two important aspects of the formula for the variance should be considered here. First, in the numerator, parentheses are placed around the deviation between each score and the mean $(X - \bar{X})$ to first calculate the deviation before performing any squaring or summing. Second, in the denominator, the sum of the squared deviations is not divided by the total number of scores (N) but instead is divided by the total number of scores minus 1 ($N - 1$). The rationale for dividing by $N - 1$ rather than by N will be discussed after we illustrate the calculations for the variance.

Table 4.3 begins the calculation of the variance for the set of nine scores. The first three columns of this table are identical to the same columns in Table 4.2—the fourth column squares each of the deviations. For example, the squared deviation for the first score is $(-1.00)^2$ or 1.00. Note that all of the squared deviations are positive numbers.

Using the squared deviations calculated in Table 4.3, the variance for the set of nine scores can be calculated using Formula 4-3:

$$\begin{aligned} s^2 &= \frac{\Sigma(X - \bar{X})^2}{N-1} \\ &= \frac{1.00 + .00 + 4.00 + \ldots + 1.00 + 9.00 + .00}{9-1} = \frac{20.00}{8} \\ &= 2.50 \end{aligned}$$

From this calculation, we may conclude that the variance in this set of data is equal to 2.50. This means that, for the sample of nine scores, the average squared deviation of a score from the mean is 2.50.

In calculating the variance, because a squared deviation is always a positive number, the sum of the squared deviations $(\Sigma(X - \bar{X})^2)$ and subsequently the variance (s^2) must also always be positive numbers. This is an important point that helps students identify calculation errors in homework and paper assignments: *Obtaining a negative value for the sum of squared deviations or the variance means you have made a mistake in your calculations.*

As a second example, let's calculate the variance for the frequency rating variable. Using the sample mean $(\bar{X})$ of 33.30 calculated earlier in this chapter, the last column in Table 4.4 shows the squared deviation for each of the 20 ratings in this sample.

TABLE 4.2 ● CALCULATION OF THE DEVIATION OF EACH SCORE FROM THE MEAN $(X - \bar{X})$, SIMPLE EXAMPLE

Score (X)	Mean ($\bar{X}$)	Score – Mean ($X - \bar{X}$)
2	3.00	–1.00
3	3.00	.00
1	3.00	–2.00
4	3.00	1.00
3	3.00	.00
1	3.00	–2.00
4	3.00	1.00
6	3.00	3.00
3	3.00	.00

TABLE 4.3 ● CALCULATION OF SQUARED DEVIATION FROM THE MEAN ($(X - \bar{X})^2$), SIMPLE EXAMPLE

Score (X)	Mean ($\bar{X}$)	Score – Mean ($X - \bar{X}$)	$(\text{Score} - \text{Mean})^2$ $(X - \bar{X})^2$
2	3.00	–1.00	1.00
3	3.00	.00	.00
1	3.00	–2.00	4.00
4	3.00	1.00	1.00
3	3.00	.00	.00
1	3.00	–2.00	4.00
4	3.00	1.00	1.00
6	3.00	3.00	9.00
3	3.00	.00	.00

Using the squared deviations, the variance for the frequency rating variable is calculated using Formula 4-3 as follows:

$$\begin{aligned} s^2 &= \frac{\Sigma(X-\bar{X})^2}{N-1} \\ &= \frac{161.29+334.89+151.29+\ldots+127.69+1{,}069.29+53.29}{20-1} = \frac{3{,}940.20}{19} \\ &= 207.38 \end{aligned}$$

For this sample of 20 participants, the variance, which is to say the average squared deviation of a frequency rating from the mean of 33.30, is equal to 207.38.

Computational Formula for the Variance

The formula for the variance presented in Formula 4-3 represents the literal definition of variance: the average squared deviation from the mean. For this reason, it is referred to as a **definitional formula,** which is a formula based on the actual or literal definition of a concept. However, using a definitional formula to analyze a set of data can be tedious (because it requires calculating the deviation of each score from the mean) and complicated (because the mean often possesses decimal places [e.g., 33.30]). Because using the definitional formula for the variance in a large or complicated data set increases the chances of making computational errors, the formula can be algebraically manipulated to create what is known as a **computational formula,** defined as a formula not based on the definition of a concept but is designed to simplify mathematical calculations.

Formula 4-4 provides the computational formula for the variance:

$$s^2 = \frac{\Sigma X^2 - \frac{(\Sigma X^2)}{N}}{N-1} \tag{4-4}$$

where ΣX^2 is the sum of squared scores, $(\Sigma X)^2$ is the sum of scores squared, and N is the total number of scores. It is critical to understand that ΣX^2 and $(\Sigma X)^2$ involve a different order of operations. ΣX^2, which is the sum of squared scores, involves first squaring each score and then summing the squared scores (i.e., first squaring, then summing). On the other hand, $(\Sigma X)^2$, which is the sum of scores squared, involves first summing a set of scores and then squaring this sum (i.e., first summing, then squaring).

TABLE 4.4 ● CALCULATION OF SQUARED DEVIATION FROM THE MEAN ($(X - \bar{X})^2$), FREQUENCY RATING VARIABLE

Frequency Rating (X)	Mean ($\bar{X}$)	Frequency Rating – Mean ($X - \bar{X}$)	(Frequency Rating – Mean)^2 ($X - \bar{X})^2$
46	33.30	12.70	161.29
15	33.30	–18.30	334.89
21	33.30	–12.30	151.29
49	33.30	15.70	246.49
23	33.30	–10.30	106.09
39	33.30	5.70	32.49
25	33.30	–8.30	68.89
27	33.30	–6.30	39.69
20	33.30	–13.30	176.89
23	33.30	–10.30	106.09
58	33.30	24.70	610.09
24	33.30	–9.30	86.49
36	33.30	2.70	7.29
20	33.30	–13.30	176.89
49	33.30	15.70	246.49
32	33.30	–1.30	1.69
45	33.30	11.70	136.89
22	33.30	–11.30	127.69
66	33.30	32.70	1069.29
26	33.30	–7.30	53.29

Table 4.5 begins the process of calculating the variance for the set of nine scores using the computational formula. The bottom of the first column contains the sum of the nine scores ($\Sigma X = 27$); the sum of scores squared ($(\Sigma X)^2$) is equal to $(27)^2$ or 729. The second column of Table 4.5 provides the squared values of each of the scores—the sum of squared scores (ΣX^2) is located at the bottom of this column ($\Sigma X^2 = 101$). The results of these calculations are then inserted into Formula 4-4 as follows:

$$s^2 = \frac{\Sigma X^2 - \frac{(\Sigma X^2)}{N}}{N-1}$$

$$= \frac{101 - \frac{729}{9}}{9-1} = \frac{101 - 81.00}{8} = \frac{20.00}{8}$$

$$= 2.50$$

It is important to note that the same value for the variance ($s^2 = 2.50$) was obtained whether we use the computational formula in Formula 4-4 or the definitional formula in Formula 4-3.

TABLE 4.5 ● CALCULATING THE SUM OF SQUARED SCORES (ΣX^2) AND SUM OF SCORES SQUARED ($(\Sigma X)^2$), SIMPLE EXAMPLE

Score (X)	(Score)^2 (X^2)
2	4
3	9
1	1
4	16
3	9
1	1
4	16
6	36
3	9
$\Sigma X = 27$	$\Sigma X^2 = 101$
$(\Sigma X)^2 = 729$	

To calculate the variance for the frequency rating variable using the computational formula, Table 4.6 calculates the sum of squared scores (ΣX^2) and the sum of scores squared ($(\Sigma X)^2$) for the 20 participants. Once these two quantities have been calculated, the variance may be calculated as follows:

$$s^2 = \frac{\Sigma X^2 - \frac{(\Sigma X)^2}{N}}{N-1}$$

$$= \frac{26{,}118 - \frac{443{,}556}{20}}{20-1} = \frac{26{,}118 - 22{,}177.80}{19} = \frac{3{,}940.20}{19}$$

$$= 207.38$$

The value of 207.38 for the variance is again identical using the definitional or the computational formula. One concern with using the computational formula is that the formula can confuse students because it does not appear to represent the concept of variance, which is based on the difference between each score and the mean $(X - \bar{X})$. However, we've included the computational formula because it simplifies the calculation of the variance, thereby reducing the chance of making computational errors.

Why Not Use the Absolute Value of the Deviation in Calculating the Variance?

To compute the variance using the definitional formula provided in Formula 4-3, the deviation between each score and the mean must be squared to eliminate negative deviations. Instead of doing all of this squaring, you may be asking yourself, "Wouldn't it be easier to eliminate the negative deviations simply by using the absolute value of each deviation? If so, you could simply calculate the average of the absolute values." The absolute value of a number, symbolized by parallel vertical lines, ignores the sign (+/–) of the number. For example, $|2 - 3.00|$, which is the absolute value of the deviation 2 – 3.00, is equal to 1.00.

TABLE 4.6 ● SUM OF SQUARED SCORES (ΣX^2) AND SUM OF SCORES SQUARED ($(\Sigma X)^2$), FREQUENCY RATING VARIABLE

Frequency Rating (X)	(Frequency Rating)^2 (X^2)	Frequency Rating (X)	(Frequency Rating)^2 (X^2)
46	2,116	58	3,364
15	225	24	576
21	441	36	1,296
49	2,401	20	400
23	529	49	2,401
39	1,521	32	1,024
25	625	45	2,025
27	729	22	484
20	400	66	4,356
23	529	26	676
		$\Sigma X = 666$	$\Sigma X^2 = 26{,}118$
		$(\Sigma X)^2 = 443{,}556$	

The logic behind using absolute values of the deviations may be intuitively appealing. However, the statistical procedures discussed in this book that involve variability require algebraic manipulations that cannot be carried out using absolute values. For this reason, it is necessary to use formulas for the variance that are based on the squaring of the deviations.

Why Divide by $N - 1$ Rather Than N in Calculating the Variance?

In calculating the variance, students often wonder why the numerator is divided by $N - 1$, rather than by N. Given that the variance is the *average* squared deviation, it would seemingly make sense to divide the sum of the squared deviations by N. One reason for dividing by $N - 1$ is to estimate the variability in a population using data collected from a sample of the population. To illustrate this concept, consider an example in which data are collected from a sample of 100 freshmen at a local university in order to represent the entire population of freshmen nationwide. It is reasonable to suspect that the smaller sample of local university students would be less diverse than the entire population. That is, the sample may not accurately represent all of the possible values for ethnicity, economic background, age, attitudes, and so on that exist in the entire population. Consequently, the amount of variability in the sample will be less than what is believed to exist in the population.

Because the amount of variability in a sample is less than the variability in the population from which the sample is drawn, the sample variance underestimates the variance in the population. As a result, the sample variance is a **biased estimate** of the population variance; a biased estimate is a statistic for a sample that systematically underestimates or overestimates the population from which the sample was drawn. What is needed, therefore, is to correct the sample variance to make it an **unbiased estimate** of the population variance, which is a statistic based on a sample that is equally likely to underestimate or overestimate the population from which the sample was drawn.

Given that the sample variance systematically underestimates the population variance, we need to increase the value of the sample variance to more accurately estimate the variability in the population. This sample variance is increased by dividing the numerator, the sum of squared deviations, by $N - 1$ rather than by N; dividing by a smaller number makes the result larger.

4.6 THE STANDARD DEVIATION (*s*)

The variance is the average squared deviation of a score from the mean. However, researchers typically want to represent the variability in a set of data using the average deviation, not the average *squared* deviation. This representation is accomplished by calculating a measure of variability known as the **standard deviation (*s*)**, defined as the square root of the variance. Mathematically, the standard deviation is the square root of the average squared deviation from the mean, and it represents the average deviation of a score from the mean.

LEARNING CHECK 2:

Reviewing What You've Learned So Far

1. Review questions
   a. Why does calculating the variance involve squaring the deviation of each score from the mean?
   b. What is the purpose of computational formulas?
   c. In calculating a measure of variability, why can't you use the absolute value of the deviation of each score from the mean rather than the squared deviation?
   d. In calculating the variance, why is the sum of squared deviations divided by $N - 1$ rather than N?
   e. What is the difference between a biased estimate and an unbiased estimate?
2. Calculate the variance (s^2) using the definitional and computational formulas for each of the following data sets.
   a. 2, 3, 3, 5, 7
   b. 5, 4, 7, 5, 10, 5, 6
   c. 10, 13, 13, 14, 15, 16, 17, 24
   d. 73, 66, 91, 84, 69, 87, 62, 79, 82, 90
   e. 11, 9, 9, 12, 10, 9, 7, 8, 9, 8, 10, 9

Definitional Formula for the Standard Deviation

The standard deviation, represented by the symbol *s*, is calculated by computing the square root of the variance. The purpose of calculating the square root is to "undo" the effect of squaring the deviations. Formula 4-5 provides the definitional formula for the standard deviation—this formula places the definitional formula for the variance (Formula 4-3) under a square root symbol:

$$s = \sqrt{\frac{\Sigma(X - \bar{X})^2}{N - 1}} \tag{4-5}$$

For the simple data set of nine scores used throughout this chapter, using the squared deviations calculated in Table 4.3, the standard deviation may be calculated as follows:

$$s = \sqrt{\frac{\Sigma(X-\bar{X})^2}{N-1}}$$
$$= \sqrt{\frac{1.00+.00+4.00+\ldots+1.00+9.00+.00}{9-1}}$$
$$= \sqrt{\frac{20}{8}} = \sqrt{2.50}$$
$$= 1.58$$

From this value of the standard deviation, we conclude that the average deviation of the nine scores in this sample from the mean of 3.00 is equal to 1.58.

For the frequency rating variable, using the calculations in Table 4.4, the standard deviation is equal to the following:

$$s = \sqrt{\frac{\Sigma(X-\bar{X})^2}{N-1}}$$
$$= \sqrt{\frac{161.29+334.89+151.29+\ldots+127.69+1{,}069.29+53.29}{20-1}}$$
$$= \sqrt{\frac{3{,}940.20}{19}} = \sqrt{207.38}$$
$$= 14.40$$

Here, we can conclude that, in this sample of 20 participants, in rating the frequency of the word *sometimes,* the average difference between a participant's frequency rating and the sample mean of 33.30% was 14.40%.

Chapter 3 mentioned that descriptive statistics are often provided within the body or text of a paper. For the frequency rating example,

> The average frequency rating for the word *sometimes* was approximately one third the distance between *Never* and *Always,* representing a 33% frequency ($M = 33.30$, $SD = 14.40$).

In this example, the symbols M and SD represent the mean and standard deviation, respectively. Both measures of central tendency and variability are reported because they provide different pieces of information about the nature and shape of the distribution of scores for a variable.

Computational Formula for the Standard Deviation

The computational formula for the standard deviation (Formula 4-6) simply places the computational formula for the variance (Formula 4-4) within the square root symbol:

$$s = \sqrt{\frac{\Sigma X^2 - \frac{(\Sigma X)^2}{N}}{N-1}} \qquad (4\text{-}6)$$

The standard deviation for the simple example of nine scores and the frequency rating variable are calculated using Formula 4-6 in Table 4.7 (the values for the sum of scores squared [$(\Sigma X)^2$] and the sum of squared scores [ΣX^2] for the two examples are found in Tables 4.5 and 4.6). Similar to the variance, note the same value for the standard deviation is obtained regardless of whether the definitional or computational formula is used.

TABLE 4.7 ● CALCULATION OF THE STANDARD DEVIATION USING THE COMPUTATIONAL FORMULA, SIMPLE EXAMPLE AND FREQUENCY RATING VARIABLE

Simple Example	Frequency Rating
$s = \sqrt{\dfrac{\Sigma X^2 - \dfrac{(\Sigma X)^2}{N}}{N-1}}$	$s = \sqrt{\dfrac{\Sigma X^2 - \dfrac{(\Sigma X)^2}{N}}{N-1}}$
$= \sqrt{\dfrac{101 - \dfrac{729}{9}}{9-1}}$	$= \sqrt{\dfrac{26118 - \dfrac{443556}{20}}{20-1}}$
$= \sqrt{\dfrac{101 - \dfrac{81.00}{8}}{8}}$	$= \sqrt{\dfrac{26118 - 22177.80}{19}}$
$= \sqrt{\dfrac{20.00}{8}} = \sqrt{2.50}$	$= \sqrt{\dfrac{3940.20}{19}} = \sqrt{207.38}$
$= 1.58$	$= 14.40$

LEARNING CHECK 3:

Reviewing What You've Learned So Far

1. Review questions
   a. What is the relationship between the variance and the standard deviation?
   b. Is it possible to get a negative value for the variance or standard deviation? Why or why not?
2. Calculate the standard deviation (s) using the definitional and computational formula for each of the following data sets.
   a. 5, 3, 9, 2, 6
   b. 11, 19, 8, 10, 9, 7, 13
   c. 7, 9, 11, 12, 13, 14, 17, 21, 23
   d. 10, 8, 12, 8, 9, 7, 23, 6, 8, 11, 10, 6
   e. 11, 9, 13, 6, 14, 12, 7, 12, 10, 8, 15, 5, 12

"Eyeball Estimating" the Mean and Standard Deviation

We've now provided formulas for the most commonly used measures of central tendency and variability: the mean and standard deviation. As you'll be calculating these two statistics in virtually every chapter for the remainder of this book, it's important for you to perform these calculations correctly. However, we feel it's equally if not more important for you to develop the ability to look at a set of data and estimate what you believe are the mean and standard deviation before punching numbers into your calculator.

As an example of "eyeball estimation," imagine a political scientist asks 15 voters the question, "How many of the 50 U.S. states require someone to pass a background check before buying a gun from a private seller (a seller who is not a firearms dealer)?" Following are 15 hypothetical responses to this question:

30 35 18 42 30 45 33 25 37 44 33 35 40 30 36

Before reaching for your pencil and calculator, we ask you to fill in the blanks in the following sentence: "*Most of the scores are between* _____ *and* _____." What percentage of the sample would you associate with the word *most?* We'll assume that "most" is more than "half" but less than "all"; for the sake of argument, let's assume "most" means about two thirds to three fourths (66%–75%) of the scores.

Imagine you say, "Most of the scores are between 30 and 40." In doing so, you've laid the foundation for estimating the mean and standard deviation. Your estimate of the mean could be located halfway between these two scores: Halfway between 30 and 40 is *35*. Your estimate of the standard deviation is the difference between the estimated mean and either of the two scores; the difference between your estimated mean of 35 and either 30 or 40 is *5* (35 − 5 = 30 and 35 + 5 = 40). Consequently, saying "Most of the scores are between 30 and 40" can be used to develop "eyeball estimates" of a mean equal to 35 and a standard deviation equal to 5.

In the above example, the calculated mean of the 15 scores is equal to 34.20 and the standard deviation is equal to 7.18. (For the sake of reference, as of mid-2017, only 18 states required background checks from private sellers [http://smartgunlaws.org/].) It is not at all important for your eyeball estimates to be 100% correct; instead, we've found these estimates help avoid the situation where students' calculations are very *in*correct. For example, in calculating the standard deviation (s), students sometimes forget the last step, which is to calculate the square root of the variance. In this example, this would result in reporting the standard deviation to be 51.60 rather than the correct 7.18. Given that the maximum value for this variable ("How many of the 50 U.S. states . . .") is 50, reporting a standard deviation of 51.60 is outside the realm of possibility. We've found "eyeball estimates" help students become more confident in assessing whether they've performed the calculations correctly as well as interpreting the calculated values of their statistics.

4.7 MEASURES OF VARIABILITY FOR POPULATIONS

The formulas for the variance and standard deviation described thus far are used for samples drawn from populations. But what if you were to collect data from the entire population? This section discusses two measures of variability for populations: the population variance and the population standard deviation.

The Population Variance (σ^2)

The variance for data collected from a population is called the **population variance** (σ^2) (σ is the lowercase Greek letter sigma), which is defined as the average squared deviation of a score from the population mean. The definitional formula for the population variance is provided in Formula 4-7:

$$\sigma^2 = \frac{\Sigma(X - \mu)^2}{N} \tag{4-7}$$

where X is a score for a variable, μ is the population mean, and N is the total number of scores.

The formula for the population variance differs from the formula for the variance s^2 (Formula 4-4) in two important ways. First, the mean in the formula is the population mean μ rather than the sample mean $\overline{X}$. Second, the sum of the squared deviations ($\Sigma(X - \mu)^2$) is divided by N rather than $N - 1$; this is because you are no longer estimating the variability in the population from data collected from a sample but instead have collected data from the entire population.

Imagine, for a moment, that the nine scores in the first column of Table 4.3 represent a population rather than a sample. Using the squared deviations for these scores calculated in the last column of Table 4.3, the population variance would be equal to the following:

$$\sigma^2 = \frac{\Sigma(X-\mu)^2}{N}$$
$$= \frac{1.00+.00+4.00+\ldots+1.00+9.00+.00}{9} = \frac{20.00}{9}$$
$$= 2.22$$

From this calculation, you would conclude that, assuming these nine scores represent the entire population, the average squared deviation of a score from the population mean of 3.00 is 2.22.

The Population Standard Deviation (σ)

Calculating the population variance involves squaring the deviation of each score from the population mean μ. In order to obtain a measure of variability that represents the average deviation (rather than the average squared deviation) from the population mean, the square root of the population variance may be calculated. This is the **population standard deviation** (σ), defined as the square root of the population variance. Mathematically, the population standard deviation is the square root of the average squared deviation from the population mean, and it represents the average deviation of a score from the population mean. Formula 4-8 provides the definitional formula for the population standard deviation.

$$\sigma = \sqrt{\frac{\Sigma(X-\mu)^2}{N}} \qquad (4\text{-}8)$$

Relying again on Table 4.3, the population standard deviation for the set of nine scores is calculated below.

$$\sigma = \sqrt{\frac{\Sigma(X-\mu)^2}{N}}$$
$$= \sqrt{\frac{1.00+.00+4.00+\ldots+1.00+9.00+.00}{9}}$$
$$= \sqrt{\frac{20}{9}} = \sqrt{2.22}$$
$$= 1.49$$

From these calculations, it may be concluded that the average deviation of the nine scores from the population mean of 3.00 is equal to 1.49.

Measures of Variability for Samples vs. Populations

The population variance and standard deviation are examples of parameters, which were introduced in Chapter 3. Parameters are numeric characteristics of populations and are distinguished from statistics, which are numeric characteristics of samples. Parameters refer to data from the entire population, whereas statistics refer to data collected from a sample of the population. This is summarized below for both a measure of central tendency (the mean) and a measure of variability (the standard deviation):

Target	Numeric Characteristic	Mean	Standard Deviation
Sample	Statistic	$\bar{X}$	s
Population	Parameter	μ	σ

LEARNING CHECK 4:

Reviewing What You've Learned So Far

1. Review questions
   a. What is the main difference between the population standard deviation and the standard deviation?
   b. Under what conditions would you calculate the population standard deviation for a set of data rather than the standard deviation?
2. Calculate the population standard deviation (σ) using the definitional and computational formula for each of the following data sets:
   a. 2, 4, 5, 6, 8
   b. 8, 6, 4, 8, 3, 7
   c. 71, 84, 65, 78, 89, 72, 60, 85
   d. 3.0, 1.5, 3.5, 3.0, 2.0, 2.5, 4.0, 1.0, 2.0
   e. 10, 8, 12, 8, 9, 7, 23, 6, 8, 11, 10, 6

Because researchers rarely collect data from entire populations, statistics such as the sample mean and standard deviation are much more likely to be calculated than population parameters. Throughout this book, when you see the words *mean, variance,* or *standard deviation,* you should assume they refer to samples rather than populations. In fact, numeric values of parameters such as μ or σ are typically values believed or hypothesized to be true rather than values based on the actual collection of data.

4.8 MEASURES OF VARIABILITY FOR NOMINAL VARIABLES

The "frequency rating" variable used in this chapter consisted of numbers such as 46, 15, and 21; the higher the number, the higher the frequency rating of the word *sometimes.* In Chapter 1, we indicated that numeric variables whose values are equally spaced along a numeric continuum are measured at the interval or ratio levels of measurement (ratio variables differ from interval variables in that ratio variables possess a true zero point, meaning that the value of zero represents the complete absence of the variable). Chapter 3 discussed two measures of central tendency used with interval or ratio variables: the mean and the median. In this chapter, we've discussed measures of variability for interval or ratio variables: the range, variance, and standard deviation. However, variables can also be measured at the nominal level of measurement, whose values differ in category or type; examples of nominal variables are gender (male, female, etc.) and state (Missouri, Pennsylvania, etc.). In Chapter 3, we noted that the mode can be used to represent the central tendency of nominal variables; a measure of variability for nominal variables is discussed below.

The Index of Qualitative Variation

As a simple example of a nominal variable, imagine you're organizing a dinner and you ask 44 people whether they would eat a chicken entree or a vegetarian entree; 21 people say "either," 7 say "chicken," 13 say "vegetarian," and 3 say "neither." Following the guidelines discussed in Chapter 2, a frequency distribution table for these data is presented below:

Type of entree	*f*	%
Either	21	48%
Chicken	7	16%
Vegetarian	13	29%
Neither	3	7%
Total	44	100%

Because "type of entree" is a nominal variable that consists of four distinct categories (either, chicken, vegetarian, neither), an appropriate measure of central tendency would be the mode, which is the score or value of a variable that appears most frequently in a set of data. Looking at the frequency distribution table, the mode type of entree is "either" because more people selected it ($f = 21$) than the other three possibilities. But looking at this frequency distribution table, we see that the four choices vary widely in their frequencies. What if we wanted to numerically represent the amount of variability in people's entree preferences?

The **index of qualitative variation (IQV)** is a measure of variability for nominal variables. The formula for the index of qualitative variation (IQV) is provided in Formula 4-9:

$$\text{IQV} = \frac{k(N^2 - \Sigma f^2)}{N^2(k-1)} \tag{4-9}$$

where k is the number of categories, N is the total sample size, and f is the frequency for each of the categories.

Using the information from the above frequency distribution table, the IQV for the type of entree example is calculated below:

$$\begin{aligned}
\text{IQV} &= \frac{k(N^2 - \Sigma f^2)}{N^2(k-1)} \\
&= \frac{4[(44)^2 - ((21)^2 + (7)^2 + (13)^2 + (3)^2)]}{(44)^2(4-1)} \\
&= \frac{4[1936 - (441 + 49 + 169 + 9)]}{1936(3)} \\
&= \frac{4(1936 - 668)}{5808} = \frac{4(1268)}{5808} = \frac{5072}{5808} \\
&= .87
\end{aligned}$$

Possible values of the IQV range from 0 to 1; as with any measure of variability, the smaller the value of the IQV, the less the amount of variability in the frequencies of the different values of the nominal variable. As an extreme example, if all 44 of the people in the sample had chosen chicken for their entree (implying no variability in people's entree preferences), the IQV would have been equal to .00. On the other hand, if an equal number of people had selected each of the four choices ($f = 11$ for all four choices), the IQV would have been equal to 1.00.

4.9 MEASURES OF VARIABILITY: DRAWING CONCLUSIONS

These early chapters of this book have discussed the importance of examining and drawing appropriate conclusions about three aspects of distributions: modality, symmetry, and variability. Measures of central tendency such as the mean, median, and mode provide a numerical description of the modality and, to a lesser degree, the symmetry of a variable by focusing on the center of the distribution and what scores have in common with each other. However, it is equally important to consider the variability in a distribution, which is the degree to which scores differ from the center of the distribution and from each other. Measures of variability such as the variance and standard deviation provide valuable information, for example, regarding the degree to which participants in a sample agree when asked the same survey question or respond in a similar manner to the same experimental manipulation.

The beginning of this chapter introduced a research study designed to measure the degree to which people differ in their interpretation of words such as *sometimes, often,* and *seldom.* Based on the variability of the frequency ratings, what did these researchers conclude regarding the degree to which people agree in their perceptions of these words? Comparing the results of their study with those conducted in other countries, they concluded the following:

> There are subtle variations in how those sharing the same language describe the intermediate points . . . so it cannot be assumed that rating scales developed in one culture can be automatically used in another, even where there is a common language. (Skevington & Tucker, 1999, p. 59)

4.10 LOOKING AHEAD

Several critical issues have been identified and discussed in the first four chapters of the book. The first is the importance of examining data before conducting statistical analyses on the data. Appropriate and meaningful conclusions cannot be drawn from statistical analyses until the accuracy of the data has been confirmed and the distribution of scores has been understood. Second, there are many different types of distributions; you have seen distributions labeled as symmetric, skewed, peaked, flat, unimodal, and bimodal. Third, distributions can be described numerically using measures of central tendency and variability. The next chapter discusses yet another type of distribution: "normal distributions." Like the other distributions discussed so far, normal distributions can be described using descriptive statistics such as measures of central tendency and variability. What makes normal distributions unique, as we will discuss in greater detail going forward, is that they possess characteristics that enable them to be the basis of a second type of statistics, known as "inferential statistics," whose goal is to test hypotheses about populations based on the information from samples.

4.11 Summary

Variability is a third aspect of distributions and refers to the amount of spread or scatter of scores in a distribution. One goal of a science such as psychology is to describe, understand, explain, and predict variability in the phenomena studied by researchers.

A ***measure of variability*** is a descriptive statistic of the amount of differences in a set of data for a variable. Common measures of variability are the range, the interquartile range, the variance, and the standard deviation.

The ***range*** is the mathematical difference between the lowest and highest scores in a set of data. The ***interquartile range*** is the range of the middle 50% of the scores for a variable, calculated by removing the highest and lowest 25% of the distribution. The ***variance*** **(s^2)** is the average squared deviation of a score from the mean. The ***standard deviation*** **(s)** is the square root of the variance, and it represents the average deviation of a score from the mean. Because the variance and standard deviation (unlike the range and interquartile range) are based on all of the scores in the distribution, they can be used in statistical analyses designed to test hypotheses about distributions.

The variance and the standard deviation may be calculated using a ***definitional formula,*** which is a formula based on the actual or literal definition of a concept, or a ***computational formula,*** defined as a formula not based on the definition of a concept but is designed to simplify mathematical calculations. Computational formulas are used because using the definitional formula in a large or complicated data set increases the chances of making computational errors.

The variance and standard deviation measure the variability in data collected from samples drawn from populations; the ***population variance*** (σ^2) (the average squared deviation of scores from the population mean) and the ***population standard deviation*** (σ) (the square root of the population variance) are calculated when data are collected from the entire population. Because data are rarely collected from entire populations, numeric values of parameters such as σ are typically based on beliefs and hypotheses rather than the actual collection of data.

In addition to measures of variability for numeric variables measured at the interval or ratio levels of measurement, measures of variability also exist for categorical variables measured at the nominal level of measurement. The ***index of qualitative variation (IQV)*** is a measure of variability for nominal variables.

4.12 Important Terms

measure of variability (p. 88)
range (p. 89)
interquartile range (p. 92)
variance (s^2) (p. 94)
definitional formula (p. 96)
computational formula (p. 96)
biased estimate (p. 99)
unbiased estimate (p. 99)
standard deviation (s) (p. 100)
population variance (σ^2) (p. 103)
population standard deviation (σ) (p. 104)
index of qualitative variation (IQV) (p. 106)

4.13 Formulas Introduced in This Chapter

Range

$$\text{Range} = \text{highest score} - \text{lowest score} \tag{4-1}$$

Interquartile Range

$$\text{Interquartile range} = (N - \frac{N}{4})\text{th score} - (\frac{N}{4} + 1)\text{th score} \tag{4-2}$$

Variance (s^2) (Definitional Formula)

$$s^2 = \frac{\Sigma(X - \bar{X})^2}{N - 1} \tag{4-3}$$

Variance (s^2) (Computational Formula)

$$s^2 = \frac{\Sigma X^2 - \frac{(\Sigma X)^2}{N}}{N - 1} \tag{4-4}$$

Standard Deviation (s) (Definitional Formula)

$$s = \sqrt{\frac{\Sigma(X - \bar{X})^2}{N - 1}} \tag{4-5}$$

Standard Deviation (s) (Computational Formula)

$$s = \sqrt{\frac{\Sigma X^2 - \frac{(\Sigma X)^2}{N}}{N - 1}} \tag{4-6}$$

Population Variance (σ^2) (Definitional Formula)

$$\sigma^2 = \frac{\Sigma(X - \mu)^2}{N} \tag{4-7}$$

Population Standard Deviation (σ) (Definitional Formula)

$$\sigma = \sqrt{\frac{\Sigma(X - \mu)^2}{N}} \tag{4-8}$$

Index of Qualitative Variation (IQV)

$$\text{IQV} = \frac{k(N^2 - \Sigma f^2)}{N^2(k-1)} \tag{4-9}$$

4.14 Using SPSS

Calculating Measures of Central Tendency and Variability: The Frequency Rating Study (4.1)

1. Define variable (name, # decimal places, label for the variable) and enter data for the variable.

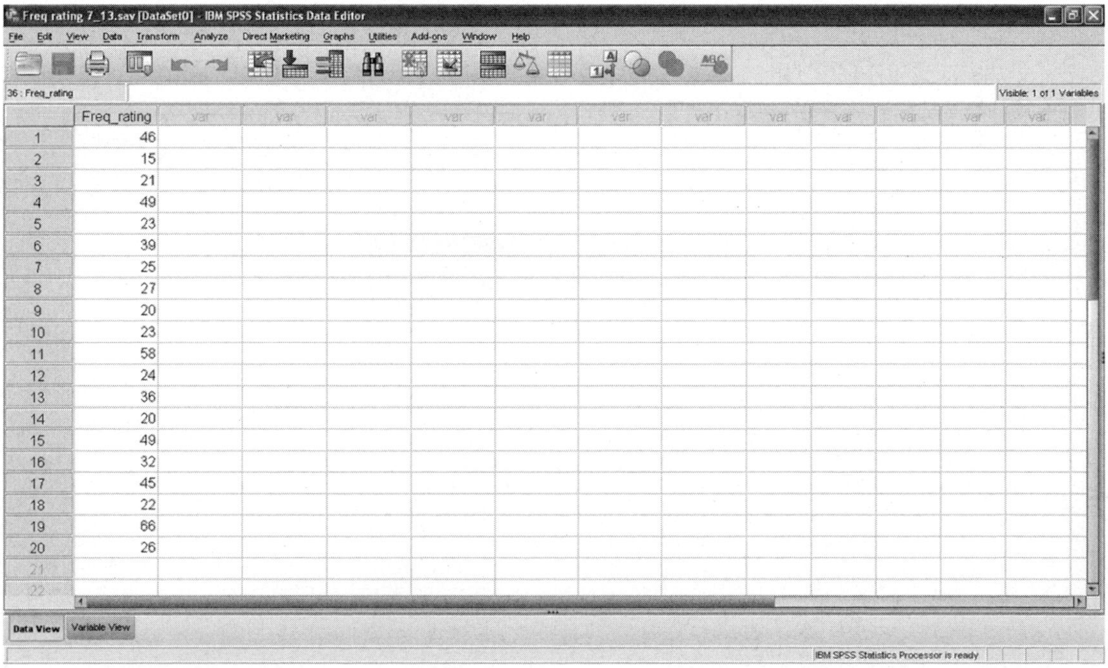

2. Select the descriptive statistics procedure within SPSS.

   How? (1) Click **Analyze** menu, (2) click **Descriptive Statistics**, and (3) click **Descriptives**.

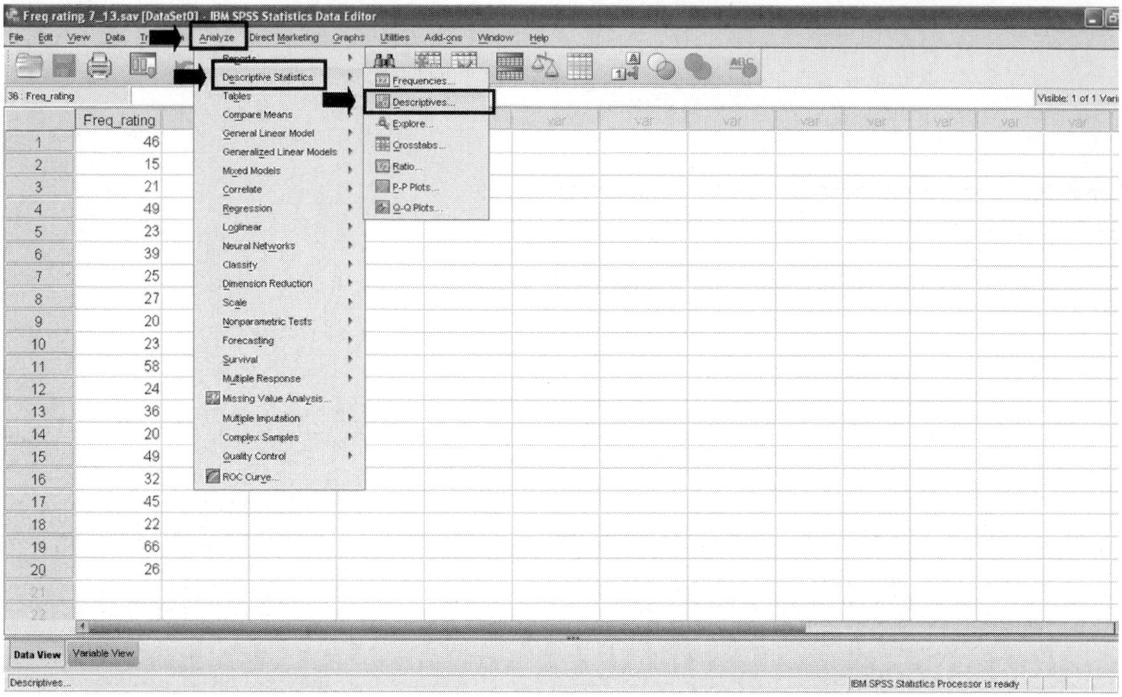

3. Select the variable to be analyzed.

How? (1) Click variable and →, and (2) click **OK**.

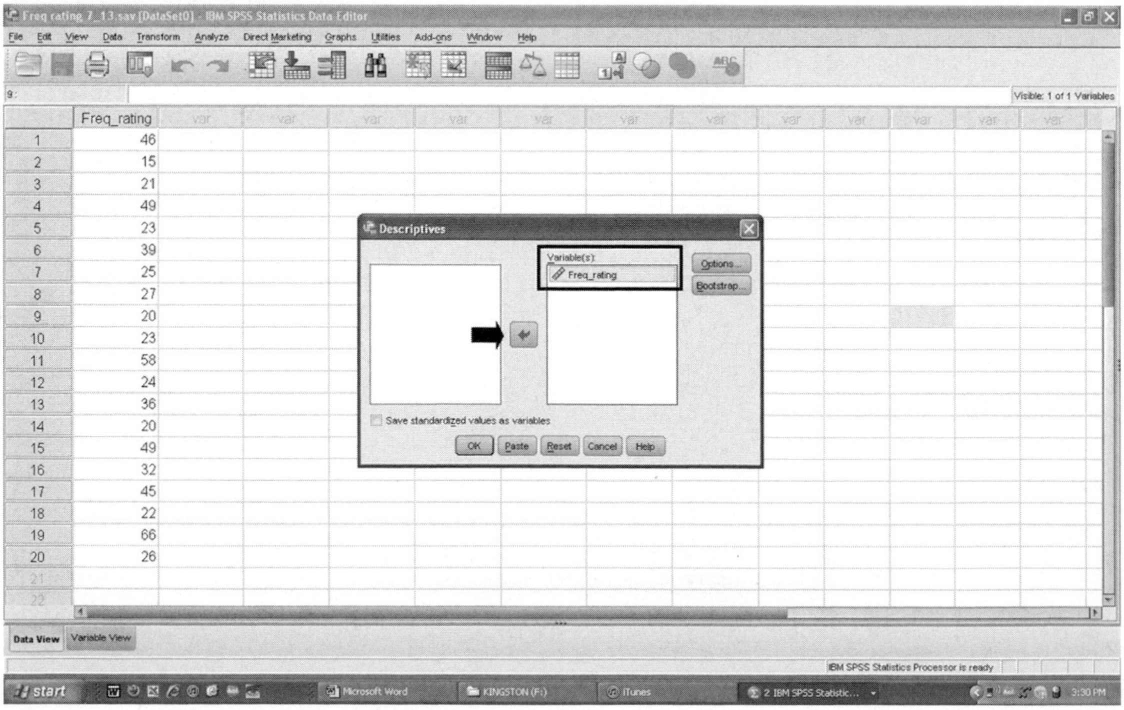

4. Examine output.

N   Mean   Standard deviation

Descriptive Statistics

	N	Minimum	Maximum	Mean	Std. Deviation
Freq_rating	20	15	66	33.30	14.401
Valid N (listwise)	20				

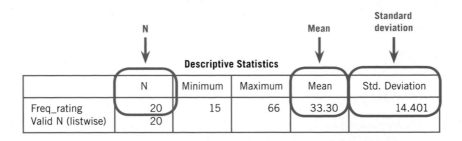

4.15 Exercises

1. Calculate the range for each of the following sets of data:
   a. 4, 6, 9
   b. 14, 17, 11, 19, 12
   c. 25, 22, 27, 30, 21, 26, 29
   d. 10, 8, 4, 16, 9, 7, 9, 13, 6, 11
   e. 73, 66, 91, 84, 69, 87, 62, 79, 82, 90
   f. 3.50, 4.21, 3.95, 2.27, 3.06, 4.58, 2.74, 3.89, 2.65, 2.03, 4.41, 3.76, 2.35
2. Calculate the range for each of the following sets of data:
   a. 9, 10, 13
   b. 5, 3, 9, 12
   c. 6, 2, 8, 12, 9, 5, 7
   d. 8, 4, 1, 6, 14, 9, 12, 5, 11, 7, 4
   e. 16.65, 12.98, 31.74, 18.80, 27.31, 29.92, 34.65, 23.68, 28.20, 20.77
3. Calculate the interquartile range for each of the following sets of data:
   a. 3, 6, 7, 12, 15, 17, 23, 28
   b. 8, 4, 1, 6, 13, 10, 12, 5
   c. 3, 8, 14, 11, 16, 7, 14, 15, 11, 9, 12, 6
   d. 15, 22, 17, 13, 31, 25, 22, 19, 26, 30, 27, 19, 23, 21, 27, 29
   e. 450, 560, 340, 510, 390, 670, 540, 420, 720, 480, 560, 510
   f. 29, 27, 26, 15, 28, 26, 32, 27, 26, 25, 30, 26, 27, 18, 23, 35
4. Calculate the interquartile range for each of the following sets of data:
   a. 1, 2, 2, 3, 4, 5, 7, 10
   b. 21, 8, 17, 7, 12, 19, 5, 12
   c. 6, 9, 5, 14, 3, 15, 19, 7, 13, 6, 8, 5
   d. 6.78, 7.81, 6.35, 9.65, 5.43, 8.62, 5.90, 7.13, 6.56, 3.27, 8.92, 4.49
   e. 78, 55, 82, 64, 93, 69, 71, 82, 59, 71, 76, 52, 89, 75, 81, 78

5. For each of the following sets of data, (1) calculate the mean of the scores ($\overline{X}$), (2) calculate the deviation of each score from the mean ($X - \overline{X}$), and (3) check to see if the sum of the deviations equals zero ($\Sigma(X - \overline{X}) = 0$).
   a. 2, 4, 6
   b. 5, 6, 13
   c. 4, 7, 8, 9
   d. 5, 2, 7, 13, 11
   e. 3, 8, 14, 11, 16, 7, 14, 15, 11

6. For each of the following sets of data, (1) calculate the mean of the scores ($\overline{X}$), (2) calculate the deviation of each score from the mean ($X - \overline{X}$), and (3) check to see if the sum of the deviations equals zero ($\Sigma(X - \overline{X}) = 0$).
   a. 7, 5, 12
   b. 6, 7, 9, 12
   c. 6, 8, 4, 11, 3, 7
   d. 5, 7, 3, 1, 4, 8, 5, 2, 7, 9, 12, 4, 15

7. For each of the sets of data in Exercise 5, calculate the variance (s^2) using the definitional formula and the computational formula.

8. For each of the sets of data in Exercise 5, calculate the population variance (σ^2) using the definitional formula. Comparing your calculations for each data set with those done in Exercise 7, which is larger, the variance or the population variance? Why?

9. For each of the sets of data in Exercise 5, calculate the standard deviation (s) and the population standard deviation (σ). For each data set, which is larger, the standard deviation or the population standard deviation?

10. (This example was discussed in Chapters 2 and 3.) A friend of yours asks 20 people to rate a movie using a 1- to 5-star rating: the higher the number of stars, the higher the recommendation. Their ratings are listed below:

Person	# Stars	Person	# Stars	Person	# Stars
1	★★★	8	★★	15	★★★
2	★★★★★	9	★★★	16	★★
3	★★	10	★★★★★	17	★★★
4	★★★★	11	★	18	★★
5	★★★	12	★★★★	19	★★★★
6	★★★	13	★★★★	20	★★★
7	★★★★	14	★★★		

   a. Calculate the variance (s^2) and standard deviation (s) of the ratings using either the definitional or the computational formulas. (Note: From Chapter 3, you may have already calculated the number of scores (N), sum of scores squared ($(\Sigma X)^2$), the sum of squared scores (ΣX^2), and the mean ($\overline{X}$).)
   b. Based on your value of the standard deviation, what would you conclude regarding the degree to which these people agree or disagree about this movie?

11. One study stated the research hypothesis, “Violence behavior in children may be reduced by teaching them conflict resolution skills” (DuRant et al., 1996). The variable “violence behavior” was measured by the number of fights in which each student was involved. Below is a frequency distribution table for the number of fights for 12 students.

# Fights	*f*	%	# Fights	*f*	%
4	1	8%	1	1	8%
3	2	17%	0	3	25%
2	5	42%	Total	12	100%

a. Calculate the variance (s^2) and standard deviation (s) of the number of fights using either the definitional or the computational formulas.

b. Based on your value of the standard deviation, what is the average difference between the number of fights a student got involved in and the mean?

12. (This example was introduced in Chapter 3.) The 10% myth study discussed in this chapter measured the beliefs of both psychology majors and non–psychology majors. The 39 non–psychology majors in this study provided the following values for the brain power variable:

a. Calculate the variance (s^2) and standard deviation (s) of the estimates using either the definitional or the computational formulas. (Note: From Chapter 3, you may have already calculated the number of scores (N), sum of scores squared ($(\Sigma X)^2$), the sum of squared scores (ΣX^2), and the mean ($\bar{X}$).)

Non–Psychology Major	Estimated Brain Power	Non–Psychology Major	Estimated Brain Power	Non–Psychology Major	Estimated Brain Power
1	15	14	10	27	15
2	45	15	20	28	25
3	10	16	50	29	10
4	40	17	25	30	20
5	5	18	45	31	10
6	10	19	40	32	5
7	50	20	25	33	15
8	45	21	10	34	10
9	10	22	5	35	5
10	10	23	15	36	60
11	35	24	5	37	10
12	5	25	40	38	5
13	15	26	30	39	15

13. (This example was introduced in Chapter 2.) An instructor administers a 27-item quiz to her class of 25 students. Each student's score on the quiz is the number of items answered correctly. These scores are listed below:

    a. Calculate the variance and standard deviation of the ratings using the computational formulas.

Person	Quiz Score	Person	Quiz Score	Person	Quiz Score
1	22	10	17	19	20
2	15	11	18	20	21
3	11	12	14	21	17
4	19	13	10	22	18
5	12	14	6	23	15
6	21	15	16	24	24
7	22	16	20	25	22
8	19	17	21		
9	23	18	19		

14. On your own, create two distributions of 10 scores that have the same mean but differ in modality: one unimodal and one bimodal. Calculate the standard deviation of the two distributions. Is there greater variability in the unimodal distribution or the bimodal distribution?

15. On your own, (a) create two distributions of 10 scores that have the same mean but differ in their amount of variability, and (b) create two distributions of 10 scores that have different means but have the same amount of variability. (c) What does this imply about the information that is necessary to accurately describe a distribution of scores for a variable?

16. What if someone told you, "There is very little variability in the scores for my variable." From this statement, would you be more likely to describe the shape of the distribution as peaked or flat?

17. What if someone told you, "There is a great deal of variability in the scores for my variable." From this statement, can you tell whether the shape of the distribution is symmetrical or skewed? Why or why not? If not, what would you do to determine the symmetry of the distribution?

18. What if someone told you, "The mean age of the participants in this sample was 35.50 and the standard deviation was 2.53." Why would you interpret the sample differently if you had been told the standard deviation was 14.71?

19. An employment survey in 2009 was conducted to find the average starting salaries of people receiving doctorate (PhD) degrees in psychology (Michalski, Kohout, Wicherski, & Hart, 2011). The mean starting salary for psychologists in assistant professor positions in universities was approximately $60,000 with a standard deviation of $11,000. Clinical psychologists in their first year of practice also report a mean of approximately $60,000 but with a standard deviation of $16,000.

    a. Which type of psychologist has more variability in income?

    b. For which type of psychologist could you more accurately predict a salary for, and why?

Answers to Learning Checks

Learning Check 1

2. a. Range = 6
   b. Range = 12
   c. Range = 19
   d. Range = .90
3. a. Interquartile range = 3
   b. Interquartile range = 8
   c. Interquartile range = 1
   d. Interquartile range = .30

Learning Check 2

2. a. $s^2 = 4.00$
   b. $s^2 = 4.00$
   c. $s^2 = 17.07$
   d. $s^2 = 105.79$
   e. $s^2 = 1.84$

Learning Check 3

2. a. $s = 2.74$
   b. $s = 4.04$
   c. $s = 5.33$
   d. $s = 4.55$
   e. $s = 3.12$

Learning Check 4

2. a. $\sigma = 2.00$
   b. $\sigma = 1.91$
   c. $\sigma = 9.58$
   d. $\sigma = .91$
   e. $\sigma = 4.36$

Answers to Odd-Numbered Exercises

1. a. Range = 5
   b. Range = 8
   c. Range = 9
   d. Range = 12

e. Range = 29

f. Range = 2.55

3. a. Interquartile range = 10

b. Interquartile range = 5

c. Interquartile range = 6

d. Interquartile range = 8

e. Interquartile range = 110

f. Interquartile range = 2

5. a.

Score (X)	Mean ($\bar{X}$)	Mean ($X - \bar{X}$)
2	4.00	–2.00
4	4.00	.00
6	4.00	2.00
		$\Sigma(X - \bar{X}) = 0$

b.

Score (X)	Mean ($\bar{X}$)	Score – Mean ($X - \bar{X}$)
4	7.00	–3.00
7	7.00	.00
8	7.00	1.00
9	7.00	2.00
		$\Sigma(X - \bar{X}) = 0$

c.

Score (X)	Mean ($\bar{X}$)	Mean ($X - \bar{X}$)
5	8.00	–3.00
6	8.00	–2.00
13	8.00	5.00
		$\Sigma(X - \bar{X}) = 0$

d.

Score (X)	Mean ($\bar{X}$)	Score – Mean ($X - \bar{X}$)
3	11.00	–8.00
8	11.00	–3.00
14	11.00	3.00
11	11.00	.00
16	11.00	5.00

Score (X)	Mean ($\bar{X}$)	Score – Mean ($X - \bar{X}$)
7	11.00	–4.00
14	11.00	3.00
15	11.00	4.00
11	11.00	.00
		$\Sigma(X - \bar{X}) = 0$

e.

Score (X)	Mean ($\bar{X}$)	Score – Mean ($X - \bar{X}$)
5	7.60	–2.60
2	7.60	–5.60
7	7.60	–.60
13	7.60	5.40
11	7.60	3.40
		$\Sigma(X - \bar{X}) = 0$

7. a. $s^2 = 4.00$
   b. $s^2 = 19.00$
   c. $s^2 = 4.67$
   d. $s^2 = 19.80$
   e. $s^2 = 18.50$

9. a. $s = 2.00$, $\sigma = 1.63$
   b. $s = 4.36$, $\sigma = 3.56$
   c. $s = 2.16$, $\sigma = 1.87$
   d. $s = 4.45$, $\sigma = 3.98$
   e. $s = 4.30$, $\sigma = 4.06$

   Note: In all data sets, the standard deviation (s) is larger than the population standard deviation (σ) because you are dividing by a smaller number ($N - 1$ rather than N).

11. a. $s^2 = 1.66$, $s = 1.29$
    b. The average difference between the number of fights of a student and the mean is 1.29 fights.

13. a. $s^2 = 19.89$, $s = 4.46$

15. a. Sample Data Set 1: 4, 4, 5, 5, 5, 5, 5, 6, 6, 6 = 5.10, $s = .74$
       Sample Data Set 2: 2, 2, 3, 4, 5, 5, 6, 6, 9, 9 = 5.10, $s = 2.51$
    b. Sample Data Set 3: 3, 3, 4, 4, 5, 5, 5, 6, 7, 7 = 4.90, $s = 1.45$
       Sample Data Set 4: 8, 8, 9, 9, 10, 10, 10, 11, 12, 12 = 9.90, $s = 1.45$
    c. Distributions with the same mean or the same standard deviation can be describing very different information; therefore, it is important to provide both measures of central tendency and variability to describe a set of data.

17. There can be a great deal of variability in both symmetrical and skewed distributions. The simplest way to determine the symmetry of a distribution is to look at a figure of the distribution.

19. a. Clinical psychologists have more variability in their starting salaries.

    b. You could more accurately predict the income for a psychologist going into an assistant professor position because there is less variability in their starting salaries.

5

NORMAL DISTRIBUTIONS

CHAPTER OUTLINE

The first chapters of this book introduced a process by which research may be conducted. This process starts by stating a research hypothesis to be tested; next, data relevant to the research hypothesis are collected. The first part of analyzing data is to examine them (using tables and figures) and describe them (using descriptive statistics such as measures of central tendency and variability). But given the goal of determining whether or not a research hypothesis is supported, data must not only be examined and described but also evaluated. The preceding chapters have emphasized the understanding of distributions because they are the basis of statistical analyses designed to test research hypotheses. The purpose of this chapter is to discuss one type of these distributions, known as normal distributions. To illustrate the characteristics and uses of normal distributions, we will use an example familiar to many college students: the SAT.

5.1 EXAMPLE: SAT SCORES

Each year, many thousands of high school seniors apply for college. One of the more anxiety-provoking aspects of this process may involve taking the SAT Reasoning Test, known by most people as the SAT. The SAT is a 3-hour, 50-minute examination divided into three sections (math, reading, and writing), each scored on a scale ranging from 200 to 800. Although educators have long disagreed regarding the appropriate role of the SAT in the college admissions process, it continues to be used by many universities.

5.2 NORMAL DISTRIBUTIONS

Imagine you have recently taken the SAT and received a score of 660 on the math section. Based on this score, how could you evaluate your performance on this section of the test? Your first step might be to compare your score with the scores of other students in the student population at large. This will require locating your score relative to the other scores within the distribution.

What Are Normal Distributions?

Evaluating a score relative to the other scores in a population may require an understanding of a specific type of distribution known as a **normal distribution**, defined as a distribution based on a population of an infinite number of scores calculated from a mathematical formula. Note that a normal distribution differs from the types of distributions discussed earlier in this book (what we have referred to as frequency distributions) in that scores in a normal distribution are generated from a mathematical formula rather than data collected from research participants. An example of a normal distribution is illustrated in Figure 5.1.

There are four distinguishing features of normal distributions. First, because they consist of an infinite number of scores generated from mathematical formulas, the left and right ends of normal distributions continue to infinity (symbolized by the Greek symbol ∞) and do not touch the horizontal (X) axis. The ends, or tails as they are sometimes called, of normal distributions do not touch the X-axis because to do so would signify the end of the distribution (a score with a frequency [f] of zero). Second, in terms of their shape, normal distributions are unimodal and symmetric and are often referred to as "bell shaped." The symmetrical nature of normal distributions implies that if you were to fold one down the middle, the left and right halves of the distribution would exactly match one another. Third, in terms of central tendency, because normal distributions are based on a population of scores, the mean of a normal distribution is the population mean μ. Similarly, in terms of variability, the standard deviation of a normal distribution is the population standard deviation σ.

Why Are Normal Distributions Important?

Normal distributions differ from frequency distributions in that frequency distributions are constructed using data collected from samples, whereas normal distributions consist of an infinite number of scores generated by mathematical formulas. Because data collected by researchers often do not closely resemble the smooth, curved shape of a normal distribution, you may wonder why normal distributions are the basis of statistical procedures designed to analyze collected data. The answer to this question is threefold.

First, normal distributions are used because many variables (e.g., height or weight) are generally believed by researchers to be normally distributed in the population. Therefore, researchers apply the characteristics of normal distributions to data collected from samples, particularly when the sample is large. The discovery that such variables are not normally distributed in a particular sample may be attributed to limitations in how the data were collected. For example, a variable may not be normally distributed in samples that are relatively small or not representative of the population.

FIGURE 5.1 ● EXAMPLE OF A NORMAL DISTRIBUTION

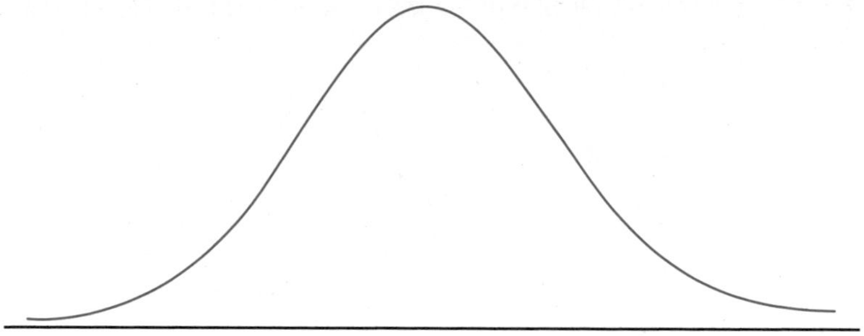

A second reason for using normal distributions, related to the first, is that many statistical procedures designed to test research hypotheses are based mathematically on the assumption that variables are normally distributed. This assumption reinforces the importance of examining the distribution of data collected from samples to assess the degree to which it resembles a normal distribution.

Last, because normal distributions are based on mathematical formulas, it is possible to determine the proportion of the distribution associated with any score. For example, because a normal distribution is symmetric, exactly 50% of the distribution is located below the mean and 50% is located above the mean. As a second example, assuming SAT scores are normally distributed in the population, one can evaluate individual scores such as the score of 660 mentioned earlier, answering questions such as, "What percentage of scores on the math section of the SAT is less than 660?"

Is There Only One Normal Distribution?

The concept of normal distributions is a very important one to researchers. However, there is one problem associated with using these distributions: There is not just one normal distribution. The mathematical formula for the normal distribution involves stating a value for the population mean (μ) and standard deviation (σ). As such, changing either of these values results in a change in the shape of the distribution.

Consider, for example, the variables height and weight. It is assumed both of these variables are normally distributed in the adult population. However, the shapes of the two distributions are not the same because they are measured using different units of measurement (inches vs. pounds). As a result, how a score is evaluated depends on how the variable is measured. For example, a score of 80 is interpreted very differently for height versus weight. For height, 80 inches (6′ 8″) is near the upper end of the distribution for adults; for weight, 80 pounds is at the lower end.

As further illustration of different types of normal distributions, Figure 5.2(a) displays two normal distributions that have different values for the population mean μ but the same standard deviation σ. Figure 5.2(b) displays two normal distributions that have the same value for μ but a different value for σ. Because there are an infinite number of possible normal distributions, how you interpret a score depends on its particular distribution.

5.3 THE STANDARD NORMAL DISTRIBUTION

To avoid confusion resulting from using normal distributions with different units of measurement, researchers rely on a single normal distribution that can be applied to any normally distributed variable, regardless of

FIGURE 5.2 ● DIFFERENT TYPES OF NORMAL DISTRIBUTIONS

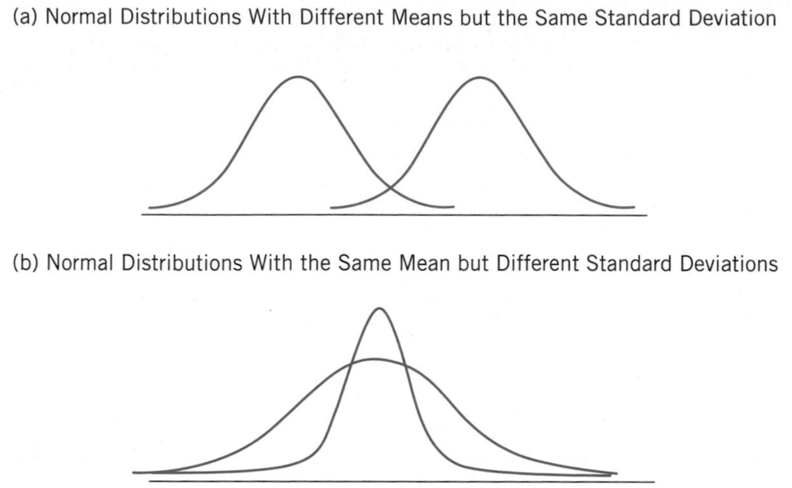

how the variable has been measured. This distribution is known as the **standard normal distribution**, which is a normal distribution measured in standard deviation units with a mean equal to 0 and a standard deviation equal to 1. Figure 5.3 is a visual illustration of the standard normal distribution.

(a) Normal Distributions With Different Means but the Same Standard Deviation
(b) Normal Distributions With the Same Mean but Different Standard Deviations

What Is the Standard Normal Distribution?

In terms of describing the standard normal distribution, it is important to understand it shares many of the features of normal distributions. There are an infinite number of scores in this distribution, with the left and right tails never touching the horizontal axis. In terms of its shape, the standard normal distribution is unimodal and symmetric.

However, the standard normal distribution is distinguished from other normal distributions in three important ways. First, scores for the standard normal distribution, known as ***z*-scores**, are measured in standard deviation units, which is the number of standard deviations the score is from the mean. This is in contrast to other normal distributions, where measurements are a function of how each particular variable is measured (such as "inches" or "pounds"). Second, the mean of the standard normal distribution is equal to zero (0) instead of the population mean μ. Third, in terms of variability, the standard deviation of the standard normal distribution is equal to 1 rather than the population standard deviation σ.

The next section will discuss the distinctive features of the standard normal distribution. In presenting these features, our discussion will focus on two ways z-scores may be interpreted: a score's distance from the mean and a score's location relative to the entire distribution.

Interpreting *z*-Scores: Distance From the Mean

Because z-scores are measured in standard deviation units, the numeric value of a z-score represents its distance from the mean in standard deviations. For example, a z-score of zero ($z = .00$) is located zero standard deviations away from the mean z-score of 0. For any z-score other than zero, the sign (+/–) of the z-score indicates its location relative to the mean. z-Scores greater than the mean are positive (+) numbers; z-scores less than the mean have negative values. Note that the sign of a z-score is typically only included when the z-score is

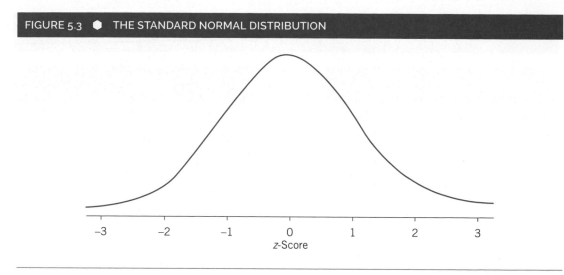

FIGURE 5.3 ● THE STANDARD NORMAL DISTRIBUTION

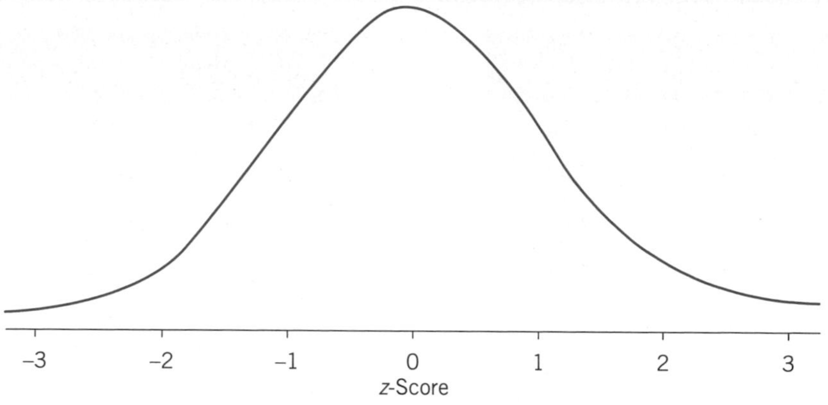

negative; for example, a *z*-score of +.35 is reported as .35. Figure 5.4 displays three *z*-scores in the standard normal distribution.

In Figure 5.4(a), the *z*-score .00 is located at the mean of the distribution. (Although there are an infinite number of *z*-scores, researchers typically round them to two decimal places.) In Figure 5.4(b), the *z*-score of 1.00 is located exactly one standard deviation to the right of (i.e., greater than) the mean. Finally, in Figure 5.4(c), the *z*-score of −1.24 is located a distance of 1.24 standard deviations below (or less than) the mean. Because the standard normal distribution is symmetric, the *z*-score of −1.24 is the same distance from the mean as the *z*-score of 1.24.

Interpreting *z*-Scores: Relative Position Within the Distribution

In addition to its distance from the mean in standard deviation units, a *z*-score may be interpreted in terms of its location relative to the entire distribution. For example, it's possible to answer questions such as, "What percentage of the standard normal distribution is less than the *z*-score of +1.00?" "What percentage is greater than a *z*-score of −2.24?" or "What percentage is between the *z*-scores −.50 and +.50?" Later in this chapter, you'll see how *z*-scores can be used to answer questions such as, "What percentage of SAT Math scores is less than a score of 660?"

To determine the area of the standard normal distribution corresponding to a particular *z*-score, we'll use the **normal curve table**, a table containing the proportion of the standard normal distribution associated with different *z*-scores. Selected parts of the normal curve table are listed in Table 5.1; the entire table may be found in the back of this book in Table 1.

FIGURE 5.4 ● THREE *z*-SCORES IN THE STANDARD NORMAL DISTRIBUTION

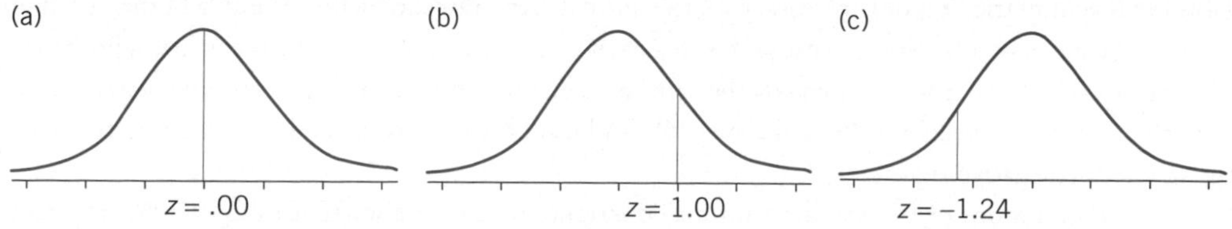

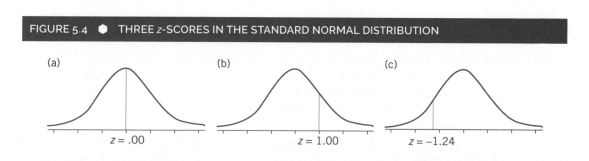

TABLE 5.1 ● PORTIONS OF THE APPENDIX (PROPORTIONS OF AREA UNDER THE STANDARD NORMAL DISTRIBUTION)

z	Area Between Mean and z	Area Beyond z	z	Area Between Mean and z	Area Beyond z
.00	.0000	.5000	1.95	.4744	.0256
.01	.0040	.4960	1.96	.4750	.0250
.02	.0080	.4920	1.97	.4756	.0244
.03	.0120	.4880	1.98	.4761	.0239
.04	.0160	.4840	1.99	.4767	.0233
.05	.0199	.4801	2.00	.4772	.0228
.06	.0239	.4761	2.01	.4778	.0222
.07	.0279	.4721	2.02	.4783	.0217
.08	.0319	.4681	2.03	.4788	.0212
.09	.0359	.4641	2.04	.4793	.0207
------------	---------------	------------	------------	---------------	------------
.85	.3023	.1977	2.95	.4984	.0016
.86	.3051	.1949	2.96	.4985	.0015
.87	.3078	.1922	2.97	.4985	.0015
.88	.3106	.1894	2.98	.4986	.0014
.89	.3133	.1867	2.99	.4986	.0014
------------	---------------	------------			
1.00	.3413	.1587	3.00	.4987	.0013
1.01	.3438	.1562	4.00	.4999	.0001
1.02	.3461	.1539			
1.03	.3485	.1515			
1.04	.3508	.1492			
------------	---------------	------------			

The first column listed in Table 5.1, "z," provides different values of z-scores starting with the mean z-score of .00. You may notice that all of the values of z in this column are positive numbers. Because the standard normal distribution is symmetric, it is only necessary for the normal curve table to include the area of the distribution to the right of the mean of .00. This is because the area of the distribution between the mean and a positive z-score (i.e., between $z = .00$ and $z = .75$) is *exactly* the same as the area between the mean and the negative z-score of the same value (i.e., between $z = .00$ and $z = -.75$). To locate the area associated with a negative z-score, you simply ignore the negative sign.

The symmetric nature of the standard normal distribution also implies that exactly 50% of z-scores are located in each of the two halves of the distribution. In other words, 50% of z-scores have positive values greater than the mean of .00, and 50% have negative values less than the mean of .00. Adding the two halves together (50% + 50% = 100%) accounts for all possible *z*-scores.

Finding the Area Between the Mean and a *z*-Score

What if you wanted to know, "How much of the standard normal distribution is located between $z = .00$ and $z = 3.00$?" Answering this question begins by examining the first column of Table 5.1—the column labeled "*z*." This column lists *z*-scores in increasing order, starting with the mean of the distribution, $z = .00$. After moving down the "*z*" column until you reach the value 3.00, move to the right until you reach the corresponding number under the column labeled "Area Between Mean and *z*." The number .4987 reveals that 49.87% (of a possible 50%) of the standard normal distribution lies *between* $z = .00$ and $z = 3.00$. This area is visually illustrated by the lighter shaded area in Figure 5.5.

As we mentioned earlier, the area of the distribution between the mean and a positive *z*-score is the same as the area between the mean and the same negative *z*-score. In other words, if we had asked, "How much of the standard normal distribution is located between $z = .00$ and $z = -3.00$?" we would have arrived at the same answer of 49.87%. Although the *z*-score may be a negative number, note that the percentage cannot be a negative number; this is because the smallest possible value of a percentage is zero (0%).

Now let's determine how much of the standard normal distribution is *greater* than the *z*-score of 3.00. Returning to Table 5.1, move from our *z*-score of 3.00 to the corresponding number in the column labeled "Area Beyond *z*." Here we find the value .0013, from which we learn that .13% (or less than 1%) of *z*-scores are greater than ("beyond") 3.00. This area is visually illustrated by the darker shaded portion located to the right of the value 3.00 in Figure 5.5. (We've found that drawing and shading figures such as the one in Figure 5.5 is a very useful way to help check your calculations of these areas.)

We can also verify our findings by noting that the two percentages (49.87% and .13%) add up to 50%, which is the entire right half of the standard normal distribution:

$$\begin{aligned}\text{Area under the right half of } &= (\text{area between mean and } z = 3.00) + \\ \text{the standard normal distribution} &\,(\text{area beyond } z = 3.00) \\ 50\% &= 49.87\% + .13\% \\ 50\% &= 50\%\end{aligned}$$

FIGURE 5.5 ● PERCENTAGE OF *z*-SCORES BETWEEN MEAN AND $z = 3.00$

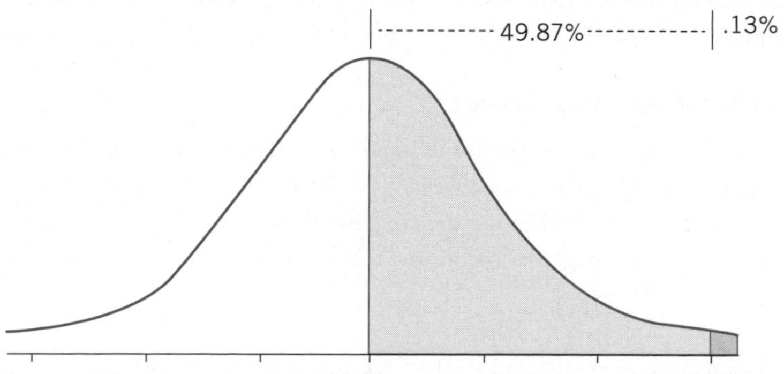

FIGURE 5.6 ● PERCENTAGE OF *z*-SCORES LESS THAN 1.00

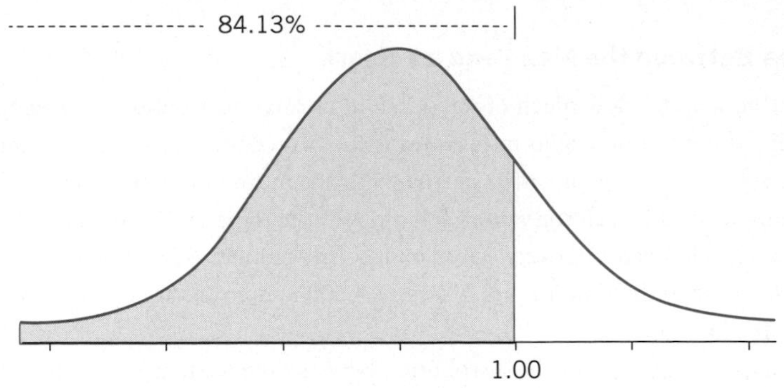

Finding the Area Less or Greater Than a *z*-Score

In addition to determining the area of the standard normal distribution between the mean *z*-score of .00 and a particular *z*-score, you can also evaluate a *z*-score in terms of its relationship to the *entire* standard normal distribution. For example, what if you asked, "What percentage of *z*-scores is less than $z = 1.00$?" This has been represented by the shaded area in Figure 5.6.

The first step in determining the percentage of *z*-scores less than $z = 1.00$ is to determine the area between the mean of .00 and the stated *z*-score. For this example, move down the "*z*" column in Table 5.1 until you reach $z = 1.00$. Under the column labeled "Area Between Mean and *z*," you find associated with $z = 1.00$ the number .3413, which implies that 34.13% of the standard normal distribution lies between the mean and $z = 1.00$. However, this only determines the area of *z*-scores less than 1.00 that are in the right half of the distribution. To find the percentage of *all* *z*-scores less than 1.00, you need to add the left half of the distribution to the percentage you just identified. Given that half, or 50%, of the standard normal distribution is less than .00, you add 50% to 34.13%:

$$\begin{aligned}\text{Area less than } z = +1.00 &= (\text{area between mean and z} = 1.00) + (\text{area less than mean})\\ &= 34.13\% + 50\%\\ &= 84.13\%\end{aligned}$$

From these calculations, we can conclude that 84.13% of the total distribution of *z*-scores is less than 1.00.

What percentage of *z*-scores in the standard normal distribution is *greater* than (to the right of) $z = 1.00$? To answer this question, turn to Table 5.1 and look under the column "Area Beyond *z*." Associated with $z = 1.00$ is the number .1587, which means that 15.87% of *z*-scores are greater than $z = 1.00$. Note that the sum of the two percentages you've just calculated (84.13 + 15.87) is equal to 100%, which is the entire distribution.

Finding the Area Between Two *z*-Scores

It is also possible to identify the percentage of the standard normal distribution that's between two *z*-scores. For example, what percentage of *z*-scores is between $z = -.85$ and $z = 1.95$? This area is shaded in Figure 5.7. Calculating the percentage of scores within this area requires three steps. The first step is to divide the area into two sections: the area between the negative *z*-score and the mean of .00 and the area between the mean and the positive *z*-score. In this example:

$$\begin{aligned}\text{Area between } z = -.85 \text{ and } z = 1.95 = &\left(\text{area between } z = -.85 \text{ and mean}\right) +\\ &\left(\text{area between mean and } z = 1.95\right)\end{aligned}$$

FIGURE 5.7 ● PERCENTAGE OF z-SCORES BETWEEN $z = -.85$ AND $z = 1.95$

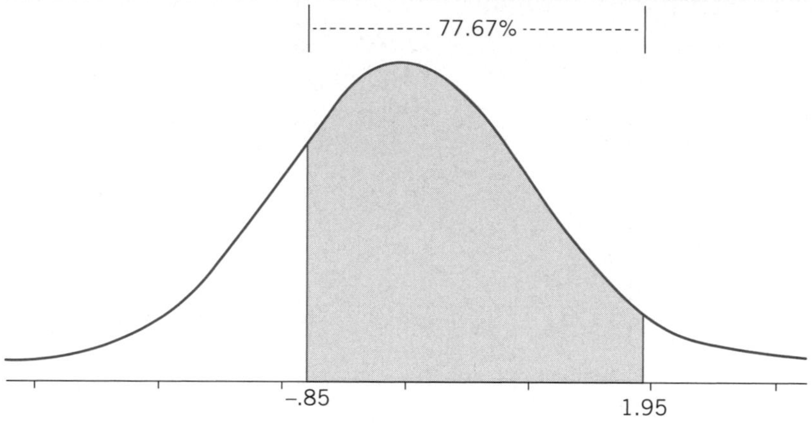

Because the distribution is symmetric, the area between the negative z-score and the mean of .00 should be treated as if it were above the mean. In this example, the area of the distribution between the mean and $z = -.85$ is the same as the area between the mean and $z = .85$.

The second step is to determine the area between the mean and each of the two z-scores. To do this, we return to the "Area Between Mean and z" column in Table 5.1. Here, the area between the mean and $z = -.85$ is 30.23%, and the area between the mean and $z = 1.95$ is 47.44%.

The third and last step is to add these two percentages together:

$$\text{Area between } z = -.85 \text{ and } z = 1.95 = 30.23\% + 47.44\%$$
$$= 77.67\%$$

On the basis of these three steps, we can conclude that 77.67% of z-scores in the standard normal distribution fall between $z = -.85$ and $z = 1.95$.

The above example shows how to calculate the area between a negative z-score and a positive z-score. Several of these areas, illustrated in Figure 5.8, are particularly important to remember. First, what percentage of z-scores is between the mean and one standard deviation in both directions ($z = \pm1.00$)? (The $\pm$ symbol means "plus or minus.") Looking at Table 5.1, under the "Area Between Mean and z" column, you find that 34.13% of z-scores are between the mean and $z = 1.00$. To cover the area both left and right of the mean ($z = \pm1.00$), you simply multiply 34.13% by 2 and conclude that 68.26%, or roughly two thirds, of the standard normal distribution lies within ±1 standard deviation of the mean. Furthermore, for $z = \pm2.00$, 95.44% (47.72% + 47.72%) of the distribution falls within 2 standard deviations of the mean. Finally, slightly less than 100% of z-scores, 99.74%, are within 3 standard deviations of the mean ($z = \pm3.00$).

In addition to these examples, it is also possible to determine the area of the distribution that is greater or less than a positive or negative z-score or that is between or beyond two different z-scores. Table 5.2 lists some simple rules regarding how the area under the standard normal distribution may be determined for stated z-scores.

FIGURE 5.8 ● PERCENTAGE OF *z*-SCORES WITHIN 1, 2, AND 3 STANDARD DEVIATIONS OF THE MEAN

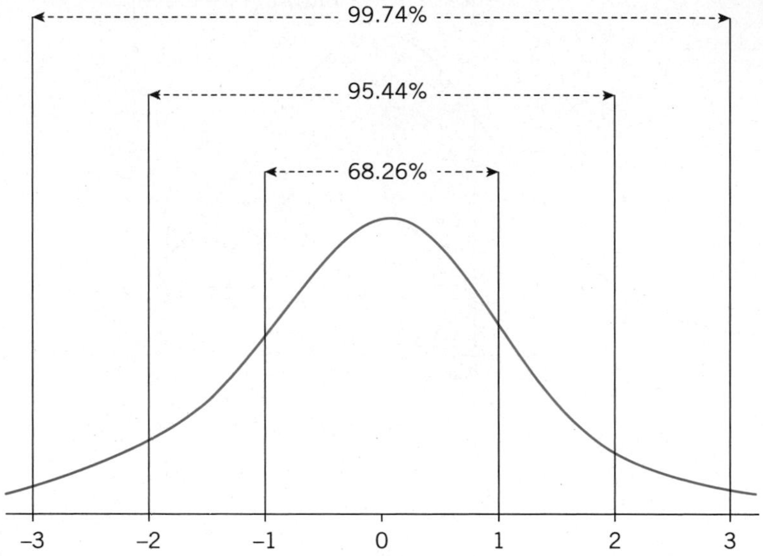

TABLE 5.2 ● RULES FOR DETERMINING AREA OF DISTRIBUTION ASSOCIATED WITH STATED *z*-SCORES

To find the percentage of *z*-scores that is . . .

- *between* the mean and a positive or negative *z*-score,
  find the "Area Between Mean and *z*"
- *greater* than a positive *z*-score,
  find the "Area Beyond *z*"
- *greater* than a negative *z*-score,
  find the "Area Between Mean and *z*" and add 50%
- *less* than a positive *z*-score,
  find the "Area Between Mean and *z*" and add 50%
- *less* than a negative *z*-score,
  find the "Area Beyond *z*"
- *between* two positive or two negative *z*-scores,
  find the "Area Between Mean and *z*" for each of the two *z*-scores and subtract the smaller area from the larger area
- *between* one positive and one negative *z*-score,
  find the "Area Between Mean and *z*" for each of the two *z*-scores and add the two areas together
- *beyond* two *z*-scores,
  calculate the area between the two *z*-scores and subtract it from 100%

LEARNING CHECK 1:

Reviewing What You've Learned So Far

1. Review questions
   a. What are the main features and characteristics of normal distributions?
   b. Why are normal distributions important to researchers?
   c. What is a problem or concern about using normal distributions?
   d. What are the main features of the standard normal distribution?
   e. What distinguishes the standard normal distribution from normal distributions?
   f. What are the characteristics of *z*-scores?
   g. What are the two main ways *z*-scores may be interpreted?
   h. Why does the normal curve table only cover half of the standard normal distribution?
   i. How do you determine the area of the standard normal distribution that is between the mean and a stated *z*-score? Less or greater than a *z*-score? Between two *z*-scores?
   j. What percentage of the standard normal distribution lies within one, two, and three standard deviations of the mean?
2. Using the normal curve table, determine the area of the standard normal distribution that is between the mean of .00 and the following *z*-scores:
   a. $z = .46$
   b. $z = 1.30$
   c. $z = 2.87$
   d. $z = -.19$
   e. $z = -1.53$
3. Using the normal curve table and the rules in Table 5.2, determine the area of the standard normal distribution that is either less (<) or greater (>) than the following *z*-scores:
   a. $z < .66$
   b. $z < 2.55$
   c. $z < -.75$
   d. $z > 1.73$
   e. $z > -2.17$
4. Using the normal curve table and the rules in Table 5.2, determine the area of the standard normal distribution that is between the following *z*-scores.
   a. $z = .25$ and $z = .50$
   b. $z = 1.33$ and $z = 2.25$
   c. $z = -.67$ and $z = -1.50$
   d. $z = -.44$ and $z = .82$
   e. $z = -1.17$ and $z = 2.60$

5.4 APPLYING *z*-SCORES TO NORMAL DISTRIBUTIONS

The previous section spent a great deal of time defining and explaining *z*-scores. However, researchers do not typically collect data in the form of *z*-scores. For example, if you ask people how tall they are, you'd get responses such as 5′ 5″, 6′ 4″, or 5′ 9″ rather than −1.33, 2.33, or .00. Fortunately, the information provided by the standard normal distribution can be applied to variables not measured in standard deviation units. This serves a valuable purpose: the ability to interpret and evaluate scores on variables. This section discusses how the standard normal distribution may be applied to variables assumed to be normally distributed in the population.

Transforming Normal Distributions Into the Standard Normal Distribution

Normal distributions and the standard normal distribution share many features, such as their bell shape and infinite number of scores. The primary way in which the two types of distributions differ is the unit of measurement. The unit of measurement for normal distributions is specific to each particular variable (i.e., inches, pounds, miles per hour); consequently, it is difficult to evaluate scores for a variable as well as compare scores across different variables. However, the unit of measurement in the standard normal distribution is common to all variables: distance from the mean in standard deviations. We can evaluate scores for variables that are assumed to be normally distributed in the population by transforming the normal distribution into the standard normal distribution—that is, transforming scores into *z*-scores.

Understanding the *z*-Score Formula

Let's return to the example of the SAT introduced earlier in this chapter. In taking the math section of the SAT, you receive a score of 660 and want to know how "good" it is—that is, how a score of 660 relates to the scores of other students. Answering this question involves applying the properties of the standard normal distribution to the distribution of SAT scores in the population.

To apply the standard normal distribution to a normal distribution, a score in the normal distribution must be transformed into a *z*-score (z) using the following formula:

$$z = \frac{X - \mu}{\sigma} \tag{5-1}$$

where X is a score for the variable, μ is the population mean, and σ is the population standard deviation.

How does Formula 5-1 transform a score into a *z*-score? To answer that question, let's first examine the numerator: $X - \mu$. Here we see that the value of a *z*-score is a function of the difference between the score and the population mean; the greater the difference, the larger the value of the *z*-score. Also, whether the score is greater (or less) than the mean determines whether the corresponding *z*-score is a positive (or negative) number. Next, the denominator for the formula is a standard deviation; dividing the numerator by a standard deviation implies that a *z*-score is measured in standard deviation units. Combining the numerator and the denominator of Formula 5-1 illustrates that a *z*-score is the difference between the score and the mean in standard deviation units.

Converting a score into a *z*-score requires two pieces of information: the population mean (μ) and standard deviation (σ) of the variable to be transformed. There are two main ways to determine the values for μ and σ. The first way is to collect data from the entire population and then calculate the mean and standard deviation from these data. However, as it is extremely difficult to collect data from an entire population, a second, and more common, method is to examine data collected from a large sample or large number of samples and then to estimate the population mean and standard deviation from these samples.

Example: Transforming SAT Scores Into *z*-Scores

Each year, the Education Testing Service (ETS) administers more than two million SAT examinations. The extremely large size of the sample increases confidence in the ETS's claim that SAT scores are normally distributed with a mean (μ) of 500 and standard deviation (σ) of 100. Therefore, SAT scores may be transformed into *z*-scores using the following formula:

$$z = \frac{X - 500}{100}$$

To illustrate how to transform scores into *z*-scores, let's say you have the SAT scores of three students: 630, 450, and 500. Substituting each score for "*X*" in the above formula, the three scores are transformed to *z*-scores as follows:

SAT = 630	SAT = 450	SAT = 500
$z = \frac{630 - 500}{100}$	$z = \frac{450 - 500}{100}$	$z = \frac{500 - 500}{100}$
$= \frac{130}{100}$	$= \frac{-150}{100}$	$= \frac{0}{100}$
$= 1.30$	$= -.50$	$= .00$

Interpreting *z*-Scores: Distance From the Mean

In the previous example, the SAT scores of 630, 450, and 500 were transformed into the *z*-scores of 1.30, −.50, and .00, respectively. As you have already learned, there are two main ways to interpret *z*-scores: the distance of the score from the mean of .00 and the location of the score relative to the entire distribution.

First, let's look at how these three SAT scores can be interpreted in terms of their distance from the mean. The first *z*-score of 1.30 indicates that an SAT score of 630 is located 1.30 standard deviations above the population mean of 500. The second *z*-score ($z = -.50$) implies that a score of 450 is .50 standard deviations below the mean. Finally, the third *z*-score of .00 means that an SAT score of 500 is equal to the population mean. In summary, an SAT score of 630 would be considered well above average, a score of 450 is below average, and a score of 500 is average performance.

Interpreting *z*-Scores: Relative Position Within Distribution

In addition to their distance from the mean, scores transformed into *z*-scores can be interpreted relative to the entire distribution. It's possible for researchers to determine the area of the distribution between specified *z*-scores (e.g., the percentage of SAT scores less than 660) by using the guidelines provided earlier in Table 5.2.

Finding the Area Between the Mean and a Score

As a first example, what if you wanted to know the percentage of SAT scores between the mean (μ) of 500 and a score of 700? The first step in answering this question is to transform the stated score (700) into a *z*-score using Formula 5-1:

$$z = \frac{700 - 500}{100} = \frac{200}{100}$$
$$= 2.00$$

The above calculation tells us that an SAT score of 700 is equal to a *z*-score of 2.00.

Now that we've calculated the *z*-score, the next step is to determine the "Area Between Mean and *z*" in the normal curve table. Looking at Table 5.1, for $z = 2.00$, we see the value .4772. Therefore, we can say that 47.72% (or almost half) of SAT scores are between 500 and 700. This is illustrated by the shaded portion of Figure 5.9.

FIGURE 5.9 ● PERCENTAGE OF SAT SCORES BETWEEN 500 AND 700

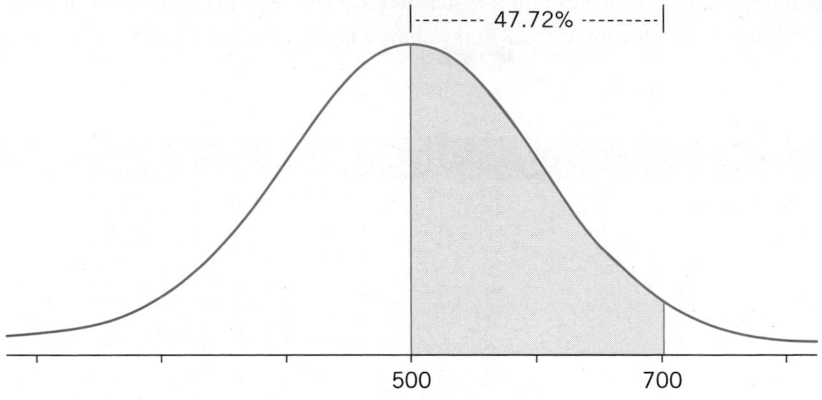

Finding the Area Less or Greater Than a Score

Earlier in this chapter, we posed the situation where you received a score of 660 on the math section of the SAT. What if you wanted to evaluate your score by determining the percentage of SAT scores less than 660? Again, you start by converting the score into the corresponding *z*-score:

$$z = \frac{660 - 500}{100} = \frac{160}{100}$$
$$= 1.60$$

Because you're determining the area less than a positive *z*-score, Table 5.2 informs you to find the "Area Between Mean and *z*" and add 50%. Looking at the Appendix, move down the "*z*" column until you reach $z = 1.60$. Under the "Area Between Mean and *z*" column, you find the value of .4452. By adding 50% to 44.52%, you can conclude that 94.52% of SAT scores are less than 660. This has been represented by the shaded area in Figure 5.10.

FIGURE 5.10 ● PERCENTAGE OF SAT SCORES LESS THAN 660

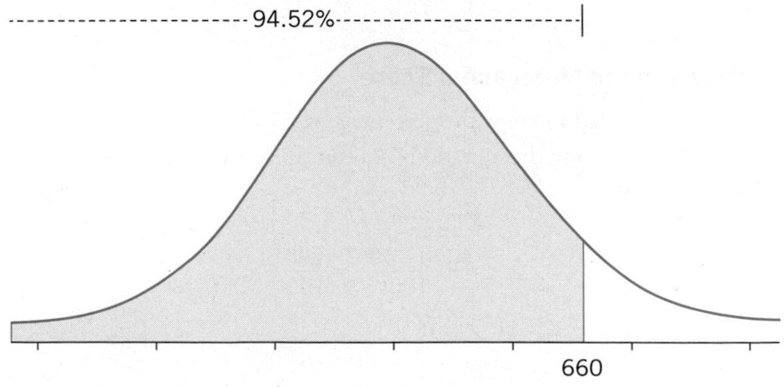

What percentage of SAT scores is *greater* than 660? Looking under the column "Area Beyond *z*," you find associated with $z = 1.60$ the number .0548, implying that 5.48% of SAT scores (100% – 94.52%) are greater than 660.

Finding the Area Between Two Scores

What if you asked, "What percentage of SAT scores is between 470 and 540?" (see Figure 5.11). You begin by converting each of these two scores to *z*-scores:

SAT = 470	SAT = 540
$z = \frac{470-500}{100}$	$z = \frac{540-500}{100}$
$= \frac{-30}{100}$	$= \frac{40}{100}$
$= -.30$	$= .40$

Because this example involves one positive and one negative *z*-score, Table 5.2 states that the next step is to divide this area into two sections: the area between the negative *z*-score and the mean of .00, as well as the area between the mean and the positive *z*-score. Ignoring the sign of the negative *z*-score, you determine the "Area Between Mean and *z*" for the two *z*-scores and add these two percentages together. In this example,

$$\begin{aligned}\text{Area between } z = -.30 \text{ and } z = .40 &= (\text{area between } z = -.30 \text{ and mean}) + \\ &\quad (\text{area between mean and } z = .40) \\ &= 11.79\% + 15.54\% \\ &= 27.33\%\end{aligned}$$

From these calculations, you would conclude that 27.33% of SAT scores are between 470 and 540.

FIGURE 5.11 ● PERCENTAGE OF SAT SCORES BETWEEN 470 AND 540

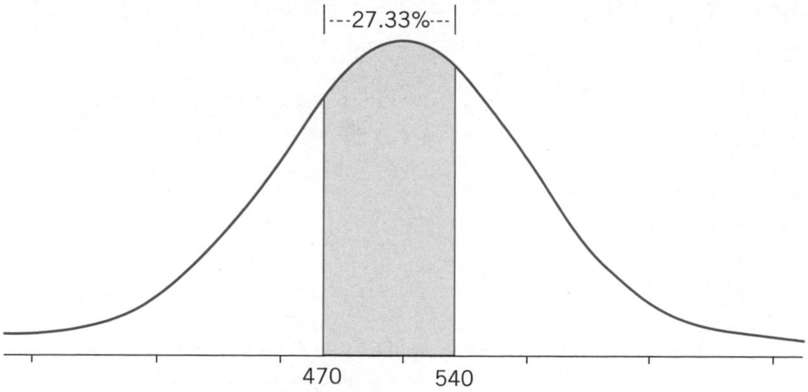

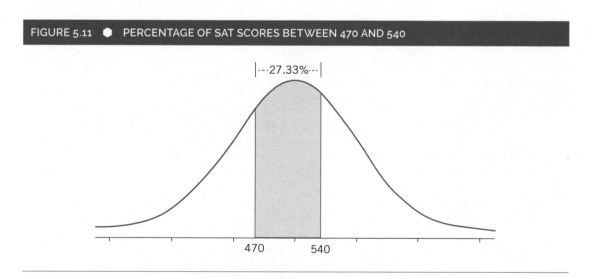

Transforming Scores Into *z*-Scores: Drawing Conclusions

The previous section discussed how the standard normal distribution may be applied to variables assumed to be normally distributed in the population. Although we've emphasized mathematical formulas and calculations, it is equally important to develop the ability to interpret and evaluate the results of these calculations. Transforming a score into a *z*-score provides the ability to draw conclusions about the score. As an example, we posed the situation where you wanted to evaluate your SAT score of 660. Transforming this score into a *z*-score revealed that 94.52% of SAT scores are less than 660, whereas only 5.48% of scores are greater than 660. Consequently, it can be concluded that only a small percentage of the population receives scores higher than 660.

Transforming scores into *z*-scores also helps researchers draw conclusions regarding whether a score is an outlier. Outliers were defined in Chapter 2 as rare, extreme scores that lie outside of the range of the majority of scores and are a concern because they may distort conclusions researchers draw about their data. Consider, for example, an SAT Math score of 760, which is transformed into a *z*-score below:

$$z = \frac{760 - 500}{100} = \frac{260}{100}$$
$$= 2.60$$

Using the normal curve table and Table 5.2, we find that an extremely large proportion (99.53%) of SAT Math scores are less than 760, with only .47% being greater than 760. Consequently, we may identify someone with an SAT score of 760 as an outlier.

LEARNING CHECK 2:

Reviewing What You've Learned So Far

1. Review questions
   a. Why might one want to apply the standard normal distribution to a normal distribution?
   b. What's the first step in applying the standard normal distribution to a normal distribution?

   (Exercises 2 and 3 are based on the following example): A study of more than 1,000 pregnant women in India found the length of pregnancy was normally distributed, with a mean (μ) of 272 days (about 9 months) and a standard deviation (σ) of 9 days (Bhat & Kushtagi, 2006).

2. Transform the following lengths of pregnancy into *z*-scores:
   a. 275 days
   b. 280 days
   c. 265 days
   d. 286 days
   e. 260 days
3. What percentage (%) of pregnancies . . .
   a. is shorter than 277 days?
   b. is shorter than 262 days?
   c. is longer than 270 days?
   d. is longer than 283 days?
   e. is between 279 and 286 days?
   f. is between 268 and 276 days?

   (Exercise 4 is based on the following example): A survey conducted in 2005 (www.macintouch.com) asked a large number of iPod owners whether their iPod had ever failed (completely stopped working). Based on their results, it was estimated that the likelihood of experiencing a failure was normally distributed, with an average likelihood of failure (μ) of 13.70%, with a standard deviation (σ) of 7.54%.

4. What percentage (%) of iPods . . .
   a. has a likelihood of failure greater than 25%?
   b. has a likelihood of failure greater than 10%?

c. has a likelihood of failure less than 20%?
d. has a likelihood of failure less than 5%?
e. has a likelihood of failure between 15% and 30%?
f. has a likelihood of failure between 11% and 18%?

5.5 STANDARDIZING FREQUENCY DISTRIBUTIONS

The previous section discussed how the standard normal distribution may be applied to normal distributions, which are distributions based on populations with a known population mean and standard deviation. The primary purpose of applying the standard normal distribution to normally distributed variables is to interpret and evaluate scores for a variable by transforming them into *z*-scores. This transformation, interpretation, and evaluation are the foundation of statistical procedures designed to test research hypotheses that we'll discuss later in this book.

In addition to transforming normal distributions, *z*-scores may also be calculated for scores within a frequency distribution, which is a distribution of scores based on data collected from a sample. For example, imagine you take a midterm exam in a class of 23 students and want to know, "How well did I do?" You might want to evaluate your score relative to the other students in the class but have no reason to believe the scores in your class are normally distributed. This section uses a published research study to show how and why researchers transform scores in a frequency distribution into *z*-scores.

Example: Measuring Politicians' Behavior

Voters have a vested interest in understanding the thoughts and behaviors of their elected officials, including the impact of politicians' personal goals on their behavior. Researchers might choose to examine, for example, whether the behavior of a state governor who wishes to be president differs from the behavior of a governor who prefers serving at a lower level of government.

One study, conducted by Dr. Rebekah Herrick from Oklahoma State University, investigated the political ambition of members of the U.S. House of Representatives (Herrick, 2001). More specifically, "Members' responses to questions about their desires for future political office were compared to their legislative behaviors to see if ambition affects their behavior" (p. 471). The example in this chapter will focus on the legislative behavior of these members.

One aspect of legislative behavior examined in this study was "floor activity," measured using two variables: the number of remarks made on the floor and the number of amendments proposed to other members' legislation. The higher the number of remarks or amendments, the higher the level of floor activity. Dr. Herrick recorded the number of remarks and amendments made by a sample of members of the House of Representatives during their 2-year term in office. To minimize space, this example will be based on the data from 23 of the 47 members in her study. (The findings reported here closely resemble the overall findings from the published study.)

Table 5.3(a) provides the number of remarks and the number of amendments for the 23 members in this example. Table 5.3(b) provides frequency distribution tables for the data presented in Table 5.3(a).

Looking at the frequency distribution tables, the two variables differ greatly in their range of possible values. The number of remarks ranges from 0 to 157; however, the number of amendments is much smaller because developing an amendment requires more time and effort than public speaking. These differences are reinforced by Table 5.4(a) and (b), which calculate the mean and standard deviation for the two variables using the formulas introduced in Chapters 3 and 4. As you can see, a comparison of the two means, $M = 65.22$ for the number of remarks and $M = 1.74$ for the number of amendments, reflects the difference in the nature of the two variables.

TABLE 5.3 ● NUMBER OF REMARKS AND NUMBER OF AMENDMENTS VARIABLES

(a) Raw Data

Member	# Remarks	# Amendments	Member	# Remarks	# Amendments
1	**49**	*0*	13	**57**	*2*
2	**82**	*1*	14	**105**	*0*
3	**66**	*3*	15	**78**	*0*
4	**25**	*0*	16	**36**	*1*
5	**44**	*0*	17	**68**	*7*
6	**113**	*13*	18	**29**	*1*
7	**52**	*1*	19	**157**	*0*
8	**61**	*1*	20	**55**	*1*
9	**70**	*2*	21	**61**	*0*
10	**54**	*2*	22	**0**	*0*
11	**84**	*1*	23	**84**	*4*
12	**70**	*0*			

(b) Frequency Distribution Tables

# Remarks	*f*	%	# Amendments	*f*	%
140+	1	4%	*7+*	2	9%
120–139	0	0%	*6*	0	0%
100–119	2	9%	*5*	0	0%
80–99	3	13%	*4*	1	4%
60–79	7	31%	*3*	1	4%
40–59	6	26%	*2*	3	13%
20–39	3	13%	*1*	7	31%
0–19	1	4%	*0*	9	39%
Total	23	100%	Total	23	100%

Dr. Herrick wanted to combine scores on the two variables into a single measure of floor activity. However, because of the difference in the distributions of the two variables, it would be inappropriate to simply add them together. For instance, although a score of 0 is an extremely low number of remarks, it is actually the modal score for the number of amendments. Consequently, to compare and combine variables with different units of measurement, the variables must first be transformed to the same unit of measurement. As she wrote, "Since the number of amendments were significantly smaller than the number of remarks, *z*-scores were calculated and used" (Herrick, 2001, p. 471).

Standardizing a Frequency Distribution

When scores in a frequency distribution are transformed into *z*-scores, the distribution of *z*-scores is called a **standardized distribution.** A standardized distribution is different from the standard normal distribution in that a standardized distribution does not assume the variable is normally distributed with a known population mean and standard deviation. The z-scores within a standardized distribution are referred to as **standardized scores.**

A frequency distribution of scores may be transformed into a standardized distribution of standardized scores using Formula 5-2:

$$z = \frac{X - \overline{X}}{s} \quad (5\text{-}2)$$

where X is a score for the variable, $\overline{X}$ is the sample mean, and s is the standard deviation.

Formula 5-2 is very similar to Formula 5-1, which is used to transform a normal distribution into the standard normal distribution. The numerator in both formulas is the deviation of a score from a mean, and both require dividing the numerator by a standard deviation. There are, however, two critical differences between the two formulas. First, creating a standardized distribution using Formula 5-2 does not assume the knowledge of a population mean (μ) or standard deviation (σ). Instead, z-scores in a standardized distribution are based on the sample mean ($\overline{X}$) and standard deviation (s). As such, you are relating the scores for a variable to the other scores in the sample rather than to scores in the population. Second, the scores in the frequency distribution are not assumed to be normally distributed. Regardless of its shape (modality, symmetry, and variability), *any* frequency distribution may be transformed into a standardized distribution.

Interpreting Standardized Scores

To illustrate how a frequency distribution may be transformed into a standardized distribution, let's transform the first three scores for the number of remarks variable into their corresponding standardized scores. Inserting the sample mean ($M = 65.22$) and standard deviation ($s = 32.16$) calculated in Table 5.4 into Formula 5-2, the three z-scores are calculated below, in which the number of remarks (49, 82, and 66) is transformed into standardized scores of −.50, .52, and .02.

Member 1 (# Remarks = 49)	Member 2 (# Remarks = 82)	Member 3 (# Remarks = 66)
$z = \frac{49 - 65.22}{32.16}$	$z = \frac{82 - 65.22}{32.16}$	$z = \frac{66 - 65.22}{32.16}$
$= \frac{-16.22}{32.16}$	$= \frac{16.78}{32.16}$	$= \frac{.78}{32.16}$
$= -.50$	$= .52$	$= .02$

Standardized scores are typically interpreted in terms of their distance from the sample mean in standard deviation units. For example, the first member's z-score of −.50 indicates that 49 remarks is .50 standard deviations, or half a standard deviation, below the sample mean of 65.22. Transforming scores into standardized scores helps researchers evaluate and interpret their data. For example, we can say that the 49 remarks made by the first member is below the average of the sample, whereas the second member's 82 remarks is above average.

Using Standardized Scores to Compare or Combine Variables

Perhaps the most common reason for creating standardized distributions is to compare and combine scores on variables measured on different units of measurement. As an example, the information from Tables 5.3 and 5.4

TABLE 5.4 ● DESCRIPTIVE STATISTICS, NUMBER OF REMARKS, AND AMENDMENTS VARIABLES

(a) Calculation of Means ($\bar{X}$)

# Remarks	# Amendments
$\bar{X} = \frac{\Sigma X}{N}$	$\bar{X} = \frac{\Sigma X}{N}$
$= \frac{49 + 82 + \ldots + 0 + 84}{23} = \frac{1500}{23}$	$= \frac{0 + 1 + \ldots + 0 + 4}{23} = \frac{40}{23}$
$= 65.22$	$= 1.74$

(b) Calculation of Standard Deviations (s)

# Remarks	# Amendments
$s = \sqrt{\frac{\Sigma(X - \bar{X})^2}{N - 1}}$	$s = \sqrt{\frac{\Sigma(X - \bar{X})^2}{N - 1}}$
$= \sqrt{\frac{(49 - 65.22)^2 + \ldots + (84 - 65.22)^2}{23 - 1}}$	$= \sqrt{\frac{(0 - 1.74)^2 + \ldots + (4 - 1.74)^2}{23 - 1}}$
$= \sqrt{\frac{263.09 + \ldots + 352.69}{22}}$	$= \sqrt{\frac{3.03 + \ldots + 5.11}{22}}$
$= \sqrt{\frac{22751.95}{22}} = \sqrt{1034.18}$	$= \sqrt{\frac{192.43}{22}} = \sqrt{8.75}$
$= 32.16$	$= 2.96$

is used to transform the scores for the first three members on the other variable in this study, the number of amendments introduced by each member:

Member 1 (# Amendments = 0)	Member 2 (# Amendments = 1)	Member 3 (# Amendments = 3)
$z = \frac{0 - 1.74}{2.96}$	$z = \frac{1 - 1.74}{2.96}$	$z = \frac{3 - 1.74}{2.96}$
$= \frac{-1.74}{2.96}$	$= \frac{-.74}{2.96}$	$= \frac{1.26}{2.96}$
$= -.59$	$= -.25$	$= .43$

Transforming scores into standardized scores provides a way of comparing scores on the two variables. Looking, for example, at the two standardized scores for the first member, the score of 49 on the number of remarks variable ($z = -.50$) is comparable to the score of 0 on the number of amendments variable ($z = -.59$). As such, we can conclude that this member's level of floor activity on both of these variables was below the average of this sample of representatives. Furthermore, because both variables are now measured on the same unit of measurement, scores on the two variables may be combined. In this example, Dr. Herrick was able to combine the number of remarks and number of amendments variables into a single measure of floor activity.

Characteristics of Standardized Distributions

This section illustrates the similarities and differences between standardized distributions and the standard normal distribution. To begin this discussion, Table 5.5(a) provides the scores and *z*-scores for the number of amendments variable in the Herrick (2001) study. Table 5.5(b) calculates the mean and standard deviation of the standardized scores for this variable.

Looking at the calculations for the mean and standard deviation, we discover that standardized distributions share several of the characteristics of the standard normal distribution. Specifically, the mean and standard deviation of both distributions are equal to 0 and 1, respectively. But is the shape of the two distributions also the same?

TABLE 5.5 ● STANDARDIZED SCORES FOR THE NUMBER OF AMENDMENTS VARIABLE

(a) Listing of Scores and *z*-Scores for Each Member

Member	# Amendments	*z*-Score
1	0	−.59
2	1	−.25
3	3	.43
4	0	−.59
5	0	−.59
6	13	3.81
7	1	−.25
8	1	−.25
9	2	.09
10	2	.09
11	1	−.25
12	0	−.59
13	2	.09
14	0	−.59
15	0	−.59
16	1	−.25
17	7	1.78
18	1	−.25
19	0	−.59
20	1	−.25
21	0	−.59
22	0	−.59
23	4	.76

(b) Descriptive Statistics of Standardized Scores

Mean	Standard Deviation
$\frac{\Sigma z}{N}$ $= \frac{-.59 + -25 + \ldots + -.59 + .76}{23} = \frac{.00}{23}$ $= .00$	$\sqrt{\frac{\Sigma\left(z - \bar{z}\right)^2}{N-1}}$ $= \sqrt{\frac{(.59 - .00)^2 + \ldots + (.76 - .00)^2}{23-1}}$ $= \sqrt{\frac{.35 + \ldots + .58}{22}} = \sqrt{\frac{22.00}{22}} = \sqrt{1.00}$ $= 1.00$

TABLE 5.6 ● FREQUENCY DISTRIBUTION TABLES, NUMBER OF AMENDMENTS VARIABLE

Scores			*z*-Scores		
Score	*f*	%	*z*-Score	*f*	%
5+	2	9%	1.78+	2	9%
4	1	4%	.76	1	4%
3	1	4%	.43	1	4%
2	3	13%	.09	3	13%
1	7	31%	–.25	7	31%
0	9	39%	–.59	9	39%
Total	23	100%	Total	23	100%

Table 5.6 presents frequency distribution tables for the number of amendments variable in both original and standardized score form. Notice that although the scores and *z*-scores differ, the numbers in the "*f*" and "%" columns are the same in both tables. As we can see from this example, standardizing a frequency distribution does *not* change the shape of the distribution. The shape of the distribution is the same, regardless of whether a variable is measured in score or standardized score form. Standardizing a distribution does *not* transform a frequency distribution into a normal distribution; if the scores for a variable are not normally distributed, they will not be normally distributed after being transformed into standardized scores.

The fact that standardizing a frequency distribution does not alter the shape of the distribution has an important implication for how standardized scores are interpreted. For example, from Table 5.1 we know that 50% of *z*-scores in the standard normal distribution lie on either side of the mean. However, in the standardized distribution in Table 5.6, only 17% (4 of 23) of the standardized scores are greater than .00. Students sometimes mistakenly believe that converting scores to standardized scores changes the shape of the distribution into a normal distribution. This is simply not the case.

Using Formula 5-2 to standardize a frequency distribution does not alter the shape of the distribution because this formula is a **linear transformation**, which is a mathematical transformation of a variable comprising addition, subtraction, multiplication, or division. Linear transformations do not change the shape of the distribution because the relative distances between scores in the distribution do not change. Another example of a linear transformation is temperature. For example, you could convert a temperature of 80° Fahrenheit to its Celsius equivalent by using the formula, Celsius $= \frac{5}{9}$ (Fahrenheit – 32). For a temperature of 80° Fahrenheit, its Celsius equivalent is $\frac{5}{9}(80-32)$, or $\frac{5}{9}(48)$, or 26.67°. If one person says today's temperature is 80° Fahrenheit and another person says it's 27° Celsius, you would not say the temperature has changed. Similarly, transforming scores to *z*-scores to standardize a frequency distribution does not change the shape of the distribution.

When Can You Apply the Normal Curve Table to a Standardized Distribution?

Given that standardizing a frequency distribution does not change the shape of the distribution, is it possible to apply the characteristics of the standard normal distribution to a standardized distribution? The answer to this question depends on the shape of the frequency distribution. The features of the standard normal distribution can be applied to standardized distributions if and when the shape of the distribution of scores is approximately normal. But this raises another question: What is meant by "approximately" normal?

Because the standard normal distribution consists of an infinite number of scores based on a mathematical formula, data collected from a sample cannot be expected to completely resemble a normal distribution. However, the larger the sample, the more likely the data will resemble a normal distribution. How large does a sample need to be to assume the data approach a normal distribution? Although there is no definitive answer to this question, some researchers have come to rely on the formula, $N = 30$. This means that a sample size of at least 30 is necessary for the normal curve to be applied to a standardized distribution.

LEARNING CHECK 3:

Reviewing What You've Learned So Far

1. Review questions
   a. What are the differences between calculating *z*-scores for normal distributions versus frequency distributions?
   b. What are common reasons for standardizing frequency distributions?
   c. What are the main characteristics of standardized distributions?
   d. Does standardizing a frequency distribution create a normal distribution? Why or why not?
   e. Under what conditions could you apply the normal curve table to a standardized distribution?
2. Using the descriptive statistics calculated in Table 5.4, calculate standardized scores for the number of remarks variable for the following members.
   a. Member 4 (# remarks = 25)
   b. Member 5 (# remarks = 44)
   c. Member 6 (# remarks = 113)
   d. Member 7 (# remarks = 61)
   e. Member 8 (# remarks = 70)
3. Imagine you take a quiz in one of your classes and learn that the class mean ($\bar{X}$) was 15.53 and the standard deviation (s) was 4.22. Calculate standardized scores for the following quiz scores.
   a. 14
   b. 18
   c. 7
   d. 17
   e. 23

Player	Batting Average	Player	Batting Average
1	**.375**	6	**.298**
2	**.250**	7	**.311**
3	**.289**	8	**.232**
4	**.321**	9	**.211**
5	**.264**		

4. You play on a softball team and at the end of the season find the batting averages of the players on your team are as follows:
   a. Calculate the mean ($\bar{X}$) and standard deviation (s) of the batting averages.
   b. Transform each batting average into a standardized score.
   c. Calculate the mean and standard deviation of the standardized scores.

5.6 LOOKING AHEAD

This chapter began by reviewing a process used to conduct research. A central part of this process is conducting statistical analyses designed to test research hypotheses. As you will see in the next chapter, these statistical analyses depend heavily on different types of distributions. The main purpose of this chapter was to introduce two critical types of distributions: normal distributions and the standard normal distribution. Two aspects of these distributions will be built upon in future chapters. First, these distributions are unimodal and symmetric, with an infinite number of scores. Second, it is possible to evaluate a score in terms of the proportion of the distribution associated with that score. However, rather than evaluating individual scores, researchers analyze data to test research hypotheses about populations. Because data are not collected from the entire population, researchers must rely on probability to evaluate the data in their sample and draw appropriate conclusions about their hypotheses. The next chapter describes the role of probability in the research process and introduces steps taken by researchers to conduct statistical analyses to test their hypotheses.

5.7 Summary

Normal distributions are distributions based on a population of an infinite number of scores calculated from mathematical formulas. Some of the features of normal distributions are that they consist of an infinite number of scores, are unimodal and symmetric, have a mean that is equal to the population mean μ, and have a standard deviation equal to the population standard deviation σ.

Normal distributions are important to researchers because many variables in populations are generally believed by researchers to be normally distributed, many statistical procedures designed to test research hypotheses are based on the assumption that variables are normally distributed, and, because normal distributions are based on mathematical formulas, it is possible to determine the proportion of the distribution associated with any score.

One problem with using normal distributions is that the shape and nature of a particular normal distribution depend on how each particular variable is measured. To address this problem, the ***standard normal distribution***, which is a normal distribution with a mean equal to 0 and a standard deviation equal to 1, may be applied to any normal distribution. Scores in the standard normal distribution, known as ***z-scores***, are measured in standard deviation units, such that a *z*-score is the number of standard deviations a score is different from the mean.

It is possible to evaluate *z*-scores in two ways: distance from the mean in standard deviation units and the percentage of the distribution associated with a stated *z*-score or scores (using a table known as the ***normal curve table***).

Scores in normally distributed variables may be evaluated by transforming the scores into *z*-scores. Scores in a frequency distribution (data collected by researchers) may also be transformed into *z*-scores known as ***standardized scores***; the distribution of standardized scores is known as a ***standardized distribution***. A common reason for creating standardized distributions is to compare and combine scores on variables measured on different scales of measurement. Because a standardized distribution does not assume the variable is normally distributed, if the scores in a frequency distribution are not normally distributed, they will not be normally distributed after being transformed into standardized scores. However, the larger the sample, the more likely it is a frequency distribution will resemble a normal distribution.

5.8 Important Terms

normal distribution (p. 120)
standard normal distribution (p. 122)
z-scores (p. 122)
normal curve table (p. 123)
standardized distribution (p. 136)
standardized scores (p. 137)
linear transformation (p. 140)

5.9 Formulas Introduced in This Chapter

z-Score

$$z = \frac{X - \mu}{\sigma} \tag{5-1}$$

Standardized Score

$$z = \frac{X - \overline{X}}{s} \tag{5-2}$$

5.10 Exercises

NOTE: In Exercises 1 to 4, it may be helpful to draw a figure such as Figure 5.5.

1. Using the normal curve table, determine the area of the standard normal distribution that is between the mean of .00 and the following *z*-scores:
   a. $z = .69$
   b. $z = 1.45$
   c. $z = 2.01$
   d. $z = -.25$
   e. $z = -1.86$
2. Using the normal curve table, determine the area of the standard normal distribution that is less than the following *z*-scores:
   a. $z = .30$
   b. $z = 1.75$
   c. $z = 2.42$
   d. $z = -.68$
   e. $z = -1.11$
3. Using the normal curve table, determine the area of the standard normal distribution that is greater than the following *z*-scores:
   a. $z = .78$
   b. $z = 1.22$
   c. $z = 2.30$
   d. $z = -.90$
   e. $z = -1.34$
4. Using the normal curve table, determine the area of the standard normal distribution that is between the following *z*-scores:
   a. $z = .30$ and $z = 1.00$
   b. $z = 1.56$ and $z = 2.45$
   c. $z = -.50$ and $z = -.80$
   d. $z = -.90$ and $z = -2.67$

e. $z = -1.34$ and $z = .10$

f. $z = -.75$ and $z = .75$

(Exercises 5 through 9 are based on the following example): IQ scores are normally distributed in the population with a mean (μ) of 100 and a standard deviation (σ) of 16. Use the normal curve table to complete these exercises.

5. Transform the following IQ scores into *z*-scores.
   a. 116
   b. 84
   c. 124
   d. 95
   e. 130

6. What percentage of IQ scores is between the mean of 100 and . . .
   a. 116
   b. 68
   c. 105
   d. 90
   e. 140

7. What percentage of IQ scores is . . .
   a. less than 122
   b. less than 77
   c. less than 104
   d. greater than 135
   e. greater than 73
   f. greater than 112

8. What percentage of IQ scores is . . .
   a. between 115 and 130
   b. between 105 and 120
   c. between 75 and 85
   d. between 80 and 95
   e. between 90 and 110
   f. between 80 and 120

9. What percentage of IQ scores is . . .
   a. less than 110 or greater than 130
   b. less than 80 or greater than 90
   c. less than 95 or greater than 105
   d. less than 84 or greater than 116

10. A researcher constructs a test to measure self-esteem. In doing so, she calculates a mean for her sample of 10 with a standard deviation of 2. Assuming that scores on this test in the larger population are normally distributed, what percentage of scores on the test is . . .
    a. greater than 12
    b. greater than 16
    c. greater than 8

d. less than 11

e. between 13 and 15

f. between 6 and 14

(Exercises 11 through 13 are based on the following example): Imagine you are the president of a toy company that builds video games. You have determined that the time needed to build these games is normally distributed with a mean of 20 minutes and a standard deviation of 5 minutes.

11. What percentage of the video games is built . . .

    a. in less than 15 minutes

    b. in less than 10 minutes

    c. in more than 20 minutes

    d. in more than 30 minutes

    e. between 26 and 32 minutes

    f. between 12 and 24 minutes

12. Five percent of the video games (the bottom 5%) take less than ____ minutes to build.

13. Ninety percent of the video games (the top 90%) are built in ____ minutes.

14. How many base hits can a baseball team expect to get in a game? Frohlich (1994) recorded the number of hits by the 28 Major League Baseball teams for all of the games played from 1989 to 1993 (each team plays 162 games a year). He found that the number of hits the teams made in the games was normally distributed, with a mean of 8.72 and a standard deviation of 1.10.

    a. In what percentage of games would you expect a team to get more than nine hits?

    b. In what percentage of the games would you expect a team to get less than six hits?

15. In another baseball-related story, between 1980 and 2000, there was a remarkable increase in the number of home runs hit in the major leagues. A recent article stated that the average number of home runs per game rose from 1.47 in 1980 to 2.34 in 2000 (Rist, 2001). A factor thought to be a contributing factor to this increase is the ball itself. One of the responsibilities of a university-based research center sponsored by Major League Baseball is to check the weight of baseballs. In a recent report, this research center found that the weight of baseballs used in the 2000 season had a mean of 5.07 ounces with a standard deviation of .06 ounces.

    a. Major League Baseball specifies that the weight of baseballs should be between 5.00 and 5.25 ounces. What percentage of the baseballs falls outside of the specified range? That is, what percentage of the 2000 baseballs was less than 5.00 ounces, and what percentage weighed more than 5.25 ounces?

    b. A sample of baseballs weighed in 1999 had a mean weight of 5.09 ounces with a standard deviation of .05 ounces. What percentage of the 1999 baseballs fell outside of the 5.00 to 5.25 range?

    c. Comparing the results of the two seasons, what conclusions might you reach regarding the weight of baseballs?

16. What if you took midterms in two different courses and happened to get the same grade of 68 on both of them. In the first course, the mean was 63 and the standard deviation was 10. In the second course, the mean was 72 and the standard deviation was 8.

    a. Calculate the z-score for your score in both courses.

    b. Relative to the other students in the courses, in which course would you consider your performance to be better? Why?

17. What if you took final examinations in two different courses and happened to get the same grade of 72 on both of them? The mean in the two courses turns out to be the same (= 60). However, the standard deviation in the two courses differs: $s = 12$ in the first course and $s = 6$ in the second.

    a. Calculate the z-score for your score in both courses.

    b. Even though your score on the two tests was the same and the mean in the two courses was the same, why are your two z-scores different from each other?

c. Relative to the other students in the courses, in which course would you consider your performance to be better? Why?

18. The mean of the Graduate Record Examination (GRE), for the verbal section, is 500 with a standard deviation of 100. Your friend scores a 529.

a. What percentage of test takers did she score above?

b. How many standard deviations above or below the mean is she?

19. Two students are comparing their recent midterm scores. The first student received a 92 and the second student received an 86. However, they are in different classes, and their tests were slightly different. The mean score in the first student's class was 82, with a standard deviation of 10 points. The mean score in the second student's class was a 75, with a standard deviation of 10. Both tests were normally distributed.

a. Convert both students' test scores to *z*-scores.

b. What percentage of test scores was below each score?

c. Which student did better on the statistics test?

20. You collect the grade point average (GPA) from 10 students:

a. Calculate the mean ($\overline{X}$) and standard deviation (s) of these GPAs.

b. Calculate the *z*-score for each GPA in order to standardize the distribution.

c. Calculate the mean and standard deviation of these *z*-scores.

d. What percentage of *z*-scores is greater than the mean? What percentage is less than the mean? Are these two percentages the same? Why or why not?

Student	GPA	Student	GPA
1	**3.45**	6	**2.55**
2	**2.40**	7	**2.30**
3	**2.80**	8	**2.80**
4	**2.55**	9	**3.05**
5	**3.80**	10	**2.30**

Answers to Learning Checks

Learning Check 1

2. a. 17.72%
   b. 40.32%
   c. 49.79%
   d. 7.53%
   e. 4.37%

3. a. 74.54%
   b. 99.46%
   c. 22.66%

d. 4.18%
e. 98.50%

4. a. 9.28%
b. 7.95%
c. 18.46%
d. 46.39%
e. 87.43%

Learning Check 2

2. a. $z = .33$
b. $z = .89$
c. $z = -.78$
d. $z = 1.56$
e. $z = -1.33$

3. a. 71.22% ($z = .56$)
b. 13.35% ($z = -1.11$)
c. 58.71% ($z = -.22$)
d. 11.12% ($z = 1.22$)
e. 15.83% ($z = .78$ and $z = 1.56$)
f. 34.01% ($z = -.44$ and $z = .44$)

4. a. 6.69% ($z = 1.50$)
b. 68.79% ($z = -.49$)
c. 79.95% ($z = .84$)
d. 12.51% ($z = -1.15$)
e. 41.71% ($z = .17$ and $z = 2.16$)
f. 35.62% ($z = -.36$ and $z = .57$)

Learning Check 3

2. a. $z = -1.25$
b. $z = -.66$
c. $z = 1.49$
d. $z = -.13$
e. $z = .15$

3. a. $z = -.36$
b. $z = .59$
c. $z = -2.02$
d. $z = .35$
e. $z = 1.77$

4. a. $\bar{X} = .284, s = .051$

b.

Player	Batting Average	z	Player	Batting Average	z
1	.375	1.78	6	.298	.27
2	.250	−.67	7	.316	.63
3	.289	.10	8	.232	1.02
4	.321	.73	9	.207	1.51
5	.264	−.39			

c. Mean of z-scores = 0, standard deviation of z-scores = 1.00.

Answers to Odd-Numbered Exercises

1. a. 25.49%
   b. 42.65%
   c. 47.78%
   d. 9.87%
   e. 46.86%

3. a. 21.77%
   b. 11.12%
   c. 1.07%
   d. 81.59%
   e. 90.99%

5. a. $z = 1.00$
   b. $z = -1.00$
   c. $z = 1.50$
   d. $z = -.31$
   e. $z = 1.88$

7. a. 91.62% ($z = 1.38$)
   b. 7.49% ($z = -1.44$)
   c. 59.87% ($z = .25$)
   d. 1.43% ($z = 2.19$)
   e. 95.45% ($z = -1.69$)
   f. 22.66% ($z = .75$)

9. a. 76.58% ($z = .63$ and $z = 1.88$)
   b. 84.13% ($z = -1.25$ and $z = -.63$)
   c. 75.66% ($z = -.31$ and $z = .31$)
   d. 31.74% ($z = -1.00$ and $z = 1.00$)

11. a. 15.87% ($z = -1.00$)
    b. 2.28% ($z = -2.00$)
    c. 50.00% ($z = .00$)
    d. 2.28% ($z = 2.00$)
    e. 10.69% ($z = 1.20$ and $z = 2.40$)
    f. 73.33% ($z = -1.60$ and $z = .80$)

13. 26.40 minutes ($z = 1.28$)

15. a. 12.10% weigh less than 5 oz ($z = -1.17$); .13% weigh more than 5.25 oz ($z = 3.00$).
    b. 3.59% weigh less than 5 oz ($z = -1.80$); .07% weigh more than 5.25 oz ($z = 3.20$).
    c. It appears baseballs that weighed outside the allowed range were being used more frequently in 2000 than in 1999.

17. a. First course: $z = 1.00$, second course: $z = 2.00$
    b. The students in the second course scored more similarly; therefore, a deviation from the mean is a greater change in relation to the other students.
    c. In relation to other students, the score in Course B is better (better than 97.72% of the class) than the score in Course A (better than 84.13% of the class).

19. a. First student: $z = 1.00$; second student: $z = 1.10$
    b. 84.13% of the class scored below the first student; 86.85% scored below the second student.
    c. The second student did better on his or her midterm, even though the score was lower.

SAGE edge™

Sharpen your skills with **SAGE edge at edge.sagepub.com/tokunaga2e**

SAGE edge for students provides a personalized approach to help you accomplish your coursework goals in an easy-to-use learning environment. Log on to access:

- eFlashcards
- Web Quizzes
- SPSS Data Files
- Video and Audio Resources

6

PROBABILITY AND INTRODUCTION TO HYPOTHESIS TESTING

CHAPTER OUTLINE

In Chapter 5, we discussed normal distributions and the standard normal distribution. One critical feature of these distributions is the ability to evaluate a score in terms of its relationship to the other scores in the distribution. Understanding these types of distributions and evaluating scores is the foundation of this chapter, which introduces the process of conducting statistical analyses to test research hypotheses. But rather than evaluate individual scores, these statistical analyses evaluate data collected from samples. Because data are typically collected from samples of a population rather than the entire population, researchers cannot test hypotheses with complete certainty but instead must determine the likelihood (or probability) of obtaining their results. Therefore, before introducing the process of hypothesis testing, the next section provides a very brief introduction to probability.

6.1 A BRIEF INTRODUCTION TO PROBABILITY

Although you may not be aware of it, you constantly come in contact with and use probability. You hear commercials saying that a particular product makes it "90% less likely" you'll be the victim of identity theft. If you play card games, you keep or discard cards based on the likelihood of making the best possible hand. At a grocery store, you look at five checkout lines and pick the one you believe is most likely to move quickly. Probability is also an area of study given a great deal of attention by researchers, with entire textbooks written on the subject. For the purposes of our discussion, however, we focus on aspects of probability relevant to the research process portrayed in this book.

What Is Probability?

At the most basic conceptual level, **probability** may be defined as the likelihood of the occurrence of a particular outcome of an event given all possible outcomes. For the sake of our discussion, an *event* may be defined as an action that takes place, and *outcomes* are the possible results of the event. For example, in the event of flipping a coin, there are two possible outcomes: heads and tails. In the event of rolling a die (half a pair of dice), there are six possible outcomes: the numbers 1 through 6.

Expressed mathematically, the probability (p) of an outcome is the number of ways an outcome can occur divided by the total number of possible outcomes:

$$p(\text{outcome}) = \frac{\text{number of ways an outcome can occur}}{\text{total number of possible outcomes}} \qquad (6\text{-}1)$$

Returning to the grocery store example presented earlier, if you randomly choose one of the five lines, the probability the chosen checkout line will be the fastest is equal to 1/5 or .20. Because a number (your one checkout line) is divided by a larger number (the five checkout lines), the probability of an outcome is a number ranging from .00 (0%) to 1.00 (100%); in this book, we will report probabilities using decimal places rather than percentages. At the extremes of the range of probabilities, a probability of .00 for an outcome means the outcome is certain not to occur, whereas a probability of 1.00 indicates the outcome is certain to occur.

Moving to a second example, for the event of flipping a coin, the probability of the outcome of "heads" is calculated by dividing the number of ways this outcome can occur (1) by the total number of possible outcomes (2: heads or tails). Using Formula 6-1, the probability of getting heads is 1/2 or .50; this can be represented by "$p(\text{heads}) = .50$." Given that the probability of the other outcome (tails) is exactly the same, we may state that "$p(\text{tails}) = .50$."

The simple event of flipping a coin illustrates two important aspects of probability. First, the sum of the probabilities of all possible outcomes of an event equals 1.00 (100%). In flipping a coin, $p(\text{heads}) + p(\text{tails}) = .50 + .50 = 1.00$. Second, according to the **addition rule**, the combined probability of mutually exclusive outcomes is the sum of their individual probabilities. Outcomes are "mutually exclusive" when they cannot occur at the same time. For example, in flipping a coin, we can get *either* heads *or* tails, but we cannot get *both* heads *and* tails. Because the two outcomes are mutually exclusive, the two probabilities can be added together, and we can say "the probability of getting either a heads or a tails from flipping a coin is .50 + .50, or 1.00." Another example of the addition rule is determining the probability of rolling a 1, 2, 3, or 4 on a single roll of a die. Because these four outcomes are mutually exclusive, their combined probability is (1/6 + 1/6 + 1/6 + 1/6) or 4/6 or 2/3 or .67; in other words, $p(1 \text{ or } 2 \text{ or } 3 \text{ or } 4) = .67$.

Why Is Probability Important to Researchers?

The previous section focused on mathematical characteristics of probability. However, conducting research is very different from waiting in grocery store lines, flipping coins, or rolling dice. How is probability a part of the research process? As we stated at the beginning of this chapter, one of the goals of research is to test hypotheses regarding what is believed to be true in the population. However, because collecting data from an entire population is difficult and impractical, researchers base their conclusions about their hypotheses on data provided by a sample of a population. Because they have not collected data from the entire population, researchers must instead rely on probability to evaluate the data collected from their samples.

The concept of **sampling error** refers to differences between statistics calculated from a sample and statistics pertaining to the population from which the sample is drawn; these differences are attributed to random, chance factors. For example, imagine the average height of adult males in the population is known to be 5′ 9″. We draw a random sample of males from the population, and in this sample we calculate a mean height of 6′ 1″. We would attribute the difference between the mean of our sample and the mean of the population to sampling error. Given the many ways in which the members of a population may differ from each other, researchers cannot be certain that any particular sample will completely resemble the population.

In addition to the difference between a sample and a population, sampling error also refers to differences between samples drawn from the same population. For example, if we were to draw 10 random samples of adult males from the population, because the samples contain different people, we may end up calculating 10 different sample means. These differences across samples may be defined as sampling error.

The possibility of sampling error exists in any study in which data are collected from a sample rather than the entire population. As a result, researchers can never be certain that their sample completely represents the population that is the basis of their research hypotheses. This implies that researchers cannot test their research hypotheses with absolute certainty but must instead rely on probability to evaluate the data collected from their samples.

Applying Probability to Normal Distributions

Research involves developing hypotheses about the relationship between variables in a population. As such, it is important to understand how probability may be applied to the distribution of values for these variables. More specifically, the principles of probability may be applied to distributions such as normal distributions and the standard normal distribution. In Chapter 5, for example, we asked questions regarding such things as the percentage of z-scores between $z = -.85$ and $z = 1.95$ and the percentage of SAT scores less than 660. As it turns out, these questions can be restated in terms of probability. For example, asking the question, "What percentage of z-scores is between $z = -.85$ and $z = 1.95$?" is the same as asking the question, "What is the probability a z-score is between $z = -.85$ and $z = 1.95$?" Similarly, our finding that 94.52% of SAT scores are less than 660 implies there is a .9452 probability that any randomly selected SAT score will be less than 660.

The ability to apply the concepts of probability to normal distributions and the standard normal distribution is critical to researchers because it enables them to evaluate data collected from samples. For instance, using our earlier example of the height of adult males, we can determine the probability of obtaining a mean of 6′ 1″ for a sample drawn from a population that has a hypothesized mean of 5′ 9″; this probability could then be used to test a hypothesis regarding the height of adult males.

Applying Probability to Binomial Distributions

In addition to normal distributions and the standard normal distribution, which involve numeric variables measured at the interval or ratio level of measurement, probability may also be applied to variables measured at the nominal level of measurement, with values consisting of two or more distinct categories. For example, a **binomial variable** is a variable consisting of exactly two categories; examples of binomial variables include gender (male, female), test response (correct, incorrect), and jury verdict (guilty, not guilty).

When probability is applied to binomial variables, the probabilities of the two categories are labeled p and q. In our example of coin flipping, $p = p(\text{heads})$ and $q = p(\text{tails})$. The sum of the probabilities of the two categories of a binomial variable is equal to 1.00; when flipping a coin, the probability of heads (p) and tails (q) is both .50, which sum to 1.00.

If a coin is flipped once, the two possible outcomes (heads or tails) each have a probability of .50. But what are the outcomes and probabilities in *two* coin flips? To answer this question, the list below provides all of the possible outcomes from two coin flips:

First Coin Flip	Second Coin Flip
Heads	Heads
Heads	*Tails*
Tails	Heads
Tails	*Tails*

Focusing on the number of heads in two coin flips, there are four possible outcomes. These outcomes have been organized into the distribution table in Table 6.1; this table is an example of a **binomial distribution**, which is the distribution of probabilities for a binomial variable. Looking at Table 6.1 and Figure 6.1, we see that in two coin flips, there is a .25 probability of getting 0 heads, a .50 probability of getting one head, and a .25 probability of getting two heads. Given that the three outcomes in the event of two coin flips are mutually exclusive, the addition rule may be applied such that the probabilities may be added together (.25 + .50 + .25), which sum to 1.00.

TABLE 6.1 ● BINOMIAL DISTRIBUTION, NUMBER OF HEADS IN TWO COIN FLIPS

# Heads	*f*	Probability
2	1	.25
1	2	.50
0	1	.25
Total	4	1.00

FIGURE 6.1 ● BINOMIAL DISTRIBUTION, NUMBER OF HEADS IN TWO COIN FLIPS

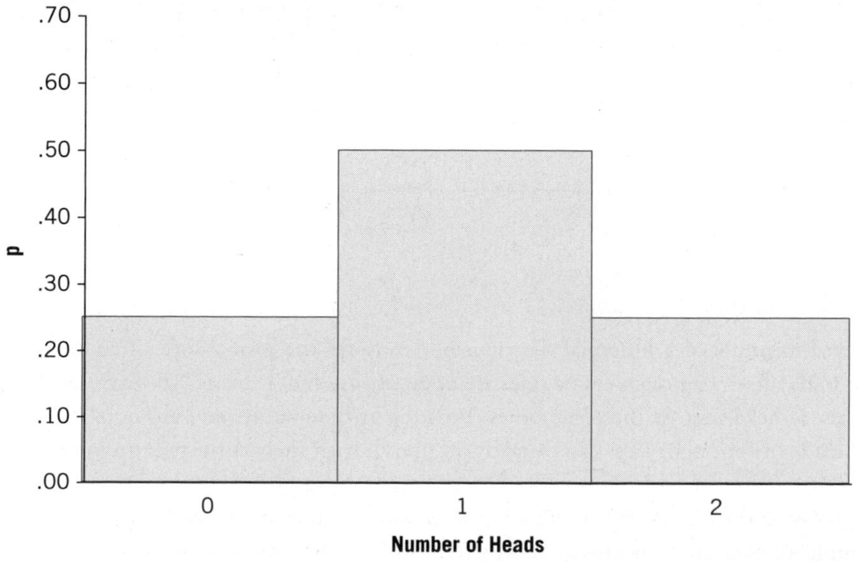

TABLE 6.2 ● POSSIBLE OUTCOMES AND BINOMIAL DISTRIBUTION TABLE, FOUR COIN FLIPS

(a) Possible Outcomes

Outcome	First Coin Flip	Second Coin Flip	Third Coin Flip	Fourth Coin Flip
1	Heads	Heads	Heads	Heads
2	Heads	Heads	Heads	*Tails*
3	Heads	Heads	*Tails*	Heads
4	Heads	Heads	*Tails*	*Tails*
5	Heads	*Tails*	Heads	Heads
6	Heads	*Tails*	Heads	*Tails*
7	Heads	*Tails*	Tails	Heads
8	Heads	*Tails*	Tails	*Tails*
9	*Tails*	Heads	Heads	Heads
10	*Tails*	Heads	Heads	*Tails*
11	*Tails*	Heads	*Tails*	Heads
12	*Tails*	Heads	*Tails*	*Tails*
13	*Tails*	*Tails*	Heads	Heads
14	*Tails*	*Tails*	Heads	*Tails*
15	*Tails*	*Tails*	*Tails*	Heads
16	*Tails*	*Tails*	*Tails*	*Tails*

(b) Binomial Distribution Table

# Heads	*f*	Probability
4	1	.06
3	4	.25
2	6	.38
1	4	.25
0	1	.06
Total	16	1.00

As a second example of a binomial distribution, consider the probabilities in a sample of four coin flips. Table 6.2(a) lists 16 possible outcomes that can occur, and Table 6.2(b) calculates the probability of the number of heads across these outcomes. Looking at these tables and at Figure 6.2, we see that the number of heads in four coin flips can range from 0 to 4, with their corresponding probabilities ranging from .06 to .38.

We've introduced the binomial distribution to illustrate the probabilities of different possible outcomes for different samples of data. In the next section, we will introduce a research situation that involves applying probability to a binomial distribution to test a research hypothesis about a population.

FIGURE 6.2 ● BINOMIAL DISTRIBUTION, NUMBER OF HEADS IN FOUR COIN FLIPS

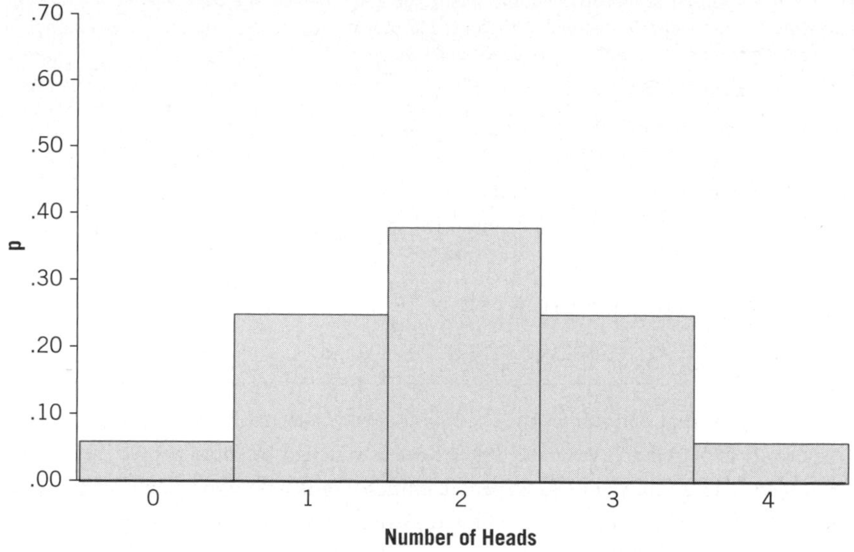

LEARNING CHECK 1:

Reviewing What You've Learned So Far

1. Review questions
   a. What is probability (conceptually and mathematically)?
   b. Why is probability important to researchers?
   c. What is the addition rule of probability?
   d. What is the relationship between sampling error and probability?
   e. How can probability be applied to distributions?
   f. What is the difference between normal distributions and binomial distributions?
2. Here is a set of scores: 7 2 6 1 9 3 5 2. If we were to randomly select one of these scores, what is the probability this score will be . . .
   a. equal to 6
   b. equal to 2
   c. greater than 4
   d. less than 8
   e. an even number
3. Using the standard normal distribution and the normal curve table, what is the probability of a *z*-score . . .
   a. greater than 1.25
   b. less than .40
   c. greater than −.50
   d. less than −1.33
   e. between .50 and 1.50

(Continued)

(Continued)

4. According to the National Center for Health Statistics, in 2005 the average birth weight of a newborn baby was approximately normally distributed with a mean of 119 ounces (7 pounds, 7 ounces) and a standard deviation of 19 ounces. What is the probability a newborn baby will weigh . . .
   a. more than 128 ounces (8 pounds)
   b. less than 112 ounces (7 pounds)
   c. more than 144 ounces (9 pounds)
   d. less than 124 ounces (7 pounds, 12 ounces)
   e. between 96 ounces (6 pounds) and 136 ounces (8.5 pounds)

6.2 EXAMPLE: MAKING HEADS OR TAILS OF THE SUPER BOWL

The Super Bowl, the game that determines the champion of the National Football League, has become an integral part of American society. Each year, the Super Bowl is watched by more people than any other televised event. Fans, gamblers, and curious onlookers spend countless hours before the game debating the merits of the two competing teams to predict which team will win. Although some factors that influence the outcome of a football game (such as physical ability and strategy) are within the control of the players and coaches, other factors such as weather and injuries also play a role. As a result, the winner of the Super Bowl cannot be predicted with 100% certainty. It is possible, however, that understanding and analyzing the potential impact of different factors may increase the probability of picking the winning team.

A coin flip is held at the beginning of every Super Bowl game to determine which team gets to decide who will receive and possess the ball at the start of the game. Assuming this decision plays a role in determining who will win the game, it's plausible that the team that wins the pregame coin flip has a greater probability of winning the game. Let's imagine we conduct a research study to test the relationship between winning the pregame coin flip and winning the game.

As discussed in Chapter 1, the first step in the research process is to *develop a research hypothesis,* which is an expected outcome or relationship between variables. Imagine that, based on an examination of the literature regarding sporting events, we believe that the team that wins the coin flip has a greater likelihood of scoring first than the team that loses the coin flip. As history has shown that the team scoring first is more likely to win the game, our research hypothesis might be stated as the following: "It is hypothesized that teams are more likely to win the Super Bowl if they win the pregame coin flip than if they lose the coin flip."

The second step in the research process is to *collect data,* which involves defining the target population, drawing a sample from this population, and determining what data to collect from the sample and how they will be collected. In the case of the Super Bowl example, the relevant population consists of the teams that have won the pregame coin flip. The example discussed below is based on a sample of the teams competing in 12 Super Bowls between 2002 and 2013 (we understand that, given the Super Bowl has only been played since 1966, it would be relatively easy to collect data from the entire population of games rather than just a sample; however, we'll use only a sample of games in order to illustrate the process of hypothesis testing). In this example, the relevant data to be collected would consist of determining whether the team that won the coin flip went on to win the game. For each of the 12 Super Bowls between 2002 and 2013, whether the team that won the coin flip went on to win or lose the game is provided in Table 6.3.

The third step in the research process is to *analyze the data that have been collected*. As we saw in Chapter 2, it's useful to examine data before proceeding with any statistical analyses. A frequency distribution table has been created in Table 6.4 for the sample of 12 Super Bowl coin flip winners. Looking at the table, the data for the Super Bowl example may be summarized as follows: "Of the 12 Super Bowls from 2002 to 2013, the team that won the coin flip won five of the games." (Simple examples like the current one do not require calculating descriptive statistics such as the mean and standard deviation discussed in Chapters 3 and 4; research situations requiring the calculation of descriptive statistics will be discussed starting in Chapter 7.)

TABLE 6.3 ● OUTCOME OF SUPER BOWL FOR TEAM WINNING COIN FLIP, 2002–2013

Super Bowl	Outcome of Game	Super Bowl	Outcome of Game
2002	**Lose**	2008	**Win**
2003	**Win**	2009	**Lose**
2004	**Lose**	2010	**Win**
2005	**Lose**	2011	**Win**
2006	**Lose**	2012	**Lose**
2007	**Lose**	2013	**Win**

TABLE 6.4 ● FREQUENCY DISTRIBUTION TABLE, OUTCOME OF SUPER BOWL FOR TEAM WINNING COIN FLIP, 2002–2013

Outcome of Game	*f*	%
Win	5	42%
Lose	7	58%
Total	12	100%

The fourth step in the research process is to *draw a conclusion regarding whether the research hypothesis has been supported* based on the results of the statistical analyses. Does finding that 5 of the 12 teams that won the coin flip went on to win the game enable us to draw a conclusion regarding whether winning the coin flip affects a team's chance of winning the Super Bowl? No—because in this example we have collected data from a sample rather than the entire population, we need to conduct statistical analyses specifically designed to evaluate data collected from samples of the population in order to test research hypotheses. The next section introduces the steps in hypothesis testing; as we will see, hypothesis testing relies heavily on probability.

6.3 INTRODUCTION TO HYPOTHESIS TESTING

Conducting statistical analyses to test research hypotheses consists of a number of steps, steps that will be used in many of the remaining chapters of this book. The purpose of this section is to use the Super Bowl example to illustrate why and how these steps are completed. The steps in hypothesis testing may be stated and summarized as follows:

- State the null and alternative hypotheses (H_0 and H_1).
  State two mutually exclusive conclusions regarding the existence of a hypothesized change, difference, or relationship in the population.
- Make a decision about the null hypothesis.
  Make a decision to reject or not reject the null hypothesis based on the probability of a calculated value of a statistic.
- Draw a conclusion from the analysis.
  Based on the decision about the null hypothesis, draw a conclusion about the hypothesized change, difference, or relationship.

- Relate the result of the analysis to the research hypothesis.
  Interpret the results of the statistical analysis in terms of whether it supports or does not support the study's research hypothesis.

Each of these steps is discussed in detail in the sections below.

State the Null and Alternative Hypotheses (H_0 and H_1)

The first step in hypothesis testing is to represent two mutually exclusive conclusions that can be made based on analyzing data collected from a sample of the population: (1) a conclusion regarding what is believed to exist in the population and (2) a mutually exclusive alternative to this conclusion. Ultimately, as a result of conducting statistical analyses, a decision is made regarding which of these conclusions is considered to be true.

Each of these two conclusions is represented by a **statistical hypothesis**: a statement about an expected outcome or relationship involving population parameters. As stated in Chapter 3, a parameter, such as the population mean μ, is a numeric characteristic of a population. Statistical hypotheses differ from research hypotheses in that research hypotheses involve concepts and are expressed using words, whereas statistical hypotheses involve mathematical terms and are expressed with numbers. Below we define the two statistical hypotheses and develop hypotheses applicable to our example of the number of winning teams in 12 Super Bowl coin flip winners.

The Null Hypothesis (H_0)

The first statistical hypothesis is the **null hypothesis (H_0)** (pronounced "H sub zero," where *sub* stands for "subscript"), defined as a statistical hypothesis that states that a hypothesized change, difference, or relationship among groups or variables does not exist in the population. The word *null* and the subscript *0* are used to designate the absence or lack of existence of a change, difference, or relationship.

In the Super Bowl example, the null hypothesis represents the conclusion that winning the coin flip does *not* affect a team's chances of winning the Super Bowl. Stating the null hypothesis requires representing this conclusion mathematically for our sample of 12 Super Bowls. If we were to draw random samples of 12 games from the population of Super Bowls, on average, how many of the 12 teams that won the coin flip would we expect to win the game? If winning the coin flip does not affect a team's chances of winning the game, assuming the probability of winning the game is .50, we would expect an average of half (or 6) of the 12 teams to win the game. Therefore, for the population of 12 Super Bowl games, the null hypothesis would be stated the following way:

$$H_0: \mu = 6$$

This null hypothesis states that, in the population of 12 Super Bowls, the mean number of wins for the teams winning the coin flip is equal to 6.

Before moving on, we want to emphasize that statistical hypotheses state an expected outcome or relationship involving population parameters rather than sample statistics. For example, the null hypothesis in the Super Bowl example does *not* state that our one sample of 12 Super Bowl coin flip winners will have 6 winning teams. Instead, the null hypothesis asserts that in the population of 12 Super Bowl coin flip winners, a population based on an infinite number of random samples, the mean number of winning teams (μ) will be equal to 6.

The Alternative Hypothesis (H_1)

The **alternative hypothesis (H_1)** (pronounced "H sub one") is a statistical hypothesis that a hypothesized change, difference, or relationship among groups or variables does exist in the population. The alternative hypothesis provides a mutually exclusive alternative to the null hypothesis; they are mutually exclusive because only one of the two hypotheses can be true.

In the Super Bowl example, the alternative hypothesis represents the conclusion that winning the coin flip does in fact affect a team's chances of winning the Super Bowl. One way to represent this conclusion is to state that winning the coin flip makes the probability of winning the game *not* equal to .50; if so, the mean number of victories by teams that win the coin flip in the population of 12 Super Bowls will *not* be equal to 6. Consequently, the alternative hypothesis could be stated as follows:

$$H_1: \mu \neq 6$$

The above alternative hypothesis implies that the population mean (μ) is either less than 6 or greater than 6.

You may be wondering, given that the research hypothesis in the Super Bowl example predicted that winning the coin flip *increases* a team's chances of winning the game, why the alternative hypothesis was not stated as H_1: $\mu > 6$. Including the > symbol rather than the ≠ symbol implies that the number of wins by teams winning the coin flip should be *greater* than the population mean of 6. However, we have stated the alternative hypothesis as H_1: $\mu \neq 6$, which means the null hypothesis will be rejected if the number of wins is either less than 6 *or* greater than 6. Including the ≠ symbol in the alternative hypothesis allows for the possibility that winning the coin flip either increases *or* decreases a team's chance of winning the game. Later in this chapter, we illustrate choices available to researchers regarding how the alternative hypothesis may be stated.

Make a Decision About the Null Hypothesis

Having stated two mutually exclusive statistical hypotheses, the next step in hypothesis testing is to decide which of the two hypotheses is supported by the data. More specifically, we will decide whether to reject the null hypothesis in favor of the alternative hypothesis, which represents the conclusion that the change, difference, or relationship does in fact exist.

We could make a decision about the null hypothesis simply by comparing the value of the statistic calculated on our sample with that of the parameter specified in the null hypothesis. In the Super Bowl example, because the statistic of five winning teams is not the same as the parameter (μ) of 6, can we conclude the alternative hypothesis is true? The answer to this question is no—as we learned from our discussion of sampling error, a sample may differ from the population simply because of random, chance factors.

Because of the existence of sampling error, it would be inappropriate to require each and every sample of 12 Super Bowl teams to contain exactly 6 winning teams in order to conclude the null hypothesis is true. Instead, the decision about the null hypothesis is made based on the likelihood, or probability, of obtaining the value of the statistic calculated from the sample. In essence, if the statistic has a low probability of occurring under the assumption that the null hypothesis is true, we will conclude that the change, difference, or relationship found in the sample is not due to random, chance factors, and the decision will be made to reject the null hypothesis in favor of the alternative hypothesis.

Defining a "Low" Probability of a Statistic: Setting Alpha (α)

In hypothesis testing, it is necessary to state how "low" the probability of a statistic must be to make the decision to reject the null hypothesis. This probability is represented by the Greek letter **alpha** (α), defined as the probability of a statistic used to make a decision whether to reject the null hypothesis.

In many academic disciplines, it is customary to define a "low" probability at .05 ($\alpha = .05$). A strategy based on this probability would be as follows:

If the probability of the statistic is less than .05, the null hypothesis will be rejected.

In the Super Bowl example, if the number of wins has a probability less than .05, we will reject the null hypothesis and conclude that winning the coin flip does in fact affect a team's chances of winning the Super Bowl. However, if this probability is *not* less than .05, we will *not* reject the null hypothesis and conclude that winning the coin flip does *not* affect a team's chances of winning the Super Bowl.

Although researchers traditionally set alpha at .05, this is by no means used by every researcher in every situation. Reading reports of research findings in journal articles, you may encounter situations where a low probability has been defined as .10 or .01. Different values for alpha may be chosen based on such things as the size of the sample or the researcher's desired level of confidence of concluding a hypothesized effect does in fact exist. The consequences of using a particular value of alpha will be further discussed later in this chapter and in subsequent chapters.

Identifying the Values of the Statistic With a Low Probability

Once alpha has been set, the next step in making the decision about the null hypothesis is to identify the values of the statistic that have the defined low probability of occurring. Calculating one of these values from the data in a sample will lead to the decision to reject the null hypothesis in favor of the alternative hypothesis.

Throughout this book, we'll discuss a variety of statistics; the specific type of statistic calculated for a particular research situation is a function of the type of variable used in the analysis. The Super Bowl example uses a binomial variable (win, lose); earlier in this chapter, we determined the probabilities for the number of heads that can occur in two or four coin flips. However, for our example of $N = 12$, determining all of the possible outcomes and the probability of each outcome would be extremely time-consuming. Fortunately, this information can be obtained by consulting existing tables of binomial distributions.

In the back of this book, Table 2 (Binomial Probabilities) presents a table of binomial probabilities for distributions of different sample sizes. Looking first at the columns of the table, these columns represent a wide range of probabilities of the two values of the variable. Consider the example of an item on a multiple-choice test that contains four alternatives (a–d); if a student were to randomly choose one of the four alternatives, the probability of answering the item correctly (p) is 1/4 or .25. If, on the other hand, the item consisted of five alternatives (a–e), the probability of answering the item correctly (p) would be 1/5 or .20. In the Super Bowl example, because we've stated that the probability (p) of winning a game is .50, we will rely on the .50 column.

The rows of the binomial probabilities table represent different distributions of probabilities for different sample sizes (N). To illustrate how to read this table, let's return to our earlier example of two coin flips. Looking under the N column, move down to the number *2* ($N = 2$). Here, we find three rows, labeled *0, 1,* and *2*; these three rows correspond to the possible number of heads that may result from two coin flips. Assuming the probability of getting a heads in a coin flip is .50, we move to the right until we reach the *.50* column. Here, the numbers *.2500, .5000,* and *.2500* correspond to the probabilities (calculated in Table 6.1) of the outcomes of 0 (.25), 1 (.50), and 2 (.25) heads.

For the Super Bowl example, we need to determine the number of wins in 12 games whose combined probability is low enough to lead to the decision to reject the null hypothesis. Looking at the binomial probabilities table, move down the N column until we reach $N = 12$. Here the 13 rows inform us of the possible outcomes for our example (0 to 12 wins in 12 games). Assuming the probability (p) of winning a game is .50, move to the column labeled .50 to determine the probability for each outcome. These probabilities are illustrated in Figure 6.3. Looking at this figure, it resembles the normal distributions discussed in Chapter 5 in two critical ways: Its shape is unimodal and symmetric, and the value in the middle of the distribution is the population mean μ.

Within hypothesis testing, distributions of values of a statistic such as the one in Figure 6.3 are divided into two parts, or "regions": the region of rejection and the region of nonrejection. The **region of rejection** represents the values of a statistic whose combined probability is low enough that obtaining one of these values results in the decision to reject the null hypothesis. The **region of nonrejection** represents the values of a statistic whose combined probability is high enough that obtaining one of these values results in the decision to *not* reject the null hypothesis. A **critical value** is the value of a statistic that separates the region of rejection from the region of nonrejection.

Figure 6.4 indicates the region of rejection, region of nonrejection, and critical values for a hypothetical distribution. If alpha is set at .05, the region of rejection contains the 5% of values of the statistic whose combined

FIGURE 6.3 ● BINOMIAL DISTRIBUTION, NUMBER OF WINS IN 12 SUPER BOWLS

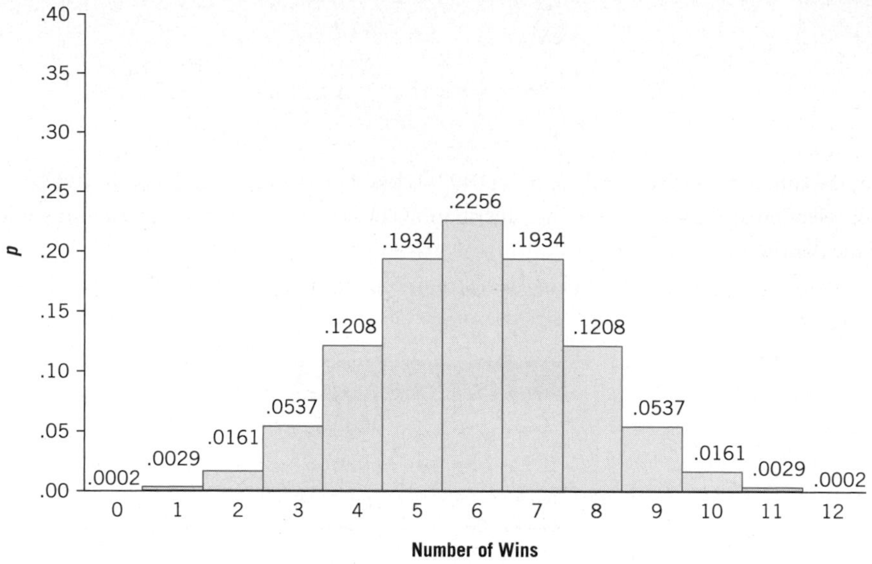

probability is low enough to lead to the rejection of the null hypothesis; the region of nonrejection contains the remaining 95% (100% – 5%) of the values of the statistic. Note that the shaded 5% region of rejection in Figure 6.4 consists of two parts: the 2½% of the distribution at each end of the distribution. The region of rejection is divided in this way when the alternative hypothesis includes the ≠ symbol (e.g., H_1: $\mu \neq 6$). Because the ≠ symbol implies the null hypothesis will be rejected if the value of the statistic is either greater than *or* less than the population parameter, the 5% region of rejection must be split into the two tails of the distribution. In the Super Bowl example, the alternative hypothesis of H_1: $\mu \neq 6$ implies the null hypothesis will be rejected if the number of wins is either less than 6 *or* greater than 6. Later in this chapter, we will discuss different ways the alternative hypothesis may be stated.

FIGURE 6.4 ● REGIONS OF REJECTION AND NONREJECTION FOR ALPHA (α) = .05

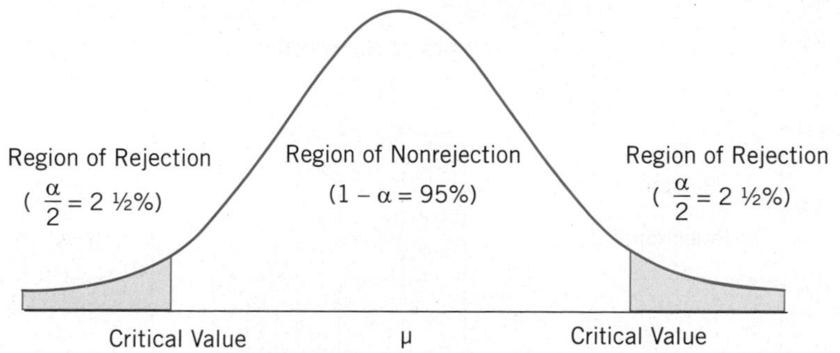

For the Super Bowl example, we need to determine the number of wins in 12 games that lie in the region of rejection. First, let's look at the probabilities of the different number of wins at the lower end of the distribution. Starting from zero wins, the probabilities of zero, one, and two wins are added together below:

# Wins	Probability
0	.0002
1	.0029
2	.0161
Sum	.0192

Note that the sum of these three probabilities (.0192) is less than the stated cutoff of .0250 (2½%). Therefore, the region of rejection at the lower end of the distribution consists of 0, 1, and 2 wins, and the critical value at this end of the distribution is 3 wins.

Turning our attention to the upper end of the distribution, the probabilities of 12, 11, and 10 wins are added together as follows:

# Wins	Probability
12	.0002
11	.0029
10	.0161
Sum	.0192

The region of rejection at the upper end of the distribution consists of 12, 11, and 10 wins, and the critical value at this end of the distribution is 9 wins.

Based on these calculations, the two critical values for the Super Bowl example may be stated as follows:

For $\alpha = .05$ and $N = 12$ games, critical values = 3 wins and 9 wins.

The above statement represents the logic, "Assuming we've defined a low probability as .05 and our sample consists of 12 Super Bowl games, 3 wins and 9 wins separate the region of rejection from the region of nonrejection." The regions of rejection and nonrejection and the critical values for the Super Bowl example are illustrated in Figure 6.5.

FIGURE 6.5 ● REGIONS OF REJECTION AND NONREJECTION, SUPER BOWL EXAMPLE ($N = 12$)

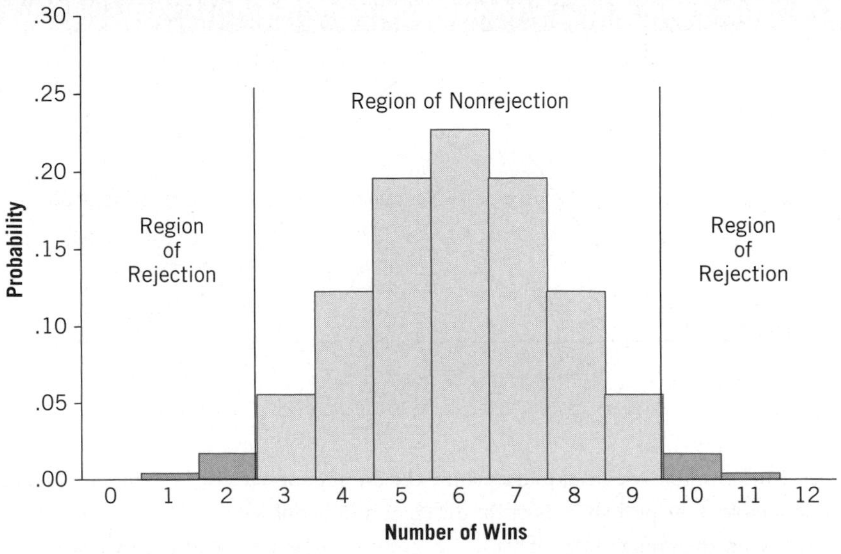

Stating a Decision Rule

Now that the critical values have been identified, it's useful to provide a rule that explicitly states the logic used to make a decision about the null hypothesis. A **decision rule** is a rule that specifies the values of the statistic that result in the decision to reject or not reject the null hypothesis. The general decision rule in hypothesis testing may be stated as follows:

> If the value of the statistic calculated for the sample lies beyond a critical value, the null hypothesis will be rejected; otherwise, the null hypothesis will not be rejected.

> The decision rule for the Super Bowl example is stated below:

> If the number of wins in 12 games is < 3 or > 9, reject H_0; otherwise, do not reject H_0.

This decision rule implies that the null hypothesis of $\mu = 6$ will be rejected if the number of wins in a sample of 12 games is either less than 3 or more than 9, as these number of wins are located in the region of rejection, which implies the number of wins has a low probability of occurring ($p < .05$). However, if the number of wins is *not* less than 3 or more than 9, the null hypothesis will *not* be rejected because the number of wins falls in the region of *non*rejection, which implies the number of wins does *not* have a low probability of occurring ($p > .05$).

Calculating a Value of a Statistic

Once the decision rule has been stated, the next step in hypothesis testing is to conduct statistical analyses on the data collected from the sample. The purpose of these analyses is to calculate a value of a statistic that will be compared with the critical values to make a decision about the null hypothesis. These statistical analyses typically involve the calculation of inferential statistics, defined in Chapter 1 as statistical procedures used to test research hypotheses.

Many of the remaining chapters of this book will introduce different inferential statistical procedures; the specific inferential statistic calculated will depend on the research hypothesis of interest, as well as the nature of the variables involved in the analysis. For example, a different statistical procedure will be used to compare the means of two groups (Chapter 9) versus comparing the means of three or more groups (Chapter 11).

Unlike future chapters, the Super Bowl example is unique in that we have already calculated the statistic relevant to the research hypothesis: the number of wins for teams that won the coin flip (5 out of 12 teams). Due to the relative simplicity of this example, the purpose of which is to introduce the steps in hypothesis testing, no additional statistical analyses or calculations are necessary.

Making the Decision Whether to Reject the Null Hypothesis

Based on the logic stated in the decision rule, the decision about the null hypothesis is made by comparing the value of a calculated statistic with the critical values. This comparison results in one of two decisions. If the value of the statistic exceeds a critical value, meaning the statistic is located in the region of rejection, the decision is made to reject the null hypothesis because the statistic has a low ($< .05$) probability of occurring. However, if the value of the statistic does not exceed a critical value, meaning the statistic is located in the region of nonrejection, the decision is made to not reject the null hypothesis because the statistic does not have a low probability of occurring—that is, the probability is greater than .05 ($p > .05$).

In the sample of 12 games in the Super Bowl example, there were five winning teams. By comparing this statistic with the critical values, the decision about the null hypothesis may be stated as the following:

> Because 5 wins is neither less than 3 nor greater than 9, the decision is made to *not* reject the null hypothesis, which implies the probability of winning 5 out of 12 games is greater than .05.

Below is a more concise way of stating the above decision:

5 wins is not < 3 or > 9; therefore, do not reject H_0 ($p > .05$).

In this example, the null hypothesis is not rejected because 5 wins in 12 games falls in the region of nonrejection, meaning that the probability of 5 wins in 12 games is greater than the .05 value of alpha (α). This probability may be stated as "$p > .05$." In essence, we've decided that winning 5 out of 12 games is not sufficiently different from the population mean of 6 and that this difference is due to random, chance factors.

Draw a Conclusion From the Analysis

The next step in hypothesis testing is to draw a conclusion based on the decision to reject or not reject the null hypothesis. Whenever a statistical analysis is conducted, drawing as specific a conclusion as possible regarding the results of the analysis reduces the likelihood that others will misunderstand or misinterpret a study's findings. In the Super Bowl example, based on our analysis, we could conclude the following:

In a sample of 12 teams that won the coin flip in the Super Bowl, 5 of the 12 went on to win the game. Consequently, the null hypothesis of $\mu = 6$ was not rejected because the probability of 5 wins in 12 games is greater than $\alpha = .05$.

Stating our conclusion in this manner informs the reader about several critical aspects of the analysis:

- The sample from whom data were collected: "a sample of 12 teams that won the coin flip in the Super Bowl"
- The value of the statistic calculated from the data: "5 of the 12 went on to win the game"
- The decision about the null hypothesis: "the null hypothesis of $\mu = 6$ was not rejected"
- The probability of the value of the statistic: "the probability . . . is greater than $\alpha = .05$"

In later chapters, as we discuss different types of inferential statistical procedures, we'll expand on this discussion regarding how to draw conclusions from statistical analyses and communicate these conclusions to others.

Relate the Result of the Analysis to the Research Hypothesis

We are now prepared to relate the results of the analysis back to the original research hypothesis. More specifically, it is important to determine whether the results of a statistical analysis support or do not support the research hypothesis.

For the Super Bowl example, the research hypothesis was stated as follows: "Teams are more likely to win the Super Bowl if they win the pregame coin flip than if they lose the coin flip." Do the results of our analysis support or not support this research hypothesis? In this analysis, we made the decision to not reject the null hypothesis, which reflects the belief that the difference between winning 5 games versus the hypothesized population mean of 6 games is due to random, chance factors. Relating this decision to the research hypothesis, we could state the following:

The results of the statistical analysis do not support the research hypothesis that teams are more likely to win the Super Bowl if they win the pregame coin flip than if they lose the coin flip.

Summary of Steps in Hypothesis Testing

Using the Super Bowl example, the steps involved in hypothesis testing are summarized in Box 6.1. As students often have difficulty when learning hypothesis testing, it's important to understand how the steps are related to each other. If you look at Box 6.1, you see such statements as, "5 wins is not < 3 or > 9; therefore, do not reject H_0 ($p > .05$)," "The null hypothesis of $\mu = 6$ was not rejected," and "The results of the statistical analysis do not support the research hypothesis." Notice that, in this example, the word *not* appears multiple times in these steps. This illustrates how the result of one step has implications for the steps that follow.

The process of hypothesis testing introduced in this chapter will be referred to and expanded on throughout the remainder of this book. Although the statistical procedures discussed in future chapters may seem very different from each other, because they share the same goal of testing hypotheses about populations based on data collected from samples, the steps used to test these hypotheses will remain the same.

SUMMARY BOX 6.1 SUMMARY OF STEPS IN HYPOTHESIS TESTING, SUPER BOWL EXAMPLE

State the null and alternative hypotheses (H_0 and H_1).

$$H_0: \mu = 6 \qquad H_1: \mu \neq 6$$

Make a decision about the null hypothesis.

Set alpha, identify the critical values, and state a decision rule.

For $\alpha = .05$, if the number of wins in 12 games is < 3 or > 9, reject H_0; otherwise, do not reject H_0.

Calculate a statistic: number of wins in 12 Super Bowls.

5 wins in 12 games

Make a decision whether to reject the null hypothesis.

5 wins is not < 3 or > 9; therefore, do not reject H_0 ($p > .05$).

Draw a conclusion from the analysis.

In a sample of 12 teams that won the coin flip in the Super Bowl, 5 of the 12 went on to win the game. Consequently, the null hypothesis of $\mu = 6$ was not rejected because the probability of 5 wins in 12 games is greater than $\alpha = .05$.

Relate the result of the analysis to the research hypothesis.

The results of the statistical analysis do not support the research hypothesis that teams are more likely to win the Super Bowl if they win the pregame coin flip than if they lose the coin flip.

LEARNING CHECK 2:

Reviewing What You've Learned So Far

1. Review questions
   a. What are the main steps in hypothesis testing?
   b. What are the differences between research hypotheses and statistical hypotheses?
   c. What are the null and the alternative hypotheses?

(Continued)

(Continued)

   d. What are the main steps in making the decision about the null hypothesis?
   e. What is the difference between the region of rejection and the region of nonrejection?
   f. Does the value of a statistic that exceeds the critical value have a low or high probability of occurring?
   g. In stating the conclusion of a statistical analysis, what information is useful to provide?

2. Imagine you decide to conduct a study to test each of the following statements. For each one, state a null and alternative hypothesis (H_0 and H_1).
   a. A survey of iPod owners conducted in 2005 estimated that the likelihood of having one's iPod completely stop working is 13.70%.
   b. The average American adult female weighs 164 pounds.
   c. In 2004, the California Student Public Interest Group (CALPIRG) published a report stating college students in California spend an average of $450 a semester on textbooks.
   d. One study estimated that the average college freshman owes $900 on his or her credit cards (Arano, 2006).
   e. A nationwide survey of college students estimated that 50% of college students have cheated at least once.
3. Returning to the Super Bowl example, imagine you draw a sample of 20 teams ($N = 20$) rather than 12. For this situation,
   a. State the null and alternative hypotheses (H_0 and H_1).
   b. Identify the critical values for $\alpha = .05$ and state a decision rule.
   c. What decision would you make about the null hypothesis if you found 16 of the 20 teams that won the coin flip also won the game?
   d. If you had found that 13 of the 20 teams that won the coin flip also won the game, would this support or not support the research hypothesis?
4. A friend of yours claims to have extrasensory perception (ESP). You test this by creating a deck of 16 cards, half of which are red and the other half green. Sitting behind a partition, you hold up each card and ask her to guess the color. Assuming the probability of guessing the color correctly is 50%, you find she correctly guesses the color on 12 of the 16 cards.
   a. State the null and alternative hypotheses (H_0 and H_1).
   b. Identify the critical values for $\alpha = .05$ and state a decision rule.
   c. Make a decision about the null hypothesis.
   d. Does the result of your analysis support or not support your friend's claim?

6.4 ISSUES RELATED TO HYPOTHESIS TESTING: AN INTRODUCTION

The previous section described the steps involved in conducting statistical analyses designed to test research hypotheses. Because the process of hypothesis testing is centered on collecting data from samples rather than populations and therefore relies on probability rather than certainty, this process raises a number of issues researchers must understand. The purpose of this section is to briefly introduce several issues related to hypothesis testing:

- the issue of "proof" in hypothesis testing,
- errors that can be made in making the decision about the null hypothesis, and
- factors that affect the decision about the null hypothesis.

The Issue of "Proof" in Hypothesis Testing

In hypothesis testing, when the decision is made to either reject or not reject the null hypothesis, does this "prove" the null hypothesis is either false or true? And regardless of which decision is made, is it appropriate to conclude a research hypothesis has been "proven" to be true or false? The issue of "proof" as it pertains to hypothesis testing may be summarized as follows:

> Hypothesis testing cannot provide proof or lack of proof for a research hypothesis; instead, it can only provide support or lack of support for a research hypothesis.

Hypothesis testing cannot prove or disprove a research hypothesis because proof implies establishing the accuracy of a statement with absolute certainty. *Proof cannot be provided for a research hypothesis because data have been collected from a sample rather than the entire population.* Regardless of what decision is made about the null hypothesis, the concept of sampling error implies the sample may not be representative of the population and that other samples of the population may not have resulted in the same decision about the null hypothesis.

Because decisions and conclusions about hypotheses are based on probability, we can never be certain that the results of statistical analyses represent what would have happened if data had been collected from the entire population. Consequently, the best we can say is that our findings either support or do not support a research hypothesis, with the word *support* implying that our findings have increased or decreased the likelihood that a particular statement is accurate.

Errors in Decision Making

Collecting data from samples rather than populations is one reason why researchers exercise restraint in drawing conclusions about their hypotheses. There is another reason why relying on probability results in carefully worded conclusions:

> It is possible that the decision made about the null hypothesis may either be correct or incorrect.

For example, when the decision is made to reject the null hypothesis, we can never completely eliminate the possibility that this was the wrong decision and that the hypothesized effect does *not* actually exist, and that the decision to reject the null hypothesis was based on chance factors. In the Super Bowl example, imagine the null hypothesis is in fact true, such that winning the coin flip does not affect the likelihood of winning the game. Even under this condition, it's still possible that all 12 of 12 teams that win the coin flip will win the game. If this were to happen, we would make the decision to reject the null hypothesis; however, this would be the incorrect decision.

It's also possible to make a second type of error: making the decision to not reject the null hypothesis when we should. In the case of the Super Bowl example, we made the decision to not reject the null hypothesis and conclude that winning the coin flip does not affect the outcome of the game. However, because we did not collect data from the entire population, we must acknowledge the possibility that the hypothesized effect may in fact exist in the population of Super Bowl games even though we did not find it in our particular sample.

We introduce these two types of errors at this point in the book to further emphasize the importance of appropriately interpreting the results of statistical analyses. Chapter 10 discusses each type of error, known as Type I error and Type II error, in greater detail, including factors that lead to these errors and how the likelihood of making each type of error can be lowered. We've chosen to discuss these errors in a later chapter because we want to first discuss additional examples of hypothesis testing in Chapters 7 and 9. However, if you or your instructor want to focus on these errors now, we encourage you to read Chapter 10 before moving on to the next chapter.

Factors Influencing the Decision About the Null Hypothesis

Although the process of making the decision to reject or not reject the null hypothesis may be fairly straightforward, a number of factors directly influence which of these two decisions are made. This section introduces three factors that influence the decision about the null hypothesis:

- the size of the sample,
- the value of alpha (the probability of the statistic used to make a decision about the null hypothesis), and
- the directionality of the alternative hypothesis.

Because each of these factors is under the control of the researcher, it is important to gain an understanding of their role in hypothesis testing.

Sample Size

The relationship between the size of the sample (i.e., the number of research participants from whom data are collected) and the decision about the null hypothesis may be summarized as follows:

The larger the sample size, the greater the likelihood of rejecting the null hypothesis.

Let's illustrate this relationship using the example of the Super Bowl. In the example described earlier, a sample size of 12 Super Bowls required relatively extreme outcomes (less than 3 or more than 9 wins) for us to decide to reject the null hypothesis. But what if the sample size had been larger? For a sample size of 30 games, how many games must the teams win to reject the null hypothesis that $\mu = 15$? Again dividing the alpha level of .05 in half, Table 6.5 uses the binomial probabilities table in the back of the book to calculate a 2½% region of rejection at each of the two ends of the distribution. (Although some of the probabilities are listed as .0000, this

TABLE 6.5 ⬢ CALCULATION OF CRITICAL VALUES FOR BINOMIAL DISTRIBUTION OF $N = 30$

Lower End of Distribution		Upper End of Distribution	
# Wins	Probability	# Wins	Probability
0	.0000	30	.0000
1	.0000	29	.0000
2	.0000	28	.0000
3	.0000	27	.0000
4	.0000	26	.0000
5	.0001	25	.0001
6	.0006	24	.0006
7	.0019	23	.0019
8	.0055	22	.0055
9	.0133	21	.0133
Sum	.0214	Sum	.0214

does *not* mean the probability is equal to zero but rather that it is less than .0001.) These calculations identify the critical values to be 10 wins or 20 wins. Therefore, for $N = 30$, the decision rule is, "If the number of wins is < 10 or > 20, reject H_0; otherwise, do not reject H_0."

The critical values for samples of $N = 12$ and $N = 30$ games are illustrated in Figure 6.6. Comparing the vertical lines that separate the regions of rejection and nonrejection, the critical values for $N = 30$ are closer to the center of the distribution than the critical values for $N = 12$. For example, when $N = 12$ (Figure 6.6(a)), the teams must win at least 10 of the 12 games (10/12 = 83%) to reject the null hypothesis. However, when $N = 30$, the teams need only win at least 21 of the 30 games (21/30 = 70%) to reject the null hypothesis (Figure 6.6(b)).

FIGURE 6.6 ● CRITICAL VALUES, BINOMIAL DISTRIBUTION FOR $N = 12$ AND $N = 30$

(a) Critical Values, $N = 12$

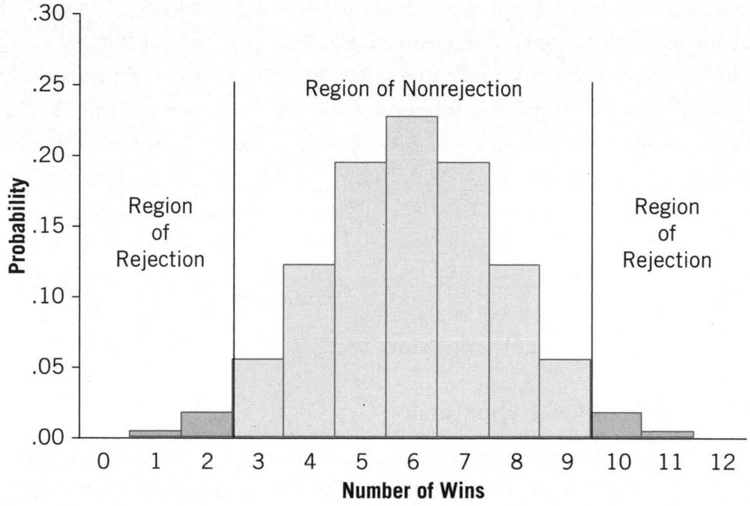

(b) Critical Values, $N = 30$

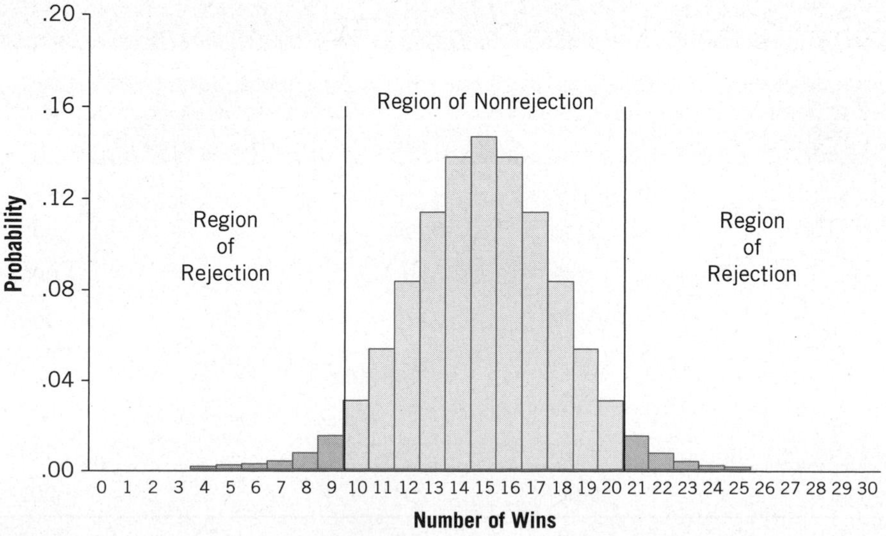

As the size of the sample increases, less extreme values of the statistic are needed to reject the null hypothesis. Consequently, the larger the sample size, the greater the likelihood of rejecting the null hypothesis.

Alpha (α)

The null hypothesis is rejected when the calculated value of a statistic has a "low" probability of occurring; this probability is represented by alpha (α). Although it is tradition in many academic disciplines to set alpha to .05, other values may be chosen. The relationship between alpha and the decision about the null hypothesis may be stated the following way:

The smaller the value of alpha, the lower the likelihood of rejecting the null hypothesis.

As we mentioned earlier in this chapter, researchers may require the probability of a statistic to be less than .01 (rather than .05) to reject the null hypothesis. One reason why alpha may be set to .01 is to require stronger, more convincing evidence before reaching the conclusion that the proposed change, difference, or relationship exists. This is analogous to what may happen in jury trials, in which a jury may be required to reach a unanimous decision rather than a two-thirds majority to find someone guilty of committing a crime.

Let's return to the Super Bowl example to illustrate the impact of using $\alpha = .01$ rather than .05 on the decision about the null hypothesis. For a sample of 30 games, Table 6.6 calculates the critical values for $\alpha = .01$ (the .01 alpha level has been divided into two halves of .005). Based on Table 6.6, for $\alpha = .01$, the decision rule may be stated as follows: "If the number of wins is < 8 or > 22, reject H_0; otherwise, do not reject H_0."

The critical values for $\alpha = .05$ and $\alpha = .01$ for $N = 30$ are illustrated in Figure 6.7. Looking at this figure, we see that the critical values are farther from the center of the distribution for $\alpha = .01$ (Figure 6.7(b)) than for $\alpha = .05$ (Figure 6.7(a)). This indicates that more extreme values of the statistic are needed to reject the null hypothesis when a low probability is defined as .01 rather than .05. Consequently, it is more difficult to reject the null hypothesis when alpha is set at .01 rather than .05.

Directionality of the Alternative Hypothesis

Hypothesis testing begins by stating two mutually exclusive statistical hypotheses, a null hypothesis (H_0) and an alternative hypothesis (H_1), that state that a hypothesized change, difference, or relationship in the population

TABLE 6.6 ● CALCULATION OF CRITICAL VALUES FOR BINOMIAL DISTRIBUTION OF $N = 30$, ALPHA = .01

Lower End of Distribution		Upper End of Distribution	
# Wins	Probability	# Wins	Probability
0	.0000	30	.0000
1	.0000	29	.0000
2	.0000	28	.0000
3	.0000	27	.0000
4	.0000	26	.0000
5	.0001	25	.0001
6	.0006	24	.0006
7	.0019	23	.0019
Sum	.0026	Sum	.0026

FIGURE 6.7 ● CRITICAL VALUES, BINOMIAL DISTRIBUTION, $\alpha = .05$ AND $\alpha = .01$ ($N = 30$)

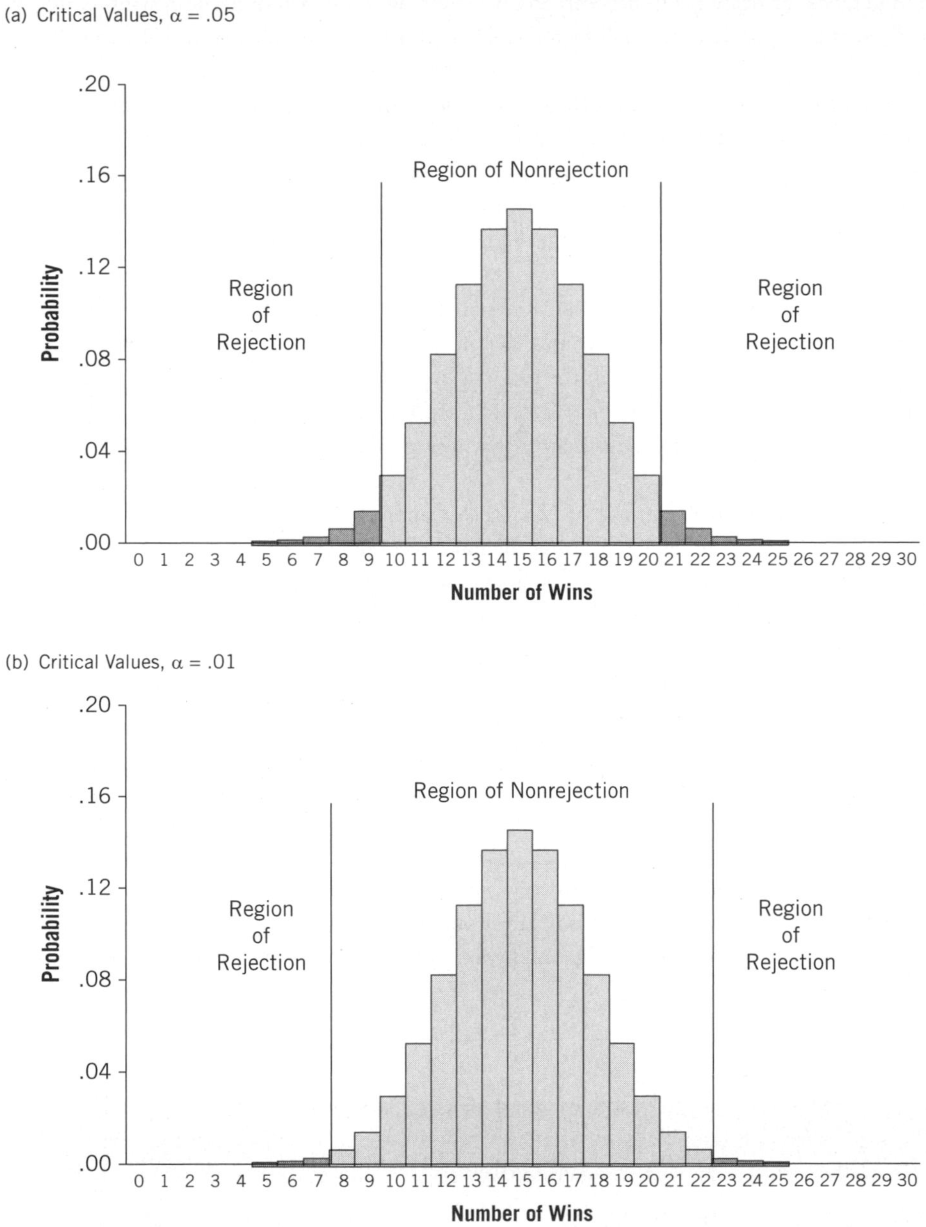

does not exist or does exist, respectively. As it turns out, the alternative hypothesis may be stated in one of two ways: nondirectional or directional. This section describes the following relationship between the directionality of the alternative hypothesis and the decision to reject the null hypothesis:

> There is a greater likelihood of rejecting the null hypothesis when the alternative hypothesis is directional than nondirectional.

In the Super Bowl example, the alternative hypothesis was stated as H_1: $\mu \neq 6$. This is an example of a **nondirectional alternative hypothesis**, which is an alternative hypothesis that does not indicate the direction of the change, difference, or relationship between groups or variables. By including the ≠ symbol, the alternative hypothesis H_1: $\mu \neq 6$ implies that the null hypothesis is rejected if the number of wins is *either* less than 6 *or* greater than 6.

A researcher may instead choose to use a **directional alternative hypothesis**, an alternative hypothesis that indicates the direction of the change, difference, or relationship by including the > or < (greater than or less than) symbol. Returning to the Super Bowl example, suppose we had reason to believe that winning the coin flip can increase (but *not* decrease) a team's chances of winning the game. If so, we may have stated the alternative hypothesis as H_1: $\mu > 6$, which implies that the null hypothesis is rejected only if the number of wins is greater than 6.

In terms of deciding which type of alternative hypothesis to use, researchers often choose nondirectional hypotheses to allow for the possibility of a change or difference in both directions from the population mean μ. In the Super Bowl example, even though the research hypothesis makes the prediction that the number of wins by teams winning the coin flip will be *greater* than the population mean of 6, we used a nondirectional alternative hypothesis to allow for the possibility that winning the coin flip could either increase *or* decrease a team's chance of winning the game.

Let's return to the Super Bowl example of $N = 30$ to demonstrate the implications of using a directional alternative hypothesis. Setting alpha at .05, using the nondirectional alternative hypothesis of H_1: $\mu \neq 15$, we identified critical values of 9 and 20. Because its region of rejection is located in both ends (or tails) of the distribution, a nondirectional alternative hypothesis is often referred to as "two-tailed." But suppose we decide to state the alternative hypothesis as H_1: $\mu > 15$. When the alternative hypothesis is directional, the region of rejection and critical value are located at only one end of the distribution ("one-tailed"). Because H_1: $\mu > 15$ includes the ">" symbol, the critical value is located at the upper end of the distribution—in Table 6.7, we find this critical value to be equal to 19. As such, the decision rule would be stated as follows: "If the number of wins is > 19, reject H_0; otherwise, do not reject H_0."

Figure 6.8 illustrates the difference between a nondirectional (two-tailed) and a directional (one-tailed) alternative hypothesis for the $N = 30$ Super Bowl example. Comparing the 5% region of rejection in Figure 6.8(b) with the 2½% region on the right end of the distribution in Figure 6.8(a), we see that the region of rejection is slightly larger when the alternative hypothesis is directional rather than nondirectional. That illustrates that a less extreme result will lead to rejection of the null hypothesis when a directional rather than a nondirectional alternative hypothesis is chosen.

TABLE 6.7 ● CALCULATION OF CRITICAL VALUES FOR BINOMIAL DISTRIBUTION OF $N = 30$, DIRECTIONAL ALTERNATIVE HYPOTHESIS (H_1: $\mu > 15$)

# Wins	Probability	# Wins	Probability
30	.0000	24	.0006
29	.0000	23	.0019
28	.0000	22	.0055
27	.0000	21	.0133
26	.0000	20	.0280
25	.0001	Sum	.0494

FIGURE 6.8 ● CRITICAL VALUES, BINOMIAL DISTRIBUTION, NONDIRECTIONAL AND DIRECTIONAL ALTERNATIVE HYPOTHESIS ($N = 30$)

(a) Critical Values, Non-Directional Alternative Hypothesis (H_1: $\mu \neq 15$)

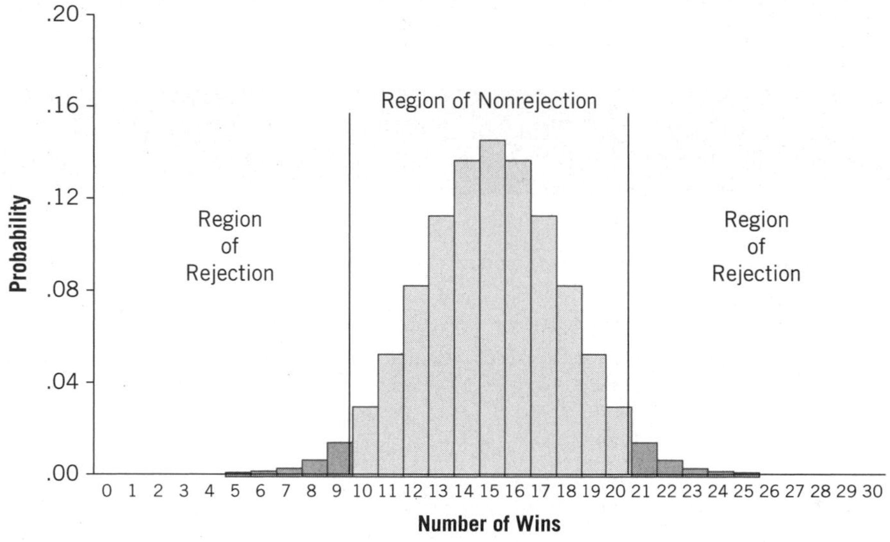

(b) Critical Values, Directional Alternative Hypothesis (H_1: $\mu > 15$)

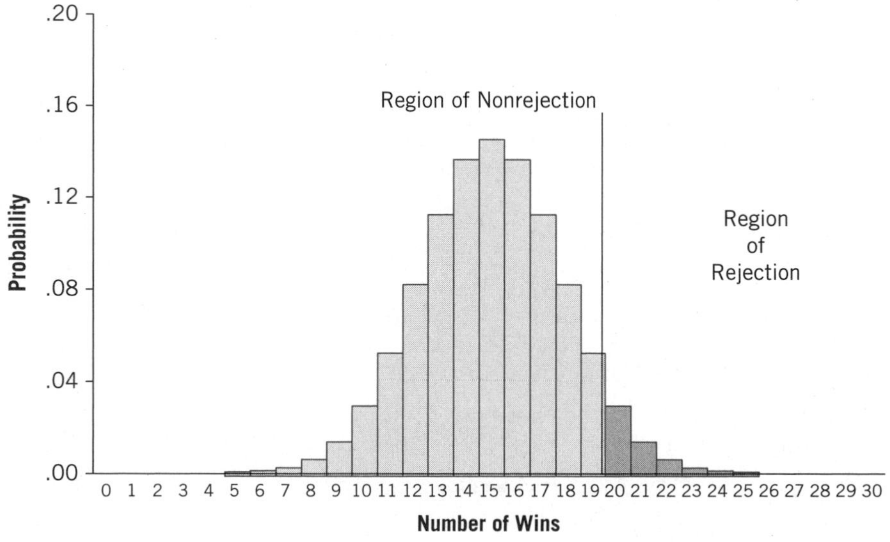

6.5 LOOKING AHEAD

The majority of this chapter focused on describing the steps, logic, and process of conducting statistical analyses to test research hypotheses; in doing so, critical concepts were defined and illustrated. It is important to gain an understanding of these steps and concepts as they will be repeated throughout many of the remaining chapters of this book. We also defined and emphasized the role of probability in conducting and interpreting the results of statistical analyses, ending the chapter with a brief introduction to issues that arise in testing hypotheses. Given the numerous steps and concepts that are part of hypothesis testing, we chose to minimize mathematical calculations in this chapter by using the relatively simple example of a binomial variable. The next chapter will be the first of several chapters that discuss testing research hypotheses by conducting statistical procedures that involve calculating and comparing sample means with population means. Although the mathematical calculations in future chapters may become increasingly complicated, keep in mind that the steps used in hypothesis testing will remain the same.

LEARNING CHECK 3:

Reviewing What You've Learned So Far

1. Review questions
   a. Within hypothesis testing, why do we say a research hypothesis has been "supported" rather than "proved"?
   b. What are the two types of errors you can make regarding the two statistical hypotheses?
   c. What are the three factors discussed in this chapter that affect the decision about the null hypothesis? In what specific ways does each of these influence this decision?
   d. What is the difference between a directional (one-tailed) and a nondirectional (two-tailed) alternative hypothesis? Under what research situations might you use one versus the other?
2. Using the Super Bowl example, imagine you draw the following three samples. Use the binomial probabilities table to determine the critical values (assume the alternative hypothesis is nondirectional [two-tailed] and alpha = .05). For each sample, also calculate the percentage of wins needed to reject the null hypothesis at the upper end of the distribution.
   a. $N = 14$
   b. $N = 16$
   c. $N = 22$
3. Returning to the Super Bowl example, imagine you draw the following three samples. Use the binomial probabilities table to determine the critical values for α .05 and α .01 (assume the alternative hypothesis is nondirectional [two-tailed]). For each sample, also calculate the percentage of wins needed to reject the null hypothesis at the upper end of the distribution.
   a. $N = 10$
   b. $N = 16$
   c. $N = 24$
4. Returning to the Super Bowl example, imagine you draw the following three samples. Use the binomial probabilities table to determine the critical values for a nondirectional alternative hypothesis and a directional alternative hypothesis (assume alpha = .05).
   a. $N = 10$ (H_1: $\mu \neq 5$ vs. H_1: $\mu > 5$)
   b. $N = 18$ (H_1: $\mu \neq 9$ vs. H_1: $\mu < 9$)
   c. $N = 26$ (H_1: $\mu \neq 13$ vs. H_1: $\mu > 13$)

6.6 Summary

Conceptually, ***probability*** may be defined as the likelihood of occurrence of a particular outcome of an event given all possible outcomes. Mathematically, the probability (p) of an outcome is the number of ways the outcome can occur divided by the total number of possible outcomes. Two important aspects of probability are that the sum of the probabilities of all possible outcomes of an event equals 1.00 (100%) and that the combined probability of mutually exclusive outcomes is the sum of their individual probabilities. This second observation is known as the ***addition rule***.

Researchers cannot be certain that their sample is completely representative of the larger population. ***Sampling error*** refers to differences between statistics calculated from a sample and statistics pertaining to the population from which the sample is drawn that are due to random, chance factors. Because of sampling error, researchers cannot test research hypotheses with absolute certainty and instead rely on probability to evaluate the data collected from their samples.

Probability may be applied to distributions such as normal distributions, the standard normal distribution, and ***binomial distributions*** (distributions of probabilities for variables consisting of exactly two categories) to determine the probability of a particular score or outcome.

Conducting statistical analyses to test research hypotheses consists of several steps. The first step is to state two ***statistical hypotheses*** (i.e., statements regarding expected outcomes or relationships involving population parameters). The first statistical hypothesis is the ***null hypothesis*** (H_0), which states that in the population, there exists no change, difference, or relationship among groups or variables. The second statistical hypothesis, the ***alternative hypothesis*** (H_1), states that the hypothesized change, difference, or relationship among groups or variables does exist in the population.

The second step in hypothesis testing is to make a decision about which of the two statistical hypotheses, the null or the alternative, is believed to be true. This step requires several steps of its own. First, ***alpha*** (α) (i.e., the probability of a statistic used to make a decision whether to reject the null hypothesis) is stated. In many academic disciplines, it is customary to set an alpha level of .05 or 5% (α .05). Setting alpha divides the distribution of values of a statistic into two regions: the ***region of rejection*** (values of the statistic whose probability is low enough to lead to the decision to reject the null hypothesis) and the ***region of nonrejection*** (values of the statistic whose probability is high enough to lead to the decision to not reject the null hypothesis). A value of the statistic that separates the regions of rejection and nonrejection is known as a ***critical value***. Second, a ***decision rule*** is stated specifying the values of the statistic that result in the decision to reject the null hypothesis. Third, statistical analyses are conducted on the data collected from the sample to calculate a value of a statistic. Fourth, the decision whether to reject the null hypothesis is then made by comparing the value of the statistic calculated from the sample with the critical values stated in the decision rule. If the value of the statistic exceeds a critical value, then the null hypothesis will be rejected; otherwise, the null hypothesis will not be rejected.

The third step in hypothesis testing is to draw a conclusion based on the decision to reject or not reject the null hypothesis, which typically involves communicating several aspects of the analysis, including the sample from whom data were collected, the value of the statistic calculated from the data, the decision about the null hypothesis, and the probability of the value of the statistic.

The fourth step in hypothesis testing is to relate the results of the analysis back to the original research hypothesis, more specifically, to determine whether the results of the analysis support or do not support the research hypothesis.

Because hypothesis testing is centered on probability rather than certainty, it raises a number of issues and concerns. First, hypothesis testing cannot provide "proof" (or a lack of proof) for a research hypothesis; instead, it can only provide "support" (or a lack of support) for a research hypothesis. Second, it is possible that the decision made about the null hypothesis may either be correct or incorrect. When the decision is made to reject the null hypothesis, we can never completely eliminate the possibility that the hypothesized effect does not actually exist and the decision to reject the null hypothesis was based on chance factors. It is also possible to make a second type of error: making the decision to not reject the null hypothesis when we should. Third, several factors (sample size, alpha, and the directionality of the alternative hypothesis) under the control of the researcher influence the decision about the null hypothesis: The larger the sample size, the greater will be the likelihood of rejecting the null hypothesis; the smaller the value of alpha, the lower will be the likelihood of rejecting the null hypothesis; and there is a greater likelihood of rejecting the null hypothesis using a ***directional (one-tailed) alternative hypothesis*** (which predicts the direction of the change, difference, or relationship) than using a ***nondirectional (two-tailed) alternative hypothesis*** (which does not predict the direction of the change, difference, or relationship).

6.7 Important Terms

probability (p. 151)
addition rule (p. 151)
sampling error (p. 152)
binomial variable (p. 152)
binomial distribution (p. 153)
statistical hypothesis (p. 158)
null hypothesis (H_0) (p. 158)
alternative hypothesis (H_1) (p. 158)
alpha (α) (p. 159)
region of rejection (p. 160)
region of nonrejection (p. 160)
critical value (p. 160)
decision rule (p. 163)
nondirectional (two-tailed) alternative hypothesis (p. 172)
directional (one-tailed) alternative hypothesis (p. 172)

6.8 Formula Introduced in This Chapter

Probability

$$p(\text{outcome}) = \frac{\text{number of ways an outcome can occur}}{\text{total number of possible outcomes}} \quad (6\text{-}1)$$

6.9 Exercises

1. You have collected the following data:

   3 8 2 6 11

   If you place these five numbers in a bag and randomly select one, what is the probability the number (X) will be . . .

   a. equal to 6
   b. less than 11
   c. greater than 3
   d. greater than 2 but less than 11

2. You have collected the following data:

   6 4 3 7 4 2 6 5 7 4

   If you randomly select one of these 10 numbers, what is the probability the number (X) will be . . .

   a. equal to 4
   b. equal to 7
   c. less than 5
   d. greater than 2
   e. greater than 4 but less than 7

3. A lottery contains 500 tickets. In this lottery, there are 25 prizes of $1, 10 prizes of $5, and 5 prizes of $25. What is the probability of . . .

   a. winning nothing ($0)
   b. winning $25
   c. winning more than $1

4. (This example was introduced in Chapter 2.) A friend of yours asks 20 people to rate a movie using a 1- to 5-star rating: the higher the number of stars, the higher the recommendation. Their ratings are listed below:

Person	# Stars	Person	# Stars	Person	# Stars
1	★★★	8	★★	15	★★★
2	★★★★★	9	★★★	16	★★
3	★★	10	★★★★	17	★★★
4	★★★★	11	★	18	★★
5	★★★	12	★★★★	19	★★★★
6	★★★	13	★★★★	20	★★★
7	★★★★	14	★★★		

a. What is the probability of this movie receiving three stars?
b. What is the probability of four or more stars?

5. For the standard normal distribution, what is the probability of having a z-score . . .
   a. greater than 1.67 ($z > 1.67$)
   b. greater than −1.67 ($z > -1.67$)
   c. less than .75 ($z < .75$)
   d. less than −2.75 ($z < -2.75$)
   e. between 1.00 and 1.25 ($1.00 < z < 1.25$)
   f. between −.25 and .25 ($-.25 < z < .25$)

6. According to the test's publishers (www.act.org), scores on the ACT college entrance examination for students graduating in 2001 were normally distributed, with $\mu = 21$ and $\sigma = 5$ (scores can range from 1–36).
   a. What is the probability of having a score greater than 30?
   b. What is the probability of having a score greater than 20?
   c. What is the probability of having a score less than 17?
   d. What is the probability of having a score less than 23?
   e. What is the probability of having a score between 18 and 25?

7. A bottling company uses a machine to fill 16-ounce bottles with orange juice. The company finds that the standard deviation of the amount of juice in these bottles is equal to $\frac{1}{4}$ ounce ($\sigma = .25$). What is the probability that a single bottle will contain . . .
   a. less than 15.60 ounces
   b. more than 16.30 ounces
   c. between 16.10 and 16.60 ounces

8. The length of time, in days, of pregnancy in healthy women is approximately normally distributed, with $\mu = 280$ days and $\sigma = 10$ days. What is the probability a woman will . . .
   a. give birth more than 1 week after her expected due date
   b. give birth more than 2 weeks before her expected due date
   c. give birth between 3 days before and 5 days after her expected due date

9. (This example was introduced in Chapter 5.) How many base hits can a baseball team expect to get in a game? Frohlich (1994) recorded the number of hits by the 28 Major League Baseball teams for all of the games played in 1989 to 1993 (each team plays 162 games a year). He found that the number of hits the teams made in the games was normally distributed, with a mean of 8.72 and a standard deviation of 1.10.
   a. What is the probability of a team getting more than 7 hits?
   b. What is the probability of a team getting less than 10 hits?
   c. What is the probability of a team getting less than 6 hits?
   d. What is the probability of a team getting between 8 and 11 hits?

NOTE: Exercises 10 to 15 use the binomial probabilities table.

10. Assuming a coin is fair, in a sample of 6 coin flips, what is the probability of getting . . .
    a. 3 heads
    b. less than 4 heads
    c. more than 2 heads and less than 5 heads

11. For a family with 5 children, assuming the probability of having a boy and having a girl are both .50 (50%), what is the probability of having . . .

a. 0 boys

b. 2 boys

c. more than 3 boys

12. Assume an equal number of people prefer the two most popular brands of cola. Under this assumption, what is the probability of the following claim being true: "9 out of 10 people prefer Brand X over Brand Y"?

13. In each of four political races, Democrats are believed to have a 60% chance of winning. If so, what is the probability that Democrats will win . . .

a. none of the elections

b. at least one election

c. the majority of the elections

14. A store advertises that there is a 90% chance their equipment will be trouble-free for a year. If you buy 6 of their products, what is the probability that . . .

a. all of the products will be trouble-free

b. half of the products will be trouble-free

15. Assuming that 35% of all marriages end in divorce, if you encountered 8 adult men, what is the probability that all of them are still married?

16. For each of the following situations, state the two competing hypotheses to be tested using words rather than mathematical symbols or formulas.

a. A company designs a program aimed at helping people stop smoking. They design a study aimed at testing the program.

b. A study wished to examine whether using seat belts affected the severity of injuries sustained by children in automobile accidents (Osberg & Di Scala, 1992).

c. A researcher hypothesizes that the more drivers use cellular phones, the greater the likelihood of getting into a traffic accident.

d. "It was hypothesized that . . . infants who spent greater amounts of time in center-based care would demonstrate more advanced exploratory behaviors than infants who did not spend as much time in center-based care" (Schuetze et al., 1999, p. 269).

e. "It is expected that achievement motivation will be a positive predictor of academic success" (Busato et al., 2000, p. 1060).

f. "The purpose of our study was to gain a better understanding of the relationship between social functioning and problem drinking. . . .We predicted that problem drinkers would endorse more social deficits than nonproblem drinkers" (Lewis & O'Neill, 2000, pp. 295–296).

17. For each of the following situations, state the two competing hypotheses to be tested using words rather than mathematical symbols or formulas.

a. Students who are taught effective learning skills will perform better on tests than students offered incentives to do well.

b. Increased use of Internet bulletin boards is associated with lower levels of television viewing.

c. Violent behavior in children may be reduced by teaching them conflict resolution skills.

d. The higher a person scores on the Graduate Management Admissions Test, the more likely the person is to succeed in graduate business school.

e. The more time children spend watching television, the more they will express a preference for unhealthy foods.

18. For each of the following situations, state a null hypothesis (H_0) and a nondirectional (two-tailed) alternative hypothesis (H_1).

a. A bank advertises that customers never have to wait more than 5 minutes in line at its branches. A researcher standing in line decides to test this advertisement.

b. The average life expectancy in this country is 78 years. A researcher wishes to see whether students, when asked how long they expect to live, give estimates different from the actual life expectancy.

c. The American Statistical Association publishes a monthly magazine sent to all of its members. Not too long ago, an article was written asking, "How many chocolate chips are there in a bag of Chips Ahoy cookies?" (Warner & Rutledge, 1999). Nabisco, the makers of these cookies, claim there are 1,000 chocolate chips in each bag. The authors of this article wish to test this claim.

d. An automobile company claims that their truck gets an average of 20 miles per gallon. A disgruntled group of truck buyers disputes this claim.

e. A company that makes batteries claims its batteries last an average of 25 hours of continuous use. A consumer group believes they have data that suggest the batteries last less than 25 hours.

19. Returning to the Super Bowl example discussed earlier in this chapter, imagine that you found 13 of 18 teams that won the coin flip went on to win the game.

a. State the null and alternative hypotheses (H_0 and H_1).

b. Identify the critical values for $\alpha = .05$ and state a decision rule.

c. Make a decision about the null hypothesis.

d. Does the result of your analysis support or not support the research hypothesis that winning the coin flip increases a team's chances of winning the game?

20. Two researchers test the same research hypothesis using the same instruments. One researcher rejects the null hypothesis but the other does not.

a. Which researcher is more likely to have had a larger sample size? Why?

b. Which researcher is more likely to have had a smaller level of alpha? Why?

c. Which researcher is more likely to have had a directional alternative hypothesis? Why?

21. A company makes a test they say can detect whether or not someone is guilty of a crime. In this situation, what are the two types of errors that could be made?

Answers to Learning Checks

Learning Check 1

2. a. $p = .13$
   b. $p = .25$
   c. $p = .50$
   d. $p = .88$
   e. $p = .38$

3. a. $p = .11$
   b. $p = .66$
   c. $p = .73$
   d. $p = .09$
   e. $p = .24$

4. a. $p = .16$
   b. $p = .07$
   c. $p = .55$

d. $p = .17$
e. $p = .03$

Learning Check 2

2. a. H_0: $\mu = .1370$; H_1: $\mu \neq .1370$
   b. H_0: $\mu = 164$; H_1: $\mu \neq 164$
   c. H_0: $\mu = 450$; H_1: $\mu \neq 450$
   d. H_0: $\mu = 900$; H_1: $\mu \neq 900$
   e. H_0: $\mu = .50$; H_1: $\mu \neq .50$
3. a. H_0: $\mu = 10$; H_1: $\mu \neq 10$
   b. For $\alpha = .05$, if the number of wins is < 6 or > 14, reject H_0; otherwise, do not reject H_0.
   c. Sixteen wins is > 14; therefore, reject H_0 ($p < .05$).
   d. Thirteen of the 20 teams would not support the research hypothesis because H_0 would not be rejected (13 wins is not < 6 or > 14).
4. a. H_0: $\mu = 8$; H_1: $\mu \neq 8$
   b. For $\alpha = .05$, if the number of correct guesses is < 4 or > 12, reject H_0; otherwise, do not reject H_0.
   c. Sixteen wins is > 15; therefore, reject H_0 ($p < .05$).
   d. The result of this analysis does not support your friend's claim to have ESP.

Learning Check 3

2. a. Critical values = 3 wins and 11 wins; 12/14 = 86%
   b. Critical values = 4 wins and 12 wins; 13/16 = 81%
   c. Critical values = 6 wins and 16 wins; 17/22 = 77%
3. a. For $\alpha = .05$, critical values = 2 wins and 8 wins (9/10 = 90%);
   for $\alpha = .01$, critical values = 1 wins and 9 wins (10/10 = 100%)
   b. For $\alpha = .05$, critical values = 4 wins and 12 wins (13/16 = 81%);
   for $\alpha = .01$, critical values = 3 wins and 13 wins (14/16 = 88%)
   c. For $\alpha = .05$, critical values = 7 wins and 17 wins (18/24 = 75%);
   for $\alpha = .01$, critical values = 6 wins and 18 wins (19/24 = 79%)
4. a. For H_1: $\mu \neq 5$, critical values = 2 wins and 8 wins;
   for H_1: $\mu > 5$, critical value = 8 wins
   b. For H_1: $\mu \neq 9$, critical values = 5 wins and 13 wins;
   for H_1: $\mu < 9$, critical value = 6 wins
   c. For H_1: $\mu \neq 13$, critical values = 8 wins and 18 wins;
   for H_1: $\mu > 13$, critical value = 17 wins

Answers to Odd-Numbered Exercises

1. a. $p = .20$
   b. $p = .80$

c. $p = .60$

d. $p = .60$

3. a. $p = .92$

   b. $p = .01$

   c. $p = .03$

5. a. $p = .0475$

   b. $p = .9525$

   c. $p = .7734$

   d. $p = .0030$

   e. $p = .0531$

   f. $p = .1974$

7. a. $p = .0548$ ($z = -1.60$)

   b. $p = .1151$ ($z = 1.20$)

   c. $p = .3364$ ($z = .40$ and $z = 2.40$)

9. a. $p = .9406$ ($z = -1.56$)

   b. $p = .8770$ ($z = 1.16$)

   c. $p = .0068$ ($z = -2.47$)

   d. $p = .7229$ ($z = -.65$ and $z = 2.07$)

11. a. $p = .0313$

    b. $p = .3125$

    c. $p = .1876$

13. a. $p = .0256$

    b. $p = .9744$

    c. $p = .4752$

15. $p = .0319$

17. a. H_0: Students taught effective learning skills will perform the same on tests as students offered incentives to do well.

       H_1: Students taught effective learning skills will perform differently on tests than students offered incentives to do well.

    b. H_0: Use of Internet bulletin boards is unrelated to television viewing.

       H_1: Use of Internet bulletin boards is related to television viewing.

    c. H_0: Children taught conflict resolution skills do not differ in violent behavior compared to children not taught conflict resolution skills.

       H_1: Children taught conflict resolution skills differ in violent behavior compared to children not taught conflict resolution skills.

    d. H_0: Scores on the Graduate Management Admissions Test are not related to the likelihood of success in graduate business school.

       H_1: Scores on the Graduate Management Admissions Test are related to the likelihood of success in graduate business school.

    e. H_0: The amount of time children spend watching television is not related to their preference for unhealthy foods.

       H_1: The amount of time children spend watching television is related to their preference for unhealthy foods.

19. a. H_0: $\mu = 9$; H_1: $\mu \neq 9$
    b. For $\alpha = .05$, if the number of wins is < 5 or > 13, reject H_0; otherwise, do not reject H_0.
    c. Thirteen wins is not < 5 or > 13; therefore, do not reject H_0 ($p > .05$).
    d. The result of this analysis does not support the research hypothesis that winning the coin flip increases a team's chances of winning the game.

21. One type of error is to reject the null hypothesis when we should not. Assuming innocence is the null hypothesis, in this case, the company concluded that the person was guilty of the crime when in fact the individual was innocent. The second type of error is to not reject the null hypothesis when we should. In this case, the company concluded that the person was innocent when in fact the person was guilty.

7

TESTING ONE SAMPLE MEAN

CHAPTER OUTLINE

Chapter 6 introduced hypothesis testing, which is the process of conducting a statistical analysis to test a research hypothesis about a population. To introduce hypothesis testing, Chapter 6 used the relatively simple example of winning the pregame coin flip at the Super Bowl. The Super Bowl example used the binomial distribution, a distribution that states with exact precision the probability of every possible outcome within the population (e.g., getting 0 heads in 2 coin flips). In contrast, the statistical procedures discussed in this chapter, as well as the chapters that follow, involve populations that cannot be fully defined or understood. This chapter examines two of these procedures, the *z*-test for one mean and the *t*-test for one mean, which test the difference between the mean of a sample and a hypothesized population mean. Our discussion will center on the findings from actual, published research studies.

7.1 AN EXAMPLE FROM THE RESEARCH: DO YOU READ ME?

What are the most critical goals of elementary school education? A team of researchers at Eastern Washington University headed by graduate student Jaclyn Reed was emphatic: "Without a doubt, reading is the most important skill that students can acquire in school. . . . Reading at high levels is associated with continued academic success, significantly reduced risk for school dropout, and higher rates of entering college and finding successful employment" (Reed, Marchand-Martella, Martella, & Kolts, 2007, p. 45).

Two fundamental aspects of reading are vocabulary and comprehension. Vocabulary pertains to the knowledge and understanding of individual words, whereas comprehension is the ability to understand why and how a passage of text is structured as well as to summarize a passage based on its meaning. In reviewing the relevant literature, the researchers wrote that "instruction that focuses on vocabulary building and text comprehension is critical for student success" (Reed et al., 2007, p. 46). Consequently, the goal of their study was to evaluate the effectiveness of a program designed to teach reading skills to fourth-grade students. On the basis of an evaluation of the literature, they hypothesized that students who went through the program would demonstrate higher vocabulary and comprehension skills than the general population of fourth graders.

The researchers collected data from students in four classrooms at one elementary school; these students received the year-long *Reading Success Level A* program, which consisted of lesson plans, exercises, workbooks, and exams. The study, which we will refer to as the reading skills study, is an example of quasi-experimental research, described in Chapter 1 as research methods using naturally formed or preexisting groups rather than employing random assignment to conditions. As the researchers noted, "Because this research project was conducted to evaluate *Reading Success Level A* for the elementary school and a control group was not available, it was not possible to assign students randomly to a control and an experimental group" (Reed et al., 2007, p. 65).

To measure students on their vocabulary and comprehension skills, the researchers used a software program known as *Read Naturally*. This program had each student read a passage of text out loud, after which the number of words that were spoken correctly and the amount of time taken to read the passage was recorded. The number of words spoken correctly was then divided by the amount of time needed to read the passage; this variable was called "number of words correct per minute" (WCPM). For example, 280 words spoken correctly in 2 minutes resulted in a WCPM score of 280 ÷ 2, or 140.

The sample in the study consisted of 93 students. However, to save space and time, the example in this chapter will use a smaller sample of 20 students ($N = 20$) designed to resemble the data from the original study. The WCPM for the 20 students are listed in Table 7.1.

To organize and illustrate the data, a grouped frequency distribution table and frequency polygon for the 20 students' WCPM is provided in Figure 7.1. Examining this figure, we see that the distribution of WCPM is roughly symmetrical, with the center of the distribution near the 141–150 and 151–160 intervals.

One step in analyzing data that have been collected is to calculate descriptive statistics. The most typical descriptive statistics are a measure of central tendency, such as the mean, and a measure of variability, such as the standard deviation. Before performing any calculations, however, it's helpful to look at the data in Table 7.1 and Figure 7.1 and come up with an "eyeball" estimate of what you believe are the mean and standard deviation. For example, given that many of the 20 scores appear to be between 140 and 160 WCPM, you could estimate the sample mean to be in the middle of the 140 to 160 range, which would be 150 WCPM. Furthermore, the

TABLE 7.1 ● WORDS CORRECT PER MINUTE (WCPM) FOR 20 STUDENTS

Student	WCPM	Student	WCPM	Student	WCPM
1	**159**	8	**126**	15	**146**
2	**140**	9	**150**	16	**174**
3	**148**	10	**136**	17	**144**
4	**156**	11	**168**	18	**119**
5	**141**	12	**153**	19	**160**
6	**178**	13	**134**	20	**145**
7	**169**	14	**161**		

difference between the estimated sample mean (150) and either of the ends of the range (140 or 160) can be an estimate of the standard deviation. In this example, our estimate of the standard deviation would be 10 WCPM.

Having created our "eyeball" estimates, we are ready to actually calculate descriptive statistics for the data in this sample. First, the calculation of the mean $(\overline{X})$ WCPM for the 20 students in the reading skills study is presented below (please refer to Chapter 3 for a review of this formula):

$$\begin{aligned}\overline{X} &= \frac{\Sigma X}{N} \\ &= \frac{159+140+148+\ldots+119+160+145}{20} = \frac{3{,}007}{20} \\ &= 150.35\end{aligned}$$

To describe the variability in this set of data, the standard deviation (s) is calculated using the definitional formula introduced in Chapter 4:

$$\begin{aligned}s &= \sqrt{\frac{\Sigma(X-\overline{X})^2}{N-1}} \\ &= \sqrt{\frac{(159-150.35)^2+(140-150.35)^2+\ldots+(160-150.35)^2+(145-150.35)^2}{20-1}} \\ &= \sqrt{\frac{74.82+107.12+\ldots+93.12+28.62}{19}} = \sqrt{\frac{4640.55}{19}} = \sqrt{224.24} \\ &= 15.63\end{aligned}$$

Looking at the descriptive statistics ($M = 150.35$, $s = 15.63$), we find that on average, students read approximately 150 words per minute, with the majority of the students having WCPM scores between 135 and 165.

Preliminary conclusions about these students' reading skills may be drawn from examining the table, figure, and descriptive statistics. However, testing the study's research hypothesis, that students completing the reading program would demonstrate higher levels of vocabulary and comprehension skills than the population of fourth graders, requires calculating a second type of statistic known as an inferential statistic.

For the reading skills study, comparing the study's sample with the population of fourth graders involves evaluating the sample mean $(\overline{X})$ in terms of its difference from a hypothesized population mean (μ). In essence, we'll evaluate the sample mean by transforming it into a statistic so that the probability of the sample mean may be determined. Although this may sound new to you, it's actually very similar to what we did in Chapter 5. In that chapter, we evaluated a score for a variable (X) based on its difference from a hypothesized population mean (μ) by transforming the score into a statistic known as a *z*-score (z); this transformation allowed us to determine the probability of the score.

FIGURE 7.1 ● GROUPED FREQUENCY DISTRIBUTION TABLE AND FREQUENCY POLYGON OF WORDS CORRECT PER MINUTE (WCPM) FOR 20 STUDENTS

(a) Grouped Frequency Distribution Table

WCPM	*f*	%
> 180	0	0%
171–180	2	10%
161–170	3	15%
151–160	4	20%
141–150	6	30%
131–140	3	15%
121–130	1	5%
< 121	1	5%
Total	20	100%

(b) Frequency Polygon

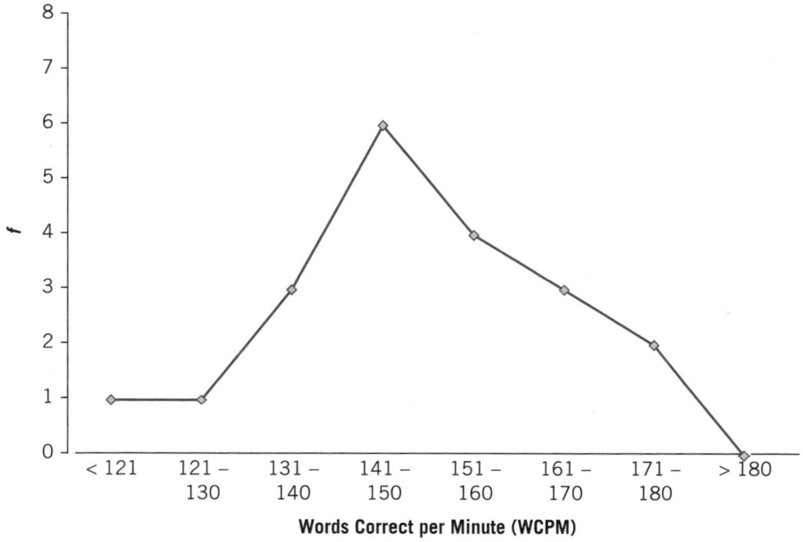

The critical difference between this chapter and Chapter 5 is that we'll use what we learned in Chapter 6 to determine whether the probability of the statistic the sample mean has been transformed into is low enough for us to decide the difference between the sample mean and the hypothesized population mean is not due to chance, random factors. Put another way, we'll use the probability of the statistic to make a decision whether to reject what is known as the null hypothesis.

In the Chapter 6 Super Bowl example, we needed to determine whether the probability of 5 wins in 12 games was low enough to reject the null hypothesis of 6 wins in 12 games (H_0: $\mu = 6$). In order to do this, we created a distribution of all of the possible wins that could occur in 12 games. That is, before we could determine the probability of our outcome, we had to determine all of the possible outcomes. For the reading skills study, determining the probability of obtaining our sample mean of 150.35 also requires a distribution—a distribution of all possible sample means. This distribution, known as the sampling distribution of the mean, is discussed in the next section.

7.2 THE SAMPLING DISTRIBUTION OF THE MEAN

The **sampling distribution of the mean** is the distribution of values of the sample mean for an infinite number of samples of size N that are randomly selected from the population. This distribution is an example of a **sampling distribution**, which is a distribution of statistics for samples randomly drawn from populations.

Let's use the example of age to illustrate both the sampling distribution of the mean and the concept of sampling distributions. We'll start by assuming the age of people who obtain their PhD degrees is normally distributed in the population with a mean of 30 years and a standard deviation of 7 years; in other words, $\mu = 30$ and $\sigma = 7$. This is illustrated in Figure 7.2(a). Now, say we draw two random samples of three people ($N = 3$) from this population and calculate the mean of the ages for each sample. Would we expect the means of the two samples to be the same? Would we expect either or both of these sample means to be equal to the population mean of 30?

The answer to both of the above questions is "perhaps, but probably not." Because the two samples are not only smaller than the population but also contain different people, we would expect the samples to differ from each other as well as from the population because of random, chance factors; these differences represent the concept of sampling error introduced in Chapter 6.

Returning to the age example, imagine we continue to draw random samples of three people from this population and calculate the mean age of each sample until we've calculated an infinite number of sample means—all of the sample means that could possibly occur within the population for a sample size of $N = 3$. The distribution of these sample means is an example of the sampling distribution of the mean, defined earlier as the distribution of sample means for an infinite number of samples of size N randomly drawn from the population. Figure 7.2(b) displays the sampling distribution of the mean for the age example.

Characteristics of the Sampling Distribution of the Mean

Like any other distribution, the sampling distribution of the mean such as the one in Figure 7.2(b) may be described in terms of its modality, symmetry, and variability. First, in terms of modality, because all of the samples are drawn from a population that has a mean equal to μ, the mean of the sample means is also expected to be equal to μ. That is, even though the samples may differ from each other, we expect the mean of the sample means to be equal to the mean of the population. In the age example, the mean of the sampling distribution of the mean is $\mu = 30$.

Next, in terms of symmetry, the sampling distribution of the mean is an approximate normal (bell-shaped) distribution, assuming the samples are sufficiently large, which is typically defined as a sample size of at least $N = 30$. For the sample means to be normally distributed implies that, even though there is variability among the sample means, we expect the majority of the sample means to be relatively close to the population mean, especially if the samples are of adequate size.

In introducing measures of variability, Chapter 3 discussed the standard deviation, which represents the average deviation of a score from the mean. However, because we are now working with sample means rather than individual scores, the variability of the sampling distribution of the mean is measured by the **standard error of the mean**, defined as the average deviation of a sample mean from the population mean. In essence, the standard error of the mean is the standard deviation of the sampling distribution of the mean.

You may wonder why the word *error* is used to describe variability in the sampling distribution of the mean. Theoretically, because all of the samples are drawn from the same population, the mean of every sample

FIGURE 7.2 ● DISTRIBUTION OF AGE IN A POPULATION AND THE SAMPLING DISTRIBUTION OF THE MEAN

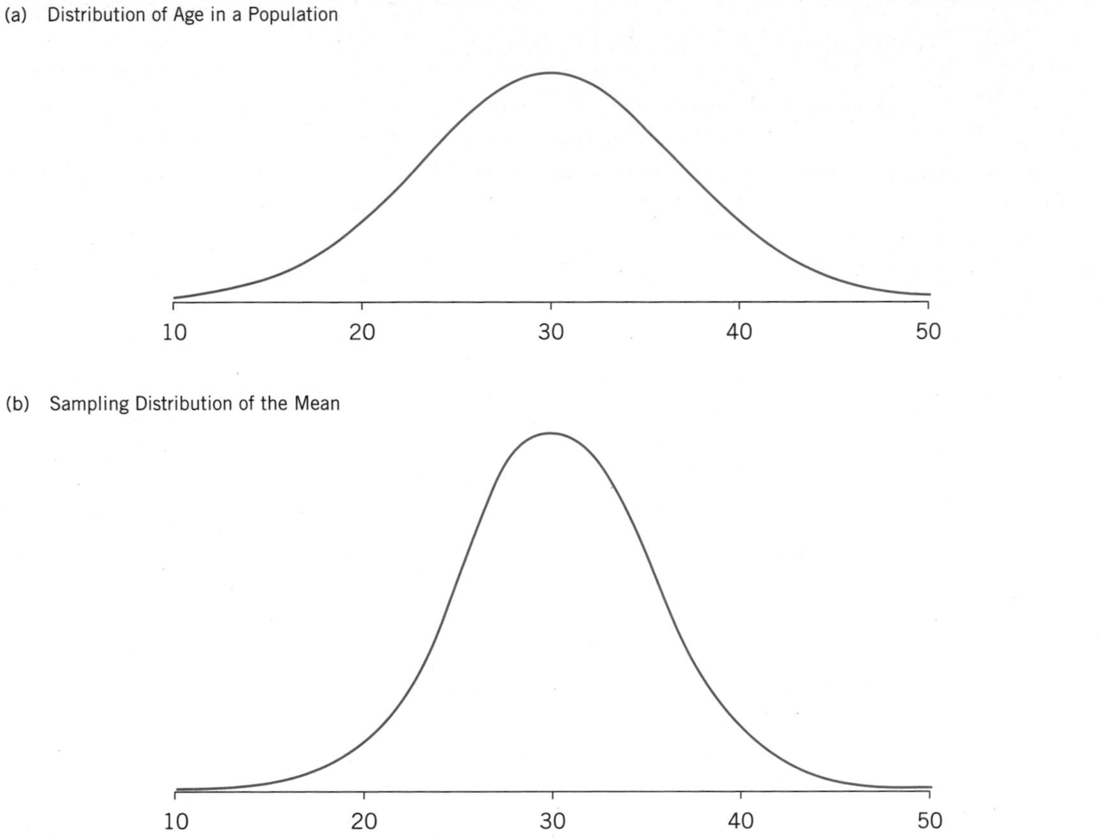

should be the same as the population mean μ. Any variability among these sample means is seen as being the result of random factors referred to as "error." So rather than use the term *standard deviation* (which measures the variability of scores for a variable), we use *standard error* to represent the variability of statistics calculated from samples.

The modality, symmetry, and variability of the sampling distribution of the mean are defined by a statistical principle known as the **central limit theorem**. This theorem states that when an infinite number of random samples are drawn from a population, the sample means are approximately normally distributed with a mean equal to the population mean μ and a standard deviation equal to the standard error of the mean, assuming the samples are of sufficient size ($N \geq 30$).

That the sampling distribution of the mean is approximately normally distributed is important to researchers. As we learned in Chapters 5 and 6, we can apply the principles of the standard normal distribution to normal distributions to determine the probability of any particular score in the distribution. Because the sampling distribution of the mean is approximately normally distributed, we can determine the probability of any particular sample mean. In our example of the reading skills study, we'll use the sampling distribution of the mean to determine the probability of obtaining a sample mean of 150.35 WCPM. The next section describes how we can calculate an inferential statistic that uses this probability to test a hypothesis regarding the difference between a sample mean and a population mean.

LEARNING CHECK 1:

Reviewing What You've Learned So Far

1. Review questions
   a. How could you create the sampling distribution of the mean for a variable?
   b. What are the main characteristics (modality, symmetry, variability) of the sampling distribution of the mean?
   c. What is measured by the standard error of the mean?
   d. What is the difference between a standard *deviation* and a standard *error*?
   e. Why is it important that the sampling distribution of the mean is normally distributed?

7.3 INFERENTIAL STATISTICS: TESTING ONE SAMPLE MEAN (σ KNOWN)

The research hypothesis in the reading skills study was that students completing the reading program would demonstrate higher levels of vocabulary and comprehension skills than the population of fourth graders. The process of hypothesis testing introduced in Chapter 6 will be used to test this research hypothesis. This process consists of four steps:

- state the null and alternative hypotheses (H_0 and H_1),
- make a decision about the null hypothesis,
- draw a conclusion from the analysis, and
- relate the result of the analysis to the research hypothesis.

Each of these steps is described below, using the reading skills study to illustrate relevant concepts and calculations. For this example, hypothesis testing centers on the calculation of an inferential statistic that evaluates the difference between a sample mean and a hypothesized population mean when the population standard deviation (σ) for the variable is known and can be stated.

State the Null and Alternative Hypotheses (H_0 and H_1)

The process of hypothesis testing begins by stating two statistical hypotheses: the null hypothesis and the alternative hypothesis; ultimately, we will make a decision regarding which hypothesis is supported by the data. The null hypothesis (H_0) implies no change, difference, or relationship exists among groups or variables in the population. In the reading skills study, if the program does not affect students' reading skills, the appropriate null hypothesis represents the conclusion that the mean WCPM of the sample is the same as the mean of the population.

To state the null hypothesis, we need a value of a hypothesized population mean (μ). In the reading skills study, the researchers noted that the *Read Naturally* software program had been administered to thousands of students and that the program's publisher reported a population mean WCPM of 124.81. Consequently, the null hypothesis for this study may be stated as

$$H_0: \mu = 124.81$$

A mutually exclusive alternative to the null hypothesis is the alternative hypothesis (H_1), which implies that a change, difference, or relationship does exist among groups or variables in the population. In the reading skills study, if the program does in fact affect students' reading skills, one appropriate alternative hypothesis reflects the conclusion that the mean WCPM of the sample is *not* the same as the population mean. This difference is represented by the following alternative hypothesis:

$$H_1: \mu \neq 124.81$$

As we learned in Chapter 6, an alternative hypothesis that includes the ≠ symbol is referred to as nondirectional or two-tailed, which implies the null hypothesis will be rejected if the mean of the sample is either greater than *or* less than the population mean.

The relationship between research hypotheses and statistical hypotheses is sometimes confusing to students. For example, the research hypothesis in the reading skills study implies the WCPM for the sample should be *greater* than the population mean. As such, the alternative hypothesis could have been stated as $H_1: \mu > 124.81$, meaning that the null hypothesis will only be rejected if the sample mean is greater than the population mean. An alternative hypothesis that includes the less than (<) or greater than (>) symbol is referred to as directional, or one-tailed. However, nondirectional alternative hypotheses such as $H_1: \mu \neq 124.81$ are often used in research to allow for the possibility of unexpected findings, particularly when there is not a large body of relevant theory or research to enable researchers to be more specific in their predictions. Using a nondirectional alternative hypothesis in the reading skills study allows for the possibility that the *Reading Success Level A* program somehow impairs students' reading skills, resulting in the sample having a WCPM that is *less* than the mean of the population.

Make a Decision About the Null Hypothesis

The primary purpose of this step is to calculate the value of an inferential statistic and to make a decision about the null hypothesis, which in this example involves deciding whether the difference between the sample mean and the population mean is not due to chance, random factors but instead is what is referred to as "statistically significant." This section describes the steps involved in this decision-making process:

- set alpha (α), identify the critical values, and state a decision rule;
- calculate a statistic: *z*-test for one mean;
- make a decision whether to reject the null hypothesis; and
- determine the level of significance.

Although the majority of these steps were introduced in Chapter 6, the last one is new and is discussed below.

Set Alpha (α), Identify the Critical Values, and State a Decision Rule

Hypothesis testing involves making one of two decisions: reject the null hypothesis or do not reject the null hypothesis. This decision is made based on the probability of the statistic that is calculated; if the statistic has a low probability of occurring, the decision is made to reject the null hypothesis. The first step in making the decision about the null hypothesis is to define a "low" probability, which is done by setting alpha (α); alpha is the probability of the statistic needed to reject the null hypothesis.

As we discussed in Chapter 6, alpha is traditionally set at .05, which implies the null hypothesis is rejected when the probability of obtaining the calculated value of a statistic is less than .05 ($p < .05$). In stating the value for alpha, it is useful to also indicate whether the alternative hypothesis is directional (one-tailed) or nondirectional (two-tailed). For the reading skills study, because the alternative hypothesis ($H_1: \mu \neq 124.81$) is nondirectional, alpha may be stated as "$\alpha = .05$ (two-tailed)."

Once alpha has been determined, the next step is to identify the values of the statistic that have a low probability of occurring; put another way, the next step is to identify the values of the statistic that result in the decision to reject the null hypothesis. This is accomplished by identifying what is known as the critical values of the statistic. In order to identify the critical values, we must first determine what particular statistic will be calculated.

In the reading skills study, the publishers of the *Read Naturally* software program reported not only a value of the population mean (μ) but also a population standard deviation (σ) of 43.26 WCPM. When both the population mean and standard deviation of a variable are known, the appropriate distribution to evaluate the difference between a sample mean and a population mean is the standard normal distribution (normal curve) introduced in Chapter 5. Scores for the standard normal distribution are referred to as *z*-scores. However, because we'll use this distribution to calculate a statistic that tests the difference between a sample mean and a population mean, we'll refer to this statistic as a *z*-statistic.

Using the standard normal distribution, identifying the critical values involves determining the values of the *z*-statistic that divide the distribution into two regions: the region of rejection (containing the values of the statistic whose probability is low enough to lead to the decision to reject the null hypothesis) and the region of nonrejection (containing the values of the statistic whose probability is high enough to lead to the decision to *not* reject the null hypothesis). Given that we have defined a "low" probability as .05 ($\alpha = .05$), we must identify the values of the standard normal distribution that fall in a 5% region of rejection.

Because the alternative hypothesis for the reading skills study is nondirectional, the 5% region of rejection is split into two halves, one at each tail of the distribution. Consequently, we divide the $\alpha = .05$ probability by 2 ($.05 \div 2 = .0250$) and find the values of the *z*-statistic associated with .0250 in the "Area Beyond *z*" column of the normal curve table (see Table 1 in the back of this book); the "Area Beyond *z*" is the percentage of *z*-statistics located in the tail (as opposed to the center) of the distribution. We first move down the "Area Beyond *z*" column until we reach the value .0250 and then move to the left until we are under the "*z*" column. Here, we find a *z*-statistic value of 1.96. Therefore, the 5% region of rejection is located in the area beyond the *z*-statistics −1.96 and 1.96; this is illustrated by the shaded areas of the distribution in Figure 7.3.

To summarize, we have thus far stated a value of alpha and identified the critical values of the *z*-statistic. In other words, we have defined a "low" probability and identified the values of the statistic that have this low probability. For the reading skills study, this may be stated as follows:

$$\text{For } \alpha = .05 \text{ (two-tailed), critical values} = \pm 1.96$$

In stating the critical values, the ± symbol (i.e., ±1.96) represents "plus or minus."

Once the critical values have been identified, a decision rule can be stated that explicitly specifies the values of the statistic that result in the rejection of the null hypothesis. For the reading skills study:

$$\text{If } z < -1.96 \text{ or } > 1.96\text{, reject } H_0\text{; otherwise, do not reject } H_0$$

This decision rule implies that the null hypothesis will be rejected if the value of the *z*-statistic calculated from the sample data is either less than −1.96 or greater than 1.96. In other words, the null hypothesis is rejected when the statistic is located in the region of rejection, which implies the statistic has a low (< .05) probability of occurring. On the other hand, if the value of the *z*-statistic is neither less than −1.96 nor greater than 1.96, the null hypothesis will *not* be rejected. The decision to not reject the null hypothesis is made when the statistic is located in the region of nonrejection, which implies the statistic does not have a low probability of occurring—that is, its probability is greater than α ($p > .05$).

Calculate a Statistic: *z*-Test for One Mean

In this research situation, we want to evaluate a sample mean in terms of its difference from a population mean ($\bar{X} - \mu$) in order to determine the probability of obtaining our value of the sample mean. For the reading skills

FIGURE 7.3 ● CRITICAL VALUES FOR z-STATISTIC, $\alpha = .05$, TWO-TAILED ALTERNATIVE HYPOTHESIS

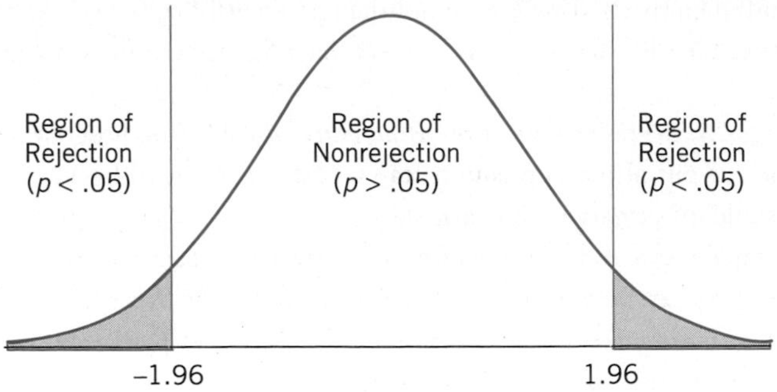

study, assuming the population mean WCPM is 124.81, is the probability of obtaining our sample mean of 150.35 low enough to reject the null hypothesis and therefore conclude the difference between the sample mean and the population mean is statistically significant? We answer this question by transforming the sample mean into a *z*-statistic.

Evaluating a sample mean by transforming it may seem new to you, but it's actually very similar to what we did in Chapter 5, where we evaluated a score in terms of the difference between the score and the population mean $(X - \mu)$ by transforming the score into a *z*-score using Formula 5-1:

$$z = \frac{X - \mu}{\sigma}$$

When the above formula is used to transform a score into a *z*-score, we can evaluate the score by determining its probability. For example, in Chapter 5, we transformed an SAT score of 660 into a *z*-score of 1.60, where we found a .0548 probability that an SAT score is greater than 660.

For the reading skills study, evaluating our sample mean $(\overline{X})$ of 150.35 involves transforming it in a manner very similar to transforming scores. One statistical procedure used to evaluate sample means, the ***z*-test for one mean**, tests the difference between a sample mean and a population mean when σ is known:

$$z = \frac{\overline{X} - \mu}{\sigma_{\overline{X}}} \qquad (7\text{-}1)$$

where $\overline{X}$ is the sample mean, μ is the population mean, and $\sigma_{\overline{X}}$ is the population standard error of the mean. As we can see, Formula 7-1 closely resembles Formula 5-1; however, the numerator in Formula 7-1 involves the difference between two means $(\overline{X} - \mu)$ rather than the difference between a score and a mean $(X - \mu)$, and the denominator reflects variability of sample means $(\sigma_{\overline{X}})$ instead of variability of scores (σ).

The first step in calculating the *z*-statistic for the *z*-test for one mean is to calculate the **population standard error of the mean** $(\sigma_{\overline{X}})$, which is the standard deviation of the sampling distribution of the mean when the population standard deviation (σ) for the variable is known. (Later in this chapter, we will discuss a standard error of the mean that's used when the population standard deviation is not known.)

The formula for the population standard error of the mean $(\sigma_{\overline{X}})$ is presented in Formula 7-2:

$$\sigma_{\overline{X}} = \frac{\sigma}{\sqrt{N}} \qquad (7\text{-}2)$$

where σ is the population standard deviation and N is the size of the sample. You may wonder why the standard error of the mean, which represents the variability of sample means, is a function of the variability of individual scores (σ) and the size of the sample (N). First, the larger the amount of variability in scores in the population, the larger the amount of variability in sample means generated from this population. As an extreme example of this principle, if all of the scores for a variable were the same (meaning there was zero variability of scores), all of the sample means would also be the same (meaning there would be zero variability of sample means). Second, the larger the samples used to calculate the sample means, the smaller the effect of random, chance factors that create variability among sample means; as a result, there should be less variability among sample means than when smaller samples are used.

In calculating the population standard error of the mean for the reading skills study, earlier we mentioned that the publishers of the *Read Naturally* program reported a population standard deviation (σ) of 43.26 WCPM. Inserting this value of σ and our sample size of $N = 20$ into Formula 7-2, $\sigma_{\overline{X}}$ for the reading skills study is calculated as follows:

$$\begin{aligned}\sigma_{\overline{X}} &= \frac{\sigma}{\sqrt{N}}\\ &= \frac{43.26}{\sqrt{20}} = \frac{43.26}{4.47}\\ &= 9.68\end{aligned}$$

It is important to remember that the standard deviation (σ) measures the variability of individual scores, whereas the standard error of the mean ($\sigma_{\overline{X}}$) estimates the variability of sample means.

Once the population standard error of the mean ($\sigma_{\overline{X}}$) has been calculated, we can calculate a value of the z-statistic. For the reading skills study, the values for the sample mean ($\overline{X} = 150.35$), population mean ($\mu = 124.81$), and population standard error of the mean ($\sigma_{\overline{X}} = 9.68$) are inserted into Formula 7-1:

$$\begin{aligned}z &= \frac{\overline{X} - \mu}{\sigma_{\overline{X}}}\\ &= \frac{150.35 - 124.81}{9.68} = \frac{25.54}{9.68}\\ &= 2.64\end{aligned}$$

Make a Decision Whether to Reject the Null Hypothesis

Now that a value for the z-statistic has been calculated, the next step in evaluating the sample mean is to make a decision whether to reject the null hypothesis. Using the decision rule stated earlier, this decision is made by comparing the value of the z-statistic calculated from the sample with the critical values. For the reading skills study, we could state this decision the following way:

> $z = 2.64$ is greater than the critical value 1.96; therefore, the decision is made to reject the null hypothesis because the probability of $z = 2.64$ is less than .05.

LEARNING CHECK 2:

Reviewing What You've Learned So Far

1. Review questions
   a. In testing one mean, the value of what parameter is stated in the null and alternative hypotheses?
   b. In testing one mean, what is implied by the null and alternative hypotheses?

(Continued)

(Continued)

c. Why are nondirectional (two-tailed) rather than directional (one-tailed) alternative hypotheses predominantly used in research studies?
d. What are the similarities and differences between the formula for the *z*-score (Chapter 5) and the *z*-test for one mean?
e. Under what conditions would one calculate the population standard error of the mean rather than the standard error of the mean?
f. What two factors influence the amount of variability in a distribution of sample means?

2. For each of the following situations, calculate the population standard error of the mean ($\sigma_{\bar{X}}$).
   a. $\sigma = 12.00$; $N = 16$
   b. $\sigma = 1.00$; $N = 9$
   c. $\sigma = 4.50$; $N = 26$
   d. $\sigma = 39.76$; $N = 50$
3. For each of the following situations, calculate the *z*-statistic (*z*).
   a. $\bar{X} = 10.00$; $\mu = 8$; $\sigma = 3$; $N = 9$
   b. $\bar{X} = 4.00$; $\mu = 7$; $\sigma = 6$; $N = 16$
   c. $\bar{X} = 3.52$; $\mu = 3.29$; $\sigma = 1.18$; $N = 21$
   d. $\bar{X} = 13.25$; $\mu = 11.87$; $\sigma = 3.42$; $N = 32$

Another, more concise, way of stating this decision is as follows:

$$z = 2.64 > 1.96 \therefore \text{ reject } H_0 \ (p < .05)$$

The $\therefore$ symbol is the mathematical symbol for *therefore.*

Statistical significance versus nonsignificance. When the null hypothesis is rejected, it is common to say that the result of the analysis is "statistically significant"; in the reading skills study, we could say that "the difference between the sample mean of 150.35 and the population mean of 124.81 is statistically significant." If, on the other hand, the decision had been made to *not* reject the null hypothesis, the result of the analysis may be referred to as "nonsignificant." Please keep in mind that "nonsignificant" is a statistical concept. You may see researchers refer to a statistic as "insignificant"; however, the word *insignificant* is not a statistical concept but is rather a value judgment that implies the statistic is not meaningful or important.

The abbreviation "n.s." may be used to indicate a *non*significant result of a statistical analysis. Similarly, the notation "$p > .05$" informs the reader that the probability of the statistic was greater than .05, meaning the probability was not low enough to reject the null hypothesis. We have found students sometimes get confused over the difference between "$p < .05$" and "$p > .05$." Remember, the null hypothesis is rejected when the value of the statistic is large enough to have a low probability, $p < .05$, of occurring.

Determine the Level of Significance

In the reading skills study, the decision was made to reject the null hypothesis because the probability of the *z*-statistic of 2.64 was less than .05. When the decision is made to reject the null hypothesis, it is customary to go one step further and determine whether the probability of the statistic is not only less than .05 ($p < .05$) but is also less than .01 ($p < .01$). We do this in order to report the results of a statistical analysis as accurately and informatively as possible. This is similar to the difference between describing someone as "a 19-year-old" rather than as a "teenager" or saying one's yearly income is "$4,500" rather than "less than $10,000."

It is important for you to understand that when researchers calculate statistics using statistical software, the software provides the exact probability of a statistic rather than simply indicating whether the probability is less or greater than .05 or .01. As a result, in reading the results of statistical analyses in journal articles, you may see something such as "p = .021"; because .021 is less than .05, this implies the null hypothesis was rejected; "p = .004" would imply that the probability of the statistic is less than .01. However, in reporting the results of analyses in tables or figures rather than in the text of a journal article, it remains customary to use levels of significance such as .05, .01, and .001. It is for that reason that we include this step within the process of hypothesis testing.

To determine whether the probability of the z-statistic for the reading skills study is less than .01, we need to identify the values of the z-statistic that correspond to the outer 1% of the distribution because these values have a combined probability less than .01. First, for a two-tailed alternative hypothesis, we divide .01 by 2, which is .0050. Next, returning to the normal curve table, we move down the "Area Beyond z" column until we reach the value .0050. Moving to the left to the "z" column, we find the z-statistic 2.58, which indicates there is less than a .01 probability of obtaining a z-statistic either less than −2.58 or greater than 2.58.

For the reading skills study, we determine whether the z-statistic meets the .01 level of statistical significance as follows:

$$z = 2.64 > 2.58 \therefore p < .01$$

Because a z-statistic value of 2.64 is greater than 2.58, we may conclude that its probability is not only less than .05 but also less than .01. Figure 7.4 illustrates that a z-statistic value of 2.64 exceeds both the α = .05 and .01 cutoffs of 1.96 and 2.58, respectively. We have found it extremely useful for students to draw the distribution with its critical values in order to determine the appropriate level of significance.

Why determine whether $p < .01$*?* As we mentioned earlier, researchers determine whether the probability of a statistic is less than .01 in order to present their results as precisely as possible. Unfortunately, researchers sometimes attach inappropriate labels to different levels of significance. A common mistake is to say that a statistic with less than a .01 probability is not simply "statistically significant" but is "*highly* significant." This is actually inappropriate because statistical significance is a dichotomy (significant vs. nonsignificant) rather than a continuum. The purpose of reporting levels of significance such as $p < .01$ (or, when appropriate, $p < .001$ or $p < .0001$) is to provide accurate descriptions of statistical analyses, not to make judgmental declarations.

When determine whether $p < .01$*?* It is only appropriate to determine whether the probability of a statistic is less than .01 when we have made the decision to reject the null hypothesis. When we do not reject the null

FIGURE 7.4 ● DETERMINING THE LEVEL OF SIGNIFICANCE FOR THE READING SKILLS STUDY

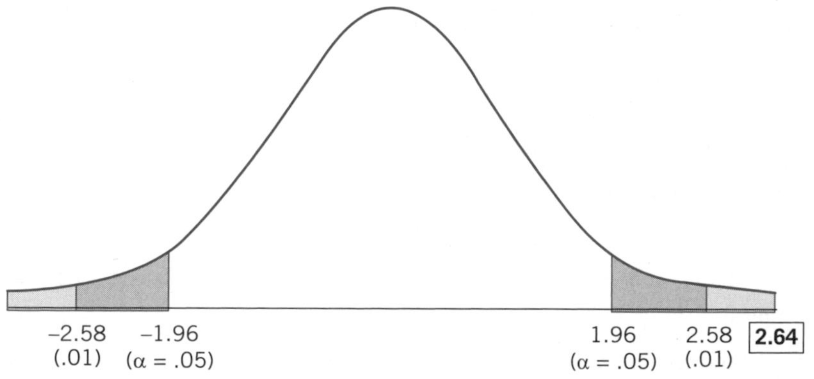

hypothesis, this implies the probability of the statistic is greater than .05 ($p > .05$). If the probability is *greater* than .05, it cannot be *less* than .01. Therefore, when we do not reject the null hypothesis, it is not necessary to determine whether $p < .01$.

What if p < *.05 but* not < *.01?* In the reading skills study, what would we have concluded if we rejected the null hypothesis but the *z*-statistic had *not* been greater than 2.58? For example, what if *z* had been 2.32 rather than 2.64? Because $z = 2.32$ is greater than the $\alpha = .05$ critical value of 1.96, we would reject the null hypothesis, implying that $p < .05$. However, because 2.32 is *less* than the .01 cutoff of 2.58, the most accurate statement we could make would be that its probability is less than .05 ($p < .05$) but *not* less than .01. Consequently, we could state

$$z = 2.32 < 2.58 \therefore p < .05 \text{ (but not} < .01)$$

This situation is illustrated in Figure 7.5. As we can see, when the calculated value of a statistic falls between the .05 and .01 critical values, its probability is less than .05 but *not* less than .01. When researchers report "$p < .05$" for an analysis, we assume they have determined that the probability of their statistic is not less than .01.

Students sometimes get confused regarding what conclusion to draw when a statistic exceeds the .05 critical value but does not exceed the .01 critical value. In the above example, even though $z = 2.32$ was greater than the $\alpha = .05$ critical value of 1.96 and as a result we rejected the null hypothesis, because 2.32 is less than the $\alpha = .01$ critical value students may draw the following conclusion: "$z = 2.32 < 2.58 \therefore$ do not reject H_0." It is important to understand that the purpose of the "Determine the Level of Significance" step is *not* to decide whether to reject the null hypothesis—we've already made this decision. Once the decision has been made to reject the null hypothesis, we do not "unmake" it at a later step.

Draw a Conclusion From the Analysis

What conclusion could we draw from the results of the analysis in the reading skills study? We could say that "the null hypothesis was rejected," but this says nothing about the purpose or nature of the analysis. Concluding that "the difference was statistically significant" is slightly better, but it doesn't indicate the specific difference to which we are referring. Saying that "the sample mean ($M = 150.35$) is significantly different from the hypothesized population mean ($\mu = 124.81$)" is better still, but perhaps there is a way to be even more specific and informative. One way we could state our conclusion regarding the reading skills study is the following:

> The number of words correct per minute (WCPM) ($M = 150.35$) in a sample of 20 fourth-grade students was significantly greater than the national normative sample mean ($\mu = 124.81$), $z = 2.64$, $p < .01$.

This brief sentence contains a great deal of information:

- The *variable* that was analyzed: "the number of words correct per minute (WCPM)"
- The *sample* from whom data were collected: "a sample of 20 fourth-grade students"
- *Descriptive statistics* of the variables: "($M = 150.35$) . . . ($\mu = 124.81$)"
- The *nature and direction* of the findings: "was significantly *greater* than the national normative sample mean" (rather than simply saying the sample mean was significantly "different" from the hypothesized population mean)
- Information about the *inferential statistic:* "$z = 2.64$, $p < .01$" (which indicates the type of statistic calculated [*z*], the calculated value of the statistic [2.64], and the level of significance of the statistic [$p < .01$])

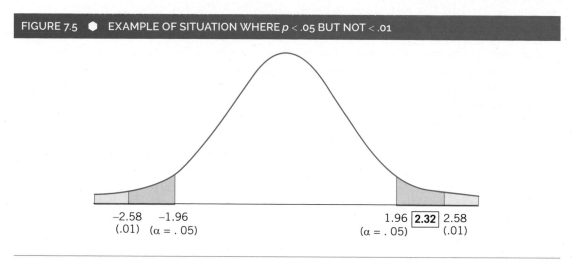

FIGURE 7.5 ● EXAMPLE OF SITUATION WHERE $p < .05$ BUT NOT $< .01$

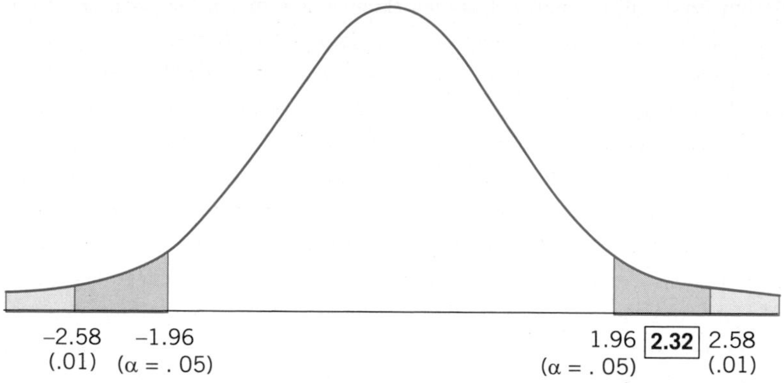

It's important to report the results of statistical analyses with as much relevant information as possible as this reduces the possibility that others will draw incomplete or inappropriate conclusions about the analysis.

Relate the Result of the Analysis to the Research Hypothesis

The final step in hypothesis testing is to interpret the results of the statistical analysis in terms of the extent to which the analysis supports or does not support a study's research hypothesis. To recall, the research hypothesis in the reading skills study was that students who went through the reading program would demonstrate higher vocabulary and comprehension skills than the general population of fourth graders. Given that the researchers found the WCPM for their sample to be significantly greater than the population mean, here is how they communicated their conclusions regarding their research hypothesis:

> This study revealed that teaching explicit, systematic reading comprehension strategies to fourth graders is likely to increase reading comprehension skills. (Reed et al., 2007, p. 64)

Notice that they did not say that their findings "proved" the effectiveness of the reading program. As we learned in Chapter 6, hypotheses regarding a population cannot be proven when data are collected from only a sample of the population.

Assumptions of the *z*-Test for One Mean

The goal of statistical procedures such as the *z*-test is to test hypotheses researchers have about populations by analyzing data collected from samples of these populations. These procedures are based on certain assumptions regarding such things as who the data have been collected from, how the variables in a research study are measured, and how scores for the variable are distributed in the population. To use these procedures appropriately, researchers must determine the extent to which their data meet these assumptions. This section discusses assumptions related to the *z*-test for one mean; these assumptions also apply to many of the statistical procedures discussed in later chapters of this book.

The first assumption related to the *z*-test for one mean is the **assumption of random sampling**, which is the assumption that the sample in a research study has been randomly selected from the population. The reason for this assumption is that distributions such as the sampling distribution of the mean and the standard normal distribution are based on random sampling. However, in conducting research, it is very difficult to meet this assumption. Imagine, for example, a researcher develops a hypothesis involving differences between American men and women. Given the millions of men and women who live in the United States, it would be extremely difficult to develop completely random samples. As a result, researchers may use inferential statistics without random sampling but, in doing so, must assess their samples in terms of how representative they are of their relevant populations and describe any limitations in their ability to apply their findings beyond their samples.

The second assumption pertains to how the variables in a research study are measured. The **assumption of interval or ratio scale of measurement** implies that the variables being analyzed are measured at the interval or ratio scale of measurement. In Chapter 1, we noted that the values of interval or ratio variables, such as WCPM, are equally spaced along a numeric continuum. This assumption is critical because the calculation of statistics such as the mean and standard deviation involves arithmetic operations such as addition and multiplication. These operations cannot be performed on variables measured at the nominal or ordinal scales of measurement, such as gender, whose values are qualitatively rather than quantitatively different from each other. One cannot, for example, add "male" and "female" to each other to calculate the "average gender" in a sample.

The third assumption is the **assumption of normality**, which is the assumption that scores for the variable are approximately normally distributed in the population. Because theoretical distributions of statistics are normally distributed, it is expected that they are based on variables that are normally distributed. However, even when this assumption is not met, research has found that statistics such as the *z*-statistic are **robust**, meaning they are able to withstand moderate violations of the assumption of normality. For example, an important aspect of the central limit theorem discussed earlier is that, assuming the samples are of sufficient size ($N \geq 30$), the sampling distribution of the mean approximates a normal distribution even if the distribution of scores for the variable is *not* normally distributed in the population. This helps the *z*-test be robust to violations of the assumption of normality.

Summary

The process of testing the mean of a sample when the standard deviation (σ) is known by conducting the *z*-test for one mean is summarized in Box 7.1, using the reading skills study as an example. The *z*-test is used when the population standard deviation σ is known. However, as it is extremely rare for researchers to collect data from an entire population, the population standard deviation is typically not known, and researchers must estimate it using the standard deviation of the sample (s). Consequently, testing the difference between a sample mean and a hypothesized population mean when the population standard deviation is not known requires a different statistical procedure. Again using a published research study, the next section introduces and discusses this procedure, known as the *t*-test for one mean.

SUMMARY BOX 7.1 CONDUCTING THE *Z*-TEST FOR ONE MEAN (READING SKILLS EXAMPLE)

State the null and alternative hypotheses (H_0 and H_1)

$$H_0: \mu = 124.81 \qquad H_1: \mu \neq 124.81$$

Make a decision about the null hypothesis

Set alpha (α), identify the critical values, and state a decision rule

If $z < -1.96$ or > 1.96, reject H_0; otherwise, do not reject H_0

Calculate a statistic: *z*-test for one mean

Calculate the population standard error of the mean ($\sigma_{\bar{X}}$)

$$\sigma_{\overline{X}} = \frac{\sigma}{\sqrt{N}} = \frac{43.26}{\sqrt{20}} = \frac{43.26}{4.47} = 9.68$$

Calculate the *z*-statistic (*z*)

$$z = \frac{\overline{X} - \mu}{\sigma_{\overline{X}}} = \frac{150.35 - 124.81}{9.68} = \frac{25.54}{9.68} = 2.64$$

Make a decision whether to reject the null hypothesis

$$z = 2.64 > 1.96 \therefore \text{ reject } H_0 \ (p < .05)$$

Determine the level of significance

$$z = 2.64 > 2.58 \therefore p < .01$$

Draw a conclusion from the analysis

In a sample of 20 fourth-grade students, the number of words correct per minute (WCPM) ($M = 150.35$) was significantly greater than the national normative sample ($\mu = 124.81$), $z = 2.64$, $p < .01$.

Relate the result of the analysis to the research hypothesis

This study revealed that teaching explicit, systematic reading comprehension strategies to fourth graders is likely to increase reading comprehension skills (Reed et al., 2007, p. 64).

LEARNING CHECK 3:
Reviewing What You've Learned So Far

1. Review questions
   a. What is the difference between a statistic that is statistically "significant" and one that is "nonsignificant"?
   b. What is the difference between a statistic referred to as "nonsignificant" and one referred to as "insignificant"?
   c. Does "$p < .05$" imply the null hypothesis was rejected or not rejected? Why?
   d. When and why would you determine whether the probability of a statistic is less than .01?
   e. Which of these would you use when the value of a statistic falls in between the .05 and .01 critical values: $p > .05$, $p < .05$, or $p < .01$?
   f. What information about a statistical analysis is typically included when communicating the results of the analysis?
   g. What are some assumptions applicable to the *z*-test for one mean?
2. For each of the following situations, calculate the *z*-statistic (*z*), make a decision about the null hypothesis (reject, do not reject), and indicate the level of significance ($p > .05$, $p < .05$, $p < .01$).
   a. $\overline{X} = 12.00$; $\mu = 6$; $\sigma_{\overline{X}} = 3.00$
   b. $\overline{X} = 3.00$; $\mu = 4$; $\sigma_{\overline{X}} = 1.00$
   c. $\overline{X} = 14.92$; $\mu = 11.76$; $\sigma_{\overline{X}} = 1.31$

(Continued)

(Continued)

3. For each of the following situations, calculate the population standard error of the mean ($\sigma_{\bar{X}}$) and the z-statistic (z), make a decision about the null hypothesis, and indicate the level of significance.
   a. $\bar{X} = 8.00$; $\mu = 5$; $\sigma = 3$; $N = 9$
   b. $\bar{X} = 3.00$; $\mu = 5$; $\sigma = 8$; $N = 16$
   c. $\bar{X} = 14.50$; $\mu = 12.75$; $\sigma = 4.41$; $N = 23$

7.4 A SECOND EXAMPLE FROM THE RESEARCH: UNIQUE INVULNERABILITY

How long do you expect to live? Each year, you may read about the number of traffic fatalities expected to occur on major holidays, or you may hear about someone who contracted a rare disease and died at a relatively young age. When you hear such forecasts or news events, you might think to yourself, "That happens to other people, not to me." One study examined the tendency people have to "distort information so that negative human outcomes are less likely to happen to us than to other people" (Snyder, 1997, p. 197).

C. R. Snyder, a researcher at the University of Kansas, set out to demonstrate this bias, called "unique invulnerability," in a classroom exercise. Dr. Snyder was interested in whether people display the unique invulnerability bias when it comes to predicting how long they expect to live. He hypothesized that, when asked to predict their age at the time of their death, people will provide estimates greater than the average life expectancy in the population.

For the sample in his study, which we will refer to as the unique invulnerability study, Dr. Snyder used 17 graduate students (4 men and 13 women). The variable of interest in this study was the estimated age of death: how old (in years) each student expected to be when he or she died. To collect the data for this study, Dr. Snyder "began by informing the class that the actuarially predicted age of death for U.S. citizens (men and women together) is 75 years. . . . After delivering this information, I asked students to write down (anonymously) their estimated ages of death . . . on a blank slip of paper" (Snyder, 1997, p. 198). The estimated age of death for the 17 students is listed in Table 7.2.

Analyzing the data from the study is facilitated by constructing a grouped frequency distribution table and frequency polygon; these are provided in Figure 7.6. A frequency polygon is used because estimated age at death is a numerical variable measured at the ratio level of measurement. Looking at this figure, we see that 82% of the students' estimates of their age at death are greater than the stated population mean of 75 years, suggesting

TABLE 7.2 ● ESTIMATED AGE OF DEATH (IN YEARS) FOR 17 STUDENTS

Student	Estimated Age of Death	Student	Estimated Age of Death	Student	Estimated Age of Death
1	**85**	7	**86**	13	**90**
2	**92**	8	**90**	14	**100**
3	**75**	9	**84**	15	**80**
4	**80**	10	**83**	16	**83**
5	**84**	11	**68**	17	**80**
6	**73**	12	**95**		

FIGURE 7.6 ● GROUPED FREQUENCY DISTRIBUTION TABLE AND FREQUENCY POLYGON OF ESTIMATED AGE OF DEATH FOR 17 STUDENTS

(a) Grouped Frequency Distribution Table

Estimated Age of Death	*f*	%
> 95	1	6%
91–95	2	12%
86–90	3	18%
81–85	5	28%
76–80	3	18%
71–75	2	12%
66–70	1	6%
< 66	0	0%
Total	17	100%

(b) Frequency Polygon

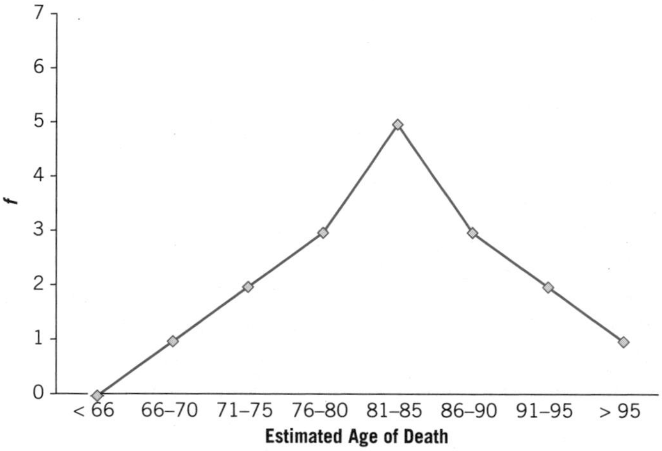

that they may indeed be displaying the unique invulnerability bias. In terms of the shape of the distribution, it appears the estimates are somewhat normally distributed, ranging from 66 years to 100 years, with the center of the distribution in the 81- to 85-year range. Before moving on, we strongly recommend you look at the table and figure and come up with an "eyeball estimate" of the sample mean and standard deviation.

Following is a numerical summary of the data for the study, achieved by calculating the sample mean ($\overline{X}$) and standard deviation (s):

$$\begin{aligned}\overline{X} &= \frac{\Sigma X}{N} \\ &= \frac{85+92+75+\ldots+80+83+80}{17} = \frac{1428}{17} \\ &= 84.00\end{aligned}$$

$$s = \sqrt{\frac{\Sigma(X-\bar{X})^2}{N-1}}$$

$$= \sqrt{\frac{(85-84.00)^2 + (92-84.00)^2 + \ldots + (83-84.00)^2 + (80-84.00)^2}{17-1}}$$

$$= \sqrt{\frac{1.00 + 64.00 + \ldots + 1.00 + 16.00}{16}} = \sqrt{\frac{1{,}026.00}{16}} = \sqrt{64.13}$$

$$= 8.01$$

Looking at the descriptive statistics, we see that the average estimated age of death of the 17 students in this sample ($M = 84.00$) is higher than the hypothesized population average of 75. The value for the standard deviation ($s = 8.01$) shows there is variability in the students' estimates, with the majority of the estimates between 76 and 92.

In the reading skills study discussed earlier in this chapter, the sample mean was transformed into a *z*-statistic that was evaluated using the standard normal distribution. The *z*-statistic is used to test the difference between a sample mean and a population mean when the population standard deviation for the variable is known. However, for the vast majority of variables studied by researchers, the population standard deviation is not known. The variables we used to illustrate *z*-scores and *z*-statistics, SAT scores and scores on the *Reading Success Level A* program, were administered to thousands of students. For the estimated age of death variable used in the unique invulnerability study, the population standard deviation is not presumed to be known. Therefore, to test the study's research hypothesis, the sample mean must be transformed into a different type of statistic that is based on a different type of distribution. This distribution, known as the *t*-distribution, is introduced in the next section.

7.5 INTRODUCTION TO THE *t*-DISTRIBUTION

As was the case in the reading skills study, the primary goal of the unique invulnerability study was to evaluate the mean of a sample based on the difference between the sample mean and the population mean. In essence, we're asking the question, "Does the difference between the sample's mean estimated age at death of 84.00 years and the hypothesized population mean of 75 years provide support for the unique invulnerability bias?" We evaluate the sample mean by transforming it into a statistic; once the statistic has been calculated, we can determine whether the probability of the statistic is low enough for us to decide the difference between the sample mean and the population mean is not due to chance, random factors but instead is statistically significant.

When the population standard deviation for the variable is not known, the sample mean is transformed into a statistic known as the *t*-statistic. The distribution of values for the *t*-statistic is known as the **Student *t*-distribution**, defined as the distribution of values of the *t*-statistic based on an infinite number of samples of size *N* randomly drawn from the population. The Student *t*-distribution was developed by W. S. Gosset, who wrote under the pen name "Student." Figure 7.7 illustrates the *t*-distribution, using a sample size of $N = 17$ as an example.

Characteristics of the Student *t*-Distribution

The Student *t*-distribution is a theoretical distribution that is created in a similar manner to the sampling distribution of the mean discussed earlier in this chapter. Using the unique invulnerability study to illustrate this distribution, imagine that the average estimated age of death in the population is 75. We draw a random sample from the population, calculate the mean estimated age of death for the sample, and transform the sample mean into a *t*-statistic. What should be the value of the *t*-statistic calculated from a sample? Similar to the *z*-statistic discussed earlier, because there should be no difference between the sample mean and the population mean, the

FIGURE 7.7 ● EXAMPLE OF THE STUDENT *t*-DISTRIBUTION

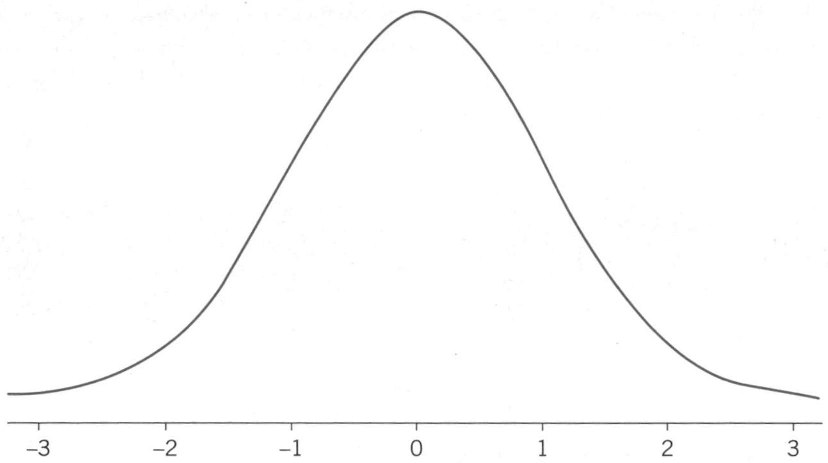

t-statistic should be zero (0). But because of the random, chance factors associated with sampling error, we know there will be variability in the values of *t*-statistics such that there is a distribution of *t*-statistics.

The Student *t*-distribution shares many of the characteristics of the standard normal distribution. First, in terms of modality, looking at the center of the distribution in Figure 7.7, we see that the mean of the Student *t*-distribution is equal to 0. Another similarity to the standard normal distribution is that the Student *t*-distribution is symmetrical, with an infinite number of values in both tails. However, unlike the standard normal distribution, there is not just one *t*-distribution but instead a *family* of *t*-distributions: different distribution of *t*-statistics for each sample size. Because *t*-statistics involve estimating the population from a sample, and this estimate is a function of the size of the sample, it makes sense that there will be a different distribution of *t*-statistics for different sample sizes. This is similar to what we saw with the binomial distribution in Chapter 6, where, for example, the distribution of the number of heads in 12 coin flips was not the same as the distribution of the number of heads in 30 coin flips.

FIGURE 7.8 ● DISTRIBUTION OF *t* FOR DIFFERENT SAMPLE SIZES

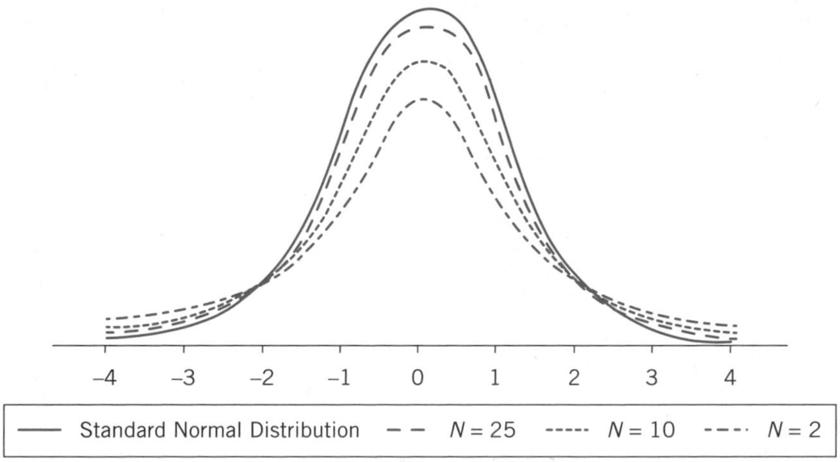

Source: Adapted from Pagno, R.R. (2009).

The three dashed lines in Figure 7.8 represent the *t*-distribution for three sample sizes: $N = 2$, $N = 10$, and $N = 25$; the solid line in this figure is the standard normal distribution. As we see, as the sample size increases, the more the *t*-distribution resembles the standard normal distribution. This is because the larger the sample, the more the sample resembles the population; the more the sample resembles the population, the more a distribution of statistics based on the sample (i.e., the *t*-distribution) resembles a distribution of statistics based on a population (i.e., the standard normal distribution).

LEARNING CHECK 4:

Reviewing What You've Learned So Far

1. Review questions
   a. What are the main characteristics of the Student *t*-distribution?
   b. In what ways is the Student *t*-distribution similar to the standard normal distribution?
   c. Why is there a different distribution of *t*-statistics for different sample sizes?
   d. What happens to the shape of the *t*-distribution as the sample size grows larger?

7.6 INFERENTIAL STATISTICS: TESTING ONE SAMPLE MEAN (σ NOT KNOWN)

As in the reading skills study, the unique invulnerability study hypothesizes a difference between a sample mean and a population mean. This difference is tested by calculating and testing an inferential statistic using the same four steps as before: state the null and alternative hypotheses (H_0 and H_1), make a decision about the null hypothesis, draw a conclusion from the analysis, and relate the result of the analysis to the research hypothesis. However, several of these steps contain differences due to the fact that, in this situation, the population standard deviation for the variable is not known. As we work through the steps, these differences will be highlighted and explained.

State the Null and Alternative Hypotheses (H_0 and H_1)

In testing one sample mean, the null hypothesis (H_0) states a value of the population mean μ that will be used to evaluate the sample mean. In this example, if the unique invulnerability bias does not exist, the estimated age of death for this sample should be the same as the actual average age of death in the population. As mentioned earlier, the average age of death in the population of U.S. citizens was claimed to be 75. Thus, the null hypothesis may be stated as

$$H_0: \mu = 75$$

The alternative hypothesis (H_1) states a mutually exclusive alternative to the null hypothesis. In the unique invulnerability study, one way to state the alternative hypothesis (H_1) reflects the belief that the estimated age of death in this sample is *not* the same as the actual average age of death in the population. This difference is represented by the following alternative hypothesis:

$$H_1: \mu \neq 75$$

As the study's research hypothesis suggests that the estimated age of death in the sample should be *greater* than the population mean, the alternative hypothesis could have been stated as the directional $H_1: \mu > 75$. But as we mentioned earlier in this chapter, using a nondirectional (two-tailed) alternative hypothesis implies the null hypothesis will be rejected if the sample mean is either greater than *or* less than the population mean. In the unique invulnerability study, this allows for the possibility that the sample may not display the unique invulnerability bias and may in fact give estimates of death that are *lower* than the average age of death in the population.

Make a Decision About the Null Hypothesis

The next step is to decide whether the difference between the sample mean and the population mean is statistically significant. The steps involved in making this decision when the population standard deviation is not known are as follows:

- calculate the degrees of freedom (*df*);
- set alpha, identify the critical values, and state a decision rule;
- calculate a statistic: *t*-test for one mean;
- make a decision whether to reject the null hypothesis; and
- determine the level of significance.

The last four steps are the same as for the *z*-test for one mean. However, when the population standard deviation is unknown, making the decision about the null hypothesis begins by calculating what are known as degrees of freedom. The next section defines and illustrates this concept.

Calculate the Degrees of Freedom (*df*)

The **degrees of freedom (*df*)** is defined as the number of values or quantities that are free to vary when a statistic is used to estimate a parameter. This step is included in this example because the standard deviation (s) is being used to estimate the population standard deviation (σ). The concept of degrees of freedom will appear throughout many of the remaining chapters of this book as different research situations and statistical procedures are discussed.

Let's illustrate the concept of degrees of freedom with a simple demonstration. Imagine a friend makes the following request of you: "Tell me any three numbers," to which you respond, "2, 35, and —". But your friend interrupts you before you can complete your response. "Stop!" he shouts. "I forgot to tell you that the three numbers must add up to 99." Once this has been said, you don't need to tell him a third number because it can only be 62. In this example, the first two numbers you chose were free to vary and could, in fact, have had any value. However, your friend's constraint, in this case the sum of 99, meant the value of the third number was no longer free to vary but was instead fixed at 62. In this example, you had two "degrees of freedom."

The concept of degrees of freedom is relevant to inferential statistics because these statistics involve estimating population parameters from data collected from samples. In the unique invulnerability example, from a sample of 17 participants ($N = 17$), the standard deviation of 8.01 is an estimate of an unknown population standard deviation. As a reminder, below is the formula for the standard deviation (s):

$$s = \sqrt{\frac{\Sigma(X - \overline{X})^2}{N - 1}}$$

We can see that the numerator of this formula involves summing deviations of scores from the mean $(X - \overline{X})$. In Chapter 4, we saw that, for any set of data, the sum of deviations $(\Sigma(X - \overline{X}))$ is equal to 0. Consequently, in a sample of 17 participants, although the first 16 deviations can be any value, the deviation of the last participant must enable the sum of the deviations to be equal to zero. Consequently, although the first 16 scores are free to vary, the 17th is not; it is predetermined by the first 16 scores.

Formula 7-3 calculates the degrees of freedom (*df*) for one sample:

$$df = N - 1 \tag{7-3}$$

where *N* is the size of the sample. In the unique invulnerability example,

$$\begin{aligned} df &= N - 1 \\ &= 17 - 1 \\ &= 16 \end{aligned}$$

Therefore, when the sample size *N* is equal to 17, there are 16 degrees of freedom. Different formulas for the degrees of freedom will be presented in later chapters of this book as different statistical procedures are introduced.

Set Alpha (α), Identify the Critical Values, and State a Decision Rule

This second step in making the decision about the null hypothesis begins by setting alpha (α), which is the probability of the statistic needed to reject the null hypothesis. As is tradition, alpha (α) will be set to .05, meaning that the null hypothesis will be rejected if the value of the calculated statistic has less than a .05 probability of occurring under the assumption that the null hypothesis is true. Furthermore, because the alternative hypothesis in this example is nondirectional (H_1: $\mu \neq 75$), alpha can be referred to as "$\alpha = .05$ (two-tailed)."

Once alpha has been determined, we need to identify the values of the *t*-statistic whose probability is low enough to lead to the decision to reject the null hypothesis. But because there are an infinite number of possible values of *t* and an infinite number of possible sample sizes, it would be impractical to have a table that lists the probability for every value of *t* for every sample size. Instead, because we are only interested in values of *t* that affect the decision about the null hypothesis, the only values of concern are the critical values, defined as the values of a statistic that separate the regions of rejection and nonrejection. A table of critical values for the *t*-distribution is provided in Table 3 in the back of this book.

To determine the critical values of the *t*-statistic, three pieces of information are needed: the degrees of freedom (*df*), alpha (α), and the directionality of the alternative hypothesis (directional [one-tailed] or nondirectional [two-tailed]). The first column of Table 7.3, labeled *df*, lists different degrees of freedom. Because there is a different *t*-distribution for each sample size, each degrees of freedom has its own critical values. The bottom row of this table includes the infinity (∞) symbol; this does not imply that a sample may actually contain an infinite number of degrees of freedom; rather, it provides the critical values for a population assumed to be normally distributed.

The eight columns to the right of the *df* column list different values of alpha (.10, .05, .01, .001) for different levels of significance. The first four of these columns are used when the alternative hypothesis (H_1) is directional (one-tailed); the last four columns are used for a nondirectional (two-tailed) alternative hypothesis.

For the unique invulnerability study, the critical values may be identified by moving down the *df* column until we reach the appropriate row for this example: $df = 16$. Next, because the alternative hypothesis is nondirectional (H_1: $\mu \neq 75$), we move to the right until we are under the heading, "Level of Significance for Two-Tailed Test." Within this set of columns, for $\alpha = .05$, we find a critical *t* value of 2.120. Therefore, for the unique invulnerability study, the critical values may be stated as follows:

$$\text{For } \alpha = .05 \text{ (two-tailed) and } df = 16, \text{ critical values} = \pm 2.120$$

The critical values and the regions of rejection and nonrejection for the unique invulnerability study are illustrated in Figure 7.9.

TABLE 7.3 ● PORTIONS OF THE TABLE OF CRITICAL VALUES OF THE *t*-STATISTIC

	Level of Significance for One-Tailed Test				Level of Significance for Two-Tailed Test			
df	.10	.05	.01	.001	.10	.05	.01	.001
1	3.078	6.314	31.821	318.310	6.314	12.706	63.657	636.619
2	1.886	2.920	6.965	22.326	2.920	4.303	9.925	31.598
3	1.638	2.353	4.541	10.213	2.353	3.182	5.841	12.941
4	1.533	2.132	3.747	7.173	2.132	2.776	4.604	8.610
5	1.476	2.015	3.365	5.893	2.015	2.571	4.032	6.859
11	1.363	1.796	2.718	4.025	1.796	2.201	3.106	4.437
12	1.356	1.782	2.681	3.930	1.782	2.179	3.055	4.318
13	1.350	1.771	2.650	3.852	1.771	2.160	3.012	4.221
14	1.345	1.761	2.624	3.787	1.761	2.145	2.977	4.140
15	1.341	1.753	2.602	3.733	1.753	2.131	2.947	4.073
16	1.337	1.746	2.583	3.686	1.746	2.120	2.921	4.015
17	1.333	1.740	2.567	3.646	1.740	2.110	2.898	3.965
18	1.330	1.734	2.552	3.610	1.734	2.101	2.878	3.922
19	1.328	1.729	2.639	3.579	1.729	2.093	2.861	3.883
20	1.325	1.725	2.528	3.552	1.725	2.086	2.845	3.850
90	1.291	1.662	2.368	3.183	1.662	1.987	2.632	3.403
120	1.289	1.658	2.358	3.232	1.658	1.980	2.617	3.373
∞	1.282	1.645	2.326	3.183	1.645	1.960	2.576	3.291

Source: Adapted from Pearson, E. S., & Hartley, H. O. (Eds.) (1966).

The next step in hypothesis testing is to state a decision rule that uses the critical values to explicitly specify the conditions under which the null hypothesis will be rejected. For the unique invulnerability study:

$$\text{If } t < -2.120 \text{ or } > 2.120, \text{ reject } H_0\text{; otherwise, do not reject } H_0$$

This decision rule implies that if the value of t calculated for the sample exceeds either of the critical values, it falls in the region of rejection and the null hypothesis will be rejected.

Calculate a Statistic: *t*-Test for One Mean

The next step is to calculate a value of the *t*-statistic using a statistical procedure known as the ***t*-test for one mean**, which tests the difference between a sample mean and a hypothesized population mean when the population standard deviation is not known:

$$t = \frac{\overline{X} - \mu}{s_{\overline{X}}} \tag{7-4}$$

where $\overline{X}$ is the sample mean, μ is the population mean, and $s_{\overline{X}}$ is the standard error of the mean.

FIGURE 7.9 ● CRITICAL VALUES FOR THE UNIQUE INVULNERABILITY STUDY (α = .05 [TWO-TAILED] AND df = 16)

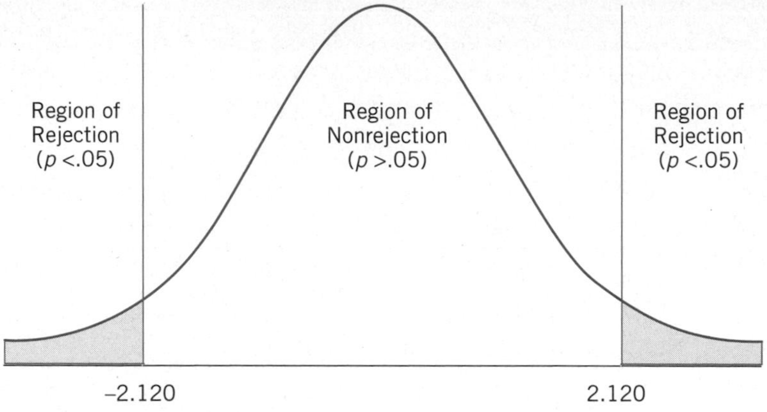

Formula 7-4 closely resembles the formula for the z-test for one mean (Formula 7-2). The numerator in both formulas is the difference between the sample mean and the population mean $(\overline{X} - \mu)$, which is the primary interest of this analysis. The denominator of Formula 7-4, however, is the **standard error of the mean ($s_{\overline{X}}$)**, which is the standard deviation of the sampling distribution of the mean when the standard deviation (s) is used to estimate an unknown population standard deviation (σ). The standard error of the mean estimates the variability of sample means in the population and helps us evaluate the mean of our sample. Formula 7-5 provides the formula for $s_{\overline{X}}$:

$$s_{\overline{X}} = \frac{s}{\sqrt{N}} \qquad (7\text{-}5)$$

where s is the standard deviation and N is the size of the sample. This formula differs from the formula for the population standard error of the mean (Formula 7-2) by having the standard deviation (s) rather than the population standard deviation (σ) in the numerator.

For the unique invulnerability study, for the sample of 17 students, the standard deviation of s = 8.01 is used to calculate the standard error of the mean as follows:

$$\begin{aligned} s_{\overline{X}} &= \frac{s}{\sqrt{N}} \\ &= \frac{8.01}{\sqrt{17}} = \frac{8.01}{4.12} \\ &= 1.94 \end{aligned}$$

Once the standard error of the mean ($s_{\overline{X}}$) has been calculated, it is inserted into Formula 7-4, along with the sample ($\overline{X}$) and population (μ) means to calculate a value of the t-statistic. For the unique invulnerability study:

$$\begin{aligned} t &= \frac{\overline{X} - \mu}{s_{\overline{X}}} \\ &= \frac{84.00 - 75}{1.94} = \frac{9.00}{1.94} \\ &= 4.63 \end{aligned}$$

LEARNING CHECK 5:
Reviewing What You've Learned So Far

1. Review questions
   a. What is the generic definition of degrees of freedom (*df*)?
   b. Under what situations are degrees of freedom (*df*) calculated?
   c. Why is there a table of critical values for the *t*-statistic rather than a single critical value?
   d. What information is needed to identify the critical value for the *t*-statistic?
   e. Under what situations would you conduct the *z*-test for one mean or the *t*-test for one mean?
   f. What are the differences between the formulas for the population standard error of the mean and the standard error of the mean?
   g. What are the similarities and differences between the formulas for the *z*-test and the *t*-test?
2. For each of the following situations, calculate the degrees of freedom and identify the critical values (assume $\alpha = .05$ [two-tailed]).
   a. $N = 9$
   b. $N = 13$
   c. $N = 18$
   d. $N = 21$
3. For each of the following situations, calculate the standard error of the mean ($s_{\bar{X}}$).
   a. $s = 3.00$; $N = 16$
   b. $s = 7.50$; $N = 25$
   c. $s = 20.00$; $N = 50$
   d. $s = 3.41$; $N = 38$
4. For each of the following situations, calculate the *t*-statistic (*t*).
   a. $\bar{X} = 14.00$; $\mu = 9$; $s = 12.00$; $N = 36$
   b. $\bar{X} = 3.00$; $\mu = 5$; $s = 6.00$; $N = 16$
   c. $\bar{X} = .56$; $\mu = .65$; $s = .18$; $N = 17$

Now that the sample mean has been transformed into a *t*-statistic, it can be evaluated to make a decision about the null hypothesis; this is discussed in the next section.

Make a Decision Whether to Reject the Null Hypothesis

Now that a value of the *t*-statistic has been calculated, we can make a decision whether to reject the null hypothesis. For the unique invulnerability study, comparing the calculated value of the *t*-statistic with the critical values leads to the following decision:

$$t = 4.63 > 2.120 \therefore \text{ reject } H_0 \ (p < .05)$$

Because the sample's *t*-statistic of 4.63 exceeds the critical value 2.120, it falls in the region of rejection, implying that its probability is sufficiently low ($p < .05$) to lead to the decision to reject the null hypothesis. This decision implies that the difference between the mean of the sample ($\bar{X} = 84.00$) and the hypothesized mean of the population ($\mu = 75$) is statistically significant, meaning it is not the result of random, chance factors.

Determine the Level of Significance

To be as precise as possible, when the null hypothesis has been rejected, we determine whether the probability of the *t*-statistic is not only less than .05 but also less than .01. For the unique invulnerability study, the .01

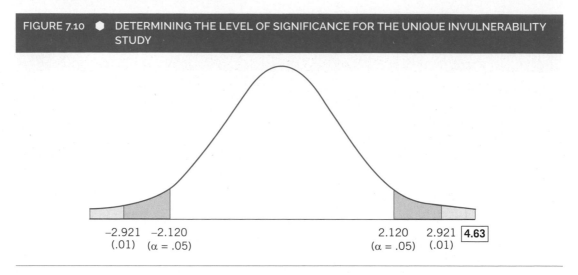

FIGURE 7.10 ● DETERMINING THE LEVEL OF SIGNIFICANCE FOR THE UNIQUE INVULNERABILITY STUDY

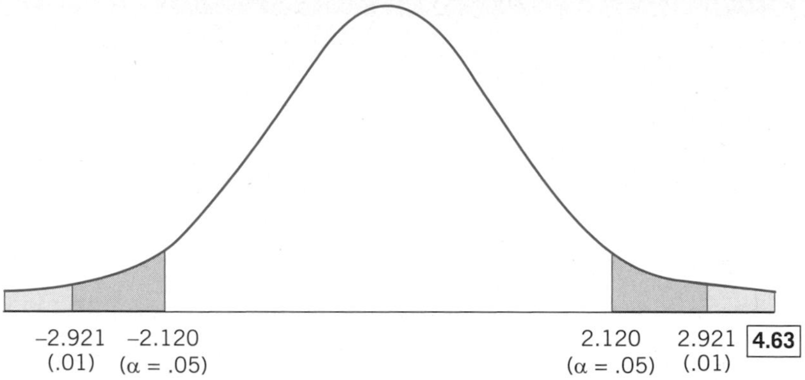

critical value is determined by returning to the table of critical values in Table 3 in the back of this book. After moving down to the *df* = 16 row, move to the right until we are under the .01 critical values of *t* for a two-tailed alternative hypothesis; here we find the critical value 2.921. Below, we compare the calculated value of *t* of 4.63 with the .01 critical value:

$$t = 4.63 > 2.921 \therefore p < .01$$

Figure 7.10 illustrates the location of *t* relative to the .05 and .01 critical values. As can be seen from this figure, because 4.63 is greater than both 2.120 and 2.921, its probability is not only less than .05 but also less than .01.

Draw a Conclusion From the Analysis

We can state the following conclusion regarding the results of the statistical analysis for the unique invulnerability study:

> The average estimated age of death (M = 84.00 years) for the 17 class members in this sample was significantly greater than the actual population average of 75 years, $t(16) = 4.63, p < .01$.

The above sentence describes the variable that was analyzed ("The average estimated age of death"), who was included in the analysis ("the 17 class members in this sample"), descriptive statistics of the variables ("(M = 84.00 years) . . . the actual population average of 75 years"), the nature and direction of the findings ("was significantly greater than the actual population average"), and information about the inferential statistic: $t(16) = 4.63, p < .01$, which indicates the type of statistic calculated (t), degrees of freedom (16), value of the statistic (4.63), and level of significance ($p < .01$).

Relate the Result of the Analysis to the Research Hypothesis

The final step in the process of hypothesis testing is to relate the results of the statistical analysis to the research hypothesis that initiated the analysis. By finding that the mean estimate of age of death in this sample was significantly greater than the actual population average, we can relate the results of this analysis to the study's research hypothesis the following way:

> The result of this analysis supports the research hypothesis that people will provide estimates of age of death greater than the average life expectancy in the population.

Dr. Snyder provided the following description of his conclusions from the study:

> These results replicate other recent classroom-demonstration findings in which students maintain self-serving positive illusions in spite of knowing about the relevant research . . . instructors should help students to find ways of abandoning biases, such as unique invulnerability, especially when such biases increase the potential for harm in students' lives. (Snyder, 1997, p. 199)

Assumptions of the *t*-Test for One Mean

Given the similarities between the two procedures, it's not surprising that the assumptions of the *z*-test for one mean (random sampling, interval or ratio scale of measurement, and normality) also apply to the *t*-test for one mean. One implication of the assumption of normality, which is the assumption that scores for the variable are approximately normally distributed in the population, is that distributions of scores for samples drawn from these populations are also expected to be approximately normal. Given we are estimating the population standard deviation from a sample, it's particularly important that the sample approximates a normal distribution. As we mentioned earlier, this is more likely to occur when the sample is sufficiently large ($N \geq 30$).

SUMMARY BOX 7.2 CONDUCTING THE *t*-TEST FOR ONE MEAN (UNIQUE INVULNERABILITY EXAMPLE)

State the null and alternative hypotheses (H_0 and H_1)

$$H_0: \mu = 75 \qquad H_1: \mu \neq 75$$

Make a decision about the null hypothesis

Calculate the degrees of freedom (*df*)

$$df = N - 1 = 17 - 1 = 16$$

Set alpha (α), identify the critical values, and state a decision rule

If $t < -2.120$ or > 2.120, reject H_0; otherwise, do not reject H_0

Calculate a statistic: *t*-test for one mean

Calculate the standard error of the mean ($s_{\bar{X}}$)

$$s_{\bar{X}} = \frac{s}{\sqrt{N}} = \frac{8.01}{\sqrt{17}} = \frac{8.01}{4.12} = 1.94$$

Calculate the *t*-statistic (*t*)

$$t = \frac{\bar{X} - \mu}{s_{\bar{X}}} = \frac{84.00 - 75}{1.94} = \frac{9.00}{1.94} = 4.63$$

Make a decision whether to reject the null hypothesis

$$t = 4.63 > 2.120 \therefore \text{reject } H_0 \ (p < .05)$$

Determine the level of significance

$$t = 4.63 > 2.921 \therefore p < .01$$

Draw a conclusion from the analysis

The average estimated age of death ($M = 84.00$ years) for the 17 class members in this sample was significantly greater than the actual population average of 75 years, $t(16) = 4.63$, $p < .01$.

Relate the result of the analysis to the research hypothesis

The result of this analysis supports the research hypothesis that people will provide estimates of age of death greater than the average life expectancy in the population.

Summary

The process of testing the mean of a sample when σ is unknown by conducting the *t*-test for one mean is summarized in Box 7.2, using the unique invulnerability study as an example. This chapter has presented several research studies that used the process of hypothesis testing first introduced in Chapter 6. The purpose of the next section is to further our discussion of another topic introduced in this earlier chapter: factors that directly affect the decision made about the null hypothesis.

LEARNING CHECK 6:

Reviewing What You've Learned So Far

1. For each of the following situations, calculate the degrees of freedom (*df*), identify the critical values (assume $\alpha = .05$ [two-tailed]), calculate the *t*-statistic (*t*), make a decision about the null hypothesis (reject, do not reject), and indicate the level of significance ($p > .05$, $p < .05$, $p < .01$).
   a. $\overline{X} = 7.00$; $\mu = 4$; $s_{\overline{X}} = 1.25$; $N = 25$
   b. $\overline{X} = 12.00$; $\mu = 10$; $s_{\overline{X}} = 1.45$; $N = 16$
   c. $\overline{X} = 1.69$; $\mu = 2.08$; $s_{\overline{X}} = .11$; $N = 19$
2. For each of the following situations, calculate the degrees of freedom (*df*), identify the critical values (assume $\alpha = .05$ [two-tailed]), calculate the standard error of the mean ($s_{\overline{X}}$), calculate the *t*-statistic (*t*), make a decision about the null hypothesis, and indicate the level of significance.
   a. $\overline{X} = 16.75$; $\mu = 12.75$; $s = 4.00$; $N = 11$
   b. $\overline{X} = 3.27$; $\mu = 2.98$; $s = 1.73$; $N = 27$
   c. $\overline{X} = 7.82$; $\mu = 10.56$; $s = 5.2$; $N = 19$

7.7 FACTORS AFFECTING THE DECISION ABOUT THE NULL HYPOTHESIS

Within hypothesis testing, one of two decisions is made: reject the null hypothesis or do not reject the null hypothesis. In this chapter, rejecting the null hypothesis implies that a sample mean is significantly different from a population mean, which may in turn imply that a study's research hypothesis is supported. In Chapter 6, we introduced three factors that directly affect the likelihood of rejecting the null hypothesis:

- sample size,
- alpha, and
- the directionality of the alternative hypothesis.

This section will use the research situations covered in this chapter to continue and expand on our earlier discussion of these factors. It is critical to understand these factors as they are, to some degree, under the control of researchers.

Sample Size

Given its repeated appearance throughout the data collection and data analysis stages of the research process, sample size clearly plays a critical role in statistical analyses. As was mentioned in Chapter 6, the relationship between sample size and the decision regarding the null hypothesis is as follows:

The larger the sample size, the greater the likelihood of rejecting the null hypothesis.

In conducting an inferential statistic such as the *t*-test for one mean, sample size influences the decision about the null hypothesis in two ways. First, keeping all other aspects of the analysis constant, increasing the size of the sample increases the numeric value of the statistic, which in turn increases the likelihood of rejecting the null hypothesis. Second, increasing the size of the sample decreases the critical values of the statistic, which also increases the likelihood of rejecting the null hypothesis.

To illustrate the relationship between sample size and the numeric value of the statistic, imagine we conduct two studies of unique invulnerability. In both studies, the sample means are the same ($M = 80.00$), the standard deviations are the same ($s = 7.00$), and the hypothesized population mean is the same ($\mu = 75$). However, the sample sizes in the two studies are different: $N = 20$ and $N = 10$, respectively. In Table 7.4 the standard error of the mean ($s_{\overline{X}}$) and the *t*-test for one mean are calculated for the two studies.

Looking at Table 7.4 using the larger sample of $N = 20$ results in a smaller value of the standard error of the mean ($s_{\overline{X}} = 1.57$ vs. 2.22), which leads to a larger value of the *t*-statistic ($t = 3.19$ vs. 2.25). Even though we have not changed the difference between the sample mean and the population mean, increasing the sample size has increased the value of the *t*-statistic, which in turn increases the likelihood of rejecting the null hypothesis.

Next, in terms of the relationship between sample size and the critical values, turn to the table of critical values for the Student *t*-distribution in Table 7.3. Moving down the *df* column, we see that as the number of degrees of freedom increases, the critical value becomes smaller. This implies that as the sample gets larger, the value of the *t*-statistic needed to reject the null hypothesis gets smaller, thereby increasing the likelihood of rejecting the null hypothesis. Why do the critical values change as a function of sample size? As we mentioned earlier in this chapter, increasing the size of a sample decreases the effect of random factors that create variability among sample means; this results in less variability among sample means and less variability among statistics such as the *t*-statistic (see Figure 7.8). Because extreme values of a statistic are less likely to occur in larger samples, a less extreme value is needed to reject the null hypothesis.

Alpha (α)

The relationship between alpha (α), the probability of a statistic needed to reject the null hypothesis, and the decision about the null hypothesis may be summarized as follows:

The larger the value of alpha, the greater the likelihood of rejecting the null hypothesis.

This is because an increase in alpha results in smaller critical values that increase the size of the region of rejection, which increases the likelihood of rejecting the null hypothesis.

Figure 7.11 illustrates the critical values for a sample size of $N = 10$ ($df = 9$) for two values of alpha: .05 and .10. Setting alpha to .10 implies that the null hypothesis will be rejected if the probability of the statistic is less than .10; this definition of a "low" probability is less stringent than $\alpha = .05$. In this figure, we see that the critical values are

TABLE 7.4 ● THE EFFECT OF SAMPLE SIZE ON THE NUMERIC VALUE OF THE *t*-TEST FOR ONE MEAN

$N = 20$	$N = 10$
$s_{\overline{X}} = \frac{7.00}{\sqrt{20}} = \frac{7.00}{4.47} = 1.57$	$s_{\overline{X}} = \frac{7.00}{\sqrt{10}} = \frac{7.00}{3.16} = 2.22$
$t = \frac{80.00 - 75}{1.57} = \frac{5.00}{1.57} = 3.19$	$t = \frac{80.00 - 75}{2.22} = \frac{5.00}{2.22} = 2.25$

FIGURE 7.11 ● RELATIONSHIP BETWEEN ALPHA (α) AND CRITICAL VALUES FOR t ($df = 9$)

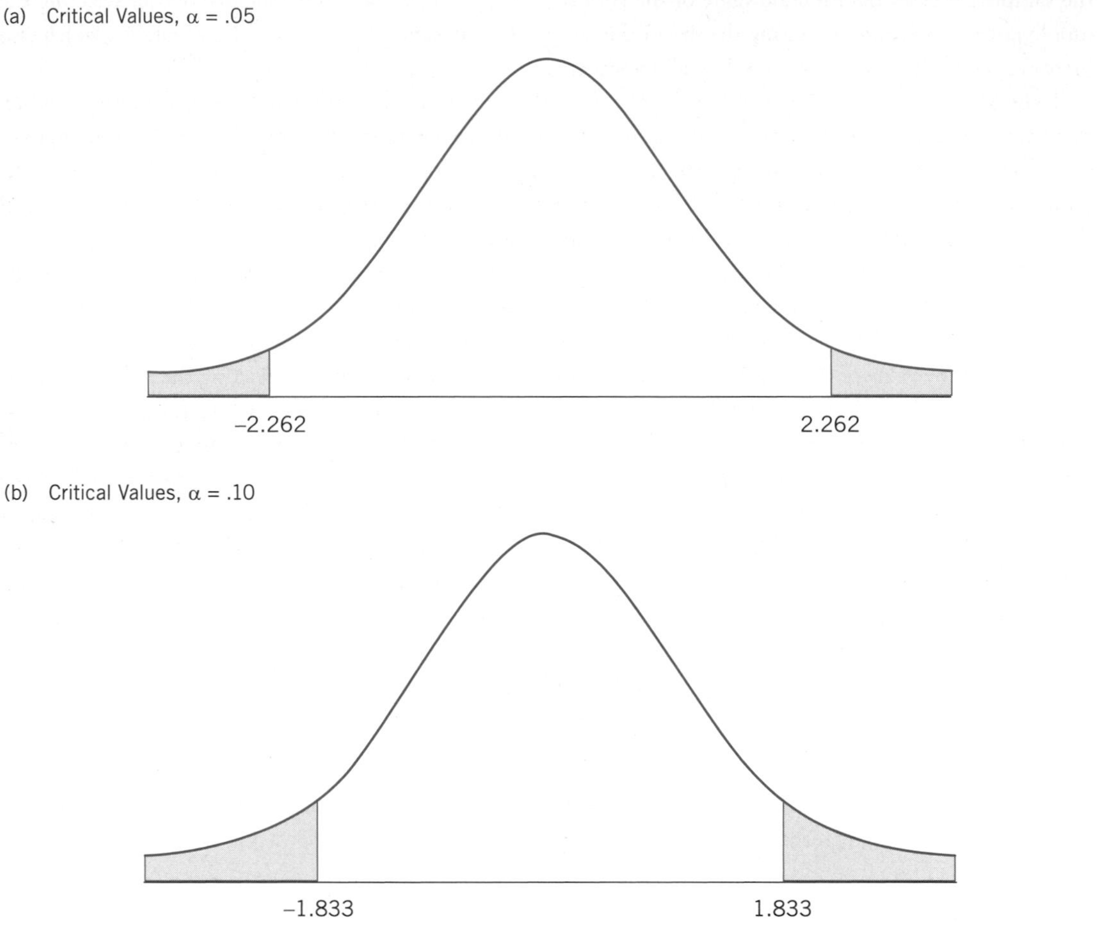

smaller for $\alpha = .10$ (±1.833) than for $\alpha = .05$ (±2.262). Because the region of rejection is larger for $\alpha = .10$, a smaller value of the statistic is needed to reject the null hypothesis, making it more likely the null hypothesis will be rejected.

Although it is tradition in academic research to set alpha to .05, researchers sometimes choose different definitions of a "low" probability. For example, researchers may believe using an alpha level of .05 is too strict, particularly in research situations where it may be difficult to have large samples. In situations such as these, the likelihood of rejecting the null hypothesis can be increased by setting alpha to .10 rather than .05. On the other hand, researchers might believe that $\alpha = .05$ is too lenient, choosing instead to reject the null hypothesis only when the statistic has a very small probability of occurring (i.e., $< .01$), thereby lowering the chances of rejecting the null hypothesis. Ultimately, it is each researcher's responsibility to set alpha at a level appropriate for the research situation at hand.

Directionality of the Alternative Hypothesis

The relationship between the directionality of the alternative hypothesis (H_1) and the decision about the null hypothesis may be stated the following way:

> There is a greater likelihood of rejecting the null hypothesis when the alternative hypothesis is directional (one-tailed) rather than nondirectional (two-tailed).

Using a directional alternative hypothesis results in a smaller critical value, which increases the likelihood of rejecting the null hypothesis when the difference between the sample mean and the population mean is in the hypothesized direction.

Critical values for a nondirectional and directional alternative hypothesis for $N = 10$ and $\alpha = .05$ are illustrated in Figure 7.12. In this figure, the positive critical value is smaller for the directional alternative hypothesis (1.833) than a nondirectional alternative hypothesis (2.262) because the 5% region of rejection is not split between the two ends of the distribution. Consequently, using a directional alternative hypothesis increases the likelihood of rejecting the null hypothesis when the result is in the predicted direction.

Using a one-tailed rather than a two-tailed alternative hypothesis may be a source of controversy, in that researchers could choose to use a one-tailed alternative hypothesis when the calculated value of their statistics is not large enough to reject the null hypothesis using a two-tailed alternative hypothesis, thereby leading to confusing conclusions regarding their statistical analyses. The decision to use a one-tailed or a two-tailed alternative hypothesis depends on several factors. The traditional and more conservative approach is to use a two-tailed alternative hypothesis, particularly if not enough is known about the topic or the population to make directional hypotheses. However, there are situations when using a one-tailed alternative hypothesis is appropriate. First, a large body of research may indicate the tendency for the predicted relationship between variables to be in only one direction. Also, it may not be possible for the relationship to exist in both directions. For example, if we conduct a study looking at the effects of a spray designed to greatly increase the length of one's hair, we may have no reason to expect the length of one's hair to *decrease* as a result of using the spray and consequently state an alternative hypothesis that only allows for an increase in length.

FIGURE 7.12 ● RELATIONSHIP BETWEEN DIRECTIONALITY OF THE ALTERNATIVE HYPOTHESIS AND CRITICAL VALUES FOR t ($df = 9$ AND $\alpha = .05$)

(a) Nondirectional (Two-Tailed) Alternative Hypothesis

(b) Directional (One-Tailed) Alternative Hypothesis

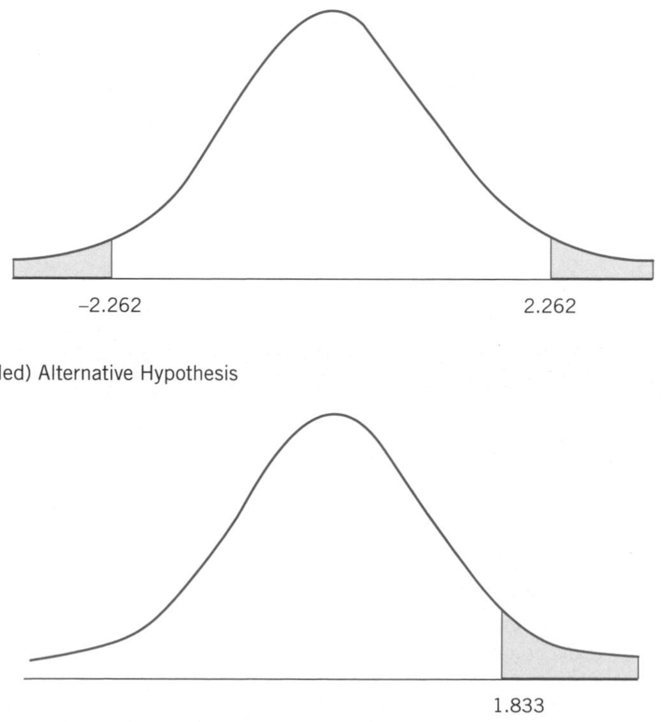

LEARNING CHECK 7:

Reviewing What You've Learned So Far

1. Review questions
   a. What are the two ways in which sample size affects the decision about the null hypothesis in testing the *t*-statistic?
   b. What is the relationship between alpha (α) and the decision about the null hypothesis?
   c. Why is it necessary to determine whether to use a one-tailed or two-tailed alternative hypothesis before conducting a statistical analysis rather than after?

At first glance, one solution would be to use a one-tailed alternative hypothesis but switch to a two-tailed alternative hypothesis when the results are in the opposite, unexpected direction. As logical as this may sound, this solution presents problems of its own because choosing this strategy essentially means using a 7½% region of rejection: 5% at one end of the distribution while still allowing for 2½% at the other end. This is roughly analogous to the captain of a softball team calling "heads" while the coin is in the air and then changing her mind after the coin hits the ground. The directionality of the alternative hypothesis should be based on sound, defensible reasoning *before* statistical analyses are conducted, without vacillating once the choice has been made.

7.8 LOOKING AHEAD

In this chapter, we have expanded our discussion of the research process in general and the process of hypothesis testing in particular. Using the *z*-test and *t*-test, we have tested the difference between a sample mean and a hypothesized population mean; later chapters will examine statistical procedures designed to test hypotheses in different research situations. In this chapter, we have also expanded on a discussion of issues relevant to hypothesis testing, such as factors that influence the decision about the null hypothesis. As we have seen, choices researchers make regarding sample size, alpha, and the directionality of the alternative hypothesis directly affect the conclusions drawn about research hypotheses. As such, hypothesis testing has been the source of concern and controversy among researchers. The next chapter discusses these concerns and presents an alternative to hypothesis testing.

7.9 Summary

To test hypotheses regarding the difference between a sample mean $(\overline{X})$ and a population mean (μ), one of two statistical techniques may be used: the ***z-test for one mean*** or the ***t-test for one mean***. The *z*-test is used when the population standard deviation (σ) for the variable being analyzed is known; the *t*-test is used when the population standard deviation is not known, and the standard deviation (*s*) is used to estimate σ.

Evaluating the difference between a sample mean and a population mean involves determining the probability of obtaining a particular value of the sample mean. To determine this probability, the ***sampling distribution of the mean***, the distribution of values of the sample mean when an infinite number of samples of size *N* are randomly selected from the population, is used. The sampling distribution of the mean is an example of a ***sampling distribution***, which is a distribution of statistics for samples randomly drawn from populations.

Three features of the sampling distribution of the mean are (1) the expected mean of the distribution is equal to the population mean (μ); (2) the distribution is an approximate normal distribution, even if the distribution of scores in the population is *not* normally distributed, assuming the samples are sufficiently large (typically defined as a sample size of at least $N = 30$); and (3) its variability is represented by the ***standard error of the mean***, which is the standard deviation of the sampling distribution of the mean. These features are defined by a statistical principle known as the ***central limit theorem***.

There are two types of standard error of the mean. The ***population standard error of the mean*** ($\sigma_{\bar{X}}$) is calculated when the population standard deviation (σ) is known; the ***standard error of the mean*** ($s_{\bar{X}}$) is calculated when the standard deviation (s) is used to estimate an unknown population standard deviation.

Making the decision about the null hypothesis for the *z*-test and the *t*-test involves many of the same steps; however, the *z*-test uses the standard normal distribution, whereas the *t*-test uses the ***Student t-distribution***, which is the distribution of values of the *t*-statistic based on an infinite number of samples of size *N* randomly drawn from the population. Using the Student *t*-distribution involves calculating the ***degrees of freedom (df)*** for a sample, which are the number of values or quantities that are free to vary when a statistic is used to estimate a parameter.

Three factors affecting the decision about the null hypothesis are sample size, alpha, and the directionality of the alternative hypothesis. In terms of sample size, the larger the sample size, the greater the likelihood of rejecting the null hypothesis. In terms of alpha, the larger the value of alpha, the greater the likelihood of rejecting the null hypothesis. Finally, there is a greater likelihood of rejecting the null hypothesis when the alternative hypothesis is directional (one-tailed) than when it is nondirectional (two-tailed). It is important to understand these factors as they are, to some degree, under the control of researchers.

7.10 Important Terms

sampling distribution of the mean (p. 187)
sampling distribution (p. 187)
standard error of the mean (p. 187)
central limit theorem (p. 188)
z-test for one mean (p. 192)
population standard error of the mean ($\sigma_{\bar{X}}$) (p. 192)
assumption of random sampling (p. 197)
assumption of interval or ratio scale of measurement (p. 198)
assumption of normality (p. 198)
robust (p. 198)
Student *t*-distribution (p. 202)
degrees of freedom (*df*) (p. 205)
t-test for one mean (p. 207)
standard error of the mean ($s_{\bar{X}}$) (p. 208)

7.11 Formulas Introduced in This Chapter

z-Test for One Mean

$$z = \frac{\bar{X} - \mu}{\sigma_{\bar{X}}} \qquad (7\text{-}1)$$

Population Standard Error of the Mean ($\sigma_{\bar{X}}$)

$$\sigma_{\bar{X}} = \frac{\sigma}{\sqrt{N}} \qquad (7\text{-}2)$$

Degrees of Freedom (*df*) for One Sample

$$df = N - 1 \qquad (7\text{-}3)$$

t-Test for One Mean

$$t = \frac{\overline{X} - \mu}{s_{\overline{X}}} \tag{7-4}$$

Standard Error of the Mean ($s_{\overline{X}}$)

$$s_{\overline{X}} = \frac{s}{\sqrt{N}} \tag{7-5}$$

7.12 Using SPSS

Testing One Sample Mean (σ Not Known): The Unique Invulnerability Study (7.6)

1. Define variable (name, # decimal places, label for the variable) and enter data for the variable.

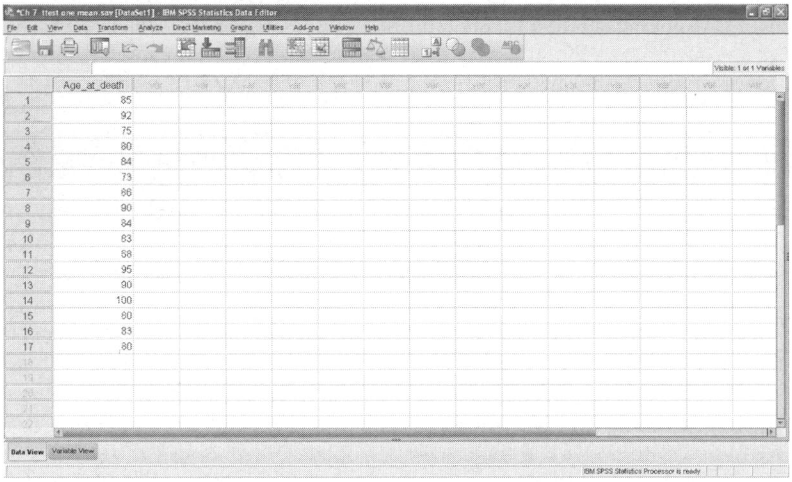

2. Select the t-test for one sample mean procedure within SPSS.

   How? (1) Click **Analyze** menu, (2) click **Compare Means**, and (3) click **One-Sample t Test**.

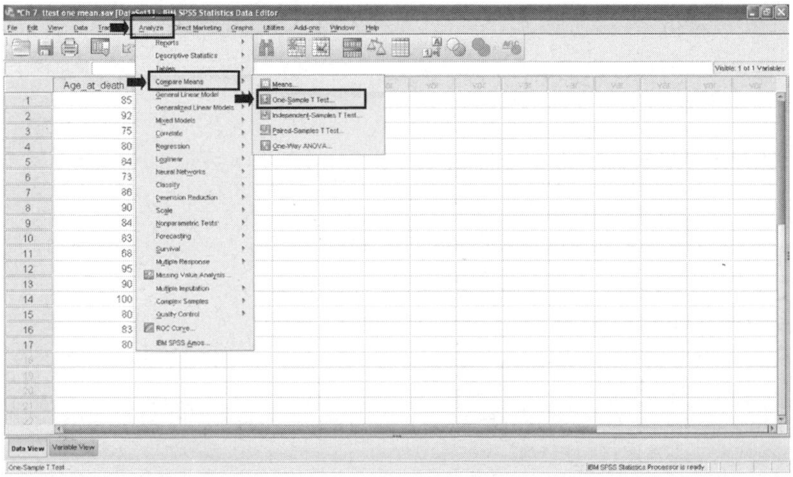

3. Select the variable to be analyzed and set the value of hypothesized population mean μ.

   How? (1) Click variable and →, (2) type value of hypothesized population mean (μ) in **Test Value** box, and (3) click **OK**.

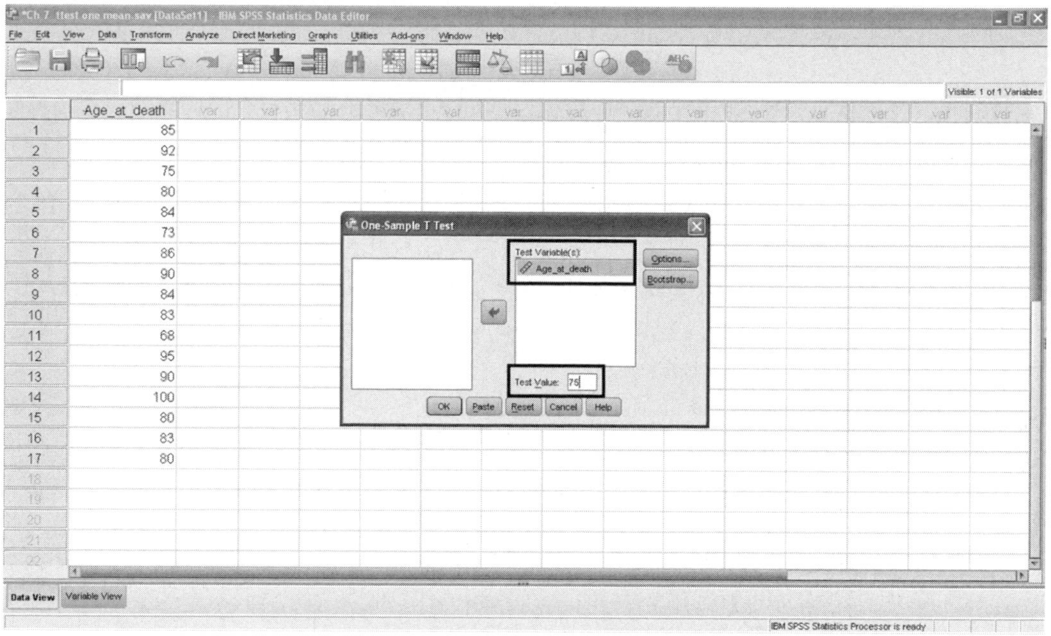

4. Examine output.

N — Mean — Standard deviation — Standard error of the mean

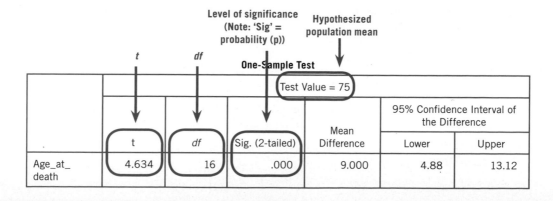

One-Sample Statistics

	N	Mean	Std. Deviation	Std. Error Mean
Age_at_death	17	84.00	8.008	1.942

t — *df* — Level of significance (Note: 'Sig' = probability (p)) — Hypothesized population mean

One-Sample Test

	Test Value = 75					
	t	*df*	Sig. (2-tailed)	Mean Difference	95% Confidence Interval of the Difference: Lower	95% Confidence Interval of the Difference: Upper
Age_at_death	4.634	16	.000	9.000	4.88	13.12

7.13 Exercises

1. For each of the following situations, calculate the population standard error of the mean ($\sigma_{\bar{X}}$).
   a. $\sigma = 8$; $N = 16$
   b. $\sigma = 12$; $N = 64$
   c. $\sigma = 2$; $N = 25$
   d. $\sigma = 3.72$; $N = 18$
   e. $\sigma = 32.86$; $N = 31$
2. For each of the following situations, calculate the population standard error of the mean ($\sigma_{\bar{X}}$).
   a. $\sigma = 18$; $N = 36$
   b. $\sigma = 9.42$; $N = 49$
   c. $\sigma = 1.87$; $N = 60$
   d. $\sigma = .91$; $N = 22$
   e. $\sigma = 21.43$; $N = 106$
3. For each of the following situations, calculate the *z*-statistic (*z*).
   a. $\bar{X} = 14.00$; $\mu = 11$; $\sigma = 6$; $N = 36$
   b. $\bar{X} = 7.00$; $\mu = 8$; $\sigma = 3$; $N = 9$
   c. $\bar{X} = 2.86$; $\mu = 2.69$; $\sigma = .40$; $N = 29$
   d. $\bar{X} = 10.12$; $\mu = 7.98$; $\sigma = 4.59$; $N = 26$
   e. $\bar{X} = 92.87$; $\mu = 101.55$; $\sigma = 20.65$; $N = 43$
4. For each of the following situations, calculate the *z*-statistic (*z*).
   a. $\bar{X} = 8.00$; $\mu = 5$; $\sigma = 6$; $N = 16$
   b. $\bar{X} = 4.00$; $\mu = 2$; $\sigma = 8$; $N = 25$
   c. $\bar{X} = 11.50$; $\mu = 9.25$; $\sigma = 5.75$; $N = 38$
   d. $\bar{X} = .95$; $\mu = .82$; $\sigma = .31$; $N = 15$
   e. $\bar{X} = 74.59$; $\mu = 81.29$; $\sigma = 13.54$; $N = 26$
5. For each of the following situations, calculate the *z*-statistic (*z*), make a decision about the null hypothesis (reject, do not reject), and indicate the level of significance ($p > .05$, $p < .05$, $p < .01$).
   a. $\bar{X} = 40.00$; $\mu = 30$; $\sigma_{\bar{X}} = 10.00$
   b. $\bar{X} = 3.70$; $\mu = 3.46$; $\sigma_{\bar{X}} = .14$
   c. $\bar{X} = 19.23$; $\mu = 15.01$; $\sigma_{\bar{X}} = 1.47$
   d. $\bar{X} = 132.65$; $\mu = 154.90$; $\sigma_{\bar{X}} = 11.79$
6. For each of the following situations, calculate the *z*-statistic (*z*), make a decision about the null hypothesis (reject, do not reject), and indicate the level of significance ($p > .05$, $p < .05$, $p < .01$).
   a. $\bar{X} = 8.00$; $\mu = 16$; $\sigma_{\bar{X}} = 4.00$
   b. $\bar{X} = 39.54$; $\mu = 34.22$; $\sigma_{\bar{X}} = 2.18$
   c. $\bar{X} = 1.19$; $\mu = .92$; $\sigma_{\bar{X}} = .17$
   d. $\bar{X} = 56.81$; $\mu = 64.45$; $\sigma_{\bar{X}} = 2.68$
7. For each of the following situations, calculate the population standard error of the mean ($\sigma_{\bar{X}}$) and the *z*-statistic (*z*), make a decision about the null hypothesis, and indicate the level of significance.
   a. $\bar{X} = 4.00$; $\mu = 2$; $\sigma = 6$; $N = 36$
   b. $\bar{X} = 24.52$; $\mu = 19.76$; $\sigma = 10.93$; $N = 23$

c. $\bar{X} = 8.11; \mu = 10.12; \sigma = 3.28; N = 19$

d. $\bar{X} = 4.54; \mu = 3.89; \sigma = 2.32; N = 33$

8. For each of the following situations, calculate the population standard error of the mean ($\sigma_{\bar{X}}$) and the *z*-statistic (*z*), make a decision about the null hypothesis, and indicate the level of significance.

   a. $\bar{X} = 12.00; \mu = 13; \sigma = 5; N = 25$

   b. $\bar{X} = 1.82; \mu = 1.53; \sigma = .67; N = 40$

   c. $\bar{X} = 6.64; \mu = 5.94; \sigma = 1.72; N = 34$

   d. $\bar{X} = 76.29; \mu = 87.71; \sigma = 30.76; N = 26$

9. Students applying to graduate schools in many disciplines are required to take the Graduate Record Examination (GRE); essentially, it is the graduate school equivalent of the SAT. Let's say an enterprising student develops a course she believes will increase students' GRE scores. She develops a sample of 25 students believed to be representative of the larger college student population and puts them through this course. Next, they take the GRE and receive a mean score of 1,075.00. Assuming the GRE has a population mean (μ) of 1,000 and a standard deviation (σ) of 200, does the course appear to significantly increase GRE scores?

   a. State the null and alternative hypotheses (H_0 and H_1) (allow for the possibility that the average GRE score may be less than 1,000).

   b. Make a decision about the null hypothesis.

      (1) Set alpha (α), identify the critical values, and state a decision rule.

      (2) Calculate a statistic: *z*-test for one mean.

      (3) Make a decision whether to reject the null hypothesis.

      (4) Determine the level of significance.

   c. Draw a conclusion from the analysis.

   d. Relate the result of the analysis to the research hypothesis.

10. In the GRE test example (Exercise 9), what if it was believed that the only possible alternative to the null hypothesis is one in which the students' GRE scores increase (i.e., they cannot decrease)?

   a. State the null and alternative hypotheses (H_0 and H_1).

   b. Make a decision about the null hypothesis.

      (1) Set alpha (α), identify the critical values, and state a decision rule. Why is the decision rule different in this situation?

      (2) Calculate a statistic: *z*-test for one mean. Is it a different value in this situation? Why or why not?

      (3) Make a decision whether to reject the null hypothesis.

      (4) Determine the level of significance.

   c. Draw a conclusion from the analysis.

   d. What are the implications of this analysis for the GRE class?

11. One study examined the personal values of 116 students studying mortuary science with the intention of becoming funeral directors (Shaw & Duys, 2005). The students completed a well-established survey measuring different work-related values, one of which was the extent to which the student valued social interaction. The researchers tested the difference between the mean social interaction score of these students ($\bar{X} = 12.91$) with an estimated population mean (μ) of 14.50 and a standard deviation (σ) of 2.87.

   a. State the null and alternative hypotheses (H_0 and H_1).

   b. Make a decision about the null hypothesis.

      (1) Set alpha (α), identify the critical values, and state a decision rule.

      (2) Calculate a statistic: *z*-test for one mean.

      (3) Make a decision whether to reject the null hypothesis.

      (4) Determine the level of significance.

   c. Draw a conclusion from the analysis.
   d. What might you say about the extent to which mortuary science students value social interaction as part of their jobs?

12. For each of the following situations, calculate the degrees of freedom (*df*) and determine the critical values of *t*.
   a. $N = 10; \alpha = .05; H_1: \mu \neq 5$
   b. $N = 20; \alpha = .05; H_1: \mu \neq 5$
   c. $N = 10; \alpha = .01; H_1: \mu \neq 5$
   d. $N = 20; \alpha = .01; H_1: \mu \neq 5$
   e. $N = 10; \alpha = .05; H_1: \mu > 5$
   f. $N = 20; \alpha = .05; H_1: \mu > 5$
   g. $N = 28; \alpha = .05; H_1: \mu \neq 5$

13. For each of the following situations, calculate the standard error of the mean ($s_{\overline{X}}$).
   a. $s = 7.00; N = 49$
   b. $s = 2.50; N = 14$
   c. $s = 8.90; N = 23$
   d. $s = 25.61; N = 54$

14. For each of the following situations, calculate the standard error of the mean ($s_{\overline{X}}$).
   a. $s = 5.00; N = 16$
   b. $s = 17.82; N = 10$
   c. $s = 2.31; N = 37$
   d. $s = 51.32; N = 21$

15. For each of the following sets of numbers, calculate the sample size (*N*), the mean ($\overline{X}$), the standard deviation (*s*), and the standard error of the mean ($s_{\overline{X}}$):
   a. 2, 3, 3, 5, 7
   b. 6, 7, 7, 9, 10, 12, 13, 16
   c. 3, 3, 5, 6, 7, 7, 8, 9, 9, 10, 11
   d. 2, 5, 7, 8, 9, 10, 13, 13, 14, 16, 18, 20
   e. 82, 69, 51, 95, 78, 65, 62, 87, 47, 80, 73, 82, 55, 61, 96

16. For each of the following sets of numbers, calculate the sample size (*N*), the mean ($\overline{X}$), the standard deviation (*s*), and the standard error of the mean ($s_{\overline{X}}$):
   a. 4, 5, 8, 9, 11, 11
   b. 27, 19, 14, 25, 19, 22, 26, 23
   c. 2.7, 3.1, 3.4, 2.1, 2.8, 3.3, 3.0, 2.9, 3.8, 3.1
   d. 12, 10, 9, 11, 9, 15, 10, 9, 8, 13, 9, 10, 7
   e. 11, 9, 20, 12, 10, 19, 7, 8, 23, 14, 34, 9, 17, 6, 14, 19

17. For each of the following situations, calculate the *t*-statistic (*t*):
   a. $\overline{X} = 12.00; \mu = 10; s_{\overline{X}} = 2.00$
   b. $\overline{X} = 6.00; \mu = 9; s_{\overline{X}} = 1.50$
   c. $\overline{X} = 4.25; \mu = 4.25; s_{\overline{X}} = .75$
   d. $\overline{X} = 5.62; \mu = 5.25; s_{\overline{X}} = .20$
   e. $\overline{X} = 9.75; \mu = 10.50; s_{\overline{X}} = 1.08$

18. For each of the following situations, calculate the *t*-statistic (*t*):

   a. $\bar{X} = 11.00; \mu = 5; s_{\bar{X}} = 3.00$
   b. $\bar{X} = 26.00; \mu = 31; s_{\bar{X}} = 2.00$
   c. $\bar{X} = 19.60; \mu = 22; s_{\bar{X}} = 3.25$
   d. $\bar{X} = 3.27; \mu = 3; s_{\bar{X}} = .13$
   e. $\bar{X} = 74.92; \mu = 65.50; s_{\bar{X}} = 5.16$

19. For each of the following situations, calculate the *t*-statistic (*t*):

   a. $\bar{X} = 4.00; \mu = 5; s = 3.00; N = 36$
   b. $\bar{X} = 1.50; \mu = 1.25; s = .75; N = 25$
   c. $\bar{X} = 16.21; \mu = 15; s = 3.23; N = 46$
   d. $\bar{X} = 8.26; \mu = 6.31; s = 3.71; N = 12$
   e. $\bar{X} = 93.70; \mu = 99.80; s = 8.16; N = 21$

20. For each of the following situations, calculate the *t*-statistic (*t*):

   a. $\bar{X} = .45; \mu = .52; s = .17; N = 56$
   b. $\bar{X} = 7.75; \mu = 6; s = 3.98; N = 40$
   c. $\bar{X} = 3.31; \mu = 4; s = 1.33; N = 35$
   d. $\bar{X} = 37.83; \mu = 40.95; s = 7.74; N = 21$
   e. $\bar{X} = 127.85; \mu = 125; s = 12.62; N = 18$

21. For each of the following situations, calculate the *t*-statistic (*t*):

   a. $\bar{X} = 20.00; \mu = 18; s_{\bar{X}} = 1.00$
   b. $\bar{X} = 20.00; \mu = 13; s_{\bar{X}} = 1.00$
   c. $\bar{X} = 12.00; \mu = 20; s_{\bar{X}} = 1.00$
   d. Looking at your answers to (a–c), how is the *t*-statistic affected by the size of the difference between the sample mean and the population mean $(\bar{X} - \mu)$?

22. For each of the following situations, calculate the degrees of freedom (*df*), identify the critical values (assume $\alpha = .05$ [two-tailed]), calculate the *t*-statistic (*t*), make a decision about the null hypothesis, and indicate the level of significance ($p > .05$, $p < .05$, $p < .01$).

   a. $\bar{X} = 2.50; \mu = 3.50; s_{\bar{X}} = .67; N = 9$
   b. $\bar{X} = 2.50; \mu = 20; s_{\bar{X}} = 2.00; N = 16$
   c. $\bar{X} = .78; \mu = .59; s_{\bar{X}} = .09; N = 14$
   d. $\bar{X} = 4.91; \mu = 2.84; s_{\bar{X}} = .60; N = 21$

23. For each of the following situations, calculate the degrees of freedom (*df*), identify the critical values (assume $\alpha = .05$ [two-tailed]), calculate the *t*-statistic (*t*), make a decision about the null hypothesis, and indicate the level of significance ($p > .05$, $p < .05$, $p < .01$).

   a. $\bar{X} = 12.71; \mu = 16.49; s_{\bar{X}} = 1.91; N = 9$
   b. $\bar{X} = 3.85; \mu = 3; s_{\bar{X}} = .42; N = 16$
   c. $\bar{X} = 24.76; \mu = 21.55; s_{\bar{X}} = 1.27; N = 12$
   d. $\bar{X} = 167.32; \mu = 187.03; s_{\bar{X}} = 6.25; N = 26$

24. For each of the following situations, calculate the degrees of freedom (*df*), identify the critical values (assume $\alpha = .05$ [two-tailed]), calculate the standard error of the mean $(s_{\bar{X}})$, calculate the *t*-statistic (*t*), make a

decision about the null hypothesis (reject, do not reject), and indicate the level of significance ($p > .05$, $p < .05$, $p < .01$).

a. $\overline{X} = 75.00$; $\mu = 65$; $s = 18.00$; $N = 9$

b. $\overline{X} = 64.00$; $\mu = 69$; $s = 6.00$; $N = 16$

c. $\overline{X} = .75$; $\mu = .63$; $s = .23$; $N = 29$

d. $\overline{X} = 78.51$; $\mu = 67.92$; $s = 18.53$; $N = 23$

25. For each of the following situations, calculate the degrees of freedom (*df*), identify the critical values (assume $\alpha = .05$ [two-tailed]), calculate the standard error of the mean $(s_{\overline{X}})$, calculate the *t*-statistic (*t*), make a decision about the null hypothesis (reject, do not reject), and indicate the level of significance ($p > .05$, $p < .05$, $p < .01$).

a. $\overline{X} = 14.01$; $\mu = 18.34$; $s = 8.82$; $N = 17$

b. $\overline{X} = 6.41$; $\mu = 5.72$; $s = 1.21$; $N = 15$

c. $\overline{X} = 78.89$; $\mu = 70$; $s = 23.54$; $N = 13$

d. $\overline{X} = 471.46$; $\mu = 500$; $s = 48.51$; $N = 25$

26. For each set of data below, calculate the sample size (*N*), the mean $(\overline{X})$, the standard deviation (*s*), the standard error of the mean $(s_{\overline{X}})$, and the *t*-statistic (*t*). Note: assume $\mu = 12$.

a. 9, 14, 17, 23

b. 11, 9, 5, 13, 6, 8, 10

c. 4, 12, 16, 8, 10, 10, 9, 13, 8

d. 5, 16, 14, 8, 15, 11, 13, 9, 12, 19, 13, 10, 17

27. For each of the following situations, draw a normal distribution (a.k.a. "bell-shaped curve"), mark the critical value(s), shade in the region of rejection, compare the stated *t*-statistic to the critical value(s), and make a decision whether to reject the null hypothesis (H_0).

a. Critical value = ±2.262; $t = 1.25$

b. Critical value = ±2.074; $t = -2.50$

c. Critical value = 1.734; $t = -2.00$

d. Critical value = −1.701; $t = -1.79$

28. The manager of a department store disputes the company's claim that the average age of customers who buy a particular brand of clothes is 16; she believes the average age is older than 16. A clerk for the store stops the next 25 people buying these clothes and asks for their age. She calculates a mean age of 18.50 years, with a standard deviation of 2.25 years. Use the data from this sample to test the manager's claim.

a. State the null and alternative hypotheses (H_0 and H_1) (allow for the possibility that the average age may be younger than 16).

b. Make a decision about the null hypothesis.

   (1) Calculate the degrees of freedom (*df*).

   (2) Set alpha (α), identify the critical values (draw the distribution), and state a decision rule.

   (3) Calculate a statistic: *t*-test for one mean.

   (4) Make a decision whether to reject the null hypothesis.

   (5) Determine the level of significance.

c. Draw a conclusion from the analysis.

d. Relate the result of the analysis to the research hypothesis.

29. In the department store example (Exercise 28), imagine that the sample had been $N = 4$ rather than $N = 25$ (for the sake of this example, keep the sample mean and standard deviation at 18.50 and 2.25, respectively).

a. State the null and alternative hypotheses (H_0 and H_1) (allow for the possibility that the average age may be younger than 16).

b. Make a decision about the null hypothesis.
   (1) Calculate the degrees of freedom (*df*).
   (2) Set alpha (α), identify the critical values, and state a decision rule. Why is the decision rule different than when $N = 25$?
   (3) Calculate a statistic: *t*-test for one mean. Why is it a different value than when $N = 25$?
   (4) Make a decision whether to reject the null hypothesis.
   (5) Determine the level of significance.

c. Draw a conclusion from the analysis.

d. What are the implications of this analysis for the store manager's claim?

30. One of the examples in this chapter discussed unique invulnerability, the belief that people will provide estimates of age at time of death that are greater than the average life expectancy in the population. In a follow-up to this article, the author again demonstrated this phenomenon, this time recognizing the fact that the actuarial age of death for people with higher levels of education is actually older than 75 and may in fact be 83 (Snyder, 1997). The estimated age of death provided by these students is presented below:

Student	Estimated Age of Death	Student	Estimated Age of Death	Student	Estimated Age of Death
1	80	7	96	13	102
2	94	8	88	14	80
3	82	9	106	15	85
4	84	10	82	16	90
5	88	11	85	17	89
6	85	12	102	#	

Use the data from this class to test whether the unique invulnerability bias occurs for people with higher levels of education.

a. Calculate the mean and standard deviation of the estimates of age of death.

b. State the null and alternative hypotheses (H_0 and H_1) (allow for the possibility that the estimated age of death may be less than the actuarial age in the population).

c. Make a decision about the null hypothesis.
   (1) Calculate the degrees of freedom (*df*).
   (2) Set alpha (α), identify the critical values (draw the distribution), and state a decision rule.
   (3) Calculate a statistic: *t*-test for one mean.
   (4) Make a decision whether to reject the null hypothesis.
   (5) Determine the level of significance.

d. Draw a conclusion from the analysis.

e. Relate the result of the analysis to the research hypothesis.

31. The American Statistical Association publishes a monthly magazine sent to all of its members. This magazine once included an article that asked, "How many chocolate chips are there in a bag of Chips Ahoy cookies?" (Warner & Rutledge, 1999). Nabisco, the makers of these cookies, claims there are (at least) 1,000 chocolate chips in each 18-ounce bag of cookies. The authors tested this claim by obtaining 42 bags of cookies, dissolving the cookies in water to separate the chocolate chips from the dough, and counting the number of chocolate chips in each bag. For the sake of simplicity, the number of chips in 18 of their bags is listed below (these 18 bags are representative of the larger sample of 42 bags):

Bag	# Chips	Bag	# Chips	Bag	# Chips
1	**1,103**	7	**1,121**	13	**1,219**
2	**1,219**	8	**1,185**	14	**1,191**
3	**1,345**	9	**1,325**	15	**1,166**
4	**1,258**	10	**1,269**	16	**1,270**
5	**1,307**	11	**1,440**	17	**1,215**
6	**1,419**	12	**1,132**	18	**1,514**

Use these data to test Nabisco's "Chips Ahoy! 1,000 Chips Challenge."

a. Calculate the mean $(\bar{X})$ and standard deviation (s) of the sample of bags of cookies.

b. State the null and alternative hypotheses (H_0 and H_1) (allow for the possibility that the number of chips may either be less or greater than that claimed by Nabisco).

c. Make a decision about the null hypothesis.

   (1) Calculate the degrees of freedom (df).

   (2) Set alpha (α), identify the critical values (draw the distribution), and state a decision rule.

   (3) Calculate a statistic: t-test for one mean.

   (4) Make a decision whether to reject the null hypothesis.

   (5) Determine the level of significance.

d. Draw a conclusion from the analysis.

e. What are the implications of this analysis for those who disbelieve Nabisco's claim?

32. A sixth-grade teacher uses a new method of teaching mathematics to her students, one she believes will increase their level of mathematical ability. To assess their mathematical ability, she administers a standardized test (the Test of Computational Knowledge [TOCK]). Scores on this test can range from a minimum of 20 to a possible maximum of 80. The TOCK scores for the 20 students are listed below:

Student	TOCK Score	Student	TOCK Score	Student	TOCK Score
1	**58**	8	**36**	15	**61**
2	**67**	9	**48**	16	**39**
3	**54**	10	**77**	17	**75**
4	**45**	11	**68**	18	**34**
5	**59**	12	**43**	19	**49**
6	**55**	13	**65**	20	**52**
7	**69**	14	**56**		

To interpret their level of performance on the test, she wishes to compare their mean with the hypothesized population mean (μ) of 50 for sixth graders in her state.

a. Calculate the mean $(\bar{X})$ and standard deviation (s) of the TOCK scores.

b. State the null and alternative hypotheses (H_0 and H_1) (allow for the possibility that the new teaching method may, for some reason, lower students' mathematical ability).

c. Make a decision about the null hypothesis.

(1) Calculate the degrees of freedom (*df*).
(2) Set alpha (α), identify the critical values (draw the distribution), and state a decision rule.
(3) Calculate a statistic: *t*-test for one mean.
(4) Make a decision whether to reject the null hypothesis.
(5) Determine the level of significance.

d. Draw a conclusion from the analysis.
e. Relate the result of the analysis to the research hypothesis.

Answers to Learning Checks

Learning Check 2

2. a. $\sigma_{\bar{X}} = 3.00$
   b. $\sigma_{\bar{X}} = .33$
   c. $\sigma_{\bar{X}} = .88$
   d. $\sigma_{\bar{X}} = 5.62$
3. a. $z = 2.00$
   b. $z = -2.00$
   c. $z = .89$
   d. $z = 2.28$

Learning Check 3

2. a. $z = 2.00$; reject H_0; $p < .05$
   b. $z = -1.00$; do not reject H_0; $p > .05$
   c. $z = 2.41$; reject H_0; $p < .05$
3. a. $\sigma_{\bar{X}} = 1.00$; $z = 3.00$; reject H_0; $p < .01$
   b. $\sigma_{\bar{X}} = 2.00$; $z = 1.00$; do not reject H_0; $p > .05$
   c. $\sigma_{\bar{X}} = .92$; $z = 1.90$; do not reject H_0; $p < .05$

Learning Check 5

2. a. $df = 8$; critical value = ± 2.306
   b. $df = 12$; critical value = ± 2.179
   c. $df = 17$; critical value = ± 2.110
   d. $df = 20$; critical value = ± 2.086
3. a. $s_{\bar{X}} = .75$
   b. $s_{\bar{X}} = 1.50$
   c. $s_{\bar{X}} = 2.83$
   d. $s_{\bar{X}} = .55$
4. a. $t = 2.50$
   b. $t = -1.33$
   c. $t = -2.06$

Learning Check 6

1. a. $df = 24$; ±2.064; $t = 2.40$; reject H_0; $p < .05$
   b. $df = 15$; ±2.131; $t = 1.38$; do not reject H_0; $p > .05$
   c. $df = 18$; ±2.101; $t = -3.55$; reject H_0; $p < .01$
2. a. $df = 10$; ±2.228; = 1.21; $t = 3.32$; reject H_0; $p < .01$
   b. $df = 26$; ±2.060; = .33; $t = .87$; do not reject H_0; $p > .05$
   c. $df = 18$; ±2.101; = 1.20; $t = -2.29$; reject H_0; $p < .05$

Answers to Odd-Numbered Exercises

1. a. $\sigma_{\overline{X}} = 2.00$
   b. $\sigma_{\overline{X}} = 1.50$
   c. $\sigma_{\overline{X}} = .40$
   d. $\sigma_{\overline{X}} = .88$
   e. $\sigma_{\overline{X}} = 5.90$
3. a. $z = 3.00$
   b. $z = -1.00$
   c. $z = 2.43$
   d. $z = 2.38$
   e. $z = -2.76$
5. a. $z = 1.00$; do not reject H_0; $p > .05$
   b. $z = 1.71$; do not reject H_0; $p > .05$
   c. $z = 2.87$; reject H_0; $p < .01$
   d. $z = -1.89$; do not reject H_0; $p > .05$
7. a. $\sigma_{\overline{X}} = 1.00; z = 2.00$; reject H_0; $p < .05$
   b. $\sigma_{\overline{X}} = 2.28$; $z = 2.09$; reject H_0; $p < .05$
   c. $\sigma_{\overline{X}} = .75; z = -2.68$; reject H_0; $p < .01$
   d. $\sigma_{\overline{X}} = .40; z = 1.63$; do not reject H_0; $p > .05$
9. a. H_0: $\mu = 1{,}000$; H_1: $\mu \neq 1{,}000$
   b. (1) If $z < -1.96$ or > 1.96, reject H_0; otherwise, do not reject H_0
      (2) $\sigma_{\overline{X}} = 40.00$; $z = 1.88$
      (3) $z = 1.88$ is not < -1.96 or > 1.96 ∴ do not reject H_0 ($p > .05$)
      (4) Not applicable (H_0 not rejected)
   c. The mean GRE score of the 25 students taking the course ($M = 1{,}075.00$) was not significantly different from the population mean of 1,000, $z = 1.88$, $p > .05$.
   d. The result of this analysis does not support the student's belief that the course will significantly increase students' GRE scores.

11. a. H_0: $\mu = 14.50$; H_1: $\mu \neq 14.50$
   (1) If $z < -1.96$ or > 1.96, reject H_0; otherwise, do not reject H_0
   (2) $\sigma_{\bar{X}} = .27; z = -5.89$
   (3) $z = -5.89 < -1.96 \therefore$ reject H_0 ($p < .05$)
   (4) $z = -5.89 < -2.58 \therefore p < .01$
   c. The mean social interaction value of the 116 mortuary students ($M = 12.91$) was significantly lower than the population mean of 14.50, $z = -5.89$, $p < .01$.
   d. The result of this analysis suggests that mortuary science students do not appear to value social interaction as part of their jobs.

13. a. $s_{\bar{X}} = 1.00$
   b. $s_{\bar{X}} = .67$
   c. $s_{\bar{X}} = 1.86$
   d. $s_{\bar{X}} = 3.49$

15. a. $N = 5, \bar{X} = 4.00, s = 2.00, s_{\bar{X}} = .89$
   b. $N = 8, \bar{X} = 10.00, s = 3.46, s_{\bar{X}} = 1.22$
   c. $N = 11, \bar{X} = 7.09, s = 2.66, s_{\bar{X}} = .80$
   d. $N = 12, \bar{X} = 11.25, s = 5.38, s_{\bar{X}} = 1.55$
   e. $N = 15, \bar{X} = 72.20, s = 15.27, s_{\bar{X}} = 3.94$

17. a. $t = 1.00$
   b. $t = -2.00$
   c. $t = .00$
   d. $t = 1.85$
   e. $t = -.69$

19. a. $t = -2.00$
   b. $t = 1.67$
   c. $t = 2.52$
   d. $t = 1.82$
   e. $t = -3.43$

21. a. $t = 2.00$
   b. $t = 7.00$
   c. $t = -8.00$
   d. The greater the difference between the sample mean and the population mean, the greater the absolute value of the *t*-statistic.

23. a. $df = 8$; ± 2.306; $t = -1.98$; do not reject H_0; $p > .05$
   b. $df = 15$; ± 2.131; $t = 2.02$; do not reject H_0; $p > .05$
   c. $df = 11$; ± 2.201; $t = 2.53$; reject H_0; $p < .05$
   d. $df = 25$; ± 2.060; $t = -3.15$; reject H_0; $p < .01$

25. a. $df = 16; \pm 2.120; s_{\bar{X}} = 2.14; t = -2.02$; do not reject H_0; $p < .05$
   b. $df = 14; \pm 2.145; s_{\bar{X}} = .31; t = 2.23$; reject H_0; $p < .05$
   c. $df = 12; \pm 2.179; s_{\bar{X}} = 6.53; t = 1.36$; do not reject H_0; $p < .05$
   d. $df = 24; \pm 2.064; s_{\bar{X}} = 9.70; t = -2.94$; reject H_0; $p < .01$

27.

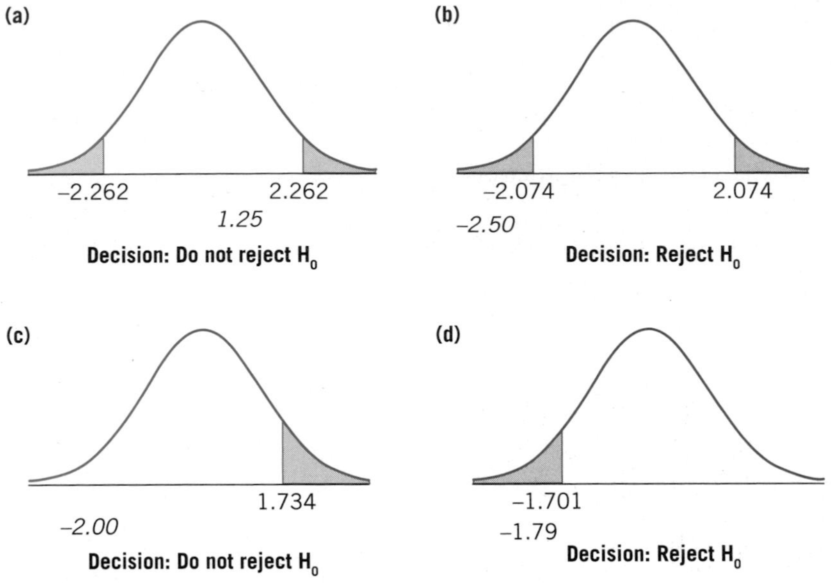

29. a. H_0: $\mu = 16$; H_1: $\mu \neq 16$

b. (1) $df = 3$

(2) If $t < -3.182$ or > 3.182, reject H_0; otherwise, do not reject H_0. The decision rule is different because the critical values are dependent on sample size—the smaller the sample, the larger the critical values.

(3) $s_{\bar{X}} = 1.13$; $t = 2.21$. The *t*-statistic is different because its value is dependent on sample size—the smaller the sample size, the smaller the value of *t*.

(4) $t = 2.21 < 3.182$ ∴ do not reject H_0 ($p > .05$)

(5) Not applicable (H_0 not rejected)

c. The mean age of the four shoppers in the sample ($M = 18.50$) is not significantly different from the company's hypothesized population mean of 16 years, $t(3) = 2.21$, $p > .05$.

d. The result of this analysis does not support the manager's claim that the average age of customers is older than 16.

31. a. $s_{\bar{X}} = 1{,}261.00$; $s = 114.01$

b. H_0: $\mu = 1{,}000$; H_1: $\mu \neq 1{,}000$

c. (1) $df = 17$

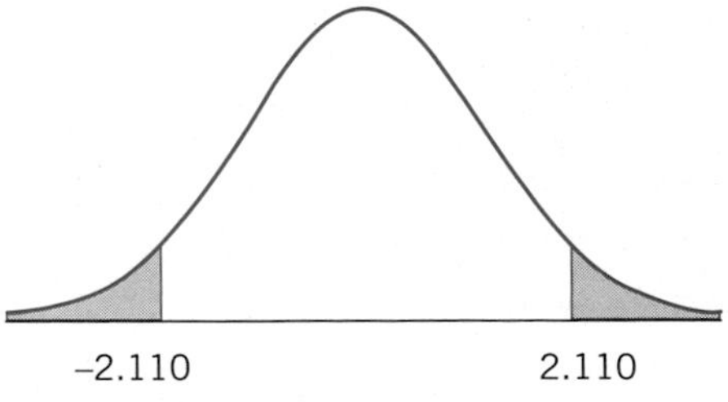

(2) If $t < -2.110$ or > 2.110, reject H_0; otherwise, do not reject H_0

(3) $s_{\bar{X}} = 26.87$; $t = 9.71$

(4) $t = 9.71 > 2.110 \therefore$ reject H_0 ($p < .05$)

(5) $t = 9.71 > 2.898 \therefore p < .01$

d. In the sample of 18 bags of cookies, the average number of chocolate chips (M = 1,261.00) was significantly greater than the claimed amount of 1,000, $t(17) = 9.71$, $p < .01$.

e. The results of this analysis support Nabisco's claim that there are at least 1,000 chocolate chips in each bag.

8

ESTIMATING THE MEAN OF A POPULATION

CHAPTER OUTLINE

Chapter 7 discussed the test of a research hypothesis regarding the difference between the mean of a sample and a hypothesized population mean (μ). However, it is important to understand there are situations in which the population mean for a variable is not known. In such cases, the goal of a research study may be to use a sample mean to develop, with a desired level of confidence, an estimate of an unknown population mean. The purpose of this chapter is to introduce **interval estimation**, defined as the estimation of a population parameter (such as μ) by specifying a range, or interval, of values within which one has a certain degree of confidence that the population parameter falls. This chapter will discuss how these intervals, known as confidence intervals, are constructed and interpreted, as well as the relationship between confidence intervals and hypothesis testing.

8.1 AN EXAMPLE FROM THE RESEARCH: SALARY SURVEY

Psychology is a diverse discipline, with areas of specialization that include clinical, developmental, social, personality, and biological psychology. Another, less well-known, area of psychology is industrial and organizational (I-O) psychology. The main goal of I-O psychology is to apply psychological theory and methods to organizations; I-O psychologists study topics such as leadership, work motivation, and employee selection.

The Society for Industrial and Organizational Psychology (SIOP), the professional organization for I-O psychologists, publishes a newsletter for its members. In a 2017 article, a research team led by Amy DuVernet reported the results of a survey in which they collected information from SIOP members (DuVernet et al., 2017). One of the goals of the study, which will be referred to as the salary survey study, was to gain information about the salaries of SIOP members. In this chapter, we'll use some of the data in the study to estimate the annual salary of a population of I-O psychologists.

To collect information from a large and representative sample, a survey was e-mailed to every member of SIOP. In this example, only a small portion of this sample, the income of 38 SIOP members who had recently (within the previous 3 years) received their master's degrees, will be used. Respondents reported their income in response to a question that asked for "total income from your primary job (in thousands of U.S. dollars)." This variable, which will be called income, is measured at the ratio level of measurement. The income for the 38 participants is listed in Table 8.1.

To gain an initial sense of the data from the salary survey study, the incomes for the sample of 38 SIOP members are organized into the grouped frequency distribution table and frequency polygon in Figure 8.1. Looking at this figure, the distribution of income for the 38 participants is somewhat normally distributed, with the center of the distribution in the \$60,001 to \$70,000 interval.

The next step in analyzing data is to calculate descriptive statistics: the sample mean $(\bar{X})$ and standard deviation (s):

$$\begin{aligned}\bar{X} &= \frac{\Sigma X}{N} \\ &= \frac{72{,}000 + 56{,}650 + \ldots + 62{,}222 + 57{,}000}{38} = \frac{2{,}451{,}336}{38} \\ &= 64{,}508.84\end{aligned}$$

$$\begin{aligned}s &= \sqrt{\frac{\Sigma(X - \bar{X})^2}{N-1}} \\ &= \sqrt{\frac{(72{,}000 - 64{,}508.84)^2 + \ldots + (57{,}000 - 64{,}508.84)^2}{38-1}} \\ &= \sqrt{\frac{56{,}117{,}446.60 + \ldots + 56{,}382{,}709.76}{37}} \\ &= \sqrt{\frac{7{,}804{,}938{,}999.07}{37}} = \sqrt{210{,}944{,}297.27} \\ &= 14{,}523.92\end{aligned}$$

TABLE 8.1 ● ANNUAL INCOME OF 38 SIOP MEMBERS WITH RECENT MASTER'S DEGREES

SIOP Member	Income	SIOP Member	Income	SIOP Member	Income
1	**72,000**	14	**67,500**	27	**53,456**
2	**56,650**	15	**45,000**	28	**67,500**
3	**60,200**	16	**76,000**	29	**83,000**
4	**93,300**	17	**60,000**	30	**68,250**
5	**65,000**	18	**85,000**	31	**31,000**
6	**80,000**	19	**60,000**	32	**42,581**
7	**66,000**	20	**73,000**	33	**52,500**
8	**46,000**	21	**83,000**	34	**61,210**
9	**86,000**	22	**65,000**	35	**47,000**
10	**39,467**	23	**48,000**	36	**75,000**
11	**67,000**	24	**60,000**	37	**62,222**
12	**87,000**	25	**80,000**	38	**57,000**
13	**65,000**	26	**64,500**		

Looking at the descriptive statistics, we see that the mean income of the 38 SIOP members was \$64,508.84. As this sample mean is in the middle of the modal income interval (\$60,001–\$70,000), the presence of a similar mean and mode indicates that the incomes in this sample are normally distributed. The standard deviation of \$14,523.92 reflects variability in income around the \$64,508.84 mean, suggesting that the majority of incomes in this sample (±1 standard deviation from the mean) are between \$50,000 and \$80,000.

8.2 INTRODUCTION TO THE CONFIDENCE INTERVAL FOR THE MEAN

At this point in examining the salary survey study, we might expect to see the steps in hypothesis testing (stating the null and alternative hypotheses, etc.) described in earlier chapters. However, a different approach is required when the population mean is not presumed to be known. Rather than test the difference between the sample mean and a known population mean, in this chapter, we'll estimate the value of an unknown population mean.

For the salary survey study, a sample mean of \$64,508.84 has been calculated. This sample mean may be thought of as an estimate of the mean income of the population of SIOP members who recently received their master's degrees. As such, the sample mean is a **point estimate**, defined as a single value used to estimate an unknown population parameter. Figure 8.2(a) illustrates the concept of the point estimate.

The sample mean of \$64,508.84 serves as a useful starting point for estimating an unknown average salary. However, based on the concepts of sampling error and the sampling distribution of the mean (Chapter 7), we know that if we were to draw additional samples of 38 SIOP members from the population, we would calculate a variety of values for the sample mean. So, rather than rely solely on a sample mean to estimate a population mean, the features of the sampling distribution of the mean can be used to surround the sample mean with a range of values that, with a stated degree of confidence, includes the population mean. This range of values is an example of a **confidence interval**, defined as a range or interval of values that has a stated probability of containing an unknown population parameter.

FIGURE 8.1 ● GROUPED FREQUENCY DISTRIBUTION TABLE AND FREQUENCY POLYGON OF INCOME FOR 38 SIOP MEMBERS

(a) Grouped Frequency Distribution Table

Income	*f*	%
> 90,000	1	3%
80,001–90,000	5	13%
70,001–80,000	6	16%
60,001–70,000	12	32%
50,001–60,000	7	18%
40,001–50,000	5	13%
30001–40,000	2	5%
< 30,000	0	0%
Total	38	100%

(b) Frequency Polygon

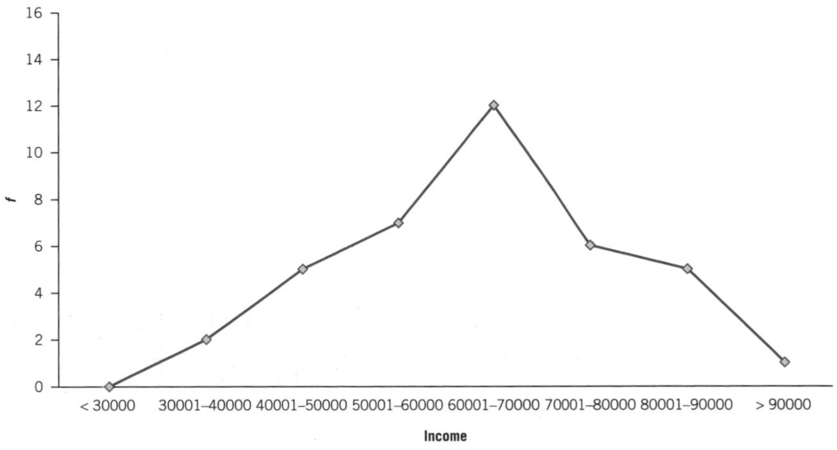

In this chapter, we'll calculate the **confidence interval for the mean**, which is an interval of values that has a stated probability of containing an unknown population mean. The concept of the confidence interval for the mean is illustrated by the gray band in Figure 8.2(b). Constructing a confidence interval for the mean implies we have a stated level of confidence that the gray band in this figure contains an unknown population mean.

Using the Sampling Distribution of the Mean to Create the Confidence Interval for the Mean

Given that we are using a sample mean to estimate a population mean, we rely on the sampling distribution of the mean to help construct the confidence interval for the mean. In Chapter 7, the sampling distribution of the mean was defined as the distribution of values of the sample mean for an infinite number of random samples of size *N* drawn from the population. Furthermore, because the confidence interval for the mean consists of a range of values around a sample mean, we are particularly interested in measuring the variability of sample means in the sampling distribution of the mean.

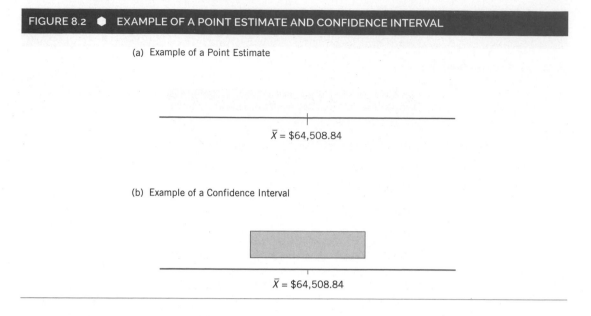

FIGURE 8.2 ● EXAMPLE OF A POINT ESTIMATE AND CONFIDENCE INTERVAL

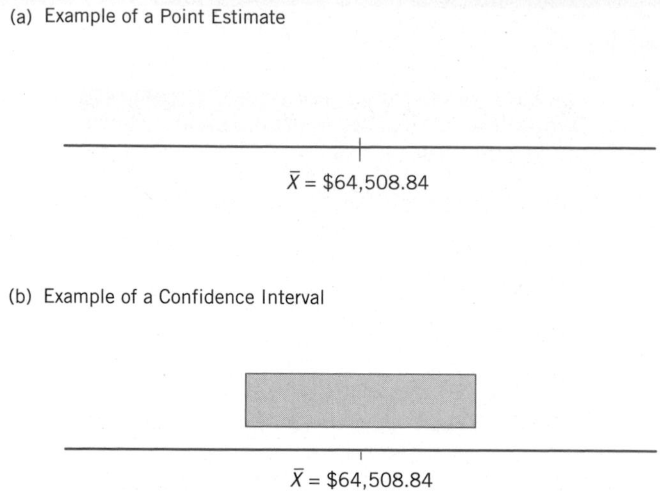

The variability of sample means is measured by a statistic known as the standard error of the mean. There are two versions of the standard error of the mean: the population standard error of the mean ($\sigma_{\bar{X}}$), which is calculated when the standard deviation in the population is known, and the standard error of the mean ($s_{\bar{X}}$), which is calculated when the standard deviation in the population is unknown and is estimated using a standard deviation (s). In this chapter, we'll create confidence intervals for the mean using both versions of the standard error of the mean.

In Chapter 7, to evaluate a sample mean, the sampling distribution of the mean and the standard error of the mean were used to transform the sample mean into one of two statistics. First, the sample mean was transformed into a z-statistic in the situation where the population standard deviation for the variable (σ) was known; the z-statistic was then evaluated using the standard normal distribution. Second, the sample mean was transformed into a t-statistic when the population standard deviation was *not* known; the t-statistic was evaluated using the Student t-distribution, with a different distribution of t-statistics for each degrees of freedom (df), defined as the number of values that are free to vary when a statistic is used to estimate a parameter.

Similar to the test of one mean discussed in Chapter 7, the confidence interval for the mean may be calculated in one of two ways, with the choice being a function of whether the population standard deviation (σ) is known or not known. In the next section, we discuss the confidence interval for the mean for a variable with an unknown population standard deviation. Later in this chapter, we'll construct a confidence interval for the mean for a variable with a known population standard deviation; we'll discuss the two confidence intervals in this order because it is extremely rare to collect enough data from a population to be able to state a population standard deviation.

8.3 THE CONFIDENCE INTERVAL FOR THE MEAN (σ NOT KNOWN)

This section discusses how to calculate and interpret the confidence interval for the mean when the population standard deviation is not known. There are three steps in calculating this confidence interval:

- state the desired level of confidence,
- calculate the confidence interval and confidence limits, and
- draw conclusions about the confidence interval.

Each of these steps is explained and completed below, using the income variable in the salary survey study.

State the Desired Level of Confidence

The size or width of the confidence interval for the mean is a function of the desired level of confidence—that is, how confident one wants to be that the interval contains the population mean. As an extreme example, how big would the confidence interval have to be if we wanted to be absolutely (which is to say 100%) confident that the interval contains the population mean? To be *absolutely* confident, the interval would have to include *all* possible values of the population mean; in other words, a 100% confidence interval would have to be infinitely wide. In our current example, this would be akin to saying, "The average income in the population of SIOP members is somewhere between 0 and infinity." Although this is true, a 100% confidence interval is uninformative due to its lack of precision.

To be of practical use, a confidence interval must have a desired level of confidence less than 100%. One traditional practice is to set the desired level of confidence at 95%, such that researchers construct a 95% confidence interval around the mean:

$$\text{Desired level of confidence} = 95\%$$

A 95% confidence interval for the mean implies that one can say with 95% confidence that the interval contains the unknown population mean. Another way of stating this is that there is a .95 probability the confidence interval contains the population mean and, conversely, a .05 probability the confidence interval does *not* contain the population mean. Later in this chapter, intervals of different levels of confidence will be calculated and compared with the 95% confidence interval.

Calculate the Confidence Interval and Confidence Limits

Once we've determined the desired level of confidence, the next step is to calculate the confidence interval. The formula for the confidence interval (CI) for the mean when the population standard deviation is unknown is presented in Formula 8-1:

$$\text{CI} = \overline{X} = \pm t(s_{\overline{X}}) \tag{8-1}$$

where $\overline{X}$ is the point estimate (i.e., the sample mean), t is the critical value of the t-statistic for the desired level of confidence, and $s_{\overline{X}}$ is the standard error of the mean. When σ is not known and is estimated using the standard deviation, we must rely on the t-statistic and the Student t-distribution to help construct the confidence interval. The inclusion of the $\pm$ symbol in Formula 8-1 indicates that the confidence interval extends from the sample mean in both the left and right directions.

Calculating the confidence interval for the mean using Formula 8-1 requires the completion of three steps:

- calculate the standard error of the mean $(s_{\overline{X}})$,
- calculate the degrees of freedom (*df*) and identify the critical value of *t*, and
- calculate the confidence interval and confidence limits.

Each of these steps will be introduced and illustrated using the income variable in the salary survey study.

Calculate the Standard Error of the Mean ($s_{\overline{X}}$)

The confidence interval for the mean is based on a sample mean, which serves as the point estimate of the population mean. However, a second piece of information needed to construct the confidence interval is a measure of variability of sample means. When the population standard deviation for a variable is not known, the variability of sample means is represented by the standard error of the mean $(s_{\overline{X}})$.

To calculate the standard error of the mean for the data from the salary survey study, the standard deviation (s) of 14,523.92 calculated earlier and the sample size ($N = 38$) are inserted into the following formula (Formula 7-5 in Chapter 7):

$$\begin{aligned} s_{\overline{X}} &= \frac{s}{\sqrt{N}} \\ &= \frac{14{,}523.92}{\sqrt{38}} = \frac{14{,}523.92}{6.16} \\ &= 2{,}356.09 \end{aligned}$$

The width of a confidence interval for the mean is a function of the amount of variability of sample means. The greater the variability of sample means, the wider an interval must be to have the desired level of confidence that the interval contains the population mean.

Calculate the Degrees of Freedom (*df*) and Identify the Critical Value of *t*

Because we are using the standard deviation to estimate the population standard deviation, the Student t-distribution is used to help construct the confidence interval for the mean. As we saw in Chapter 7, using this distribution requires the calculation of the degrees of freedom (df) for the sample. Using Formula 7-5, the degrees of freedom for the sample in the salary survey study is calculated as follows:

$$\begin{aligned} df &= N - 1 \\ &= 38 - 1 \\ &= 37 \end{aligned}$$

In a sample of 38 scores, the first 37 scores are free to vary; that is, there are 37 degrees of freedom.

Once the degrees of freedom have been determined, the next step is to identify the appropriate critical value of the t-statistic for the desired level of confidence. For the salary survey study, what value of t is associated with a 95% confidence interval? As a 95% confidence interval implies there is a .05 probability the interval does *not* contain the population mean, we can identify the critical value for the confidence interval by identifying the values of t that have less than a .05 probability of occurring. Consequently, stating a desired level of confidence is similar to stating a significance level of alpha (α); within the context of hypothesis testing, alpha is the probability of a statistic needed to reject a null hypothesis.

Stating a 95% level of confidence implies we must identify the critical value for the $\alpha = .05$ (two-tailed) region of rejection. For the salary survey study, we need to find the critical value of t for $df = 37$ and $\alpha = .05$ (two-tailed). Turning to the table of critical t values in Table 3 at the back of the book, we start at the df column. Moving down this column, we don't find a row for $df = 37$; instead, we find rows for $df = 30$ and $df = 40$. To find the critical value for the salary survey example, the $df = 30$ row is chosen; it would be inappropriate to use the $df = 40$ row because this would imply the sample is larger than it really is. Within the $df = 30$ row, the critical value of t for $\alpha = .05$ (two-tailed) is identified by moving to the right until we are under the .05 column under the "Level of significance for two-tailed test" label; here, we find the critical value 2.042. Therefore, for the data from the salary survey study:

For a 95% confidence interval and $df = 37$, critical value of $t = 2.042$

Calculate the Confidence Interval and Confidence Limits

Inserting the information obtained from the earlier steps (desired level of confidence [95%], point estimate $[\overline{X} = 64{,}508.84]$, standard error of the mean $[s_{\overline{X}} = 2{,}356.09]$, and critical value of t [2.042]) into Formula 8-1, the 95% CI for the mean for the salary survey study is calculated as follows:

$$95\%\ \text{CI} = \overline{X} = \pm t(s_{\overline{X}})$$
$$= 64{,}508.84 \pm 2.042(2{,}356.09)$$
$$= 64{,}508.84 \pm 4{,}811.14$$

By subtracting and adding (±) 4,811.14 from the point estimate of 64,508.84, **confidence limits**, defined as the lower and upper boundaries of the confidence interval, may be calculated. The lower and upper confidence limits for the income variable are as follows:

$$\text{Lower limit} = 64{,}508.84 - 4{,}811.14 = 59{,}697.70 \qquad \text{Upper limit} = 64{,}508.84 + 4{,}811.14 = 69{,}319.98$$

Therefore, the 95% confidence interval for the income variable ranges from 59,697.70 to 69,319.98, with the point estimate (the sample mean of 64,508.84) located in the center of the interval. The confidence interval for the mean may be represented the following way:

$$95\%\ \text{CI} = (59{,}697.70,\ 69{,}319.98)$$

where the lower and upper confidence limits are placed within parentheses and separated with a comma.

Draw a Conclusion About the Confidence Interval

The third and final step is to draw a conclusion about the confidence interval that has been calculated. This conclusion for the salary survey study may be stated as follows:

> Based on a sample of 38 SIOP members who recently received their master's degrees, there is a .95 probability that the interval of $59,697.70 to $69,319.98 contains the mean annual income for the population of recent master's degrees recipients in I-O psychology.

This sentence provides useful information about the confidence interval:

- The *sample* used to create the confidence interval: "38 SIOP members who recently received their master's degrees"
- The *confidence interval*: ".95 probability that the interval of $59,697.70 to $69,319.98"
- The *variable* being measured: "mean annual income"
- The *population whose mean is being estimated*: "recent master's degrees recipients in I-O psychology"

Income and employment surveys of the type conducted in the salary survey study are useful for a variety of reasons. For example, an estimate of the mean income for the population may assist students in developing realistic expectations regarding their future employment. In one study, students were asked to estimate the starting salaries of those completing doctoral degrees in clinical psychology (Gallucci, 1997). Compared with the results of a salary survey, students overestimated starting salaries by as much as $20,000. Consequently, creating a confidence interval may help college students make more informed choices regarding their choice of academic major and career goals.

Probability of an Interval or the Probability of a Population Mean?

For the salary survey study, we concluded there is a .95 probability that the interval of $59,697.70 to $69,319.98 contains the mean income for the population of recent master's degrees recipients in I-O psychology. Let's use the sampling distribution of the mean to help explain what is inferred by this probability. One feature of this distribution is that, assuming the samples are sufficiently large (typically defined as a sample size of $N \geq 30$), 95% of the sample means fall within ±1.96 standard errors of the population mean μ. (This is analogous to the standard normal distribution, in which 95% of *z*-scores are between ±1.96.) This feature of the sampling distribution of the mean is illustrated for a hypothetical variable in Figure 8.3(a), where the unshaded area in the middle of the distribution represents the 95% of sample means located in the area $\mu \pm 1.96\ \sigma_{\overline{X}}$.

FIGURE 8.3 ● UNDERSTANDING CONCLUSIONS DRAWN REGARDING THE CONFIDENCE INTERVAL FOR THE MEAN

(a) Sampling Distribution of the Mean, Hypothetical Variable

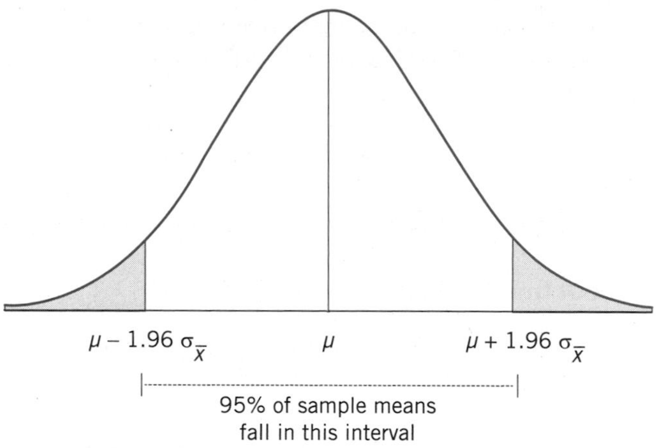

(b) Confidence Intervals, Hypothetical Variable

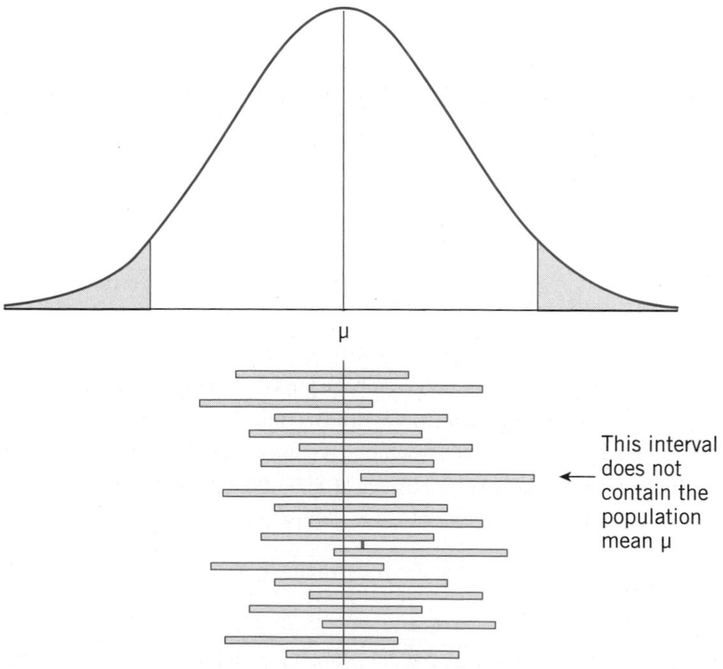

Still working with the hypothetical variable in Figure 8.3(a), suppose we drew a large number of random samples of sufficient size, and for each sample we calculated the confidence interval for the mean. The confidence intervals for these samples are graphically represented by the bars in Figure 8.3(b). Because 95% of the sample means are located within ±1.96 standard errors of the population mean μ, it stands to reason that 95% of the confidence intervals will contain the population mean μ. Consequently, for any single sample, there is a .95 probability the confidence interval calculated for that sample will contain the population mean.

It is useful to make a distinction between the "probability of a confidence interval" and the "probability of a population mean." When interpreting a confidence interval, one may be tempted to draw a conclusion regarding the probability that the population mean falls in the interval. In the salary survey study, this would be represented by stating, "There is a .95 probability that the mean income for the population is between $59,697.70 and $69,319.98." This is actually an incorrect statement, however. The .95 probability refers to the probability that the confidence interval contains an unknown population mean—it does *not* refer to the probability that the population mean is in the interval.

Let's look at Figure 8.3(b) to explain the distinction between the probability of a population mean and the probability of a confidence interval. In this figure, we see that the population mean μ, represented by the vertical line below the μ symbol, remains the same across all of the samples. In other words, the population mean does not change—it is either located within an interval or it is not. Consequently, it does not make sense to talk about the "probability" of the population mean being in an interval. However, the intervals do change and the probability that an interval contains the population mean varies from sample to sample. Again, a confidence interval for the mean is the probability the interval contains the population mean, *not* the probability that the population mean is in the interval.

Summary

Box 8.1 summarizes the steps in calculating the confidence interval for the mean when the population standard deviation σ is not known, using the salary survey study as the example. In the next section, we discuss the relatively rare situation in which the confidence interval for the mean is calculated for a variable with a known population standard deviation (σ).

SUMMARY BOX 8.1 CALCULATING THE CONFIDENCE INTERVAL FOR THE MEAN (σ NOT KNOWN) (SALARY SURVEY STUDY EXAMPLE)

State the desired level of confidence

Desired level of confidence = 95%

Calculate the confidence interval and confidence limits

Calculate the standard error of the mean ($s_{\bar{X}}$)

$$s_{\bar{X}} = \frac{s}{\sqrt{N}} = \frac{14{,}523.92}{\sqrt{38}} = \frac{14{,}523.92}{6.16} = 2{,}356.09$$

Calculate the degrees of freedom (*df*) and identify the critical value of *t*

Calculate the degrees of freedom (*df*)

$$df = N - 1 = 38 - 1 = 37$$

(Continued)

(Continued)

Identify the critical value

For a 95% confidence interval and $df = 37$, critical value of $t = 2.042$

Calculate the confidence interval and confidence limits

Calculate the confidence interval

$$95\%\ \text{CI} = \overline{X} \pm t\,(s_{\overline{X}}) = 64{,}508.84 \pm 2.042(2{,}356.09)$$
$$= 64{,}508.84 \pm 4{,}811.14$$

Calculate the confidence limits

$$\text{Lower limit} = \overline{X} - t(s_{\overline{X}}) = 64{,}508.84 - 4{,}811.14 = 59{,}697.70$$
$$\text{Upper limit} = \overline{X} + t(s_{\overline{X}}) = 64{,}508.84 + 4{,}811.14 = 69{,}319.98$$

Draw a conclusion about the confidence interval

Based on a sample of 38 SIOP members who recently received their master's degrees, there is a .95 probability that the interval of $59,697.70 to $69,319.98 contains the mean annual income for the population of recent master's degrees recipients in I-O psychology.

LEARNING CHECK 1:
Reviewing What You've Learned So Far

1. Review questions
   a. For what types of situations might one calculate the confidence interval for the mean rather than the *t*-test for one mean?
   b. What is the difference between a point estimate and the confidence interval for the mean?
   c. Why is a 100% confidence interval of relatively little use?
   d. "There is a .95 probability that the population mean is between 3 and 6." Why is this conclusion about a confidence interval inappropriate?
2. For each of the following situations, calculate a 95% confidence interval for the mean (σ not known), beginning with the step, "Calculate the degrees of freedom (*df*) and identify the critical value of *t*."
   a. $N = 11, \overline{X} = 3.00, s_{\overline{X}} = .50$
   b. $N = 22, \overline{X} = 7.50, s_{\overline{X}} = 1.25$
   c. $N = 26, \overline{X} = 16.42, s_{\overline{X}} = 2.27$
3. For each of the following sets of numbers, calculate a 95% confidence interval for the mean (σ not known); before calculating the confidence interval, the sample mean ($\overline{X}$) and standard deviation (s) must be calculated.
   a. 3, 6, 4, 2, 5
   b. 9, 13, 19, 6, 20, 15, 11
   c. 59, 54, 61, 72, 50, 66, 48, 70, 53, 63
4. A team of researchers interested in reducing alcohol-related problems in college fraternity members asked 159 members to report the number of drinks consumed on a typical occasion (Larimer et al., 2001). The mean

number of drinks in their sample was 4.93, with a standard deviation of 2.73. Calculate a 95% confidence interval for this set of data.

a. State the desired level of confidence.
b. Calculate the confidence interval and confidence limits.
   (1) Calculate the standard error of the mean ($s_{\overline{X}}$).
   (2) Calculate the degrees of freedom (*df*) and identify the critical value of *t*.
   (3) Calculate the confidence interval and confidence limits.
c. Draw a conclusion about the confidence interval.

8.4 THE CONFIDENCE INTERVAL FOR THE MEAN (σ KNOWN)

This section illustrates how to calculate the confidence interval for the mean for the uncommon situation where the population standard deviation (σ) for a variable is known. Let's return to the example of SAT scores introduced in Chapter 5, where the math section of the SAT has a stated population standard deviation (σ) of 100. Imagine a high school counselor wants to develop an estimate of the mean SAT score for the population of students attending her high school. She obtains the SAT scores from a sample of 12 students from her school and calculates a sample mean of 550.00 ($\overline{X} = 550.00$). (We don't need to calculate the standard deviation (*s*) for the sample because it's not needed to calculate the confidence interval mean when σ is known.)

Calculating the confidence interval for the mean when σ is known involves the same three steps as when σ is not known:

- state the desired level of confidence,
- calculate the confidence interval and confidence limits, and
- draw a conclusion about the confidence interval.

However, even though the steps are the same, there are a few differences in completing these steps when σ is known; these differences are noted below.

State the Desired Level of Confidence

The first step requires stating the desired level of confidence that the confidence interval contains the unknown population mean. In this example, the desired level of confidence will once again be set at the traditional 95%.

$$\text{Desired level of confidence} = 95\%$$

A desired level of confidence of 95% implies that there is a .95 probability that the calculated confidence interval contains an unknown population mean.

Calculate the Confidence Interval and Confidence Limits

The formula for the CI for the mean when σ is known is provided in Formula 8-2:

$$\text{CI} = \overline{X} \pm z(\sigma_{\overline{X}}) \quad (8\text{-}2)$$

where $\overline{X}$ is the point estimate (i.e., the sample mean), z is the critical value of the z-statistic for the desired level of confidence, and $\sigma_{\overline{X}}$ is the population standard error of the mean. There are two critical differences between Formula 8-2 and the formula for the confidence interval for the mean when σ is not known (Formula 8-1). First, the population standard error of the mean($\sigma_{\overline{X}}$)is used to estimate the variability of sample means rather than the standard error of the mean ($s_{\overline{X}}$).Second, the critical value is based on the standard normal distribution of z-statistics rather than the Student t-distribution.

Using Formula 8-2 to calculate the confidence interval for the mean involves three steps:

- calculate the population standard error of the mean ($\sigma_{\overline{X}}$),
- identify the critical value of z, and
- calculate the confidence interval and confidence limits.

Each of these three steps is described below, using the SAT example to illustrate the necessary calculations.

Calculate the Population Standard Error of the Mean ($\sigma_{\overline{X}}$)

When σ is known, the variability of sample means is measured by the population standard error of the mean ($\sigma_{\overline{X}}$). Using the SAT's stated population standard deviation of $\sigma = 100$ and the sample size for this example ($N = 12$), $\sigma_{\overline{X}}$ is calculated for the SAT example using Formula 7-2 from Chapter 7:

$$\begin{aligned}\sigma_{\overline{X}} &= \frac{\sigma}{\sqrt{N}} \\ &= \frac{100}{\sqrt{12}} = \frac{100}{3.46} \\ &= 28.90\end{aligned}$$

Identify the Critical Value of z

The next step is to identify the critical value of a distribution associated with the desired level of confidence. When σ is known, the appropriate distribution is the standard normal distribution—this is the same distribution used in Chapter 7 to test a single mean when σ is known.

In the SAT example, a stated level of confidence of 95% requires us to identify the values of the z-statistic associated with the middle 95% of the standard normal distribution. In Chapter 7, these values were found to be ± 1.96. Therefore, the critical value of z may be stated as follows:

For a 95% confidence interval, critical value of $z = 1.96$

Note that, unlike the income variable in the salary survey study, we did not calculate the degrees of freedom (df) for the sample; this is because we aren't estimating the population standard deviation from the sample's standard deviation.

Calculate the Confidence Interval and Confidence Limits

Using the desired level of confidence (95%), point estimate ($\overline{X} = 550.00$), population standard error of the mean ($\sigma_{\overline{X}} = 28.90$), and critical value of z (1.96), we are ready to calculate the confidence interval for the mean for the SAT example using Formula 8-2:

$$\begin{aligned}95\%\ \text{CI} &= \overline{X} \pm z(\sigma_{\overline{X}}) \\ &= 550.00 \pm 1.96(28.90) \\ &= 550.00 \pm 56.64\end{aligned}$$

Next, the lower and upper confidence limits for this 95% confidence interval may be calculated:

$$\text{Lower limit} = 550.00 - 56.64 = 493.36 \qquad \text{Upper limit} = 550.00 + 56.64 = 606.64$$

The 95% confidence interval for the SAT example may be represented by the following:

$$95\%\ \text{CI} = (493.36, 606.64)$$

Draw a Conclusion About the Confidence Interval

A conclusion regarding the confidence interval for the hypothetical SAT example may be described in the following way:

> Based on a sample of 12 high school students, there is a .95 probability that the interval of 493.36 to 606.64 contains the mean SAT score for the population of students attending the counselor's high school.

In drawing conclusions about the confidence interval for the mean, it's useful to explicitly identify the sample used to develop the confidence interval ("12 high school students"), the confidence interval (".95 probability that the interval of 493.36 to 606.64"), the variable being measured ("SAT score"), and the population whose mean is being estimated ("the population of students attending the counselor's high school").

Summary

Box 8.2 summarizes the steps in calculating the confidence interval for the mean when the population standard deviation σ is known, using the SAT variable as the example. The main purpose of the confidence interval for the mean is to construct a range of values for a variable that has a stated probability of containing an unknown population mean. As such, the smaller the interval, the more precisely the population mean can be estimated. In the next section, we discuss two factors under the control of researchers that influence the width of a confidence interval for the mean.

8.5 FACTORS AFFECTING THE WIDTH OF THE CONFIDENCE INTERVAL FOR THE MEAN

In Chapter 7, we discussed factors that affect the decision to reject the null hypothesis. Similarly, factors influence the width of a confidence interval for the mean, which in turn influences a researcher's ability to estimate an unknown population mean with a desired degree of precision. In this section, we'll use the income variable from the SIOP salary survey study to illustrate two of these factors:

- sample size and
- the desired level of confidence.

As with hypothesis testing, it's important to understand the role these factors play in the development of a confidence interval as they are to a certain degree under the control and discretion of researchers.

SUMMARY BOX 8.2 CALCULATING THE CONFIDENCE INTERVAL FOR THE MEAN (σ KNOWN) (SAT EXAMPLE)

State the desired level of confidence

Desired level of confidence = 95%

Calculate the confidence interval and confidence limits

Calculate the population standard error of the mean ($\sigma_{\overline{X}}$)

$$\sigma_{\overline{X}} = \frac{\sigma}{\sqrt{N}} = \frac{100}{\sqrt{12}} = \frac{100}{3.46} = 28.90$$

Identify the critical value of z

For a 95% confidence interval, critical value of $z = 1.96$

Calculate the confidence interval and confidence limits

Calculate the confidence interval

$$95\% \text{ CI} = \overline{X} \pm z(\sigma_{\overline{X}}) = 550.00 \pm 1.96(28.90)$$
$$= 550.00 \pm 56.64$$

Calculate the confidence limits

$$\text{Lower limit} = \overline{X} - z(\sigma_{\overline{X}}) = 550.00 - 56.64 = 493.36$$
$$\text{Upper limit} = \overline{X} + z(\sigma_{\overline{X}}) = 550.00 + 56.64 = 606.64$$

Draw a conclusion about the confidence interval

Based on a sample of 12 high school students, there is a .95 probability that the interval of 493.36 to 606.64 contains the mean SAT math score for the population of students attending the counselor's high school.

LEARNING CHECK 2:

Reviewing What You've Learned So Far

1. Review questions
   a. How does the calculation of the confidence interval for the mean change depending on whether the population standard deviation σ is known?
2. For each of the following situations, calculate a 95% confidence interval for the mean (σ known), beginning with the step, "Identify the critical value of z."
   a. $\overline{X} = 4.00$, $\sigma_{\overline{X}} = 1.25$
   b. $\overline{X} = 14.50$, $\sigma_{\overline{X}} = 2.16$
   c. $\overline{X} = 97.34$, $\sigma_{\overline{X}} = 3.61$
3. For each of the following sets of numbers, calculate a 95% confidence interval for the mean (σ known); before calculating the confidence interval, the sample mean ($\overline{X}$) and standard deviation (s) must be calculated.

a. 7, 5, 8, 7, 4, 6 (assume $\sigma = 1.00$)
b. 23, 19, 21, 22, 20, 18, 22, 21 (assume $\sigma = 2.75$)
c. 3.05, 2.72, 3.69, 3.11, 2.38, 2.90, 3.37, 2.56, 3.73, 3.02, 2.66 (assume $\sigma = .50$)

4. Although the link between obesity and one's physical condition is well established, a research team stated that less was known about the relationship between obesity and one's mental state of mind (Fontaine, Cheskin, & Barofsky, 1996). These researchers asked 312 obese people to complete a survey measuring social functioning. The researchers reported a sample mean ($\bar{X}$) on social functioning of 77.10 and a population standard deviation (σ) of 22.30. Calculate a 95% confidence interval for the mean for the social functioning variable.
   a. State the desired level of confidence.
   b. Calculate the confidence interval and confidence limits.
      (1) Calculate the population standard error of the mean ($\sigma_{\bar{X}}$).
      (2) Identify the critical value of *z*.
      (3) Calculate the confidence interval and confidence limits.
   c. Draw a conclusion about the confidence interval.

Sample Size

The relationship between the size of a sample and the width of the confidence interval for the mean may be summarized as follows:

The larger the sample size, the narrower the confidence interval for the mean.

In other words, the larger the sample size, the more precisely one is able to estimate a population mean.

To illustrate the impact of sample size on confidence intervals, suppose the sample from the salary survey study had consisted of 100 master's degree recipients rather than 38. Table 8.2 shows the calculation of the confidence interval for the mean for the two sample sizes; to focus on the effect of sample size, we'll assume both samples have the same standard deviation (s = 14,523.92). Looking at the bottom row of this table, we find that the confidence interval for N = 100 ($61,622.94, $67,394.74) is narrower than the confidence interval for N = 38 ($59,697.70, $69,319.98), which means we can more precisely estimate the population mean with the larger sample.

TABLE 8.2 ● THE EFFECT OF SAMPLE SIZE ON THE CONFIDENCE INTERVAL FOR THE MEAN

Step	$N = 38$	$N = 100$
Calculate $s_{\bar{X}}$	$s_{\bar{X}} = \frac{14{,}523.92}{\sqrt{38}} = 2{,}356.09$	$s_{\bar{X}} = \frac{14{,}523.92}{\sqrt{100}} = 1{,}452.39$
Calculate *df*	$df = \mathbf{38} - 1 = 37$	$df = \mathbf{100} - 1 = 99$
Identify critical value of *t*	2.042	1.987
Calculate confidence interval	$\text{CI} = 64{,}508.84 \pm 2.042(2{,}356.09)$ $= 64{,}508.84 \pm 4{,}811.14$	$\text{CI} = 64{,}508.84 \pm 1.987(1{,}452.39)$ $= 64{,}508.84 \pm 2{,}885.90$
Calculate confidence limits	($59,697.70, $69,319.98)	($61,622.94, $67,394.74)

To further illustrate the effect of sample size on the width of a confidence interval, Figure 8.4 illustrates confidence intervals for the SIOP income variable for five different sample sizes ($N = 10, 20, 38, 100$, and 300). Comparing the width of the gray bands in this figure, we find that the bands become narrower as the sample size increases. As the sample size increases, the population mean can be estimated with greater precision.

The relationship between sample size and the width of a confidence interval may be explained by the effect of sampling error on the relationship between samples and populations. In Chapter 6, we defined sampling error as the difference between a statistic calculated from a sample (i.e., the sample mean) and the corresponding population parameter (i.e., the population mean) due to random, chance factors. Because larger samples more closely approximate the population than smaller samples, larger samples have less sampling error. This implies that the distribution of sample means for larger samples is more closely clustered around the population mean, with less variability among sample means. Looking at the calculations in Table 8.2 we see that the larger ($N = 100$) sample has a smaller value for the standard error of the mean ($s_{\overline{X}}$), which in turn narrows the width of the confidence interval. In summary, the larger the sample, the more confidence a researcher has that the sample mean estimates the unknown population mean μ and the smaller the confidence interval around the sample mean.

Estimating the Sample Size Needed for a Desired Confidence Interval

The previous section illustrated the impact of sample size on confidence intervals. As researchers may wish to estimate a population mean with a stated degree of precision, it is possible to determine the minimum sample size needed to attain a confidence interval of a desired width. The relationship between a desired interval width and necessary sample size may be described the following way:

The smaller the desired width of a confidence interval, the larger the necessary sample size.

To illustrate the determination of sample size for a confidence interval, imagine it is now several years since the last SIOP income survey was conducted. The researchers again set out to estimate the mean income of master's degree recipients; however, they now have the goal of creating a confidence interval with a stated degree of

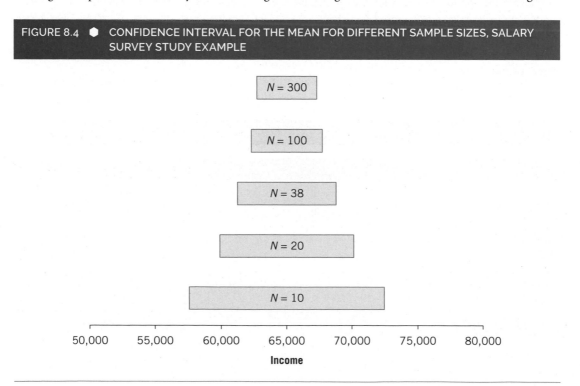

FIGURE 8.4 ● CONFIDENCE INTERVAL FOR THE MEAN FOR DIFFERENT SAMPLE SIZES, SALARY SURVEY STUDY EXAMPLE

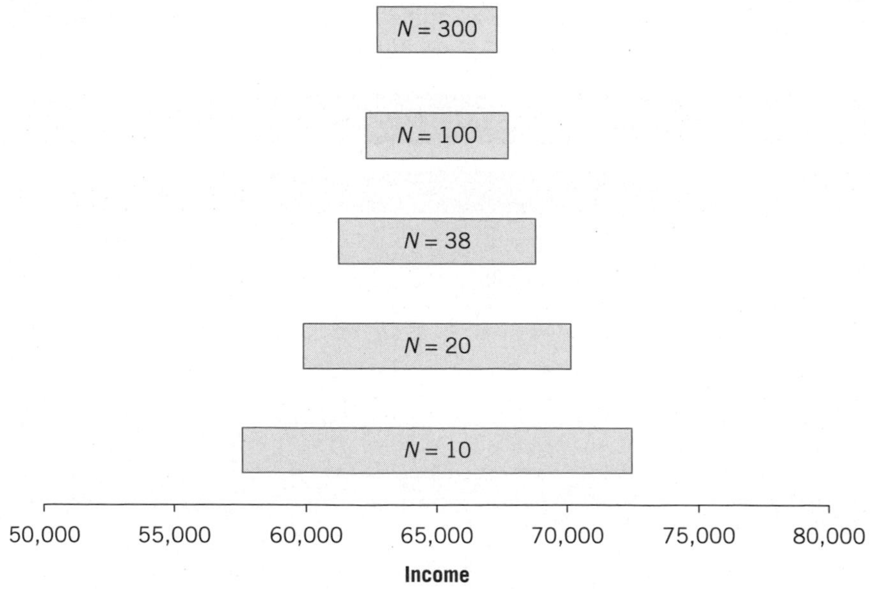

precision: a width of \$4,000. In other words, they want to surround their point estimate (the sample mean $\bar{X}$) with an interval that stretches from \$2,000 below the sample mean to \$2,000 above the sample mean. How large must the sample be to achieve this result?

The estimated sample size for the confidence interval for the mean of a desired width may be calculated using Formula 8-3:

$$N' = \left(\frac{z(\hat{\sigma})}{\text{CI} \div 2} \right)^2 \tag{8-3}$$

where N' is the necessary sample size, z is the critical value of the z-statistic for a desired level of confidence, $\hat{\sigma}$ is an estimate of the population standard deviation, and CI is the desired width of confidence interval.

To calculate the necessary sample size for the income variable, the first piece of information needed is the z-statistic associated with a stated confidence level. Assuming the researchers wish to have the traditional 95% confidence that the interval contains the population mean, the critical value of z is 1.96. Next, we need an estimate of the population standard deviation for the variable ($\hat{\sigma}$). As this estimate is often obtained from the results of existing or previous research, the researchers could use as their estimate the standard deviation (s) from the 2016 income survey (14,523.92). Finally, in terms of the denominator of Formula 8-3, it was proposed earlier that they wanted the confidence interval to have a width of \$4,000; therefore, CI = 4,000.

Based on these three pieces of information ($z = 1.96$, $\hat{\sigma} = 14{,}523.92$, CI = 4,000), the necessary sample size (N') for the new salary survey study is calculated as follows:

$$\begin{aligned} N' &= \left(\frac{z(\hat{\sigma})}{\text{CI} \div 2} \right)^2 \\ &= \left(\frac{1.96(14{,}523.92)}{4{,}000 \div 2} \right)^2 \\ &= \left(\frac{28{,}466.88}{2{,}000} \right)^2 = (14.23)^2 \\ &= 202.49 \end{aligned}$$

Rounding the result of these calculations, the researchers need a sample size of at least 203 participants to construct a 95% confidence interval with a width of \$4,000.

To illustrate the relationship between desired confidence interval width and necessary sample size, Table 8.3 lists the necessary sample size for 95% confidence intervals of five different widths for the SIOP income variable (12,000, 8,000, 4,000, 2,000, and 1,000). Looking at this table, you see that the smaller the desired width of a confidence interval, the larger the necessary sample size. In other words, the greater the desired precision of the estimate of the population mean, the larger the sample must be to attain this precision.

It must be emphasized that calculating a necessary sample size does not guarantee that the confidence interval contains the population mean; it simply determines the sample size necessary to achieve the desired width of the interval. Nonetheless, determining minimum sample sizes can be informative to researchers. For example, as a result of these calculations, a researcher may decide that he or she lacks the resources to collect the amount of data needed to attain a desired level of precision and may consequently change the goals of the study.

Level of Confidence

In addition to sample size, a researcher's desired level of confidence that the interval contains the population mean also affects the width of a confidence interval. The relationship between level of confidence and the width of the confidence interval can be stated as follows:

The higher the desired level of confidence, the wider the confidence interval.

TABLE 8.3 ● ESTIMATED SAMPLE SIZE, 95% CONFIDENCE INTERVALS OF DIFFERENT WIDTHS, SALARY SURVEY STUDY EXAMPLE

Desired Width of Interval	Estimated Sample Size (N')
12,000	22.51
8,000	50.65
4,000	202.49
2,000	810.36
1,000	3,241.45

LEARNING CHECK 3:

Reviewing What You've Learned So Far

1. Review questions
   a. What happens to the width of a confidence interval for the mean as sample size increases?
   b. What are the two ways sample size affects the width of the confidence interval for the mean?
   c. Why might a researcher want to estimate the sample size needed for a confidence interval of a desired width?
2. Counselors at two colleges (College Blue and College Gold) want to estimate the average grade point average (GPA) of students attending their respective colleges. The College Blue counselor collects GPAs from 10 students ($N = 10$); however, the College Gold counselor collects GPAs from 30 students ($N = 30$). In analyzing their data, by an amazing coincidence, they calculate the same mean and standard deviation ($\overline{X} = 2.81$, $s = .90$).
   a. Calculate the 95% confidence interval for the mean (σ not known) for the two colleges.
   b. Which college has the narrower confidence interval? Why?
3. Earlier in this chapter, a 95% confidence interval for the mean (σ known) was calculated for the SAT example ($N = 12$). Assuming the sample mean ($\overline{X} = 550.00$) and population standard deviation ($\sigma = 100$) stay the same . . .
   a. What is the confidence interval for the mean for $N = 25$ students rather than 12?
   b. What is the confidence interval for the mean for $N = 50$ students rather than 12?
   c. What is the confidence interval for the mean for $N = 100$ students rather than 12?
   d. What is the effect of increasing sample size on the width of the confidence interval?
4. In the SAT example, for $N = 12$, the 95% confidence interval had a width of about 115 points (the difference between the lower limit of 493.36 and the higher limit of 606.64). If we again use 100 as the estimate of the population standard deviation ($\hat{\sigma}$) . . .
   a. How large should the sample be for a desired width of 80 points rather than 115?
   b. How large should the sample be for a desired width of 40 points rather than 115?
   c. How large should the sample be for a desired width of 20 points rather than 115?

In other words, the greater the desired probability that the confidence interval contains the population mean, the wider the interval must be to attain this probability.

All of our examples thus far have calculated a 95% confidence interval for the mean. This is the traditional approach, as it corresponds to the traditional value of .05 for alpha (α), the level of significance used in hypothesis testing. However, researchers can also determine confidence intervals that are either higher or lower than 95%. For example, one researcher may want the probability that the interval contains the population mean to be higher than 95%, even if this wider interval implies a loss of precision in estimating the population mean.

However, another researcher may choose a level of confidence lower than 95%, even if a narrower interval lowers the researcher's confidence that the interval contains the population mean. In this section, we'll calculate and interpret confidence intervals of different levels of confidence.

One alternative to a 95% confidence interval is a 99% confidence interval, which is an interval that has a .99 probability of containing an unknown population mean. Consequently, a 99% confidence interval has a higher probability of containing the population mean than does a 95% confidence interval. Another common alternative is a 90% confidence interval, which is an interval with a .90 probability of containing the unknown population mean. A 90% confidence interval has a lower probability of containing the population mean than does a 95% or 99% confidence interval.

In Table 8.4, 99% and 90% confidence intervals are constructed for the income variable in the SIOP salary survey study. Looking at the bottom row of this table, we see that the 99% confidence interval ($58,029.55, $70,988.13) is wider than the 90% confidence interval ($60,510.53, $68,507.15). Looking at the steps in the calculations, the 99% confidence interval is wider because the critical value of t is larger (2.750) than for the 90% interval (1.697). A larger critical value is needed for a 99% confidence interval in order to encompass a larger percentage of the possible values of the population mean.

Choosing Between Different Levels of Confidence: Probability Versus Precision

The three bars in Figure 8.5 illustrate the 90%, 95%, and 99% confidence intervals for the SIOP income variable. Comparing the three confidence intervals, we see that the 99% confidence interval is wider than the 95% interval, which in turn is wider than the 90% interval. In other words, the higher the desired level of confidence, the larger the confidence interval.

The choice of level of confidence is ultimately a trade-off between probability and precision. The higher the level of confidence, the greater the probability the interval contains the population mean. However, this heightened probability results in a corresponding loss of precision. For example, comparing a 99% and 90% confidence interval, although a 99% confidence interval has a greater probability of containing the population mean, it is wider than a 90% confidence interval, meaning that it estimates the population mean with less precision.

The choice of level of confidence may depend on the other factor discussed earlier: sample size. The relationship between level of confidence and sample size may be summarized as follows:

> The larger the sample size, the smaller the effect of level of confidence on the width of the confidence interval.

TABLE 8.4 ● THE EFFECT OF LEVEL OF CONFIDENCE ON THE CONFIDENCE INTERVAL FOR THE MEAN

Step	99% Confidence	90% Confidence
Calculate $s_{\overline{X}}$	$s_{\overline{X}} = \frac{14,523.92}{\sqrt{38}}$ $= 2,356.09$	$s_{\overline{X}} = \frac{14,523.92}{\sqrt{38}}$ $= 2,356.09$
Calculate df	$df = 38 - 1 = 37$	$df = 38 - 1 = 37$
Identify critical value of t	2.750	1.697
Calculate confidence interval	$CI = 64,508.84 \pm 2.750(2,356.09)$ $= 64,508.84 \pm 6,479.29$	$CI = 64,508.84 \pm 1.697(2,356.09)$ $= 64,508.84 \pm 3,998.31$
Calculate confidence limits	($58,029.55, $70,988.13)	($60,510.53, $68,507.15)

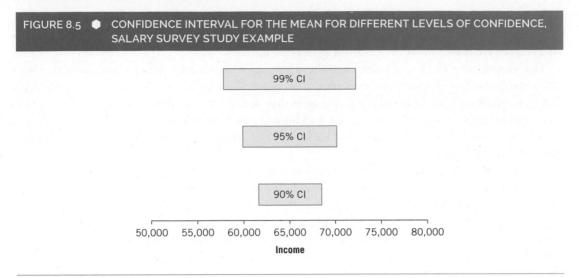

FIGURE 8.5 ● CONFIDENCE INTERVAL FOR THE MEAN FOR DIFFERENT LEVELS OF CONFIDENCE, SALARY SURVEY STUDY EXAMPLE

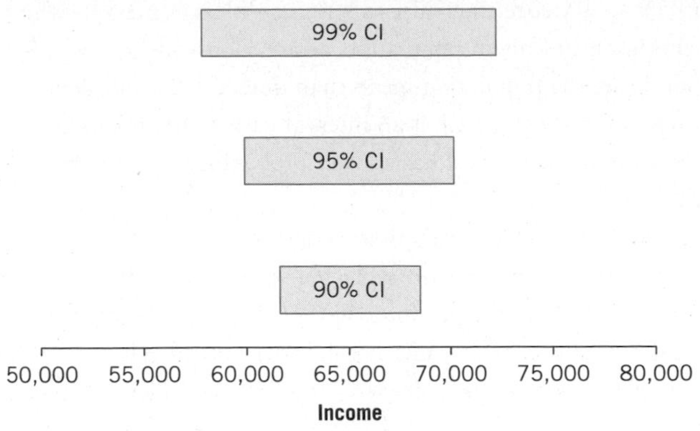

Table 8.5 displays 90%, 95%, and 99% confidence intervals for the income variable for three sample sizes: $N = 20$, 38, and 100 (some of these confidence intervals have been calculated earlier in this chapter). Keeping the sample mean and standard deviation the same, we see that as the sample size increases, the level of confidence has less of an impact on the width of the confidence interval. For the largest sample size ($N = 100$), there is little difference between the 90%, 95%, and 99% confidence intervals. On the other hand, a much bigger difference exists between the three confidence intervals for the smallest sample ($N = 20$). Put another way, the smaller the sample, the greater the trade-off between probability and precision.

8.6 INTERVAL ESTIMATION AND HYPOTHESIS TESTING

Throughout this chapter, we've identified many similarities between the confidence interval for the mean and the test of one mean discussed in Chapter 7. Both involve many of the same pieces of information, such as hypothesized population means, sample means, sampling distributions, and probability. However, they differ in their primary purpose. The purpose of hypothesis testing is to make a decision about the null hypothesis; in testing one mean, we decide whether a sample mean is significantly different from a stated value of a population mean μ. In contrast, the purpose of the confidence interval for the mean is to estimate an unknown population mean. In this section, we discuss concerns about hypothesis testing and the relationship between interval estimation and hypothesis testing.

TABLE 8.5 ● COMBINED EFFECT OF SAMPLE SIZE AND LEVEL OF CONFIDENCE ON WIDTH OF THE CONFIDENCE INTERVAL, SALARY SURVEY STUDY EXAMPLE

Level of Confidence	$N = 20$		$N = 36$		$N = 100$	
	Lower	Upper	Lower	Upper	Lower	Upper
90%	51764.39	61142.77	53023.16	59884.00	54437.77	58469.39
95%	50777.20	62129.96	52325.76	60581.40	54043.59	58863.57
99%	48694.32	64212.84	50894.56	62012.60	53261.28	59645.88

LEARNING CHECK 4:
Reviewing What You've Learned So Far

1. Review questions
   a. What happens to the size of a confidence interval if the level of confidence is changed from 95% to 99%? From 95% to 90%?
   b. What are relative strengths and weaknesses of confidence intervals of different levels of confidence?
2. Using our earlier example of SAT ($N = 12, \bar{X} = 550.00, \sigma = 100$)...
   a. Calculate the 90% and 99% confidence interval for the mean (σ known).
   b. Which confidence interval is wider, and why?
3. As part of an examination of baseballs used in Major League Baseball games (Rist, 2001), a research team weighs a sample of 80 baseballs manufactured in 2000. They find the mean weight to be 5.11 ounces, with a standard deviation of .06 ounces.
   a. Calculate the 90%, 95%, and 99% confidence interval for the mean (σ not known).
   b. Would you say the three confidence intervals are very similar or very different? Why?

Concerns About Hypothesis Testing

Since the 1950s, researchers have expressed concerns about the uses, and potential misuses, of hypothesis testing. For example, in 1996, one researcher, Frank Schmidt, wrote that "my conclusion is that we must abandon the statistical significance test" (Schmidt, 1996, p. 116). Another writer, Geoffrey Loftus, supported this conclusion in an article entitled, "Psychology will be a much better science when we change the way we analyze data" (Loftus, 1996). More recently, Geoff Cumming (2014) wrote, "I conclude from the arguments and evidence I have reviewed that best research practice is not to use NHST [null hypothesis significance testing] at all" (p. 26). The following sections focus on two concerns about hypothesis testing raised by researchers such as Drs. Schmidt, Loftus, and Cumming that highlight differences between hypothesis testing and interval estimation:

- the interpretation of statistical significance and nonsignificance and
- the influence of sample size on the decision about the null hypothesis.

Keep in mind that these concerns are not inherent in hypothesis testing but are due to how researchers interpret the results of statistical analyses.

The Interpretation of Statistical Significance and Nonsignificance

Hypothesis testing centers on two mutually exclusive statistical hypotheses: the null hypothesis and the alternative hypothesis. Based on the probability of a calculated value of a statistic such as the *t*-statistic, one of two decisions is made: reject or do not reject the null hypothesis. When the null hypothesis is rejected, we describe the result of the analysis as "statistically significant"; if the null hypothesis is not rejected, the result is referred to as "nonsignificant."

Dichotomizing statistical analyses as "significant" or "nonsignificant" leads to the possibility that researchers may misinterpret the results of their analyses. For example, there is the tendency to diminish or ignore nonsignificant findings, due in part to the tendency of academic journals to publish studies with significant rather than nonsignificant results. Consequently, readers may mistakenly believe "that if a difference or relationship is not statistically significant, then it is zero" (Schmidt, 1996, p. 126). For example, imagine you compare two *t*-statistics, $t = 2.10$ and $t = .57$, with a critical value of 2.120. Because both *t*-statistics are less than 2.120, you would not reject the null hypothesis in either situation. However, given that the *t*-statistic of 2.10 is barely below the critical value of 2.120, it would be inappropriate to consider it to be the same as the *t*-statistic of .57.

The Influence of Sample Size on Statistical Analyses

An overreliance on the "significant" versus "nonsignificant" dichotomy to evaluate statistical analyses is also problematic because researchers may fail to take into account factors that affect the decision about the null hypothesis, one of which is sample size. As we mentioned in Chapters 6 and 7, the larger the sample size, the greater the likelihood of rejecting the null hypothesis. Consequently, researchers independently studying the same topic may make different decisions regarding the null hypothesis simply because of differences in sample size; this can lead to confusion regarding whether a hypothesized relationship does in fact exist.

Joseph Rossi, a researcher at the University of Rhode Island, was interested in the concept of "spontaneous recovery" (Rossi, 1997). To illustrate this concept, imagine you're given a set of material to learn and are tested on your recall of this material. As you might expect, you remember some, but not all, of the material. Next, you're given a second set of material to learn that's similar to the first set. Given that the two sets of material are similar, spontaneous recovery occurs when you suddenly remember information in the first set that you did not remember earlier.

Dr. Rossi analyzed 40 published research studies that tested for the existence of spontaneous recovery; he found that half of the studies found a statistically significant spontaneous recovery effect and half did not. In examining these studies, Dr. Rossi found a critical difference between the studies that found a significant effect and those that did not: the studies that found a significant spontaneous recovery effect had larger samples. As he concluded, "These results suggest that the inconsistency among spontaneous recovery studies may have been due to the emphasis reviewers and researchers placed on the level of [statistical] significance" (Rossi, 1997, p. 183).

Interval Estimation as an Alternative to Hypothesis Testing

We've discussed two concerns related to the decision to reject the null hypothesis within hypothesis testing. The first concern is that, because this decision categorizes the results of a statistical analysis as either "significant" or "nonsignificant," researchers may incorrectly conclude that a hypothesized effect either completely exists or completely does not exist. The second concern is that researchers may fail to take into consideration the critical role of sample size in the decision to reject the null hypothesis.

Because the decision making that is an integral part of hypothesis testing has the potential to lead to confusion and misunderstanding, researchers such as Drs. Schmidt and Loftus have suggested replacing hypothesis testing with interval estimation. For example, rather than make a decision as to whether the difference between a sample mean and a hypothesized population mean is statistically significant, researchers could instead use the sample mean to develop a confidence interval for the mean to estimate the population mean. Interval estimation is seen as being of particular value when researchers are not certain of the value of the population mean.

The concerns about hypothesis testing discussed in this chapter are valid and important. Nevertheless, hypothesis testing will be used as the primary method for testing research hypotheses throughout this book. As researcher Raymond Nickerson (2000) pointed out, there are several reasons for retaining hypothesis testing. First, in relatively complex research situations, it may not be possible to do interval estimation. Second, like hypothesis testing, the width of confidence intervals is affected by such things as sample size and alpha; as such, the evaluation and interpretation of confidence intervals are subject to some of the same concerns as hypothesis testing.

One strategy employed by researchers to address the concerns related to hypothesis testing is to calculate and report statistics that estimate the size or magnitude of the hypothesized effect being tested in a statistical analysis. These statistics, known as measures of effect size, are used to supplement the decision regarding the null hypothesis. Measures of effect size, which we introduce in Chapter 10, are valuable in that they are not affected by sample size.

Ultimately, we believe many of the perceived flaws and weaknesses of hypothesis testing are not inherent to the statistical procedures themselves but are rather a function of how researchers use and interpret these procedures. It is up to researchers to appropriately interpret and communicate the results of their statistical analyses. For example, rather than focusing solely on the decision about the null hypothesis, researchers must understand and discuss how factors such as sample size and alpha may have affected this decision.

8.7 LOOKING AHEAD

The purpose of this chapter was to introduce interval estimation and confidence intervals, which are statistical procedures designed to estimate unknown population parameters rather than test research hypotheses. Although there are many similarities between interval estimation and hypothesis testing, there remain several fundamental differences. We ended the chapter with a discussion of concerns about hypothesis testing, concerns that will be relevant for the remainder of this book. In the next chapter, we again introduce a statistical procedure designed to test research hypotheses. However, unlike Chapter 7, in which we were testing one mean, in the next chapter, we will test hypotheses regarding the difference between two means.

8.8 Summary

Interval estimation is the estimation of a population parameter by specifying a range, or interval, of values within which one has a certain degree of confidence the population parameter falls.

A ***point estimate*** is a single value used to estimate an unknown population parameter. A ***confidence interval*** is a range or interval of values that has a stated probability of containing an unknown population parameter. The ***confidence interval for the mean*** is a range of values for a variable that has a stated probability of containing an unknown population mean. ***Confidence limits*** are the lower and upper boundaries of the confidence interval.

Confidence intervals for the mean may be calculated when the population standard deviation (σ) is not known or it is known. When the population standard deviation is not known, the confidence interval for the mean is developed using the t-statistic and the Student t-distribution. When the population standard deviation is known, the confidence interval for the mean is developed using the z-statistic and the standard normal distribution.

Two factors influence the width or size of a confidence interval for the mean. The first factor is sample size, such that the larger the sample size, the narrower the width of the confidence interval. It is possible to determine the minimum sample size needed to attain a confidence interval of a desired width; the smaller the desired width of a confidence interval, the larger the necessary sample size. The second factor is the desired level of confidence, such that the higher the desired level of confidence, the wider the confidence interval.

Although there are many similarities between confidence intervals and the test of one mean (Chapter 7), there are also some critical differences between hypothesis testing and interval estimation. Researchers have expressed concerns regarding two aspects of hypothesis testing: The first is the interpretation of statistical significance and nonsignificance, such that dichotomizing statistical analyses as either "significant" or "nonsignificant" may lead to inappropriate interpretations of the results of statistical analyses. The second concern pertains to the influence of sample size on the decision about the null hypothesis. Because interval estimation does not require researchers to make a decision about the null hypothesis, researchers have suggested that hypothesis testing be replaced with or supplemented by interval estimation, particularly when researchers are not certain of the value of the population mean.

8.9 Important Terms

interval estimation (p. 233)
point estimate (p. 234)
confidence interval (p. 234)
confidence interval for the mean (p. 235)
confidence limits (p. 239)

8.10 Formulas Introduced in This Chapter

Confidence Interval for the Mean (σ Not Known)

$$CI = \bar{X} \pm t(s_{\bar{X}}) \quad (8\text{-}1)$$

Confidence Interval for the Mean (σ Known)

$$\mathrm{CI} = \bar{X} \pm z(\sigma_{\bar{X}}) \tag{8-2}$$

Estimated Sample Size for the Confidence Interval for the Mean

$$N' = \left(\frac{z(\hat{\sigma})}{\mathrm{CI} \div 2} \right)^2 \tag{8-3}$$

8.11 Using SPSS

Calculating the confidence interval for the mean (The Salary Survey Study) (8.1)

1. Define variable (name, # decimal places, label for the variable) and enter data for the variable.

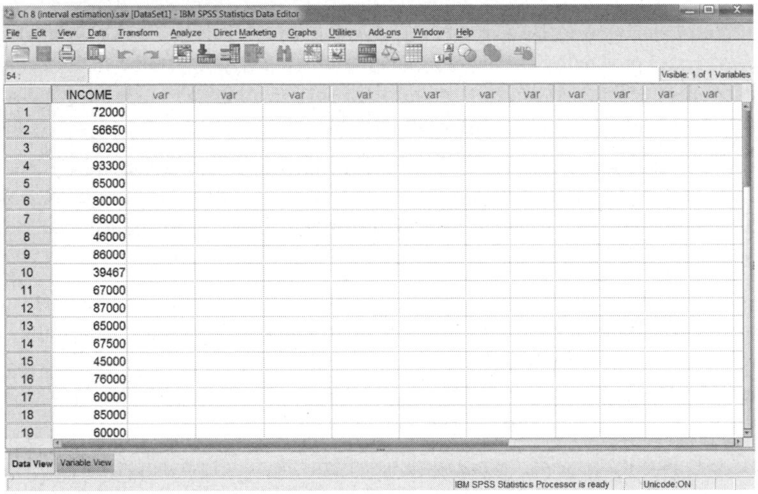

2. Select the *t*-test for one sample mean procedure within SPSS.

   How? (1) Click **Analyze** menu, (2) click **Compare Means**, and (3) click **One-Sample T Test**.

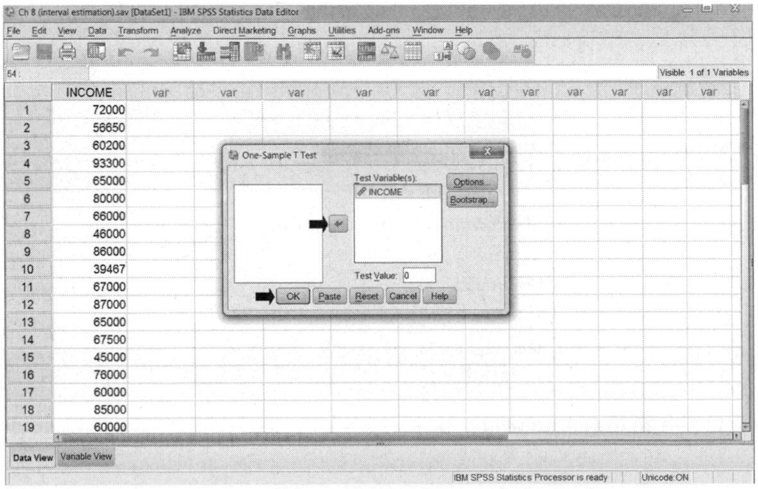

3. Select the variable to be analyzed.

   How? (1) Click variable and →, and (2) click **OK**.

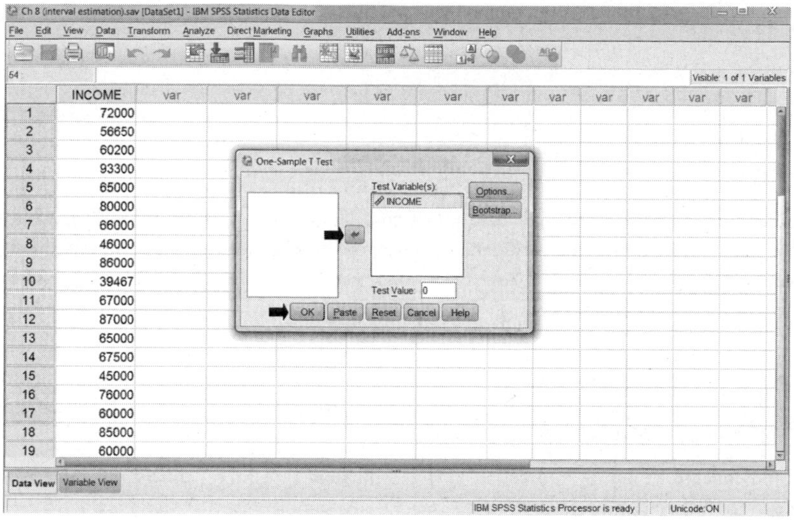

4. Examine output.

T - Test

N — Mean — Standard deviation — Standard error of the mean

One-Sample Statistics

	N	Mean	Std. Deviation	Std. Error Mean
INCOME	38	64508.84	14523.922	2356.091

degrees of freedom — Lower limit — Upper limit

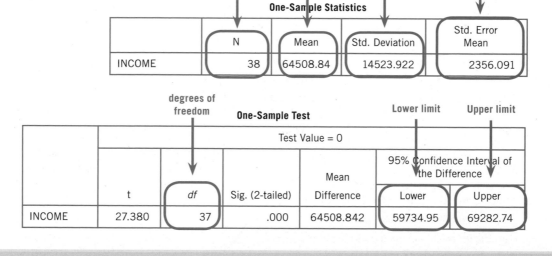

One-Sample Test

	Test Value = 0					
					95% Confidence Interval of the Difference	
	t	*df*	Sig. (2-tailed)	Mean Difference	Lower	Upper
INCOME	27.380	37	.000	64508.842	59734.95	69282.74

8.12 Exercises

1. For each of the following situations, calculate a 95% confidence interval for the mean (σ not known), beginning with the step, "Calculate the degrees of freedom (*df*) and identify the critical value of *t*."

   a. $N = 15, \overline{X} = 6.00, s_{\overline{X}} = 1.50$

   b. $N = 24, \overline{X} = 23.40, s_{\overline{X}} = 1.73$

c. $N = 13, \bar{X} = 8.50, s_{\bar{X}} = .73$

d. $N = 19, \bar{X} = 3.37, s_{\bar{X}} = .11$

2. For each of the following situations, calculate a 95% confidence interval for the mean (σ not known), beginning with the step, "Calculate the degrees of freedom (*df*) and identify the critical value of *t*."

   a. $N = 11, \bar{X} = 3.00, s_{\bar{X}} = .13$

   b. $N = 20, \bar{X} = 17.83, s_{\bar{X}} = 2.37$

   c. $N = 17, \bar{X} = 1.56, s_{\bar{X}} = .14$

   d. $N = 30, \bar{X} = 69.71, s_{\bar{X}} = 4.76$

3. For each of the following sets of numbers, calculate a 95% confidence interval for the mean (σ not known); before going through the steps in calculating the confidence interval, the sample mean ($\bar{X}$) and standard deviation (*s*) must first be calculated.

   a. 7, 2, 5, 9, 6, 6

   b. 4, 8, 13, 6, 7, 11, 15, 10

   c. 6, 25, 20, 7, 10, 9, 21, 14, 11, 15

   d. 89, 92, 87, 84, 90, 88, 91, 80, 87, 93, 85

4. For each of the following sets of numbers, calculate a 95% confidence interval for the mean (σ not known); before going through the steps in calculating the confidence interval, the sample mean ($\bar{X}$) and standard deviation (*s*) must first be calculated.

   a. 12, 11, 16, 9, 14, 10

   b. 3.50, 2.25, 3.30, 4.75, 2.60, 4.00, 3.80

   c. 24, 31, 28, 26, 19, 33, 22, 17, 25

   d. 7, 4, 13, 8, 6, 10, 9, 5, 17, 10, 7, 12

5. For each of the following sets of numbers, calculate a 95% confidence interval for the mean (σ not known); before going through the steps in calculating the confidence interval, the sample mean ($\bar{X}$) and standard deviation (*s*) must first be calculated. Next, draw a conclusion about each confidence interval.

   a. 22, 29, 30, 25, 21, 19, 17

   b. 10, 1, 7, 4, 5, 5, 6, 2

   c. 2.25, 1.48, 3.31, 1.90, 2.82, 3.07, 1.98, 1.54, 2.09, 2.56, 3.81

   d. 15, 29, 10, 14, 23, 9, 19, 12, 21, 34, 11, 5, 17, 13, 25

6. The exercises in Chapter 7 included a study designed to measure the number of chocolate chips in a bag of Chips Ahoy cookies (Warner & Rutledge, 1999). From this example, for the sample of 18 bags of cookies, the mean ($\bar{X}$) and standard deviation (*s*) of the number of chips were found to be 1,261.00 and 114.01, respectively. Imagine you want to estimate the average number of chocolate chips in the population of Chips Ahoy bags of cookies. Calculate a 95% confidence interval for the mean for this set of data.

   a. State the desired level of confidence.

   b. Calculate the confidence interval and confidence limits.

      (1) Calculate the standard error of the mean ($s_{\bar{X}}$).

      (2) Calculate the degrees of freedom (*df*) and identify the critical value of *t*.

      (3) Calculate the confidence interval and confidence limits.

   c. Draw a conclusion about the confidence interval.

7. In an earlier SIOP salary survey (Katkowski & Medsker, 2001), 73 of the respondents (33 males, 40 females) received their master's degrees in the 10-year period from 1991 to 2000. Listed below are descriptive statistics of the income of the two genders:

Calculate two 95% confidence intervals, one for males and one for females. For each gender . . .

a. State the desired level of confidence.

b. Calculate the confidence interval and confidence limits.

(1) Calculate the standard error of the mean ($s_{\bar{X}}$).

(2) Calculate the degrees of freedom (*df*) and identify the critical value of *t*.

(3) Calculate the confidence interval and confidence limits.

c. Draw a conclusion about the confidence interval.

Gender	*N*	Mean ($\bar{X}$)	*SD* (*s*)
Male	33	$70,727	$41,845
Female	40	$63,200	$30,182

8. How long does it take an ambulance to respond to a request for emergency medical aid? One of the goals of one study was to estimate the response time of ambulances using warning lights (Ho & Lindquist, 2001). They timed a total of 67 runs in a small rural county in Minnesota. They calculated the mean response time to be 8.51 minutes, with a standard deviation of 6.64 minutes. Calculate a 95% confidence interval for the mean for this set of data.

a. State the desired level of confidence.

b. Calculate the confidence interval and confidence limits.

(1) Calculate the standard error of the mean ($s_{\bar{X}}$).

(2) Calculate the degrees of freedom (*df*) and identify the critical value of *t*.

(3) Calculate the confidence interval and confidence limits.

c. Draw a conclusion about the confidence interval.

9. For each of the following situations, calculate a 95% confidence interval for the mean (σ known), beginning with the step, "Identify the critical value of *z*."

a. $\bar{X} = 7.00, \sigma_{\bar{X}} = 1.00$

b. $\bar{X} = 65.50, \sigma_{\bar{X}} = 2.18$

c. $\bar{X} = 26.40, \sigma_{\bar{X}} = 1.05$

d. $\bar{X} = 112.00, \sigma_{\bar{X}} = 10.13$

10. For each of the following situations, calculate a 95% confidence interval for the mean (σ known), beginning with the step, "Identify the critical value of *z*."

a. $\bar{X} = 50.00, \sigma_{\bar{X}} = 3.00$

b. $\bar{X} = 1.58, \sigma_{\bar{X}} = .12$

c. $\bar{X} = 7.34, \sigma_{\bar{X}} = 1.87$

d. $\bar{X} = 32.56, \sigma_{\bar{X}} = 3.70$

11. For each of the following sets of numbers, calculate a 95% confidence interval for the mean (σ known); before going through the steps in calculating the confidence interval, the sample mean ($\bar{X}$) and standard deviation (*s*) must first be calculated. Next, draw a conclusion about each confidence interval.

a. 9, 2, 8, 16, 7, 4, 12 (assume σ = 5.00)

b. 1.50, .75, 3.22, 1.25, .78 (assume σ = .30)

c. 116, 123, 97, 108, 112, 101, 119, 104 (assume σ = 15)

12. For each of the following sets of numbers, calculate a 95% confidence interval for the mean (σ known); before going through the steps in calculating the confidence interval, the sample mean ($\overline{X}$) and standard deviation (s) must first be calculated. Next, draw a conclusion about each confidence interval.

    a. 4, 7, 3, 6, 2, 5, 2, 4, 3 (assume $\sigma = 1.50$)

    b. 43, 34, 48, 31, 39, 32, 45, 40, 46, 37, 42 (assume $\sigma = 6.25$)

    c. .18, .15, .20, .16, .14, .18, .22, .17, .26, .13, .15, .09, .23, .13 (assume $\sigma = .06$)

13. An exercise in Chapter 7 referred to a program designed to improve scores on the Graduate Record Examination (GRE). The class had 25 students and they scored a mean of 1075.00 on the GRE. Assuming the population standard deviation for the GRE (σ) is 200, estimate the mean GRE score for the population of students who take this program by calculating the 95% confidence interval for the mean.

    a. State the desired level of confidence.

    b. Calculate the confidence interval and confidence limits.

        (1) Calculate the population standard error of the mean ($\sigma_{\overline{X}}$).

        (2) Identify the critical value of z.

        (3) Calculate the confidence interval and confidence limits.

    c. Draw a conclusion about the confidence interval.

14. William and Meagan are working on their senior projects, both of which involve estimating the average alcohol consumption of students on their college campuses. Imagine they achieve the same mean and standard deviation ($\overline{X} = 2.60$ drinks per week, $s = 1.10$), but the sample sizes of the two studies differ (William: $N = 25$, Meagan: $N = 100$).

    a. Calculate the 95% confidence interval for the mean (σ not known) for each of the two studies.

    b. Compare the width of the confidence intervals and explain why they differ.

15. In the Chips Ahoy cookie example (Exercise 6), a 95% confidence interval of (1204.30, 1317.70) was constructed for $N = 18$ bags of cookies. Assuming the mean and standard deviation remained the same ($\overline{X} = 1261.00$, $s = 114.01$)...

    a. What is the confidence interval for the mean for $N = 36$ bags rather than 18?

    b. What is the confidence interval for the mean for $N = 200$ bags rather than 18?

    c. What happens to the confidence interval as the sample gets larger (18 to 36 to 200)?

16. In the ambulance response time example (Exercise 8), a 95% confidence interval was constructed for $N = 67$ ambulance runs. Assuming the mean and standard deviation remained the same ($\overline{X} = 8.51$, $s = 6.64$)...

    a. What is the confidence interval for the mean for $N = 125$ ambulance runs rather than 67?

    b. What is the confidence interval for the mean for $N = 33$ ambulance runs rather than 67?

    c. How do these two confidence intervals compare with the 95% confidence interval of (6.89, 10.13) calculated in Exercise 8?

17. In the Chips Ahoy cookie example (Exercise 6), a 95% confidence interval with a width of approximately 100 (the difference between the lower limit of 1,204.30 and the higher limit of 1,317.70) was constructed when the sample consisted of 18 bags of cookies ($N = 18$). Using the standard deviation of 114.01 as the estimate of the population standard deviation ($\hat{\sigma}$) . . .

    a. How large should the sample be for a desired 95% confidence interval width of 50 chocolate chips?

    b. How large should the sample be for a desired 95% confidence interval width of 30 chocolate chips?

    c. How large should the sample be for a desired 95% confidence interval width of 10 chocolate chips?

18. In the ambulance response time example (Exercise 8), a 95% confidence interval with a width of approximately 3¼ minutes was constructed based on a sample of 67 runs ($N = 67$). Using the standard deviation of 6.64 as the estimate of the population standard deviation ($\hat{\sigma}$) . . .

    a. How large should the sample be for a desired 95% confidence interval width of 2 minutes?

    b. How large should the sample be for a desired 95% confidence interval width of 1 minute?

    c. How large should the sample be for a desired 95% confidence interval width of ½ minute (30 seconds)?

19. Calculate the 90% confidence interval for the first three situations in Exercise 1 (a–d).

20. Calculate the 99% confidence interval for the first three situations in Exercise 1 (a–d).

21. Calculate the 90% and 99% confidence interval for the Chips Ahoy example (Exercise 6).

22. Calculate the 90% and 99% confidence interval for the ambulance response time example (Exercise 8).

23. A pediatrician wanted to estimate the average temperature of children who come to her for treatment. She records the temperature of 23 children and calculates a mean of 98.83° and a standard deviation of 2.11°.
    a. Calculate the 90%, 95%, and the 99% confidence interval for the mean (σ not known) for her data.
    b. Which of the three confidence intervals is the widest? Which is the narrowest?

Answers to Learning Checks

Learning Check 1

2. a. $df = 10$; critical value = 2.228; 95% CI = 3.00 ± 1.11; confidence limits = (1.89, 4.11)
   b. $df = 21$; critical value = 2.080; 95% CI = 7.50 ± 2.60; confidence limits = (4.90, 10.10)
   c. $df = 25$; critical value = 2.060; 95% CI = 16.42 ± 4.68; confidence limits = (11.74, 21.10)

3. a. $\bar{X} = 4.00$; $s = 1.58$; $s_{\bar{X}} = .71$; $df = 4$; critical value = 2.776; 95% CI = 4.00 ± 1.97; confidence limits = (2.03, 5.97)
   b. $\bar{X} = 14.50$; $s = 5.12$; $s_{\bar{X}} = 1.93$; $df = 6$; critical value = 2.447; 95% CI = 14.50 ± 4.72; confidence limits = (8.56, 18.01)
   c. $\bar{X} = 97.34$; $s = 8.29$; $s_{\bar{X}} = 2.62$; $df = 9$; critical value = 2.262; 95% CI = 97.34 ± 5.93; confidence limits = (53.67, 65.53)

4. a. Desired level of confidence = 95%
   b. $s_{\bar{X}} = .22$; $df = 158$; critical value = 1.980; 95% CI = 4.93 ± .44; confidence limits = (4.49, 5.37)
   c. There is a .95 probability the interval of 4.49 to 5.38 contains the mean number of drinks consumed in the population of fraternity members.

Learning Check 2

2. a. Critical value = 1.96; 95% CI = 4.00 ± 2.45; confidence limits = (1.55, 6.45)
   b. Critical value = 1.96; 95% CI = 14.50 ± 4.23; confidence limits = (10.27, 18.73)
   c. Critical value = 1.96; 95% CI = 97.34 ± 7.08; confidence limits = (90.26, 104.42)

3. a. $\bar{X} = 6.17$; = .41; critical value $\sigma_{\bar{X}} = 1.96$; 95% CI = 6.17 ± .80; confidence limits = (5.36, 6.97); we are 95% confident that the interval of 5.36 to 6.97 contains the population mean.
   b. $\bar{X} = 20.75$; = .97; critical value $\sigma_{\bar{X}} = 1.96$; 95% CI = 20.75 ± 1.90; confidence limits = (18.85, 22.65); we are 95% confident that the interval of 18.85 to 22.65 contains the population mean.
   c. $\bar{X} = 3.02$; = .15; critical value $\sigma_{\bar{X}} = 1.96$; 95% CI = 3.02 ± .29; confidence limits = (2.72, 3.31); we are 95% confident that the interval of 2.72 to 3.31 contains the population mean.

4. a. Desired level of confidence = 95%
   b. $\sigma_{\bar{X}} = 1.26$; critical value = 1.96; 95% CI = 77.10 ± 2.47; confidence limits = (74.63, 79.57)
   c. There is a .95 probability the interval of 74.63 to 79.57 contains the mean level of social functioning in the population of people seeking treatment for obesity.

Learning Check 3

2. a. College Blue:
   (1) Desired level of confidence = 95%
   (2) $s_{\bar{X}} = .28$; $df = 9$; critical value = 2.262; 95% CI = 2.81 ± .63; confidence limits = (2.18, 3.44)
   College Gold:
   (1) Desired level of confidence = 95%
   (2) $s_{\bar{X}} = .16$; $df = 29$; critical value = 1.045; 95% CI = 2.81 ± .33; confidence limits = (2.48, 3.14)
   b. The confidence interval for College Gold is narrower and more precise because of its larger sample size, which decreases the standard error of the mean and lowers the critical value.
3. a. $N = 25$
   (1) Desired level of confidence = 95%
   (2) $\sigma_{\bar{X}} = 20.00$; critical value = 1.960; 95% CI = 550.00 ± 39.20; confidence limits = (510.80, 589.20)
   b. $N = 50$
   (1) Desired level of confidence = 95%
   (2) $\sigma_{\bar{X}} = 14.14$; critical value = 1.960; 95% CI = 550.00 ± 27.71; confidence limits = (522.29, 577.71)
   c. $N = 100$
   (1) Desired level of confidence = 95%
   (2) $\sigma_{\bar{X}} = 10.00$; critical value = 1.960; 95% CI = 550.00 ± 19.60; confidence limits = (530.40, 569.60)
   d. Increasing the sample size narrows the width of the confidence interval. For example, increasing the sample size from $N = 25$ to $N = 100$ cuts the width of the confidence interval in half.
4. a. $z = 1.96$; $\hat{\sigma} = 100.0$; CI = 80; $N' = 24.01$, sample size of 25 is needed
   b. $z = 1.96$; $\hat{\sigma} = 100.0$; CI = 40; $N' = 96.04$, sample size of 97 is needed
   c. $z = 1.96$; $\hat{\sigma} = 100.0$; CI = 20; $N' = 384.16$, sample size of 385 is needed

Learning Check 4

2. a. 90%:
   (1) Desired level of confidence = 90%
   (2 $\sigma_{\bar{X}} = 28.90$; critical value = 1.64; 90% CI = 550.00 ± 47.40; confidence limits = (502.60, 597.40)
   99%:
   (1) Desired level of confidence = 99%
   (2) $\sigma_{\bar{X}} = 28.90$; critical value = 2.58; 99% CI = 550.00 ± 74.56; confidence limits = (475.44, 624.56)
   b. The 99%: confidence interval is wider than the 90% confidence interval—in this example, it is 50 points wider. The 99% interval is wider to have a greater level of confidence that the interval contains the population mean.
3. a. 90%:
   (1) Desired level of confidence = 90%
   (2) $s_{\bar{X}} = .007$; $df = 79$; critical value = 1.671; 90% CI = 5.11 ± .012; confidence limits = (5.098, 5.122)
   95%:
   (1) Desired level of confidence = 95%
   (2) $s_{\bar{X}} = .007$; $df = 79$; critical value = 2.000; 95% CI = 5.11 ± .014; confidence limits = (5.096, 5.124)
   99%:
   (1) Desired level of confidence = 99%
   (2) $s_{\bar{X}} = .007$; $df = 79$; critical value = 2.660; 99% CI = 5.11 ± .019; confidence limits = (5.091, 5.129)
   b. The three confidence intervals appear to be very similar in width; this is in part due to the small standard deviation ($s = .06$), which indicates that there is very little variability in the weights of baseballs. The lack of variability increases the precision with which the population mean may be estimated.

Answers to Odd-Numbered Exercises

1. a. $df = 14$; critical value = 2.145; 95% CI = 6.00 ± 3.22; confidence limits = (2.78, 9.22)
   b. $df = 23$; critical value = 2.069; 95% CI = 23.40 ± 3.58; confidence limits = (19.82, 26.98)
   c. $df = 12$; critical value = 2.179; 95% CI = 8.50 ± 1.59; confidence limits = (6.91, 10.09)
   d. $df = 18$; critical value = 2.101; 95% CI = 3.37 ± .23; confidence limits = (3.14, 3.60)

3. a. $\bar{X} = 5.83$; $s = 2.32$; $s_{\bar{X}} = .95$; $df = 5$; critical value = 2.571; 95% CI = 5.83 ± 2.44; confidence limits = (3.39, 8.28)
   b. $\bar{X} = 9.25$; $s = 3.69$; $s_{\bar{X}} = 1.31$; $df = 7$; critical value = 2.365; 95% CI = 9.25 ± 3.10; confidence limits = (6.15, 12.35)
   c. $\bar{X} = 13.80$; $s = 6.41$; $s_{\bar{X}} = 2.03$; $df = 9$; critical value = 2.262; 95% CI = 13.80 ± 4.49; confidence limits = (9.21, 18.39)
   d. $\bar{X} = 87.82$; $s = 3.82$; $s_{\bar{X}} = 1.15$; $df = 10$; critical value = 2.228; 95% CI = 87.82 ± 2.56; confidence limits = (85.26, 90.38)

5. a. $\bar{X} = 23.29$; $s = 4.92$; $s_{\bar{X}} = 1.86$; $df = 6$; critical value = 2.447; 95% CI = 23.29 ± 4.55; confidence limits = (18.73, 27.84); we are 95% confident that the interval of 18.73 to 27.84 contains the population mean.
   b. $\bar{X} = 5.00$; $s = 2.83$; $s_{\bar{X}} = 1.00$; $df = 7$; critical value = 2.365; 95% CI = 5.00 ± 2.37; confidence limits = (2.64, 7.37); we are 95% confident that the interval of 2.64 to 7.37 contains the population mean.
   c. $\bar{X} = 2.44$; $s = .75$; $s_{\bar{X}} = .23$; $df = 10$; critical value = 2.228; 95% CI = 2.44 ±.51; confidence limits = (1.92, 2.95); we are 95% confident that the interval of 1.92 to 2.95 contains the population mean.
   d. $\bar{X} = 17.13$; $s = 8.02$; $s_{\bar{X}} = 2.07$; $df = 14$; critical value = 2.145; 95% CI = 17.13 ± 4.44; confidence limits = (12.69, 21.57); we are 95% confident that the interval of 12.69 to 21.57 contains the population mean.

7. Males:
   a. Desired level of confidence = 95%
   b. $s_{\bar{X}} = 7{,}290.07$; $df = 32$; critical value = 2.042; 95% CI = 70,727.00 ± 14,886.32; confidence limits = (55,840.68, 85,613.32)
   c. There is a .95 probability the interval of $55,840.68 to $85,613.32 contains the mean income for the population of males who recently received their master's degrees in I-O psychology.

   Females:
   a. Desired level of confidence = 95%
   b. $s_{\bar{X}} = 4{,}775.63$; $df = 39$; critical value = 2.042; 95% CI = 63,200.00 ± 9,751.84; confidence limits = (53,448.16, 72,951.84)
   c. There is a .95 probability the interval of $53,448.16 to $72,951.84 contains the mean income for the population of females who recently received their master's degrees in I-O psychology.

9. a. Critical value = 1.96; 95% CI = 7.00 ± 1.96; confidence limits = (5.04, 8.96)
   b. Critical value = 1.96; 95% CI = 65.50 ± 4.27; confidence limits = (61.23, 69.77)
   c. Critical value = 1.96; 95% CI = 26.40 ± 2.06; confidence limits = (24.34, 28.46)
   d. Critical value = 1.96; 95% CI = 112.00 ± 19.85; confidence limits = (92.15, 131.85)

11. a. $\bar{X} = 8.29$; $\sigma_{\bar{X}} = 1.89$; critical value = 1.96; 95% CI = 8.29 ± 3.70; confidence limits = (4.58, 11.99); we are 95% confident that the interval of 4.58 to 11.99 contains the population mean.
    b. $\bar{X} = 1.50$; $\sigma_{\bar{X}} = .13$; critical value = 1.96; 95% CI = 1.50 ± .25; confidence limits = (1.25, 1.75); we are 95% confident that the interval of 1.25 to 1.75 contains the population mean.
    c. $\bar{X} = 110.00$; $\sigma_{\bar{X}} = 5.30$; critical value = 1.96; 95% CI = 110.00 ± 10.39; confidence limits = (99.61, 120.39); we are 95% confident that the interval of 99.61 to 120.39 contains the population mean.

13. a. Desired level of confidence = 95%
    b. $\sigma_{\bar{X}} = 40.00$; critical value = 1.96; 95% CI = 1,075.00 ± 78.40; confidence limits = (996.60, 1,153.40)
    c. There is a .95 probability the interval of 996.60 to 1,153.40 contains the mean GRE score for the population of students who take the program.

15. a. $N = 36$
   (1) $s_{\bar{X}} = 19.00$; $df = 35$; critical value = 2.042; 95% CI = 1,261.00 ± 38.80; confidence limits = (1,222.20, 1,299.80)
   b. $N = 200$
   (1) $s_{\bar{X}} = 8.06$; $df = 199$; critical value = 1.980; 95% CI = 1,261.00 ± 15.96; confidence limits = (1,245.04, 1,276.96)
   c. As the sample size increases, the width of the confidence gets smaller. We are able to estimate the population mean with greater precision.
17. a. $z = 1.96$; $\hat{\sigma} = 114.01$; CI = 25; $N' = 79.92$, a sample size of 80 is needed.
   b. $z = 1.96$; $\hat{\sigma} = 114.01$; CI = 15; $N' = 222.01$, a sample size of 223 is needed.
   c. $z = 1.96$; $\hat{\sigma} = 114.01$; CI = 5; $N' = 1{,}997.20$, a sample size of 1,998 is needed.
19. a. $s_{\bar{X}} = 1.50$; $df = 14$; critical value = 1.761; 90% CI = 6.00 ± 2.64; confidence limits = (3.36, 8.64)
   b. $s_{\bar{X}} = 1.73$; $df = 23$; critical value = 1.714; 90% CI = 23.40 ± 2.97; confidence limits = (20.43, 26.37)
   c. $s_{\bar{X}} = .73$; $df = 12$; critical value = 1.782; 90% CI = 8.50 ± 1.30; confidence limits = (7.20, 9.80)
   d. $s_{\bar{X}} = .11$; $df = 18$; critical value = 1.734; 90% CI = 3.37 ± .19; confidence limits = (3.18, 3.56)
21. a. 90% confidence interval
   (1) $s_{\bar{X}} = 26.89$; $df = 17$; critical value = 1.740; 90% CI = 1,261.00 ± 46.79; confidence limits = (1,214.21, 1,307.79)
   b. 99% confidence interval
   (1) $s_{\bar{X}} = 26.89$; $df = 17$; critical value = 2.898; 99% CI = 1,261.00 ± 77.93; confidence limits = (1,183.07, 1,338.93)
23. a. 90% confidence interval
   (1) $s_{\bar{X}} = .44$; $df = 22$; critical value = 1.717; 90% CI = 98.83 ± .76; confidence limits = (98.07, 99.59)
   95% confidence interval
   (2) $s_{\bar{X}} = .44$; $df = 22$; critical value = 2.074; 95% CI = 98.83 ± .91; confidence limits = (97.92, 99.74)
   99% confidence interval
   (3) $s_{\bar{X}} = .44$; $df = 22$; critical value = 2.819; 99% CI = 98.83 ± 1.24; confidence limits = (97.59, 100.07)
   b. The 99% confidence interval is the widest; the 90% confidence interval is the narrowest (the greater the level of desired confidence, the wider the interval has to be to contain the population mean).

9

TESTING THE DIFFERENCE BETWEEN TWO MEANS

CHAPTER OUTLINE

(Continued)

This chapter returns to a discussion of the process of calculating inferential statistics to test research hypotheses. Chapter 7 introduced this process with the simplest example, the test of one mean, which evaluated the difference between a sample mean ($\bar{X}$) and a hypothesized population mean (μ). This chapter, on the other hand, discusses inferential statistics that evaluate the difference between the means of two samples drawn from two populations. Although the calculations in this chapter are slightly more complicated than those in Chapter 7, throughout this chapter you'll see many critical similarities between the test of one mean and the difference between two means.

9.1 AN EXAMPLE FROM THE RESEARCH: YOU CAN JUST WAIT

Like most college towns, Berkeley has an inordinate number of coffee houses. The author of this book has spent a great deal of time in these businesses preparing lectures, grading exams, and, of course, drinking a lot of coffee. It is for this reason that this book's author became introduced to a sign next to a restroom door:

> Remember: How long a minute is depends on which side of the door you're on.

Does it ever seem to you that people move slower when they know you're waiting for them than when they don't know you're there? Well, perhaps it isn't your imagination.

Two sociologists at Pennsylvania State University, Barry Ruback and Daniel Juieng, were interested in studying territorial behavior, defined as "marking, occupying, or defending a location in order to indicate presumed rights to the particular place" (Ruback & Juieng, 1997, p. 821). Although you may think of territorial behavior as protecting our homes from burglars, the researchers studied this behavior in public places. For example, imagine you're in a busy library working on a photocopy machine when someone walks up and, without saying a word to you, makes it apparent he also wishes to use it. Do you (a) speed up to finish more quickly, (b) continue to work in the same manner as if no one were waiting, or (c) deliberately slow down and actually take longer than if no one were there?

The theory of territorial behavior states that people sometimes select choice (c). In describing this behavior, Ruback and Juieng (1997) proposed that when a person possesses a limited resource that is desired by others, the person will maintain possession of the resource to defend it from "intruders" and "would be territorial even when they had completed their task at the location and the territory no longer served any function to them" (p. 823).

The researchers chose to test their beliefs in a familiar setting: a shopping mall. Picture yourself driving in a crowded parking lot when you see someone leave the mall and walk to his car. You drive over and wait for him to leave. And you wait . . . and wait . . . and wait. Based on the theory of territorial behavior, the research hypothesis in the Ruback and Juieng (1997) study was that even if people no longer need a parking space, they will take longer to leave the space when another driver is waiting than when no such "intruder" is present.

To collect their data, the researchers watched drivers leave parking spaces at a large shopping mall. As such, this study, which we will refer to as the parking lot study, is an example of observational research. As defined in Chapter 1, observational research involves the systematic and objective observation of naturally occurring events, with little or no intervention on the part of the researcher.

The researchers in the parking lot study were interested in measuring the amount of time taken by drivers to leave a parking space. Using a stopwatch, they "started timing the moment the departing shopper opened the driver's side car door and stopped timing when the car had completely left the parking space" (Ruback & Juieng, 1997, p. 823). They also recorded whether or not another driver was waiting to use the parking space. If there was another driver waiting, the departing driver was defined as having an "intruder"; if not, the departing driver had "no intruder" present. Therefore, this study involved two variables. Driver group, the independent variable, consisted of two groups: Intruder and No Intruder. Because the Driver group variable consists of distinct categories or groups, it is measured at the nominal level of measurement. Time, the dependent variable, was the amount of time in seconds taken to leave the parking space. Because Time is a numeric variable with a true zero point (0 seconds), it is measured at the ratio level of measurement.

The original study consisted of 200 drivers. To save space, the example in this chapter will use a smaller sample of 30 drivers, equally divided between the Intruder and No Intruder groups. (Although the data in this example differ from the original study, the results of the analysis mirror those reached by the researchers.) The time in seconds taken by the 15 drivers in each group to leave their parking spaces is presented in Table 9.1(a); these data have been organized into grouped frequency distribution tables (Table 9.1(b)). An examination of these tables shows that the departure times of those in the Intruder group are generally longer than those in the No Intruder group. For example, the modal interval for the Intruder group is 31 to 40 seconds as opposed to 21 to 30 seconds for the No Intruder group. However, the shape of the distribution for both groups is somewhat normal, with each group having a range of about 40 seconds from the quickest driver to the slowest.

The next step in analyzing the collected data is to calculate descriptive statistics of the dependent variable for each level of the independent variable (Table 9.2). This table includes additions and changes to the notational system used in this book. For example, subscripts have been added to the mathematical symbols to distinguish each group's descriptive statistics—in the parking lot study, the sample sizes for the Intruder and No Intruder groups are symbolized by N_1 and N_2 rather than simply N. Also, the subscript i is used instead of a number to represent a group without specifying any particular one. For example, the symbol $\bar{X}_i$ represents the mean of either of the two groups.

Previous chapters have created figures to illustrate the distribution of scores for a variable. For example, bar charts were used in Chapter 2 to display the frequencies for the values of a variable measured at the nominal or ordinal level of measurement, and histograms and frequency polygons were used for interval or ratio variables. This chapter introduces figures used to illustrate descriptive statistics such as measures of central tendency and variability. Just as there are different types of figures for variables, there are different ways of displaying descriptive statistics, the choice being a function of the nature of the variables.

When the independent variable is measured at the nominal or ordinal level of measurement and the dependent variable is measured at the interval or ratio level of measurement, descriptive statistics are typically displayed using a bar graph. A **bar graph** is a figure in which bars are used to represent the mean of the dependent variable for each level of the independent variable. For the parking lot study, the bar graph in Figure 9.1 displays the descriptive statistics for the dependent variable Time for the Intruder and No Intruder groups.

The height of the bars in the bar graph in Figure 9.1 represents the mean of the dependent variable (Time) for each of the two groups. The sample mean serves as an estimate of the mean of the population from which the sample was drawn. However, from our discussion of sampling error in Chapters 6 and 7, we know that the means of samples drawn from the population vary as the result of random, chance factors; this variability is estimated using a statistic known as the standard error of the mean. To illustrate the variability of sample means, the T-shaped lines extending above and below the mean in each bar in Figure 9.1 measure one standard error of the mean above and below the sample mean ($\pm 1\, s_{\bar{X}}$).

TABLE 9.1 ● THE TIME (IN SECONDS) FOR DRIVERS IN THE INTRUDER AND NO INTRUDER GROUPS

(a) Raw Data

Intruder			
Driver	Time	Driver	Time
1	23	9	35
2	62	10	45
3	28	11	37
4	55	12	43
5	31	13	39
6	51	14	41
7	33	15	40
8	48		

No Intruder			
Driver	Time	Driver	Time
1	54	9	35
2	19	10	25
3	46	11	34
4	20	12	28
5	42	13	31
6	21	14	29
7	38	15	30
8	23		

(b) Grouped Frequency Distribution Tables

Intruder		
Time	*f*	%
> 60	1	7%
51–60	2	13%
41–50	4	27%
31–40	6	40%
21–30	2	13%
≤ 20	0	0%
Total	15	100%

No Intruder		
Time	*f*	%
> 60	0	0%
51–60	1	7%
41–50	2	13%
31–40	4	27%
21–30	6	40%
≤ 20	2	13%
Total	15	100%

Using Formula 7-5 from Chapter 7, the standard error of the mean $s_{\overline{X}}$ for the two groups is calculated as follows:

Intruder

$$s_{\overline{X}} = \frac{s_1}{\sqrt{N_1}} = \frac{10.42}{\sqrt{15}} = \frac{10.42}{3.87} = 2.69$$

No Intruder

$$s_{\overline{X}} = \frac{s_2}{\sqrt{N_2}} = \frac{10.08}{\sqrt{15}} = \frac{10.08}{3.87} = 2.60$$

TABLE 9.2 ● DESCRIPTIVE STATISTICS OF TIME FOR DRIVERS IN THE INTRUDER AND NO INTRUDER GROUPS

(a) Mean ($\overline{X}_i$)

Intruder	No Intruder
$\overline{X}_1 = \frac{\Sigma X}{N} = \frac{23+62+\ldots+41+40}{15} = \frac{611}{15} = 40.73$	$\overline{X}_2 = \frac{\Sigma X}{N} = \frac{54+19+\ldots+29+30}{15} = \frac{475}{15} = 31.67$

(b) Standard Deviation (s_i)

$s_1 = \sqrt{\dfrac{\Sigma(X - \bar{X})^2}{N-1}}$	$s_2 = \sqrt{\dfrac{\Sigma(X - \bar{X})^2}{N-1}}$
$= \sqrt{\dfrac{(23 - 40.73)^2 + \ldots + (40 - 40.73)^2}{15-1}}$	$= \sqrt{\dfrac{(54 - 31.67)^2 + \ldots + (30 - 31.67)^2}{15-1}}$
$= \sqrt{\dfrac{314.47 + \ldots + .54}{14}} = \sqrt{108.50}$	$= \sqrt{\dfrac{498.78 + \ldots + 2.78}{14}} = \sqrt{101.52}$
$= 10.42$	$= 10.08$

In Figure 9.1, for the Intruder group, the area covered by the T-shaped lines extends from 40.73 ± 2.69. From our discussion of confidence intervals in Chapter 8, we know that the range represented by the T-shaped line represents an interval or range with a stated probability of containing the mean of the population on the dependent variable.

Looking at Table 9.2 and Figure 9.1, we see that the mean time for the Intruder group ($M = 40.73$) is 9.06 seconds greater than that for the No Intruder group ($M = 31.67$). This difference in departure time provides initial support for the hypothesis that people will take longer to leave a parking space when an intruder is present than when there is no intruder. However, to formally test the study's research hypothesis, the next step is to calculate an inferential statistic to conclude the difference between the two sample means is not due to chance but instead is statistically significant.

FIGURE 9.1 ● BAR GRAPH OF TIME FOR DRIVERS IN THE INTRUDER AND NO INTRUDER GROUPS

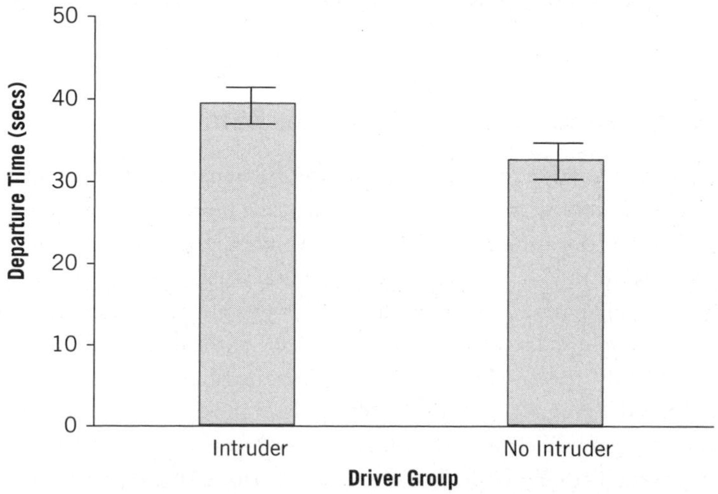

In Chapter 7, testing the difference between a sample mean and a population mean ($\bar{X}-\mu$) involved determining the probability of obtaining the value of the sample mean. To address the question, "What is the probability of obtaining a sample mean of 150.35 WCPM assuming the mean in the population is 124.81 WCPM?", we relied on the sampling distribution of the mean, defined as the distribution of all possible values of the sample mean when an infinite number of samples of size N are randomly selected from the population. However, as the goal in this chapter is to test the difference between two sample means ($\bar{X}_1 - \bar{X}_2$), we now need to determine the probability of obtaining our particular difference between the two sample means. In the parking lot study, we need to answer the question, "What is the probability of obtaining our difference in departure time of 9.06 seconds between the Intruder and No Intruder groups?" To answer this question, we need a different type of sampling distribution: a distribution of differences between sample means. This distribution is introduced and illustrated in the next section.

9.2 THE SAMPLING DISTRIBUTION OF THE DIFFERENCE

The **sampling distribution of the difference** is the distribution of all possible values of the difference between two sample means when an infinite number of pairs of samples of size N are randomly selected from two populations. It, like the sampling distribution of the mean (Chapter 7), is an example of sampling distribution, which is a distribution of statistics for samples randomly drawn from populations. The sampling distribution of the difference is used to determine the probability of obtaining any particular difference between two sample means. As such, we will rely on this distribution to test the difference between the departure times of the Intruder and No Intruder groups in the parking lot study.

To illustrate the sampling distribution of the difference, imagine you have two populations that do *not* differ on some variable, such that the two population means μ_1 and μ_2 are equal. You randomly draw samples from these two populations, calculate the mean for each sample, and then calculate the difference between the two sample means ($\bar{X}_1 - \bar{X}_2$). What should be the difference between the two means? Because we've stated that the two populations do not differ, we expect the difference between the two sample means to be zero. However, because the concept of sampling error implies there will be variability among sample means drawn from the populations, there will also be variability among differences between sample means. Consequently, the difference between the two sample means may not be equal to zero. Furthermore, if we were to repeat this process of randomly drawing samples and calculating the difference between sample means an infinite number of times, we could create a distribution of these differences. This distribution is known as the sampling distribution of the difference.

Characteristics of the Sampling Distribution of the Difference

Like any other distribution, the sampling distribution of the difference may be characterized in terms of its modality, symmetry, and variability. In terms of its modality, the mean of the sampling distribution of the difference is equal to 0 under the assumption that if the two population means are equal, the average difference between sample means should be zero. Next, in terms of its symmetry, the sampling distribution of the difference is approximately normal, assuming the two samples are of sufficient size (typically defined as $N \geq 30$); like the t-distribution discussed in Chapter 7, the shape of the distribution changes as a function of the size of the samples. Finally, the variability of the sampling distribution of the difference is measured by the **standard error of the difference** ($s_{\bar{X}_1-\bar{X}_2}$), defined as the standard deviation of the sampling distribution of the difference. The standard error of the difference represents the variability of differences between two sample means (the formula used to calculate the standard error of the difference will be presented later in this chapter).

The characteristics of the sampling distribution of the difference enable researchers to determine the probability of obtaining any particular difference between the means of two samples. For the parking lot study, determining the probability of obtaining our difference of 9.06 seconds between the Intruder ($M = 40.73$) and No Intruder ($M = 31.67$) groups will allow us to test the study's research hypothesis that people will take longer to leave a parking space when another driver is waiting than when no such "intruder" is present. In order to determine the probability of the difference between two sample means, it must be transformed into a statistic in a manner very similar to what we encountered in Chapter 7. The next section describes the process of calculating and evaluating an inferential statistic that tests the difference between two sample means.

LEARNING CHECK 1

Reviewing What You've Learned So Far

1. Review questions
   a. What changes to this book's notational system are introduced as a function of having two groups rather than one?
   b. What are the differences between visual displays of variables such as bar charts and visual displays of descriptive statistics such as bar graphs?
   c. What is the main purpose and characteristics of the sampling distribution of the difference?
   d. What is the difference between the sampling distribution of the mean (Chapter 7) and the sampling distribution of the difference?
   e. What is the difference between the standard error of the mean (Chapter 7) and the standard error of the difference?
2. Construct a bar graph for each of the following situations (assume the independent variable is Gender and the dependent variable is Score):
   a. Males ($N = 6$, $M = 3.00$, $s = 1.79$); Females ($N = 6$, $M = 5.00$, $s = 1.41$)
   b. Males ($N = 9$, $M = 13.00$, $s = 1.73$); Females ($N = 9$, $M = 11.00$, $s = 2.06$)
   c. Males ($N = 12$, $M = .64$, $s = .27$); Females ($N = 12$, $M = .45$, $s = .25$)
3. For each of the following situations, (a) calculate the mean, standard deviation, and standard error of the mean for each group and (b) construct a bar graph (assume the independent variable is Type of Pet and the dependent variable is Obedience).
   a. Dog: 1, 7, 1, 0, 1
      Cat: 6, 10, 5, 1, 8
   b. Dog: .76, .80, .67, .42, .56, .78, .49, .31, .24, .69
      Cat: .92, .51, .48, .65, .32, .71, .93, .26, .37, .14
   c. Dog: 13, 7, 9, 11, 12, 7, 8, 10, 6, 7, 5, 10, 9, 8
      Cat: 5, 4, 8, 10, 7, 6, 8, 4, 6, 9, 3, 5, 6, 7

9.3 INFERENTIAL STATISTICS: TESTING THE DIFFERENCE BETWEEN TWO SAMPLE MEANS

This section describes the steps involved in calculating and evaluating an inferential statistic that tests the difference between two sample means. In the parking lot study, do drivers take longer to leave a parking space when another driver is waiting than when there is no intruder? To test this hypothesis, we'll follow the steps introduced in earlier chapters:

- state the null and alternative hypotheses (H_0 and H_1),
- make a decision about the null hypothesis,

- draw a conclusion from the analysis, and
- relate the result of the analysis to the research hypothesis.

In discussing each of these steps, we will note both similarities and differences between the research situations discussed in this chapter versus those in Chapter 7, which involved one sample mean rather than the difference between two sample means.

State the Null and Alternative Hypotheses (H_0 and H_1)

The process of hypothesis testing begins by stating the null hypothesis (H_0), which reflects the conclusion that no change, difference, or relationship exists among groups or variables in the population. When testing the difference between two sample means, the null hypothesis states that the means of the two populations are equal. For the parking lot study, the null hypothesis is that the mean amount of time taken to leave a parking space is the same for drivers with or without an intruder. This absence of difference between the two populations is reflected in the following null hypothesis:

$$H_0\text{: } \mu_{\text{Intruder}} = \mu_{\text{No Intruder}}$$

The alternative hypothesis (H_1), which is mutually exclusive from the null hypothesis, states that in the population there does exist a change, difference, or relationship between groups or variables. For the parking lot study, one way to state the alternative hypothesis is to reflect the conclusion that the mean departure time in the two populations is *not* equal to one another. This conclusion is represented by the following alternative hypothesis:

$$H_1\text{: } \mu_{\text{Intruder}} \neq \mu_{\text{No Intruder}}$$

The use of the "not equals" ($\neq$) symbol in the above alternative hypothesis implies that the mean departure time of drivers in the No Intruder group in the population may *either* be greater than *or* less than the mean departure time of drivers in the Intruder group in the population. As such, this alternative hypothesis is considered "nondirectional" or "two-tailed." From the study's research hypothesis, which predicts that people will take a *greater* amount of time to leave in the presence of an intruder than when there is no intruder, it may have been reasonable for the researchers to propose the directional (one-tailed) alternative hypothesis H_1: $\mu_{\text{Intruder}} > \mu_{\text{No Intruder}}$. However, they chose the traditional approach of a nondirectional alternative hypothesis to allow for the possibility of a statistically significant difference in the opposite direction of what was anticipated.

Make a Decision About the Null Hypothesis

Once the null and alternative hypotheses have been stated, the next step is to make the decision whether to reject the null hypothesis. This involves completing the steps introduced in Chapter 7:

- calculate the degrees of freedom (*df*);
- set alpha (α), identify the critical values, and state a decision rule;
- calculate a statistic: *t*-test for independent means;
- make a decision whether to reject the null hypothesis; and
- determine the level of significance.

We complete each of these steps for the parking lot study below, highlighting changes new to this chapter.

Calculate the Degrees of Freedom (*df*)

The degrees of freedom (*df*) is defined as the number of values or quantities that are free to vary when a statistic is used to estimate a parameter. In testing the mean of one sample in Chapter 7, the degrees of freedom was equal to $N - 1$ (the number of scores in the sample minus 1). However, because samples have now been drawn from two populations rather than one, the number of degrees of freedom has changed. Formula 9-1 presents the formula for the degrees of freedom for the difference between two sample means:

$$df = (N_1 - 1) + (N_2 - 1) \tag{9-1}$$

where N_1 and N_2 are the sample sizes of the two groups.

For the parking lot study, where both groups have a sample size of 15 ($N_i = 15$), the degrees of freedom are calculated as follows:

$$\begin{aligned} df &= (N_1 - 1) + (N_2 - 1) \\ &= (15 - 1) + (15 - 1) = 14 + 14 \\ &= 28 \end{aligned}$$

By combining the degrees of freedom for the two samples, there are a total of 28 degrees of freedom ($df = 28$) for the data in the parking lot study.

Set Alpha (α), Identify the Critical Values, and State a Decision Rule

The second step in making the decision to reject the null hypothesis consists of three parts. The first part is to set alpha (α), which is the probability of the inferential statistic needed to reject the null hypothesis. The second part is to use the degrees of freedom and stated value of alpha to identify the critical values of the statistic we are about to calculate; the critical values determine the values of the statistic that result in the decision to reject the null hypothesis. The third part is to state a decision rule that explicitly states the logic to be followed in making the decision whether to reject the null hypothesis.

In this chapter, alpha will be set to the traditional value of .05, implying the null hypothesis will be rejected if the probability of the calculated value of the statistic is less than .05. Furthermore, because the alternative hypothesis in the parking lot study (H_1: $\mu_{\text{Intruder}} \neq \mu_{\text{No Intruder}}$) is nondirectional, alpha may be stated as "$\alpha = .05$ (two-tailed)."

To identify the critical values of the statistic, we must first determine which particular inferential statistic will be calculated. In Chapter 7, when the population standard deviation for a variable was unknown, the sample mean was transformed into a *t*-statistic, which was part of the Student *t*-distribution. In this chapter, we have a similar situation in that the population standard deviations for the two groups are assumed to be unknown. As such, we will transform the difference between the two sample means into a *t*-statistic and again rely on the Student *t*-distribution.

To demonstrate how to identify the critical value of *t*, we return to the table of critical values for the *t*-distribution provided in Table 3 in the back of this book. Three pieces of information are needed to identify the critical value: the degrees of freedom (*df*), alpha (α), and the directionality of the alternative hypothesis (one-tailed or two-tailed). For the parking lot study, because we've calculated the degrees of freedom to be 28 ($df = 28$), we move down the *df* column of Table 3 until we reach the number 28. Next, because our alternative hypothesis is nondirectional, we move to the right to the set of columns labeled "Level of significance for two-tailed test." Within this set of columns, we move to the column for the stated value of .05 for alpha (α). Using these three pieces of information, we find a critical *t* value of 2.048. Therefore, for the parking lot study, the critical values may be stated as the following:

For $\alpha = .05$ (two-tailed) and $df = 28$, critical values $= \pm 2.048$.

The critical values and regions of rejection and nonrejection for the parking lot study are illustrated in Figure 9.2. This figure demonstrates that values of the *t*-statistic that are either less than −2.048 or greater than 2.048 have a "low" probability of occurring, that is, a probability less than .05.

Once the critical values have been identified, it is helpful to explicitly state a decision rule that specifies the logic to be followed in making the decision to reject or not reject the null hypothesis. For the parking lot study, the following decision rule is stated:

$$\text{If } t < -2.048 \text{ or } > 2.048, \text{ reject } H_0; \text{ otherwise, do not reject } H_0.$$

In this situation, the null hypothesis will be rejected if the value of *t* we calculate from our data is either less than −2.048 or greater than 2.048 because such a value is located in the region of rejection, meaning it has a low (< .05) probability of occurring.

Calculate a Statistic: *t*-Test for Independent Means

The next step in hypothesis testing is to calculate a value of an inferential statistic. As mentioned earlier, in testing the difference between two sample means, we will calculate a value of a *t*-statistic. Formula 9-2 provides the formula for the ***t*-test for independent means**, defined as an inferential statistic that tests the difference between the means of two samples drawn from two populations:

$$t = \frac{\overline{X}_1 - \overline{X}_2}{s_{\overline{X}_1 - \overline{X}_2}} \tag{9-2}$$

where $\overline{X}_1$ and $\overline{X}_2$ are the means for the two groups and $(s_{\overline{X}_1 - \overline{X}_2})$ is the standard error of the difference. This statistic is called the *t*-test for "independent" means to indicate that the data were collected from two samples drawn from two different populations, and that scores in one sample are unrelated to scores in the other sample. (Later in this chapter, we'll discuss research situations in which the two scores are related to each other because both are collected from the same participant.) For the parking lot study, "independence" implies that the No Intruder and Intruder drivers are from two different populations of drivers.

If you compare Formula 9-2 with Chapter 7's Formula 7-4 (the *t*-test for one mean), you see the two formulas are very similar. First, the numerator of both formulas involves the difference between means, which in both cases is the primary issue of interest in the analysis. However, the numerator of Formula 9-2 includes

FIGURE 9.2 ● CRITICAL VALUES AND REGIONS OF REJECTION AND NONREJECTION FOR THE PARKING LOT STUDY

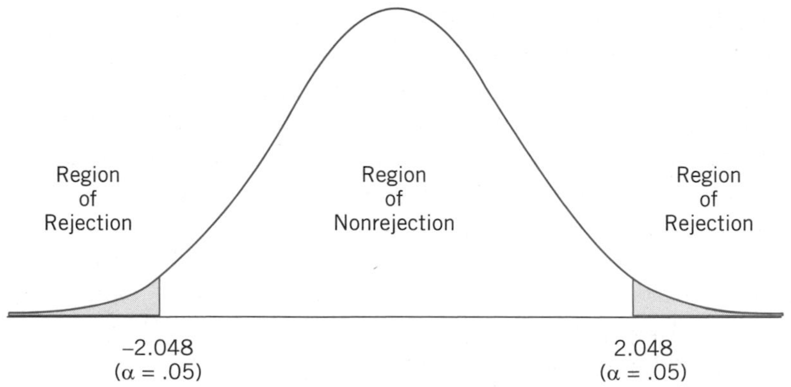

the difference between two sample means rather than the difference between a sample mean and a population mean. Second, although the denominator of both formulas contains a measure of variability, the denominator of Formula 9-2 measures the variability of differences between sample means rather than the variability of sample means.

The first step in calculating the *t*-statistic is to calculate the standard error of the difference $(s_{\overline{X}_1-\overline{X}_2})$ using Formula 9-3:

$$s_{\overline{X}_1-\overline{X}_2}=\sqrt{\frac{s_1^2}{N_1}+\frac{s_2^2}{N_2}} \tag{9-3}$$

where s_1 and s_2 are the standard deviations for the two groups and N_1 and N_2 are the sample sizes of the two groups. The formula for the standard error of the difference is very similar to the formula for the standard error of the mean $(s_{\overline{X}})$ discussed in Chapter 7, the critical difference being that we must now take into account the variability in two samples rather than one.

Using the standard deviations calculated in Table 9.2, the standard error of the difference for the parking lot study is calculated as follows:

$$\begin{aligned}s_{\overline{X}_1-\overline{X}_2}&=\sqrt{\frac{s_1^2}{N_1}+\frac{s_2^2}{N_2}}\\&=\sqrt{\frac{(10.42)^2}{15}+\frac{(10.08)^2}{15}}=\sqrt{7.24+6.77}=\sqrt{14.01}\\&=3.74\end{aligned}$$

In calculating the standard error of the difference, students sometimes get confused by the fact that the symbol $s_{\overline{X}_1-\overline{X}_2}$ includes a minus sign (–) but the formula contains the plus sign (+). The formula involves addition (+) because in order to estimate the variability of differences between two sample means, we need to combine (add together) the variability of the two samples. We are evaluating the difference between sample means, *not* the difference between sample standard deviations.

Inserting our value of the standard error of the difference and the Intruder and No Intruder sample means into Formula 9-2, a value of the *t*-statistic may now be calculated:

$$\begin{aligned}t&=\frac{\overline{X}_1-\overline{X}_2}{s_{\overline{X}_1-\overline{X}_2}}\\&=\frac{40.73-31.67}{3.74}=\frac{9.06}{3.74}\\&=2.42\end{aligned}$$

As you can see from these calculations, we've transformed the difference between the two sample means (40.73 –31.67) into a *t*-statistic. Furthermore, the calculated value for the *t*-statistic ($t=2.42$) is a positive number. This is because the mean of the first group, $M_{\text{Intruder}}=40.73$, is greater than the mean of the second group, $M_{\text{No Intruder}}=31.67$. Because the designation of the first and second groups is arbitrary, the *t*-statistic would have been –2.42 if the No Intruder group had been the first group and the Intruder group the second.

Make a Decision Whether to Reject the Null Hypothesis

The next step is to make a decision whether to reject the null hypothesis by comparing the value of the *t*-statistic calculated from the data with the identified critical values. If the *t*-statistic exceeds one of the critical values, it falls in the region of rejection, and the decision is made to reject the null hypothesis. This decision implies that

LEARNING CHECK 2

Reviewing What You've Learned So Far

1. Review questions
   a. In testing the difference between two sample means, what conclusions are represented by the null and alternative hypotheses (H_0 and H_1)?
   b. What is the difference between the degrees of freedom (*df*) for the test of one sample mean (Chapter 7) and the difference between two sample means?
   c. What are the main differences between the formulas for the *t*-test for one sample mean (Chapter 7) and the *t*-test for independent means?
   d. What is the main difference between the formulas for the standard error of the mean (Chapter 7) and the standard error of the difference?
2. State the null and alternative hypotheses (H_0 and H_1) for the following research questions:
   a. Do children from single-parent families have higher self-esteem than children from intact families?
   b. Does tap water taste any different than bottled water?
   c. Who is more likely to intervene when a crime takes place: someone walking alone or someone walking with other people?
3. For each of the following situations, calculate the degrees of freedom (*df*) and identify the critical value of *t* (assume $\alpha = .05$).
   a. $N_1 = 7, N_2 = 7, H_1: \mu_1 \neq \mu_2$
   b. $N_1 = 11, N_2 = 11, H_1: \mu_1 \neq \mu_2$
   c. $N_1 = 8, N_2 = 8, H_1: \mu_1 > \mu_2$
4. For each of the following situations, calculate the standard error of the difference ($s_{\overline{X}_1 - \overline{X}_2}$).
   a. $N_1 = 9, s_1 = 3.00, N_2 = 9, s_2 = 6.00$
   b. $N_1 = 16, s_1 = 8.00, N_2 = 16, s_2 = 12.00$
   c. $N_1 = 24, s_1 = 3.38, N_2 = 24, s_2 = 3.76$
5. For each of the following situations, calculate the standard error of the difference ($s_{\overline{X}_1 - \overline{X}_2}$) and the *t*-test for independent means.
   a. $N_1 = 6, \overline{X}_1 = 9.00, s_1 = 3.00, N_2 = 6, \overline{X}_2 = 6.00, s_2 = 2.00$
   b. $N_1 = 15, \overline{X}_1 = 53.62, s_1 = 13.54, N_2 = 15, \overline{X}_2 = 49.12, s_2 = 11.87$
   c. $N_1 = 36, \overline{X}_1 = 7.63, s_1 = 2.90, N_2 = 36, \overline{X}_2 = 10.50, s_2 = 3.38$

the difference between the means of the two groups is statistically significant, meaning it is unlikely to occur as the result of chance factors.

For the parking lot study, the decision about the null hypothesis may be stated the following way:

$$t = 2.42 > 2.048 \therefore \text{ reject } H_0 \ (p < .05)$$

Because our calculated value of *t* of 2.42 is greater than the critical value 2.048, it falls in the region of rejection and the decision is made to reject the null hypothesis because its probability is less than .05; in other words, $p < .05$. As a result, we've decided that the difference between the two groups in the amount of time taken to leave their parking spaces ($M_{\text{Intruder}} = 40.73$ and $M_{\text{No Intruder}} = 31.67$) is statistically significant.

Determine the Level of Significance

If and when the null hypothesis is rejected, it is appropriate to determine whether the probability of the *t*-statistic is not only less than .05 but also less than .01. In Chapter 7, we discussed that when researchers calculate a statistic using statistical software, they often report the exact probability of the statistic (i.e., "$p = .017$") rather than use cutoffs such as "$p < .05$." However, in this textbook, we include the step of determining whether the probability of a statistic meets the .01 level of significance because "$p < .01$" is commonly included in tables and figures of statistics in journal articles.

To determine whether the probability of the t-statistic for the parking lot study is less than .01, we return to the table of critical values in Table 3. Moving down to the $df = 28$ row, we move to the right until we are under the .01 column under "Level of significance for two-tailed test." Here, we find the critical value 2.763. Comparing the calculated t-statistic of 2.42 with the $\alpha = .01$ critical value of 2.763, we reach the following conclusion:

$$t = 2.42 < 2.763 \therefore p < .05 \text{ (but not} < .01)$$

Figure 9.3 illustrates the location of the value of t from this example relative to the .05 and .01 critical values. Because the calculated t-statistic of 2.42 was greater than the .05 critical value of 2.048, it falls in the region of rejection and we made the decision to reject the null hypothesis. However, because 2.42 is *less* than the .01 critical value of 2.763, its probability is less than .05 but *not* less than .01. Consequently, we would report "$p < .05$" as the level of significance. As a reminder, when the value of a statistic doesn't exceed the .01 critical value, this does *not* mean the null hypothesis is not rejected; that decision has already been made. It simply means that the probability of the statistic is not less than .01.

Draw a Conclusion From the Analysis

Given that we've made the decision to reject the null hypothesis, what conclusion can be drawn about the difference between the departure times of the Intruder and No Intruder groups? One way to report the results of the analysis is the following:

The mean departure time for the 15 drivers in the Intruder group ($M = 40.73$ s) is significantly greater than the mean departure time for the 15 drivers in the No Intruder group ($M = 31.67$ s), $t(28) = 2.42$, $p < .05$.

Note that this single sentence provides the following information:

- the dependent variable ("The mean departure time"),
- the two samples ("15 drivers . . . 15 drivers"),
- the independent variable ("Intruder group . . . No Intruder group"),
- descriptive statistics ("$M = 40.73$ s . . . $M = 31.67$ s"),

FIGURE 9.3 ● DETERMINING THE LEVEL OF SIGNIFICANCE FOR THE PARKING LOT STUDY

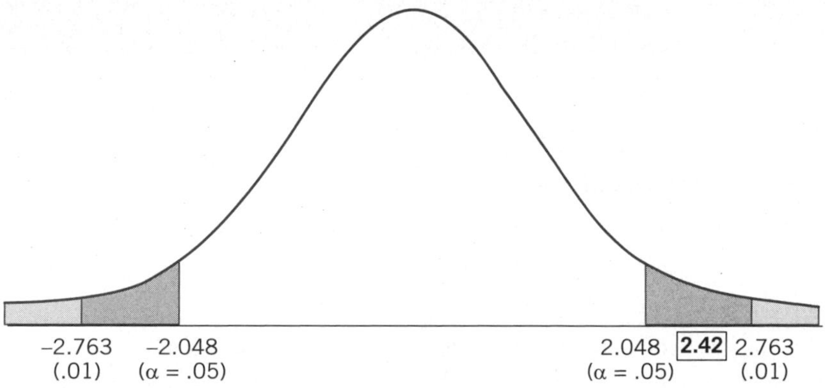

- the nature and direction of the findings ("significantly greater than"), and
- information about the inferential statistic ("$t(28) = 2.42, p < .05$"), which indicates the inferential statistic calculated (t), the degrees of freedom (28), the value of the statistic (2.42), and the level of significance ($p < .05$).

Relate the Result of the Analysis to the Research Hypothesis

It's critical to relate the result of the statistical analysis back to the research hypothesis it was designed to test. In the parking lot study, does the finding that drivers took a significantly longer amount of time to leave a parking space in the presence of an intruder support or not support the study's research hypothesis? Here is what the authors of the study had to say:

> The present series of studies is consistent with prior findings that people display territorial defense in public territories. . . . What is new about the present research is that it suggests people sometimes display territorial behavior merely to keep others from possessing the space even when it no longer has any value to them. (Ruback & Juieng, 1997, p. 831)

Note that the researchers did not say that their study "proved" their hypothesis but rather that it "*suggests* people *sometimes* display territorial behavior." As was discussed in Chapter 6, research hypotheses cannot be proven because data are collected from a sample rather than the entire population.

The process of testing the difference between two sample means is summarized in Box 9.1, using the parking lot study as an example. In closing, you may remember an old saying: "A watched pot never boils." Research suggests this may reflect how people behave in social situations. Sometimes a 'watched' person really does take longer to move!

Assumptions of the *t*-Test for Independent Means

The goal of statistical procedures such as the *t*-test is to test hypotheses researchers have about populations by analyzing data collected from samples of these populations. These procedures make certain mathematical assumptions about the distribution of scores for variables in the population. To appropriately use these procedures, researchers must determine whether the data in their samples meet these assumptions. This section discusses assumptions related to the *t*-test for independent means and strategies available to researchers if their data do not meet these assumptions.

SUMMARY BOX 9.1 TESTING THE DIFFERENCE BETWEEN TWO SAMPLE MEANS (PARKING LOT STUDY EXAMPLE)

State the null and alternative hypotheses (H_0 and H_1)

$$H_0: \mu_{Intruder} = \mu_{No\ Intruder} \quad H_1: \mu_{Intruder} \neq \mu_{No\ Intruder}$$

Make a decision about the null hypothesis

Calculate the degrees of freedom (df)

$$df = (N_1 - 1) + (N_2 - 1) = (15 - 1) + (15 - 1) = 14 + 14 = 28$$

Set alpha (α), identify the critical values, and state a decision rule

If $t < -2.048$ or > 2.048, reject H_0; otherwise, do not reject H_0

Calculate a statistic: *t*-test for independent means

Calculate the standard error of the difference $(s_{\bar{X}_1-\bar{X}_2})$

$$s_{\bar{X}_1-\bar{X}_2}=\sqrt{\frac{s_1^2}{N_1}+\frac{s_2^2}{N_2}}=\sqrt{\frac{(10.42)^2}{15}+\frac{(10.08)^2}{15}}=\sqrt{14.01}=3.74$$

Calculate the *t*-statistic (t)

$$t=\frac{\bar{X}_1-\bar{X}_2}{s_{\bar{X}_1-\bar{X}_2}}=\frac{40.73-31.67}{3.74}=\frac{9.06}{3.74}=2.42$$

Make a decision whether to reject the null hypothesis

$$t=2.42>2.048 \therefore \text{ reject } H_0 \ (p<.05)$$

Determine the level of significance

$$t=2.42<2.763 \therefore p<.05 \text{ (but not } <.01)$$

Draw a conclusion from the analysis

The mean departure time for the 15 drivers in the Intruder group (M = 40.73 s) is significantly greater than the mean departure time for the 15 drivers in the No Intruder group (M = 31.67 s), $t(28) = 2.42$, $p < .05$.

Relate the result of the analysis to the research hypothesis

"The present series of studies is consistent with prior findings that people display territorial defense in public territories. . . . What is new about the present research is that it suggests people sometimes display territorial behavior merely to keep others from possessing the space even when it no longer has any value to them" (Ruback & Juieng, 1997, p. 831).

The first assumption related to the *t*-test for independent means is the assumption of normality, which is the assumption that scores for the dependent variable in each of the two populations are approximately normally distributed; that is, the shape of the two distributions resembles normal (bell-shaped) curves. The implication of this assumption is that distributions of scores for samples drawn from these populations are also expected to be approximately normal. The assumption of normality is common to many of the statistical procedures discussed in this book.

The second assumption applicable to the *t*-test for independent means is **homogeneity of variance**, which is the assumption that the variance of scores in the two populations is the same. Assuming this assumption is true, the variance in two samples drawn from the two populations should also be the same. Figure 9.4(a) illustrates two distributions that meet the assumptions of normality and homogeneity of variance.

A violation of homogeneity of variance might take place if the variance in one sample is much less or greater than the variance in the other sample; this is illustrated in Figure 9.4(b). When the homogeneity of variance assumption is violated, it's possible that the two samples are not representative of their populations, which in turn raises doubt about the results of statistical analyses conducted on the samples.

Violating the assumptions of normality and homogeneity of variance increases the possibility of making the wrong decision regarding the null hypothesis for a statistic such as the *t*-test. For example, a researcher may decide to reject the null hypothesis and conclude the difference between the two sample means is significant when the groups do not actually differ in the populations from which the samples were drawn. (These types of errors in decision making are discussed in greater detail in the next chapter.)

Statistical tests have been developed to determine whether the assumptions of normality and homogeneity of variance are met in a set of data. However, we will not discuss these tests for two reasons: They are

FIGURE 9.4 ● ILLUSTRATION OF THE HOMOGENEITY OF VARIANCE ASSUMPTION

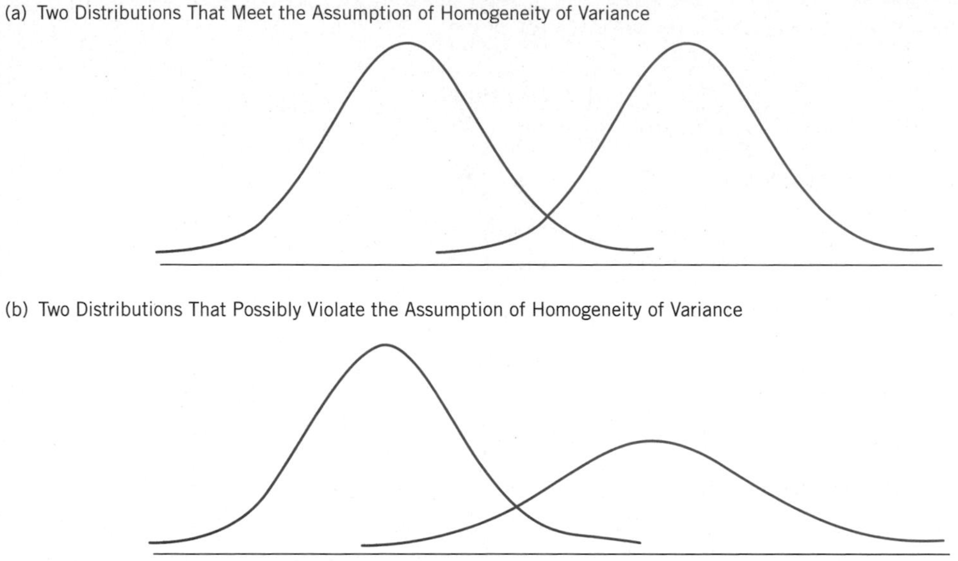

beyond the scope of this textbook, and the need to conduct these tests has been called into question (i.e., Zimmerman, 2004). As we mentioned in Chapter 7, research has found that statistics such as the *t*-statistic are "robust," meaning they can withstand moderate violations of their mathematical assumptions. To withstand a violation of an assumption implies that, even if the data from a sample violate an assumption, the decision made regarding the null hypothesis is the same decision that would have been made had the assumption not been violated. For example, imagine that two populations do not differ on a variable that is normally distributed in the populations. Even if data from samples of these populations happen to violate the assumptions, a robust statistic such as the *t*-test will lead to the correct decision, which in this case is to not reject the null hypothesis. This is particularly true when the two samples are of equal size and are sufficiently large (i.e., $N_i \geq 30$).

Even when a statistic is considered robust, researchers may wish to address possible violations of assumptions by altering how they analyze their data. One way to do this is to use different formulas to calculate and evaluate the *t*-test (Welch, 1938); we will not present these formulas for the same reasons mentioned earlier. A second way to address extreme violations of assumptions is to employ an alternative statistical procedure, one that is not based on these assumptions. For example, rather than calculate the *t*-test for independent means, the data can be analyzed using a statistical procedure known as the Mann-Whitney *U*. Examples of these alternative types of statistical procedures are introduced in Chapter 14 of this textbook.

Summary

The parking lot study compared the means of two groups calculated from samples of equal size ($N_i = 15$). Having sample sizes equal to each other is preferable because it helps ensure that the two samples are treated equally in analyzing the data. However, as it is possible that the two groups may not have the same sample size, the next section describes how to test the difference between two sample means that are based on unequal sample sizes.

LEARNING CHECK 3

Reviewing What You've Learned So Far

1. Review questions
   a. What are the two mathematical assumptions related to the *t*-test for independent means?
   b. What does it mean to violate the assumption of heterogeneity of variance?
   c. What does it mean for a statistic to be robust?
   d. What can researchers do when they believe assumptions have been violated?
2. How much do you think the average adult woman weighs? Let's imagine that a researcher, in response to available evidence from both the media and society, entertains the assumption that men believe that women weigh less than they actually do. Consequently, the researcher hypothesizes that men will give lower estimates of the average woman's weight than will women. To test this hypothesis, the researcher asks samples of men and women, "How much (in pounds) do you believe the average adult woman weighs?" The researcher reports the following descriptive statistics regarding the mean estimate of weight: Men: $N = 13$, $M = 135.62$, $s = 19.78$, Women: $N = 13$, $M = 137.69$, $s = 12.68$.
   a. State the null and alternative hypotheses (H_0 and H_1).
   b. Make a decision about the null hypothesis.
      (1) Calculate the degrees of freedom (*df*).
      (2) For $\alpha = .05$ (two-tailed), identify the critical values and state a decision rule.
      (3) Calculate the *t*-test for independent means.
      (4) Make a decision whether to reject the null hypothesis.
      (5) Determine the level of significance.
   c. Draw a conclusion from the analysis.
   d. Relate the result of the analysis to the research hypothesis.

9.4 INFERENTIAL STATISTICS: TESTING THE DIFFERENCE BETWEEN TWO SAMPLE MEANS (UNEQUAL SAMPLE SIZES)

The example in this section tests the difference between the means of two samples with different sample sizes. With unequal *N*s, the calculation of the *t*-test is slightly more complicated because the formula for the standard error of the difference ($s_{\bar{X}_1-\bar{X}_2}$) must be modified.

An Example From the Research: Kids' Motor Skills and Fitness

Childhood obesity is a serious concern in this country as it has been linked to psychological problems, such as lowered self-esteem and depression, and physical problems such as diabetes and high blood pressure. A variety of interventions have been developed to combat childhood obesity, some of which have the goal of changing children's lifestyles and eating habits. However, two researchers at the University of Northern Iowa, Oksana Matvienko and Iradge Ahrabi-Fard, wondered whether a brief intervention focusing on "the development of motor skills applied in popular sports and games is an effective approach for increasing physical activity particularly among young and preadolescent children" (Matvienko & Ahrabi-Fard, 2010, p. 299).

The researchers developed a short, 4-week intervention aimed at developing the motor skills of kindergarten and first-grade students. As part of an after-school program, students received classroom lessons on topics such as human anatomy and nutrition, played exercises and games designed to increase their physical strength and endurance, and learned different motor skills such as kicking balls and rope jumping.

The researchers obtained the permission of four elementary schools to participate in the study; students at two of the schools received the intervention, and students at the other two schools did not. This study is an example of quasi-experimental research, which involves comparing preexisting groups rather than randomly assigning participants to conditions. We will refer to the independent variable in this study as Group, which consists of two levels: Intervention and Control.

Students in both the Intervention and Control groups were measured on different physical activities. In this example, we'll focus on the number of times the student was able to successfully jump rope in 30 seconds; the dependent variable in this example will be called Jumps. To investigate how long the effect of the intervention may last, the number of jumps for each student was measured 4 months after the intervention.

Because the number of students attending each of the schools differed, the number of students in the Intervention and Control groups was not equal to each other. For this example, the number of students in the Intervention and Control groups will be 16 and 11, respectively. Data reflecting the results of this study for the two groups are presented in Table 9.3.

The descriptive statistics for the number of jumps variable are calculated in Table 9.5; the means of the two groups are displayed in Figure 9.5. Looking at this table and figure, we see that students in the Intervention group averaged a higher number of jumps in 30 seconds ($M = 27.31$) than did students in the Control group ($M = 11.91$).

Inferential Statistics: Testing the Difference Between Two Sample Means (Unequal Sample Sizes)

As in the parking lot study, the next step in analyzing the motor skills study is to calculate an inferential statistic to test the study's research hypothesis. Except for one difference, the steps followed in testing the difference between two sample means are the same whether or not the two sample sizes are equal. The difference is that with unequal sample sizes, calculating the standard error of the difference ($s_{\bar{X}_1-\bar{X}_2}$) is slightly more complicated.

State the Null and Alternative Hypotheses (H_0 and H_1)

In this example, the null hypothesis is that the average number of jumps for students who receive the intervention is the same as for students who do not receive the intervention; this represents the conclusion that the intervention does not affect students' physical fitness. The null hypothesis for the motor skills study is stated as follows:

$$H_0: \mu_{\text{Intervention}} = \mu_{\text{Control}}$$

TABLE 9.3 ● NUMBER OF JUMPS FOR STUDENTS IN THE INTERVENTION AND CONTROL GROUPS

(a) Raw Data

Intervention			
Student	Jumps	Student	Jumps
1	**37**	9	**30**
2	**16**	10	**12**
3	**26**	11	**44**
4	**20**	12	**17**
5	**39**	13	**35**
6	**11**	14	**27**
7	**22**	15	**38**
8	**34**	16	**29**

Control			
Student	Jumps	Student	Jumps
1	**16**	9	**8**
2	**6**	10	**3**
3	**33**	11	**6**
4	**2**		
5	**13**		
6	**7**		
7	**11**		
8	**26**		

(b) Grouped Frequency Distribution Tables

Intervention		
Jumps	f	%
> 40	1	6%
36–40	3	19%
31–35	2	13%
26–30	4	25%
21–25	1	6%
16–20	3	19%
11–15	2	13%
6–10	0	0%
≤ 5	0	0%
Total	16	100%

Control		
Jumps	f	%
> 40	0	0%
36–40	0	0%
31–35	1	9%
26–30	1	9%
21–25	0	0%
16–30	1	9%
11–15	2	18%
6–10	4	36%
≤ 5	2	18%
Total	11	100%

TABLE 9.4 ● DESCRIPTIVE STATISTICS OF JUMPS FOR STUDENTS IN THE INTERVENTION AND CONTROL GROUPS

(a) Mean ($\overline{X}_i$)

Intervention

$$\overline{X}_1 = \frac{\Sigma X}{N}$$
$$= \frac{37 + 16 + \ldots + 38 + 29}{16} = \frac{437}{16}$$
$$= 27.31$$

Control

$$\overline{X}_2 = \frac{\Sigma X}{N}$$
$$= \frac{16 + 6 + \ldots + 3 + 6}{11} = \frac{131}{11}$$
$$= 11.91$$

(b) Standard Deviation (s_i)

$$s_1 = \sqrt{\frac{\Sigma(X - \overline{X})^2}{N - 1}}$$
$$= \sqrt{\frac{(37 - 27.31)^2 + \ldots + (29 - 27.31)^2}{16 - 1}}$$
$$= \sqrt{\frac{93.85 + \ldots + 2.85}{15}} = \sqrt{103.70}$$
$$= 10.18$$

$$s_2 = \sqrt{\frac{\Sigma(X - X)^2}{N - 1}}$$
$$= \sqrt{\frac{(16 - 11.91)^2 + \ldots + (6 - 11.91)^2}{11 - 1}}$$
$$= \sqrt{\frac{16.74 + \ldots + 34.92}{10}} = \sqrt{94.89}$$
$$= 9.74$$

In this example, a nondirectional (two-tailed) alternative hypothesis will be used, stating that the mean number of jumps for the two groups of students in the population is *not* equal:

$$H_1: \mu_{\text{Intervention}} \neq \mu_{\text{Control}}$$

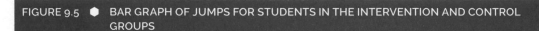

FIGURE 9.5 ● BAR GRAPH OF JUMPS FOR STUDENTS IN THE INTERVENTION AND CONTROL GROUPS

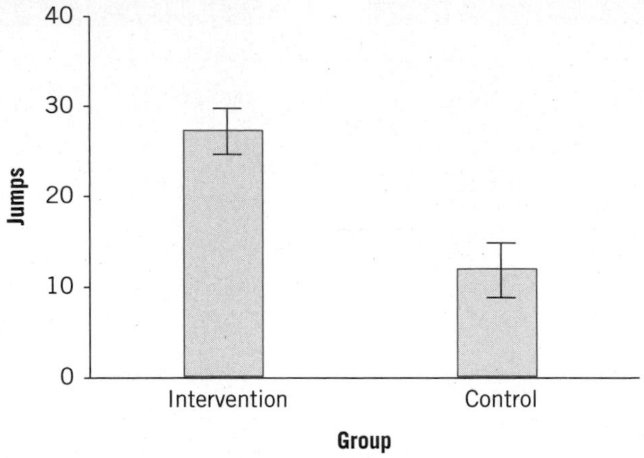

Make a Decision About the Null Hypothesis

The first step in making the decision whether to reject the null hypothesis is to *calculate the degrees of freedom (df)*. Inserting the sample sizes of $N_1 = 16$ and $N_2 = 11$ for the Intervention and Control groups into Formula 9-1, the degrees of freedom for the motor skills study are

$$\begin{aligned} df &= (N_1 - 1) + (N_2 - 1) \\ &= (16 - 1) + (11 - 1) = 15 + 10 \\ &= 25 \end{aligned}$$

The next step is to set alpha (α), identify the critical values, and state a decision rule. In this example, alpha will once again be defined as "$\alpha = .05$ (two-tailed)." Next, the critical values are identified by moving down the *df* column in Table 3 until we reach the $df = 25$ row. For $\alpha = .05$ (two-tailed), we find a critical value of 2.060. Consequently,

For $\alpha = .05$ (two-tailed) and $df = 25$, critical value $= \pm 2.060$.

Based on our values of alpha and the critical values, we can state a decision rule used to make the decision about the null hypothesis. For the data in the motor skills study:

If $t < -2.060$ or > 2.060, reject H_0; otherwise, do not reject H_0.

The next step in making a decision about the null hypothesis is to *calculate a statistic*—in this case, the *t*-test for independent means. Looking back at Formula 9-2, the first part of calculating the *t*-statistic is to calculate the standard error of the difference ($s_{\bar{X}_1 - \bar{X}_2}$). This is the one aspect of the analysis that differs depending on whether the sample sizes are equal. Formula 9-4 presents the formula for the standard error of the difference for unequal sample sizes:

$$s_{\bar{X}_1 - \bar{X}_2} = \sqrt{\frac{(N_1 - 1)s_1^2 + (N_2 - 1)s_2^2}{N_1 + N_2 - 2}\left(\frac{1}{N_1} + \frac{1}{N_2}\right)} \tag{9-4}$$

where N_1 and N_2 are the sample sizes of the two groups, and s_1 and s_2 are the standard deviations of the two groups. Notice that this formula is more complicated algebraically than Formula 9-3 (the standard error of the difference with equal sample sizes) in that the sample size for each group (N_1 and N_2) must be represented multiple times. The standard error of the difference for the motor skills example is calculated as follows:

$$\begin{aligned} s_{\bar{X}_1-\bar{X}_2} &= \sqrt{\frac{(N_1-1)s_1^2+(N_2-1)s_2^2}{N_1+N_2-2}\left(\frac{1}{N_1}+\frac{1}{N_2}\right)} \\ &= \sqrt{\frac{(16-1)(10.18)^2+(11-1)(9.74)^2}{16+11-2}\left(\frac{1}{16}+\frac{1}{11}\right)} \\ &= \sqrt{\frac{(15)(103.63)+(10)(94.87)}{25}(.06+.09)} \\ &= \sqrt{\frac{2,503.15}{25}(.15)} = \sqrt{15.02} \\ &= 3.88 \end{aligned}$$

Once the standard error of the difference has been calculated, we can calculate a value of the t-statistic using Formula 9-2:

$$\begin{aligned} t &= \frac{\bar{X}_1-\bar{X}_2}{s_{\bar{X}_1-\bar{X}_2}} \\ &= \frac{27.31-11.91}{3.88} = \frac{15.40}{3.88} \\ &= 3.97 \end{aligned}$$

Next, we make a decision whether to reject the null hypothesis. In this example:

$$t = 3.97 > 2.060 \therefore \text{ reject } H_0 \ (p < .05)$$

Here, because the calculated t-statistic of 3.97 is greater than the critical value 2.060, the null hypothesis is rejected because 3.97 lies in the region of rejection at the right end of the t-distribution. This decision implies that the mean number of jumps for the two groups (27.31 for the Intervention group and 11.91 for the Control group) are significantly different from each other.

Because the decision has been made to reject the null hypothesis, it is appropriate to *determine the level of significance*. Returning to the table of critical values in Table 3, for $df = 25$ and a .01 (two-tailed) probability we find a critical value of 2.787. Comparing the calculated t-statistic with this critical value, we draw the following conclusion:

$$t = 3.97 > 2.787 \therefore p < .01$$

Because the t-statistic of 3.97 is greater than the .01 critical value of 2.787, its probability is not only less than .05 but also less than .01 (see Figure 9.6). Therefore, the probability of obtaining the observed difference in the mean number of jumps for the Intervention and Control groups is less than .01.

Draw a Conclusion From the Analysis

What conclusions could we make on the basis of this analysis? Following the format of the earlier examples, the following statement could be made:

> The average number of rope jumps in 30 seconds is significantly greater for the 16 students who received the intervention ($M = 27.31$) than for the 11 students in the control group who did not receive the intervention ($M = 11.91$), $t(25) = 3.97$, $p < .01$.

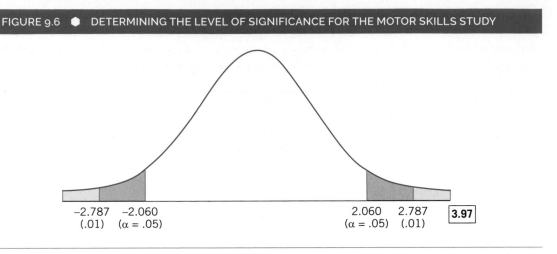

FIGURE 9.6 ● DETERMINING THE LEVEL OF SIGNIFICANCE FOR THE MOTOR SKILLS STUDY

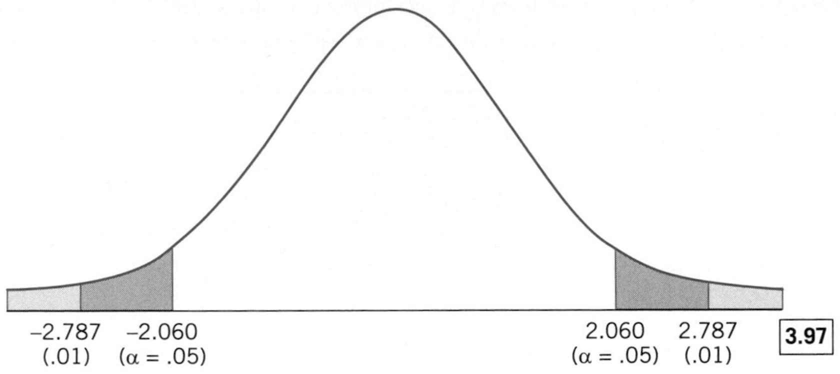

Relate the Result of the Analysis to the Research Hypothesis

The purpose of the motor skills study was to evaluate the effectiveness of a brief intervention aimed at improving the physical activity levels of kindergarten and first-grade students. What are the implications of our statistical analysis for evaluating the effectiveness of this intervention? Here is what the authors of the study said:

> This finding suggests that programs emphasizing the enhancement of basic motor skills that children apply in a variety of games and sports may be an effective approach to increasing overall activity and fitness levels of young children. (Matvienko & Ahrabi-Fard, 2010, p. 303)

The process of testing the difference between two sample means with unequal samples is summarized in Box 9.2.

Assumptions of the *t*-Test for Independent Means (Unequal Sample Sizes)

Earlier in this chapter, we introduced two assumptions related to the *t*-test for independent means: the assumption of normality, which is the assumption that the distribution of scores in the two populations is approximately normal, and the assumption of homogeneity of variance, in which the variance of scores in the two populations is the same. The data collected by researchers should meet these assumptions to draw appropriate conclusions from the results of statistical analyses. This is of particular concern when the sample sizes of the two groups are not equal to each other, as unequal sample sizes exacerbate the effects of differences in the shape and variance of the distributions of the two samples.

Statistics such as the *t*-statistic have been found to be robust, meaning that even if the data from a sample violate an assumption, the decision made regarding the null hypothesis is the same that would have been made if the assumption had not been violated. However, researchers have found that the robustness of statistics such as the *t*-statistic is lessened when the sample sizes of the groups are not equal to each other, especially when the variances are also unequal. The combination of unequal sample sizes and unequal variances may result in the need to consider the alternatives to the *t*-test for independent means discussed earlier in this chapter.

Summary

The examples discussed thus far in this chapter have compared participants in different conditions or groups. For example, the parking lot study compared the departure times of drivers who either had or did not have an intruder, and the motor skills study measured the jumping rope ability of students who either received or did not receive the after-school intervention. As such, these studies are examples of **between-subjects research designs**, defined as research designs in which each research participant appears in only one level or category of the independent variable. The word *between* indicates these designs involve testing differences between different groups of participants. The next section introduces a different type of research design, one in which each participant appears in all levels or categories of the independent variable, and as a result the researcher is testing differences *within* each participant.

SUMMARY BOX 9.2 TESTING THE DIFFERENCE BETWEEN TWO SAMPLE MEANS (UNEQUAL SAMPLE SIZES) (MOTOR SKILLS EXAMPLE)

State the null and alternative hypotheses (H_0 and H_1)

$$H_0: \mu_{Intervention} = \mu_{Control} \quad H_1: \mu_{Intervention} \neq \mu_{Control}$$

Make a decision about the null hypothesis

Calculate the degrees of freedom (*df*)

$$df = (N_1 - 1) + (N_2 - 1) = (16 - 1) + (11 - 1) = 15 + 10 = 25$$

Set alpha (α), identify the critical values, and state a decision rule

If $t < -2.060$ or > 2.060, reject H_0; otherwise, do not reject H_0

Calculate a statistic: *t*-test for independent means

Calculate the standard error of the difference (unequal sample sizes) ($s_{\bar{X}_1 - \bar{X}_2}$).

$$s_{\bar{X}_1-\bar{X}_2} = \sqrt{\frac{(N_1-1)s_1^2 + (N_2-1)s_2^2}{N_1+N_2-2}\left(\frac{1}{N_1}+\frac{1}{N_2}\right)}$$

$$= \sqrt{\frac{(16-1)(10.18)^2 + (11-1)(9.74)^2}{16+11-2}\left(\frac{1}{16}+\frac{1}{11}\right)}$$

$$= \sqrt{\frac{2{,}503.15}{25}(.15)} = \sqrt{100.13(.15)} = \sqrt{15.02} = 3.88$$

Calculate the *t*-statistic (*t*)

$$t = \frac{\bar{X}_1 - \bar{X}_2}{s_{\bar{X}_1-\bar{X}_2}} = \frac{27.31 - 11.91}{3.88} = \frac{15.40}{3.88} = 3.97$$

Make a decision whether to reject the null hypothesis

$$t = 3.97 > 2.060 \therefore \text{reject } H_0 \ (p < .05)$$

Determine the level of significance

$$t = 3.97 > 2.787 \therefore p < .01$$

Draw a conclusion from the analysis

The average number of rope jumps in 30 seconds is significantly greater for the 16 students who received the intervention ($M = 27.31$) than for the 11 students in the control group who did not receive the intervention ($M = 11.91$), $t(25) = 3.97$, $p < .01$.

Relate the result of the analysis to the research hypothesis

"This finding suggests that programs emphasizing the enhancement of basic motor skills that children apply in a variety of games and sports may be an effective approach to increasing overall activity and fitness levels of young children" (Matvienko & Ahrabi-Fard, 2010, p. 303).

LEARNING CHECK 4

Reviewing What You've Learned So Far

1. Review questions
   a. How does the calculation of the *t*-test for the difference between two sample means change when the sample sizes of the two groups are unequal?
2. For each of the following situations, calculate the standard error of the difference ($s_{\bar{X}_1-\bar{X}_2}$) using the formula for unequal sample sizes.
   a. $N_1 = 4$, $s_1 = 2.00$, $N_2 = 5$, $s_2 = 3.00$
   b. $N_1 = 20$, $s_1 = 6.00$, $N_2 = 10$, $s_2 = 4.00$
   c. $N_1 = 11$, $s_1 = 8.64$, $N_2 = 17$, $s_2 = 11.75$
3. To assist parents whose children have been diagnosed as having the eating disorder anorexia nervosa, one study evaluated the impact of a program designed to provide parents access to information and support groups (Carlton & Pyle, 2007). At the end of the program, parents who participated in the program ($N = 29$) and a group of parents of children with anorexia who did not participate ($N = 53$) completed a survey asking them to what extent they know what to feed their child at home (the higher the score, the higher the level of knowledge). The researchers reported the following descriptive statistics: participated ($M = 4.68$, $s = 1.57$) and did not participate ($M = 2.77$, $s = 2.02$).
   a. State the null and alternative hypotheses (H_0 and H_1).
   b. Make a decision about the null hypothesis.
      (1) Calculate the degrees of freedom (*df*).
      (2) For $\alpha = .05$ (two-tailed), identify the critical values and state a decision rule.
      (3) Calculate a value for the *t*-test for independent means (when calculating the standard error of the difference ($s_{\bar{X}_1-\bar{X}_2}$), be sure to use the formula for unequal sample sizes).
      (4) Make a decision whether to reject the null hypothesis.
      (5) Determine the level of significance.
   c. Draw a conclusion from the analysis.
   d. What conclusions might the researchers draw regarding the impact of the program on parents' knowledge of anorexia nervosa?

9.5 INFERENTIAL STATISTICS: TESTING THE DIFFERENCE BETWEEN PAIRED MEANS

This section will again use a research study to introduce a statistical procedure that tests the difference between two means. However, this study differs from the earlier examples in that each research participant appears in both levels of the independent variable rather than just one. It is an example of a **within-subjects research design**, which is a research design in which each participant appears in all levels or categories of the independent variable. Rather than testing the difference *between* different groups of participants, within-subjects research designs test differences *within* the same participant.

Research Situations Appropriate for Within-Subjects Research Designs

There are several types of research situations where within-subjects research designs are used. First, these designs are used to examine differences within a person regarding different situations or stimuli. An example of this situation would be to have people taste two types of ice cream and rate both types in terms of their flavor. One research study that used a within-subjects design for this purpose looked at police officers' beliefs regarding eyewitnesses' ability to provide accurate information about a crime (Kebbell & Milne, 1998). A sample of police officers were asked how often they believed eyewitnesses provided accurate information regarding four different aspects of a crime: the *person* who committed the crime, the *action* taken by the perpetrator, the *object* or target of the crime, and the *surroundings* in which the crime took place. Comparing these officers' ratings of these four aspects, the researchers found that officers believed eyewitnesses were more likely to provide useful information regarding the action involved in the crime than about the person committing the crime, the object of the crime, or the surroundings. That is, they believed witnesses are better able to describe the actions of a criminal than the actual criminal.

A second use of within-subjects designs involves collecting data on the same variable across repeated administrations. **Longitudinal research** is a research design in which the same information is collected from a research participant over two or more administrations to assess changes occurring within the person over time. An example of a longitudinal study, conducted by the author of this textbook (Tokunaga, 1985), examined the relationship between experiencing the death of a close friend or relative and changes in one's own fear of death during the initial bereavement period. People who had experienced this type of loss were contacted and asked to complete a survey measuring death-related fears every 4 months over a 12-month period.

Another type of longitudinal research, the **pretest-posttest research design**, involves collecting data from participants before and after the introduction of an intervention or experimental manipulation to determine whether the intervention or manipulation is associated with changes in the dependent variable. The research study described below is an example of a pretest-posttest design.

An Example From the Research: Web-Based Family Interventions

Given the large number of families in which both parents work, concern has been expressed regarding parents' ability to be aware of their children's well-being. A team of researchers lead by Diane Deitz set out to develop and evaluate a program designed to increase parents' knowledge of problems such as childhood anxiety and depression (Deitz, Cook, Billings, & Hendrickson, 2009). As the researchers wrote, "The prevalence of mental disorders in youth is substantial; however, these disorders often go unrecognized by parents and those closest to them . . . [therefore] the purpose of the project was to test a web-based program providing working parents with the knowledge and skills necessary for prevention and early intervention of mental health problems in youth" (p. 488).

In the study, the researchers developed a web-based program in which parents could work through a series of online modules covering such things as information about different mental disorders and treatment options for these disorders. As part of their study, the researchers hypothesized that "parents receiving the web-based program would exhibit significant gains in . . . knowledge of mental health issues in youth" (Deitz et al., 2009, p. 489).

The participants in the study, referred to as the web-based intervention study, were working parents with at least one child living at home. Before starting the program, each parent answered a test of 32 true/false items assessing their knowledge of depression, anxiety, treatment options, and parenting. Approximately 3 weeks later, after completing the program, parents completed the test a second time. In the study, the dependent variable, Knowledge, is the number of items each parent answered correctly. The independent variable, Time, consists of two levels: the test score before starting the program (Pretest) and after completing the program (Posttest). The purpose of the study was to see whether parents' knowledge of mental health issues in youth were higher after completing the program than at the start of the program.

The findings of the study will be illustrated using a sample of 20 parents. The parents' scores on the Knowledge variable both before (Pretest) and after (Posttest) the program are presented in Table 9.7(a). Notice that each of the research participants appears in both levels of the independent variable. For example, before the program was introduced, the first parent answered 16 items correctly; after the program, this same parent received a score of 24.

Table 9.8 calculates the mean and standard deviation for the Knowledge variable for the Pretest and Posttest time periods. From examining the descriptive statistics and the bar graph in Figure 9.7, we see that the mean Knowledge scores among these parents increased from the Pretest ($M = 15.55$) to Posttest ($M = 21.15$).

Inferential Statistics: Testing the Difference Between Paired Means

To test the study's research hypothesis, the next step is to calculate an inferential statistic to determine whether the difference between the two means is statistically significant. As in the earlier examples, the inferential statistic will be a *t*-test. However, having each research participant appear in both levels of the independent variable alters the steps used to test the difference between the two means.

Calculate the Difference Between the Paired Scores

In a between-subjects design, scores in the different groups are independent of each other. For example, in the parking lot study, the departure time for the first driver in the Intruder group is completely unrelated to the departure time of the first driver in the No Intruder group. However, in a pretest-posttest research design, the scores in the two groups are related to each other. For example, in the web-based intervention study, the first pretest and posttest knowledge scores of 16 and 24 in Table 9.5 were produced by the same person.

TABLE 9.5 ● KNOWLEDGE OF ANXIETY AND DEPRESSION FOR PARENTS AT PRETEST AND POSTTEST

(a) Raw Data

	Time	
Parent	Pretest	Posttest
1	**16**	**24**
2	**23**	**23**
3	**15**	**19**
4	**12**	**17**
5	**21**	**27**
6	**17**	**17**
7	**21**	**31**
8	**20**	**17**
9	**13**	**20**
10	**19**	**27**

	Time	
Parent	Pretest	Posttest
11	**12**	**22**
12	**14**	**19**
13	**17**	**28**
14	**8**	**9**
15	**12**	**20**
16	**16**	**22**
17	**10**	**15**
18	**14**	**27**
19	**12**	**21**
20	**19**	**18**

(b) Grouped Frequency Distribution Tables

	Pretest	
Knowledge	*f*	%
> 30	0	0%
26–30	0	0%
21–25	3	15%
16–20	7	35%
11–15	8	40%
6–10	2	10%
≤ 5	0	0%
Total	20	100%

	Posttest	
Knowledge	*f*	%
> 30	1	5%
26–30	4	20%
21–25	5	25%
16–20	8	40%
11–15	1	5%
6–10	1	5%
≤ 5	0	0%
Total	20	100%

TABLE 9.6 ● DESCRIPTIVE STATISTICS OF KNOWLEDGE FOR PARENTS AT THE PRETEST AND POSTTEST TIME PERIODS

(a) Mean ($\overline{X}_i$)

Pretest	Posttest
$\overline{X}_1 = \frac{\Sigma X}{N}$	$\overline{X}_2 = \frac{\Sigma X}{N}$
$= \frac{16+23+\ldots+12+19}{20} = \frac{311}{20}$	$= \frac{24+23+\ldots+21+18}{20} = \frac{423}{20}$
$= 15.55$	$= 21.15$

(b) Standard Deviation (s_i)

$s_1 = \sqrt{\frac{\Sigma(X-\overline{X})^2}{N-1}}$	$s_2 = \sqrt{\frac{\Sigma(X-\overline{X})^2}{N-1}}$
$= \sqrt{\frac{(16-15.55)^2+\ldots+(19-15.55)^2}{20-1}}$	$= \sqrt{\frac{(24-21.15)^2+\ldots+(18-11.91)^2}{20-1}}$
$= \sqrt{\frac{.20+\ldots+11.90}{19}} = \sqrt{16.47}$	$= \sqrt{\frac{8.12+\ldots+9.92}{19}} = \sqrt{27.29}$
$= 4.06$	$= 5.22$

Having the same research participant appear in all levels of an independent variable has important consequences for how the data are analyzed. When we discussed sampling error in Chapters 6 and 7, we noted that samples may vary from each other as the result of random, chance factors. One of the factors that causes variability across samples is having different people in the different samples. Using a pretest-posttest design eliminates this source of random variability. Therefore, analyzing data collected using this type of design requires explicit recognition that the scores in the two groups are related to each other and may be paired together. To indicate this pairing, we calculate the difference between each participant's two scores.

For the web-based intervention study, the difference between the knowledge scores at the posttest and pretest for each of the 20 parents, represented by the symbol D, is calculated in the last column of Table 9.7(a). Notice that some of these differences are negative numbers. The presence of negative numbers simply indicates that a participant's posttest score is greater than his or her pretest score. For example, the first participant in Table 9.7 had a pretest score of 16 and a posttest score of 24, resulting in a difference of −8. Because they will be needed later in the analysis, Table 9.7(b) calculates the mean ($\overline{X}_D$) and standard deviation (s_D) of the difference scores.

The consequence of pairing the two scores is that rather than having two scores for each participant, we now have just one score—the difference score (D). As a result, the steps in hypothesis testing for paired means are essentially identical to those presented in Chapter 7, in which we compared the mean of one sample against a hypothesized population mean. These steps are illustrated below.

State the Null and Alternative Hypotheses (H_0 and H_1)

The null hypothesis in a pretest-posttest research design reflects the belief that the two means in the population are equal to each other. In the web-based intervention study, the null hypothesis is that there is no difference between pretest and posttest knowledge scores. This absence of difference is reflected in the following null hypothesis:

$$H_0\text{: } \mu_D = 0$$

where the subscript D stands for "difference." Stating that the mean difference between the two populations is equal to zero is simply another way of stating that the two population means are equal to each other ($\mu_{\text{Pretest}} = \mu_{\text{Posttest}}$).

FIGURE 9.7 ● BAR GRAPH OF KNOWLEDGE FOR PARENTS AT THE PRETEST AND POSTTEST TIMES

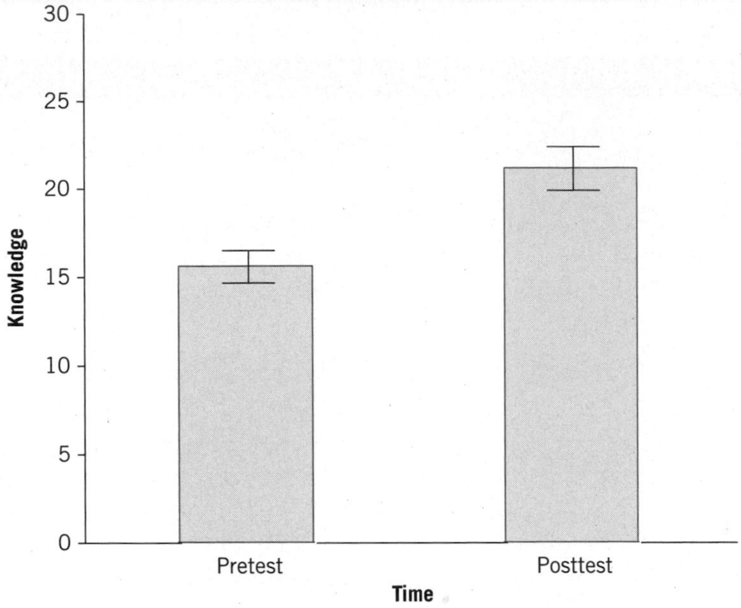

TABLE 9.7 ● DIFFERENCE (*D*) SCORES OF KNOWLEDGE FOR PARENTS AT PRETEST AND POSTTEST

(a) Difference (*D*) Scores

Parent	Pretest	Posttest	Difference (*D*)
1	16	24	**−8**
2	23	23	**0**
3	15	19	**−4**
4	12	17	**−5**
5	21	27	**−6**
6	17	17	**0**
7	21	31	**−10**
8	20	17	**3**
9	13	20	**−7**
10	19	27	**−8**

Parent	Pretest	Posttest	Difference (*D*)
11	12	22	**−10**
12	14	19	**−5**
13	17	28	**−11**
14	8	9	**−1**
15	12	20	**−8**
16	16	22	**−6**
17	10	15	**−5**
18	14	27	**−13**
19	12	21	**−9**
20	19	18	**1**

(b) Descriptive Statistics

$$\bar{X}_D = \frac{\sum D}{N_D} = \frac{-8+0+-4+\ldots+-13+-9+1}{20} = \frac{-112}{20} = -5.60$$

$$s_D = \sqrt{\frac{\sum(D-\bar{X}_D)^2}{N_D-1}} = \sqrt{\frac{(-8-(-5.60))^2+\ldots+(1-(-5.60))^2}{20-1}} = \sqrt{\frac{5.76+\ldots+43.56}{19}} = \sqrt{18.88} = 4.35$$

As in the previous examples, the alternative hypothesis may be directional or nondirectional. In the web-based intervention study, the following nondirectional alternative hypothesis will be used:

$$H_1: \mu_D \neq 0$$

Therefore, in this example, the null hypothesis will be rejected if the posttest knowledge scores are either significantly less *or* significantly greater than pretest knowledge scores.

Make a Decision About the Null Hypothesis

The first step in making the decision to reject the null hypothesis is to calculate the degrees of freedom (*df*). Formula 9-5 presents the formula for the degrees of freedom for paired means:

$$df = N_D - 1 \qquad (9\text{-}5)$$

where N_D is the number of difference scores. This degrees of freedom differs from the degrees of freedom for the difference between two sample means because we're now working with one set of scores (the difference [*D*] scores) rather than two. For the web-based intervention study, because there are 20 parents, the degrees of freedom are

$$\begin{aligned} df &= N_D - 1 \\ &= 20 - 1 \\ &= 19 \end{aligned}$$

The next step is to set alpha (α), identify the critical values, and state a decision rule. Assuming alpha is set to .05, the nondirectional alternative hypothesis in this example ($H_1: \mu_D \neq 0$) leads us to state that "$\alpha = .05$ (two-tailed)." To determine the critical values, we move down the *df* column of Table 3 until we reach $df = 19$ and then to the right until we are under the .05 heading within the "Level of significance for two-tailed test" columns. For the web-based intervention study:

For $\alpha = .05$ (two-tailed) and $df = 19$, critical value $= \pm 2.093$

We can now state a decision rule regarding the conditions under which the null hypothesis will be rejected. Given the critical values we've just identified:

If $t < -2.093$ or > 2.093, reject H_0; otherwise, do not reject H_0.

The next step is to calculate an inferential statistic. As in this chapter's other examples, we'll calculate a *t*-test. However, in this situation we'll calculate the ***t*-test for dependent means (*t*)** to test the difference between two means based on the same participant or paired participants. The word *dependent* is used to indicate that the two means are from one sample drawn from one population, as opposed to the two samples and two populations that are the basis of the *t*-test for independent means discussed earlier in this chapter.

The formula for the *t*-test for dependent means is presented in Formula 9-6:

$$t = \frac{\overline{X}_D - \mu_D}{s_{\overline{D}}} \qquad (9\text{-}6)$$

where $\overline{X}_D$ is the mean of the difference scores, μ_D is the hypothesized population mean of the difference scores, and $s_{\overline{D}}$ is the standard error of the difference scores. Except for the *D* in the subscripts, Formula 9-6 is virtually identical to the formula for the *t*-test of one mean (Formula 7-4) presented in Chapter 7.

The three pieces of information needed to calculate the *t*-test for dependent means are located in different places. First, the value for $\overline{X}_D$ is found in the descriptive statistics (Table 9.9(b)). For the web-based intervention study, $\overline{X}_D = -5.60$. Second, the population mean μ_D is located in the null hypothesis. Because of how the null hypothesis in this example has been stated (H_0: $\mu_D = 0$), μ_D is equal to 0. The third piece of information, the standard error of the difference scores $(s_{\overline{D}})$, is calculated using Formula 9-7:

$$s_{\overline{D}} = \frac{s_D}{\sqrt{N}} \tag{9-7}$$

where s_D is the standard deviation of the difference scores and N is the sample size. Obtaining the standard deviation of the difference scores (s_D) for the 20 parents in the web-based intervention study from Table 9.9(b), the standard error of the difference scores $(s_{\overline{D}})$ is as follows:

$$\begin{aligned} s_{\overline{D}} &= \frac{s_D}{\sqrt{N}} \\ &= \frac{4.35}{\sqrt{20}} = \frac{4.35}{4.47} \\ &= .97 \end{aligned}$$

Placing these three pieces of information into Formula 9-7, we may now calculate a value of the *t*-statistic:

$$\begin{aligned} t &= \frac{\overline{X}_D - \mu_D}{s_{\overline{D}}} \\ &= \frac{-5.60 - 0}{.97} = \frac{-5.60}{.97} \\ &= -5.77 \end{aligned}$$

The value of the *t*-statistic for the web-based intervention study is a negative number because the first mean ($M_{\text{Pretest}} = 15.55$) is less than the second mean ($M_{\text{Posttest}} = 21.15$).

Now that we've transformed the mean of the difference scores into a *t*-statistic, we're ready to make a decision whether to reject the null hypothesis. Comparing the *t*-statistic of −5.77 calculated from this example with the critical values, we make the following decision:

$$t = -5.77 < -2.093 \;\therefore\; \text{reject } H_0 \; (p < .05)$$

Rejecting the null hypothesis in this study implies that the knowledge scores at the pretest and posttest are significantly different from each other.

Because the null hypothesis has been rejected, it's appropriate to determine the level of significance by determining whether the probability of our *t*-statistic is less than .01. In Table 3, for $\alpha = .01$ (two-tailed) and $df = 19$, we find the critical value −2.861. Comparing the calculated *t*-statistic for this example with the .01 critical value:

$$t = -5.77 < -2.861 \;\therefore\; p < .01$$

Because the *t*-statistic of −5.77 is less than the .01 critical value of −2.861, the level of significance for this set of data is $p < .01$ (see Figure 9.8).

Draw a Conclusion From the Analysis

For the web-based intervention study, we've found that the 20 parents' mean knowledge scores at the posttest are significantly greater than at the pretest. To present the results of this analysis in a more informative way, we could say the following:

> The average knowledge scores for the 20 parents were significantly higher after completing the web-based intervention program ($M = 21.15$) than before beginning the program ($M = 15.55$), $t(19) = -5.77$, $p < .01$.

Relate the Result of the Analysis to the Research Hypothesis

One purpose of the web-based intervention study was to determine whether the program could provide working parents knowledge that may help them and their children with mental health–related problems. The researchers discuss the implication of finding a significant increase in knowledge from the pretest to the posttest below:

> These findings indicate that the program can be an effective intervention for improving parents' knowledge of children's mental health problems and boost their confidence in handling such issues. . . . The study findings lend support to the growing literature on the utility of offering web-based programs to improve the health of the general population. (Deitz et al., 2009, p. 492)

Assumptions of the *t*-Test for Dependent Means

As with the *t*-test for independent means, the appropriate use of the *t*-test for dependent means requires that certain mathematical assumptions be met. One of the fundamental assumptions of statistics such as the *t*-test is the assumption of normality: that the data being analyzed are normally distributed. Given that the *t*-test for dependent means involves analyzing difference (*D*) scores, it's not surprising that this statistic is based on the assumption that difference scores in the population are normally distributed. The larger the sample size, the greater the likelihood of meeting this assumption.

Summary

The steps in testing the difference between paired means are summarized in Box 9.3. Looking at this table, we see a great amount of similarity between these steps and those used to test a single mean in Chapter 7. Research studies may differ in their purpose, hypotheses, variables of interest, or method of data collection; however, the analysis of data follows a common sequencing and logic.

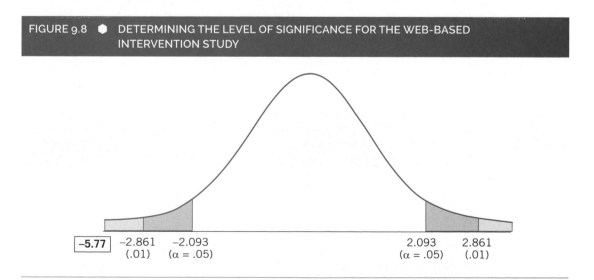

FIGURE 9.8 ● DETERMINING THE LEVEL OF SIGNIFICANCE FOR THE WEB-BASED INTERVENTION STUDY

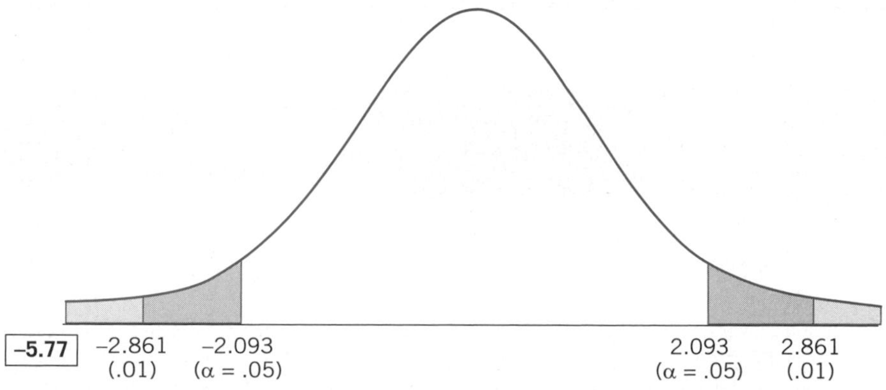

SUMMARY BOX 9.3 TESTING THE DIFFERENCE BETWEEN PAIRED MEANS (WEB-BASED INTERVENTION EXAMPLE)

State the null and alternative hypotheses (H_0 and H_1)

$$H_0: \mu_D = 0 \qquad H_1: \mu_D \neq 0$$

Make a decision about the null hypothesis

Calculate the degrees of freedom (*df*)

$$df = N_D - 1 = 20 - 1 = 19$$

Set alpha (α), identify the critical values, and state a decision rule

If $t < -2.093$ or > 2.093, reject H_0; otherwise, do not reject H_0

Calculate a statistic: *t*-test for dependent means

Calculate the standard error of the difference scores ($s_{\overline{D}}$)

$$s_{\overline{D}} = \frac{s_D}{\sqrt{N}} = \frac{4.35}{\sqrt{20}} = \frac{4.35}{4.47} = .97$$

Calculate the *t*-statistic (*t*)

$$t = \frac{\overline{X}_D - \mu_D}{s_{\overline{D}}} = \frac{-5.60 - 0}{.97} = \frac{-5.60}{.97} = -5.77$$

Make a decision whether to reject the null hypothesis

$$t = -5.77 < -2.093 \therefore \text{ reject } H_0 \ (p < .05)$$

Determine the level of significance

$$t = -5.77 < -2.861 \therefore p < .01$$

Draw a conclusion from the analysis

The average knowledge scores for the 20 parents were significantly higher after completing the web-based intervention program ($M = 21.15$) than before beginning the program ($M = 15.55$), $t(19) = -5.77$, $p < .01$.

Relate the result of the analysis to the research hypothesis

"These findings indicate that the program can be an effective intervention for improving parents' knowledge of children's mental health problems and boost their confidence in handling such issues. . . . The study findings lend support to the growing literature on the utility of offering web-based programs to improve the health of the general population" (Deitz et al., 2009, p. 492).

LEARNING CHECK 5

Reviewing What You've Learned So Far

1. Review questions
   a. What is the main difference between between-subjects and within-subjects research designs?
   b. For what types of research situations might you use a within-subjects design?
   c. In analyzing a pretest-posttest research design, why do we calculate the difference between the scores in the two groups?
   d. When would you calculate the *t*-test for dependent means rather than the *t*-test for independent means?

2. You conduct a class project in which you divide classmates into seven pairs; each pair consists of a shy student and an outgoing student. You have each pair of students work on a puzzle but do not allow them to talk to each other. After completing the puzzle, you ask each student to indicate how much he or she enjoyed working on the puzzle:

Pair	Shy	Outgoing
1	8	4
2	4	6
3	8	2
4	6	6
5	3	7
6	6	3
7	5	4

   a. Calculate the difference score (D) for each pair of students.
   b. Calculate the mean ($\bar{X}_D$), standard deviation (s_D), and standard error ($s_{\bar{D}}$), of the difference scores.

3. For each of the following situations, calculate the standard error of the difference scores (s_D) and the t-test for dependent means.
   a. $N_D = 9, \bar{X}_D = 2.00, s_{\bar{D}} = 1.75$
   b. $N_D = 12, \bar{X}_D = 1.50, s_{\bar{D}} = 3.50$
   c. $N_D = 21, \bar{X}_D = 12.63, s_{\bar{D}} = 24.82$

4. As adults get older, their lives may become increasingly sedentary. Given concerns about physical and psychological problems associated with a lack of physical activity, one study investigated whether some office tasks such as typing, working on a computer, and reading may be performed at a satisfactory level while walking on a treadmill rather than sitting (John, Bassett, Thompson, Fairbrother, & Baldwin, 2009). One part of the study involved having 20 adults take a typing test two times: once while sitting at a desk and once while walking on a treadmill. The average number of words per minute (WPM) typed correctly for the two conditions was as follows: sitting ($M = 40.20, s = 10.20$) and treadmill ($M = 36.90, s = 11.10$). Furthermore, the researchers reported a mean difference score ($\bar{X}_D$) of 3.30 and a standard deviation of the difference scores (s_D) of 4.70. Test the difference in typing speed for the two conditions.
   a. State the null and alternative hypotheses (H_0 and H_1).
   b. Make a decision about the null hypothesis.
      (1) Calculate the degrees of freedom (df).
      (2) For $\alpha = .05$ (two-tailed), identify the critical values and state a decision rule.
      (3) Calculate a value for the t-test for dependent means.
      (4) Make a decision whether to reject the null hypothesis.
      (5) Determine the level of significance.
   c. Draw a conclusion from the analysis.
   d. What conclusions might the researchers draw regarding whether the task of typing may be performed at a satisfactory level while walking on a treadmill rather than sitting?

9.6 LOOKING AHEAD

The main purpose of the present chapter was to further your understanding of the process of hypothesis testing using situations slightly more complicated than those presented in earlier chapters. As you move further along in this book, you'll encounter research situations of growing complexity. Although the statistical procedures may become more challenging, the steps in hypothesis testing will remain essentially the same. The main purpose of hypothesis testing is to make one of two decisions about the null hypothesis: reject or not reject. This decision is based on the probability of obtaining a calculated value of an inferential statistic when the null hypothesis is true. Relying on probability creates the possibility that the decision made about the null hypothesis may be in error. The next chapter discusses these errors in greater detail, as well as what researchers can do to minimize the possibility and impact of making these errors.

9.7 Summary

To test the difference between two sample means, we create the ***sampling distribution of the difference,*** which is the distribution of all possible values of the difference between two sample means when an infinite number of pairs of samples of size N are randomly selected from two populations. Three characteristics of the sampling distribution of the difference are that its mean is equal to zero (0), the distribution is approximately normal in shape, and the variability of this distribution is measured by the ***standard error of the difference*** ($s_{\bar{X}_1-\bar{X}_2}$), defined as the standard deviation of the sampling distribution of the difference. These characteristics of the sampling distribution of the difference enable researchers to determine the probability of obtaining any particular difference between the means of two samples.

The ***t-test for independent means,*** an inferential statistic that tests the difference between the means of two samples drawn from two populations, is used to test a hypothesis about the difference between two population means. The *t*-test may be calculated when the sample sizes for the two groups are either equal or unequal; with unequal sample sizes, the calculation of the *t*-test is slightly more complicated because the formula for the standard error of the difference must be modified.

The *t*-test for independent means is used in ***between-subjects research designs,*** in which each research participant appears in only one level or category of the independent variable, and these designs involve testing differences between different groups of participants. The ***t-test for dependent means*** is used in ***within-subjects designs,*** in which each participant appears in all levels or categories of the independent variable involve collecting information from each participant more than once. An example of within-subjects designs is ***longitudinal research,*** which involves collecting the same information from a research participant over two or more administrations to assess changes occurring within the person over time. One example of longitudinal research is the ***pretest-posttest research design,*** which involves collecting data from participants twice—before and after the introduction of an intervention or experimental manipulation—to determine whether the intervention or manipulation is associated with changes in the dependent variable.

9.8 Important Terms

bar graph (p. 267)
sampling distribution of the difference (p. 270)
standard error of the difference ($s_{\bar{X}_1-\bar{X}_2}$) (p. 270)
t-test for independent means (p. 274)
homogeneity of variance (p. 279)
between-subjects research designs (p. 287)
within-subjects research designs (p. 288)
longitudinal research (p. 289)
pretest-posttest research design (p. 289)
t-test for dependent means (*t*) (p. 293)

9.9 Formulas Introduced in This Chapter

Degrees of Freedom (*df*), Difference Between Two Sample Means

$$df = (N_1 - 1) + (N_2 - 1) \quad (9\text{-}1)$$

***t*-Test for Independent Means**

$$t = \frac{\bar{X}_1 - \bar{X}_2}{s_{\bar{X}_1-\bar{X}_2}} \quad (9\text{-}2)$$

Standard Error of the Difference ($s_{\bar{X}_1-\bar{X}_2}$),

$$s_{\bar{X}_1-\bar{X}_2} = \sqrt{\frac{s_1^2}{N_1} + \frac{s_2^2}{N_2}} \quad (9\text{-}3)$$

Standard Error of the Difference (Unequal Sample Sizes)

$$s_{\bar{X}_1-\bar{X}_2} = \sqrt{\frac{(N_1-1)s_1^2 + (N_2-1)s_2^2}{N_1+N_2-2}\left(\frac{1}{N_1}+\frac{1}{N_2}\right)} \quad (9\text{-}4)$$

Degrees of Freedom (*df*), Difference Between Paired Means

$$df = N_D - 1 \quad (9\text{-}5)$$

***t*-Test for Dependent Means**

$$t = \frac{\bar{X}_D - \mu_D}{s_{\bar{D}}} \quad (9\text{-}6)$$

Standard Error of Difference Scores ($s_{\bar{D}}$)

$$s_{\bar{D}} = \frac{s_D}{\sqrt{N}} \quad (9\text{-}7)$$

9.10 Using SPSS

Testing the Difference Between Two Sample Means: The Parking Lot Study (9.1)

1. Define independent and dependent variables (name, # decimals, labels for the variables, labels for values of the independent variable) and enter data for the variables.

   NOTE: Numerically code values of the independent variable (i.e., 1 = Intruder, 2 = No Intruder) and provide labels for these values in the **Values** box within **Variable View**.

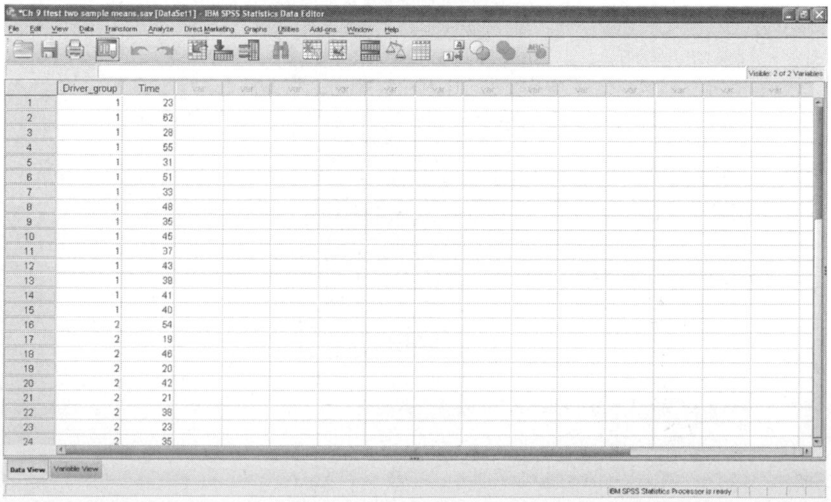

2. Select the *t*-test for independent means procedure within SPSS.

   How? (1) Click **Analyze menu**, (2) click **Compare Means**, and (3) click **Independent-Samples T Test**.

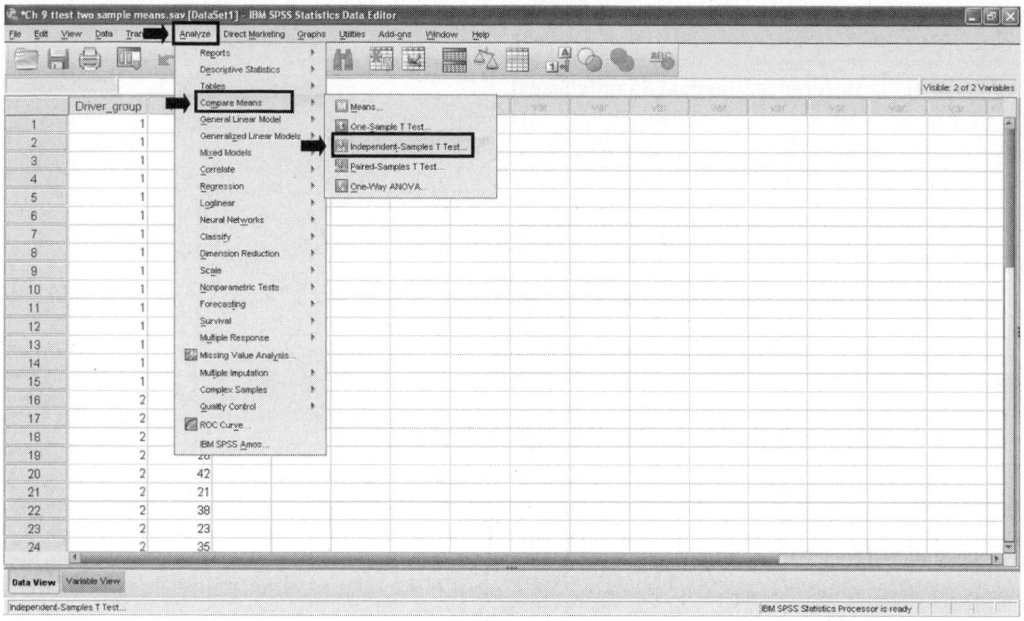

3. Identify the dependent variable, the independent variable, and the values of the independent variable.

   How? (1) Click dependent variable and → **Test Variable**, (2) click independent variable and → **Grouping Variable**, (3) click **Define Groups** and type the values for the independent variable, (4) click **Continue**, and (5) click **OK**.

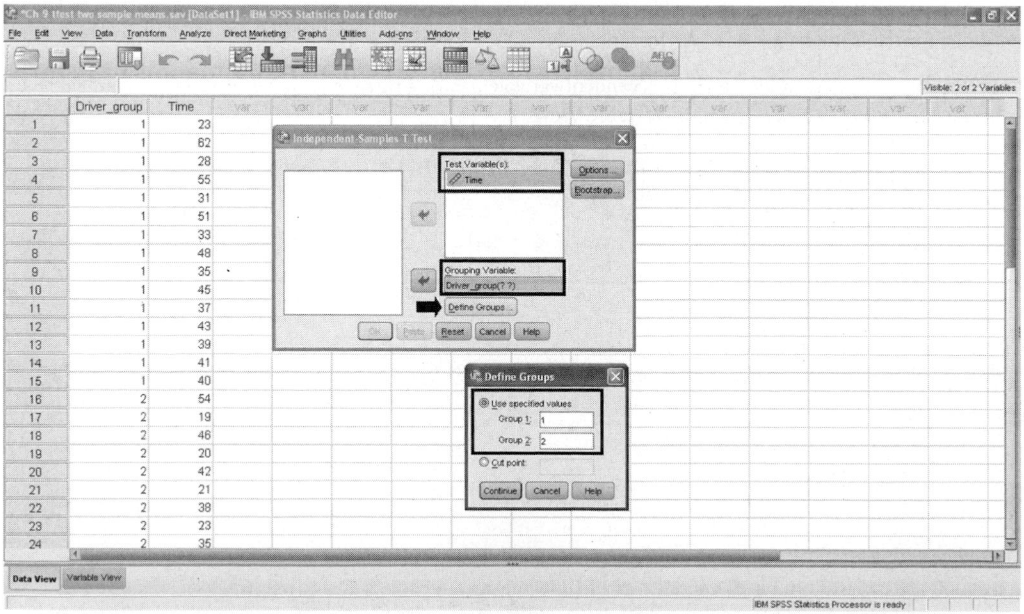

4. Examine output.

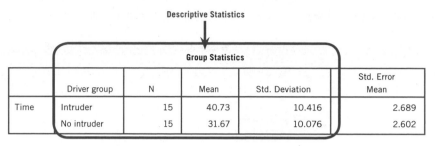

Descriptive Statistics

Group Statistics

	Driver group	N	Mean	Std. Deviation	Std. Error Mean
Time	Intruder	15	40.73	10.416	2.689
	No intruder	15	31.67	10.076	2.602

Independent Samples Test

		Levene's Test for Equality of Variances		t-test for Equality of Means					95% Confidence Interval of the Difference	
		F	Sig	t	*df*	Sig (2-tailed)	Mean Difference	Std. Error Difference	Lower	Upper
Time	Equal variances assumed	.003	.959	2.423	28	.022	9.067	3.742	1.402	16.731
	Equal variances not assumed			2.423	27.97	.022	9.067	3.742	1.402	16.732

t ***df*** **Level of significance (Note: 'Sig.'= probability (p))** **Standard error of the difference**

Testing the Difference Between Paired Means: The Web-Based Intervention Study (9.5)

1. Define the two levels of the independent variable (name, # decimals, labels for the variables) and enter data for each level.

   NOTE: Each participant has scores on two variables—each variable represents one of the two levels of the independent variable (i.e., pretest, posttest).

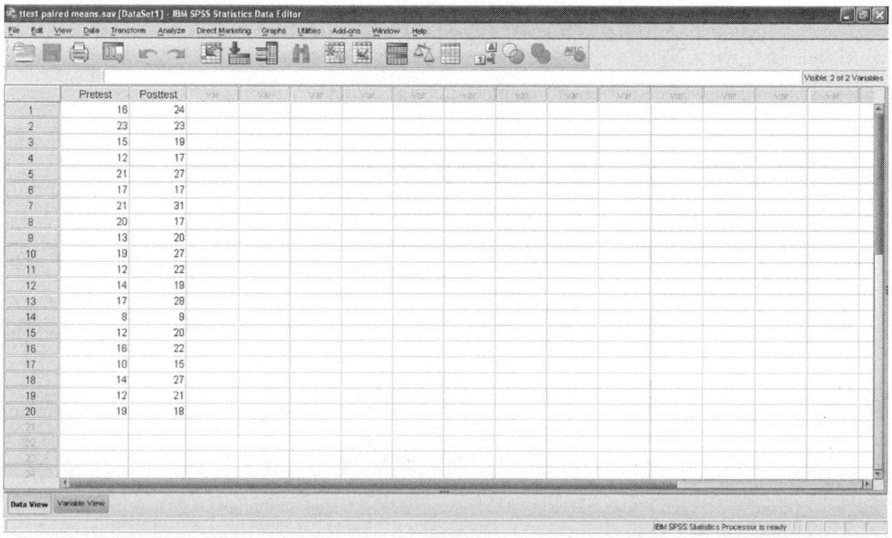

	Pretest	Posttest
1	16	24
2	23	23
3	15	19
4	12	17
5	21	27
6	17	17
7	21	31
8	20	17
9	13	20
10	19	27
11	12	22
12	14	19
13	17	28
14	8	9
15	12	20
16	16	22
17	10	15
18	14	27
19	12	21
20	19	18

2. Begin the *t*-test for dependent means procedure within SPSS.

   How? (1) Click **Analyze menu**, (2) click **Compare Means**, and (3) click **Paired-Samples T Test**.

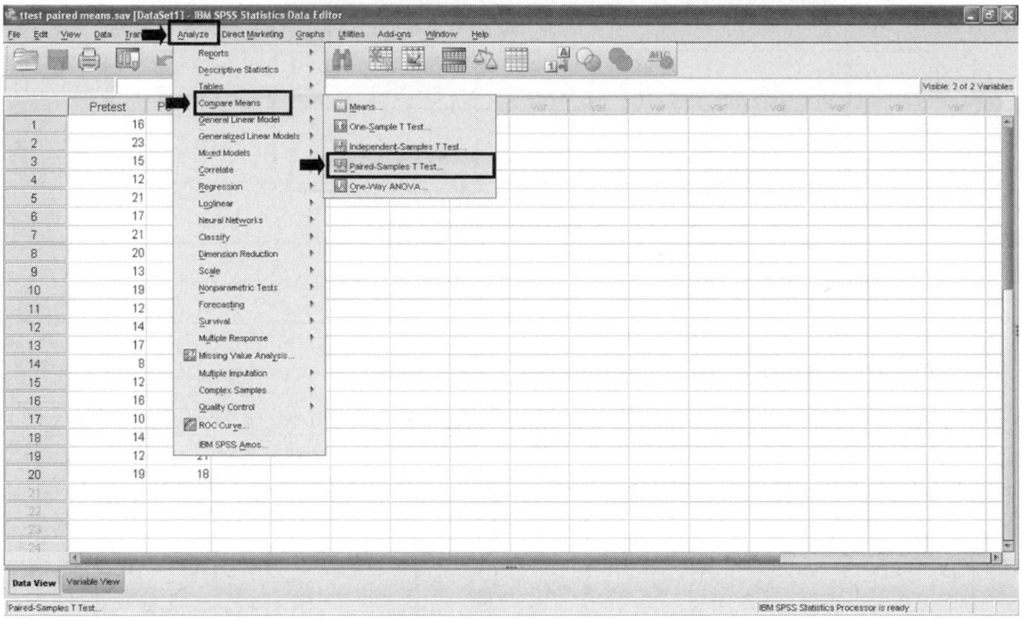

3. Identify the paired variables.

   How? (1) Click the two variables and → **Paired Variables** and (2) click **OK**.

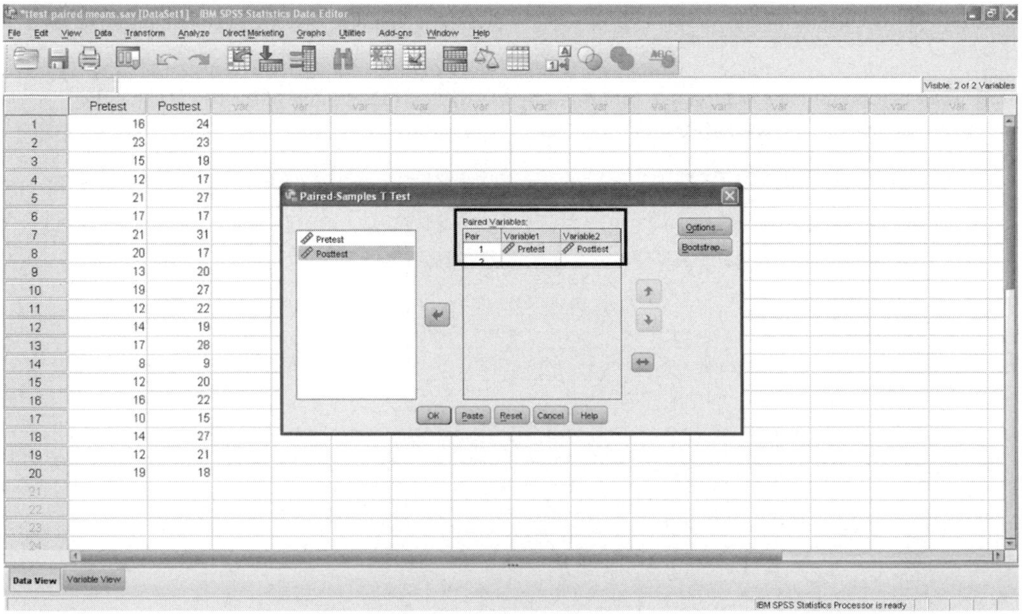

4. Examine output.

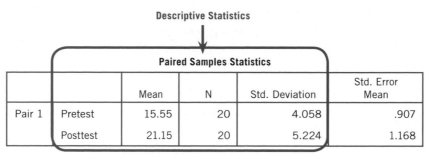

Descriptive Statistics

Paired Samples Statistics

		Mean	N	Std. Deviation	Std. Error Mean
Pair 1	Pretest	15.55	20	4.058	.907
	Posttest	21.15	20	5.224	1.168

Difference score (D) statistics

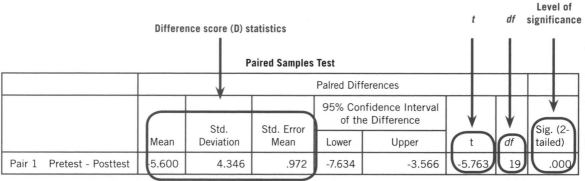

t **df** **Level of significance**

Paired Samples Test

	Palred Differences							
	Mean	Std. Deviation	Std. Error Mean	95% Confidence Interval of the Difference Lower	95% Confidence Interval of the Difference Upper	t	*df*	Sig. (2-tailed)
Pair 1 Pretest - Posttest	-5.600	4.346	.972	-7.634	-3.566	-5.763	19	.000

9.11 Exercises

1. Construct a bar graph for each of the following (assume the independent variable is group and the dependent variable is time):
   a. Group A ($N = 5$, $M = 4.00$, $s = 1.58$); Group B ($N = 5$, $M = 6.00$, $s = 2.12$)
   b. Group A ($N = 8$, $M = 26.00$, $s = 2.56$); Group B ($N = 8$, $M = 23.00$, $s = 2.33$)
   c. Group A ($N = 11$, $M = 1.77$, $s = .29$); Group B ($N = 11$, $M = 1.53$, $s = .22$)
   d. Group A ($N = 18$, $M = 63.59$, $s = 23.70$); Group B ($N = 18$, $M = 71.42$, $s = 20.91$)
2. Construct a bar graph for each of the following (assume the independent variable is group and the dependent variable is time):
   a. Group A ($N = 21$, $M = 14.05$, $s = 3.63$); Group B ($N = 21$, $M = 12.33$, $s = 3.26$)
   b. Group A ($N = 28$, $M = 6.79$, $s = 3.11$); Group B ($N = 28$, $M = 7.93$, $s = 2.36$)
   c. Group A ($N = 16$, $M = 52.56$, $s = 23.77$); Group B ($N = 16$, $M = 60.38$, $s = 21.91$)
   d. Group A ($N = 25$, $M = 5.76$, $s = 2.14$); Group B ($N = 25$, $M = 4.43$, $s = 2.27$)
3. For each of the following, (a) calculate the mean, standard deviation, and standard error of the mean for each group and (b) construct a bar graph (assume the independent variable is condition and the dependent variable is test score).
   a. Experimental: 3, 7, 4, 1, 10, 4, 6
      Control: 10, 7, 12, 4, 8, 9

b. Experimental: 15, 12, 10, 14, 17, 15, 18, 19

Control: 22, 12, 17, 19, 20, 21, 16, 17

c. Experimental: 3.89, 3.04, 3.95, 2.91, 2.72, 3.70, 3.16, 3.21, 2.86

Control: 2.96, 3.38, 2.82, 2.07, 3.22, 2.56, 2.44, 3.11, 2.68

d. Experimental: 2, 4, 3, 5, 1, 4, 3, 5, 4, 1, 4, 5, 3

Control: 5, 2, 5, 1, 4, 3, 5, 2, 1, 4, 5, 1, 4

4. State the null and alternative hypotheses (H_0 and H_1) for each of the following research questions:

a. Are the average starting salaries for clinical psychologists in private practice the same as or different from that of psychological researchers in business or the government?

b. In Chapter 3, we looked at psychology majors' and non–psychology majors' belief in the myth that we only use 10% of our brains. Do the two groups differ in this belief?

c. Are men paid more than women for doing the same job?

d. Do Republicans and Democrats similarly support a national health insurance program or does one group favor this more than the other?

5. For each of the following, calculate the degrees of freedom (*df*) and determine the critical values of *t* (assume $\alpha = .05$).

a. $N_1 = 21, N_2 = 21, H_1: \mu_1 \neq \mu_2$

b. $N_1 = 14, N_2 = 14, H_1: \mu_1 \neq \mu_2$

c. $N_1 = 4, N_2 = 4, H_1: \mu_1 > \mu_2$

d. $N_1 = 32, N_2 = 32, H_1: \mu_1 < \mu_2$

6. For each of the following, calculate the degrees of freedom (*df*) and determine the critical values of *t* (assume $\alpha = .05$).

a. $N_1 = 5, N_2 = 5, H_1: \mu_1 \neq \mu_2$

b. $N_1 = 12, N_2 = 12, H_1: \mu_1 < \mu_2$

c. $N_1 = 9, N_2 = 9, H_1: \mu_1 \neq \mu_2$

d. $N_1 = 16, N_2 = 16, H_1: \mu_1 > \mu_2$

7. For each of the following, calculate the standard error of the difference ($s_{\bar{X}_1-\bar{X}_2}$).

a. $N_1 = 35, s_1 = 1.50, N_2 = 35, s_2 = 3.25$

b. $N_1 = 4, s_1 = 4.30, N_2 = 4, s_2 = 2.10$

c. $N_1 = 23, s_1 = 8.10, N_2 = 23, s_2 = 7.50$

d. $N_1 = 20, s_1 = 1.20, N_2 = 20, s_2 = 1.75$

8. For each of the following, calculate the standard error of the difference ($s_{\bar{X}_1-\bar{X}_2}$).

a. $N_1 = 10, s_1 = 2.00, N_2 = 10, s_2 = 3.00$

b. $N_1 = 19, s_1 = 1.73, N_2 = 19, s_2 = 1.48$

c. $N_1 = 27, s_1 = 24.91, N_2 = 27, s_2 = 27.02$

d. $N_1 = 50, s_1 = 12.29, N_2 = 50, s_2 = 10.63$

9. For each of the following, calculate the *t*-test for independent means.

a. $\bar{X}_1 = 3.49, \bar{X}_2 = 3.14, s_{\bar{X}_1-\bar{X}_2} = .31$

b. $\bar{X}_1 = 13.27, \bar{X}_2 = 16.45, s_{\bar{X}_1-\bar{X}_2} = 1.52$

c. $\bar{X}_1 = .76, \bar{X}_2 = .91, s_{\bar{X}_1-\bar{X}_2} = .09$

d. $\bar{X}_1 = 1.52, \bar{X}_2 = 1.36, s_{\bar{X}_1-\bar{X}_2} = .05$

10. For each of the following, calculate the *t*-test for independent means.

a. $\bar{X}_1 = 7.00, \bar{X}_2 = 11.00, s_{\bar{X}_1-\bar{X}_2} = 1.17$

b. $\overline{X}_1 = 65.56, \overline{X}_2 = 60.92, s_{\overline{X}_1-\overline{X}_2} = 2.88$

c. $\overline{X}_1 = 137.73, \overline{X}_2 = 114.09, s_{\overline{X}_1-\overline{X}_2} = 10.71$

d. $\overline{X}_1 = 73.24, \overline{X}_2 = 81.53, s_{\overline{X}_1-\overline{X}_2} = 4.39$

11. For each of the following, calculate the standard error of the difference ($s_{\overline{X}_1-\overline{X}_2}$) and the t-test for independent means.

    a. $N_1 = 6, \overline{X}_1 = 18.50, s_1 = 2.00, N_2 = 6, \overline{X}_2 = 19.00, s_2 = 2.50$

    b. $N_1 = 13, \overline{X}_1 = 36.23, s_1 = 4.17, N_2 = 13, \overline{X}_2 = 29.59, s_2 = 6.01$

    c. $N_1 = 29, \overline{X}_1 = 7.80, s_1 = 1.25, N_2 = 29, \overline{X}_2 = 7.17, s_2 = 1.63$

    d. $N_1 = 38, \overline{X}_1 = 21.14, s_1 = 4.38, N_2 = 38, \overline{X}_2 = 23.29, s_2 = 3.91$

12. For each of the following, calculate the standard error of the difference ($s_{\overline{X}_1-\overline{X}_2}$) and the t-test for independent means.

    a. $N_1 = 11, \overline{X}_1 = 12.00, s_1 = 2.00, N_2 = 11, \overline{X}_2 = 10.00, s_2 = 3.00$

    b. $N_1 = 21, \overline{X}_1 = 39.85, s_1 = 5.23, N_2 = 21, \overline{X}_2 = 44.16, s_2 = 4.60$

    c. $N_1 = 33, \overline{X}_1 = 4.37, s_1 = 1.07, N_2 = 33, \overline{X}_2 = 4.92, s_2 = .94$

    d. $N_1 = 57, \overline{X}_1 = 53.98, s_1 = 6.96, N_2 = 57, \overline{X}_2 = 50.74, s_2 = 6.03$

13. When you're interviewing for a job, is your behavior influenced by beliefs the interviewer has about you? Researchers Ridge and Reber (2002) observed the interactions of 54 men, each of whom was interviewing a woman for a job. Right before conducting their interviews, half of the men were told the woman whom they were interviewing was attracted to them. The researchers hypothesized that "these same men would elicit relatively more flirtatious behavior from women than would men holding no such belief" (p. 2). The interviews were videotaped, and the number of flirtatious behaviors performed by each woman was counted. The researchers reported the following descriptive statistics regarding the mean number of flirtatious behaviors: attraction condition ($N = 27$, $M = 37.44$, $s = 5.21$) and no attraction condition ($N = 27$, $M = 34.59$, $s = 4.54$).

    a. State the null and alternative hypotheses (H_0 and H_1).

    b. Make a decision about the null hypothesis.

        (1) Calculate the degrees of freedom (df).

        (2) Set alpha (α), identify the critical values, and state a decision rule.

        (3) Calculate a value for the t-test for independent means.

        (4) Make a decision whether to reject the null hypothesis.

        (5) Determine the level of significance.

    c. Draw a conclusion from the analysis.

    d. Relate the result of the analysis to the research hypothesis.

14. An advertising agency is interested in learning how to fit its commercials to the interests and needs of the viewing audience. It asked samples of 41 men and 41 women to report the average amount of television watched daily. The men reported a mean television time of 1.70 hours per day with a standard deviation of .70. The women reported a mean of 2.05 hours per day with a standard deviation of .80. Use these data to test the manager's claim that there is a significant gender difference in television viewing.

    a. State the null and alternative hypotheses (H_0 and H_1).

    b. Make a decision about the null hypothesis.

        (1) Calculate the degrees of freedom (df).

        (2) Set alpha (α), identify the critical values (draw the distribution), and state a decision rule.

        (3) Calculate a value for the t-test for independent means.

        (4) Make a decision whether to reject the null hypothesis.

        (5) Determine the level of significance.

c. Draw a conclusion from the analysis.

d. What are the implications of this analysis for the advertising agency?

15. A pair of researchers was interested in studying men's preferences for a potential mate and how these preferences may change as men get older (Alterovitz & Mendelsohn, 2009). The researchers believed that men prefer women who are younger than themselves; furthermore, they hypothesized that the preferred difference in age is greater for older men than for younger men. As part of a larger study, they examined personal advertisements placed by men who were either 20 to 34 years old or 40 to 54 years old; in looking at these ads, the researchers calculated the difference between the man's age and the man's preferred age of a potential partner. The descriptive statistics for the difference in age variable for the two age groups are as follows:

20 to 34 years old: $N_1 = 25$, $\overline{X}_1 = 1.04$, $s_1 = 2.72$

40 to 54 years old: $N_1 = 25$, $\overline{X}_1 = 4.98$, $s_1 = 3.80$

a. State the null and alternative hypotheses (H_0 and H_1).

b. Make a decision about the null hypothesis.

(1) Calculate the degrees of freedom (*df*).

(2) Set alpha (α), identify the critical values (draw the distribution), and state a decision rule.

(3) Calculate a value for the *t*-test for independent means.

(4) Make a decision whether to reject the null hypothesis.

(5) Determine the level of significance.

c. Draw a conclusion from the analysis.

d. Relate the result of the analysis to the research hypothesis.

16. A third-grade teacher is interested in comparing the effectiveness of two styles of instruction in language comprehension: imagery, in which the students are asked to picture a situation involving the word, and repetition, in which the students repeat the definition of the word multiple times. After 6 weeks of instruction, she gives the students a language comprehension test. The following scores are the number of correct answers for each student. Determine whether the two styles of instruction differ in their effectiveness.

Imagery: 12, 13, 11, 11, 13, 13, 15, 12, 9, 12

Repetition: 6, 11, 10, 12, 9, 10, 11, 12, 10, 8

a. For each group, calculate the sample size (N), mean ($\overline{X}$), and standard deviation (s).

b. State the null and alternative hypotheses (H_0 and H_1).

c. Make a decision about the null hypothesis.

(1) Calculate the degrees of freedom (*df*).

(2) Set alpha (α), identify the critical values (draw the distribution), and state a decision rule.

(3) Calculate a value for the *t*-test for independent means.

(4) Make a decision whether to reject the null hypothesis.

(5) Determine the level of significance.

d. Draw a conclusion from the analysis.

e. What are the implications of this analysis for the teacher?

17. Imagine you believe that men and women have different beliefs regarding the average cost of a wedding. More specifically, because they are traditionally more involved in the planning of weddings and therefore have a more complete understanding of their costs, you hypothesize that young women will provide a higher estimate of the cost of the average wedding than will men. To test your hypothesis, you ask a sample of college students (13 women and 13 men), "How much (in thousands) do you believe the average wedding costs?" The estimates (in thousands of dollars) of these students are below:

Women: 20, 9, 14, 11, 25, 18, 12, 33, 24, 10, 30, 14, 22

Men: 15, 6, 19, 16, 33, 4, 21, 13, 10, 7, 24, 5, 20

a. For each group, calculate the sample size (N), mean ($\overline{X}$), and standard deviation (s).

b. State the null and alternative hypotheses (H_0 and H_1).

c. Make a decision about the null hypothesis.

(1) Calculate the degrees of freedom (df).

(2) Set alpha (α), identify the critical values (draw the distribution), and state a decision rule.

(3) Calculate a value for the t-test for independent means.

(4) Make a decision whether to reject the null hypothesis.

(5) Determine the level of significance.

d. Draw a conclusion from the analysis.

e. Relate the result of the analysis to the research hypothesis.

18. For each of the following, calculate the standard error of the difference ($s_{\overline{X}_1-\overline{X}_2}$) (be sure to use the formula for unequal sample sizes).

a. $N_1 = 8, s_1 = 2.30, N_2 = 12, s_2 = 2.00$

b. $N_1 = 25, s_1 = 5.15, N_2 = 22, s_2 = 7.80$

c. $N_1 = 13, s_1 = 4.50, N_2 = 19, s_2 = 5.89$

d. $N_1 = 50, s_1 = 21.00, N_2 = 45, s_2 = 17.50$

19. For each of the following, calculate the standard error of the difference ($s_{\overline{X}_1-\overline{X}_2}$) (be sure to use the formula for unequal sample sizes).

a. $N_1 = 9, s_1 = 4.00, N_2 = 7, s_2 = 2.00$

b. $N_1 = 10, s_1 = 8.50, N_2 = 20, s_2 = 7.80$

c. $N_1 = 33, s_1 = 1.87, N_2 = 41, s_2 = 5.89$

d. $N_1 = 21, s_1 = 16.21, N_2 = 19, s_2 = 3.56$

20. A team of researchers stated that effective methods of therapy help people access and process their emotions (Watson & Bedard, 2006). They looked at the level of emotional processing brought about by two different types of therapy for clients diagnosed with depression. The first group ($N = 17$) took part in cognitive-behavioral therapy (CBT). The second group ($N = 21$) took part in process-experiential therapy (PET). The Experiencing Scale measured their level of emotional processing. The CBT group scored a mean of 2.73 on the scale, with a standard deviation of .46. The PET group scored a mean of 3.04, with a standard deviation of .42. Use the steps of hypothesis testing to determine whether one therapeutic technique brings about more emotional processing than the other.

a. State the null and alternative hypotheses (H_0 and H_1).

b. Make a decision about the null hypothesis.

(1) Calculate the degrees of freedom (df).

(2) Set alpha (α), identify the critical values (draw the distribution), and state a decision rule.

(3) Calculate a value for the t-test for independent means (when calculating the standard error of the difference [$s_{\overline{X}_1-\overline{X}_2}$], be sure to use the formula for unequal sample sizes).

(4) Make a decision whether to reject the null hypothesis.

(5) Determine the level of significance.

c. Draw a conclusion from the analysis.

d. Does one of the therapy methods appear to be more effective than the other?

21. A team of audiologists was interested in examining whether their patients' satisfaction with their hearing aids was related to how long they had used hearing aids (Williams, Johnson, & Danhauer, 2009). They divided their patients into two categories, new users ($N = 30$) and experienced users ($N = 34$), and asked them to indicate how satisfied they were with their hearing aids; the higher the score, the greater the satisfaction. The new users reported a mean satisfaction of 26.90 on the scale (standard deviation = 3.96), and the experienced users reported a mean satisfaction of 28.03 (standard deviation = 5.04). Use the steps of hypothesis testing to test the difference in satisfaction between new and experienced users.
   a. State the null and alternative hypotheses (H_0 and H_1).
   b. Make a decision about the null hypothesis.
      (1) Calculate the degrees of freedom (*df*).
      (2) Set alpha (α), identify the critical values (draw the distribution), and state a decision rule.
      (3) Calculate a value for the *t*-test for independent means (when calculating the standard error of the difference [$s_{\overline{X}_1-\overline{X}_2}$], be sure to use the formula for unequal sample sizes).
      (4) Make a decision whether to reject the null hypothesis.
      (5) Determine the level of significance.
   c. Draw a conclusion from the analysis.
   d. Is patients' satisfaction with hearing aids related to how long they have used hearing aids?

22. Imagine a researcher asks a sample of five people to drive two types of cars and rate each of them on a 1 to 20 scale. Listed below are the data she collected:

Person	Type A	Type B
1	20	10
2	10	11
3	11	4
4	23	12
5	7	0

   a. Calculate the difference score (*D*) for each of the five participants.
   b. Calculate the mean ($\overline{X}_D$), standard deviation ($s_{\overline{D}}$), and standard error ($s_{\overline{D}}$) of the difference scores.

23. For each of the following, calculate the standard error of the difference scores ($s_{\overline{D}}$) and the *t*-test for dependent means.
   a. $N_D = 25, \overline{X}_D = 10.00, s_{\overline{D}} = 8.00$
   b. $N_D = 5, \overline{X}_D = 6.80, s_{\overline{D}} = 4.71$
   c. $N_D = 16, \overline{X}_D = 2.12, s_{\overline{D}} = 2.17$
   d. $N_D = 20, \overline{X}_D = .80, s_{\overline{D}} = 3.10$

24. For each of the following, calculate the standard error of the difference scores ($s_{\overline{D}}$) and the *t*-test for dependent means.
   a. $N_D = 10, \overline{X}_D = 12.00, s_{\overline{D}} = 9.00$
   b. $N_D = 19, \overline{X}_D = 3.52, s_{\overline{D}} = 6.14$
   c. $N_D = 8, \overline{X}_D = 2.89, s_{\overline{D}} = 4.86$
   d. $N_D = 30, \overline{X}_D = 1.44, s_{\overline{D}} = 8.32$

25. As people in our society live longer, there is a growing need for older adults to maintain a healthy lifestyle, part of which is their physical fitness. One study evaluated a program designed to increase older adults' level of fitness using

weight and strength training (Doll, 2009). In the study, eight adults (mean age = 75.60 years) received training on a weight machine over an 8-week period, doing exercises such as bench presses and abdominal crunches. Before and after beginning the training, each adult was measured on a number of functional tasks, one of which was the number of times they could lift a 5-pound weight over their heads. The descriptive statistics for the number of overhead lifts were the following: pretest ($M = 45.60$, $s = 15.00$) and posttest ($M = 53.90$, $s = 18.60$). Furthermore, the mean difference score ($\overline{X}_D$) was −8.30; the standard deviation of the difference scores ($s_{\overline{D}}$) was 9.84. Test the difference in overhead lifts between the pretest and posttest.

a. State the null and alternative hypotheses (H_0 and H_1).
b. Make a decision about the null hypothesis.
   (1) Calculate the degrees of freedom (*df*).
   (2) Set alpha (α), identify the critical values (draw the distribution), and state a decision rule.
   (3) Calculate a value for the *t*-test for dependent means.
   (4) Make a decision whether to reject the null hypothesis.
   (5) Determine the level of significance.
c. Draw a conclusion from the analysis.
d. What conclusions might the researchers draw regarding the effectiveness of a weight and strength training program designed for older adults?

26. Another part of the study discussed in Exercise 25 had a second sample of nine older adults take part in an 8-week calisthenics program in which they did squats, hamstring curls, and lunges. As the program was designed to help these adults perform everyday activities more easily, before and after the program, each adult was asked how difficult it was for them to get in and out of a bathtub (the higher the number, the greater the difficulty). The descriptive statistics for this bathtub-related difficulty variable were the following: pretest ($M = 4.30$, $s = 2.00$) and posttest ($M = 2.90$, $s = .90$). Furthermore, the mean difference score ($\overline{X}_D$) was 1.40; the standard deviation of the difference scores ($s_{\overline{D}}$) was 1.86. Test the difference in bathtub-related difficulty between the pretest and posttest.

    a. State the null and alternative hypotheses (H_0 and H_1).
    b. Make a decision about the null hypothesis.
       (1) Calculate the degrees of freedom (*df*).
       (2) Set alpha (α), identify the critical values (draw the distribution), and state a decision rule.
       (3) Calculate a value for the *t*-test for dependent means.
       (4) Make a decision whether to reject the null hypothesis.
       (5) Determine the level of significance.
    c. Draw a conclusion from the analysis.
    d. What conclusions might the researchers draw regarding the effectiveness of a calisthenics program designed for older adults?

27. A group of friends decided to work together to decrease their cigarette smoking. They hypothesize that they can decrease their level of smoking by doing such things as giving one another encouragement, using carrots and candy to replace the physical action of smoking, and using deep breathing and counting through cravings. Each friend keeps track of how many cigarettes he or she smokes each day. The first set of data is each person's baseline (before beginning the plan), and the second set of data is after 4 weeks of using the tools. Complete the steps below to test their hypothesis.

Person	Before	After
1	8	0
2	25	13

(Continued)

(Continued)

Person	Before	After
3	17	0
4	5	1
5	12	14
6	20	10

a. For each of the two time periods, calculate the sample size (N_i), the mean ($\bar{X}_i$), and the standard deviation (s_i).

b. Calculate the mean ($\bar{X}_D$) and the standard deviation of the difference scores (s_D).

c. State the null and alternative hypotheses (H_0 and H_1).

d. Make a decision about the null hypothesis.

(1) Calculate the degrees of freedom (*df*).

(2) Set alpha (α), identify the critical values (draw the distribution), and state a decision rule.

(3) Calculate a value for the *t*-test for dependent means.

(4) Make a decision whether to reject the null hypothesis.

(5) Determine the level of significance.

e. Draw a conclusion from the analysis.

f. Relate the result of the analysis to the research hypothesis.

28. A researcher hypothesizes that an afternoon dose of caffeine lessens the amount of time a person sleeps at night. The participants are instructed to report their average hours of sleep per night for a week without any caffeine in the afternoons and their average amount of sleep for a week in which they drink two caffeinated beverages between 2 and 3 p.m. Using their reported data, run through the steps of hypothesis testing to test the research hypothesis.

Person	Without Caffeine	With Caffeine
1	9.00	6.00
2	6.00	5.50
3	8.00	8.00
4	7.00	6.25
5	7.50	5.00
6	7.00	5.00
7	6.00	7.00
8	9.00	8.00
9	8.00	7.00
10	7.00	6.00

a. For each of the two caffeine conditions, calculate the sample size (N_i), the mean ($\bar{X}_i$), and the standard deviation (s_i).

b. Calculate the mean ($\bar{X}_D$) and the standard deviation of the difference scores (s_D).

c. State the null and alternative hypotheses (H_0 and H_1).

d. Make a decision about the null hypothesis.

(1) Calculate the degrees of freedom (*df*).

(2) Set alpha (α), identify the critical values (draw the distribution), and state a decision rule.

(3) Calculate a value for the t-test for dependent means.

(4) Make a decision whether to reject the null hypothesis.

(5) Determine the level of significance.

e. Draw a conclusion from the analysis.

f. Relate the result of the analysis to the research hypothesis.

Answers to Learning Checks

Learning Check 1

2.

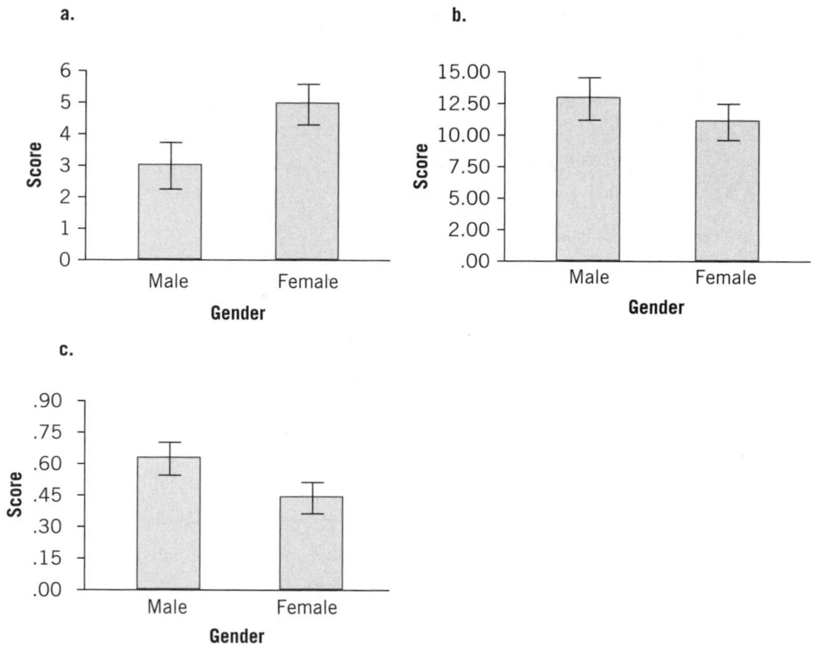

3. a. Dog: $\bar{X} = 2.00$, $s = 2.83$, $s_{\bar{X}} = 1.27$; Cat: $\bar{X} = 6.00$; $s = 3.39$, $s_{\bar{X}} = 1.52$

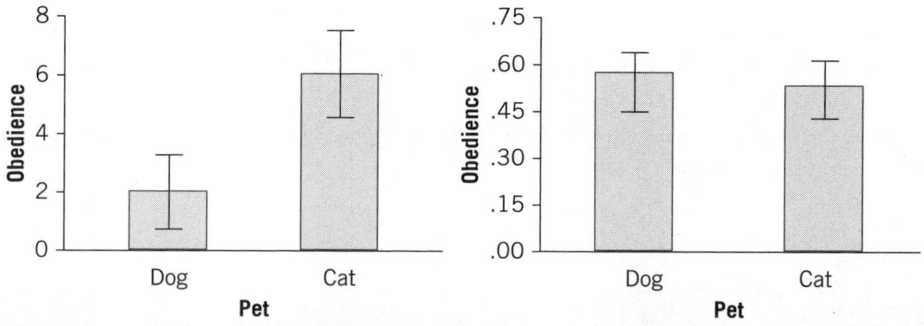

b. Dog: $\overline{X} = .57, s = .20, s_{\overline{X}} = .06$; Cat: $\overline{X} = .53, s = .27, s_{\overline{X}} = .09$

c. Dog: $\overline{X} = 8.71, s = 2.30, s_{\overline{X}} = .62$; Cat: $\overline{X} = 6.29, s = 2.02, s_{\overline{X}} = .54$

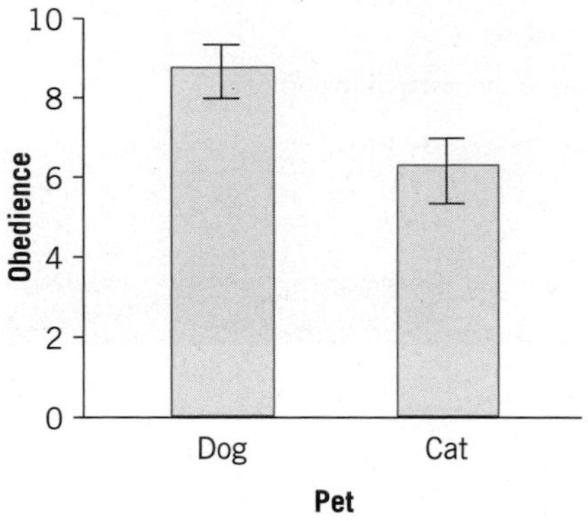

Learning Check 2

2. a. H_0: $\mu_{\text{Single parent}} = \mu_{\text{Intact}}$; H_1: $\mu_{\text{Single parent}} \neq \mu_{\text{Intact}}$

   b. H_0: $\mu_{\text{Tap water}} = \mu_{\text{Bottled water}}$; H_1: $\mu_{\text{Tap water}} \neq \mu_{\text{Bottled water}}$

   c. H_0: $\mu_{\text{Alone}} = \mu_{\text{With others}}$; H_1: $\mu_{\text{Alone}} \neq \mu_{\text{With others}}$

3. a. $df = 12$; critical value = ± 2.179

   b. $df = 20$; critical value = ± 2.086

   c. $df = 14$; critical value = 1.761

4. a. $s_{\overline{X}_1 - \overline{X}_2} = 2.24$

   b. $s_{\overline{X}_1 - \overline{X}_2} = 3.61$

   c. $s_{\overline{X}_1 - \overline{X}_2} = 1.03$

5. a. $s_{\overline{X}_1 - \overline{X}_2} = 1.47; t = 2.04$

   b. $s_{\overline{X}_1 - \overline{X}_2} = 4.65; t = .97$

   c. $s_{\overline{X}_1 - \overline{X}_2} = .74; t = -3.88$

Learning Check 3

2. a. H_0: $\mu_{\text{Men}} = \mu_{\text{Women}}$; H_1: $\mu_{\text{Men}} \neq \mu_{\text{Women}}$

   b. (1) $df = 24$

      (2) If $t < -2.064$ or > 2.064, reject H_0; otherwise, do not reject H_0

      (3) $s_{\overline{X}_1 - \overline{X}_2} = 6.52; t = -.32$

      (4) $t = -.32$ is not < -2.064 or > 2.064 $\therefore$ do not reject H_0 ($p > .05$)

      (5) Not applicable (H_0 not rejected)

   c. The average estimate of a woman's weight in the sample of 13 men ($M = 135.62$) and 13 women ($M = 137.69$) was not significantly different, $t(24) = -.32, p > .05$.

   d. The results of this analysis do not support the research hypothesis that men will give lower estimates of the average woman's weight than will women.

Learning Check 4

2. a. $s_{\overline{X}_1 - \overline{X}_2} = 1.76$

   b. $s_{\overline{X}_1 - \overline{X}_2} = 2.11$

c. $s_{\bar{X}_1-\bar{X}_2} = 4.13$

3. a. H_0: $\mu_{Participate} = \mu_{Not\ participate}$; H_1: $\mu_{Participate} \neq \mu_{Not\ participate}$
   b. (1) $df = 80$
      (2) If $t < -2.000$ or > 2.000, reject H_0; otherwise, do not reject H_0
      (3) $s_{\bar{X}_1-\bar{X}_2} = .42; t = 4.55$
      (4) $t = 4.55 > 2.000$ ∴ reject H_0 ($p < .05$)
      (5) $t = 4.55 > 2.660$ ∴ $p < .01$
   c. The average level of knowledge regarding what to feed their children at home was significantly higher for the sample of 29 parents who participated in the program ($M = 4.68$) than the sample of 53 parents who did not participate ($M = 2.77$), $t(80) = 4.55$, $p < .01$.
   d. The results of this analysis suggest that the program may provide information for parents of children with anorexia nervosa.

Learning Check 5

2. a.

Pair	Shy	Outgoing	Difference (*D*)
1	8	4	**4**
2	4	6	**–2**
3	8	2	**6**
4	6	6	**0**
5	3	7	**–4**
6	6	3	**3**
7	5	4	**1**

   b. $\bar{X}_D\ s_{\bar{D}} = 1.14; s_{\bar{D}} = 3.48; = 1.32$

3. a. $s_{\bar{D}} = .50; t = 6.00$
   b. $s_{\bar{D}} = .38, t = 3.95$
   c. $s_{\bar{D}} = 5.42, t = 2.33$

4. a. H_0: $\mu_D = 0$; H_1: $\mu_D \neq 0$
   b. (1) $df = 19$
      (2) If $t < -2.093$ or > 2.093, reject H_0; otherwise, do not reject H_0
      (3) $s_{\bar{D}} = 1.05; t = 3.14$
      (4) $t = 3.14 > 2.093$ ∴ reject H_0 ($p < .05$)
      (5) $t = 3.14 > 2.861$ ∴ $p < .01$
   c. The average number of words per minute (WPM) typed correctly for a sample of 20 adults was significantly greater when tested while sitting ($M = 40.20$) than while they were walking on a treadmill ($M = 36.90$), $t(19) = 3.14$, $p < .01$.
   d. The results of this analysis suggest that typing is not a task that may be performed at a satisfactory level while walking on a treadmill rather than sitting.

Answers to Odd-Numbered Exercises

1.

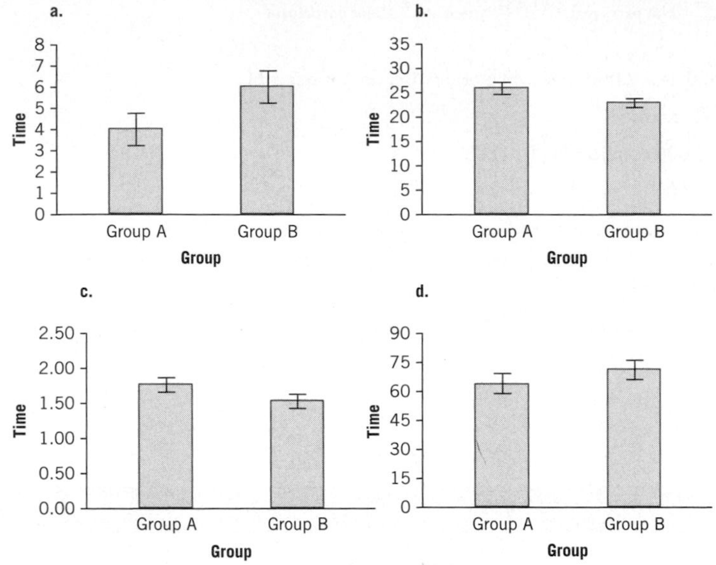

3. a. Experimental: $\bar{X} = 5.00$, $s = 2.94$, $s_{\bar{X}} = 1.11$; Control: $\bar{X} = 9.00$, $s = 3.06$, $s_{\bar{X}} = 1.16$

b. Experimental: $\bar{X} = 15.00$, $s = 3.02$, $s_{\bar{X}} = 1.07$; Control: $\bar{X} = 18.00$; $s = 3.21$, $s_{\bar{X}} = 1.13$

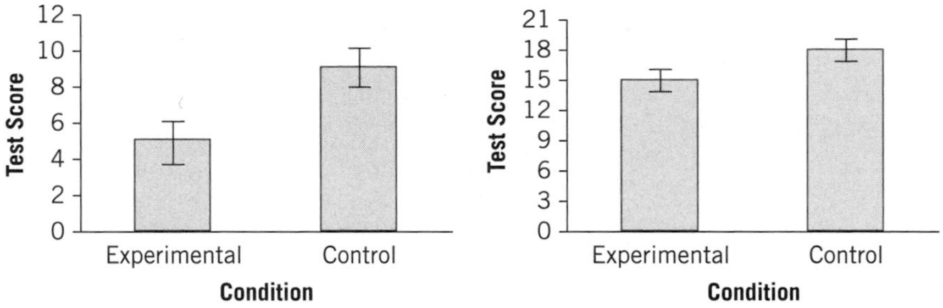

c. Experimental: $\bar{X} = 3.27$; $s = .46$, $s_{\bar{X}} = .15$; Control: $\bar{X} = 2.80$; $s = .41$, $s_{\bar{X}} = .14$

d. Experimental: $\bar{X} = 3.38$; $s = 1.39$, $s_{\bar{X}} = .39$; Control: $\bar{X} = 3.23$; $s = 1.64$, $s_{\bar{X}} = .46$

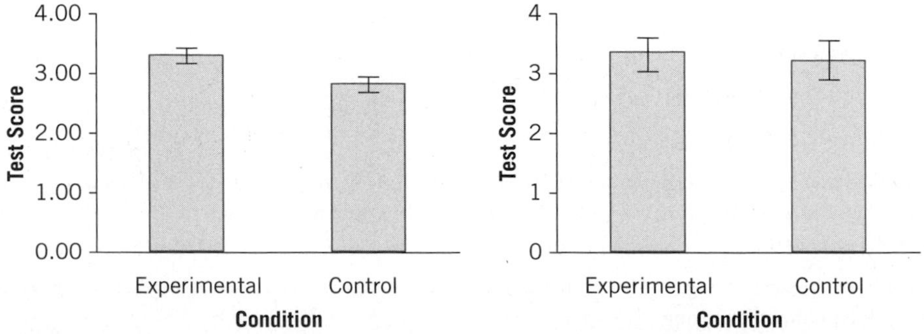

5. a. $df = 40$; critical value = ±2.021
   b. $df = 26$; critical value = ±2.056
   c. $df = 6$; critical value = 1.943
   d. $df = 62$; critical value = −1.671

7. a. $s_{\bar{X}_1-\bar{X}_2} = .60$
   b. $s_{\bar{X}_1-\bar{X}_2} = 2.39$
   c. $s_{\bar{X}_1-\bar{X}_2} = 2.30$
   d. $s_{\bar{X}_1-\bar{X}_2} = .47$

9. a. $t = 1.13$
   b. $t = -2.09$
   c. $t = -1.67$
   d. $t = 3.20$

11. a. $s_{\bar{X}_1-\bar{X}_2} = 1.31; t = -.38$
    b. $s_{\bar{X}_1-\bar{X}_2} = 2.03; t = 3.27$
    c. $s_{\bar{X}_1-\bar{X}_2} = .37; t = 1.70$
    d. $s_{\bar{X}_1-\bar{X}_2} = .95; t = -2.26$

13. a. H_0: $\mu_{\text{Attraction}} = \mu_{\text{No attraction}}$; H_1: $\mu_{\text{Attraction}} \neq \mu_{\text{No attraction}}$
    b. (1) $df = 52$
       (2) If $t < -2.009$ or > 2.009, reject H_0; otherwise, do not reject H_0
       (3) $s_{\bar{X}_1-\bar{X}_2} = 1.33; t = 2.14$
       (4) $t = 2.14 > 2.009$ ∴ reject H_0 ($p < .05$)
       (5) $t = 2.14 < 2.678$ ∴ $p < .05$ (but not $< .01$)
    c. The average number of women's flirtatious behaviors was significantly higher in the sample of 13 interviewers in the attraction group ($M = 37.44$) than the 13 interviewers in the no attraction group ($M = 34.59$), $t(24) = 2.14$, $p < .05$.
    d. The results of this analysis support the research hypothesis that interviewers who believe that the women they interview are attracted to them elicit more flirtatious behavior from these women than interviewers who do not hold this belief.

15. a. H_0: $\mu_{20-34} = \mu_{40-54}$; H_1: $\mu_{20-34} \neq \mu_{40-54}$
    b. (1) $df = 48$
       (2) If $t < -2.021$ or > 2.021, reject H_0; otherwise, do not reject H_0

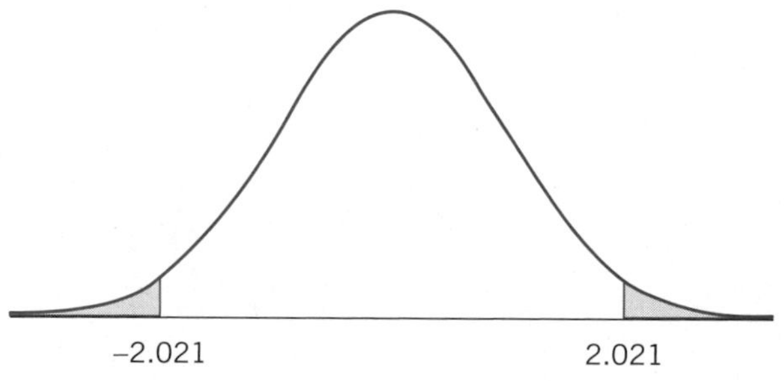

       (3) $s_{\bar{X}_1-\bar{X}_2} = .94; t = -4.22$
       (4) $t = -4.19 < -2.021$ ∴ reject H_0 ($p < .05$)
       (5) $t = -4.19 < -2.704$ ∴ $p < .01$

c. The average difference in age between themselves and their preferred female partners was significantly higher for the sample of 25 men aged 40 to 54 years ($M = 4.98$) than the sample of 25 men aged 20 to 34 years ($M = 1.04$), $t(48) = -4.19, p < .01$.

d. The results of this analysis support the research hypothesis that the preferred difference in age between themselves and their preferred partners is greater for older men than for younger men.

17. a. Women: $N = 13, \bar{X} = 18.62, s = 7.81$

Men: $N = 13, \bar{X} = 14.85, s = 8.55$

b. H_0: $\mu_{Women} = \mu_{Men}$; H_1: $\mu_{Women} \neq \mu_{Men}$

c. (1) $df = 24$

(2) If $t < -2.064$ or > 2.064, reject H_0; otherwise, do not reject H_0

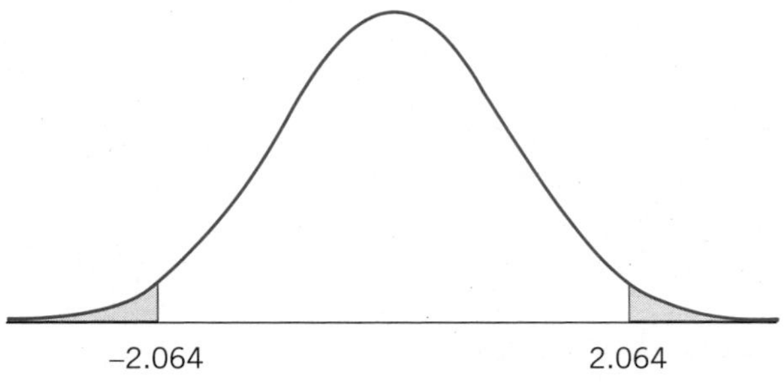

(3) $s_{\bar{X}_1-\bar{X}_2} = 3.21; t = 1.17$

(4) $t = 1.17$ is not < -2.064 or > 2.064 $\therefore$ do not reject H_0 ($p > .05$)

(5) Not applicable (H_0 not rejected)

d. The average estimates of the cost of a wedding (in thousands of dollars) for a sample of 13 women ($M = 18.62$) and a sample of 13 men ($M = 14.85$) were not significantly different, $t(24) = 1.17, p > .05$.

e. The result of this analysis does not support the research hypothesis that young women will provide a higher estimate of the cost of the average wedding than will men.

19. a. $s_{\bar{X}_1-\bar{X}_2} = 1.65$

b. $s_{\bar{X}_1-\bar{X}_2} = 3.11$

c. $s_{\bar{X}_1-\bar{X}_2} = 1.02$

d. $s_{\bar{X}_1-\bar{X}_2} = 3.08$

21. a. H_0: $\mu_{New\ user} = \mu_{Experienced\ user}$; H_1: $\mu_{New\ user} \neq \mu_{Experienced\ user}$

b. (1) $df = 62$

(2) If $t < -2.000$ or > 2.000, reject H_0; otherwise, do not reject H_0

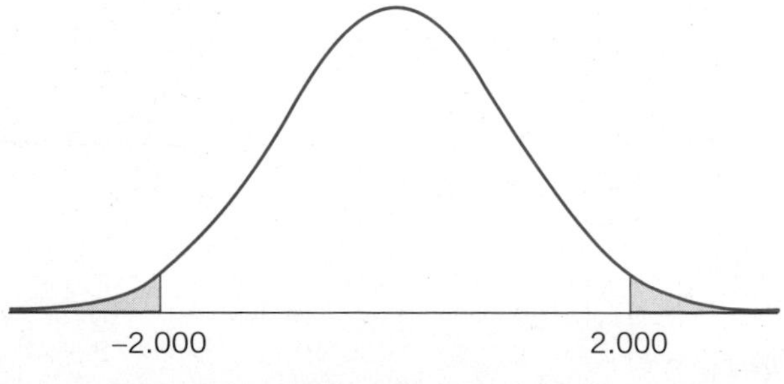

(3) $s_{\bar{X}_1 - \bar{X}_2} = 1.12; t = -1.01$

(4) $t = -1.01$ is not < -2.000 or > 2.000 ∴ do not reject H_0 ($p > .05$)

(5) Not applicable (H_0 not rejected)

c. The average level of satisfaction with their hearing aids of the sample of 30 new users ($M = 26.90$) and a sample of 34 experienced users ($M = 28.03$) was not significantly different, $t(62) = -1.01, p > .05$.

d. The result of this analysis indicates that these patients' satisfaction with their hearing aids is not related to how long they've used their hearing aids.

23. a. $s_{\bar{D}} = 1.60; t = 6.25$

b. $s_{\bar{D}} = 2.11; t = 3.22$

c. $s_{\bar{D}} = .54; t = 3.93$

d. $s_{\bar{D}} = .69; t = 1.16$

25. a. H_0: $\mu_D = 0$; H_1: $\mu_D \neq 0$

b. (1) $df = 7$

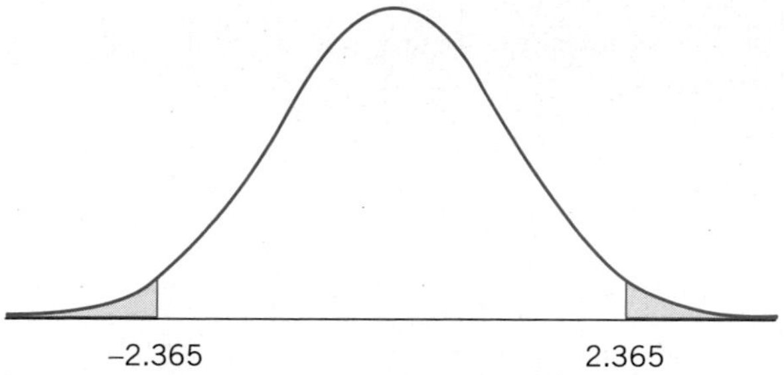

(2) If $t < -2.365$ or > 2.365, reject H_0; otherwise, do not reject H_0

(3) $s_{\bar{D}} = 3.48; t = -2.39$

(4) $t = -2.39 < -2.365$ ∴ reject H_0 ($p < .05$)

(5) $t = -2.39 > -3.499$ ∴ $p < .05$ (but not $< .01$)

e. The average number of overhead lifts for a sample of eight older adults in the weight and strength training program significantly increased over the 8-week period from the pretest ($M = 45.60$) to the posttest ($M = 53.90$), $t(7) = -2.39, p < .05$.

f. The results of this analysis suggest that weight and strength training may increase older adults' levels of physical fitness.

27. a. Before: $N_1 = 6, \bar{X}_1 = 14.50, s_1 = 7.56$

After: $N_2 = 6, \bar{X}_2 = 6.33, s_2 = 6.71$

b. $\bar{X}_D = 8.17; = 6.59$

c. H_0: $\mu_D = 0$; H_1: $\mu_D \neq 0$

d. (1) $df = 5$

(2) If $t < -2.571$ or > 2.571, reject H_0; otherwise, do not reject H_0

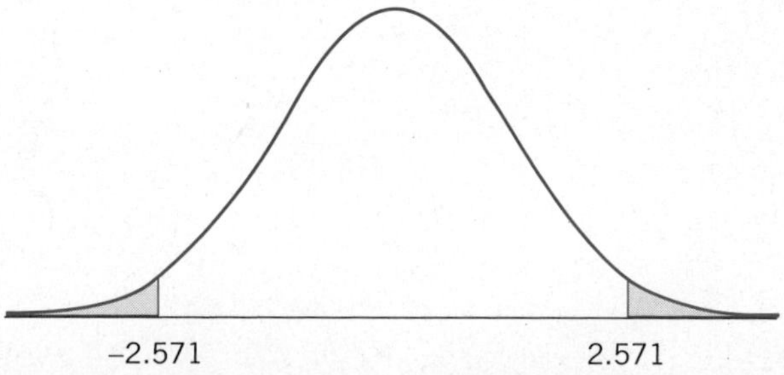

(3) $s_{\bar{D}} = 2.69; t = 3.04$

(4) $t = 3.04 > 2.571$ ∴ reject H_0 ($p < .05$)

(5) $t = 3.04 < 4.032$ ∴ $p < .05$ (but not $< .01$)

e. The average number of cigarettes smoked per day for a sample of six smokers significantly decreased from before ($M = 14.50$) to 4 weeks after ($M = 6.33$) using the tools, $t(5) < 3.04$, $p < .05$.

f. Using tools such as group support, substituting food for cigarettes, and deep breathing may be effective means of reducing cigarette smoking.

10

ERRORS IN HYPOTHESIS TESTING, STATISTICAL POWER, AND EFFECT SIZE

CHAPTER OUTLINE

In the last few chapters, we've discussed several statistical procedures used to test research hypotheses. Hypothesis testing is centered on making one of two decisions, reject or do not reject the null hypothesis, based on the probability of a statistic that has been calculated. Making the decision whether to reject the null hypothesis implies either that a hypothesized difference or relationship exists in the population or it does not. However, because this decision is based on probability rather than certainty, the possibility remains that whatever decision is made about the null hypothesis may either be correct or incorrect. In this chapter, two types of errors in hypothesis testing will be defined and described. More important, we will discuss strategies used by researchers to minimize the occurrence and consequences of these errors. To facilitate this discussion, we will use what may at first glance seem to be an unlikely analogy of hypothesis testing: the American justice system.

10.1 HYPOTHESIS TESTING VS. CRIMINAL TRIALS

As part of conducting research, the process of hypothesis testing follows a number of well-defined steps:

(1) A researcher, on the basis of an evaluation of a literature, states a research hypothesis regarding the relationship between variables and collects data to be analyzed.

(2) At the start of a statistical analysis, a null hypothesis, one that states a hypothesized relationship does not exist, is presumed to be true.

(3) The researcher statistically analyzes the collected data.

(4) On the basis of the analysis of the data, the researcher makes one of two decisions: reject or do not reject the null hypothesis. If the probability of the value of the statistic calculated from the data is sufficiently low, the null hypothesis is rejected; otherwise, the null hypothesis is not rejected.

Note that a decision to "not reject the null hypothesis" is not the same as "accept the null hypothesis." As discussed in Chapter 6, a statistical analysis cannot "prove" a hypothesized relationship does not exist in a population because data have been collected only from a limited sample of the population. Consequently, the purpose of an analysis isn't to prove the null hypothesis is true—it's to determine whether there is sufficient evidence to reject the null hypothesis.

The procedure for conducting criminal trials within the American justice system in many ways resembles the steps taken in hypothesis testing, even though the two processes occur in different settings and have different objectives:

(1) A law enforcement agency, on the basis of its evaluation of a crime that has taken place, collects evidence regarding an identified suspect, who may eventually be brought to trial.

(2) At the start of a trial, the accused person is presumed to be innocent.

(3) The attorneys present their evidence to the jury, which evaluates it.

(4) On the basis of its evaluation of the evidence, the jury makes one of two decisions: guilty or not guilty. In order to reject the presumption of innocence, the jury must decide that the evidence is convincing beyond a reasonable doubt. If the jury decides the reasonable doubt standard has been met, the presumption of innocence is rejected, and the person is found to be guilty; otherwise, the person is found to be not guilty.

It is important to note that a verdict of "not guilty" is not the same as deciding the person is "innocent." The purpose of a criminal trial isn't to prove the person's innocence—it's to determine whether there is sufficient evidence to conclude the person is guilty.

As we can see, the goal of both hypothesis testing and criminal trials is to analyze and evaluate collected evidence to make one of two decisions: The initial assumption is true or the initial assumption is not true. In both situations, because this decision is based on probability rather than certainty, it's possible to make the wrong decision. The next section discusses a research study illustrating the two types of incorrect decisions that may occur in the justice system, which will lead to a discussion of two types of errors that can be made in hypothesis testing.

10.2 AN EXAMPLE FROM THE RESEARCH: TRUTH OR CONSEQUENCES

One piece of evidence used to evaluate suspects in criminal investigations is the result of a psychophysiological detection device. The most well-known example of this type of device is the polygraph examination, sometimes referred to as the lie detector test. A polygraph examination involves attaching tubes and electrodes to a suspect's body to record breathing, skin response, and heart rate. Next, a trained test examiner asks the suspect a series of questions and then evaluates the suspect's physiological responses to these questions. For example, the examiner may ask the suspect where he or she was the day the crime was committed, during which any changes in the suspect's heart rate are noted. Ultimately, the examiner may make a decision whether or not the suspect should be considered guilty of the crime. It is important to understand that polygraph examinations cannot objectively prove someone's guilt; instead, the test examiner's decision is based on the likelihood an innocent person would have reacted the way the suspect did to the questions.

Can polygraph examinations accurately identify criminals? This has been the subject of debate for decades. One questioning technique used in polygraph examinations is the guilty knowledge test (GKT). The GKT is a multiple-choice examination measuring physiological responses to facts known to be relevant to a crime. For example, a suspect may be asked, "What kind of gun was used to shoot Mr. Doe? Was it a .22-caliber rifle, a 12-gauge shotgun, a .38-caliber revolver, or a 9-mm handgun?" The suspect's physiological response to the correct choice is compared to his or her responses to the incorrect choices. As noted by Eitan Elaad and his colleagues, "Innocent subjects, who are unable to distinguish relevant from irrelevant alternatives, are not expected to respond differentially to the relevant and irrelevant items . . . [but] if a subject's physiological responses are consistently greater to the relevant item than to the irrelevant ones, knowledge about a crime is inferred" (Elaad, Ginton, & Jungman, 1992, p. 757).

The goal of the GKT is to identify guilty suspects as accurately as possible. Consequently, Mr. Elaad and his research team conducted a study examining the accuracy of the GKT in actual crime investigations (Elaad et al., 1992). Part of this study consisted of giving GKT examinations to 68 people interrogated by police during criminal investigations. What is unique about this study is that, after taking these examinations, 38 of the 68 suspects confessed to having actually committed the crimes. As a result, this study was able to assess the accuracy of the GKT by comparing the actual innocence or guilt of the suspect with the decision (not guilty vs. guilty) made by the GKT.

Correct Decisions and Mistakes in Criminal Trials

In the GKT study, two pieces of information were recorded for each suspect: (1) the suspect's actual innocence or guilt (this study assumed that a suspect who did not confess to the crime and was eventually found not guilty during the trial was in fact innocent) and (2) the GKT's decision of whether the suspect should be considered not guilty or guilty. The four possible outcomes that could occur as a result of combining the "actual" with the "decision" are illustrated in Figure 10.1.

The top row of Figure 10.1 corresponds to suspects who were actually innocent of the crime. Within this row, Cell 1 represents innocent subjects whom the GKT classified as not guilty; this cell represents a correct GKT decision. Cell 2, on the other hand, comprises people innocent of the crime that the GKT defined as guilty. This cell represents an incorrect GKT decision, a mistake made by the justice system in which an innocent person could be sent to jail.

FIGURE 10.1 ● CORRECT AND INCORRECT DECISIONS USING THE GUILTY KNOWLEDGE TEST (GKT)

		GKT Decision	
		Not Guilty	Guilty
Suspect Actuality	Innocent	1 **Correct decision**	2 *Incorrect decision*
	Guilty	3 *Incorrect decision*	4 **Correct decision**

The bottom row of Figure 10.1 contains suspects actually guilty of the crime. On the right side of this row (Cell 4) are guilty suspects identified by the GKT as guilty; this cell represents a correct GKT decision. However, Cell 3 on the left represents people who confessed to committing the crime but were categorized by the GKT as not guilty. This is an incorrect decision, a mistake made by the justice system in which a guilty person may go free rather than be punished.

In summary, Figure 10.1 represents four possible outcomes that can occur using the GKT to make decisions regarding a person's guilt. Two of these outcomes represent correct decisions: classifying an innocent person as not guilty (Cell 1) and classifying a guilty person as guilty (Cell 4). As the goal of the justice system is to protect the innocent while convicting the guilty, we would want the proportion of people in these two cells to be as large as possible.

The other two outcomes, deciding an innocent person is guilty (Cell 2) and deciding a guilty person is not guilty (Cell 3), are incorrect decisions. We would certainly want to minimize the possibility of making either of these mistakes; however, do we consider them equally undesirable? Simply put, which do we as a society believe is worse: sending an innocent person to jail or letting a guilty person go free? Historically, a large number of laws and judicial rulings have been based on the belief that a greater miscarriage of justice has occurred if an innocent person is incarcerated than if a criminal escapes punishment. As a result, although we want the proportion of decisions in Cells 2 and 3 to be as small as possible, it is preferable to label a guilty person as not guilty (Cell 3) than to decide an innocent person is guilty (Cell 2).

In terms of the GKT study's findings regarding the accuracy of the GKT in criminal investigations, Figure 10.2 provides the number of people in each of the four possible outcomes. Looking first at the innocent subjects (Cells 1 and 2), we see that 37 of the 38 innocent subjects (97%) were correctly categorized by the GKT as not guilty, with only 1 of the 38 (3%) mistakenly labeled as guilty. Therefore, the GKT was found to be highly accurate in identifying innocent suspects. However, looking at the bottom row of Figure 10.2, the GKT was not as precise in identifying guilty suspects, with the 30 guilty suspects almost equally likely to be labeled not guilty (14/30 or 47%) as guilty (16/30 or 53%). Comparing Cells 2 and 3, the GKT was much more likely to lead to the incorrect decision of letting a guilty person go free (Cell 3) than of sending an innocent person to jail (Cell 2).

FIGURE 10.2 ● ACCURACY OF GKT TEST (ELAAD ET AL., 1992)

		GKT Decision: Not Guilty	GKT Decision: Guilty	Total
Suspect Actuality	Innocent	1 **37**	2 *1*	38
	Guilty	3 *14*	4 **16**	30

10.3 TWO ERRORS IN HYPOTHESIS TESTING: TYPE I AND TYPE II ERROR

The GKT study illustrates correct and incorrect decisions that can be made in the American justice system. Just as mistakes can be made regarding a suspect's innocence or guilt, in hypothesis testing, it is possible to commit errors in concluding whether a hypothesized effect exists in the population. Two types of errors that can be made in hypothesis testing, Type I and Type II error, are defined and described below.

Type I Error: The Risk in Rejecting the Null Hypothesis

Hypothesis testing starts with the assumption that the null hypothesis is true, which implies that the hypothesized effect or relationship does *not* exist in the population. This is similar to the process of criminal trials, which start with the assumption that the accused has not committed a crime. The top row in Figure 10.3 portrays the situation where the null hypothesis is in fact true; within this row, Cell 1 represents the situation where the null hypothesis is true and the decision is made to *not* reject the null hypothesis. In other words, the hypothesized effect does not exist in the population, and we conclude that it does not exist. This represents a correct decision, similar to the GKT deciding that an innocent suspect is not guilty.

Cell 2, on the other hand, represents a **Type I error**, which is an error that occurs when the null hypothesis is true but the decision is made to reject the null hypothesis. That is, the hypothesized effect does not exist in the population, but we conclude that it does exist. Type I error may also be expressed as "rejecting a true null hypothesis," "rejecting the null hypothesis when we shouldn't," or "concluding an effect exists when it actually does not."

To illustrate Type I error, let's first return to our Super Bowl example in Chapter 6, in which we tested a hypothesis regarding whether winning the pregame coin flip affects a team's chances of winning the Super Bowl. In this study, a Type I error would occur if winning the pregame coin flip does not affect a team's chances of winning the game, but we decide that it does. In the parking lot study from Chapter 9, a Type I error would happen if the two groups of drivers in the population, Intruder and No Intruder, do not differ in the time taken to leave a parking space, yet we conclude that they do differ. Within the GKT study, the equivalent of a Type I error occurred when the GKT concluded a person was guilty when he or she was in fact innocent.

FIGURE 10.3 ● CORRECT DECISIONS AND ERRORS IN HYPOTHESIS TESTING

		Decision About Null Hypothesis	
		Do not reject H_0	Reject H_0
Reality in Population	H_0 is true	1 **Correct decision**	2 *Type I error*
	H_1 is true	3 *Type II error*	4 **Correct decision**

What Is the Probability of Making a Type I Error?

In analyzing a set of data, what is the chance we will reject the null hypothesis when in fact we shouldn't? Given that a Type I error occurs only when the decision is made to reject the null hypothesis, the probability of making this type of error is the same as the probability of rejecting the null hypothesis. Given that alpha (α) is the probability of a statistic needed to reject the null hypothesis, the probability of making a Type I error is equal to α:

$$p(\text{Type I error}) = \alpha \quad (10\text{-}1)$$

If alpha is set to the traditional .05, the probability of making a Type I error is .05. Consequently, in making the decision to reject the null hypothesis, there is a .05 probability of concluding an effect exists when it in fact does not.

What Is the Probability of Not Making a Type I Error?

If the probability of making a Type I error is equal to the probability of rejecting the null hypothesis, it stands to reason that the probability of *not* making a Type I error is equal to the probability of *not* rejecting the null hypothesis. If the probability of making a Type I error is equal to alpha (α), the probability of not making this error is equal to $1 - \alpha$:

$$p(\text{not making Type I error}) = 1 - \alpha$$

Assuming $\alpha = .05$, the probability of not making a Type I error is $1 - .05$ or .95. Therefore, in making the decision to reject the null hypothesis, there is a .95 probability of correctly concluding an effect does not exist.

Note that there is a much greater probability of not making a Type I error (.95) than making a Type I error (.05). In hypothesis testing, researchers assume the null hypothesis is true unless the data provide convincing evidence to the contrary. The assumption that an effect does not exist is rejected only when the evidence is very strong. Again, this closely parallels the situation in criminal trials, in which the presumption of innocence is rejected only when the evidence is convincing beyond a reasonable doubt.

Why Is Type I Error a Concern?

It is important for researchers to avoid making Type I errors because these errors may result in the communication of incorrect information. Science is built on the accumulation of research such that researchers base their studies on others' findings. Consequently, stating that a relationship or effect exists when it does not may adversely affect others' thinking, time, energy, and resources.

Imagine, for example, you're a researcher for a pharmaceutical company evaluating a drug used to fight lung cancer. You most likely base your research on the results of published research studies regarding the effectiveness of the drug. If it turned out that the findings of some of these studies were incorrect in that the drug was in fact not effective, you might lose a great deal of time and effort as well as need to change how you study the topic. Furthermore, Type I error not only affects researchers but also may have negative consequences for a broader audience. Returning to the drug example, learning that a drug is not as effective as previously believed may affect the lives of people relying on the drug to combat their ailments.

Why Does Type I Error Occur?

Type I errors occur as the result of random, chance factors. To illustrate how and why these errors occur, let's return to the example in Chapter 6 of 30 Super Bowl games (Table 6.6 and Figure 6.6). In this example, we developed a distribution of the number of possible wins for 30 teams that win the pregame coin flip under the assumption that the null hypothesis is true. Using this distribution, we developed the decision rule that if the teams that won the coin flip won either less than 10 games or more than 20 games, the null hypothesis would be rejected. Consequently, even when it's true that winning the coin flip does not influence the likelihood of winning the game, if, say, 23 teams win the games, we may incorrectly make the decision to reject the null hypothesis.

LEARNING CHECK 1:

Reviewing What You've Learned So Far

1. Review questions
   a. How is making the decision whether to reject the null hypothesis similar to making the decision regarding guilt in a criminal trial?
   b. What are the different ways of describing a Type I error?
   c. If you conducted a study comparing two groups, what would be a Type I error?
   d. Why is the probability of Type I error equal to alpha (α)?
   e. Why is the probability of *not* making a Type I error equal to 1 – alpha ($1 - \alpha$)?
   f. Why is Type I error a concern?
   g. Why do researchers allow for the possibility of making Type I errors?
2. For each of the situations below, describe what would be considered a Type I error.
   a. A restaurant owner wants to see if his customers can tell if they are drinking alcoholic or nonalcoholic beer.
   b. A dentist tests two types of anesthesia on her patients to see which type is better at alleviating pain.
   c. A consumer protection agency wants to know if computer users are less likely to get infected if they pay for virus protection software rather than download free software.

Given the possible undesirable effects of Type I error, you may wonder why researchers allow this mistake to happen. Wouldn't it be more reasonable to create a situation where the null hypothesis can never be incorrectly rejected? Eliminating the possibility of Type I error is undesirable for one basic reason: The only way to never make a Type I error is to never reject the null hypothesis, no matter how low the probability of the value of a calculated statistic. A similar dilemma exists in the justice system, where the only way to never convict an innocent person is to never convict anyone, no matter how convincing the evidence.

Creating a situation in which the null hypothesis could never be rejected (or suspects could never be convicted) would render hypothesis testing (or the justice system) ineffective and obsolete. Thus, the process of hypothesis testing must allow for the possibility of making a Type I error. Later in this chapter, we will discuss methods used by researchers to minimize the possibility of Type I error.

Type II Error: The Risk in Not Rejecting the Null Hypothesis

In addition to rejecting the null hypothesis when it is in fact true, a second type of error can be made in hypothesis testing: not rejecting the null hypothesis when the hypothesized effect does exist in the population. Returning to Figure 10.3, the bottom row in this figure illustrates the situation where the alternative hypothesis (H_1) is true, meaning that the hypothesized effect does in fact exist in the population. Starting this time with the cell on the right half of this row, Cell 4 represents the situation where the alternative hypothesis is true and the decision is made to reject the null hypothesis. This is a correct decision, similar to the GKT test correctly identifying a guilty suspect in a criminal investigation.

Unfortunately, it's possible to make a wrong decision regarding the alternative hypothesis. Cell 3 represents a **Type II error**, which is an error that occurs when the decision is made to not reject the null hypothesis even though the alternative hypothesis is true. That is, the hypothesized effect exists in the population, but in our data we conclude that it does not exist. Other ways of describing a Type II error include "failing to reject a false null hypothesis," "not rejecting the null hypothesis when we should," and "concluding an effect does not exist when it does."

Returning to the Super Bowl example in Chapter 6, a Type II error would occur if winning the pregame coin flip does in fact affect a team's chances of winning the game, but from our analysis of a set of data, we conclude that it does not. In the Chapter 9 parking lot study, a Type II error would be if Intruder and No Intruder drivers in the population differ in the time taken to leave a parking space, yet in our study we conclude they do not differ. In the GKT study, a Type II error occurred when the GKT concluded a person was not guilty but the person later confessed to committing the crime.

What Is the Probability of Making a Type II Error?

The probability of making a Type II error, of not rejecting the null hypothesis when we should, is represented by the Greek symbol beta (β):

$$p(\text{Type II error}) = \beta \qquad (10\text{-}2)$$

Mathematically, what is the probability of making a Type II error? If the probability of making a Type I error is typically .05 ($\alpha = .05$), it might seem reasonable to assume that the probability of a Type II error (β) is equal to .95. However, β is *not* equal to $1 - \alpha$; this is because Type I and Type II errors take place under different situations.

Looking at Figure 10.3, we see that Type I error occurs in situations where the null hypothesis (H_0) is true, whereas Type II error occurs in situations where it is the alternative hypothesis (H_1) that is true. Using the GKT study (Figure 10.2) as an illustration, if we were to randomly pick one of the 38 innocent people, the probability of making what would be considered a Type I error, labeling an innocent suspect as guilty, is 1/38 or .03. However, looking at the bottom left cell of this figure, the probability of making a Type II error, labeling a guilty suspect as innocent, is not .97 (37/38) but rather .47 (14/30).

Unlike the GKT study, in which the suspects could be definitively identified as being either innocent or guilty, in a statistical analysis it is extremely difficult to determine the exact probability of making a Type II error. This difficulty is due to the inherent nature of how the alternative hypothesis is stated. One way to explain this is by providing an example that compares the null hypothesis (where the precise value of the population mean is specified) and the alternative hypothesis (where the precise value of the population mean is not specified).

In the reading skills study example in Chapter 7, the null hypothesis was stated as H_0: $\mu = 124.81$. Because the null hypothesis states a specific value for the population mean (124.81), we can create one sampling distribution of the mean with $\mu = 124.81$ in the center of this distribution and use this distribution to identify the values of the sample mean that have a probability less than alpha (α). As a result, we can determine the probability of making a Type I error, of rejecting the null hypothesis when it is true.

Turning to the probability of making a Type II error, the alternative hypothesis for the reading skills study was stated as H_1: $\mu \neq 124.81$. Consequently, the alternative hypothesis simply implies μ is some value *other* than 124.81; rejecting the null hypothesis could imply the population mean is 120.54, 132.98, 126.39, or any other possible value. Given the infinite possible values of μ, we are unable to determine the probability of either accepting or failing to accept the alternative hypothesis for any value of the sample mean. In short, the probability of making a Type II error in a statistical analysis is extremely difficult to determine because the value of a parameter such as μ is unknown when the alternative hypothesis is true.

What Is the Probability of Not Making a Type II Error?

If the probability of making a Type II error is represented symbolically by β, the probability of *not* making a Type II error is represented by $1 - \beta$:

$$p(\text{not making a Type II error}) = 1 - \beta$$

Given our earlier discussion of the difficulty in determining the exact probability of making a Type II error, it's not surprising that it's equally difficult to determine the probability of *not* making this error.

The probability of not making a Type II error has been given a special name, **statistical power**, defined as the probability of rejecting the null hypothesis when the alternative hypothesis is true. In other words, the statistical power of an analysis is the probability of detecting an effect when it does in fact exist in the population. Because rejecting the null hypothesis and detecting effects typically supports a study's research hypothesis, it is in a researcher's best interests to maximize the statistical power of an analysis. In Section 10.4 of this chapter, we discuss strategies used by researchers to maximize statistical power.

Why Is Type II Error a Concern?

Type II error, concluding an effect doesn't exist when in fact it does, is as important a concern to researchers as Type I error. However, it's important for a very different reason: A Type II error may result in the noncommunication of correct information. A Type II error occurs when the null hypothesis isn't rejected and a researcher concludes a difference or relationship doesn't exist; this conclusion implies a study's research hypothesis has not been supported. Because research studies whose hypotheses haven't been supported do not typically get published in academic journals or presented at conferences, research findings where the null hypothesis has not been rejected might not be communicated to other people. Returning to the drug example introduced earlier, a Type II error would occur if a drug is in fact effective but the researcher concludes it isn't. In such a situation, people's lives would be adversely affected in that they might not get access to a drug that could help them. However, they may not become aware that this error has occurred because the drug will not be produced or sold.

Why Does Type II Error Occur?

Type II errors occur when research studies do not find differences or relationships that actually exist in the population. As with Type I error, this can take place as the result of random, chance factors. Also, the effect being studied may be very small and difficult to detect. However, Type II error may also result from how researchers conduct their studies. For example, the sample in the study may have been small or unrepresentative of the population, or the methods used to measure or manipulate variables may have been inappropriate or flawed in some way.

As you can see, it is difficult to identify the precise reasons why the null hypothesis is not rejected. However, in conducting their studies, researchers have control over factors that increase the likelihood of rejecting the null hypothesis. Later in this chapter, methods used by researchers to decrease the probability of making a Type II error, and thereby increase the statistical power of a statistical analysis, will be described and evaluated.

Type I and Type II Error: Summary

To summarize, statistical analyses designed to test research hypotheses begin with two statistical hypotheses: the null hypothesis (H_0) and the alternative hypothesis (H_1). In the population, one of two actualities may be true: The null hypothesis is true or the alternative hypothesis is true. In conducting a statistical analysis, one of two decisions is made: reject or do not reject the null hypothesis.

By combining the two actualities with the two decisions, one of four possible outcomes may occur whenever a decision is made regarding the null hypothesis. The top row of Figure 10.4 represents the two outcomes that can occur when the null hypothesis is true. Here, deciding to not reject a true null hypothesis (Cell 1) is a correct decision, but rejecting a true null hypothesis (Cell 2) is an incorrect decision known as a Type I error. The probability (p) of making a Type I error is equal to alpha (α), typically set at .05; the probability of not making this error, $1 - \alpha$, is typically equal to .95. Type I error is a concern because the communication of incorrect information negatively affects others' thinking, time, energy, and resources. Type I errors occur as the result of random, chance factors, but researchers must allow for the possibility of making this error because the only way it may be eliminated is to never reject the null hypothesis regardless of the evidence.

FIGURE 10.4 ● SUMMARY OF TYPE I AND TYPE II ERRORS

Decision About Null Hypothesis

Reality in Population	Do not reject H_0	Reject H_0
H_0 is true	1 Correct decision $p = 1 - \alpha$	2 *Type I error* $p = \alpha$
H_1 is true	3 *Type II error* $p = \beta$	4 Correct decision $p = 1 - \beta$ *(statistical power)*

LEARNING CHECK 2:

Reviewing What You've Learned So Far

1. Review questions
   a. What are the different ways of describing a Type II error?
   b. If you conducted a study comparing two groups, what would be a Type II error?
   c. What is the difficulty in calculating the probability of Type II error?
   d. What is the concept of statistical power? What is its relationship to Type II error?

e. Why is Type II error a concern?
f. Why do Type II errors occur?

2. For each of the situations below, identify an example of a Type II error.
   a. A snowboarder tries two different kinds of wax to see which one lasts longer.
   b. A high school counselor wants to see whether students who pay for test preparation services are more likely to get into college than students who take free practice exams.
   c. A driver trying to save money uses both regular and premium gasoline and wants to know whether the type of gasoline affects her car's performance.
3. For each of the following hypotheses, create a chart that shows the four possible outcomes that could occur as a result of hypothesis testing. There should be two correct decisions and two incorrect decisions. Label the type of error for each of the incorrect decisions.
   a. People who drive electric vehicles live closer to their workplaces than people who drive gasoline-driven vehicles.
   b. People who drive after playing driving-related video games are more likely to get into automobile accidents than people who drive after playing stationary video games.

The two outcomes depicted in the bottom row of Figure 10.4 take place when the alternative hypothesis is true. It is possible to commit a Type II error (Cell 3), which is failing to reject a false null hypothesis. The exact probability of committing a Type II error, represented by beta (β), is extremely difficult to determine because of the inherently imprecise nature of the alternative hypothesis. Type II errors are a concern because they may result in the noncommunication of correct information that is potentially useful to others. Type II errors occur as the result of random, chance factors as well as how researchers conduct their studies, and the probability of making this error is affected by a number of factors discussed later in this chapter. Finally, Cell 4 represents a correct decision, in which the null hypothesis is rejected when the alternative hypothesis is true. The probability of this outcome (1 – β), which is the probability of not making a Type II error, is the statistical power of an analysis.

10.4 CONTROLLING TYPE I AND TYPE II ERROR

This section discusses methods used by researchers to minimize the occurrence of Type I and Type II errors, as well as concerns and drawbacks related to these methods. To demonstrate these methods, we will revisit the parking lot study discussed in Chapter 9. Box 10.1 lists the steps used to test the difference between the departure times of the Intruder and No Intruder groups. As we will see, these steps change depending on the particular method used to control for Type I or Type II error.

SUMMARY BOX 10.1 SUMMARY OF HYPOTHESIS TESTING, PARKING LOT STUDY (CHAPTER 9)

State the null and alternative hypotheses (H_0 and H_1)

$$H_0: \mu_{\text{Intruder}} = \mu_{\text{No Intruder}} \qquad H_1: \mu_{\text{Intruder}} \neq \mu_{\text{No Intruder}}$$

Make a decision about the null hypothesis

Calculate the degrees of freedom (*df*)

$$df = (N_1 - 1) + (N_2 - 1) = (15 - 1) + (15 - 1) = 28$$

(Continued)

(Continued)

Set alpha (α), identify the critical values, and state a decision rule

If $t < -2.048$ or > 2.048, reject H_0; otherwise, do not reject H_0

Calculate a statistic: t-test for independent means

Calculate the standard error of the difference ($s_{\bar{X}_1-\bar{X}_2}$)

$$s_{\bar{X}_1-\bar{X}_2} = \sqrt{\frac{s_1^2}{N_1} + \frac{s_2^2}{N_2}} = \sqrt{\frac{(10.42)^2}{15} + \frac{(10.08)^2}{15}} = 3.74$$

Calculate the t-statistic (t)

$$t = \frac{\bar{X}_1 - \bar{X}_2}{s_{\bar{X}_1-\bar{X}_2}} = \frac{40.73 - 31.67}{3.74} = 2.42$$

Make a decision whether to reject the null hypothesis

$t = 2.42 > 2.048 \therefore$ reject H_0 ($p < .05$)

Determine the level of significance

$t = 2.42 < 2.763 \therefore p < .05$ (but not $< .01$)

Controlling Type I Error

Given that a Type I error can occur when the decision is made to reject the null hypothesis, the probability of making this error can be reduced by making it more difficult to reject the null hypothesis. More specifically, Type I error may be controlled by lowering the value of alpha (α), the probability of a statistic needed to reject the null hypothesis. For example, rather than using the traditional $\alpha = .05$ probability cutoff, a researcher could choose to reject the null hypothesis only when the probability of a calculated statistic is less than .01 ($p < .01$).

Table 10.2 uses the parking lot study to demonstrate how to reduce Type I error by lowering alpha. In this table, the difference between the Intruder and No Intruder groups is tested under two conditions: $\alpha = .05$ and $\alpha = .01$. Comparing the steps in the two conditions, changing alpha doesn't affect how the null and alternative hypotheses (H_0 and H_1) are stated or how the inferential statistic (t) is calculated. However, lowering alpha from .05 to .01 increases the critical value of the statistic, in this example from 2.048 to 2.763, which makes the region of rejection smaller. Consequently, when α is equal to .01, a more extreme value of the statistic is needed to reject the null hypothesis than when $\alpha = .05$.

Looking at Table 10.1, we see that the null hypothesis in the parking lot study is not rejected when α is equal to .01, even though it was rejected when $\alpha = .05$. Reducing the probability of making a Type I error by lowering alpha means we are adopting a more conservative approach, only rejecting the null hypothesis when the calculated value of a statistic has an extremely low probability of occurring.

Concerns About Controlling Type I Error

Reducing Type I error by lowering the value of alpha is a relatively simple matter. However, there are several concerns about this method. First, statistical significance is traditionally defined using the $\alpha = .05$ cutoff. Researchers may be reluctant to set α to .01 rather than .05 because doing so may lead to confusion regarding the interpretation of the term *statistical significance.*

TABLE 10.1 ● CONTROLLING FOR TYPE I ERROR BY LOWERING ALPHA (α = .05 VS. .01)

Step	α = .05	α = .01
State H_0 and H_1	H_0: $\mu_{Intruder} = \mu_{No\ Intruder}$ H_1: $\mu_{Intruder} \neq \mu_{No\ Intruder}$	Same
Make decision about H_0		
Calculate *df*	*df* = 28	Same
Set α, identify critical values, and state a decision rule	If $t < -2.048$ or > 2.048, reject H_0; otherwise, do not reject H_0	If $t < -2.763$ or > 2.763, reject H_0; otherwise, do not reject H_0
Calculate a statistic	$s_{\bar{X}_1 - \bar{X}_2} = 3.74$ $t = 2.42$	Same Same
Make a decision whether to reject H_0	$t = 2.42 > 2.048$ $\therefore$ reject H_0 ($p < .05$)	$t = 2.42 < 2.763$ $\therefore$ do not reject H_0 ($p > .05$)

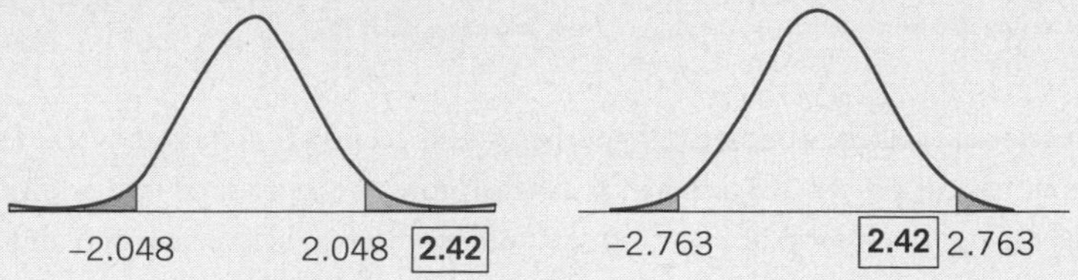

A second concern about controlling Type I error is that reducing the probability of Type I error increases the probability of Type II error. That is, the more difficult we make it to conclude an effect exists, the less likely we are to detect the effect when it does in fact exist. Using the justice system analogy, the harder we make it to decide a person is guilty, the less likely we are to detect those who are in fact guilty.

Figure 10.5 illustrates the trade-off between Type I and Type II error. The figure in the upper-left corner, Figure 10.5(a), represents a sampling distribution of a hypothetical statistic under the assumption the null hypothesis H_0 is true; the shaded area in this figure represents the probability of Type I error when $\alpha = .05$. Moving down, we find Figure 10.5(b), which represents a distribution of the statistic under the assumption that the alternative hypothesis H_1 is true; this figure is a symbolic rather than an actual representation of this distribution because, as mentioned earlier, the mean of this distribution cannot be defined. The shaded area in Figure 10.5(b) represents the probability of making a Type II error using the same $\alpha = .05$ cutoff as in Figure 10.5(a).

What is the consequence of reducing Type I error by lowering α? In the distribution in the upper-right hand corner (Figure 10.5(c)), the shaded area represents the probability of making a Type I error when $\alpha = .01$ rather than .05. Comparing Figures 10.5(a) and 10.5(c), lowering α from .05 to .01 reduces the probability of Type I error. However, comparing the shaded areas in the two bottom distributions, Figure 10.5(b) and 10.5(d), we see that lowering alpha from .05 to .01 increases the probability of Type II error.

Controlling Type II Error

A Type II error occurs when the decision is made to not reject the null hypothesis when the alternative hypothesis is true. Reducing Type II error, thereby increasing the statistical power of an analysis, involves increasing the likelihood of rejecting the null hypothesis. As discussed in several chapters of this book, a number of factors influence the likelihood of rejecting the null hypothesis: the size of the sample, the value of alpha (α), and the

FIGURE 10.5 ● TRADE-OFF BETWEEN TYPE I AND TYPE II ERROR ($\alpha = .05$ VS. $\alpha = .01$)

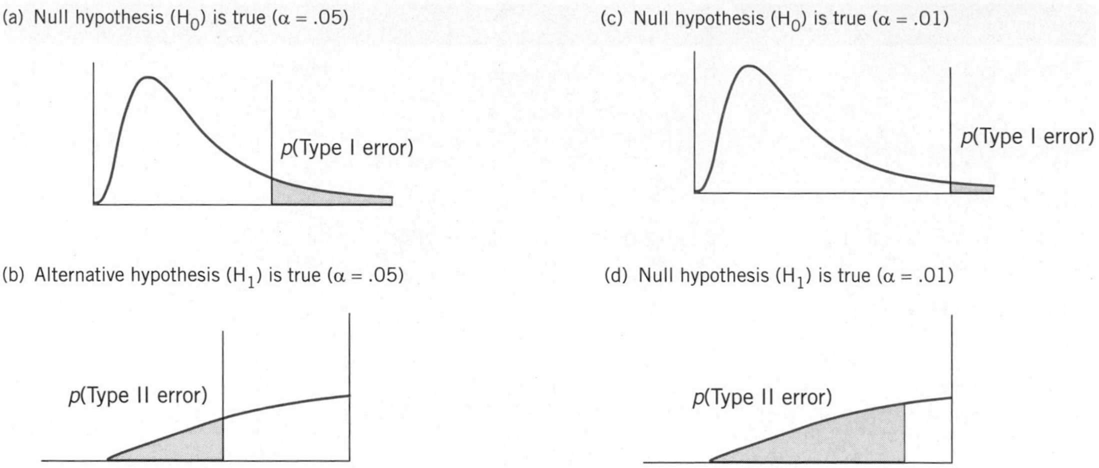

Source: Adapted from Keppel, G., Saufley, W. H., & Tokunaga, H. (1992).

directionality of the alternative hypothesis. This section illustrates how these factors may be changed to increase the statistical power of an analysis; furthermore, we introduce two additional factors that may be altered to increase the likelihood of rejecting the null hypothesis: between-group and within-group variability.

Increasing Sample Size

Perhaps the most commonly used method to increase the likelihood of rejecting the null hypothesis (i.e., increase the statistical power of an analysis) is to increase the size of a sample. In Table 10.3, the parking lot study is altered by doubling the sample size from $N = 15$ to $N = 30$. Keeping all other aspects of the study the same, increasing the sample size alters a number of aspects of the analysis. For example, a larger sample size results in smaller critical values (±2.009 vs. ±2.048); the smaller the critical values, the greater the likelihood of rejecting the null hypothesis. Also, the larger sample results in a bigger calculated value of the *t*-statistic (3.39 vs. 2.42); the larger the calculated value of a statistic, the greater the likelihood of rejecting the null hypothesis.

Although it may be simple to say, "If you want to reject the null hypothesis, use a big sample," the statistical advantages of a large sample must be balanced against pragmatic concerns. Increasing the sample size increases the amount of time, energy, and resources needed to collect data. As such, the answer to the question, "How large should my sample be?" may depend on the size of the hypothesized effect being studied. The smaller the effect, the larger the sample must be to detect it.

Raising Alpha (α)

The probability of making a Type II error may also be decreased by raising alpha (α). Table 10.3 alters the parking lot study by raising alpha from .05 to .10, implying the null hypothesis is rejected when the probability of the calculated statistic is only less than .10 (rather than the more stringent .05). In this table, raising alpha lowers the critical value used to make the decision to reject the null hypothesis (±1.701 vs. ±2.048), which in turn increases the size of the region of rejection. Because a smaller value of the statistic is needed to reject the null hypothesis when $\alpha = .10$ than when $\alpha = .05$, the likelihood the null hypothesis will be rejected has been increased.

There are several drawbacks of raising alpha from .05 to .10 to lower the probability of Type II error. First, using a value of alpha other than the traditional value of .05 may create uncertainty over the interpretation of the term *statistical significance.* Second, raising alpha from .05 to .10 increases the probability of making a Type I

TABLE 10.2 ● CONTROLLING FOR TYPE II ERROR BY INCREASING SAMPLE SIZE

Step	$N = 15$	$N = 30$
State H_0 and H_1	H_0: $\mu_{Intruder} = \mu_{No\ Intruder}$ H_1: $\mu_{Intruder} \neq \mu_{No\ Intruder}$	Same
Make decision about H_0		
Calculate *df*	$df = 28$	$df = 58$
Set α, identify critical values, and state a decision rule	If $t < -2.048$ or > 2.048, reject H_0; otherwise, do not reject H_0	If $t < -2.009$ or > 2.009, reject H_0; otherwise, do not reject H_0
Calculate a statistic	$s_{\bar{X}_1-\bar{X}_2} = 3.74$ $t = 2.42$	$s_{\bar{X}_1-\bar{X}_2} = \mathbf{2.67}$ $\mathbf{t = 3.39}$
Make a decision whether to reject H_0	$t = 2.42 > 2.048$ ∴ reject H_0 ($p < .05$)	$\mathbf{t = 3.39 > 2.009}$ ∴ reject H_0 ($p < .05$)

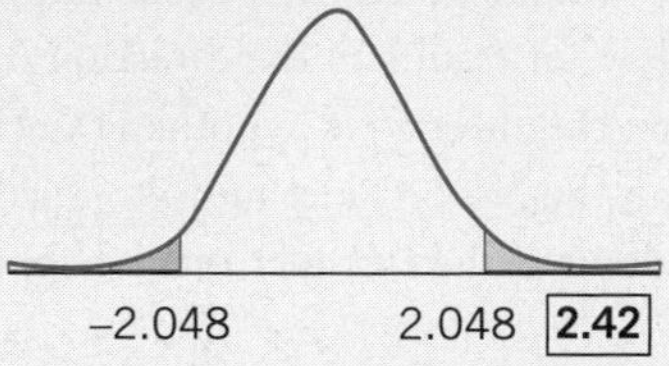

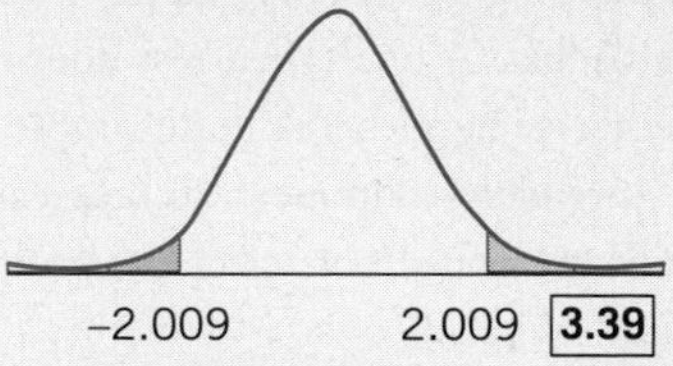

TABLE 10.3 ● CONTROLLING FOR TYPE II ERROR BY RAISING ALPHA (α = .10 VS. .05)

Step	$\alpha = .05$	$\alpha = .10$
State H_0 and H_1	H_0: $\mu_{Intruder} = \mu_{No\ Intruder}$ H_1: $\mu_{Intruder} \neq \mu_{No\ Intruder}$	Same
Make decision about H_0		
Calculate *df*	$df = 28$	Same
Set α, identify critical values, and state a decision rule	If $t < -2.048$ or > 2.048, reject H_0; otherwise, do not reject H_0	If $t < -1.701$ or > 1.701, reject H_0; otherwise, do not reject H_0
Calculate a statistic	$s_{\bar{X}_1-\bar{X}_2} = 3.74$ $t = 2.42$	Same Same
Make a decision whether to reject H_0	$t = 2.42 > 2.048$ ∴reject H_0 ($p < .05$)	$t = 2.42 > 1.701$ ∴ reject H_0 ($p < .10$)

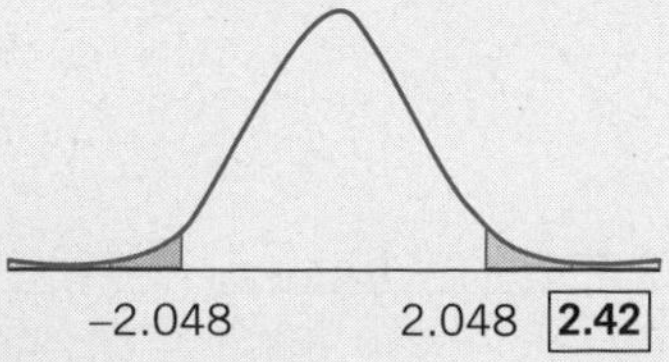

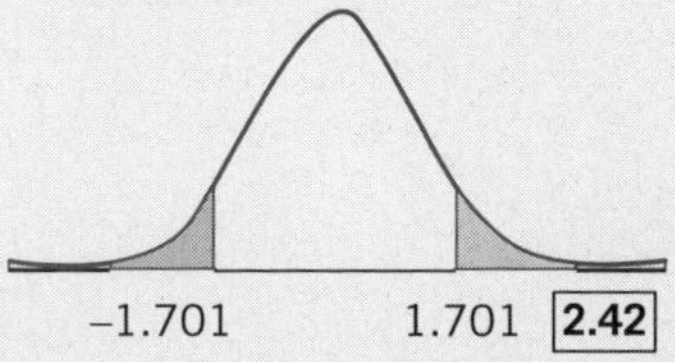

error. In other words, increasing the likelihood of deciding an effect exists also increases the possibility of concluding the effect exists when it does not. Given our concerns about Type I error, researchers do not typically increase the power of an analysis by raising alpha.

Using a Directional Alternative Hypothesis (H_1)

Another, somewhat controversial, way to increase the probability of rejecting the null hypothesis is to use a directional (one-tailed) rather than a nondirectional (two-tailed) alternative hypothesis. In the parking lot study, the nondirectional alternative hypothesis of H_1: $\mu_{Intruder} \neq \mu_{No\ Intruder}$ was used to allow for the possibility that the departure time for the Intruder group may be longer *or* shorter than that for the No Intruder group. However, given the research hypothesis that the Intruder group should take longer to leave than the No Intruder group, what if we were to use a directional hypothesis such as H_1: $\mu_{Intruder} > \mu_{No\ Intruder}$?

Table 10.4 analyzes the parking lot study using both a nondirectional and directional alternative hypothesis. The critical value is smaller for a directional hypothesis than for a nondirectional hypothesis (1.701 vs. 2.048), with the .05 region of rejection at one end of the *t*-distribution rather than split between the two ends, making it more likely to reject the null hypothesis.

Using a directional alternative hypothesis to decrease the probability of making a Type II error is similar to changing a two-tailed alpha from .05 to .10. However, there is one important distinction between the two approaches: A one-tailed test does not allow for results in the direction opposite from that specified by the alternative hypothesis. For example, stating the alternative hypothesis as H_1: $\mu_{Intruder} > \mu_{No\ Intruder}$ requires the Intruder group mean to be greater than the No Intruder mean to reject the null hypothesis. In this situation, we would not reject the null hypothesis if the Intruder mean happened to be much *less* than the No Intruder mean.

TABLE 10.4 ● CONTROLLING FOR TYPE II ERROR BY USING A DIRECTIONAL ALTERNATIVE HYPOTHESIS

Step	H_1: $\mu_{Intruder} \neq \mu_{No\ Intruder}$	H_1: $\mu_{Intruder} > \mu_{No\ Intruder}$
State H_0 and H_1	H_0: $\mu_{Intruder} = \mu_{No\ Intruder}$ H_1: $\mu_{Intruder} \neq \mu_{No\ Intruder}$	Same H_1: $\mu_{Intruder} > \mu_{No\ Intruder}$
Make decision about H_0		
Calculate *df*	$df = 28$	Same
Set α, identify critical values, and state a decision rule	If $t < -2.048$ or > 2.048, reject H_0; otherwise, do not reject H_0	If $t > 1.701$, reject H_0; otherwise, do not reject H_0
Calculate a statistic	$s_{\bar{X}_1 - \bar{X}_2} = 3.74$ $t = 2.42$	Same Same
Make a decision whether to reject H_0	$t = 2.42 > 2.048$ $\therefore$ reject H_0 ($p < .05$)	$t = 2.42 > 1.701$ $\therefore$ reject H_0 ($p < .05$)

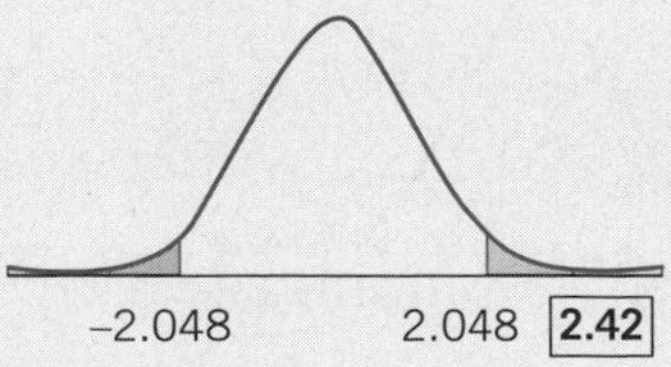

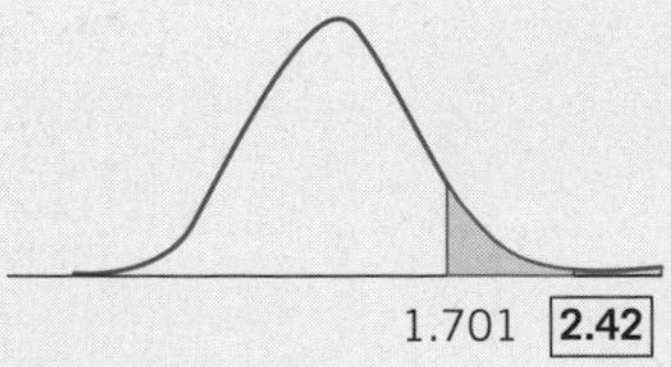

Using a directional alternative hypothesis may be justified when it is reasonable to believe the hypothesized effect may only occur in one direction. When this is not the case, however, using a directional alternative hypothesis simply to increase the likelihood of rejecting the null hypothesis is a questionable practice.

Increasing Between-Group Variability

The purpose of this section is to introduce another factor that may be altered to increase the likelihood of rejecting the null hypothesis: between-group variability. This factor will be illustrated using the formula for the *t*-test for independent means (Formula 9-2):

$$t = \frac{\overline{X}_1 - \overline{X}_2}{s_{\overline{X}_1 - \overline{X}_2}}$$

We've seen that the larger the value of the *t*-statistic calculated from a set of data, the more likely we are to reject the null hypothesis. One factor affecting the value of the *t*-statistic is the numerator of this formula: the difference between means $(\overline{X}_1 - \overline{X}_2)$. The greater this difference, the larger the calculated value of *t* and therefore the greater the likelihood of rejecting the null hypothesis.

As the numerator of the formula for the *t*-test involves the difference between group means, it may be referred to as **between-group variability**, defined as differences among the means of the groups that comprise an independent variable. The greater the amount of between-group variability, the greater the likelihood of rejecting the null hypothesis.

Table 10.5 illustrates the effect of increasing between-group variability on the decision to reject the null hypothesis. In the original parking lot study, the means of the Intruder and No Intruder groups were 40.73

TABLE 10.5 ● CONTROLLING FOR TYPE II ERROR BY INCREASING BETWEEN-GROUP VARIABILITY

Step	Intruder ($M = 40.73$) No Intruder ($M = 31.67$)	Intruder ($M = 45.73$) No Intruder ($M = 26.67$)
State H_0 and H_1	H_0: $\mu_{\text{Intruder}} = \mu_{\text{No Intruder}}$ H_1: $\mu_{\text{Intruder}} \neq \mu_{\text{No Intruder}}$	Same
Make decision about H_0		
Calculate *df*	$df = 28$	Same
Set α, identify critical values, and state a decision rule	If $t < -2.048$ or > 2.048, reject H_0; otherwise, do not reject H_0	Same
Calculate a statistic	$s_{\overline{X}_1 - \overline{X}_2} = 3.74$ $t = 2.42$	Same $t = 5.09$
Make a decision whether to reject H_0	$t = 2.42 > 2.048$ ∴ reject H_0 ($p < .05$)	$t = 5.09 > 2.048$ ∴ reject H_0 ($p < .05$)
	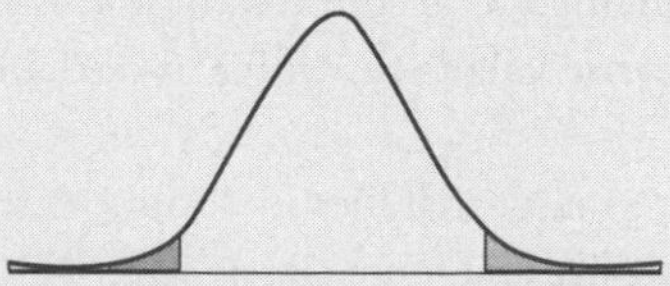 −2.048 2.048 **2.42**	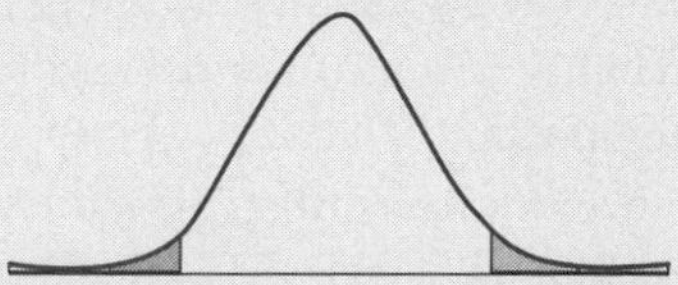 2.048 2.048 **5.09**

and 31.67, respectively. In Table 10.6, this difference has been increased by adding 5 seconds to the Intruder group mean and subtracting 5 seconds from the No Intruder mean. Keeping all other aspects of the analysis the same, increasing between-group variability increases the calculated value of the *t*-statistic (5.09 vs. 2.42), which increases the likelihood the null hypothesis will be rejected.

How can between-group variability in a study be increased? This may be accomplished by using research methods and manipulations that are effective and sensitive to differences between groups. For example, in the parking lot study, we could have one of the researchers act as the intruding driver rather than using actual drivers. The researcher could behave in a way that enhanced the level of intrusion, such as honking the horn or talking to the departing driver, which may increase the difference between the departing times of the Intruder and No Intruder groups. Returning to the drug example introduced earlier in this chapter, in testing the effectiveness of the drug, you could decide to administer it only to people who have no medical conditions that might interfere with the drug's ability to fight lung cancer.

Decreasing Within-Group Variability

Another factor affecting the likelihood of rejecting the null hypothesis is within-group variability. This factor is illustrated in the denominator of the formula for the *t*-test for two means: the standard error of the difference ($s_{\overline{X}_1-\overline{X}_2}$). Below is the formula for the standard error of the difference (Formula 9-3):

$$s_{\overline{X}_1-\overline{X}_2} = \sqrt{\frac{s_1^2}{N_1} + \frac{s_2^2}{N_2}}$$

Looking at this formula, the calculated value of $s_{\overline{X}_1-\overline{X}_2}$ is a function of the standard deviations of the two groups ($s_1$and s_2). The smaller the standard deviations, the smaller the standard error of the difference; the smaller the standard error of the difference, the larger the value of the *t*-test and the greater the likelihood of rejecting the null hypothesis.

The standard deviations represent the variability of scores within each of the groups. Therefore, the denominator of the formula for a statistic such as the *t*-test may be referred to as **within-group variability**, defined as the variability of scores within the groups that comprise an independent variable. The smaller the within-group variability in a study, the greater the likelihood of rejecting the null hypothesis.

Table 10.6 modifies the parking lot study, this time decreasing within-group variability by dividing each of the two standard deviations by 2. Again keeping all other aspects of the analysis the same, decreasing within-group variability increases the value of the *t*-statistic (4.84 vs. 2.42), thereby increasing the likelihood of rejecting the null hypothesis.

To answer the question regarding how researchers can reduce within-group variability in a study, we must first consider what within-group variability represents. How much variability should there be for the scores for a group? Theoretically, because all of the members of a group are presumed to be the same, there should be zero or no variability. For example, all of the drivers who are intruded upon should take exactly the same amount of time to leave their parking spaces. Any variability in the departure times of the intruded-upon drivers cannot be explained or accounted for and is referred to as *error*. Within-group variability is sometimes referred to as **error variance**, which is variability that cannot be explained or accounted for.

There are several ways to reduce within-group variability in a study. Perhaps the simplest way is to increase the size of the sample. As a sample grows in size and more closely approximates the population, the amount of variability among scores decreases. But error may also be reduced by collecting data in a systematic, standardized

TABLE 10.6 ● CONTROLLING FOR TYPE II ERROR BY DECREASING WITHIN-GROUP VARIABILITY

Step	Intruder (s = 10.42) No Intruder (s = 10.08)	Intruder (s = 5.21) No Intruder (s = 5.04)
State H_0 and H_1	H_0: $\mu_{Intruder} = \mu_{No\ Intruder}$ H_1: $\mu_{Intruder} \neq \mu_{No\ Intruder}$	Same
Make decision about H_0		
Calculate *df*	$df = 28$	Same
Set α, identify critical values, and state a decision rule	If $t < -2.048$ or > 2.048, reject H_0; otherwise, do not reject H_0	Same
Calculate a statistic	$s_{\bar{X}_1 - \bar{X}_2} = 3.74$ $t = 2.42$	$s_{\bar{X}_1 - \bar{X}_2} = 1.87$ $t = 4.84$
Make a decision whether to reject H_0	$t = 2.42 > 2.048$ $\therefore$ reject H_0 ($p < .05$)	$t = 4.84 > 2.048$ $\therefore$ reject H_0 ($p < .05$)
	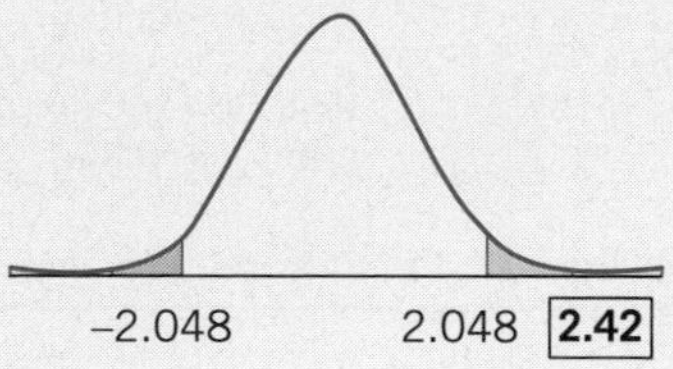	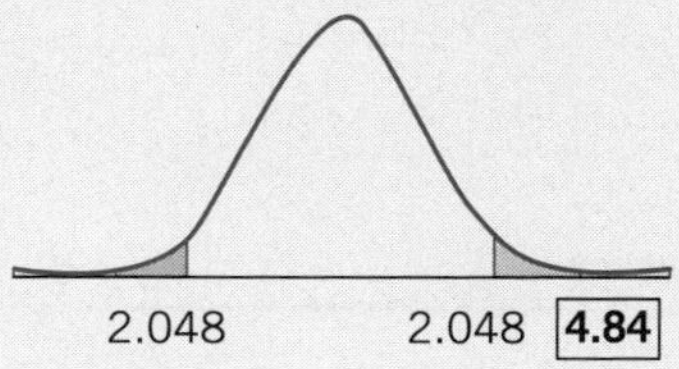

manner. For example, in the parking lot study, error variance was reduced by having all of the data collectors measure the variable "departure time" the same way: the difference between when the departing drivers opened their door and when the car had completely left the parking space.

Controlling Type I and Type II Error: Summary

Box 10.2 summarizes the methods available to researchers to control for Type I and Type II error. It is important to note that these methods differ in both the concerns they raise and the demands they place on researchers. For example, raising alpha or using a directional alternative hypothesis is a relatively simple method to decrease Type II error; however, researchers may be reluctant to use these methods because they are contrary to traditional practices. Consequently, researchers might instead focus on how data are collected rather than how they are analyzed, perhaps by increasing the size of the sample or determining ways to maximize between-group variability or minimize within-group variability, even though this may add to the time and resources necessary to conduct a study.

Thus far in this chapter, we've defined errors that can occur in making the decision to reject the null hypothesis, the probability of making these errors, why these errors occur, and methods used to reduce the occurrence of these errors. The next section of this chapter introduces a strategy used by researchers to appropriately interpret the results of their statistical analyses regardless of the decision about the null hypothesis.

SUMMARY BOX 10.2 METHODS CONTROLLING FOR TYPE I AND TYPE II ERROR

Type of Error	Probability of Making Error	Method for Controlling	Concern About Controlling
Type I (rejecting H_0 when H_0 is true)	α (typically .05)	Lower alpha (i.e., .05 to .01)	Traditional use of $\alpha = .05$
			Increases p (Type II error)
Type II (not rejecting H_0 when H_1 is true)	Difficult to calculate	Increase sample size	Pragmatic concerns (time, energy, cost)
		Raise alpha (α) (i.e., .05 to .10)	Traditional use of $\alpha = .05$
			Increases p (Type I error)
		Use directional (one-tailed) H_1	Ability to hypothesize direction of relationship
			Does not allow for result in opposite direction
		Increase between-group variability	Determining method to collect data
		Decrease within-group variability	Determining method to collect data

LEARNING CHECK 3:

Reviewing What You've Learned So Far

1. Review questions
   a. What is the main method used to control Type I error? What are concerns with this method?
   b. What are some methods for controlling Type II error? What are concerns with each of these methods?
   c. What is the relationship between the probability of making Type I and Type II errors?
   d. What happens to the critical values and the region of rejection when you control Type I error by lowering alpha or control Type II error by raising alpha?
   e. What happens to the critical values and the region of rejection when you control Type II error by using a directional alternative hypothesis?
   f. What methods for increasing the statistical power of an analysis require the least amount of effort from a researcher? What methods require the greatest amount of effort?
   g. What is the difference between between-group variability and within-group variability?
2. Imagine two studies are independently conducted regarding the same topic. Somehow, both studies find the same means and standard deviations for two groups: $\overline{X}_1 = 7.00$, $s_1 = 2.00$, and $\overline{X}_2 = 5.00$, $s_2 = 1.00$. However, the two studies differ in the sample sizes of the two groups: $N_i = 4$ in the first study and $N_i = 9$ in the second study.
   a. Calculate the standard error of the difference ($s_{\overline{X}_1 - \overline{X}_2}$) and the t-test for independent means for each study.
   b. How does increasing the sample size from $N_i = 4$ to $N_i = 9$ affect the calculated value of the t-statistic, as well as the probability of making a Type II error?

3. Two researchers set out to test the same research hypothesis. Both studies are based on the same sample size ($N_i = 23$), and somehow both studies end up with the same standard deviations for the two groups ($s_1 = 5.00$ and $s_2 = 6.00$). However, the experimental manipulation is more effective for the second researcher than for the first researcher, which results in a bigger difference between the means of the two groups: $\bar{X}_1 = 10.00$ and $\bar{X}_2 = 12.00$ for the first researcher and $\bar{X}_1 = 9.00$ and $\bar{X}_2 = 13.00$ for the second researcher.
   a. Calculate the standard error of the difference ($s_{\bar{X}_1 - \bar{X}_2}$) and the *t*-test for independent means for each study.
   b. How does the increased between-group variability in the second study affect the calculated values of the *t*-statistic, as well as the probability of making a Type II error?

10.5 MEASURES OF EFFECT SIZE

The previous section discussed a number of factors that directly influence the decision to reject the null hypothesis. The decision about the null hypothesis affects how researchers interpret the results of their statistical analyses and how they draw conclusions regarding their research hypotheses. This section illustrates a situation that may develop when researchers do not take into account the impact of one of these factors, sample size, on the decision to reject the null hypothesis. Next, strategies used by researchers to address this situation are presented and illustrated.

Interpreting Inferential Statistics: The Influence of Sample Size

Because sample size directly influences the decision regarding the null hypothesis, it also affects conclusions drawn from statistical analyses. In this section, the Chapter 9 parking lot study will be used to illustrate the role of sample size in interpreting and communicating the results of statistical analyses. When we discussed this study in Chapter 9, the decision was made to reject the null hypothesis and conclude that the departure time of the Intruder group ($M = 40.73$ s) was significantly greater than that of the No Intruder group ($M = 31.67$ s), $t(28) = 2.42$, $p < .05$.

Table 10.7 presents a modified version of the data from the parking lot study. In this new version, the original 15 scores in each of the two groups have been duplicated. For example, the last 15 scores for the Intruder group are the same as the first 15 scores. This example is similar to what we did earlier in showing how Type II error can be reduced by increasing sample size; however, this time we'll discuss how sample size influences both the calculations and interpretation of inferential statistics.

Table 10.7(b) provides the mean and standard deviation of the departure times for the two groups. It is important to notice that the two means ($M = 40.73$ and $M = 31.67$) are exactly the same as in the original example; even though the size of the sample has increased, the size of difference between the groups has not. On the other hand, the standard deviations of the two groups (10.24 and 9.90) for $N_i = 30$ are smaller than when $N_i = 15$ (10.42 and 10.08), reinforcing our earlier statement that within-group variance may be reduced by increasing sample size.

Table 10.8 tests the difference between the two means for the modified data. Comparing this table with Box 10.1, we see that moving from $N_i = 15$ to $N_i = 30$ results in a number of differences in the analysis. For example, the degrees of freedom have increased from 28 to 58, the critical value has decreased from 2.048 to 2.009, and the standard error of the difference ($s_{\bar{X}_1 - \bar{X}_2}$) has decreased from 3.74 to 2.60. Most notably, the larger sample size results in a larger value of the *t*-statistic ($t = 3.49$ vs. $t = 2.42$), and the level of significance has changed from $p < .05$ to $p < .01$. So, even though the difference between the two groups did not change, increasing the sample size changed the level of significance from $p < .05$ to $p < .01$. This highlights a critical issue regarding inferential statistics such as the *t*-test:

> The statistical significance of inferential statistics is a direct function of sample size: The larger the sample size, the more likely an inferential statistic will be statistically significant.

TABLE 10.7 ● PARKING LOT STUDY DATA, SAMPLE SIZE DOUBLED

(a) Raw Data

Intruder				No Intruder			
Driver	Time	Driver	Time	Driver	Time	Driver	Time
1	**23**	16	**23**	1	**54**	16	**54**
2	**62**	17	**62**	2	**19**	17	**19**
3	**28**	18	**28**	3	**46**	18	**46**
4	**55**	19	**55**	4	**20**	19	**20**
5	**31**	20	**31**	5	**42**	20	**42**
6	**51**	21	**51**	6	**21**	21	**21**
7	**33**	22	**33**	7	**38**	22	**38**
8	**48**	23	**48**	8	**23**	23	**23**
9	**35**	24	**35**	9	**35**	24	**35**
10	**45**	25	**45**	10	**25**	25	**25**
11	**37**	26	**37**	11	**34**	26	**34**
12	**43**	27	**43**	12	**28**	27	**28**
13	**39**	28	**39**	13	**31**	28	**31**
14	**41**	29	**41**	14	**29**	29	**29**
15	**40**	30	**40**	15	**30**	30	**30**

(b) Descriptive Statistics

	Intruder	No Intruder
Sample size (N_i)	30	30
Mean ($\bar{X}_i$)	40.73	31.67
Standard deviation (s_i)	10.24	9.90

The Inappropriateness of "Highly Significant"

The difference in level of significance in the two versions of the parking lot study may lead to differences in how the two analyses are interpreted. Researchers sometimes distinguish between different levels of significance by referring to $p < .05$ findings as "significant" and $p < .01$ findings as "highly significant." Furthermore, the phrase "highly significant" may be used to suggest that analyses at the .01 level of significance are somehow "better" than findings at the .05 level.

In Chapter 7, we mentioned that the phrase "highly significant" is inappropriate because statistical significance is a dichotomy (significant vs. nonsignificant) rather than a continuum. In hypothesis testing, we either reject or do not reject the null hypothesis; describing a result as "highly significant" implies the null hypothesis wasn't just rejected but it was *really* rejected. It is important for you to understand that although a statistic may have different levels of significance, one cannot say that one statistic is "more" significant than another.

Furthermore, the analysis of the modified parking lot study data provides another reason why the phrase "highly significant" is improper: The level of significance is a direct function of sample size. In doubling the sample size in Table 10.7, the level of significance changed from $p < .05$ to $p < .01$ *even though we did not change*

TABLE 10.8 ● ANALYSIS, MODIFIED PARKING LOT STUDY DATA

State the null and alternative hypotheses (H_0 and H_1)

$$H_0: \mu_{Intruder} = \mu_{No\ Intruder} \qquad H_1: \mu_{Intruder} \neq \mu_{No\ Intruder}$$

Make a decision about the null hypothesis

Calculate the degrees of freedom (*df*)

$$df = (N_1 - 1) + (N_2 - 1) = (30 - 1) + (30 - 1) = 58$$

Set alpha (α), identify the critical values, and state a decision rule

If $t < -2.009$ or > 2.009, reject H_0; otherwise, do not reject H_0

Calculate a statistic: *t*-test for independent means

Calculate the standard error of the difference ($s_{\bar{X}_1 - \bar{X}_2}$)

$$s_{\bar{X}_1-\bar{X}_2} = \sqrt{\frac{s_1^2}{N_1} + \frac{s_2^2}{N_2}} = \sqrt{\frac{(10.42)^2}{30} + \frac{(9.90)^2}{30}} = 2.60$$

$$t = \frac{\bar{X}_1 - \bar{X}_2}{s_{\bar{X}_1 \bar{X}_2}} = \frac{40.73 - 31.67}{2.60} = 3.49$$

Calculate the *t*-statistic (*t*)

Make a decision whether to reject the null hypothesis

$$t = 3.49 > 2.009 \therefore \text{ reject } H_0 \ (p < .05)$$

Determine the level of significance

$$t = 3.49 > 2.768 \therefore p < .01$$

the difference between the means of the two groups. This implies that two researchers studying the same effect may find the same difference between groups but a different level of significance simply by having different sample sizes. Just because a statistic has a probability less than .01 doesn't necessarily mean the effect or relationship is greater than for a statistic with a probability less than .05. Although significance levels of .01 or .001 may provide precise descriptions of a statistical analysis, they may not reflect the size or magnitude of the effect that is being studied.

At this point, it's understandable if you feel like you've received two opposing messages. On one hand, we've said that larger samples are desirable because they increase the probability of rejecting the null hypothesis. On the other hand, we've also said that larger samples may be problematic if they lead researchers to misinterpret the results of their analyses. There are two points to remember in resolving this apparent contradiction. First, samples should be as large as possible because the larger the sample, the greater the confidence in *any* decision made regarding the null hypothesis. That is, not only does a large sample increase the probability of rejecting the null hypothesis, it also increases confidence about a decision to *not* reject the null. In other words, if an analysis has a high likelihood of rejecting the null hypothesis and the null hypothesis isn't rejected, we may begin to conclude that the hypothesized effect may not exist.

Second, levels of significance such as .01 and .001 should be included to describe the results of an analysis as precisely as possible; it's up to researchers to avoid drawing inappropriate conclusions regarding the size of the effect being studied. To assist researchers in drawing appropriate conclusions about the relative size of an effect, what's needed is a statistic that estimates the size of the effect without being affected by sample size. The next section introduces these statistics, which are known as measures of effect size.

Effect Size as Variance Accounted for

To create a statistic that measures the size of an effect but is not influenced by sample size, let's return to the concept of variance introduced in Chapter 4. Statistically, variance measures the amount of variability in a distribution of scores for a variable. Conceptually, variance represents differences in the phenomena studied by researchers. People differ in such things as behavior, attitudes, and beliefs. One of the goals of theory and research is to explain these differences; put another way, the goal is to account for variance.

The notion of "accounting for variance" may be introduced with a simple example. Imagine there's a perfect relationship between people's heights and their weights, such that knowing a person's height enables us to predict with perfect accuracy how much the person weighs. This perfect relationship might imply that everyone who is 5′ 7″ tall weighs exactly the same amount, such as 155 pounds, and everyone who is 6′ 2″ tall weighs exactly 210 pounds. Put another way, this perfect relationship allows us to completely account for differences—that is to say, explain the variance—in people's weights by their differences in height.

Given the presumed perfect relationship between the two variables, how much of the variance in people's weights can be accounted for by differences in height? The answer is all, or 100%, of the variance. On the other hand, what if there was absolutely no relationship between the two variables? A complete lack of relationship implies that none, or 0%, of the variance in weight can be accounted for by differences in height. In between these two extremes of 0% and 100% lie relationships of varying strengths.

Measure of Effect Size for the Difference Between Two Sample Means: r^2

A **measure of effect size** is a statistic that measures the magnitude of the relationship between variables. In this section, we introduce several measures of effect size that are appropriate for the first research situation described in Chapter 9: testing the difference between two sample means. First, we discuss a measure of effect size that represents the percentage of variance in a variable that is accounted for by another variable, ranging from .00 to 1.00 (0% to 100%). Next, we'll discuss a second measure of effect size that measures the effect in terms of the difference between the means of the two groups.

When testing the difference between two sample means, one measure of effect size is the ***r* squared** (r^2) statistic, defined as the percentage of variance in one variable that is accounted for by another variable. The formula for r^2 is presented as follows:

$$r^2 = \frac{t^2}{t^2 + df} \tag{10-3}$$

where t is the calculated value of the t-test for two sample means and df is the degrees of freedom for two sample means. Using Box 10.1, let's calculate r^2 for the original Chapter 9 parking lot study:

$$\begin{aligned} r^2 &= \frac{t^2}{t^2 + df} \\ &= \frac{(2.42)^2}{(2.42)^2 + 28} = \frac{5.86}{33.86} \\ &= .17 \end{aligned}$$

Interpreting Measures of Effect Size

How can the r^2 value of .17 for the parking lot study be interpreted? Because this measure of effect size is the percentage of variance in one variable accounted for by another variable, we could conclude the following:

> 17% of the differences in drivers' departure times may be accounted for by whether or not they are intruded upon by another driver.

But what does it mean to account for 17% of the variance in a variable? Is that "a lot"? Is that "good"? Jacob Cohen, a psychologist and statistician at New York University, devoted much of his 30-year career to examining the size of effects found in research studies. In 1997, his important contributions to this area of research were honored by the American Psychological Association, which presented him with a lifetime achievement award.

Categorizing effect sizes as "small," "medium," and "large," Cohen (1988) provided the following guidelines for the r^2 statistic for the difference between two sample means:

A "small" effect produces an r^2 of .01.

A "medium" effect produces an r^2 of .06.

A "large" effect produces an r^2 of .15 or greater.

Using Cohen's guidelines, the r^2 of .17 for the parking lot study would be considered an effect of large magnitude. It is important to distinguish words such as *large,* which describe the magnitude of the effect, from words such as *meaningful* and *important,* which are subjective value judgments.

You may be surprised and perhaps discouraged by these guidelines regarding effect size. For example, even an effect considered large ($r^2 = .17$) fails to account for the wide majority ($1 - .17 = 83\%$) of the variance in a variable. Put another way, $r^2 = .17$ implies that 83% of the differences in drivers' departure times is *not* explained by the presence or absence of an intruding driver. The relatively small amount of accounted for variance is a reflection of the complexity of phenomena studied by social scientists. People vary in such things as intelligence, personality, and driving behavior for a wide variety of reasons; it is not surprising that any one variable can explain only a small amount of these differences.

Presenting Measures of Effect Size

One purpose of measures of effect size is to supplement the results of inferential statistical analyses such as the *t*-test. Therefore, these measures are typically provided at the end of the reporting of an inferential statistic. For example, for the Chapter 9 parking lot study, we could write the following:

> The mean departure time for the 15 drivers in the Intruder group ($M = 40.73$ s) is significantly greater than the mean departure time for the 15 drivers in the No Intruder group ($M = 31.67$ s), $t(28) = 2.42$, $p < .05$, $r^2 = .17$. The difference between the two groups of drivers, using Cohen's (1988) guidelines, comprises a large effect.

Many of the statistical procedures discussed in the remainder of this book will illustrate how to report and communicate different measures of effect size.

Reasons for Calculating Measures of Effect Size

There are a number of reasons for calculating and reporting measures of effect size. Perhaps most important, measures of effect size are not affected by sample size. To illustrate this point, we return to the modified parking lot study data in Table 10.7 and Table 10.8, where we found that increasing the sample size from $N_i = 15$ to $N_i = 30$ changed the level of significance from $p < .05$ to $p < .01$ even though the difference between the means of the two groups was the same. Below we calculate r^2 for the modified ($N_i = 30$) data:

$$\begin{aligned} r^2 &= \frac{t^2}{t^2 + df} \\ &= \frac{(3.49)^2}{(3.49)^2 + 58} = \frac{12.18}{70.18} \\ &= .17 \end{aligned}$$

Comparing this value for r^2 with the one calculated earlier for $N_i = 15$, we see that the two r^2 values are identical ($r^2 = .17$). As opposed to inferential statistics such as the *t*-test, measures of effect size are not affected by sample size.

A second reason for calculating measures of effect size such as r^2 is that, because they are measured on the same .00 to 1.00 scale and are unaffected by sample size, they are comparable across different samples and different studies. Because inferential statistics such as the *t*-test are affected by sample size, the results of analyses conducted in two different studies cannot be compared to each other, even when they are from the same population. For example, the same *t* value of 2.02 may be statistically significant in a study involving 250 participants but nonsignificant in a sample of 25.

A third reason for using measures of effect size is that they aid in the interpretation of statistically significant findings. It is possible for an analysis to be statistically significant yet have a very small measure of effect size. For example, we may read the following: $t(158) = 2.76$, $p < .01$, $r^2 = .05$. Rather than focusing solely on the level of significance ($p < .01$), the small value for r^2 informs us that only a small percentage of the variance in the variable is accounted for. The relationship between an inferential statistic and its corresponding measure of effect size is sometimes referred to as the difference between "statistical significance" and "practical significance." Whereas statistical significance focuses on the decision to reject the null hypothesis, practical significance addresses the degree to which the effect accounts for differences in the phenomenon being examined.

Finally, measures of effect size not only help interpret statistically significant findings but also help us understand analyses in which the null hypothesis is *not* rejected. For example, how might we interpret the following analysis: $t(18) = 2.05$, $p > .05$, $r^2 = .19$? Although the null hypothesis was not rejected ($p > .05$), the relatively large r^2 value of .19 suggests the effect may in fact be large. Just as a statistically significant finding may have a small effect size, a nonsignificant finding may have a large effect size. The large value for r^2 in this example suggests the researcher may wish to reexamine the effect, perhaps collecting data from a larger sample to increase the likelihood of rejecting the null hypothesis.

A Second Measure of Effect Size for the Difference Between Two Sample Means: Cohen's *d*

As a measure of effect size, the r^2 statistic represents the percentage (%) of variance in a variable accounted for by another variable. As we will see in later chapters of this book, the r^2 statistic is particularly useful as it can be used in a variety of research situations. However, other measures of effect size may be used for a specific research situation. In this section, we introduce a measure of effect size that measures an effect in terms of the difference between the means of two groups.

Cohen's *d* statistic is an estimate of the magnitude of the difference between the means of two groups measured in standard deviation units. In other words, it measures the size of the treatment effect in terms of the number of standard deviations the means of the two groups differ from each other. As such, Cohen's *d* is similar to the *z*-score, introduced in Chapter 5, which measures a score in standard deviation units (the number of standard deviations a score is different from the mean).

There exist a number of different formulas for Cohen's *d;* these formulas differ depending on whether the sample sizes or standard deviations of the two groups are equal or unequal. Formula 10-4 presents a formula for Cohen's *d* that may be used when the sample sizes of the two groups are equal (Rosenthal & Rosnow, 1991):

$$d = \frac{2t}{\sqrt{df}} \qquad (10\text{-}4)$$

where *t* is the calculated value of the *t*-test for two sample means and *df* is the degrees of freedom for two sample means.

Below we calculate Cohen's d for the original Chapter 9 parking lot study (Box 10.1):

$$d = \frac{2t}{\sqrt{df}}$$
$$= \frac{2(2.42)}{\sqrt{28}} = \frac{4.84}{5.29}$$
$$= .91$$

Cohen's d value of .91 for the parking lot study may be interpreted as the difference between the means of the two groups in standard deviations. More specifically, it indicates that the mean departure times of the Intruder and No Intruder groups are .91 standard deviations different from each other. (Note that if t is a negative number, the calculated value of Cohen's d will also be negative; however, it is always reported as a positive value because it represents the number of standard deviations the groups differ from each other.)

As with the r^2 statistic, Cohen (1988) provided guidelines for interpreting values of Cohen's d:

A "small" effect produces a Cohen's d of .20.

A "medium" effect produces a Cohen's d of .50.

A "large" effect produces a Cohen's d of .80 or greater.

Using these guidelines, the Cohen's d of .91 for the parking lot study would be considered an effect of large magnitude. Note that this is the same conclusion reached from the r^2 value of .17 calculated from the same data.

LEARNING CHECK 4:

Reviewing What You've Learned So Far

1. Review questions
   a. Why is it inappropriate to use terms such as *highly significant* to describe analyses significant at the $p < .01$ or .001 level?
   b. Why is it useful to include a measure of effect size when reporting the results of a statistical analysis such as the t-test?
   c. Conceptually, what does it mean for one variable to account for the variance of another variable?
   d. What are Cohen's (1988) guidelines for interpreting different effect sizes?
   e. Why are measures of effect size comparable across different research studies?
   f. What is the difference between statistical significance and practical significance?
   g. A researcher conducts an analysis in which the null hypothesis is not rejected, yet a large measure of effect size is calculated. What advice might you give the researcher, and why?
   h. What is the difference between the r^2 statistic and Cohen's d measure of effect size?
2. For each of the situations below, calculate the r^2 and Cohen's d measures of effect size and specify whether each is a "large," "medium," or "small" effect as outlined in the chapter.
   a. $df = 28$; $t = 2.08$
   b. $df = 10$; $t = 1.47$
   c. $df = 80$; $t = 2.14$
3. Below are sets of data for two studies. Note that the data for the second study double the data from the first study.

(Continued)

(Continued)

Original data:	Boy:	6, 7, 4, 5, 9, 6
	Girl:	3, 5, 1, 7, 4, 6
Doubled data:	Boy:	6, 7, 4, 5, 9, 6, 6, 7, 4, 5, 9, 6
	Girl:	3, 5, 1, 7, 4, 6, 3, 5, 1, 7, 4, 6

Conduct the following analyses for the two sets of data:

a. For each group, calculate the sample size (N_i), mean ($\bar{X}_i$), and standard deviation (s_i).
b. State the null and alternative hypothesis (H_0 and H_1).
c. Make a decision about the null hypothesis.
   (1) Calculate the degrees of freedom (*df*).
   (2) For $\alpha = .05$ (two-tailed), identify the critical values and state a decision rule.
   (3) Calculate a value for the *t*-test for independent means.
   (4) Make a decision whether to reject the null hypothesis.
   (5) Determine the level of significance.
   (6) Calculate r^2 and Cohen's *d* and describe size of effect (small, medium, large).
d. Compare the results of the analysis of the original and doubled data in terms of level of significance and measure of effect size.

10.6 LOOKING AHEAD

This chapter began by discussing errors that may occur in making the decision to reject the null hypothesis. Next, methods used by researchers to minimize the occurrence of these errors were described and evaluated. For conceptual and pragmatic reasons, completely eliminating errors in hypothesis testing is neither possible nor desirable. Instead, it is important to understand the consequences of making these errors. Both the research process in general and hypothesis testing in particular are influenced by a researcher's beliefs, motivations, and skills. Much of the concerns about hypothesis testing are not directed at the statistical procedures themselves but rather how these procedures are used and perhaps misused by researchers. Consequently, it is important to learn to draw appropriate conclusions from statistical analyses. This becomes particularly true when researchers conduct studies requiring relatively complex statistical analyses. The next chapter, for example, discusses a statistical procedure used to compare differences between the means of three or more groups that comprise one independent variable.

10.7 Summary

Within hypothesis testing, the decision to reject or not reject the null hypothesis is based on probability rather than certainty. Consequently, the possibility remains that whatever decision is made about the null hypothesis may either be correct or incorrect. There are two types of errors that can be committed in making the decision to reject the null hypothesis: A ***Type I error*** occurs when the null hypothesis is true but the decision is made to reject the null hypothesis; a ***Type II error*** occurs when the decision is made to not reject the null hypothesis even though the alternative hypothesis is true.

In terms of Type I error, the probability of making this error is equal to alpha (α), typically .05; the probability of not making a Type I error is equal to $1 - \alpha$, typically .95. Type I error is a concern because the communication of incorrect information negatively affects others' thinking, time, energy, and resources. Type I errors occur as the result of random, chance factors, and the probability of making this error may be reduced by making the null hypothesis more difficult to reject (e.g., setting alpha to

.01 rather than .05). However, decreasing the likelihood of making a Type I error increases the likelihood of making a Type II error. Ultimately, researchers must allow for the possibility of making this error because the only way it may be eliminated is to never reject the null hypothesis regardless of the evidence.

In terms of Type II error, the probability of committing a Type II error is represented by beta (β), which is difficult to calculate because of the imprecise nature of the alternative hypothesis. Type II errors are a concern because they may result in the non-communication of correct information that is potentially useful to others. Type II errors occur as the result of how researchers conduct their studies, and the probability of not making a Type II error is what is known as ***statistical power***. Reducing Type II error, thereby increasing the statistical power of an analysis, involves increasing the likelihood of rejecting the null hypothesis. Factors such as the sample size, alpha, directionality of the alternative hypothesis, between-group variability, and within-group variability may be altered to increase statistical power.

One limitation of inferential statistics such as the *t*-test is that the statistical significance of these statistics is a function of sample size. To draw conclusions about the relative magnitude of a hypothesized effect, researchers can calculate a ***measure of effect size***, which is a statistic that measures the magnitude of the relationship between variables. Two measures of effect size related to the difference between two sample means are r^2, which is the percentage of variance in one variable that is accounted for by another variable, and ***Cohen's d***, which is an estimate of the magnitude of the difference between the means of two groups measured in standard deviation units. Some reasons for calculating measures of effect size include that they are not affected by sample size, they are comparable across different samples and different studies, and they aid in the interpretation of statistically significant and nonsignificant findings.

10.8 Important Terms

Type I error (p. 323)
Type II error (p. 326)
statistical power (p. 327)
between-group variability (p. 335)
within-group variability (p. 336)
error variance (p. 336)
measure of effect size (p. 342)
r squared (r^2) (p. 342)
Cohen's *d* (p. 344)

10.9 Formulas Introduced in This Chapter

Probability of Type I Error

$$p(\text{Type I error}) = \alpha \qquad (10\text{-}1)$$

Probability of Type II error

$$p(\text{Type II error}) = \beta \qquad (10\text{-}2)$$

r Squared (r^2)

$$r^2 = \frac{t^2}{t^2 + df} \qquad (10\text{-}3)$$

Cohen's *d*

$$d = \frac{2t}{\sqrt{df}} \qquad (10\text{-}4)$$

10.10 Exercises

1. For each of the following hypotheses, create a chart that shows the four possible outcomes that could occur as a result of hypothesis testing. There should be two correct decisions and two incorrect decisions. Label the type of error for each of the incorrect decisions.
   a. A new reading program in the elementary schools is hypothesized to improve reading comprehension.
   b. People who take at least two vacations each year are happier, on average, than people who do not take at least two vacations per year.
   c. A college student cramming for finals hypothesizes that caffeine improves his ability to concentrate.
2. If we reject the null hypothesis and we were wrong to do so, what type of error have we made? Give an example of a research situation in which this could occur.
3. If we fail to reject the null hypothesis but should have rejected it, what type of error have we made? Provide an example of a research situation in which this could occur.
4. For each of the situations below, describe what would be considered a Type I and Type II error.
   a. A consumer group wants to see if people can tell whether they are drinking tap water or bottled water.
   b. A college instructor wishes to see whether his students prefer to work on assignments individually or in groups.
   c. A teacher evaluating a program designed to improve math skills gives her students a test before and after the program.
   d. A college student conducts searches on two different Internet search engines to see if they differ in terms of the number of relevant results provided.
5. For each of the situations below, describe what would be considered a Type I and Type II error.
   a. A political scientist compares older and younger adults' attitudes toward the death penalty.
   b. A movie critic wants to know if people prefer to see movies in 3-D or stereoscopic (not 3-D).
   c. An owner of a grocery store tests shoppers' ability to tell the difference between brand-name cereal and generic cereal.
   d. A teacher wants to know if boys and girls differ in their career goals.
6. Below are two hypothetical situations:

   Situation 1: $N_i = 30$, $\alpha = .05$ (two-tailed test)

   Situation 2: $N_i = 30$, $\alpha = .01$ (two-tailed test)

   a. Find the critical values for each of the two situations.
   b. In which situation is there less of a chance of making a Type I error? Why?
   c. What is the effect of changing alpha from .05 to .01 on the probability of making a Type II error?
7. Below are two hypothetical situations:

   Study A: $N_1 = 20, \overline{X}_1 = 14.50, s_1 = 2.50$

   $N_2 = 20, \overline{X}_2 = 12.75, s_2 = 3.25$

   Study B: $N_1 = 65, \overline{X}_1 = 14.50, s_1 = 2.50$

   $N_2 = 65, \overline{X}_2 = 12.75, s_2 = 3.25$

   a. Calculate the standard error of the difference ($s_{\overline{X}_1 - \overline{X}_2}$) and the *t*-test for independent means for each study.
   b. How does increasing the sample size from $N_i = 20$ to $N_i = 65$ affect the calculated values of the standard error of the difference and the *t*-statistic?
   c. How does increasing the sample size affect the probability of making a Type II error?
8. Imagine two studies examining the same research question find the same means and standard deviations for two groups: $\overline{X}_1 = 10.00, s_1 = 3.00$, and $\overline{X}_2 = 12.00, s_2 = 2.00$. However, the two studies differ in sample sizes: $N_i = 9$ in the first study and $N_i = 16$ in the second study.

a. Calculate the standard error of the difference ($s_{\bar{X}_1-\bar{X}_2}$) and the *t*-test for independent means for each study.

b. How does increasing the sample size from $N_i = 9$ to $N_i = 16$ affect the calculated values of the standard error of the difference and the *t*-statistic, as well as the probability of making a Type II error?

9. Two researchers separately testing the same research hypothesis end up with the same sample size ($N_i = 12$) and standard deviations for the two groups ($s_1 = .50$ and $s_2 = .75$). However, the experimental manipulation is more effective for the second researcher, which results in a bigger difference between the means of the two groups: $\bar{X}_1 = 2.50$ and $\bar{X}_2 = 2.15$ (first researcher), as well as $\bar{X}_1 = 2.70$ and $\bar{X}_2 = 1.95$ (second researcher).

a. Calculate the standard error of the difference ($s_{\bar{X}_1-\bar{X}_2}$) and the *t*-test for independent means for each study.

b. How does the increased between-group variability in the second study affect the calculated value of the *t*-statistic, as well as the probability of making a Type II error?

10. A study compares the effectiveness of a new (Drug A) and an old medicine (Drug B) and finds no difference in effectiveness: $t(68) = 1.34$, $p > .05$. The researchers run the study a second time: Both studies are based on the same sample size ($N_i = 35$), and somehow both studies end up with the same standard deviations for the two groups. However, in the second study, the size of the dosage of Drug A was increased, which increased the difference between the two groups (between-group variability):

Drug A: $\bar{X}_1 = 21.00$

Drug B: $\bar{X}_1 = 10.50$

($s_{\bar{X}_1-\bar{X}_2}$) = 3.60

a. Calculate the *t*-test for independent means.

b. What is the effect of increasing between-group variability on the likelihood of rejecting the null hypothesis?

c. What is the effect of increasing between-group variability on the likelihood of making a Type II error?

11. Two research studies compare two groups on the same dependent variable. Both studies are based on the same sample size ($N_i = 17$), and somehow both studies end up with the same means for the two groups: $\bar{X}_1 = 77.62$ and $\bar{X}_2 = 68.05$. However, in collecting her data, the second researcher provides clearer instructions to the research participants; as a result, the scores in the second study are less influenced by random, chance factors. The standard deviations for the two groups are $s_1 = 11.23$ and $s_2 = 10.86$ for the first study and $s_1 = 9.47$ and $s_2 = 8.59$ for the second study.

a. Calculate the standard error of the difference ($s_{\bar{X}_1-\bar{X}_2}$) and the *t*-test for independent means for each study.

b. How does the decreased within-group variability in the second study affect the calculated value of the *t*-statistic, as well as the probability of making a Type II error?

12. A researcher wishes to test a drug that claims to improve memory. The researcher sets up an experiment where a control group receives a placebo (i.e., a sugar pill) and an experimental group receives the memory pill. The following data are the number of sentences each participant correctly recalls after listening to a two-paragraph story. Determine if there is a significant difference between the two groups.

Placebo: 10, 3, 4, 6, 9, 1, 7

Experimental: 12, 5, 6, 10, 4, 7, 8

a. For each group, calculate the sample size (N_i), mean ($\bar{X}_i$), and standard deviation (s_i).

b. State the null and alternative hypothesis (H_0 and H_1).

c. Make a decision about the null hypothesis.

(1) Calculate the degrees of freedom (*df*).

(2) Set alpha (α), identify the critical values, and state a decision rule.

(3) Calculate a statistic: the *t*-test for independent means.

(4) Make a decision whether to reject the null hypothesis.

(5) Determine the level of significance.

d. Draw a conclusion from the analysis.

13. The researcher in Exercise 12 notices the large amount of variability within each of the two groups. Investigating, he finds the research assistants counting the number of correctly remembered sentences differed in what they counted as a correct sentence. Some experimenters counted them as correct if the participant remembered the general idea of the sentence, whereas others only counted it as correct if every single word was recalled in the correct order. To decrease this source of error, he handed out instructions on what should be counted as "correct." Using the following new data and the steps of hypothesis testing, determine if there is a significant difference between the groups.

Placebo: 7, 4, 5, 6, 5, 7, 6

Experimental: 9, 6, 7, 8, 5, 8, 9

a. For each group, calculate the sample size (N_i), mean ($\overline{X}_i$), and standard deviation (s_i).
b. State the null and alternative hypothesis (H_0 and H_1).
c. Make a decision about the null hypothesis.
   (1) Calculate the degrees of freedom (*df*).
   (2) Set alpha (α), identify the critical values, and state a decision rule.
   (3) Calculate a statistic: the *t*-test for independent means.
   (4) Make a decision whether to reject the null hypothesis.
   (5) Determine the level of significance.
d. Draw a conclusion from the analysis.
e. What type of error (Type I or Type II) was committed the first time the study was conducted?

14. For each of the situations below, calculate both the r^2 and Cohen's *d* measure of effect size and specify whether each is a "small," "medium," or "large" effect as outlined in the chapter.

a. $df = 18, t = 2.23$
b. $df = 30, t = 1.90$
c. $df = 20, t = 2.09$
d. $df = 80, t = 2.09$
e. $df = 74, t = 3.49$

15. For each of the situations below, calculate both the r^2 and Cohen's *d* measure of effect size and specify whether each is a "small," "medium," or "large" effect as outlined in the chapter.

a. $df = 12, t = 2.07$
b. $df = 24, t = 1.90$
c. $df = 24, t = 2.51$
d. $df = 62, t = 2.96$
e. $df = 32, t = 1.31$

16. A researcher is investigating the birth-order theory that middle siblings are greater risk takers than are first-born siblings. The following statistics are obtained for the study as a measure of risk taking:

Middle siblings: $N_1 = 50, \overline{X} = 23.75, s_1 = 3.40$
First born: $N_2 = 50, \overline{X} = 21.80, s_2 = 3.56$

a. Calculate the standard error of the difference ($s_{\overline{X}_1 - \overline{X}_2}$) and the *t*-test for independent means.
b. Is this difference statistically significant? At what level of significance?
c. Calculate r^2 and Cohen's *d* and specify whether it is a "small," "medium," or "large" effect.
d. Report the result of this analysis in APA format.

17. Below are sets of data for two studies: The data for the second study double the data from the first study.

Original data: Group A: 20, 14, 18, 12, 24, 14, 22

Group B: 15, 18, 16, 13, 10, 17, 8

Doubled data: Group A: 20, 14, 18, 12, 24, 14, 22, 20, 14, 18, 12, 24, 14, 22

Group B: 15, 18, 16, 13, 10, 17, 8, 15, 18, 16, 13, 10, 17, 8

Conduct the following analyses for the two sets of data.

a. For each group, calculate the sample size (N_i), mean ($\overline{X}_i$), and standard deviation (s_i).

b. State the null and alternative hypothesis (H_0 and H_1).

c. Make a decision about the null hypothesis.

(1) Calculate the degrees of freedom (*df*).

(2) Set alpha (α), identify the critical values, and state a decision rule.

(3) Calculate a value for the *t*-test for independent means.

(4) Make a decision whether to reject the null hypothesis.

(5) Determine the level of significance.

(6) Calculate r^2 and Cohen's *d* and describe size of effect (small, medium, large).

d. Compare the results of the analysis of the original and doubled data in terms of level of significance and measure of effect size.

Answers to Learning Checks

Learning Check 1

2. a. His customers think they are drinking alcoholic beer when they are actually drinking nonalcoholic beer.

b. The dentist decides the two types of anesthesia differ in their ability to alleviate pain when they actually don't.

c. The consumer protection agency concludes computer users are less likely to get infected if they pay for virus protection software when in fact they are not.

Learning Check 2

2. a. The snowboarder decides the two different kinds of wax last the same amount of time when in fact they don't.

b. The high school counselor decides that whether or not one pays for test preparation services isn't related to the likelihood of getting into college when in fact it is.

c. The driver decides the type of gasoline doesn't affect her car's performance when in fact it does.

3. a.

	Do Not reject H_0	Reject H_0
H_0 is true	**Correct decision:** Conclude the two types of drivers live the same distance from work and they really do	**Type I error:** Conclude the two types of drivers do *not* live the same distance from work but they really do
H_1 is true	**Type II error:** Conclude the two types of drivers live the same distance from work and they really do *not*	**Correct decision:** Conclude the two types of drivers do not live the same distance from work and they really do not

b.

	Do Not Reject H_0	Reject H_0
H_0 is true	**Correct decision:** Conclude the two types of video games are not related to automobile accidents and they really are not	**Type I error:** Conclude the two types of video games are related to automobile accidents and they really are *not*
H_1 is true	**Type II error:** Conclude the two types of video games are *not* related to automobile accidents and they really are	**Correct decision:** Conclude the two types of video games are related to automobile accidents and they really are

Learning Check 3

2. a. First study: $s_{\bar{X}_1-\bar{X}_2} = 1.12; t = 1.79$
   Second study: $s_{\bar{X}_1-\bar{X}_2} = .74; t = 2.70$
   b. Increasing the sample size increases the value of the *t*-statistic and reduces the probability of making a Type II error.
3. a. First study: $s_{\bar{X}_1-\bar{X}_2} = 1.63; t = -1.23$
   Second study: $s_{\bar{X}_1-\bar{X}_2} = 1.63; t = -2.45$
   b. Increasing between-group variability increases the value of the *t*-statistic and reduces the probability of making a Type II error.

Learning Check 4

2. a. $r^2 = .13$; Cohen's $d = .79$; medium effect
   b. $r^2 = .18$; Cohen's $d = .93$; large effect
   c. $r^2 = .05$; Cohen's $d = .48$; small effect
3. Original data
   a. Boy: $N_1 = 6, \bar{X}_1 = 6.17, s_1 = 1.72$
   Girl: $N_2 = 6, \bar{X}_2 = 4.33, s_2 = 2.16$
   b. H_0: $\mu_{Boy} = \mu_{Girl}$; H_1: $\mu_{Boy} \neq \mu_{Girl}$
   c. (1) $df = 10$
   (2) If $t < -2.228$ or > 2.228, reject H_0; otherwise, do not reject H_0
   (3) $s_{\bar{X}_1-\bar{X}_2} = 1.13; t = 1.63$
   (4) $t = 1.63 < 2.228 \therefore$ do not reject H_0 ($p > .05$)
   (5) Not applicable (H_0 not rejected)
   (6) $r^2 = .21$, $d = 1.03$; large effect

   Doubled data
   a. Boy: $N_1 = 12, \bar{X}_1 = 6.17, s_1 = 1.64$
   Girl: $N_2 = 12, \bar{X}_2 = 4.33, s_2 = 2.06$
   b. H_0: $\mu_{Boy} = \mu_{Girl}$; H_1: $\mu_{Boy} \neq \mu_{Girl}$
   c. (1) $df = 22$
   (2) If $t < -2.074$ or > 2.074, reject H_0; otherwise, do not reject H_0
   (3) $s_{\bar{X}_1-\bar{X}_2} = .75; t = 2.45$
   (4) $t = 2.45 > 2.074 \therefore$ reject H_0 ($p < .05$)
   (5) $t = 2.45 < 2.819 \therefore p < .05$ (but not $< .01$)
   (6) $r^2 = .21$, $d = 1.03$; large effect
   d. The null hypothesis was not rejected for the original data ($t(10) = 1.63, p > .05$) but was rejected for the doubled data ($t(22) = 2.45, p < .05$). However, the measures of effect size for the two sets of data were the same ($r^2 = .15, d = 1.03$), showing that effect size is not influenced by sample size.

Answers to Odd-Numbered Exercises

1. a. Decision re: null hypothesis

	Do Not Reject H_0	Reject H_0
H_0 is true	**Correct decision:** Conclude the program does not improve comprehension and it really does not	**Type I error:** Conclude the program improves comprehension but it really does *not*
H_1 is true	**Type II error:** Conclude the program does *not* improve comprehension but it really does	**Correct decision:** Conclude the program improves comprehension and it really does

b. Decision re: null hypothesis

	Do Not Reject H_0	Reject H_0
H_0 is true	**Correct decision:** Conclude vacationing people are not happier than nonvacationing people and they really are not	**Type I error:** Conclude vacationing people are happier than nonvacationing people but they really are *not*
H_1 is true	**Type II error:** Conclude vacationing people are *not* happier than nonvacationing people but they really are	**Correct decision:** Conclude vacationing people are happier than nonvacationing people and they really are

c. Decision re: null hypothesis

	Do Not Reject H_0	Reject H_0
H_0 is true	**Correct decision:** Conclude caffeine does not improve attention and it really does not	**Type I error:** Conclude caffeine improves attention but it really does *not*
H_1 is true	**Type II error:** Conclude caffeine does *not* improve attention but it really does	**Correct decision:** Conclude caffeine improves attention and it really does

3. Type II error

5. a. Type I error: deciding older and younger adults' attitudes differ when they actually don't.

   Type II error: deciding older and younger adults' attitudes don't differ when they actually do.

   b. Type I error: deciding people's movie-watching preferences differ when they actually don't.

   Type II error: deciding people's movie-watching preferences don't differ when they actually do.

   c. Type I error: deciding shoppers can tell the difference between the two types of cereals when they actually can't.

   Type II error: deciding shoppers can't tell the difference between the two types of cereals when they actually can.

   d. Type I error: deciding boys and girls differ in their career goals when they actually don't.

   Type II error: deciding boys and girls don't differ in their career goals when they actually do.

7. a. Study A: $s_{\bar{X}_1-\bar{X}_2} = .92; t = 1.90$

   Study B: $s_{\bar{X}_1-\bar{X}_2} = .51; t = 3.43$

   b. Increasing the sample size reduces the standard error of the difference and increases the value of the *t*-statistic.

   c. Increasing the sample size reduces the probability of making a Type II error.

9. a. First study: $s_{\bar{X}_1-\bar{X}_2} = .26; t = 1.25$

   Second study: $s_{\bar{X}_1-\bar{X}_2} = .26; t = 2.88$

   b. Increasing between-group variability increases the value of the t-statistic and reduces the probability of making a Type II error.

11. a. First study: = 3.79; $t = 2.53$

    Second study: = 3.10; $t = 3.09$

    b. Decreasing within-group variability increases the value of the t-statistic and reduces the probability of making a Type II error.

13. a. Placebo: $N_1 = 7, \bar{X}_1 = 5.71, s_1 = 1.11$

    Experimental: $N_2 = 7, \bar{X}_2 = 7.43, s_2 = 1.51$

    b. H_0: $\mu_{Placebo} = \mu_{Experimental}$; H_1: $\mu_{Placebo} \neq \mu_{Experimental}$

    c. (1) $df = 12$

       (2) If $t < -2.179$ or > 2.179, reject H_0; otherwise, do not reject H_0

       (3) $s_{\bar{X}_1-\bar{X}_2} = .71; t = -2.42$

       (4) $t = -2.42 < -2.179$ ∴ reject H_0 ($p < .05$)

       (5) $t = -2.42 > -3.055$ ∴ $p < .05$ (but not $< .01$)

    d. The number of sentences recalled by the Experimental group ($M = 7.43$) was significantly greater than the Placebo group ($M = 5.71$), $t(12) = -2.42, p < .05$.

    e. The first study is most likely to have made a Type II error, failing to spot an effect that is there. By decreasing within-group variability, the probability of making this error was decreased.

15. a. $r^2 = .26$; Cohen's $d = 1.20$; large effect

    b. $r^2 = .13$; Cohen's $d = .78$; medium effect

    c. $r^2 = .21$; Cohen's $d = 1.02$; large effect

    d. $r^2 = .12$; Cohen's $d = .75$; medium effect

    e. $r^2 = .05$; Cohen's $d = .46$; small effect

17. Original data

    a. Group A: $N_1 = 7, \bar{X}_1 = 17.71, s_1 = 4.54$

       Group B: $N_2 = 7, \bar{X}_2 = 13.86, s_2 = 3.72$

    b. H_0: $\mu_{Group\ A} = \mu_{Group\ B}$; H_1: $\mu_{Group\ A} \neq \mu_{Group\ B}$

    c. (1) $df = 12$

       (2) If $t < -2.179$ or > 2.179, reject H_0; otherwise, do not reject H_0

       (3) $s_{\bar{X}_1-\bar{X}_2} = 2.22; t = 1.73$

       (4) $t = 1.73 < 2.179$ ∴ do not reject H_0 ($p > .05$)

       (5) Not applicable (H_0 not rejected)

       (6) $r^2 = .20, d = 1.00$; large effect

    Doubled data

    a. Group A: $N_1 = 14, \bar{X}_1 = 17.71, s_1 = 4.36$

       Group B: $N_2 = 14, \bar{X}_2 = 13.86, s_2 = 3.57$

    b. H_0: $\mu_{Group\ A} = \mu_{Group\ B}$; H_1: $\mu_{Group\ A} \neq \mu_{Group\ B}$

    c. (1) $df = 26$

       (2) If $t < -2.056$ or > 2.056, reject H_0; otherwise, do not reject H_0

(3) $s_{\bar{X}_1-\bar{X}_2} = 1.51; t = 2.55$

(4) $t = 2.55 > 2.056 \therefore$ reject H_0 ($p < .05$)

(5) $t = 2.55 < 2.779 \therefore p < .05$ (but not $< .01$)

(6) $r^2 = .20$, $d = 1.00$; large effect

d. The null hypothesis was not rejected for the original data ($t(14) = 1.73$, $p > .05$) but was rejected for the doubled data ($t(26) = 2.55$, $p < .05$). However, the measures of effect size for the two sets of data were the same ($r^2 = .20$, $d = 1.00$), showing that effect size is not influenced by sample size.

11

ONE-WAY ANALYSIS OF VARIANCE (ANOVA)

CHAPTER OUTLINE

This chapter returns to the discussion of a statistical procedure designed to test research hypotheses. This statistical procedure, the one-way analysis of variance (ANOVA), is used to compare differences between the means of three or more groups that comprise one independent variable. Conceptually, the one-way ANOVA is a simple extension of the *t*-test (Chapter 9) that tests the difference between the means of two groups. However, having more than two groups makes the calculation and interpretation of the analysis somewhat more complicated. In discussing the one-way ANOVA, this chapter incorporates many of the issues presented thus far in this book, such as descriptive statistics, inferential statistics, errors in hypothesis testing, and measures of effect size.

11.1 AN EXAMPLE FROM THE RESEARCH: IT'S YOUR MOVE

What are effective ways of learning a complicated new topic? A team of researchers at Erasmus University Rotterdam in the Netherlands, led by Anique de Bruin, focused their attention on the process of "principled understanding," through which the learning of basic principles helps people develop strategies for solving problems within a specific domain (de Bruin, Rikers, & Schmidt, 2007). The researchers believed that learning principles allows the person to "apply the principle to diverse situations, and thereby revise and expand knowledge of the skill in which the principle is based" (p. 190).

The researchers examined principled understanding using the example of learning how to play chess. They emphasized that learning the *rules* of chess is distinct from learning the *principles* of chess. "The basic rules cover the factual knowledge that is necessary to learn the principles of the domain . . . [whereas] the principles describe the procedures necessary to reach certain subgoals in the game . . . [in other words], the rules explain how to *play* a chess game, whereas the principles instruct how to *win* a chess game" (de Bruin et al., 2007, p. 190).

The main goal of the study, which we will refer to as the chess study, was to compare three strategies used by beginning players to develop a principled understanding of chess. These strategies differed in the extent to which players had to explain their thinking to others. The first strategy, *observation,* consisted of having people watch chess games being played without explaining what they were looking at. The second strategy, *prediction,* also involved having people watch chess games. However, these people were also asked to make verbal predictions regarding what they believed should be a player's next move. The third strategy, *explanation,* not only involved observing and predicting but also required players to explain the reasoning behind their predictions. The researchers believed that the more people have to verbalize and explain their predictions, the more they have to assess their comprehension of the principles, which in turn leads to higher levels of principled understanding.

In the chess study, a sample of college students who had never played chess were randomly assigned to one of three conditions (Observation, Prediction, and Explanation). Given the length of chess games, rather than have students watch entire games being played, the researchers had students watch scenarios in a computer software program in which only three chess pieces were on the board: a white rook and king and the black king. In these scenarios, the pieces were placed in different positions on the board and students watched a researcher play against the computer. The researcher moved the white pieces with the goal of winning the game by checkmating the computer's black king. The students first watched a series of scenarios while employing one of the three assigned strategies (Observation, Prediction, or Explanation). Next, each student played five scenarios with the goal of winning as many of the five games as possible.

In the published report of the findings, de Bruin et al. (2007) described the study's research hypothesis in the following way: "We hypothesized that participants in the [explanation] condition would . . . checkmate the King more often . . . than the other two conditions . . . [we also] hypothesized that the prediction condition would outperform the observation condition" (p. 192).

In our example, we will mirror the study's findings using a sample of 42 students equally divided into the Observation, Prediction, and Explanation conditions. The number of checkmates in five games for each student is listed in Table 11.1(a). Frequency distribution tables, constructed for each group in Table 11.1(b), suggest that students in the Explanation condition had a greater number of checkmates than did students in the Observation and Prediction conditions. For example, 50% of the students in the Explanation condition had four or five checkmates; on the other hand, the majority of students in the Observation (64%) and Prediction (79%) conditions had only zero or one checkmates.

The next step is to calculate descriptive statistics (mean [$\overline{X}_i$] and standard deviation [s_i]) of the dependent variable for each group (Table 11.2). Figure 11.1 presents a bar graph comparing the group means, with the T-shaped lines in each bar representing the distance of one standard error of the mean above and below the mean (±1). The standard errors have been calculated using Formula 7-5 in Chapter 7.

Looking at the descriptive statistics and bar graph, we see that the mean number of checkmates for the Explanation group (M = 3.00) is the highest of the three groups. Contrary to the research hypothesis, however, the mean of the Prediction group (M = .86) is less than that of the Observation group (M = 1.36). Although these differences indicate initial support for the research hypothesis that the Explanation group would show the highest level of principled understanding, the data does not appear to support the other hypothesis that principled understanding is higher when people make verbal predictions rather than simply watching. However, it is important to keep in mind that any conclusions about the research hypotheses are premature; later in this chapter, we will discuss statistical analyses designed to test differences between groups.

The next step in analyzing the data is to calculate an inferential statistic that tests the differences between the means of the groups. The following section introduces both a family of statistical procedures used to test differences between group means and a statistic specifically used to compare the means of three or more groups that make up one independent variable.

11.2 INTRODUCTION TO ANALYSIS OF VARIANCE (ANOVA)

As we mentioned at the start of this chapter, a research situation containing an independent variable that has three or more groups is an extension of the situations discussed in Chapter 9, which involved comparing the means of two groups. Therefore, we'll introduce the statistical procedure used in this chapter by reexamining the t-test for independent means.

Review: Testing the Difference Between the Means of Two Samples

In Chapter 9, the difference between the means of two samples was evaluated using the t-test for independent means:

$$t = \frac{\overline{X}_1 - \overline{X}_2}{s_{\overline{X}_1 - \overline{X}_2}}$$

TABLE 11.1 ● THE NUMBER OF CHECKMATES FOR STUDENTS IN THE OBSERVATION, PREDICTION, AND EXPLANATION CONDITIONS OF THE CHESS STUDY

(a) Raw Data

Observation		Prediction		Explanation	
Student	# Checkmates	Student	# Checkmates	Student	# Checkmates
1	**0**	1	**3**	1	**1**
2	**5**	2	**0**	2	**5**
3	**0**	3	**1**	3	**2**
4	**0**	4	**0**	4	**1**
5	**2**	5	**1**	5	**5**
6	**1**	6	**2**	6	**4**
7	**5**	7	**1**	7	**1**
8	**1**	8	**0**	8	**5**
9	**2**	9	**2**	9	**5**
10	**0**	10	**0**	10	**1**
11	**0**	11	**1**	11	**5**
12	**1**	12	**0**	12	**4**
13	**0**	13	**0**	13	**1**
14	**2**	14	**1**	14	**2**

(b) Frequency Distribution Tables

Observation			Prediction			Explanation		
# Checkmates	*f*	%	# Checkmates	*f*	%	# Checkmates	*f*	%
5	2	14%	5	0	0%	5	5	36%
4	0	0%	4	0	0%	4	2	14%
3	0	0%	3	1	7%	3	0	0%
2	3	21%	2	2	14%	2	2	14%
1	3	21%	1	5	36%	1	5	36%
0	6	43%	0	6	43%	0	0	0%
Total	14	100%	Total	14	100%	Total	14	100%

As we learned in Chapter 10, the calculated value of a *t*-statistic is a function of two factors corresponding to the numerator and denominator of this formula. The numerator, the difference between the means of the two groups $(\overline{X}_1 - \overline{X}_2)$ was referred to as between-group variability. Between-group variability was defined as differences among the means of the different groups that comprise an independent variable. The denominator, the standard error of the difference $(s_{\overline{X}_1 - \overline{X}_2})$, represents within-group variability, defined as the variability of the dependent variable within the groups that comprise an independent variable. In this chapter, we'll discuss the relationship between between-group and within-group variability.

TABLE 11.2 ● DESCRIPTIVE STATISTICS OF NUMBER OF CHECKMATES FOR THE THREE CONDITIONS IN THE CHESS STUDY

Mean ($\bar{X}_i$)	Standard Deviation (s_i)
$\bar{X}_1 = \frac{\Sigma X}{N}$	$s_i = \sqrt{\frac{\Sigma(X - \bar{X})^2}{N - 1}}$
a. Observation	
$\bar{X}_1 = \frac{0+5+0+\ldots+1+0+2}{14}$ $= \frac{19}{14}$ $= 1.36$	$s_1 = \sqrt{\frac{(0-1.36)^2 + \ldots + (2-1.36)^2}{14-1}}$ $= \sqrt{3.02}$ $= 1.74$
b. Prediction	
$\bar{X}_2 = -\frac{3+0+\ldots+0+0+1}{14}$ $= \frac{12}{14}$ $= .86$	$s_2 = \sqrt{\frac{(3-.86)^2 + \ldots + (1-.86)^2}{14-1}}$ $= \sqrt{.90}$ $= .95$
c. Explanation	
$\bar{X}_3 = \frac{1+5+2+\ldots+4+1+2}{14}$ $= \frac{42}{14}$ $= 3.00$	$s_3 = \sqrt{\frac{(1-3.00)^2 + \ldots + (2-3.00)^2}{14-1}}$ $= \sqrt{3.38}$ $= 1.84$

FIGURE 11.1 ● BAR GRAPH OF NUMBER OF CHECKMATES FOR THE THREE CONDITIONS IN THE CHESS STUDY

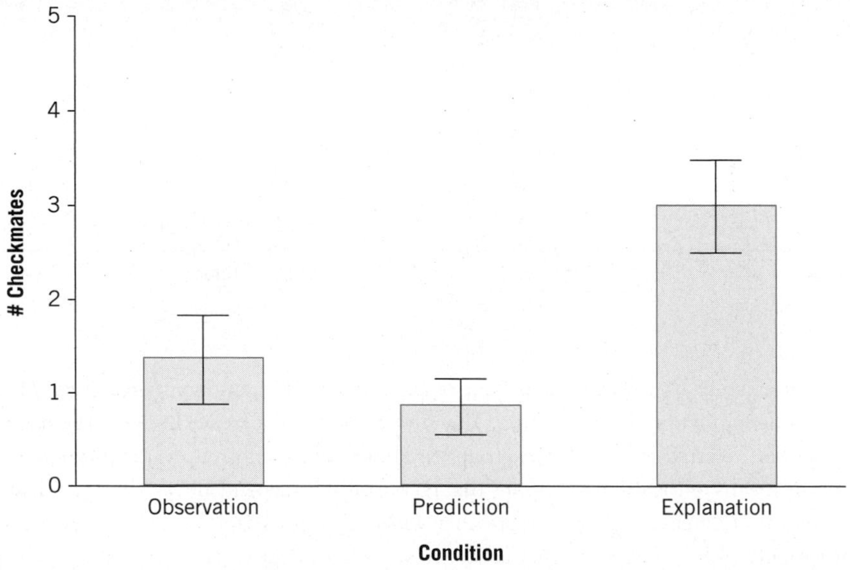

Understanding Between-Group and Within-Group Variability

The numerator of the formula for the *t*-test, the difference between the means of the two groups $(\overline{X}_1 - \overline{X}_2)$, represents between-group variability. Why might groups in a study differ from each other? For example, imagine a researcher develops a program designed to increase students' reading skills. After one group of students are put through the program, they're given a test and their scores are compared to a group that did not go through the program.

Assuming the test scores of the two groups are not the same, how can we account for this difference? The difference between groups could be due to the hypothesized effect being tested—in this case, participating or not participating in the program. However, the difference may also be due to random, chance factors, as well as other factors the researcher did not take into account. In Chapter 10, we referred to variability that cannot be explained or accounted for as "error." Taking all this into consideration, we may state that between-group variability consists of two parts: the hypothesized "effect" and error.

Turning our attention to the denominator of the formula for the *t*-test, within-group variability is represented by the standard error of the difference $(s_{\overline{X}_1-\overline{X}_2})$. Formula 9-1 (see below) provides the formula for the standard error of the difference:

$$s_{\overline{X}_1-\overline{X}_2} = \sqrt{\frac{s_1^2}{N_1} + \frac{s_2^2}{N_2}}$$

One part of this formula is the standard deviation of each group (s_i); the standard deviation measures the variability of scores within each group. Based on what we learned in Chapter 10, because all members of a group are assumed to be the same, there should theoretically be no variability within each group; that is, the standard deviation for each group should be zero (0). Consequently, any differences within a group are considered to be unexplained or unaccounted for. In other words, within-group variability may be defined as error.

To summarize, testing the difference between group means involves the relationship between two types of variability: between-group variability and within-group variability. Between-group variability (differences between groups) consists of the hypothesized effect and error, whereas within-group variability (differences within groups) is error. A statistic such as the *t*-statistic is calculated by dividing between-group variability by within-group variability in the following way:

$$\begin{aligned} t &= \frac{\overline{X}_1 - \overline{X}_2}{s_{\overline{X}_1-\overline{X}_2}} \\ &= \frac{\text{between} - \text{group variability}}{\text{within} - \text{group variability}} \\ &= \frac{\text{difference between groups}}{\text{difference within groups}} \\ &= \frac{\text{effect} + \text{error}}{\text{error}} \end{aligned}$$

Relating Between-Group and Within-Group Variability

The calculated value of the *t*-statistic is determined by the relationship of between-group and within-group variability. In essence, the amount of between-group variability (the numerator of the formula) must be sufficiently greater than the amount of within-group variability (the denominator of the formula) to reject the null hypothesis and conclude there is a statistically significant difference between group means. Stated another way, the hypothesized effect must be large enough to overcome the amount of error that exists in the data.

Figure 11.2 illustrates the relationship between between-group variability and within-group variability. Figure 11.2(a) demonstrates the effect of within-group variability on the overlap between distributions. Looking at the two sets of distributions in Figure 11.2(a), the difference between groups (between-group variability) is the same in the two situations: $((\overline{X}_1 - \overline{X}_2) = (40 - 20) = 20)$. However, the distributions in the right half of Figure 11.2(a) are flatter and wider than the distributions on the left, indicating a greater amount of within-group variability. The greater the amount of within-group variability, the greater the overlap between the two distributions; this overlap lowers the ability to conclude that groups differ from each other.

Figure 11.2(b) illustrates the consequence of within-group variability on the ability to conclude that groups differ. The two sets of distributions both reflect situations where there is a large amount of within-group variability. Comparing the distributions on the left with the distributions on the right, we see that to reduce the amount of overlap between distributions, the difference between groups must increase, in this case from $\overline{X}_1 - \overline{X}_2 = (40 - 20) = 20$ to $\overline{X}_1 - \overline{X}_2 = (50 - 10) = 40$. This shows that the greater the amount of within-group variability, the greater the between-group variability must be to determine that groups differ from each other. Put another way, the more error there is in a set of data, the greater the effect must be in order to be detected.

Introduction to Analysis of Variance (ANOVA) and the *F*-Ratio

Statistical procedures that test differences between group means involve relating between-group variability and within-group variability. In Chapter 4, we measured variability using a statistic known as variance. The *t*-test is a member of a larger family of statistical procedures known as the analysis of variance. The **analysis of variance (ANOVA)** is a family of statistical procedures used to test differences between two or more group means. It is called the "analysis" of variance because it involves analyzing different types of variance, such as between-group and within-group variance. This chapter will discuss one particular type of analysis of variance: the one-way ANOVA. The **one-way ANOVA** is a statistical procedure used to test differences between the means of three or more groups that comprise a single independent variable.

FIGURE 11.2 ● THE RELATIONSHIP BETWEEN BETWEEN-GROUP AND WITHIN-GROUP VARIABILITY

(a) Effect of Increased Within-Group Variability on Overlap Between Distributions

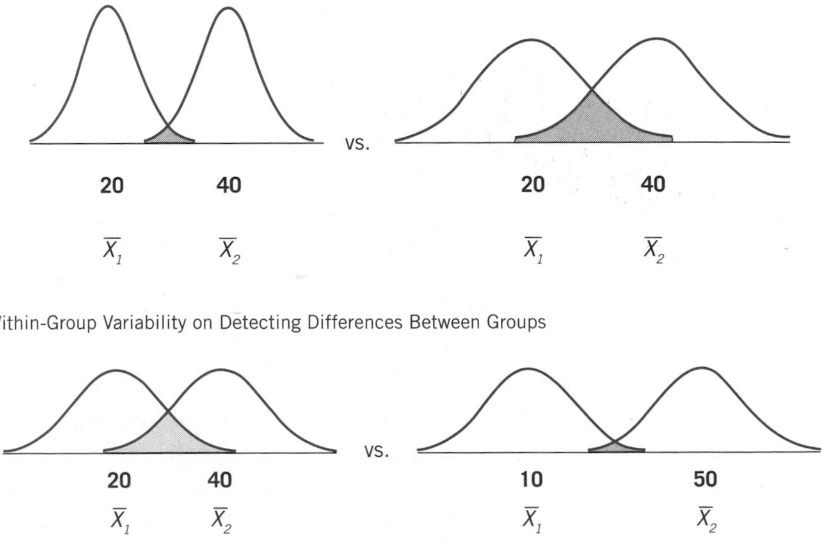

Conceptually, the one-way ANOVA is a simple extension of the *t*-test, in that both statistical procedures involve dividing between-group variance by within-group variance. However, rather than calculating a *t*-statistic, in the one-way ANOVA, we will calculate an ***F*-ratio**, defined as a statistic used to test differences between group means. The word *ratio* indicates that the *F*-ratio is calculated by *dividing* between-group variance by within-group variance:

$$F = \frac{\text{between-group variance}}{\text{within-group variance}}$$

Characteristics of the *F*-Ratio Distribution

Figure 11.3 provides an illustration of a distribution of values for the *F*-ratio. Keep in mind that the *F*-ratio, like the *t*-statistic, cannot be expressed using only one distribution. This is because the shape of the distribution changes as a function of sample size and the number of groups that comprise an independent variable.

Unlike the other theoretical distributions we've examined thus far in this book, the *F*-ratio distribution is not a bell-curve shape with both positive and negative values. Instead, it is positively skewed, starting with the value 0 and extending to the right to infinity (∞).

The *F*-ratio distribution contains only positive values because calculating *F*-ratios involves calculating variances, whose values can only be positive numbers. Here, for example, is the formula for the sample variance (s^2), defined in Chapter 4 as the average squared deviation from the mean:

$$s^2 = \frac{\Sigma(X - \overline{X})^2}{N - 1}$$

The numerator of this formula is based on squared deviations of scores from the mean $((X - \overline{X})^2)$. Because squared deviations can only be positive numbers, the *F*-ratio can only be a positive number.

In terms of the modality of the *F*-ratio distribution, looking at Figure 11.3, we see that the mode of the *F*-ratio distribution is approximately equal to 1. To explain this, recall that a statistic such as the *F*-ratio may be characterized by the formula (effect + error) ÷ error.

If, as stated in the null hypothesis, the groups do not differ, this implies there is no (meaning zero) "effect." Consequently, the expected value of the *F*-ratio is (0 + error) ÷ error, which is error ÷ error—any number divided by itself is equal to 1.

FIGURE 11.3 ● EXAMPLE OF THE *F*-RATIO DISTRIBUTION

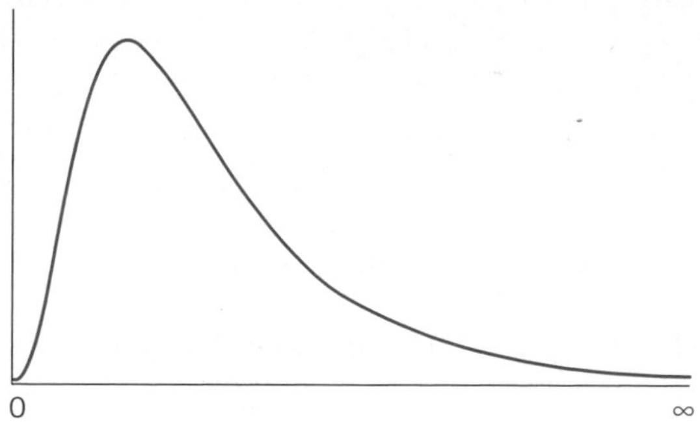

LEARNING CHECK 1:

Reviewing What You've Learned So Far

1. Review questions
   a. Conceptually, what are between-group and within-group variability? What do they consist of or represent?
   b. In testing the difference between group means, how are between-group and within-group variability related to each other? What has to exist for the null hypothesis to be rejected?
   c. In testing the difference between group means, what is the implication of a large amount of within-group variability (error) on the amount of between-group variability (effect) needed to reject the null hypothesis?
   d. What does *ANOVA* stand for, and why?
   e. In testing the difference between means, why is the inferential statistic called the *F*-ratio?
   f. What is the shape of the theoretical distribution of *F*-ratios? Why is it this shape?
   g. What is the modality of the theoretical distribution of *F*-ratios?

11.3 INFERENTIAL STATISTICS: ONE-WAY ANALYSIS OF VARIANCE (ANOVA)

Having introduced the analysis of variance, we turn our attention to testing a research hypothesis regarding differences between group means by calculating and evaluating the *F*-ratio for the one-way ANOVA. This process of hypothesis testing consists of the same steps outlined in earlier chapters:

- state the null and alternative hypotheses (H_0 and H_1),
- make a decision about the null hypothesis,
- draw a conclusion from the analysis, and
- relate the result of the analysis to the research hypothesis.

As we move through these four steps, differences between the one-way ANOVA and the *t*-test will be identified and discussed.

State the Null and Alternative Hypotheses (H_0 and H_1)

In the parking lot study in Chapter 9, the null hypothesis used to test the difference between the means of two samples was stated as H_0: $\mu_{\text{Intruder}} = \mu_{\text{No Intruder}}$. But what is the null hypothesis in an analysis involving more than two groups? In the chess study, the null hypothesis may be stated as H_0: $\mu_{\text{Observation}} = \mu_{\text{Prediction}} = \mu_{\text{Explanation}}$, which implies that the mean number of checkmates for the three groups in the population is equal to each other. Another, more generic, way of stating the null hypothesis for the one-way ANOVA is the following:

$$H_0\text{: all } \mu\text{s are equal}$$

Stating that all of the population means (μs) are equal is simply another way of stating that there are no differences among the means of the groups.

If the null hypothesis is stated as "all μs are equal," does it therefore follow that the alternative hypothesis (H_1) is "all μs are *not* equal"? The answer is no. Stating the alternative hypothesis this way, which in the chess study is the same as stating H_1: $\mu_{\text{Observation}} \neq \mu_{\text{Prediction}} \neq \mu_{\text{Explanation}}$, is actually incorrect because it requires every mean to be different from every other mean for the null hypothesis to be rejected. In fact, all that's needed to reject the null hypothesis is for just *one* of the means to be different from the others, as this would imply that the null hypothesis ("all μs are equal") is not supported by the data. Therefore, the correct form of the alternative hypothesis is as follows:

$$H_1\text{: not all } \mu\text{s are equal}$$

To illustrate the difference between the incorrect "all μs are *not* equal" and the correct "*not* all μs are equal," consider the following null hypothesis: "All people like coconut." An incorrect alternative hypothesis for this null hypothesis would be, "All people do *not* like coconut." This is incorrect because the only way to reject the null hypothesis would be if no one likes coconut. Instead, the correct alternative hypothesis is "*Not* all people like coconut"; this is correct because all that's needed to reject the null hypothesis, "All people like coconut," is to find one person who doesn't like coconut.

Make a Decision About the Null Hypothesis

Once the null and alternative hypotheses have been stated, we move on to making the decision whether to reject the null hypothesis that all of the means are equal. As we have already observed, this decision is made using the following steps:

- calculate the degrees of freedom (*df*);
- set alpha (α), identify the critical value, and state a decision rule;
- calculate a statistic: *F*-ratio for the one-way ANOVA;
- make a decision whether to reject the null hypothesis;
- determine the level of significance; and
- calculate a measure of effect size.

Except for the addition of the last step, these are the same steps used in Chapters 7 and 9. As discussed in Chapter 10, measures of effect size are statistics used to estimate the magnitude of the hypothesized effect in the population. These measures provide useful information in interpreting the results of an inferential statistical procedure and will be included throughout the remainder of this book.

Calculate the Degrees of Freedom (*df*)

The first step in making the decision to reject the null hypothesis is to calculate the degrees of freedom in the set of data. Unlike the *t*-test, the *F*-ratio for a one-way ANOVA contains not one but two degrees of freedom: one for between-group variance (df_{BG}) and one for within-group variance (df_{WG}). Because a research study may have any number of groups as well as any number of scores within a group, these two degrees of freedom are distinct from each other and must be calculated separately.

First, the degrees of freedom for between-group variance (df_{BG}) in the one-way ANOVA is equal to the number of groups that comprise the independent variable, minus 1:

$$df_{BG} = \#\text{ groups} - 1 \qquad (11\text{-}1)$$

To understand the reasoning behind df_{BG}, recall from Chapter 7 that degrees of freedom was defined as the number of values or quantities that are free to vary when a statistic is used to estimate a parameter. Because between-group variance involves differences between group means, we are using these means ($\overline{X}_i$) to estimate the population mean μ. In the case of the chess study, if we were to combine the means of the three groups to estimate μ, how many of the group means are free to vary? The answer to this question is two; this is because once the first two group means have been calculated, the third one is fixed. For the chess study:

$$\begin{aligned} df_{BG} &= \# \text{ groups} - 1 \\ &= 3 - 1 \\ &= 2 \end{aligned}$$

Next, the degrees of freedom for within-group variance (df_{WG}) in the one-way ANOVA is equal to

$$df_{WG} = \Sigma(N_i - 1) \qquad (11\text{-}2)$$

where N_i is the number of scores in each group. The formula for df_{WG} is a simple extension of the formula for the degrees of freedom for the *t*-test for two sample means ($df = (N_1 - 1) + (N_2 - 1)$). The degrees of freedom for each group is the number of scores in the group minus 1 ($N_i - 1$). Calculating df_{WG} involves combining the degrees of freedom of the groups by summing $N - 1$ across the number of groups. In the chess study, because each group contains 14 scores, the degrees of freedom is equal to the following:

$$\begin{aligned} df_{WG} &= \Sigma(N_i - 1) \\ &= (14 - 1) + (14 - 1) + (14 - 1) = 13 + 13 + 13 \\ &= 39 \end{aligned}$$

To summarize, for the chess study, the two degrees of freedom are $df_{BG} = 2$ and $df_{WG} = 39$.

Set Alpha (α), Identify the Critical Value, and State a Decision Rule

Alpha (α) is the probability of a statistic necessary to reject the null hypothesis. Alpha for the one-way ANOVA is traditionally set at .05 ($\alpha = .05$). In Chapters 7 and 9, stating alpha for the *t*-test also required indicating whether alpha was one-tailed or two-tailed. However, this distinction isn't necessary for an ANOVA because, as we've already observed, *F*-ratios can only be positive numbers.

The next step in our calculations is to identify the critical value of the *F*-ratio that separates the region of rejection from the region of nonrejection. As with the *t*-test, the critical value changes depending on the size of the sample. Furthermore, given that an independent variable may consist of any number of groups, there is a different distribution of *F*-ratios and different critical values for every combination of sample sizes and number of groups.

Table 4 in the back of this book provides a table of critical values for the *F*-ratio; a portion of this table is listed in Table 11.3. The table consists of a series of columns under the label "Degrees of Freedom for Numerator" and a series of rows next to the label "Degrees of Freedom for Denominator." These columns correspond to the numerator and the denominator of the *F*-ratio. For a one-way ANOVA, the degrees of freedom for the numerator is df_{BG} and the degrees of freedom for the denominator is df_{WG}. In the chess study example, the degrees of freedom are $df_{BG} = 2$ and $df_{WG} = 39$.

To identify the critical value for the chess study, we turn to the table of critical values in Table 4. Based on our knowledge that $df_{BG} = 2$, we move to the 2 column under "Degrees of Freedom for Numerator." Next, because $df_{WG} = 39$, we move down the rows of the table looking for a row corresponding to 39 for the "Degrees of Freedom for Denominator." Here we find that the table does not provide critical values for $df_{WG} = 39$ but only for $df = 30$ and $df = 40$. For the chess study, we use the smaller of these two values: $df = 30$. It would be inappropriate to use $df = 40$ as this would imply the sample was larger than it actually was.

TABLE 11.3 ● EXAMPLE OF CRITICAL VALUES OF THE *F*-RATIO

Degrees of Freedom for Denominator	Degrees of Freedom for Numerator											
	1	2	3	4	5	6	7	8	9	10	20	∞
1	**161.4**	**199.5**	**215.7**	**224.6**	**230.2**	**234.0**	**236.8**	**238.9**	**240.5**	**241.9**	**248.0**	**254.3**
	4052	4999	5403	5625	5764	5159	5928	5981	6021	6056	6208	6366
2	**18.51**	**19.00**	**19.16**	**19.25**	**19.30**	**19.33**	**19.36**	**19.37**	**19.38**	**19.39**	**19.44**	**19.50**
	98.49	99.00	99.17	99.25	99.30	99.33	99.36	99.37	99.39	99.40	99.45	99.50
3	**10.13**	**9.55**	**9.28**	**9.12**	**9.01**	**8.94**	**8.88**	**8.84**	**8.81**	**8.78**	**8.66**	**8.53**
	34.12	30.82	29.46	28.71	28.24	27.91	27.67	27.49	27.34	27.23	26.69	26.12
4	**7.71**	**6.94**	**6.59**	**6.39**	**6.26**	**6.16**	**6.09**	**6.04**	**6.00**	**5.96**	**5.80**	**5.63**
	21.20	18.00	16.69	15.98	15.52	15.21	14.98	14.80	14.66	14.54	14.02	13.46
5	**6.61**	**5.79**	**5.41**	**5.19**	**5.05**	**4.95**	**4.88**	**4.82**	**4.78**	**4.74**	**4.56**	**4.36**
	16.26	13.27	12.06	11.39	10.97	10.67	10.45	10.29	10.15	10.05	9.55	9.02
6	**5.99**	**5.14**	**4.76**	**4.53**	**4.39**	**4.28**	**4.21**	**4.15**	**4.10**	**4.06**	**3.87**	**3.67**
	13.74	10.92	9.78	9.15	8.75	8.47	8.26	8.10	7.98	7.87	7.39	6.33

Source: Based on Pearson, E. S., & Hartley, H. O. (Eds.) (1966).

For the combination of $df_{BG} = 2$ and $df_{WG} = 30$, we find two critical values: a boldfaced **3.32** and 5.39 immediately below it. The boldfaced number is the critical value for $\alpha = .05$, and the number below it is the critical value for $\alpha = .01$. Therefore, the critical value for the chess study may be stated as follows:

$$\text{For } \alpha = .05 \text{ and } df = 2, 39, \text{ critical value} = 3.32$$

The critical value and regions of rejection and nonrejection for the chess study are illustrated in Figure 11.4.

Once the critical value has been identified, the next step is to state a decision rule that specifies the values of the *F*-ratio resulting in the rejection of the null hypothesis. For the chess study, the following decision rule may be stated:

$$\text{If } F > 3.32, \text{ reject } H_0\text{; otherwise, do not reject } H_0$$

The region of rejection for this example is represented by the shaded area of the distribution in Figure 11.4.

Calculate a Statistic: *F*-Ratio for the One-Way ANOVA

The next step in hypothesis testing is to calculate a value of an inferential statistic. Because calculating the *F*-ratio for the one-way ANOVA involves calculating variances, it's useful to review the formula for the sample variance (s^2), which was defined in Chapter 4 as the average squared deviation of a score from its mean:

$$s^2 = \frac{\Sigma(X - \bar{X})^2}{N - 1} = \frac{SS}{df}$$

The numerator of this formula, the sum of squared deviations ($\Sigma(X - \bar{X})^2$), is also known as the Sum of Squares (*SS*). The denominator of this formula ($N - 1$) is the degrees of freedom (*df*) for a set of scores.

Calculating the *F*-ratio for a one-way ANOVA involves modifying the above formula to calculate two types of variance: between-group variance (MS_{BG}) and within-group variance (MS_{WG}). Variances calculated within an ANOVA have a specific label, Mean Square (*MS*), because the literal definition of variance is the average squared, or *m*ean *s*quared, deviation. The remainder of this section demonstrates how these two variances and the *F*-ratio are calculated.

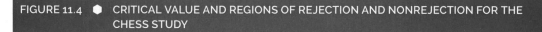

FIGURE 11.4 ● CRITICAL VALUE AND REGIONS OF REJECTION AND NONREJECTION FOR THE CHESS STUDY

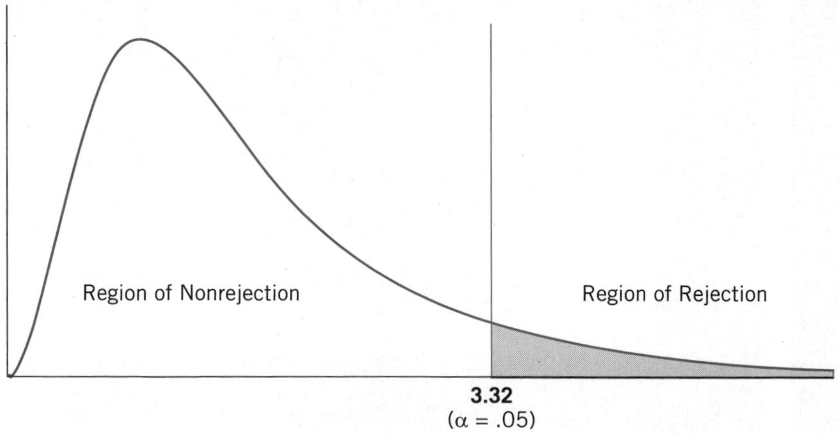

Calculate between-group variance (MS_{BG}). Between-group variance refers to differences among the groups comprising an independent variable; it measures the hypothesized effect being tested. In the *t*-test, between-group variance was simply the difference between the means of two groups $(\bar{X}_1 - \bar{X}_2)$. However, in the one-way ANOVA, between-group variance is based on the difference between the mean of each group and the mean for the total sample $(\bar{X}_i - \bar{X}_T)$.

For a simple illustration of between-group variance in the one-way ANOVA, imagine a situation where, in comparing the means of two groups, you find they do not differ: $\bar{X}_1 = 10$ and $\bar{X}_2 = 10$. If you were to combine the data from the two groups and calculate the mean of the total sample $(\bar{X}_T)$, you'd find $\bar{X}_T$ is also equal to 10. This situation is illustrated in Figure 11.5(a). Looking at the figure, the lack of difference between group means is reflected in the lack of difference between the group means and the total mean $(\bar{X}_1 - \bar{X}_2 = 10 - 10 = 0$ and $\bar{X}_2 - \bar{X}_T = 10 - 10 = 0)$.

Figure 11.5(b) illustrates the situation where the means of two groups do differ from each other $(\bar{X}_1 = 5$ and $\bar{X}_2 = 15)$. With $\bar{X}_T$ again equal to 10, this figure shows that the difference between group means is represented by the differences between group means and the total mean $(\bar{X}_1 - \bar{X}_2 = -5$ and $\bar{X}_2 - \bar{X}_T = 5)$.

Based on differences between group means and the total mean, MS_{BG} for the one-way ANOVA is calculated using Formula 11-3:

$$MS_{BG} = \frac{SS_{BG}}{df_{BG}} = \frac{N_i \Sigma(\bar{X}_i - \bar{X}_T)^2}{\#\text{groups} - 1} \qquad (11\text{-}3)$$

where N_i is the number of scores in each group, $\bar{X}_i$ is the mean for each group, and $\bar{X}_T$ is the mean for the total sample.

Although Formula 11-3 may look confusing, it's actually a simple extension of the formula for the variance (Formula 4-3) presented earlier in Chapter 4:

$$s^2 = \frac{\Sigma(X - \bar{X})^2}{N - 1}$$

The difference between the two formulas was that in Chapter 4, variance was based on the difference between a score and the mean of the scores $(X - \bar{X})$ and the number of scores in each group (N). In this chapter, variance is based on the difference between a group mean and the mean of the group means $(\bar{X}_i - \bar{X}_T)$ and the number of groups (# groups).

FIGURE 11.5 ● BETWEEN-GROUP VARIANCE AS DIFFERENCES BETWEEN GROUP MEANS ($\bar{X}_i$) AND THE TOTAL MEAN ($\bar{X}_T$)

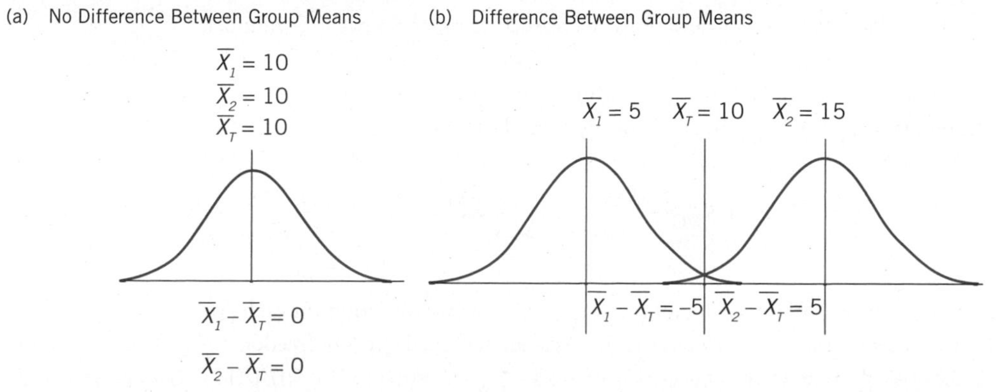

The first step in calculating MS_{BG} is to calculate the mean of the total sample ($\overline{X}_T$). When all of the groups have the same sample size (N_i), the mean of the total sample in a one-way ANOVA is the mean of the group means:

$$\overline{X}_T = \frac{\Sigma\overline{X}_i}{\#\text{groups}} \quad (11\text{-}4)$$

where $\overline{X}_i$ is the mean for each group. For the chess study, using the group means in Table 11.2, $\overline{X}_T$ is calculated below:

$$\begin{aligned}\overline{X}_T &= \frac{\Sigma\overline{X}_i}{\#\text{groups}}\\ &= \frac{1.36+.86+3.00}{3} = \frac{5.22}{3}\\ &= 1.74\end{aligned}$$

One mistake students make in calculating $\overline{X}_T$ is that they forget to divide the summed means (i.e., 1.36 + .86 + 3.00) by the number of groups (i.e., 3). Remember: $\overline{X}_T$ is the mean of the means, *not* the sum of the means.

Once $\overline{X}_T$ has been determined, MS_{BG} may be calculated. As part of calculating MS_{BG}, the degrees of freedom (df_{BG}) were calculated earlier in identifying the critical value: the number of groups minus 1. Because the chess study contained three groups, df_{BG} is equal to 2. Furthermore, because each group consisted of 14 scores, N_i is equal to 14. Using the group means from Table 11.2, we find that MS_{BG} is equal to

$$\begin{aligned}MS_{BG} &= \frac{SS_{BG}}{df_{BG}} = \frac{N_i\Sigma(\overline{X}_i - \overline{X}_T)^2}{\#\text{groups}-1}\\ &= \frac{14[(1.36-1.74)^2+(.86-1.74)^2+(3.00-1.74)^2]}{3-1}\\ &= \frac{14[(-.38)^2+(-.88)^2+(1.26)^2]}{2}\\ &= \frac{14(.14+.77+1.59)}{2} = \frac{14(2.50)}{2} = \frac{35.00}{2}\\ &= 17.50\end{aligned}$$

Students sometimes make computational errors in calculating MS_{BG}; one common mistake is to forget to square the difference between a group mean and the total mean, such as (1.36 – 1.74) or (.86 – 1.74). Keep in mind that variance is based on *squared* deviations and as such cannot be a negative number.

Calculate within-group variance (MS_{WG}). Within-group variance (MS_{WG}), the variance of scores within groups for the one-way ANOVA, is calculated using Formula 11-5:

$$MS_{WG} = \frac{SS_{WG}}{df_{WG}} = \frac{(N_i-1)\Sigma s_i^2}{\Sigma(N_i-1)} \quad (11\text{-}5)$$

where N_i is the number of scores in each group and s_i is the standard deviation for each group.

Starting with the denominator of Formula 11-5, we calculated the degrees of freedom (df_{WG}) earlier in order to determine the critical value. In the chess study example, df_{WG} is equal to (14 – 1) + (14 – 1) + (14 – 1) or 39.

Next, using the standard deviations calculated in Table 11.2, MS_{WG} for the chess study is calculated in the following way:

$$\begin{aligned} MS_{WG} &= \frac{SS_{WG}}{df_{WG}} = \frac{(N_i - 1)\Sigma s_i^2}{\Sigma(N_i - 1)} \\ &= \frac{(14-1)[(1.74)^2 + (.95)^2 + (1.84)^2]}{(14-1)+(14-1)+(14-1)} \\ &= \frac{13[3.03 + .90 + 3.39]}{13+13+13} = \frac{13(7.32)}{39} = \frac{95.16}{39} \\ &= 2.44 \end{aligned}$$

In calculating MS_{WG}, it's important to square the standard deviations before adding them together (i.e., $(1.74)^2 + (.95)^2 + (1.84)^2$). Remember, variance is based on *squared* standard deviations rather than simply standard deviations. Also, to avoid calculating the two degrees of freedom (df_{BG} and df_{WG}) incorrectly, remember that the degrees of freedom for between-group variance (df_{BG}) is based on the number of groups, whereas the degrees of freedom for within-group variance (df_{WG}) is based on the number of scores in each group.

Calculate the F-ratio (F). In a one-way ANOVA, the *F*-ratio (F) is calculated using Formula 11-6:

$$F = \frac{MS_{BG}}{MS_{WG}} \tag{11-6}$$

where MS_{BG} is between-group variance and MS_{WG} is within-group variance. In the chess study example:

$$\begin{aligned} F &= \frac{MS_{BG}}{MS_{WG}} \\ &= \frac{17.50}{2.44} \\ &= 7.17 \end{aligned}$$

Create an ANOVA summary table. To present the calculation of the *F*-ratio, researchers may construct what is known as an ANOVA summary table. An **ANOVA summary table** is a table that summarizes the calculations of an analysis of variance (ANOVA).

We have constructed three ANOVA summary tables in Table 11.4. Table 11.4(a) includes the symbols and notation used to designate the *SS, df, MS,* and *F*-ratio. The middle summary table (Table 11.4(b)) contains formulas and calculations for the different parts of the one-way ANOVA. Table 11.4(c) could be used to report the results for the chess study. The first column in each table uses the label "Source" to describe the hypothesized explanation or cause of each type of variance. In Table 11.4(c), the word *Condition* refers to the independent variable in this particular study, and the word *Error* is used to describe variance that is not accounted for.

You may have noticed that the bottom row of each ANOVA summary table is labeled "Total". This row corresponds to the total sample and consists of the total Sum of Squares (SS_T) and degrees of freedom (df_T). In a one-way ANOVA, the total Sum of Squares is the sum of the between-group and within-group Sums of Squares ($SS_{BG} + SS_{WG}$); in this example, $SS_T = 35.00 + 95.16 = 130.16$. The total degrees of freedom is calculated in a similar manner, where $df_T = df_{BG} + df_{WG}$; in the chess study, $df_T = 2 + 39 = 41$. Although not used in the calculation of the *F*-ratio, SS_T will be used later in this chapter to calculate a measure of effect size associated with the *F*-ratio.

TABLE 11.4 ● SUMMARY TABLES FOR THE ONE-WAY ANOVA

a. Notation and Symbols

Source	*SS*	*df*	*MS*	*F*
Between-group (BG)	SS_{BG}	df_{BG}	MS_{BG}	*F*
Within-group (WG)	SS_{WG}	df_{WG}	MS_{WG}	
Total (T)	SS_T	df_T		

b. Formulas

Source	*SS*	*df*	*MS*	*F*
Between-group (BG)	$N_i\Sigma(\bar{X}_i - \bar{X}_T)^2$	# groups – 1	$\frac{SS_{BG}}{df_{BG}}$	$\frac{MS_{BG}}{MS_{WG}}$
Within-group (WG)	$(N_i - 1)\Sigma s_i^2$	$\Sigma(N_i - 1)$	$\frac{SS_{WG}}{df_{WG}}$	
Total (T)	$SS_{BG} + SS_{WG}$	$df_{BG} + df_{WG}$		

c. Chess Study Example

Source	*SS*	*df*	*MS*	*F*
Condition	35.00	2	17.50	7.17
Error	95.16	39	2.44	
Total	130.16	41		

LEARNING CHECK 2:

Reviewing What You've Learned So Far

1. Review questions
   a. If you were conducting a one-way ANOVA involving four groups (Group 1 to Group 4), what are the two ways you could state the null hypothesis (H_0)?
   b. Which of these is the correct way to state the alternative hypothesis: H_1: all μs are not equal *or* H_1: not all μs are equal? Why is one way correct and the other way incorrect?
   c. Why does hypothesis testing in the one-way ANOVA involve two degrees of freedom?
   d. Why is df_{BG} = # groups = 1 and $df_{BG} = \Sigma(N_i - 1)$?
   e. How is stating the critical value and decision rule for the *F*-ratio different from the *t*-test?
   f. In measuring between-group and within-group variance, what does the symbol *MS* represent?
2. For each of the following, calculate the degrees of freedom (*df*) and identify the critical value of *F* (assume $\alpha = .05$).
   a. # groups = 4, $N_i = 11$
   b. # groups = 5, $N_i = 6$
   c. # groups = 6, $N_i = 8$

3. For each of the following, calculate between-group variance (MS_{BG}).
   a. $N_i = 9, \overline{X}_1 = 4.00, \overline{X}_2 = 9.00, \overline{X}_3 = 2.00$
   b. $N_i = 10, \overline{X}_1 = 27.00, \overline{X}_2 = 39.00, \overline{X}_3 = 33.00, \overline{X}_4 = 37.00$
   c. $N_i = 7, \overline{X}_1 = 3.75, \overline{X}_2 = 4.25, \overline{X}_3 = 3.85, \overline{X}_4 = 2.90, \overline{X}_5 = 5.25$
4. For each of the following, calculate within-group variance (MS_{WG}).
   a. $N_i = 18, s_1 = 2.00, s_2 = 4.00, s_3 = 5.00$
   b. $N_i = 5, s_1 = 12.00, s_2 = 9.00, s_3 = 11.00, s_4 = 8.00$
   c. $N_i = 11, s_1 = .27, s_2 = .51, s_3 = .46, s_4 = .34, s_5 = .40$
5. For each of the following, calculate the F-ratio (F) and create an ANOVA summary table.
   a. $N_i = 12, \overline{X}_1 = 3.50, s_1 = 1.25, \overline{X}_2 = 4.50, s_2 = 1.15, \overline{X}_3 = 4.00, s_3 = .80$
   b. $N_i = 8, \overline{X}_1 = 2.74, s_1 = .49, \overline{X}_2 = 2.26, s_2 = .38, \overline{X}_3 = 2.90, s_3 = .58, \overline{X}_4 = 2.03, s_4 = .45$

Make a Decision Whether to Reject the Null Hypothesis

The decision regarding the null hypothesis is made by comparing the calculated value of the F-ratio with the critical value; if the calculated value is greater than the critical value, the null hypothesis is rejected. For the chess study, this decision is stated as follows:

$$F = 7.17 > 3.32 \therefore \text{ reject } H_0 \ (p < .05)$$

Because the F-ratio value of 7.17 is greater than the critical value of 3.32, it falls in the region of rejection, and the decision is made to reject the null hypothesis. This decision implies that the mean number of checkmates in the three conditions is not all equal to each other, with at least one condition being different from the others.

Determine the Level of Significance

If and when the decision is made to reject the null hypothesis, it's useful to provide a precise assessment of the probability of the calculated F-ratio by comparing it with the .01 critical value. For $df_{BG} = 2$ and $df_{WG} = 39$, the .01 critical value in Table 4 is equal to 5.39. For the chess study, the level of significance may be stated as follows:

$$F = 7.17 > 5.39 \therefore p < .01$$

The level of significance for the chess study is illustrated in Figure 11.6. As we can see, the calculated F-ratio of 7.17 exceeds both the $\alpha = .05$ and .01 critical values. This implies that its probability is not only less than .05 but is also less than .01.

Calculate a Measure of Effect Size (R^2)

As introduced in Chapter 10, a measure of effect size is an index or estimate of the size or magnitude of a hypothesized effect. More specifically, the effect size may be measured as the percentage of variance in the dependent variable that is accounted for by the independent variable, ranging from .00 to 1.00 (0% to 100%). In contrast to inferential statistics such as the t-test or F-ratio, measures of effect size are useful in interpreting statistical analyses because they are not affected by sample size.

There are several measures of effect size appropriate for a one-way ANOVA, each of which has advantages and disadvantages. For the purposes of our discussion, we will use the measure of effect size known as $\boldsymbol{R^2}$. The R^2 statistic is the percentage of variance in the dependent variable associated with differences between the groups that comprise the independent variable. The R^2 statistic is comparable to the r^2 measure of effect size for Chapter 9's t-test.

The measure of effect size R^2 for the one-way ANOVA is calculated using Formula 11-7:

$$R^2 = \frac{SS_{BG}}{SS_T} \tag{11-7}$$

where SS_{BG} is the sum of squares for the between-group variance and SS_T is the sum of squares for the total amount of variance. SS_{BG} corresponds to variance in the dependent variable attributed to differences between groups and SS_T represents the total amount of variance in the dependent variable. By dividing SS_{BG} by SS_T, we're calculating a percentage of the total amount of variance in the dependent variable accounted for by differences between groups.

Using the values for SS_{BG} and SS_T provided in the ANOVA summary table in Table 11.4(c), we can calculate R^2 for the independent variable Condition in the chess study as follows:

$$\begin{aligned} R^2 &= \frac{SS_{BG}}{SS_T} \\ &= \frac{35.00}{130.16} \\ &= .27 \end{aligned}$$

Because a measure of effect size such as R^2 is meant to supplement the results of inferential statistical analyses, the results of the analysis in the chess study may be reported as

$$F(2, 39) = 7.17, p < .01, R^2 = .27$$

The R^2 in this example may be described in the following way: "27% of the variance in the number of checkmates is explained by the different strategies used by students." To interpret the value of R^2, we can use Cohen's (1988) rules of thumb that we previously encountered in Chapter 10:

A "small" effect produces an R^2 of .01.

A "medium" effect produces an R^2 of .06.

A "large" effect produces an R^2 of .15 or greater.

Interpreted in this manner, the R^2 of .27 for the chess study would be considered a relatively large effect.

FIGURE 11.6 ● DETERMINING THE LEVEL OF SIGNIFICANCE FOR THE CHESS STUDY

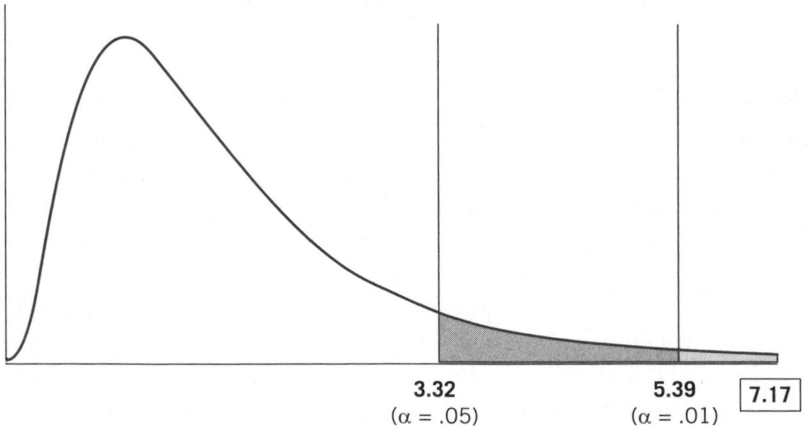

Draw a Conclusion From the Analysis

What conclusion can be drawn from rejecting the null hypothesis in a one-way ANOVA? One way to report the results of the analysis in the chess study is the following:

> The mean number of checkmates for the 14 students in each of the Observation ($M = 1.36$), Prediction ($M = .86$), and Explanation ($M = 3.00$) conditions was not all equal to each other, $F(2, 39) = 7.17$, $p < .01$, $R^2 = .27$.

Notice that this sentence contains a great deal of relevant information about the analysis:

- the dependent variable ("number of checkmates"),
- the sample size for each group ("14 students"),
- the independent variable ("Observation . . . Prediction . . . and Explanation . . . conditions"),
- descriptive statistics ("$M = 1.36$ . . . $M = .86$ . . . $M = 3.00$"),
- the nature and direction of the findings ("were not all equal to each other"), and
- information about the inferential statistic ("$F(2, 39) = 7.17$, $p < .01$, $R^2 = .27$"), which includes the inferential statistic calculated (F), the degrees of freedom (2, 39), the value of the statistic (7.17), the level of significance ($p < .01$), and the measure of effect size ($R^2 = .27$).

Relate the Result of the Analysis to the Research Hypothesis

The last step in hypothesis testing is to relate the result of the analysis to the study's research hypothesis. The research hypothesis in the chess study was that the number of checkmates would be greater for students in the Explanation condition than for students in the Prediction condition, which in turn would be greater than for students in the Observation condition. At this point in the analysis, can we draw a conclusion regarding the extent to which this research hypothesis has been supported? The answer, quite simply, is no.

Rejecting the null hypothesis in the one-way ANOVA does *not* indicate the specific nature and direction of any differences between the means of the groups. Looking back at the null (H_0) and alternative (H_1) hypothesis, we see the alternative hypothesis is stated as "not all μs are equal." Consequently, when the null hypothesis is rejected, the most specific conclusion that can be made is that the means are not all equal to each other—we *cannot* state which of the group means are significantly different from the others. As a result, we are not yet able to draw conclusions regarding whether a study's research hypothesis has been supported.

To test a study's research hypotheses, it is necessary to conduct additional, more specific analyses. For example, one of the hypotheses in the chess study was that the Explanation strategy is more effective than Prediction. To test this hypothesis, we need to conduct an analysis that involves only these two groups, excluding the Observation condition. Section 11.5 of this chapter describes these analyses, known as "analytical comparisons." Because we have not yet conducted analytical comparisons, the most appropriate statement we can make at this point regarding a study's research hypothesis is the following:

> At this point in the analysis, we are unable to determine whether the research hypothesis in the study is supported until analytical comparisons are conducted.

Later in this chapter, we introduce and illustrate the process of conducting and interpreting analytical comparisons.

Summary

The process of conducting a one-way ANOVA is summarized in Box 11.1, using the chess study as an example. As mentioned above, additional analyses known as analytical comparisons must be conducted to draw conclusions regarding the extent of support for a study's research hypothesis. However, before moving on to a discussion of analytical comparisons, a second example of a one-way ANOVA will first be presented and analyzed.

SUMMARY BOX 11.1 CONDUCTING THE ONE-WAY ANOVA (CHESS STUDY EXAMPLE)

State the null and alternative hypotheses (H_0 and H_1)

H_0: all μs are equal $\qquad$ H_1: not all μs are equal

Make a decision about the null hypothesis

Calculate the degrees of freedom (*df*)

$$df_{BG} = \#\text{ groups} - 1 = 3 - 1 = 2$$

$$df_{WG} = \Sigma(N_i - 1) = (14 - 1) + (14 - 1) + (14 - 1) = 39$$

Set alpha (α), identify the critical value, and state a decision rule

If $F > 3.32$, reject H_0; otherwise, do not reject H_0

Calculate a statistic: *F*-ratio for one-way ANOVA

Calculate between-group variance (MS_{BG})

$$MS_{BG} = \frac{SS_{BG}}{df_{BG}} = \frac{N_i\Sigma(\bar{X}_i - \bar{X}_T)^2}{\#groups - 1}$$
$$= \frac{14[(1.36 - 1.74)^2 + (.86 - 1.74)^2 + (3.00 - 1.74)^2]}{3 - 1}$$
$$= \frac{14(.14 + .77 + 1.59)}{2} = \frac{14(2.50)}{2} = \frac{35.00}{2} = 17.50$$

Calculate within-group variance (MS_{WG})

$$MS_{WG} = \frac{SS_{WG}}{df_{WG}} = \frac{(N_i - 1)\Sigma s_i^2}{\Sigma(N_i - 1)}$$
$$= \frac{(14 - 1)[(1.74)^2 + (.95)^2 + (1.84)^2]}{(14 - 1) + (14 - 1) + (14 - 1)}$$
$$= \frac{13[3.03 + .90 + 3.39]}{13 + 13 + 13} = \frac{13(7.32)}{39} = \frac{95.16}{39} = 2.44$$

Calculate the *F*-ratio (*F*)

$$F = \frac{MS_{BG}}{MS_{WG}} = \frac{17.50}{2.44} = 7.17$$

Make a decision whether to reject the null hypothesis

$$F = 7.17 > 3.32 \therefore \text{ reject } H_0\ (p < .05)$$

Determine the level of significance

$$F = 7.17 > 5.39 \therefore p < .01$$

Calculate a measure of effect size (R^2)

$$R^2 = \frac{SS_{BG}}{SS_T} = \frac{35.00}{130.16} = .27$$

Draw a conclusion from the analysis

The mean number of checkmates for the 14 students in the Observation (M = 1.36), Prediction (M = .86), and Explanation (M = 3.00) conditions is not all equal to each other, $F(2, 39) = 7.17$, $p < .01$, $R^2 = .27$.

Relate the result of the analysis to the research hypothesis

At this point in the analysis, we are unable to determine whether the research hypothesis in the study is supported until analytical comparisons are conducted.

LEARNING CHECK 3:

Reviewing What You've Learned So Far

1. Review questions
   a. What does R^2 in a one-way ANOVA represent?
   b. What is the most specific conclusion that can be drawn when the null hypothesis for a one-way ANOVA has been rejected?
   c. Does rejecting the null hypothesis for a one-way ANOVA support or not support a research hypothesis?
2. Although research has examined factors women take into consideration when making birth control decisions, relatively few studies have looked at men. As male oral contraceptives were starting to become available, Jaccard, Hand, Ku, Richardson, and Abella (1981) conducted a study to investigate males' concerns regarding these contraceptives. In the study, male college students were asked to rate the importance of one of three aspects in deciding whether they would use an oral contraceptive: effectiveness as a birth control device, convenience (time and effort needed to use the device), and potential risks to their own health—the higher the rating, the greater the importance. Jaccard et al. hypothesized that males would rate risks to their health as more important than either the contraceptive's effectiveness or convenience. Imagine the following descriptive statistics are reported for the ratings of the three aspects: effectiveness (N = 9, M = 3.18, s = 1.21), convenience (N = 9, M = 3.43, s = 1.14), and health risks (N = 9, M = 4.51, s = 1.03).
   a. State the null and alternative hypotheses (H_0 and H_1).
   b. Make a decision about the null hypothesis.
      (1) Calculate the degrees of freedom (df).
      (2) Set alpha (α), identify the critical value, and state a decision rule.
      (3) Calculate the F-ratio for the one-way ANOVA and create a summary table.
      (4) Make a decision whether to reject the null hypothesis.
      (5) Determine the level of significance.
      (6) Calculate a measure of effect size (R^2).
   c. Draw a conclusion from the analysis.
   d. Relate the result of the analysis to the research hypothesis.

11.4 A SECOND EXAMPLE: THE PARKING LOT STUDY REVISITED

This section presents a second example of conducting a one-way ANOVA. There are two main purposes of this example. The first purpose is to more quickly illustrate the process of conducting a one-way ANOVA than in the first example. Because the first example explained both why the specific calculations were being made as well as how to perform the calculations, we think it would be useful to provide a second example that focuses

TABLE 11.5 ● DESCRIPTIVE STATISTICS, PARKING LOT STUDY (CHAPTER 9)

	Driver Group	
	Intruder	No intruder
N_i	15	15
Mean ($\bar{X}_i$)	40.73	31.67
Standard deviation (s_i)	10.42	10.08

primarily on the calculations. The second purpose of this example is to compare the one-way ANOVA with the *t*-test. Because they both involve comparing the means of groups, it's useful for you to understand the relationship between these two statistical procedures.

To accomplish these two purposes, we will use the parking lot study from Chapter 9. This study involved testing the difference in the amount of time taken to leave a parking space by two groups of drivers: those intruded upon and those not intruded upon by another driver. Table 11.5 summarizes the descriptive statistics for the data from the parking lot study.

When this example was analyzed in Chapter 9 using the *t*-test, we reached the following conclusion: "The mean departure time for the 15 drivers in the Intruder group ($M = 40.73$ s) is significantly greater than the mean departure time for the 15 drivers in the No Intruder group ($M = 31.67$ s), $t(28) = 2.42$, $p < .05$."

In this section, we'll show how the parking lot study may be analyzed using a one-way ANOVA, following the same four steps used in analyzing the chess study:

- state the null and alternative hypotheses (H_0 and H_1),
- make a decision about the null hypothesis,
- draw a conclusion from the analysis, and
- relate the result of the analysis to the research hypothesis.

State the Null and Alternative Hypotheses (H_0 and H_1)

The null hypothesis for the one-way ANOVA states that the means of the groups in the population are all equal to each other, whereas the alternative hypothesis states that the means are not all equal. In the parking lot study, these two statistical hypotheses may be stated as follows:

$$H_0\text{: all } \mu\text{s are equal}$$
$$H_1\text{: not all } \mu\text{s are equal}$$

Make a Decision About the Null Hypothesis

The decision whether to reject the null hypothesis for the parking lot study is made using the same steps as in the earlier example of the chess study. Each of these steps is summarized below.

Calculate the Degrees of Freedom (*df*)

The two degrees of freedom for the one-way ANOVA, df_{BG} and df_{WG}, correspond to between-group and within-group variance. Using the information in Table 11.6, the two degrees of freedom are calculated as follows:

$$df_{BG} = \#\ groups - 1 = 2 - 1 = 1$$

$$df_{WG} = \Sigma(N_i - 1) = (15 - 1) + (15 - 1) = 28$$

The degrees of freedom for the parking lot study may be reported as 1, 28.

Set Alpha (α), Identify the Critical Value, and State a Decision Rule

Assuming $\alpha = .05$ and using the degrees of freedom calculated above, the critical value of the *F*-ratio is identified using Table 4. Because in this example $df_{BG} = 1$ and $df_{WG} = 28$, we locate the 1 column under the label "Degrees of Freedom for Numerator" and then move down this column until we reach the row corresponding to 28 degrees of freedom for the denominator df_{WG}. For this combination of alpha and degrees of freedom, we find a critical value of 4.20. Therefore,

$$\text{For } \alpha = .05 \text{ and } df = 1, 28, \text{ critical value} = 4.20$$

For the parking lot study, the decision rule for the *F*-ratio may be stated as follows:

$$\text{If } F > 4.20, \text{ reject } H_0\text{; otherwise, do not reject } H_0$$

If the *F*-ratio calculated from the data exceeds 4.20, it falls in the region of rejection and will lead to the rejection of the null hypothesis.

Calculate a Statistic: *F*-Ratio for the One-Way ANOVA

Three steps are used to calculate the *F*-ratio for the one-way ANOVA: calculate between-group variance (MS_{BG}), calculate within-group variance (MS_{WG}), and calculate the *F*-ratio (*F*). Each of these steps is completed below for the parking lot study.

Calculate between-group variance (MS_{BG}). The first step in calculating the *F*-ratio for the one-way ANOVA is to calculate between-group variance (MS_{BG}). To calculate MS_{BG}, we must first calculate $\bar{X}_T$, which is the mean for the total sample of data. Using the group means provided in Table 11.5, $\bar{X}_T$ is calculated below:

$$\bar{X}_T = \frac{\Sigma \bar{X}_i}{\#\text{groups}} = \frac{40.73 + 31.67}{2} = \frac{72.40}{2} = 36.20$$

Once $\bar{X}_T$ has been calculated, we can calculate MS_{BG}:

$$MS_{BG} = \frac{N_i \Sigma(\bar{X}_i - \bar{X}_T)^2}{\#\text{groups} - 1}$$

$$= \frac{15[(40.73 - 36.20)^2 + (31.67 - 36.20)^2]}{2 - 1}$$

$$= \frac{15[(4.53)^2 + (-4.53)^2]}{1}$$

$$= \frac{15(20.52) + (20.52)}{1} = \frac{15(41.04)}{1} = \frac{615.60}{1}$$

$$= 615.60$$

Calculate within-group variance (MS_{WG}). The second step in calculating the *F*-ratio is to calculate within-group variance (MS_{WG}). Again using the information in Table 11.5, we find that

$$\begin{aligned} MS_{WG} &= \frac{(N_i - 1)\Sigma s_i^2}{\Sigma(N_i - 1)} \\ &= \frac{(15-1)[(10.42)^2 + (10.08)^2]}{(15-1)+(15-1)} \\ &= \frac{14[108.58 + 101.61]}{14+14} = \frac{14(210.19)}{28} = \frac{2942.66}{28} \\ &= 105.10 \end{aligned}$$

Calculate the F*-ratio* (F). Once the two variances have been calculated, the next step is to calculate a value for the *F*-ratio by dividing MS_{BG} by MS_{WG}. For the parking lot study, we find the following:

$$\begin{aligned} F &= \frac{MS_{BG}}{MS_{WG}} \\ &= \frac{615.60}{105.10} \\ &= 5.86 \end{aligned}$$

An ANOVA summary table for this analysis is presented in Table 11.5.

Make a Decision Whether to Reject the Null Hypothesis

We make the decision whether to reject the null hypothesis by comparing the *F*-ratio calculated from the data with the critical value. For the parking lot study, the following decision is made:

$$F = 5.86 > 4.20 \therefore \text{reject } H_0 \ (p < .05)$$

Because the calculated *F*-ratio of 5.86 is greater than the critical value of 4.20, we reject the null hypothesis and conclude that the means are not all equal to each other.

Determine the Level of Significance

Because we've rejected the null hypothesis, it is appropriate to determine whether the probability of the *F*-ratio is less than .01. Comparing the *F* value of 5.86 ($df = 1, 28$) with the .01 critical value of 7.64 found in Table 4, the following conclusion may be made:

$$F = 5.86 < 7.64 \therefore p < .05 \text{ (but not } < .01)$$

The level of significance for this example is illustrated in Figure 11.7. Looking at the figure, we see that the *F*-ratio of 5.86 falls between the $\alpha = .05$ (4.20) and .01 (7.64) critical values. Therefore, the appropriate level of significance for this analysis is $p < .05$.

TABLE 11.6 ● SUMMARY TABLE, ONE-WAY ANOVA, PARKING LOT STUDY

Source	*SS*	*df*	*MS*	*F*
Driver group	615.60	1	615.60	5.86
Error	2942.66	28	105.10	
Total	3558.26	29		

FIGURE 11.7 ● DETERMINING THE LEVEL OF SIGNIFICANCE FOR THE PARKING LOT STUDY

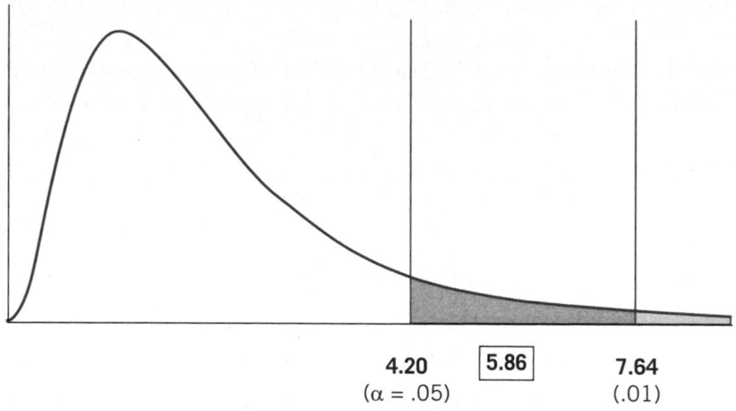

Calculate a Measure of Effect Size (R^2)

To supplement the information provided by the results of the one-way ANOVA, a measure of effect size (R^2) is calculated. Obtaining values of SS_{BG} and SS_T from the ANOVA summary table in Table 11.6, R^2 for the parking lot study is calculated below:

$$R^2 = \frac{SS_{BG}}{SS_T} = \frac{615.60}{3358.26} = .17$$

From the value of R^2, it may be concluded that 17% of the variance in drivers' departure times is accounted for by the presence or absence of another driver. Using Cohen's (1988) rules of thumb presented earlier in this chapter, we see this is a relatively large effect.

Draw a Conclusion From the Analysis

What conclusion can be drawn from the analysis in this example? In the earlier chess study example, the independent variable Condition consisted of more than two groups. Consequently, the most specific conclusion we could draw from rejecting the null hypothesis was that the means of the three groups were not all equal to one another. However, because the parking lot study's independent variable (Driver group) consists of only two groups, rejecting the null hypothesis enables us to state a more specific conclusion regarding the nature and direction of the difference between groups:

> The mean departure time for the 15 drivers in the Intruder group ($M = 40.73$ s) is significantly greater than the mean departure time for the 15 drivers in the No Intruder group ($M = 31.67$ s), $F(1, 28) = 5.86$, $p < .05$, $R^2 = .17$.

The Relationship Between the *t*-Test and the *F*-Ratio

The parking lot study has now been analyzed with two different statistical procedures: the *t*-test and the one-way ANOVA. Although both analyses resulted in the decision to reject the null hypothesis, the statistics do not look the same: $t(28) = 2.42$ and $F(1, 28) = 5.86$. Knowing that the *t*-test and one-way ANOVA both involve testing differences between group means, what is the relationship between the *t*-statistic and the *F*-ratio?

From Formula 11-8, we see that the squared value for the *t*-statistic is equal to the value of the *F*-ratio:

$$t^2 = F \tag{11-8}$$

In other words, for the same set of data, the *t*-statistic and the *F*-ratio are algebraically equivalent. To equate the *t*-statistic and *F*-ratio, the calculated value of the *t*-statistic is squared to convert any negative *t*-statistics to positive numbers.

The values of t^2 and F for the parking lot study are compared below:

$$t^2 = F$$

$$(2.42)^2 = 5.86$$

$$5.86 = 5.86$$

This example illustrates that either the *t*-test or the one-way ANOVA can be used to test the difference between the means of two groups. This choice may be based on such things as personal preference, ease of calculation, and familiarity. The important thing to remember is that the conclusions drawn from both methods are the same and that we don't get different results by using one statistical procedure instead of the other.

11.5 ANALYTICAL COMPARISONS WITHIN THE ONE-WAY ANOVA

This section returns to the chess study analyzed earlier in this chapter. In that example, the most specific conclusion we could draw from rejecting the null hypothesis was that the mean number of checkmates of the three groups of learners (Observation, Prediction, Explanation) was not all equal to one another. Because the result of a one-way ANOVA doesn't indicate which groups differ or not differ from each other, we must conduct additional analyses in order to test the study's research hypotheses. Because these analyses involve comparing groups that are part of a larger study, they are known as analytical comparisons.

Analytical comparisons are comparisons between groups that are part of a larger research design. This section will discuss how to calculate two types of analytical comparisons: planned comparisons and unplanned comparisons. These two types of analytical comparisons differ more in terms of their rationale and evaluation than in how they are conducted.

Planned Versus Unplanned Comparisons

The heart of many studies is the research hypothesis, which specifies the nature and direction of expected relationships between variables. When researchers design a study, they typically have a plan regarding the comparisons between groups needed to test their research hypotheses. These comparisons are known as **planned comparisons**, defined as comparisons that researchers build into the research design prior to the data collection process. Planned comparisons are sometimes referred to as "a priori" comparisons, which literally means "prior to."

The purpose of planned comparisons is to test research hypotheses. For example, one of the research hypotheses in the chess study was that the number of checkmates would be greater for the Explanation condition than the Prediction condition. Therefore, even before the actual research began, the researcher anticipated the importance of comparing the means of these two groups, excluding the Observation condition.

In addition to planned comparisons, there is a second type of analytical comparison known as **unplanned comparisons**, which are comparisons that have not been built into the research design prior to the data collection process. The decision to conduct unplanned comparisons is typically made after the null hypothesis has been rejected for the F-ratio for the one-way ANOVA. These types of analytical comparisons are also known as "post hoc" comparisons, meaning "after the fact."

There are two common reasons for conducting unplanned comparisons. First, the research study may be examining a topic that has not been extensively studied, making it difficult to state research hypotheses before data collection has been conducted. Second, the results of initial statistical analyses may suggest something the researcher did not anticipate when designing the study, leading the researcher to conduct additional analyses to investigate these unexpected results.

The following sections will discuss between-group and within-group variance in analytical comparisons, as well as how to conduct planned and unplanned comparisons for the one-way ANOVA. Although both types of comparisons involve the calculation of an F-ratio, they differ in terms of how the F-ratio is evaluated to make the decision about the null hypothesis.

Between-Group and Within-Group Variance in Analytical Comparisons

Because analytical comparisons are part of a larger research design involving differences between group means, conducting these comparisons requires the calculation of an F-ratio. We will represent the F-ratio in an analytical comparison by the symbol F_{comp}. Similar to the method for calculating the F-ratio for the one-way ANOVA, calculating F_{comp} consists of dividing between-group variance by within-group variance.

Between-group variance is different in an analytical comparison than it is in the one-way ANOVA. This is because we are comparing a subset of the groups, rather than all of the groups, that comprise the independent variable. As a result, the "effect" we are evaluating is not the same. For example, although the chess study consisted of three groups (Observation, Prediction, and Explanation), the "effect" we may be interested in testing is the difference between Explanation and Prediction. Consequently, between-group variance in an analytical comparison is not MS_{BG}, which is based on all of the groups, but instead is represented by $MS_{\text{BG}_{\text{comp}}}$, which is based only on the groups being compared.

Within-group variance represents an estimate of the amount of "error" in a set of data. In calculating F_{comp}, we could calculate within-group variance using only the groups involved in the analytical comparison. For example, in comparing the Explanation and Prediction conditions, we could calculate within-group variance using the data for only these two groups. However, in a set of data, we have an estimate of error based not just on some of the data but on *all* of the data: MS_{WG} from the one-way ANOVA. Because it provides the best estimate of the amount of error in a set of data, MS_{WG} is the measure of within-group variance used in conducting analytical comparisons.

To summarize, the F-ratio for an analytical comparison, F_{comp}, is calculated by dividing $MS_{\text{BG}_{\text{comp}}}$ (between-group variance ["effect"] based on the difference between the two groups being compared) by MS_{WG} (within-group variance ["error"] based on all of the groups):

$$F_{\text{comp}} = \frac{MS_{\text{BG}_{\text{comp}}}}{MS_{\text{WG}}} \tag{11-9}$$

Below we illustrate how to calculate and evaluate F_{comp} for both planned and unplanned comparisons.

Conducting Planned Comparisons

This section illustrates how to conduct planned comparisons for the difference between the means of two groups. (For those who would like to explore the topic in greater depth and detail than this book can accommodate, more complicated planned comparisons are discussed in Keppel, Saufley, and Tokunaga [1992] and Keppel and Wickens [2004].) You will recall that researchers in the chess study hypothesized that "participants in the [explanation] condition would . . . checkmate the King more often . . . than the other two conditions" (de Bruin et al., 2007, p. 192). To test this research hypothesis, we'll conduct a planned comparison between the Explanation and Prediction conditions.

Although planned comparisons may be a new concept, they are conducted following the same steps as the one-way ANOVA, each of which is discussed below:

- state the null and alternative hypotheses (H_0 and H_1),
- make a decision about the null hypothesis,
- draw a conclusion from the analysis, and
- relate the result of the analysis to the research hypothesis.

State the Null and Alternative Hypotheses (H_0 and H_1)

The statistical hypotheses for a planned comparison are very similar to those in earlier examples. In testing the difference between the number of checkmates for the Explanation and Prediction groups, the null and alternative hypotheses may either be stated as

$$H_0: \mu_{\text{Explanation}} = \mu_{\text{Prediction}} \qquad H_1: \mu_{\text{Explanation}} \neq \mu_{\text{Prediction}}$$

or in the more generic form:

$$H_0: \text{all } \mu\text{s are equal} \qquad H_1: \text{not all } \mu\text{s are equal}$$

Make a Decision About the Null Hypothesis

The decision about the null hypothesis is made following the same steps as before. Each of these steps is discussed below.

Calculate the degrees of freedom (df). There are two degrees of freedom for a planned comparison that correspond to the numerator and denominator of F_{comp}:

$$df_{\text{BG}_{\text{comp}}} = \#\text{ groups} - 1 \qquad df_{\text{WG}} = \Sigma(N_i - 1)$$

In terms of the numerator, the degrees of freedom is the number of groups being compared minus 1. Because planned comparisons involve two groups, the between-group degrees of freedom ($df_{\text{BG}_{\text{comp}}}$) is equal to 2 – 1, or 1. However, the degrees of freedom for the denominator (df_{WG}) are based on *all* of the groups in the study and are associated with MS_{WG} from the one-way ANOVA. Looking at the ANOVA summary table for the chess study in Table 11.4(c), we see that $df_{\text{WG}} = 39$. To summarize, the degrees of freedom for the Explanation versus Prediction comparison is

$$\begin{aligned} df_{\text{BG}_{\text{comp}}} &= \#\text{ groups} - 1 \\ &= 2 - 1 \\ &= 1 \end{aligned} \qquad \begin{aligned} df_{\text{WG}} &= \Sigma(N_i - 1) \\ &= (14 - 1) + (14 - 1) + (14 - 1) \\ &= 39 \end{aligned}$$

The two degrees of freedom for this analytical comparison may be represented as $df = 1, 39$.

Set alpha (α), identify the critical value, and state a decision rule. Alpha, the probability of F_{comp} needed to reject the null hypothesis, can be set to the traditional value of .05 ($\alpha = .05$). The critical value for an *F*-ratio, whether for a one-way ANOVA or a planned comparison, may be found in Table 4. Because the degrees of freedom for the Explanation versus Prediction comparison are $df_{BG_{comp}} = 1$ and $df_{WG} = 39$, we identify the critical value by locating the 1 column under "Degrees of Freedom for Numerator" and moving down to the 30 row for the "Degrees of Freedom for Denominator." Here, we find a critical value of 4.17 for $\alpha = .05$. Therefore,

$$\text{For } \alpha = .05 \text{ and } df = 1, 39, \text{ critical value} = 4.17$$

Now that the critical value has been identified, a decision rule may be stated regarding the conditions leading to rejection of the null hypothesis that the two means are equal. For the Explanation versus Prediction planned comparison:

$$\text{If } F_{comp} > 4.17, \text{ reject } H_0; \text{ otherwise, do not reject } H_0$$

Calculate a statistic: F-ratio for analytical comparisons (F_{comp}). The steps used to calculate F_{comp} involve two of the three steps used to calculate the F-ratio for the one-way ANOVA: calculate between-group variance and calculate the F-ratio.

The between-group variance for an analytical comparison ($MS_{BG_{comp}}$) is calculated using the following formula:

$$MS_{BG_{comp}} = \frac{N_i(\bar{X}_i - \bar{X}_j)^2}{2} \qquad (11\text{-}10)$$

where N_i is the number of scores for each group, $\bar{X}_i$ is the mean for the first group involved in the comparison, and $\bar{X}_j$ is the mean for the other group in the comparison. Notice that the difference between the two means $(\bar{X}_i - \bar{X}_j)$ must be squared in order to make $MS_{BG_{comp}}$ a positive number.

Obtaining the means of the Explanation and Prediction groups from Table 11.2, each of which is based on a sample size of $N_i = 14$, $MS_{BG_{comp}}$ is calculated as follows:

$$\begin{aligned} MS_{BG_{comp}} &= \frac{N_i(\bar{X}_i - \bar{X}_j)^2}{2} \\ &= \frac{14(3.00 - .86)^2}{2} \\ &= \frac{14(2.14)^2}{2} = \frac{14(4.58)}{2} = \frac{64.12}{2} \\ &= 32.06 \end{aligned}$$

Once $MS_{BG_{comp}}$ has been calculated, we are ready to calculate F_{comp} by dividing $MS_{BG_{comp}}$ (between-group variance) by MS_{WG} (within-group variance) using Formula 11-9. The value for MS_{WG} may be obtained from the summary table for the one-way ANOVA. Looking at Table 11.4(c), a MS_{WG} value of 2.44 was calculated for the chess study. Consequently, F_{comp} for the analytical comparison between the Explanation and Prediction groups may now be calculated:

$$\begin{aligned} F_{comp} &= \frac{MS_{BG_{comp}}}{MS_{WG}} \\ &= \frac{32.06}{2.44} \\ &= 13.14 \end{aligned}$$

Make a decision whether to reject the null hypothesis. The decision to reject the null hypothesis is made by comparing the calculated value of F_{comp} with the critical value for the analytical comparison. For the Explanation versus Prediction comparison, we may state that

$$F_{\text{comp}} = 13.14 > 4.17 \therefore \text{ reject } H_0 \ (p < .05)$$

By making the decision to reject the null hypothesis, we have decided that the mean number of checkmates for the Explanation and Prediction conditions is significantly different from each other.

Determine the level of significance. Because we've made the decision to reject the null hypothesis, it's appropriate to determine whether the probability of F_{comp} is less than .01. For this example, using the .01 critical value of 7.56 for $df = 1, 39$, we determine that

$$F_{\text{comp}} = 13.14 > 7.56 \therefore p < .01$$

Because the value of 13.14 for F_{comp} exceeds the .01 critical value of 7.56, its probability is not only less than .05 but also less than .01.

Calculate a measure of effect size (R^2_{comp}). Just as we did with the one-way ANOVA, we can calculate a measure of effect size for analytical comparisons. Calculating R^2 for an analytical comparison (R^2_{comp}) is a simple modification of the formula for R^2 presented earlier:

$$R^2_{\text{comp}} = \frac{SS_{BG_{\text{comp}}}}{SS_T} \qquad (11\text{-}11)$$

where $SS_{\text{BG}_{\text{comp}}}$ is the between-group sum of squares for the comparison and SS_{T} is the sum of squares for the total amount of variance.

In calculating R^2_{comp}, the value for $SS_{\text{BG}_{\text{comp}}}$ is the same as the value for $MS_{\text{BG}_{\text{comp}}}$:

$$\begin{aligned} MS_{\text{BG}_{\text{comp}}} &= \frac{SS_{\text{BG}_{\text{comp}}}}{df_{\text{BG}_{\text{comp}}}} \\ &= \frac{SS_{\text{BG}_{\text{comp}}}}{1} \\ &= SS_{\text{BG}_{\text{comp}}} \end{aligned}$$

As indicated by the calculations above, $SS_{\text{BG}_{\text{comp}}}$ and $MS_{\text{BG}_{\text{comp}}}$ are the same. For the Explanation versus Prediction analytical comparison, $SS_{\text{BG}_{\text{comp}}}$ is equal to the $MS_{\text{BG}_{\text{comp}}}$ value of 32.06 calculated earlier.

The next step in calculating R^2_{comp} is to locate the value for SS_{T}, which is found in the summary table for the one-way ANOVA. For the chess study (Table 11.4(c)), $SS_{\text{T}} = 130.16$. Now, inserting our values of $SS_{\text{BG}_{\text{comp}}}$ and SS_{T} into Formula 11-11, R^2_{comp} for the Explanation versus Prediction analytical comparison is calculated as follows:

$$\begin{aligned} R^2_{\text{comp}} &= \frac{SS_{\text{BG}_{\text{comp}}}}{SS_{\text{T}}} \\ &= \frac{32.06}{130.16} \\ &= .25 \end{aligned}$$

Using Cohen's (1988) guidelines, the difference in the number of checkmates between the Explanation and Prediction conditions represents a relatively large effect.

Draw a Conclusion From the Analysis

The following conclusion may be drawn from the Explanation versus Prediction comparison:

> The mean number of checkmates for the 14 students in the Explanation condition ($M = 3.00$) is significantly greater than the number of checkmates for the 14 students in the Prediction condition ($M = .86$), $F(1, 39) = 13.14$, $p < .01$, $R^2 = .25$.

Note that, like the *t*-test, we can indicate the specific nature and direction of differences between the groups because analytical comparisons involve only two groups.

Relate the Result of the Analysis to the Research Hypothesis

In light of the chess study's research hypotheses, what is the implication of finding a significantly greater number of checkmates for students in the Explanation condition than we found in the Prediction condition? The researchers answered this question in the following manner:

> The results of the study suggest that predicting the next move combined with self-explaining the predictions positively contributed to the development of principled understanding of a chess endgame. Participants in the prediction and self-explanation condition showed better understanding of the principles that underlie the KRK [king rook king] endgame than the prediction only condition. (de Bruin et al., 2007, p. 202)

It is worth noting that the researchers did not imply that their research hypothesis had been "proven." Instead, they wrote that their findings "suggest" that having students explain their reasoning leads to higher levels of principled understanding.

Box 11.2 summarizes the steps involved in conducting planned comparisons, using the Explanation versus Prediction comparison as an example. The next section discusses how to conduct the second type of analytical comparison known as unplanned comparisons.

Conducting Unplanned Comparisons

Whereas planned comparisons are designed at the beginning of the study and are guided by research hypotheses, the decision to conduct unplanned (post hoc) comparisons is not made until after the *F*-ratio for the one-way ANOVA has been found to be statistically significant. This section will first discuss concerns related to unplanned comparisons, followed by a presentation of one type of unplanned comparison in which each group is compared with each of the other groups.

SUMMARY BOX 11.2 CONDUCTING PLANNED COMPARISONS (EXPLANATION VS. PREDICTION IN THE CHESS STUDY EXAMPLE)

State the null and alternative hypotheses (H_0 and H_1)

$H_0: \mu_{Explanation} = \mu_{Prediction}$ $H_1: \mu_{Explanation} \neq \mu_{Prediction}$

Make a decision about the null hypothesis

Calculate the degrees of freedom (*df*)

$$df_{BG_{comp}} = \#\ groups - 1 = 2 - 1 = 1$$

$$df_{WG} = \Sigma(N_i - 1) = (14 - 1) + (14 - 1) + (14 - 1) = 39$$

(Continued)

(Continued)

Set alpha (α), identify the critical value, and state a decision rule

$$\text{If } F_{comp} > 4.17, \text{ reject } H_0; \text{ otherwise, do not reject } H_0$$

Calculate a statistic: F-ratio for analytical comparison (F_{comp})

Calculate between-group variance ($MS_{BG_{comp}}$)

$$MS_{BG_{comp}} = \frac{N_i(\bar{X}_i - \bar{X}_j)^2}{2} = \frac{14(3.00 - .86)^2}{2} = \frac{14(2.14)^2}{2}$$

$$= \frac{14(4.58)}{2} = \frac{64.12}{2} = 32.06$$

Calculate the F-ratio (F_{comp})

$$F_{comp} = \frac{MS_{BG_{comp}}}{MS_{WG}} = \frac{32.06}{2.44} = 13.14$$

Make a decision whether to reject the null hypothesis

$$F_{comp} = 13.14 > 4.17 \therefore \text{ reject } H_0 \ (p < .05)$$

Determine the level of significance

$$F_{comp} = 13.14 > 7.56 \therefore p < .01$$

Calculate a measure of effect size (R^2_{comp})

$$R^2_{comp} = \frac{SS_{BG_{comp}}}{SS_T} = \frac{32.06}{130.16} = .25$$

Draw a conclusion from the analysis

The mean number of checkmates for the 14 students in the Explanation condition ($M = 3.00$) is significantly greater than the number of checkmates for the 14 students in the Prediction condition ($M = .86$), $F(1, 39) = 13.14$, $p < .01$, $R^2 = .25$.

Relate the result of the analysis to the research hypothesis

"The results of the study suggest that predicting the next move combined with self-explaining the predictions positively contributed to the development of principled understanding of a chess endgame. Participants in the prediction and self-explanation condition showed better understanding of the principles that underlie the KRK [king rook king] endgame than the prediction only condition" (de Bruin et al., 2007, p. 202).

Concerns Regarding Unplanned Comparisons

Two types of situations may result in conducting unplanned comparisons. First, a research study may be of an exploratory nature, and researchers consequently may not be able to formulate research hypotheses regarding specific differences between groups. Second, initial statistical analyses may have unexpected findings, resulting in the need for further analyses.

Conducting analytical comparisons in these types of situations raises two main concerns. First, because unplanned comparisons are conducted in response to the results of statistical analyses rather than specified hypotheses, the results of these comparisons may be a function of the characteristics and idiosyncrasies of the samples. In other words, the fact that significant differences between groups are found in one study doesn't imply that they will be found in another study. This observation is particularly relevant when the samples are small or perhaps not representative of the larger populations.

A second concern about unplanned comparisons involves the probability of committing a Type I error over a set of analyses. Type I error, as we learned in Chapter 10, consists of rejecting the null hypothesis when it is in fact true, thereby concluding an effect exists when in fact it does not. **Familywise error**, defined as the probability of making at least one Type I error across a set of comparisons, is a concern when multiple unplanned comparisons are conducted to determine the source of a significant *F*-ratio. The probability of making a Type I error in any individual comparison is equal to the stated alpha level, typically .05. However, the more comparisons made within a set of data, the greater the probability of making at least one Type I error.

Let's use a hypothetical example to illustrate the concept of familywise error as well as to demonstrate how unplanned comparisons are conducted and evaluated. Imagine we conduct a study consisting of five groups, each having nine participants ($N_i = 9$). In analyzing the data for this hypothetical study, the following ANOVA summary table is created:

Source	*SS*	*df*	*MS*	*F*
Between-group	365.60	4	91.40	3.52
Within-group	1,040.00	40	26.00	
Total	1,405.60	44		

In making a decision about the null hypothesis, for $\alpha = .05$ and $df = 4, 40$, we find a critical value of 2.61 in Table 4. Because the calculated *F*-ratio of 3.52 is greater than the critical value of 2.61, the null hypothesis is rejected.

Because the only conclusion we can draw from rejecting the null hypothesis is that the means of the five groups are not all equal to each other, analytical comparisons could be conducted to determine significant differences among the groups. In the absence of specific hypotheses regarding which groups should be included in these comparisons, we decide to make all possible comparisons between the groups. In other words, we conduct unplanned comparisons in which each group is compared with each of the other groups. For the five groups in our study, there are 10 possible unplanned comparisons:

$\overline{X}_1$ vs. $\overline{X}_2$ $\quad$ $\overline{X}_1$ vs. $\overline{X}_3$ $\quad$ $\overline{X}_1$ vs. $\overline{X}_4$ $\quad$ $\overline{X}_1$ vs. $\overline{X}_5$

$\overline{X}_2$ vs. $\overline{X}_3$ $\quad$ $\overline{X}_2$ vs. $\overline{X}_4$ $\quad$ $\overline{X}_2$ vs. $\overline{X}_5$

$\overline{X}_3$ vs. $\overline{X}_4$ $\quad$ $\overline{X}_3$ vs. $\overline{X}_5$

$\overline{X}_4$ vs. $\overline{X}_5$

In this situation, making 10 comparisons increases the likelihood of familywise error (i.e., making at least one Type I error among these comparisons). The greater the number of comparisons made in a set of data, the greater the likelihood that groups will differ from each other as a function of random, chance factors.

Methods of Controlling Familywise Error

A number of different methods may be used to control familywise error, a full discussion of which is beyond the scope of this book. (See Keppel & Wickens [2004] for a detailed description of these methods.) For our purposes, it's important for you to understand two aspects of these methods. First, methods for controlling familywise error lower the probability of making a Type I error across a set of analytical comparisons by making it more difficult to reject the null hypothesis in each individual comparison. Second, the choice of a specific method depends on a variety of factors, such as the goals of the research study, the number and types of analytical comparisons to be conducted, and the extent to which the researcher wishes to reduce the probability of making a Type I error.

To illustrate several of the methods used to control familywise error, imagine a researcher conducts a study consisting of four groups: three different types of treatments (Treatments I, II, and III) and a control group that doesn't receive any type of treatment (Control). Furthermore, the researcher is only interested in comparing each of the three treatments with the control group (Treatment I vs. Control, Treatment II vs. Control, Treatment III vs. Control). In a situation like this, which involves comparing each group with a single comparison group, the **Dunnett test** (Dunnett, 1955) would be the appropriate method for controlling familywise error.

Another type of research situation may involve combining groups before conducting analytical comparisons. For example, suppose a study consists of three groups (A, B, and C). The researcher may want to compare each group with each other (A vs. B, A vs. C, B vs. C). These are examples of **simple comparisons**, defined as analytical comparisons between two groups. However, the researcher may also want to compare each group with the average of the other two groups (A vs. the average of B and C, B vs. the average of A and C, C vs. the average of A and B). These are examples of **complex comparisons**, which are analytical comparisons involving more than two groups. In this research situation, the **Scheffé test** (Scheffé, 1953) may be used. The Scheffé test is a statistical procedure used to control familywise error in situations in which a researcher conducts all possible simple and complex comparisons.

Compared to these first two examples, the most typical research situation involving unplanned comparisons consists of comparing each group with each of the other groups—in other words, conducting all possible simple comparisons (but no complex comparisons) in a set of data. In this situation, the Tukey test (Tukey, 1953) is often used. The next section illustrates how the Tukey test can be used to control familywise error.

Controlling Familywise Error: The Tukey Test

The **Tukey test** (Tukey, 1953), sometimes referred to as the Tukey Honestly Significant Difference (HSD) test, is a statistical procedure used to limit familywise error when conducting all possible simple comparisons between groups. The Tukey test limits familywise error by increasing the critical value for each comparison, thereby making it more difficult to reject the null hypothesis.

In terms of the steps followed in conducting unplanned comparisons using the Tukey test, it's important to note these steps are (with one exception) *exactly* the same as for planned comparisons:

- state the null and alternative hypotheses (H_0 and H_1),
- make a decision about the null hypothesis,
- draw a conclusion from the analysis, and
- relate the result of the analysis to the research hypothesis.

The only difference between planned and unplanned comparisons occurs within the step, "Make a decision about the null hypothesis." Rather than beginning this step with "Set alpha (α) and identify the critical value," we instead start with "Set the desired probability of familywise error (α_{FW}) and identify the critical value (F_T)." This is illustrated below using the hypothetical study of five groups with $N_i = 9$ introduced earlier.

Set the desired probability of familywise error (αFW), identify the critical value (F_T), and state a decision rule. The first step in making a decision about the null hypothesis using the Tukey test is to set the probability of familywise error. Typically, this probability is set to .05, meaning there is a combined .05 probability of making a Type I error across the entire set of unplanned comparisons. This probability, symbolized by $\alpha_{FW} = .05$, is very different from planned comparisons, in which each individual comparison has a .05 probability of Type I error ($\alpha = .05$).

Once α_{FW} has been set, the critical value used to evaluate F_{comp} (F_T) is calculated using the following formula:

$$F_T = \frac{(q_T)^2}{2} \qquad (11\text{-}12)$$

where q_T is a statistic known as the Studentized Range statistic (Table 5 in the back of this book). Looking at Table 5, three pieces of information are needed to determine q_T: k (the number of groups that make up the independent variable), df_{error} (df_{WG} in a one-way ANOVA), and α_{FW} (the probability of familywise error).

In the hypothetical example described above ($F(4, 40) = 3.52, p < .05$), k is equal to 5, df_{error} is equal to 40, and α_{FW} has been set at the traditional $\alpha = .05$. Using Table 5, we identify q_T by locating the $k = 5$ column and then moving down until we reach the $df_{error} = 40$ row. For $\alpha_{FW} = .05$, we find q_T is equal to 4.04. Entering $q_T = 4.04$ into Formula 11-12:

$$\begin{aligned} F_T &= \frac{(q_T)^2}{2} \\ &= \frac{(4.04)^2}{2} = \frac{16.32}{2} \\ &= 8.16 \end{aligned}$$

Therefore, the critical value for all 10 unplanned comparisons in this example is 8.16.

Once the critical value F_T has been calculated, we can state a decision rule used to decide whether the result of an unplanned comparison is statistically significant. For the hypothetical example, the following decision rule may now be stated:

If $F_{comp} > 8.16$ reject H_0; otherwise, do not reject H_0

To reject the null hypothesis for any of the 10 unplanned comparisons, the value of F_{comp} in the comparison must be greater than 8.16.

Once the decision rule is stated, we are ready to move onto the next two parts of the step, which are "Calculate a statistic: F-ratio for analytical comparison (F_{comp})" and "Make a decision whether to reject the null hypothesis." Here, there are two aspects of unplanned comparisons to understand. First, using a procedure such as the Tukey test does *not* affect the value of the F-ratio for the analytical comparison—F_{comp} is calculated *exactly* the same way regardless of whether an analytical comparison is planned or unplanned. Second, making the decision to reject the null hypothesis is the same for planned and unplanned comparisons: If F_{comp} is greater than the critical value, the null hypothesis is rejected. However, comparing the Tukey critical value (F_T) with the critical value for planned comparisons highlights the consequence of conducting unplanned comparisons.

For the hypothetical example described earlier, the critical value for unplanned comparisons (F_T) is 8.16. If, on the other hand, these comparisons had been planned, the critical value for = .05 and df = 1, 40 would have been only 4.08. Rejecting the null hypothesis for unplanned comparisons requires a larger value of F_{comp} than for planned comparisons. Using a method to control familywise error such as the Tukey test makes it more difficult to reject the null hypothesis, thereby lowering the probability of Type I error.

In summary, methods such as the Dunnett, Scheffé, and Tukey tests are designed to reduce the possibility of concluding an effect exists in the population when it does not. By making it more difficult to reject the null hypothesis, we take a conservative approach, one that minimizes the effects of chance factors that may exist in a particular sample of data.

LEARNING CHECK 4:

Reviewing What You've Learned So Far

1. Review questions
   a. What is the purpose of conducting analytical comparisons?
   b. What are two types of analytical comparisons? In what research situations would you use one versus the other?
   c. What is between-group and within-group variance in an analytical comparison?
   d. What are concerns researchers have about unplanned comparisons?
   e. What is familywise error? How can it be controlled?
2. Earlier in this chapter, a study was described examining males' concerns regarding oral contraceptives (Jaccard et al., 1981). The study hypothesized that males would rate risks to their health as more important to them than either the contraceptive's effectiveness or convenience. Below is the ANOVA summary table for the study:

   To test one of the study's research hypotheses, you conduct a planned comparison of Health Risks ($N = 9$, $M = 4.51$) versus Effectiveness ($N = 9$, $M = 3.18$). Complete the following steps for this planned comparison.
   a. State the null and alternative hypotheses (H_0 and H_1).
   b. Make a decision about the null hypothesis.
      (1) Calculate the degrees of freedom (*df*).
      (2) Set alpha (α), identify the critical value, and state a decision rule.
      (3) Calculate the *F*-ratio for an analytical comparison (F_{comp}).
      (4) Make a decision whether to reject the null hypothesis.
      (5) Determine the level of significance.
      (6) Calculate a measure of effect size (R^2_{comp}).
   c. Draw a conclusion from the analysis.
   d. Relate the result of the analysis to the research hypothesis.

Source	*SS*	*df*	*MS*	*F*
Aspect of contraception	9.00	2	4.50	3.54
Error	30.56	24	1.27	
Total	39.56	26		

3. Imagine that a researcher conducts a study consisting of five groups, each of which consists of seven participants ($N_i = 7$). The researcher conducts a one-way ANOVA and rejects the null hypothesis. Because her research addresses a new topic, the researcher decides to compare each group with each of the other groups, controlling familywise error using the Tukey test.
   a. How many unplanned comparisons will be made in this situation?
   b. What is the value of q_T?
   c. Assuming that the probability of familywise error is set at .05 ($\alpha_{FW} = .05$), what is the critical value for each comparison (F_T)?

11.6 LOOKING AHEAD

This chapter illustrates both the advantages and disadvantages of increasing the complexity of a research study. Compared with the *t*-test discussed in Chapter 9, research studies analyzed with the one-way ANOVA and analytical comparisons are somewhat more difficult to conduct and analyze, but they provide the opportunity to test theories and research hypotheses in a more complex, comprehensive, and efficient way. The next chapter extends the discussion of comparing group means by discussing studies that have not one but two independent variables. As we will see, conducting this type of study not only allows researchers to examine more than one effect in the same study but also provides the opportunity to see how these effects may combine or interact with each other.

11.7 Summary

The ***analysis of variance (ANOVA)*** is a family of statistical procedures designed to test differences between two or more group means. It is called the "analysis" of variance because it involves analyzing different types of variance such as between-group and within-group variance.

The ***one-way ANOVA*** tests the differences between the means of two or more groups that comprise a single independent variable.

Testing differences between group means using a one-way ANOVA involves calculating a statistic known as the ***F-ratio,*** which is calculated by dividing between-group variance (MS_{BG}) by within-group variance (MS_{WG}). The theoretical distribution of F-ratios is positively skewed with a mean approximately equal to 1. To organize and summarize the calculations of a one-way ANOVA, researchers construct a table known as an ***ANOVA summary table***.

Researchers often calculate a measure of effect size as an index or estimate of the size or magnitude of the hypothesized effect. Within the one-way ANOVA, one measure of effect size is the R^2 statistic, which is the percentage of variance in the dependent variable associated with differences between the groups that comprise the independent variable.

Rejecting the null hypothesis in the one-way ANOVA does not indicate the specific nature and direction of the differences between the means of the groups. Therefore, to determine which of the groups are different or not different from one another, researchers conduct ***analytical comparisons***, which are comparisons between groups that are part of a larger research design.

There are two types of analytical comparisons. ***Planned (a priori) comparisons*** are comparisons the researcher builds into the research design before data collection has begun; the purpose of these comparisons is to test research hypotheses. The second type, ***unplanned (post hoc) comparisons***, are comparisons that have not been built into the research design prior to the data collection process; the decision to conduct these comparisons is made after the F-ratio for the one-way ANOVA has been calculated and the null hypothesis has been rejected. Although both types of analytical comparisons involve the calculation of an F-ratio (F_{comp}), they differ in terms of how this F-ratio is evaluated to make the decision about the null hypothesis.

Two main concerns have been expressed about unplanned comparisons. First, because they are based on the results of statistical analyses rather than theory and hypotheses, the results and interpretation of unplanned comparisons may be a function of the characteristics and idiosyncrasies of the sample. Second, the more comparisons that are made within a set of data, the greater the probability of making at least one Type I error (i.e., rejecting the null hypothesis when it is in fact true). ***Familywise error*** is the probability of committing a Type I error over a set of analyses.

Researchers use a number of different methods used to control familywise error, including the ***Dunnett test***, the ***Scheffé test***, and the ***Tukey test.*** These methods lower the probability of making a Type I error across a set of analytical comparisons by making it more difficult to reject the null hypothesis in each individual comparison. The choice of method depends on such factors as the goals of the research study, the number and types of analytical comparisons to be conducted, and the extent to which the researcher wishes to reduce the probability of making Type I errors across a set of comparisons.

11.8 Important Terms

analysis of variance (ANOVA) (p. 362)
one-way ANOVA (p. 362)
F-ratio (p. 363)
ANOVA summary table (p. 371)
R^2 (p. 373)

analytical comparisons (p. 382)
planned (a priori) comparisons (p. 382)
unplanned (post hoc) comparisons (p. 382)
familywise error (p. 389)
Dunnett test (p. 390)
simple comparisons (p. 390)
complex comparisons (p. 390)
Scheffé test (p. 390)
Tukey test (p. 390)

11.9 Formulas Introduced in This Chapter

Degrees of Freedom for Between-Group Variance, One-Way ANOVA (df_{BG})

$$df_{\text{BG}} = \#\text{ groups} - 1 \tag{11-1}$$

Degrees of Freedom for Within-Group Variance, One-Way ANOVA (df_{WG})

$$df_{\text{WG}} = \Sigma(N_i - 1) \tag{11-2}$$

Between-Group Variance, One-Way ANOVA (MS_{BG})

$$MS_{\text{BG}} = \frac{SS_{\text{BG}}}{df_{\text{BG}}} = \frac{N_i\Sigma(\bar{X}_i - \bar{X}_T)^2}{\#\text{groups} - 1} \tag{11-3}$$

Total Sample Mean, One-Way ANOVA ($\bar{X}_T$)

$$\bar{X}_T = \frac{\Sigma\bar{X}_i}{\#\text{groups}} \tag{11-4}$$

Within-Group Variance, One-Way ANOVA (MS_{WG})

$$MS_{\text{WG}} = \frac{SS_{\text{WG}}}{df_{\text{WG}}} = \frac{(N_i - 1)\ \Sigma s_i^2}{\Sigma(N_i - 1)} \tag{11-5}$$

***F*-Ratio, One-Way ANOVA (F)**

$$F = \frac{MS_{\text{BG}}}{MS_{\text{WG}}} \tag{11-6}$$

Measure of Effect Size, One-Way ANOVA (R^2)

$$R^2 = \frac{SS_{\text{BG}}}{SS_{\text{T}}} \tag{11-7}$$

Relationship Between *t*-Test and *F*-Ratio for the One-Way ANOVA

$$t^2 = F \tag{11-8}$$

F-Ratio, Analytical Comparison (F_{comp})

$$F_{\text{comp}} = \frac{MS_{\text{BG}_{\text{comp}}}}{MS_{\text{WG}}} \qquad (11\text{-}9)$$

Between-Group Variance, Analytical Comparison ($MS_{\text{BG}_{\text{comp}}}$)

$$MS_{\text{BG}_{\text{comp}}} = \frac{N_i(\bar{X}_i - \bar{X}_j)^2}{2} \qquad (11\text{-}10)$$

Measure of Effect Size, Analytical Comparison (R^2_{comp})

$$R^2_{\text{comp}} = \frac{SS_{\text{BG}_{\text{comp}}}}{SS_{\text{T}}} \qquad (11\text{-}11)$$

Critical Value, Tukey Test for Unplanned Comparisons (F_T)

$$F_T = \frac{(q_T)^2}{2} \qquad (11\text{-}12)$$

11.10 Using SPSS

One-Way Analysis of Variance (ANOVA): The Chess Study (11.1)

1. Define independent and dependent variables (name, # decimals, labels for the variables, labels for values of the independent variable) and enter data for the variables.

   NOTE: Numerically code values of the independent variable (i.e., 1 = Observation, 2 = Prediction, 3 = Explanation) and provide labels for these values in the **Values** box within **Variable View.**

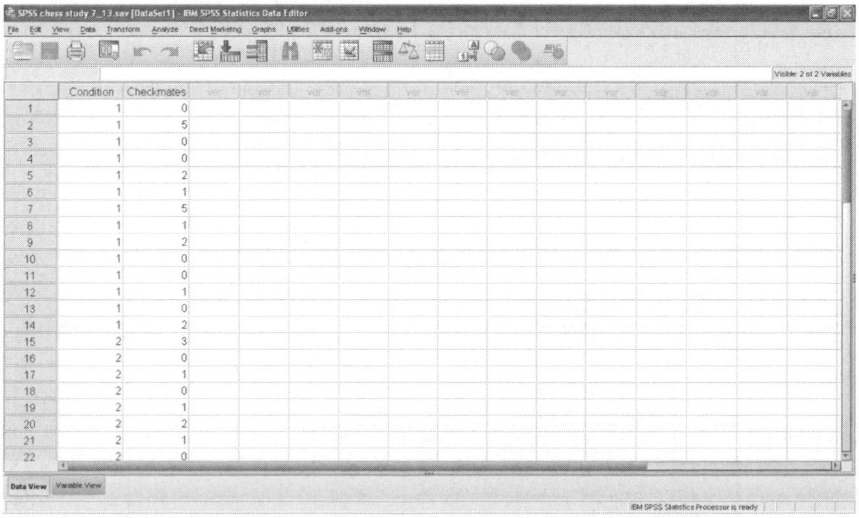

	Condition	Checkmates
1	1	0
2	1	5
3	1	0
4	1	0
5	1	2
6	1	1
7	1	5
8	1	1
9	1	2
10	1	0
11	1	0
12	1	1
13	1	0
14	1	2
15	2	3
16	2	0
17	2	1
18	2	0
19	2	1
20	2	2
21	2	1
22	2	0

2. Select the One-way ANOVA procedure within SPSS.

   How? (1) Click **Analyze** menu, (2) click **Compare Means**, and (3) click **One-Way ANOVA**.

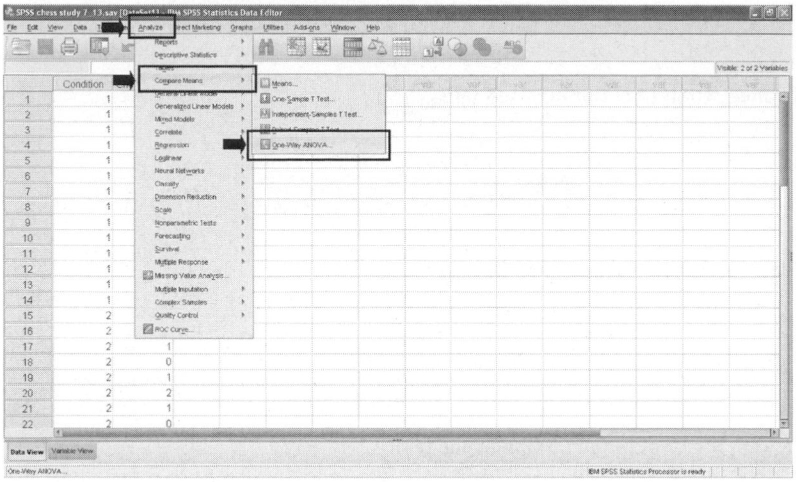

3. Identify the dependent variable and the independent variable, and ask for descriptive statistics.

   How? (1) Click dependent variable and → **Dependent List**, (2) click independent variable and → **Factor**, (3) click **Options** and click **Descriptive**, (4) click **Continue**, and (5) click **OK**.

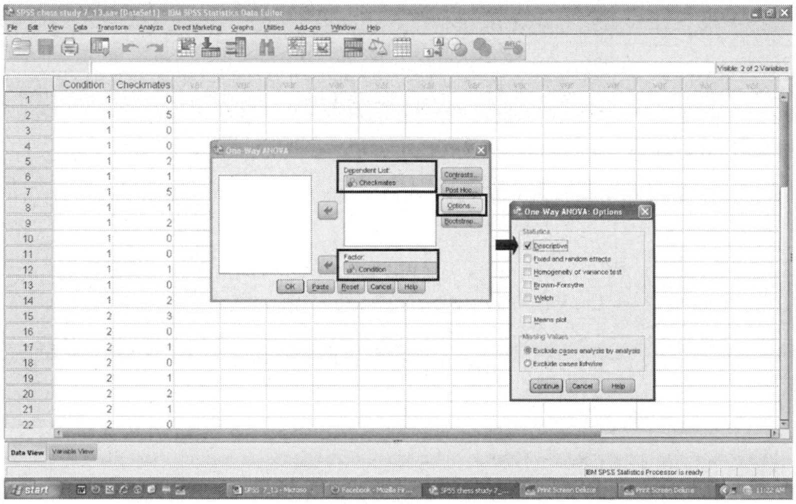

4. Examine output.

Descriptives

Descriptive statistics

Checkmates

	N	Mean	Std. Deviation	Std. Error	95% Confidence Interval for Mean Lower Bound	95% Confidence Interval for Mean Upper Bound	Minimum	Maximum
1 Observation	14	1.36	1.737	.464	.35	2.36	0	5
2 Prediction	14	.86	.949	.254	.31	1.41	0	3
3 Explanation	14	3.00	1.840	.492	1.94	4.06	1	5
Total	42	1.74	1.781	.275	1.18	2.29	0	5

ANOVA

Level of significance

ANOVA summary table

Checkmates

	Sum of Squares	df	Mean Square	F	Sig.
Between Groups	35.190	2	17.595	7.229	.002
Within Groups	94.929	39	2.434		
Total	130.119	41			

Analytical Comparisons: Planned Comparisons—Explanation vs. Prediction (11.5)

1. Within the one-way ANOVA procedure

   How? (1) Click **Analyze** menu, (2) click **Compare Means**, (3) click **One-Way ANOVA**, (4) click **Contrasts**, (5) assign 1 and –1 coefficients to the two groups being compared and assign 0 to group(s) not being compared, (6) click **Continue**, and (7) click **OK**.

   NOTE: Because 1 = Observation, 2 = Prediction, and 3 = Explanation, the Observation group is assigned a coefficient = 0, the Prediction group is assigned a coefficient = 1, and the Explanation group is assigned a coefficient = –1.

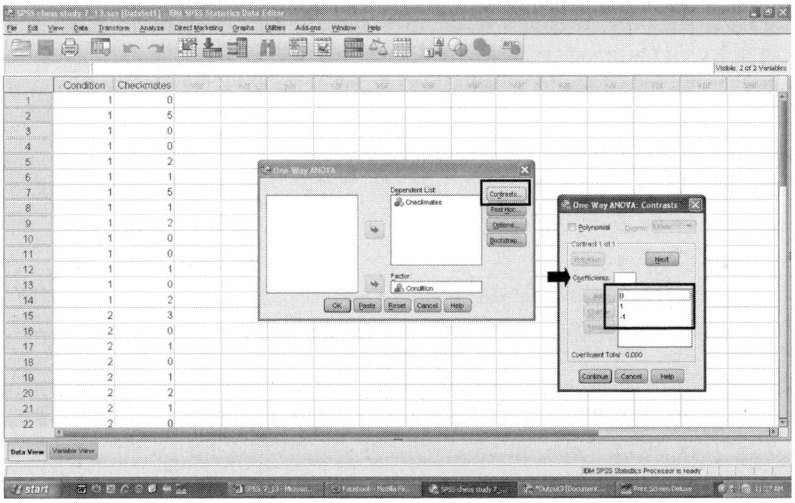

2. Examine output.

Identification of groups being compared

Contrast Coefficients

	Condition		
Contrast	1 Observation	2 Prediction	3 Explanation
1	0	1	-1

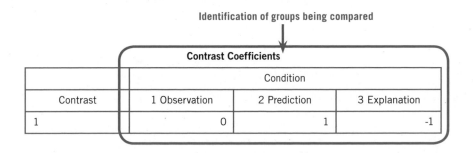

(Continued)

(Continued)

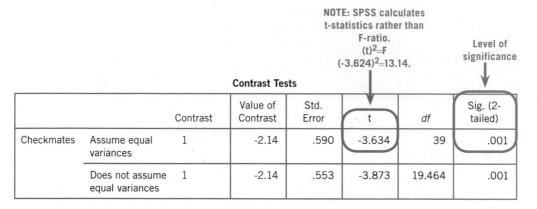

NOTE: SPSS calculates t-statistics rather than F-ratio. $(t)^2 = F$ $(-3.624)^2 = 13.14$.

Level of significance

Contrast Tests

		Contrast	Value of Contrast	Std. Error	t	*df*	Sig. (2-tailed)
Checkmates	Assume equal variances	1	-2.14	.590	-3.634	39	.001
	Does not assume equal variances	1	-2.14	.553	-3.873	19.464	.001

11.11 Exercises

1. Researchers Latané and Rodin (1969) were interested in studying helping behavior. In one of their experiments, they set up a situation in which the participant is brought into the waiting room either by himself or herself or with two or four confederates. The researcher says she will be right back, but as she walks into the other room, she pretends to sprain her ankle. The number of seconds from her cry until the participant offers help is recorded as the dependent variable. The following data are representative of their findings:

   Alone: 26, 25, 30, 20, 32

   Two confederates: 30, 33, 29, 40, 36

   Four confederates: 32, 39, 35, 41, 44

   a. Calculate the sample size (N_i), mean ($\overline{X}_i$), and standard deviation (s_i) for each group.

   b. Create a bar graph for the data with the number of confederates on the X-axis and the mean number of seconds on the Y-axis.

2. For each of the following, calculate the degrees of freedom (*df*) and identify the critical value of *F* (assume $\alpha = .05$).

   a. # groups = 3, $N_i = 10$

   b. # groups = 5, $N_i = 26$

   c. # groups = 4, $N_i = 15$

   d. There are five groups, with 20 participants in each group.

3. For each of the following, calculate the degrees of freedom (*df*) and identify the critical value of *F* (assume $\alpha = .05$).

   a. # groups = 3, $N_i = 8$

   b. # groups = 6, $N_i = 5$

   c. # groups = 4, $N_i = 9$

   d. There are three groups, with five participants in each group.

4. For each of the following, calculate between-group variance (MS_{BG}).

   a. $N_i = 13, \overline{X}_1 = 3.00, \overline{X}_2 = 5.00, \overline{X}_3 = 10.00$

   b. $N_i = 11, \overline{X}_1 = 19.00, \overline{X}_2 = 13.00, \overline{X}_3 = 21.00, \overline{X}_4 = 15.00$

   c. $N_i = 15, \overline{X}_1 = 1.13, \overline{X}_2 = 1.97, \overline{X}_3 = 2.60, \overline{X}_4 = 2.16, \overline{X}_5 = 3.82$

   d. $N_i = 9, \overline{X}_1 = 57.96, \overline{X}_2 = 64.68, \overline{X}_3 = 61.72, \overline{X}_4 = 68.53, \overline{X}_5 = 62.04, \overline{X}_6 = 56.09$

5. For each of the following, calculate between-group variance (MS_{BG}).
   a. $N_i = 20, \overline{X}_1 = 5.50, \overline{X}_2 = 3.75, \overline{X}_3 = 4.25$
   b. $N_i = 8, \overline{X}_1 = 14.85, \overline{X}_2 = 18.33, \overline{X}_3 = 19.92, \overline{X}_4 = 12.61$
   c. $N_i = 9, \overline{X}_1 = 40.81, \overline{X}_2 = 38.74, \overline{X}_3 = 33.29, \overline{X}_4 = 34.76, \overline{X}_5 = 41.15$
   d. $N_i = 6, \overline{X}_1 = 3.97, \overline{X}_2 = 4.56, \overline{X}_3 = 4.05, \overline{X}_4 = 3.31, \overline{X}_5 = 3.74, \overline{X}_6 = 2.91$
6. For each of the following, calculate within-group variance (MS_{WG}).
   a. $N_i = 7, s_1 = 1.00, s_2 = 2.00, s_3 = 6.00$
   b. $N_i = 6, s_1 = 13.00, s_2 = 16.00, s_3 = 12.00, s_4 = 15.00$
   c. $N_i = 9, s_1 = 23.72, s_2 = 18.50, s_3 = 19.97, s_4 = 16.39, s_5 = 20.06$
   d. $N_i = 23, s_1 = 4.53, s_2 = 4.18, s_3 = 3.70, s_4 = 3.46, s_5 = 3.03, s_6 = 2.71$
7. For each of the following, calculate within-group variance (MS_{WG}).
   a. $N_i = 15, s_1 = .75, s_2 = .48, s_3 = .63$
   b. $N_i = 19, s_1 = 4.52, s_2 = 4.86, s_3 = 4.28, s_4 = 4.47$
   c. $N_i = 22, s_1 = 17.76, s_2 = 14.92, s_3 = 16.20, s_4 = 20.01, s_5 = 19.57$
   d. $N_i = 12, s_1 = 2.38, s_2 = 1.91, s_3 = 1.37, s_4 = 3.06, s_5 = 2.87, s_6 = 2.54$
8. For each of the following, calculate the F-ratio (F) and create an ANOVA summary table.
   a. $N_i = 20, \overline{X}_1 = 20.00, s_1 = 8.00, \overline{X}_2 = 25.00, s_2 = 10.00, \overline{X}_3 = 15.00, s_3 = 9.00$
   b. $N_i = 5, \overline{X}_1 = 40.00, s_1 = 10.70, \overline{X}_2 = 35.00, s_2 = 8.00, \overline{X}_3 = 50.00, s_3 = 12.50, \overline{X}_4 = 55.00, s_4 = 12.00$
   c. $N_i = 11, \overline{X}_1 = 7.00, s_1 = 2.50, \overline{X}_2 = 12.00, s_2 = 4.00, \overline{X}_3 = 16.00,$
      $s_3 = 5.50, \overline{X}_4 = 9.00, s_4 = 3.75, \overline{X}_5 = 11.00, s_5 = 4.00$
9. For each of the following, calculate the F-ratio (F) and create an ANOVA summary table.
   a. $N_i = 10, \overline{X}_1 = 14.00, s_1 = 2.50, \overline{X}_2 = 12.50, s_2 = 2.10, \overline{X}_3 = 17.00, s_3 = 3.00$
   b. $N_i = 16, \overline{X}_1 = 7.50, s_1 = 1.70, \overline{X}_2 = 8.00, s_2 = 3.00, \overline{X}_3 = 6.75, s_3 = 2.50, \overline{X}_4 = 8.10, s_4 = 1.89$
   c. $N_i = 14, \overline{X}_1 = 65.24, s_1 = 10.43, \overline{X}_2 = 71.52, s_2 = 9.86, \overline{X}_3 = 75.06,$
      $s_3 = 12.79, \overline{X}_4 = 69.62, s_4 = 11.94, \overline{X}_5 = 61.91, s_5 = 11.43$
10. Using the means and standard deviations calculated in Exercise 1 for the helping behavior study, determine if the number of seconds until the participant helps the person in distress is the same for the three experimental conditions.
    a. State the null and alternative hypotheses (H_0 and H_1).
    b. Make a decision about the null hypothesis.
       (1) Calculate the degrees of freedom (df).
       (2) Set alpha (α), identify the critical value, and state a decision rule.
       (3) Calculate the F-ratio (F) and create an ANOVA summary table.
       (4) Make a decision whether to reject the null hypothesis.
       (5) Determine the level of significance.
       (6) Calculate a measure of effect size (R^2).
    c. Draw a conclusion from the analysis.
11. To evaluate the effectiveness of four different teaching methods, algebra students are randomly assigned to one of four methods (A, B, C, or D) and then given a test. The descriptive statistics for each of the four methods are provided below. Determine whether there are any differences in the test scores of the four methods.

    Method A: $N_1 = 20, \overline{X}_1 = 76.00, s_1 = 10.80$

    Method B: $N_2 = 20, \overline{X}_2 = 75.00, s_2 = 9.50$

Method C: $N_3 = 20, \overline{X}_3 = 85.00, s_3 = 11.00$

Method D: $N_4 = 20, \overline{X}_4 = 87.00, s_4 = 10.80$

a. State the null and alternative hypotheses (H_0 and H_1).
b. Make a decision about the null hypothesis.
   (1) Calculate the degrees of freedom (*df*).
   (2) Set alpha (α), identify the critical value, and state a decision rule.
   (3) Calculate the *F*-ratio (*F*) and create an ANOVA summary table.
   (4) Make a decision whether to reject the null hypothesis.
   (5) Determine the level of significance.
   (6) Calculate a measure of effect size (R^2).
c. Draw a conclusion from the analysis.

12. Although men and women have been found to perform differently on tests of mental ability, less is known about possible reasons for these differences. A researcher hypothesizes that one's beliefs play a role, such that women who believe they should not perform as well on these tests as men will in fact not perform well. To test this hypothesis, a group of women are given a written test of spatial abilities. Three different instructions were included with this test: (1) Women perform better on the test than men, (2) men perform better on the test than women, or (3) women and men perform equally well on the test. It was hypothesized that women who were told that women would perform better than men would score higher than women who were told that men would score better or that women and men would perform equally as well. Here are descriptive statistics of the test scores: Women better: $N = 11$, $M = 6.91$, $s = 2.63$; Men better: $N = 11$, $M = 7.45$, $s = 2.07$; Women and men equal: $N = 11$, $M = 9.82$, $s = 2.23$.

    a. State the null and alternative hypotheses (H_0 and H_1).
    b. Make a decision about the null hypothesis.
       (1) Calculate the degrees of freedom (*df*).
       (2) Set alpha (α), identify the critical value, and state a decision rule.
       (3) Calculate the *F*-ratio (*F*) and create an ANOVA summary table.
       (4) Make a decision whether to reject the null hypothesis.
       (5) Determine the level of significance.
       (6) Calculate a measure of effect size (R^2).
    c. Draw a conclusion from the analysis.
    d. Relate the result of the analysis to the research hypothesis.

13. Past research has examined relationships between people's backgrounds and their personality. For example, Eysenck (1982) found a relationship between blood type and introversion, and Pellegrini (1973) concluded that astrological signs were related to level of femininity. Gupta (1992) examined the relationship between season of birth and impulsivity (the lack of ability or desire to control one's behavior). Gupta hypothesized that people born in the seasons with more extreme temperatures (winter and summer) are more impulsive than those born in the more mild seasons (spring and autumn). She asked a sample of adults to report their season of birth and then administered a personality measure of impulsivity. Imagine the following descriptive statistics are reported for the test scores: Winter: $N = 6$, $M = 8.31$, $s = 1.74$; Spring: $N = 6$, $M = 6.52$, $s = 1.87$; Summer: $N = 6$, $M = 9.09$, $s = 2.35$; Fall: $N = 6$, $M = 5.86$, $s = 1.95$.

    a. State the null and alternative hypotheses (H_0 and H_1).
    b. Make a decision about the null hypothesis.
       (1) Calculate the degrees of freedom (*df*).
       (2) Set alpha (α), identify the critical value, and state a decision rule.
       (3) Calculate the *F*-ratio (*F*) and create an ANOVA summary table.
       (4) Make a decision whether to reject the null hypothesis.

(5) Determine the level of significance.

(6) Calculate a measure of effect size (R^2).

c. Draw a conclusion from the analysis.

d. Relate the result of the analysis to the research hypothesis.

14. Employers seek ways to improve the performance of their employees. Gardner, Van Dyne, and Pierce (2004) hypothesized that performance is influenced by organizational self-esteem, defined as an employee's evaluation of his or her personal adequacy as an organizational member. More specifically, they hypothesized that the higher one's organizational self-esteem, the higher will be one's job performance. The data below (representative of the study's findings) represent the employees' job performance level on a scale from 1 (low performance) to 5 (high performance). Determine whether job performance varies as a function of organizational self-esteem.

Organizational Self-esteem

Low	Medium	High
3.0	4.0	5.0
2.5	3.5	4.0
3.0	4.0	4.5
3.0	3.0	4.0
2.0	4.0	5.0
3.5	4.0	4.5

a. Calculate the sample size (N_i), mean ($\overline{X}_i$), and standard deviation (s_i) for each group.

b. State the null and alternative hypotheses (H_0 and H_1).

c. Make a decision about the null hypothesis.

(1) Calculate the degrees of freedom (*df*).

(2) Set alpha (α), identify the critical value, and state a decision rule.

(3) Calculate the *F*-ratio (*F*) and create an ANOVA summary table.

(4) Make a decision whether to reject the null hypothesis.

(5) Determine the level of significance.

(6) Calculate a measure of effect size (R^2).

d. Draw a conclusion from the analysis.

e. Relate the result of the analysis to the research hypothesis.

15. A researcher studying the effects of different learning techniques upon memory hypothesizes people learn better when training is distributed over time rather than massed all at once. She conducts an experiment in which 32 participants are randomly assigned to one of four conditions. Giving each subject several lessons to learn, she systematically varies the period of time between each lesson: 0 minutes between lessons (a massed learning condition), 10 minutes, 20 minutes, or 30 minutes. After the participants receive all of the lessons, they complete a 25-item test measuring comprehension of the material. Their scores on the test are given below—conduct a one-way ANOVA on these data.

a. Calculate the sample size (N_i), mean ($\overline{X}_i$), and standard deviation (s_i) for each group.

b. State the null and alternative hypotheses (H_0 and H_1).

c. Make a decision about the null hypothesis.

(1) Calculate the degrees of freedom (*df*).

(2) Set alpha (α), identify the critical value, and state a decision rule.

(3) Calculate the F-ratio (F) and create an ANOVA summary table.

(4) Make a decision whether to reject the null hypothesis.

(5) Determine the level of significance.

(6) Calculate a measure of effect size (R^2).

d. Draw a conclusion from the analysis.

e. Relate the result of the analysis to the research hypothesis.

Time Between Lessons

0 min	10 min	20 min	30 min
4	10	14	21
5	11	12	18
11	9	8	16
4	6	5	13
9	4	15	20
3	5	11	14
7	4	7	12
5	5	6	10

16. What is the relationship between the t-test and an ANOVA? In Chapter 9, we looked at two different types of instruction in reading comprehension (imagery and repetition). A t-test for independent means found a significant difference between the effectiveness of the two types, $t(18) = 2.85$, $p < .05$. To see the relationship between the t-test and the ANOVA, calculate the F-ratio for the one-way ANOVA for the same data and compare the results.

Imagery: $N = 10, \overline{X}_1 = 12.10, s_1 = 1.60$

Repetition: $N = 10, \overline{X}_2 = 9.90, s_2 = 1.85$

a. State the null and alternative hypotheses (H_0 and H_1).

b. Make a decision about the null hypothesis.

(1) Calculate the degrees of freedom (df).

(2) Set alpha (α), identify the critical value, and state a decision rule.

(3) Calculate the F-ratio (F) and create an ANOVA summary table.

(4) Make a decision whether to reject the null hypothesis.

(5) Determine the level of significance.

(6) Calculate a measure of effect size (R^2).

c. Draw a conclusion from the analysis.

d. Compare your value of the F-ratio with the t-test value of 2.85 to confirm their mathematical equivalence.

17. An important issue in court cases is the accuracy of eyewitness testimony. Behavioral scientists have suggested eyewitnesses can be influenced by how a question is phrased. A researcher conducts a study where 15 people watch a film of an accident in which Car A runs a stop sign and hits Car B at a speed of 20 miles per hour. After watching the film, she asks each of the 15 people to estimate Car A's speed at the moment of impact (they do not know the actual speed). However, the question is phrased three different ways. Five people are asked, "How fast was Car A going at the time of the *accident* with Car B?" Another five are asked, "How fast was Car A going when it *hit* Car B?" The last five are asked, "How fast was Car A going when it *smashed* into Car B?" The researcher hypothesizes

estimates of speed will vary as a function of the wording of the question, with more extreme wordings leading to higher estimates.

Wording of Question

"Accident"	"Hit"	"Smashed"
18	27	28
23	22	24
17	24	30
19	19	22
21	25	27

a. Calculate the sample size (N_i), mean ($\overline{X}_i$), and standard deviation (s_i) for each group.
b. State the null and alternative hypotheses (H_0 and H_1).
c. Make a decision about the null hypothesis.
   (1) Calculate the degrees of freedom (*df*).
   (2) Set alpha (α), identify the critical value, and state a decision rule.
   (3) Calculate the *F*-ratio (*F*) and create an ANOVA summary table.
   (4) Make a decision whether to reject the null hypothesis.
   (5) Determine the level of significance.
   (6) Calculate a measure of effect size (R^2).
d. Draw a conclusion from the analysis.

18. Looking back at Exercise 17, the most specific conclusion that could be drawn from rejecting the null hypothesis was that the estimates of speed for the three types of wording were not all equal to each other. However, the researcher hypothesized that more extreme wordings would lead to higher estimates of speed such that participants would report a higher speed when the word *smashed* was used rather than *accident.* Conduct the appropriate analytical comparison to test this specific hypothesis.
   a. State the null and alternative hypotheses (H_0 and H_1).
   b. Make a decision about the null hypothesis.
      (1) Calculate the degrees of freedom (*df*).
      (2) Set alpha (α), identify the critical value, and state a decision rule.
      (3) Calculate the *F*-ratio for an analytical comparison (F_{comp}).
      (4) Make a decision whether to reject the null hypothesis.
      (5) Determine the level of significance.
      (6) Calculate a measure of effect size (R^2_{comp}).
   c. Draw a conclusion from the analysis.
   d. Relate the result of the analysis to the research hypothesis.

19. The hypothetical study in Exercise 15 examined the effects of different learning techniques upon memory. Given the hypothesis is that people learn better when training is distributed over time rather than massed all at once, conduct an analytical comparison testing the difference in comprehension test scores for the 0-minute versus 10-minute conditions.
   a. State the null and alternative hypotheses (H_0 and H_1).
   b. Make a decision about the null hypothesis.

(1) Calculate the degrees of freedom (*df*).

(2) Set alpha (α), identify the critical value, and state a decision rule.

(3) Calculate the *F*-ratio for an analytical comparison (F_{comp}).

(4) Make a decision whether to reject the null hypothesis.

(5) Determine the level of significance.

(6) Calculate a measure of effect size (R^2_{comp}).

c. Draw a conclusion from the analysis.

d. Relate the result of the analysis to the research hypothesis.

20. The chess study discussed in this chapter (de Bruin et al., 2007) hypothesized that the number of checkmates should be greater for students asked to make predictions while learning to play chess than students who simply observe games of chess being played. Using the ANOVA summary table in Table 11.4(c), conduct an analytical comparison of the number of checkmates for the Prediction (N = 14, M = .86) versus Observation (N = 14, M = 1.36) conditions.

a. State the null and alternative hypotheses (H_0 and H_1).

b. Make a decision about the null hypothesis.

(1) Calculate the degrees of freedom (*df*).

(2) Set alpha (α), identify the critical value, and state a decision rule.

(3) Calculate the *F*-ratio for an analytical comparison (F_{comp}).

(4) Make a decision whether to reject the null hypothesis.

(5) Determine the level of significance.

(6) Calculate a measure of effect size (R^2_{comp}).

c. Draw a conclusion from the analysis.

d. Relate the result of the analysis to the research hypothesis.

21. An experimenter ran a study that consisted of four groups, each with 16 participants. This research study was exploratory research, and the experimenter did not have any specific hypotheses or expectations about the results of the analyses. He found a significant *F*-ratio for the one-way ANOVA and now would like to compare each group to each of the other groups to locate the differences.

a. How many unplanned comparisons will he need to conduct?

b. What are the concerns about conducting so many comparisons?

c. What is the value of q_T?

d. Assuming familywise error is set at the traditional value of .05 (α_{FW} = .05), what is the critical value for each of the unplanned comparisons (F_T)?

22. Imagine the *F*-ratio for a one-way ANOVA is rejected for a study involving six groups, each with nine participants. A researcher plans on comparing each group with every other group, controlling familywise error using the Tukey test.

a. How many unplanned comparisons will be conducted?

b. What is the value of df_{error}?

c. What is the value of q_T?

d. Assuming familywise error is set at .05 (α_{FW} = .05), what is the critical value for each comparison (F_T)?

Answers to Learning Checks

Learning Check 2

2 a. $df_{BG} = 3$, $df_{WG} = 40$, critical value = 2.84
b. $df_{BG} = 4$, $df_{WG} = 25$, critical value = 2.76
c. $df_{BG} = 5$, $df_{WG} = 42$, critical value = 2.45

3. a. $\bar{X}_T = 5.00$, $SS_{BG} = 234.00$, $df_{BG} = 2$, $MS_{BG} = 117.00$
b. $\bar{X}_T = 34.00$, $SS_{BG} = 840.00$, $df_{BG} = 3$, $MS_{BG} = 280.00$
c. $\bar{X}_T = 4.00$, $SS_{BG} = 20.37$, $df_{BG} = 4$, $MS_{BG} = 5.09$

4. a. $SS_{WG} = 765.00$, $df_{WG} = 51$, $MS_{WG} = 15.00$
b. $SS_{WG} = 1640.00$, $df_{WG} = 16$, $MS_{WG} = 102.50$
c. $SS_{WG} = 8.20$, $df_{WG} = 50$, $MS_{WG} = .16$

5. a.

Source	*SS*	*df*	*MS*	*F*
Between-group	6.00	2	3.00	2.56
Within-group	38.72	33	1.17	
Total	44.72	35		

Source	*SS*	*df*	*MS*	*F*
Between-group	4.00	3	1.33	5.78
Within-group	6.44	28	.23	
Total	10.44	31		

Learning Check 3

2. a. H_0: all μs are equal; H_1: not all μs are equal
b. (1) $df_{BG} = 2$, $df_{WG} = 24$
(2) If $F > 3.40$, reject H_0; otherwise, do not reject H_0

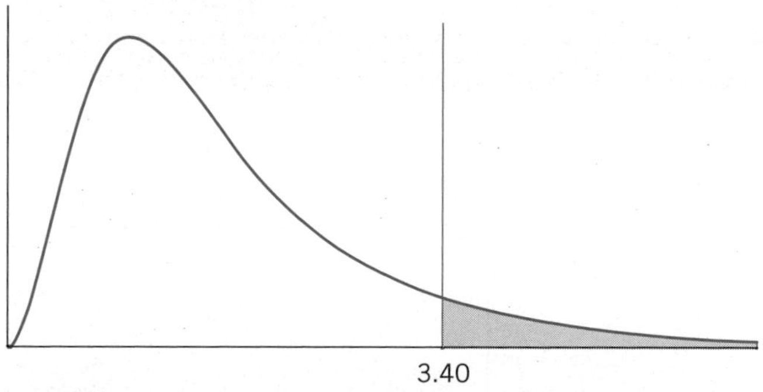

(3)

Source	*SS*	*df*	*MS*	*F*
Aspect of contraception	9.00	2	4.50	3.54
Error	30.56	24	1.27	
Total	39.56	26		

(4) $F = 3.54 > 3.40$ ∴ reject H_0 ($p < .05$)

(5) $F = 3.54 < 5.61$ ∴ $p < .05$ (but not $< .01$)

(6) $R^2 = .23$

c. The mean importance ratings of three aspects of oral contraceptives in samples of nine males (Effectiveness, $M = 3.18$; Convenience, $M = 3.43$; Health Risks, $M = 4.51$) were not all equal to each other, $F(2, 24) = 3.54$, $p < .05$, $R^2 = .23$.

d. We are unable to determine whether the research hypothesis in the study is supported until analytical comparisons are conducted.

Learning Check 4

2. a. H_0: $\mu_{\text{Health Risks}} = \mu_{\text{Convenience}}$; H_1: $\mu_{\text{Health Risks}} \neq \mu_{\text{Convenience}}$

b. (1) $df_{BG_{comp}} = 1$, $df_{WG} = 24$

(2) If $F_{comp} > 4.26$, reject H_0; otherwise, do not reject H_0

(3) $MS_{BG_{comp}} = 7.96$; $MS_{WG} = 1.27$; $F_{comp} = 6.27$

(4) $F_{comp} = 6.27 > 4.26$ ∴ reject H_0 ($p < .05$)

(5) $F_{comp} = 6.27 < 7.82$ ∴ $p < .05$ (but not $< .01$)

(6) $= .20$

c. In rating the importance of different aspects of oral contraceptives, the samples of nine males rated health risks ($M = 4.51$) as being significantly more important than its effectiveness ($M = 3.18$), $F(1, 24) = 6.27$, $p < .05$, $R^2 = .20$.

d. The result of this analytical comparison supports the research hypothesis that males would rate risks to their health as more important to them than the contraceptive's effectiveness.

3. a. Ten comparisons

b. $df_{error} = 30$; $q_T = 4.10$

c. $F_T = 8.41$

Answers to Odd-Numbered Exercises

1. a. Alone: $N = 5$, $\bar{X} = 26.60$, $s = 4.67$

Two confederates: $N = 5$, $\bar{X} = 33.60$, $s = 4.51$

Four confederates: $N = 5$, $\bar{X} = 38.20$, $s = 4.76$

b.

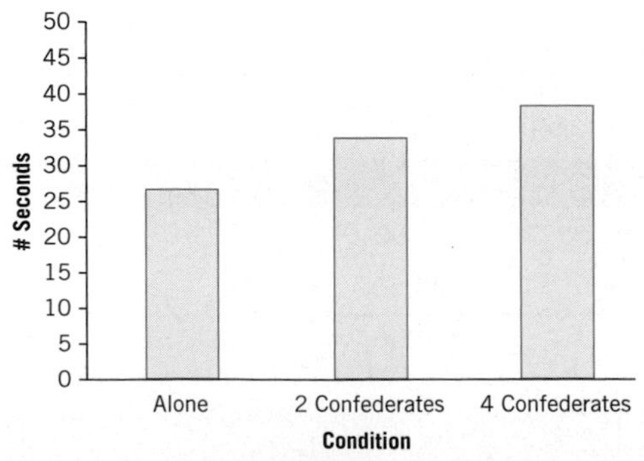

3. a. $df_{BG} = 2$, $df_{WG} = 21$, critical value = 3.47
   b. $df_{BG} = 5$, $df_{WG} = 24$, critical value = 2.84
   c. $df_{BG} = 3$, $df_{WG} = 32$, critical value = 2.92
   d. $df_{BG} = 2$, $df_{WG} = 12$, critical value = 3.89

5. a. $\bar{X}_T = 4.50, SS_{BG} = 32.40, df_{BG} = 2, MS_{BG} = 16.20$
   b. $\bar{X}_T = 16.43, SS_{BG} = 263.04, df_{BG} = 3, MS_{BG} = 87.68$
   $\bar{X}_T = 37.75, SS_{BG} = 456.57, df_{BG} = 4, MS_{BG} = 114.14$
   $\bar{X}_T = 3.76, SS_{BG} = 10.08, df_{BG} = 5, MS_{BG} = 2.02$

7. a. $SS_{WG} = 16.66, df_{WG} = 42, MS_{WG} = .40$
   b. $SS_{WG} = 1482.30, df_{WG} = 72, MS_{WG} = 20.59$
   c. $SS_{WG} = 33{,}260.85, df_{WG} = 105, MS_{WG} = 316.77$
   d. $SS_{WG} = 387.64, df_{WG} = 66, MS_{WG} = 5.87$

9. a.

Source	*SS*	*df*	*MS*	*F*
Between-group	105.00	2	52.50	8.02
Within-group	176.94	27	6.55	
Total	281.94	29		

b.

Source	*SS*	*df*	*MS*	*F*
Between-group	1,502.34	4	375.59	2.92
Within-group	8,356.14	65	128.56	
Total	9,858.48	69		

c.

Source	*SS*	*df*	*MS*	*F*
Between-group	18.40	3	6.13	1.13
Within-group	325.65	60	5.43	
Total	344.05	63		

11. a. H_0: all μs are equal; H_1: not all μs are equal
    b. (1) $df_{BG} = 3$, $df_{WG} = 76$
    (2) If $F > 2.76$, reject H_0; otherwise, do not reject H_0

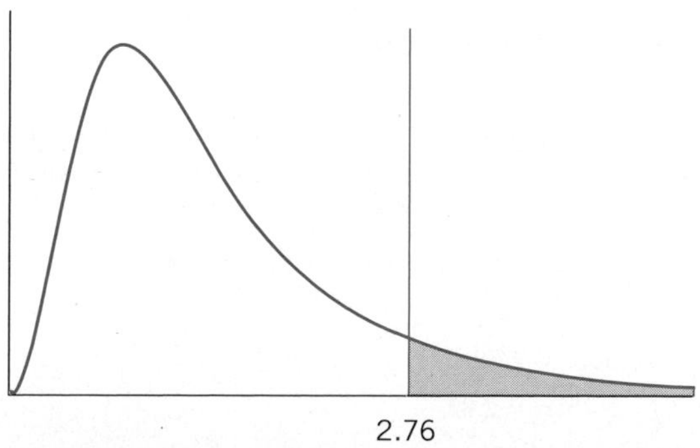

(3)

Source	*SS*	*df*	*MS*	*F*
Teaching method	2,254.80	3	751.60	6.76
Error	8,446.07	76	111.13	
Total	10,700.87	79		

(4) $F = 6.76 > 2.76$ ∴ reject H_0 ($p < .05$)

(5) $F = 6.76 > 4.13$ ∴ $p < .01$

(6) $R^2 = .21$

c. This analysis found that the mean examination scores for the 20 students in the four different teaching methods (Method A, $M = 76.00$; Method B, $M = 75.00$; Method C, $M = 85.00$; and Method D, $M = 87.00$) were not all equal, $F(3, 76) = 6.76$, $p < .01$, $R^2 = .21$.

13. a. H_0: all μs are equal; H_1: not all μs are equal

b. (1) $df_{BG} = 3$, $df_{WG} = 20$

(2) If $F > 3.10$, reject H_0; otherwise, do not reject H_0

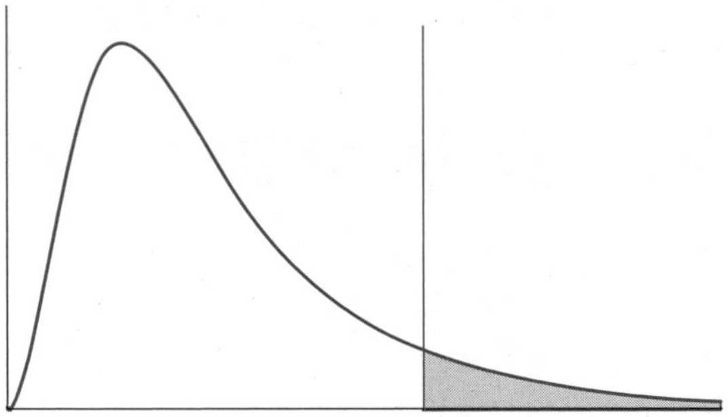

(3)

Source	*SS*	*df*	*MS*	*F*
Season of birth	40.92	3	13.64	3.44
Error	79.25	20	3.96	
Total	120.17	23		

(4) $F = 3.44 > 3.10$ ∴ reject H_0 ($p < .05$)

(5) $F = 3.44 < 4.94$ ∴ $p < .05$ (but not $< .01$)

(6) $R^2 = .34$

c. The mean levels of impulsivity in samples of six people in the four seasons of birth (winter, $M = 8.31$; spring, $M = 6.52$; summer, $M = 9.09$; fall, $M = 5.86$) were not all equal to each other, $F(3, 20) = 3.44$, $p < .05$, $R^2 = .34$.

d. We are unable to determine whether the research hypothesis in the study is supported until analytical comparisons are conducted.

15. a. 0 minutes: $N_1 = 8$, $\overline{X}_1 = 6.00$, $s_1 = 2.78$

10 minutes: $N_2 = 8$, $\overline{X}_2 = 6.75$, $s_2 = 2.82$

20 minutes: $N_3 = 8, \overline{X}_3 = 9.75, s_3 = 3.77$

30 minutes: $N_4 = 8, \overline{X}_4 = 15.50, s_4 = 3.93$

b. H_0: all μs are equal; H_1: not all μs are equal

c. (1) $df_{BG} = 3$, $df_{WG} = 28$

(2) If $F > 2.95$, reject H_0; otherwise, do not reject H_0

(3)

Source	*SS*	*df*	*MS*	*F*
Time between lessons	446.96	3	148.99	13.15
Error	317.31	28	11.33	
Total	764.37	31		

(4) $F = 13.15 > 2.95$ ∴ reject H_0 ($p < .05$)

(5) $F = 13.15 > 4.57$ ∴ $p < .01$

(6) $R^2 = .58$

d. The mean comprehension test scores of eight employees with different amounts of time between lessons (0 minutes, $M = 6.00$; 10 minutes, $M = 6.75$; 20 minutes, $M = 9.75$; 30 minutes, $M = 15.50$) were not all equal to each other, $F(3, 28) = 13.15, p < .01, R^2 = .58$.

e. We are unable to determine whether the research hypothesis in the study is supported until analytical comparisons are conducted.

17. a. "Accident": $N = 5, \overline{X} = 19.60, s = 2.41$

"Hit": $N = 5, \overline{X} = 23.40, s = 3.05$

"Smashed": $N = 5, \overline{X} = 26.20, s = 3.19$

b. H_0: all μs are equal; H_1: not all μs are equal

c. (1) $df_{BG} = 2$, $df_{WG} = 12$

(2) If $F > 3.88$, reject H_0; otherwise, do not reject H_0

(3)

Source	*SS*	*df*	*MS*	*F*
Wording	109.75	2	54.88	6.51
Error	101.16	12	8.43	
Total	210.91	14		

(4) $F = 6.51 > 3.88$ ∴ reject H_0 ($p < .05$)

(5) $F = 6.51 < 6.93$ ∴ $p < .05$ (but not < .01)

(6) $R^2 = .52$

d. This analysis reveals that the estimates of the car's speed at the time of the accident for the five people exposed to different wording of the question (accident, $M = 19.60$; hit, $M = 23.40$; smashed, $M = 26.20$) were not all equal to each other, $F(2, 12) = 6.51, p < .05, R^2 = .52$.

19. a. $H_0: \mu_{0\ min} = \mu_{10\ min}$; $H_1: \mu_{0\ min} \neq \mu_{10\ min}$

b. (1) $df_{BG_{comp}} = 1$, $df_{WG} = 28$

(2) If $F_{comp} > 4.20$, reject H_0; otherwise, do not reject H_0

(3) $MS_{BG_{comp}} = 2.25$; $MS_{WG} = 11.33$; $F_{comp} = .20$

(4) $F_{comp} = .20 < 4.20$ ∴ do not reject H_0 ($p > .05$)

(5) Not applicable (H_0 not rejected)

(6) $R^2 = .01$

c. The mean comprehension test scores of eight employees with 0 minutes between lessons ($M = 6.00$) and 10 minutes between lessons ($M = 6.75$) did not significantly differ, $F(1, 28) = .20$, $p > .05$, $R^2 = .003$.

d. The results of this analytical comparison do not support the research hypothesis that people learn better when training is distributed over time rather than massed all at once.

21. a. Six comparisons

b. Familywise error (the probability of making at least one Type I error across a set of comparisons) increases as the number of comparisons made increases.

c. $q_T = 3.74$

d. $F_T = 6.99$

12

TWO-WAY ANALYSIS OF VARIANCE (ANOVA)

CHAPTER OUTLINE

Several chapters in this book have discussed statistical procedures used to test differences between groups. In Chapter 9, we compared two groups of drivers (Intruder and No Intruder) in how long they took to leave a shopping mall parking space. Chapter 11 compared the effectiveness of three strategies (Observation, Prediction, and Explanation) used to learn to play chess. These research studies, which consisted of a single independent variable, provided useful information regarding a research topic. However, studying the effects of independent variables one at a time has conceptual, statistical, and pragmatic limitations. In this chapter, we introduce the analysis of studies consisting of two independent variables.

12.1 AN EXAMPLE FROM THE RESEARCH: VOTE—OR ELSE!

Throughout your life, people have attempted to influence your behavior. When you were younger, your parents may have tried to get you in the habit of regularly brushing your teeth. If so, what strategy did they employ? Did they tell you what a nice smile you'd have if you brushed regularly? Or did they frighten you with horror stories about cavities and visits to the dentist? A group of researchers headed by Howard Lavine of the State University of New York at Stony Brook made the following observation about how fear can affect behavior: "The use of fear appeals has long played a central role in attempts to change and shape attitudes by means of persuasive messages" (Lavine et al., 1999, p. 337).

Lavine and his fellow researchers evaluated the effectiveness of two types of messages persuading people to vote in elections. The first type of message emphasized rewards one may receive by voting, whereas the second type focused on negative consequences for failing to vote. But rather than simply proposing that one type of message is more persuasive than the other, the researchers hypothesized that the effectiveness of a particular type of message depends on characteristics of the person receiving the message. For some voters, they suggested, a rewarding message might be more effective than a threatening message, whereas for other types of voters, a threatening message may be more persuasive.

The characteristic of voters the researchers chose to examine was authoritarianism, defined as the extent to which people "perceive a great deal of threat from their immediate and broader environment . . . [and are] predisposed to view the world as dangerous" (Lavine et al., 1999, p. 338). The researchers hypothesized that the effectiveness of a particular type of message depends on the person's level of authoritarianism. As they wrote,

> Our key prediction was that . . . high authoritarian recipients would perceive the threat message as more valid and persuasive than the reward message. Because high authoritarians perceive the world as a dangerous and threatening place, they should resonate to a message highlighting the potential negative consequences of not voting. . . . We also explored the possibility that a threat-related persuasive message would lead to less persuasion among low authoritarian recipients, relative to the reward message. (Lavine et al., 1999, p. 340)

To summarize, the researchers hypothesized that the effectiveness of a type of message depends on the person's level of authoritarianism. More specifically, they predicted that a threatening message would be more persuasive than one emphasizing rewards for people with a high level of authoritarianism, but a message focusing on rewards would be more effective than a threatening message for people with low authoritarianism.

The sample in the study consisted of "86 voting-eligible (i.e., > 17 years old and American citizens) students (n = 34 men; n = 52 women) at the University of Minnesota" (Lavine et al., 1999, p. 341). The critical criterion for inclusion in the study was that the person be old enough to vote.

In conducting the study, the first independent variable was the type of message (rewarding or threatening) used to persuade people to vote. Research participants in the study were randomly assigned to receive one of two booklets. One version of the booklet contained statements emphasizing rewards gained from voting, such as "Voting allows one to play an active role in the direction taken by your government." The other version of the booklet described threats or punishments that may occur from failing to vote, such as "Not voting allows others to take away your right to express your values."

The second independent variable was the person's level of authoritarianism. To measure this variable, participants completed a 10-item scale that included statements designed to gauge their level of authoritarianism, such as "Our country will be destroyed someday if we do not smash the perversions eating away at our moral fiber and traditional beliefs." On the basis of their responses, participants were classified as either "low" or "high" in authoritarianism; the more the person agreed with the statements, the more he or she was categorized as "high."

The dependent variable in the study was the perceived quality of the message conveyed by each type of booklet. To gauge perceptions of quality, the participants completed a 12-item questionnaire in which they indicated their level of agreement with statements related to the booklet's quality and effectiveness, such as "I found the material to be convincing." The average rating across the 12 items was calculated; the higher the average rating, the higher the perceived message quality.

In summary, the study contained two independent variables: Message Type, which consisted of two types (Reward and Threat), and Authoritarianism, which comprised two levels (Low and High). The dependent variable in the study, Message Quality, was a continuous variable ranging from 1.00 to 7.00. This study is the first one in this book that includes more than one independent variable. The characteristics of this type of research design, known as a factorial research design, are discussed in the next section.

12.2 INTRODUCTION TO FACTORIAL RESEARCH DESIGNS

In Chapters 9 and 11, each research study was an example of a **single-factor research design**, which is a research design consisting of one independent variable. In a single-factor research design, each research participant is associated with one level of an independent variable. For example, in the parking lot study (Chapter 9), drivers were randomly assigned to either the Intruder or No Intruder group; in the chess study (Chapter 11), students received one of three strategies: Observation, Prediction, or Explanation.

Unlike previous chapters, the study in this chapter, which will be referred to as the voting message study, contains two independent variables: Message Type and Authoritarianism. Consequently, each participant belongs to one *combination* of the two independent variables: Reward/Low, Reward/High, Threat/Low, or Threat/High. Combining the levels of the two independent variables, the voting message study may be represented the following way:

		Message Type	
		Reward	Threat
Authoritarianism	Low		
	High		

The voting message study is an example of a **factorial research design**, which is a research design consisting of all possible combinations of two or more independent variables. This section discusses two critical features of factorial research designs: the ability to test interaction effects and the ability to test main effects.

Testing Interaction Effects

A critical reason for including two or more independent variables in the same study is the opportunity to determine whether the variables combine, or interact, to influence the dependent variable. That is, factorial research designs allow researchers to test interaction effects between variables. An **interaction effect** occurs when the effect of one independent variable on the dependent variable changes at the different levels of another independent variable. Put another way, an interaction effect exists when the effect of one independent variable depends on the level of another independent variable.

In the voting message study, the researchers hypothesized an interaction effect between message type and authoritarianism such that the effect of message type (reward vs. threat) on perceived message quality depends on the participant's level of authoritarianism (low vs. high). For people with a low level of authoritarianism, the reward message was hypothesized to be more persuasive than was the threat message; however, for people with a high level of authoritarianism the threat message was predicted to be more persuasive than the reward message.

The ability to examine interaction effects is not attainable from single-factor designs. For this reason, examining interaction effects represents a unique and important characteristic of factorial research designs. Below, a hypothetical research study is used to illustrate both the presence and the absence of an interaction effect.

An Example of Interaction Effect

What is the effect of physical exercise on one's state of mind? More specifically, for people who exercise on a regular basis, what happens to their psychological well-being when they're not able to work out? In a study headed by Gregory Mondin at the University of Wisconsin, a group of regular exercisers were asked to refrain from exercising for a 5-day period (Mondin et al., 1996). The researchers found that the level of anxiety in the research subjects increased from the first day (Monday) to the third day (Wednesday), but then dropped from Wednesday to Friday as the period of exercise deprivation ended.

Let's say you wonder whether the changes in anxiety level is the same for all regular exercisers. For example, your examination of research might suggest that the gender of the exerciser should be taken into account. It could be hypothesized, for example, that women are better able to maintain a steady mood state during periods of exercise deprivation than are men, whose level of anxiety rises and falls more sharply during the same period. On the basis of such an assumption, you might hypothesize that the effect of exercise deprivation on anxiety level depends on the gender of the exerciser. If so, you would be hypothesizing an interaction effect between Day and Gender.

A table of hypothetical means for this study is provided in Figure 12.1(a). Looking at this figure, we see that combining three levels of Day (Monday, Wednesday, Friday) with two levels of Gender (Male, Female) results in six possible combinations of the two variables. We will use the term **cell** to refer to a combination of independent variables; the exercise deprivation study therefore consists of a total of six cells. The numbers in the cells of this table are **cell means**, defined as the mean of the dependent variable for a particular combination of independent variables. For example, in Figure 12.1(a), the upper-left cell mean of 2.20 is the mean anxiety level (measured on a 1 [low] to 5 [high] scale) on Monday for the male exercisers.

Figure 12.1(b) is a line graph of the cell means for the six Day × Gender combinations. The cell means have been portrayed using a line graph because the independent variable "Day," located along the horizontal (X) axis, is a continuous variable with the same distance (2 days) between the three values (Monday, Wednesday, Friday). In the line graph, the means for the three levels of Day are plotted separately for each level of Gender (Male, Female), which is the other independent variable. If Day had been a categorical rather than a continuous variable, it would not be appropriate to "connect the dots" using a line graph, and we would have instead used a bar chart to represent the cell means.

Looking at the two lines in Figure 12.1(b), we see that changes in anxiety level during the period of exercise deprivation are not the same for the two genders. For women, the level of anxiety changes only slightly during this period. The level of anxiety for men, however, rises sharply from Monday to Wednesday before dropping back down on Friday. This hypothetical study is an illustration of an interaction effect. This is because the effect of one independent variable (Day) on the dependent variable (Anxiety Level) is not the same at the different levels of the other independent variable (Gender). That is, the effect of Day depends on the level of Gender.

Figure 12.2 presents other examples of possible interaction effects of Day and Gender. As we can see from these figures, there are a wide variety of possible interaction effects. Although graphs such as those in Figures 12.1 and 12.2 are extremely useful in inspecting data for the presence of interaction effects, the simple fact that the

FIGURE 12.1 ● EXAMPLE OF AN INTERACTION EFFECT (DAY BY GENDER)

(a) Table of Cell Means

		Day		
		Monday	Wednesday	Friday
Gender	Male	2.20	3.60	2.40
	Female	2.00	2.40	2.10

(b) Line Graph of Cell Means

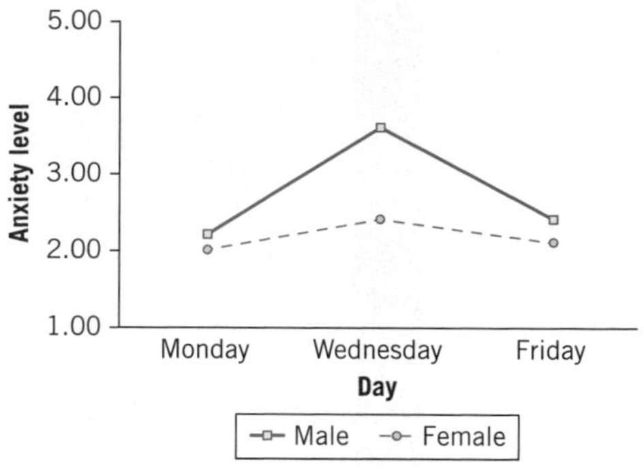

lines in a graph may not be the same does not necessarily mean there is a statistically significant interaction effect. Because of chance factors that operate in every study, we cannot draw a conclusion about an interaction effect until appropriate statistical analyses have been conducted.

An Example of No Interaction Effect

The hypothetical data for the exercise study provided in Figure 12.3 illustrate a research situation where no interaction effect is present between two variables. Looking at the cell means and the line graph, we observe that the changes in anxiety level across days are the same for men as for women. For example, the level of anxiety for the male participants increased from 2.30 to 3.10 from Monday to Wednesday (a difference of .80 [3.10 – 2.30]) and decreased from 3.10 to 2.20 from Wednesday to Friday (a difference of .90 [3.10 – 2.20]). Similar changes were found for the female participants (2.90 – 2.10 = .80 and 2.90 – 2.00 = .90).

Unlike the research situations in Figure 12.2, the lines of the two groups in Figure 12.3 are parallel. The parallel lines reflect the fact that the effect of the independent variable (Day) is the same at the different levels of the other independent variable (Gender)—the changes across days are the same for men and women. When the effect of one variable does not depend on the levels of the other variable, there is no interaction effect between variables.

Figure 12.4 presents other examples of two variables that do not interact. Similar to the presence of interaction, the absence of an interaction effect can take a wide variety of shapes. As we mentioned earlier, whether or not an interaction effect is present or absent is ultimately determined by conducting statistical analyses rather than by simply examining a table or figure.

Testing Main Effects

In addition to the ability to test for interaction effects between independent variables, a second critical aspect of factorial research designs is the ability to examine **main effects**, defined as the effect of an independent variable on the dependent variable within a factorial research design; the main effect of an independent variable is separate from the effects of the other independent variables. In the exercise deprivation study, for example, we can determine whether anxiety levels change as a function of Day, regardless of the research participant's gender. We can also test whether men and women differ in their level of anxiety, regardless of the day of the week in which they do not exercise. These effects of Day and Gender, distinct from any interaction effect between them, are known as main effects.

FIGURE 12.2 ⬢ OTHER EXAMPLES OF AN INTERACTION EFFECT (DAY BY GENDER)

(a)

		Day		
		Mon	Wed	Fri
Gender	Male	2.10	3.60	2.10
	Female	3.60	2.10	3.60

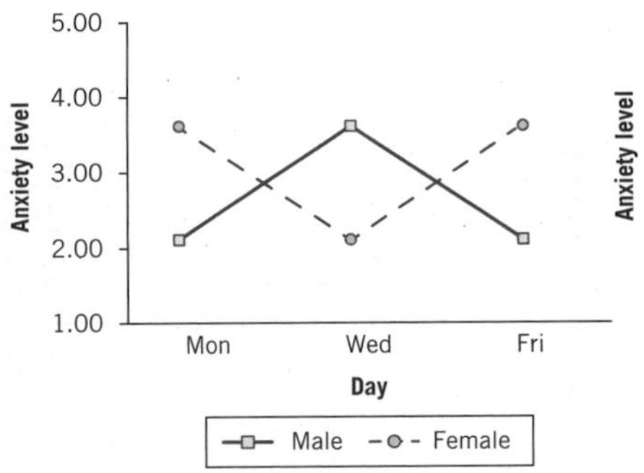

(b)

		Day		
		Mon	Wed	Fri
Gender	Male	2.80	3.20	3.60
	Female	2.80	2.40	2.00

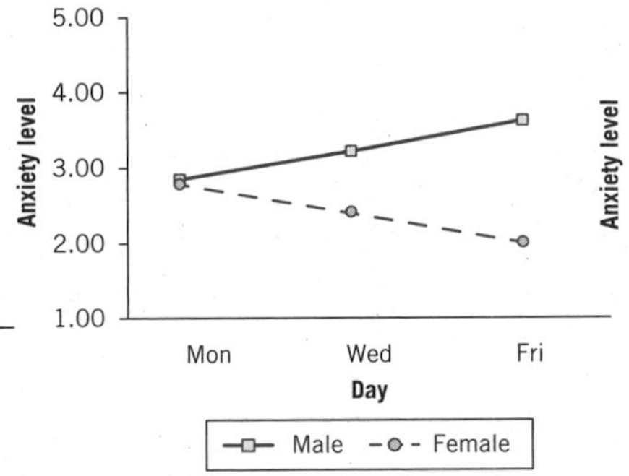

(c)

		Day		
		Mon	Wed	Fri
Gender	Male	2.80	3.80	3.80
	Female	2.80	1.70	2.70

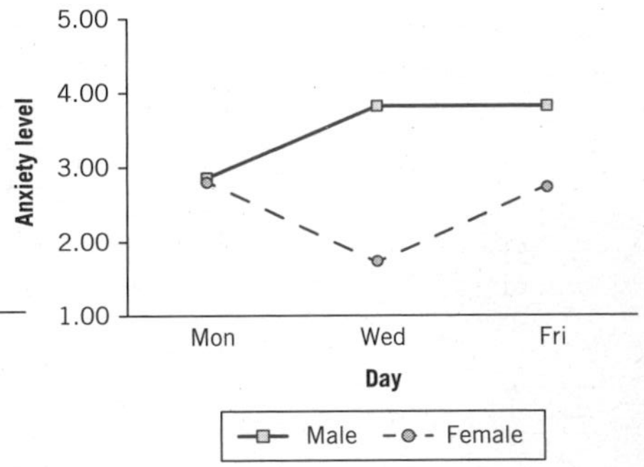

FIGURE 12.3 ● EXAMPLE OF NO INTERACTION EFFECT (DAY BY GENDER)

(a) Table of Cell Means

		Day		
		Monday	Wednesday	Friday
Gender	Male	2.30	3.10	2.20
	Female	2.10	2.90	2.00

(b) Line Graph of Cell Means

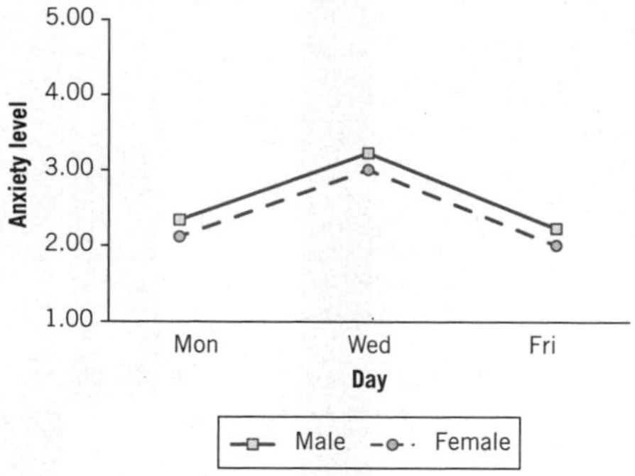

Based on the table of cell means in Figure 12.1(a), Table 12.1 illustrates two main effects for the exercise deprivation study. The boldfaced means in the margins of the table represent the two main effects of Day and Gender, respectively. These means are called **marginal means**, defined as the means in a table that represent a main effect.

When each cell is based on the same number of scores, the marginal means for an independent variable are calculated by averaging the results over the different levels of the other independent variable. For example, assuming the exercise deprivation study contained an equal number of males and females, the marginal means for the main effect of Day are calculated by computing the mean of the males and females for each of the 3 days:

$$\bar{X}_{\text{Monday}} = \frac{2.20 + 2.00}{2} = \frac{4.20}{2} = 2.10 \qquad \bar{X}_{\text{Wednesday}} = \frac{3.60 + 2.40}{2} = \frac{6.00}{2} = 3.00 \qquad \bar{X}_{\text{Friday}} = \frac{2.40 + 2.10}{2} = \frac{4.50}{2} = 2.25$$

By combining the means of the two genders for each day, we are in fact acting as though the independent variable Gender does not exist.

The marginal means for the main effect of Gender are determined by calculating the average anxiety level across the 3 days separately for the male and female exercisers:

$$\bar{X}_{\text{Male}} = \frac{2.20 + 3.60 + 2.40}{3} = \frac{8.20}{3} = 2.73 \qquad \bar{X}_{\text{Female}} = \frac{2.00 + 2.40 + 2.10}{3} = \frac{6.50}{3} = 2.17$$

It is important to understand that main effects are equivalent to the information that would be obtained by conducting separate single-factor research designs for each of the independent variables. For example, the three marginal means of 2.10, 3.00, and 2.25 constitute a single-factor research design in which we can determine whether anxiety levels differ as a function of Day. Similarly, the marginal means for males ($M = 2.73$) and females ($M = 2.17$) represent a single-factor research design for Gender. Therefore, rather than having to conduct two separate studies with two different samples to test the effects of Day and Gender on anxiety level, in a factorial research design, both effects can be tested within the same study.

FIGURE 12.4 ● OTHER EXAMPLES OF NO INTERACTION EFFECT (DAY BY GENDER)

(a)

		Day		
		Mon	Wed	Fri
Gender	Male	2.20	2.80	3.40
	Female	2.10	2.70	3.30

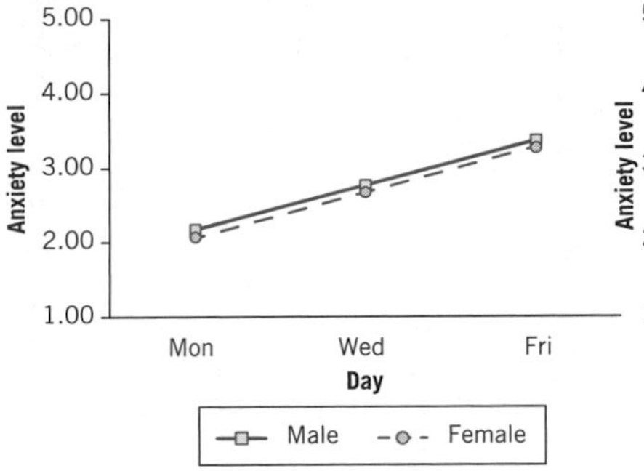

(b)

		Day		
		Mon	Wed	Fri
Gender	Male	3.20	3.20	3.20
	Female	2.50	2.50	2.50

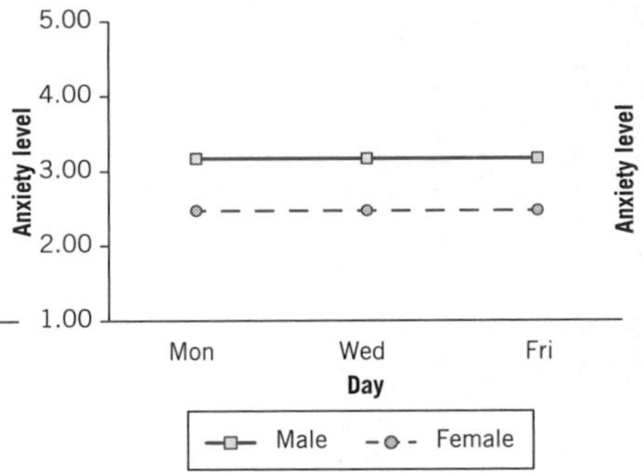

(c)

		Day		
		Mon	Wed	Fri
Gender	Male	2.30	2.30	3.80
	Female	2.60	2.60	4.10

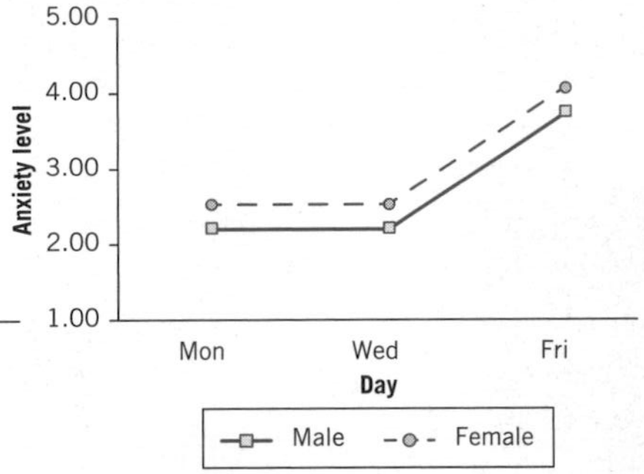

TABLE 12.1 ● EXAMPLES OF MAIN EFFECTS (DAY AND GENDER)

		Day			
		Monday	Wednesday	Friday	
Gender	Male	2.20	3.60	2.40	**2.73**
	Female	2.00	2.40	2.10	**2.17**
		2.10	**3.00**	**2.25**	

The Relationship Between Main Effects and Interaction Effects

Factorial research designs contain both main effects and interaction effects, any of which may be present or absent in any given research study. This section discusses two aspects of the relationship between main effects and interaction effects: (1) The presence or absence of main effects provides no indication of whether an interaction effect is present or absent, and vice versa, and (2) whether and how one interprets main effects depends on the presence or absence of interaction effects.

To illustrate these two aspects of the relationship between main effects and interaction effects, Figure 12.5 provides hypothetical data for eight different research designs. In a research design with two independent variables, there are eight possible combinations of present or absent main effects and interaction effect. In this figure, we will assume that any difference between means represents the presence of an effect; in reality, however, we would need to conduct statistical analyses on the data to determine whether this is true.

The top four line graphs (Figures 12.5(a–d)) all have an absent interaction effect, which is represented by two parallel lines. However, looking at the tables of means, we see these four scenarios differ in whether either or both of the main effects are present. For example, in Figure 12.5(a), neither of the two main effects is present; in Figure 12.5(d), both main effects are present.

In the bottom four line graphs (Figures 12.5(e–h)), in which an interaction effect is present, it is possible to have the same four pairings of present or absent main effects as in Figures 12.5(a–d). As a whole, these eight scenarios illustrate a critical aspect of the relationship between main effects and interaction effects:

> The presence or absence of main effects provides no indication of whether an interaction effect is either present or absent, and vice versa.

That is to say, whether or not a particular effect is present has no direct influence on the other effects.

The second aspect of the relationship between main effects and interaction effects pertains to the order in which they are examined. Even though main effects and interaction effects are independent of each other, there is typically a certain order to be followed in interpreting the results of a factorial research design:

> Whether and how one interprets main effects depends on the presence or absence of the interaction effect.

If an interaction effect is present, this implies that the effects of one independent variable depend on the level of the other independent variable. As a result, we cannot assess the effect of one independent variable in isolation of the other but must rather consider both of them simultaneously. Even if one or both of the main effects are present, the interaction effect takes precedence; attention is focused on the cell means to understand the source and nature of the interaction.

FIGURE 12.5 ● COMBINATIONS OF ABSENT AND PRESENT MAIN EFFECTS AND INTERACTION EFFECT

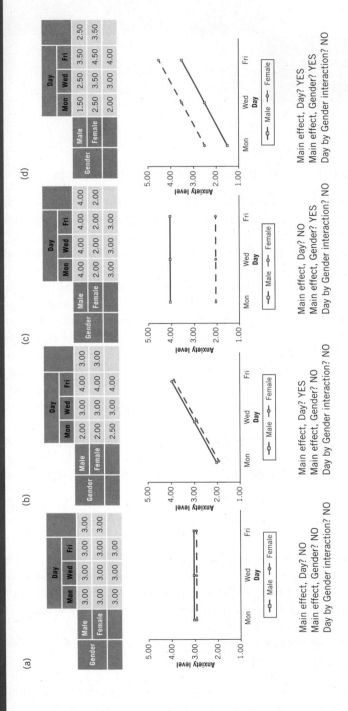

(a)

		Day			
		Mon	Wed	Fri	
Gender	Male	3.00	3.00	3.00	3.00
	Female	3.00	3.00	3.00	3.00
		3.00	3.00	3.00	

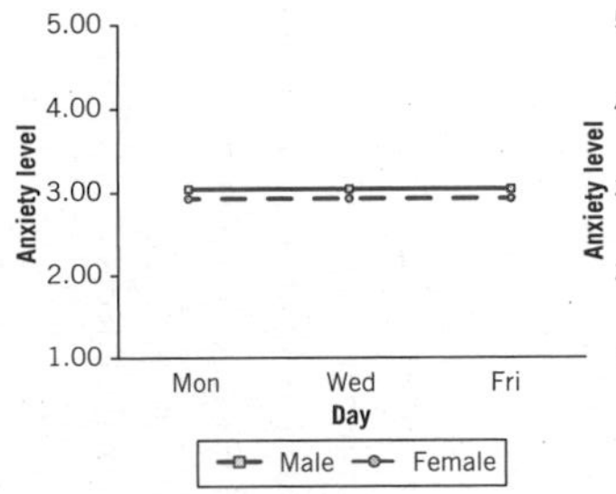

Main effect, Day? NO
Main effect, Gender? NO
Day by Gender interaction? NO

(b)

		Day			
		Mon	Wed	Fri	
Gender	Male	2.00	3.00	4.00	3.00
	Female	2.00	3.00	4.00	3.00
		2.50	3.00	4.00	

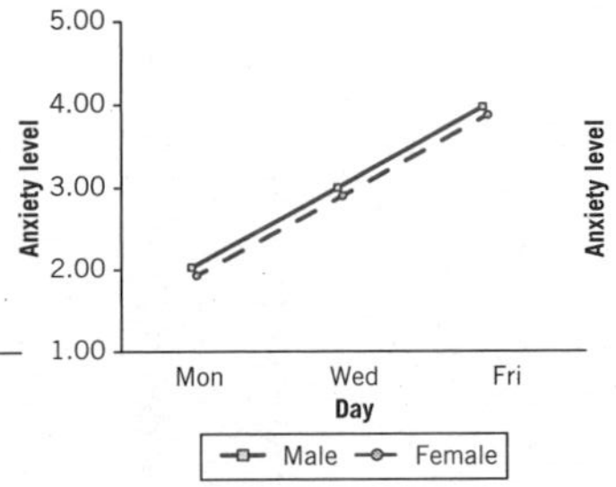

Main effect, Day? YES
Main effect, Gender? NO
Day by Gender interaction? NO

(c)

		Day			
		Mon	Wed	Fri	
Gender	Male	4.00	4.00	4.00	4.00
	Female	2.00	2.00	2.00	2.00
		3.00	3.00	3.00	

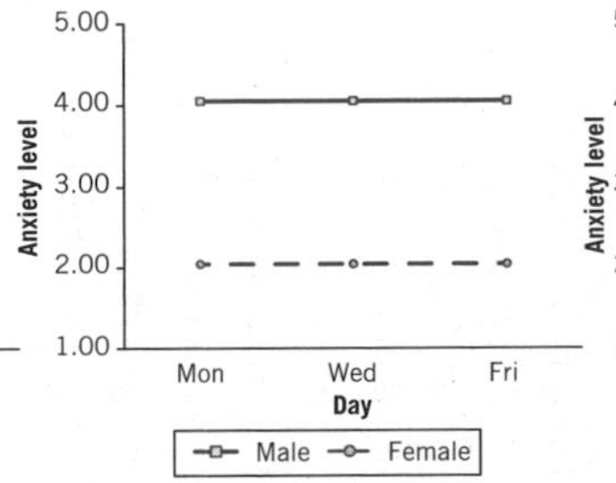

Main effect, Day? NO
Main effect, Gender? YES
Day by Gender interaction? NO

(d)

		Day			
		Mon	Wed	Fri	
Gender	Male	1.50	2.50	3.50	2.50
	Female	2.50	3.50	4.50	3.50
		2.00	3.00	4.00	

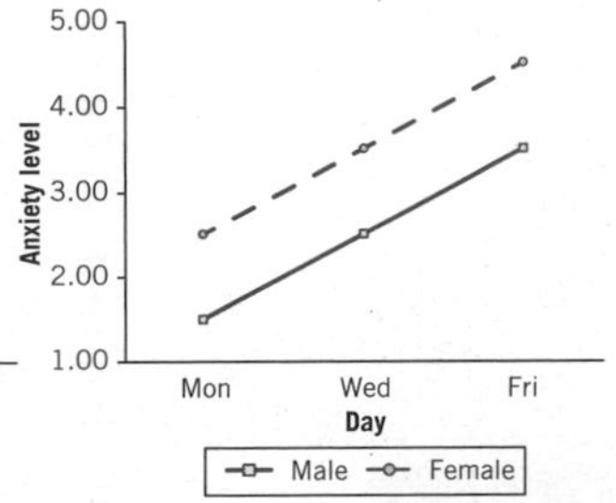

Main effect, Day? YES
Main effect, Gender? YES
Day by Gender interaction? NO

FIGURE 12.5 (Continued)

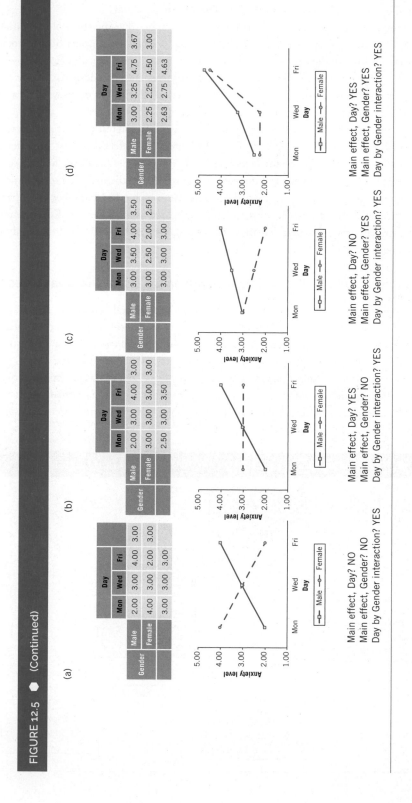

(a)

		Day			
		Mon	Wed	Fri	
Gender	Male	2.00	3.00	4.00	3.00
	Female	4.00	3.00	2.00	3.00
		3.00	3.00	3.00	

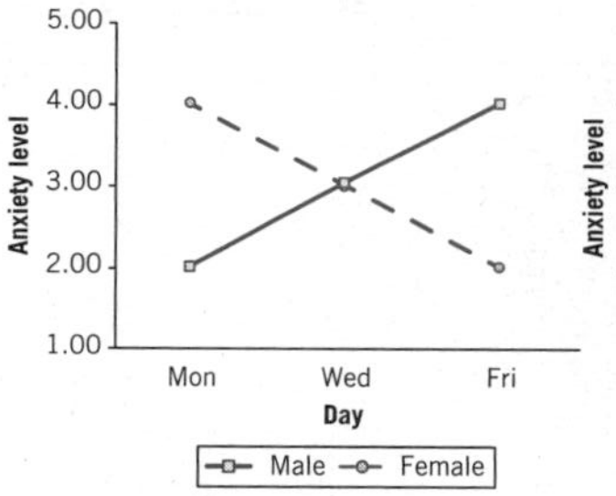

Main effect, Day? NO
Main effect, Gender? NO
Day by Gender interaction? YES

(b)

		Day			
		Mon	Wed	Fri	
Gender	Male	2.00	3.00	4.00	3.00
	Female	3.00	3.00	3.00	3.00
		2.50	3.00	3.50	

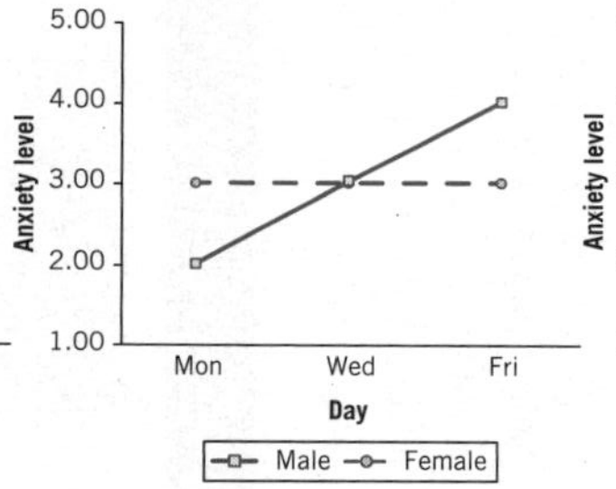

Main effect, Day? YES
Main effect, Gender? NO
Day by Gender interaction? YES

(c)

		Day			
		Mon	Wed	Fri	
Gender	Male	3.00	3.50	4.00	3.50
	Female	3.00	2.50	2.00	2.50
		3.00	3.00	3.00	

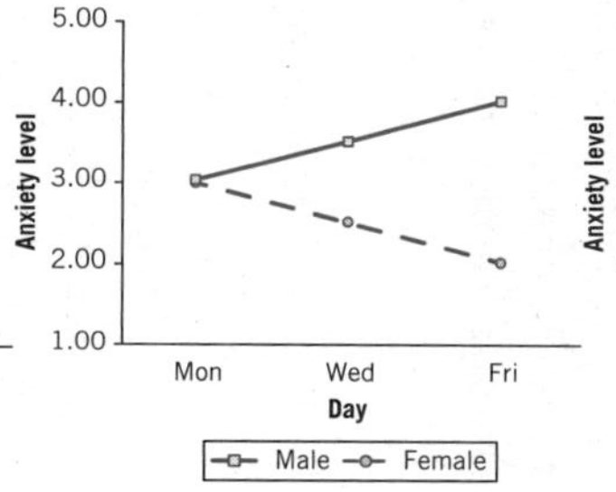

Main effect, Day? NO
Main effect, Gender? YES
Day by Gender interaction? YES

(d)

		Day			
		Mon	Wed	Fri	
Gender	Male	3.00	3.25	4.75	3.67
	Female	2.25	2.25	4.50	3.00
		2.63	2.75	4.63	

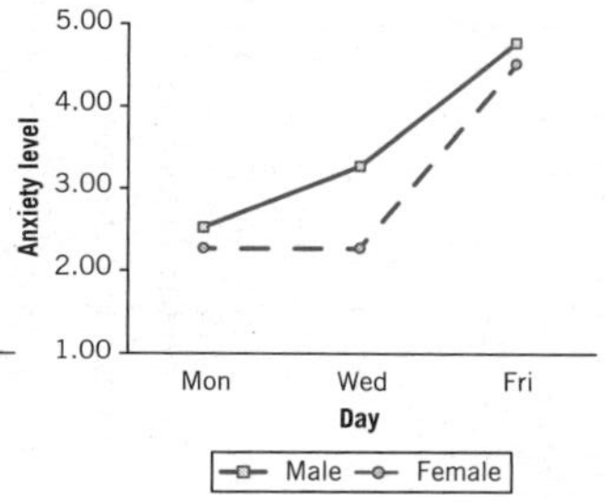

Main effect, Day? YES
Main effect, Gender? YES
Day by Gender interaction? YES

Let's first imagine there was an interaction effect between Day and Gender in the exercise deprivation study. If so, an appropriate answer to the question, "Does anxiety level change across days?" would be, "It depends on the exerciser's gender." Similarly, the question, "Do male and female exercisers differ in their anxiety levels?" would be answered by saying, "It depends on the day of exercise deprivation."

But what if an interaction effect is *not* present? Because this implies that the effect of one variable does not change across the different levels of the other variable, we would not need to consider both factors at the same time. Instead, we could think of the two factors as being distinct from each other, allowing us to shift our focus from the interaction effect to the main effects. In essence, we may now treat this design as if it were two separate single-factor designs, one for each main effect. When an interaction effect is not present, we turn our attention to the marginal means that represent the main effects.

Figure 12.6 illustrates one process for considering and analyzing factorial research designs. Note that the first question addresses the presence of an interaction effect. If an interaction effect is present, attention turns to the cell means to determine the nature of the interaction effect. If, on the other hand, the interaction effect is not present, we focus instead on the main effects. If a main effect is present, the marginal means are examined to determine which groups within an independent variable differ from each other. However, if neither of the main effects is present, the analysis of the design stops; no other comparisons are necessary. Keep in mind, however, that this is a basic strategy; the particular demands and research hypotheses contained in a study must always be taken into account.

Factorial Designs vs. Single-Factor Designs: Advantages and Disadvantages

Compared with single-factor designs, factorial research designs have several conceptual and pragmatic advantages. Conceptually, factorial research designs allow researchers to determine whether an interaction effect exists between independent variables. Pragmatically, these designs provide researchers with the opportunity to examine multiple effects in a single study. Each of these advantages is discussed below.

A critical conceptual benefit of including two or more independent variables is the opportunity to determine whether these variables interact to influence the dependent variable. This is particularly important because although a variable may have an effect on the dependent variable, this effect rarely occurs in isolation. In the

FIGURE 12.6 ● BASIC ANALYSIS STRATEGY FOR FACTORIAL RESEARCH DESIGNS

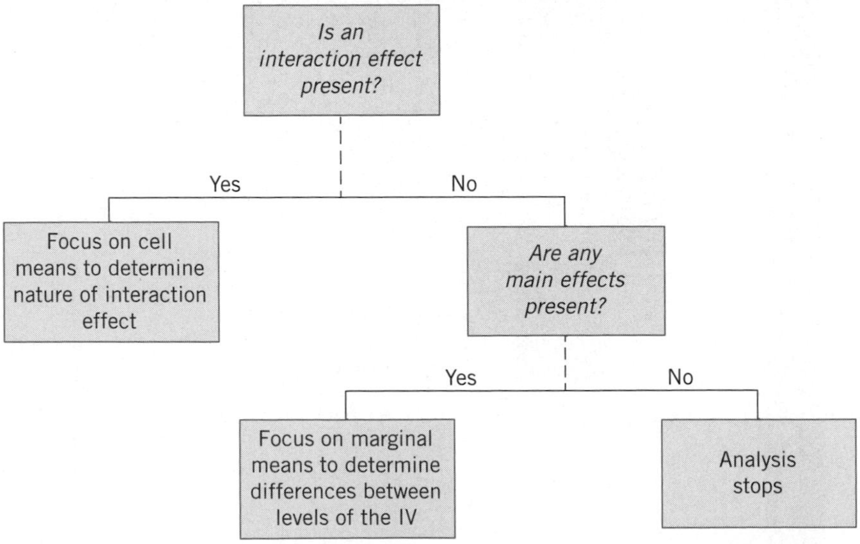

real world, a variable typically operates at the same time as other variables. Consequently, the most appropriate way to represent the complexity of phenomena studied by researchers is to include multiple variables in the same study and determine the presence of interaction effects.

Factorial research designs allow researchers not only to determine whether interaction effects are present among independent variables, but also to test the significance of each independent variable separately from the others. This leads to a second, pragmatic, advantage of factorial research designs over single-factor designs: the ability to study multiple main effects in one study. The ability to study multiple effects in one study results in savings of time and energy. To understand this, imagine that in conducting the exercise deprivation study described earlier, we had 10 participants in each of the six combinations of Day and Gender (see Figure 12.7). If so, we would test the effects of Day and Gender with a total of 60 participants. On the other hand, if we had used two separate single-factor research designs to study the effects of Day and Gender, we would have needed a total of 120 participants (60 participants in each of the two studies).

As you can see, factorial research designs have distinct advantages over their single-factor counterparts. However, two disadvantages of factorial research designs are an increase in the complexity of the calculations needed to analyze the data and an increase in the complexity of interpreting the results of the statistical analyses.

Compared with a single-factor design, analyzing factorial research designs requires analyzing the main effect of each independent variable as well as the interaction effect between independent variables. The interpretation of factorial research designs is complicated by the fact that the presence or absence of interaction effects must be determined before examining any main effects that may be present. Also, when an interaction effect is present, the researcher must appropriately analyze and interpret the interaction effect in light of the study's research hypothesis.

FIGURE 12.7 ● ILLUSTRATION OF SAMPLE SIZE IN A FACTORIAL DESIGN (10 PARTICIPANTS IN EACH COMBINATION)

		Day			
		Monday	Wednesday	Friday	
Gender	Male	10	10	10	**30**
	Female	10	10	10	**30**
		20	**20**	**20**	**60**

LEARNING CHECK 1:

Reviewing What You've Learned So Far

1. Review questions
   a. What are differences between the single-factor research design and factorial research designs?
   b. What does it mean for there to be (or *not* be) an interaction effect between independent variables?
   c. Looking at a figure of cell means, what would indicate an interaction effect is present?
   d. What is the difference between cell means and marginal means? What is represented by each of these two types of means?
   e. What is the relationship between the presence or absence of main effects and the presence or absence of interaction effects?
   f. Why does a determination of an interaction effect precede an examination of main effects?
   g. Compared with the single-factor research design, what are the advantages and disadvantages of factorial research designs?

(Continued)

(Continued)

2. Create a line graph for each of the following tables of cell means (assume the scores have a possible range of 0 to 10 and put Factor A along the X [horizontal] axis).

a.

		Factor A	
		a_1	a_2
Factor B	b_1	4	8
	b_2	4	4

b.

		Factor A	
		a_1	a_2
Factor B	b_1	7	3
	b_2	5	1

c.

		Factor A			
		a_1	a_2	a_3	a_4
Factor B	b_1	6	5	8	3
	b_2	4	3	6	1

d.

		Factor A		
		a_1	a_2	a_3
Factor B	b_1	3	3	3
	b_2	4	4	4
	b_3	6	2	8

3. In each of the following figures, is an interaction effect present or not present?

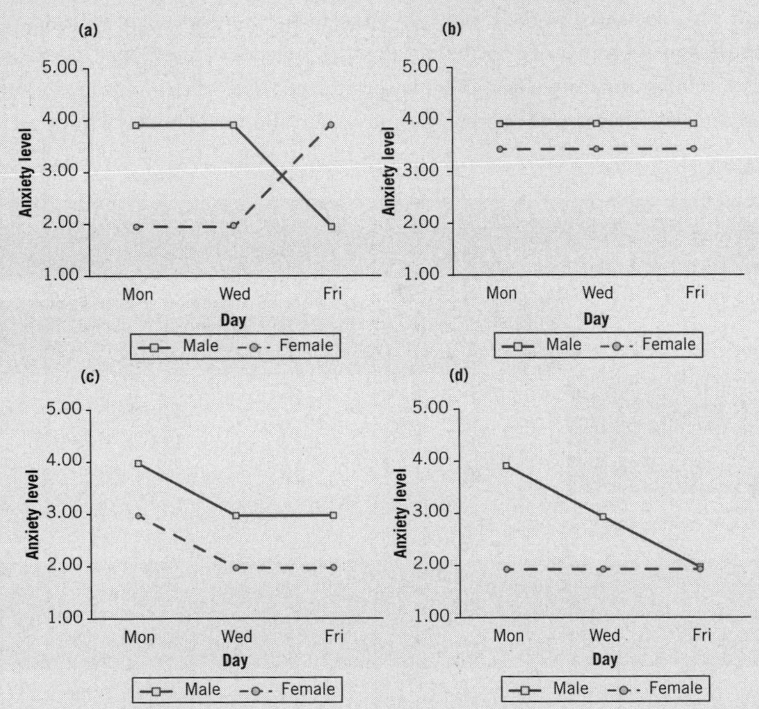

4. For the figures in Exercise 3, develop a table of cell means and marginal means and determine whether the main effects and the interaction effect are present or absent.

Testing and interpreting the effects in a factorial research design is somewhat more complicated than analyzing single-factor designs. Hopefully, this will not discourage you from using factorial research designs in your own research. The goal of this book is to help you understand the costs and benefits of different research designs and to provide you the information and skills necessary to analyze these designs effectively.

12.3 THE TWO-FACTOR (A × B) RESEARCH DESIGN

Now that the concepts of factorial research designs, main effects, and interaction effects have been introduced, let's return to the voting message study introduced earlier. This study is an example of the simplest factorial research design, one consisting of two independent variables.

A factorial research design consisting of two independent variables is referred to as a **two-factor research design** or an A × B research design (read "A-by-B"), in which "A" and "B" are represented by the numbers of levels in each of the two independent variables. For example, because Message Type (Reward vs. Threat) and Authoritarianism (Low vs. High) consist of two levels, the voting message study can be referred to as a "2 × 2 factorial design." Note that the total number of combinations in a factorial design may be calculated by multiplying the number of levels of the independent variables. In the voting message study, there are (2 × 2) or four combinations of the two independent variables.

Notational System for Components of the Two-Factor Research Design

By including a second independent variable, the two-factor research design necessitates additions to this book's notational system. First, the two independent variables will be labeled Factor A and Factor B. For the voting message study, Message Type will be called Factor A and Authoritarianism will be labeled Factor B. It doesn't matter which independent variable is given which label; however, it is important to remember each variable's label to prevent errors in later calculations. Next, the lowercase letters a and b are used to indicate the number of levels of each factor. In this example, Message Type (Factor A) consists of $a = 2$ levels (a_1 = Reward and a_2 = Threat), and Authoritarianism (Factor B) also consists of two levels (b_1 = Low and b_2 = High). The following table illustrates the designation of factors and levels of factors:

		Message type (Factor A)	
		Reward (a_1)	Threat (a_2)
Authoritarianism (Factor B)	Low (b_1)	$a_1\ b_1$	$a_2\ b_1$
	High (b_2)	$a_1\ b_2$	$a_2\ b_2$

The number of scores in each combination of the two independent variables is represented by the symbol N_{AB}. For the voting message study, we will assign eight participants to each of the four combinations, making N_{AB} equal to 8 ($N_{AB} = 8$). Table 12.2 summarizes the notational symbols for the A × B two-factor research design and illustrates these symbols using the voting message study.

Analyzing the Two-Factor Research Design: Descriptive Statistics

Although the original voting message study contained a sample of 86 participants, the study's findings will be reproduced in this chapter using a smaller sample of 32 participants, with 8 participants assigned to each of the four combinations ($N_{AB} = 8$). The ratings of message quality for the eight participants in each of the four combinations are presented in Table 12.3(a).

As was the case in analyzing single-factor designs, the first step in analyzing the two-factor research design is to calculate descriptive statistics. Table 12.3(b) provides descriptive statistics of the ratings of message quality for each combination in the voting message study. This table introduces the notation for cell means ($\overline{X}_{AB}$) and standard deviations (s_{AB}); the letters AB in the subscript indicate that these statistics represent a particular combination of Factors A and B.

Notational System for Means in the Two-Factor Research Design

To properly summarize the data and to prepare for the calculation of inferential statistics, the cell means in Table 12.3 have been used to create the tables in Table 12.4. Table 12.4(a) introduces further additions to this

TABLE 12.2 ● NOTATIONAL SYMBOLS FOR THE TWO-FACTOR RESEARCH DESIGN

Aspect of Design	Symbol(s)	Voting Message Study Example
Factors (Independent variables)	Factor A	Factor A = Message type
	Factor B	Factor B = Authoritarianism
# of levels within each factor	a, b	$a = 2$, $b = 2$
Level within a factor	a_i	a_1 = Reward, a_2 = Threat
	b_i	b_1 = Low, b_2 = High
# of combinations of factors	$a \times b$	$2 \times 2 = 4$
Combination of factors	$a_i\ b_i$	$a_1\ b_1$ = Reward/Low
		$a_1\ b_2$ = Reward/High
		$a_2\ b_1$ = Threat/Low
		$a_2\ b_2$ = Threat/High
# of scores for each combination	N_{AB}	$N_{AB} = 8$

book's notational system. In this table, the symbols $\overline{X}_A$ and $\overline{X}_B$ are used to represent the marginal means of the dependent variable for the two main effects (Factor A and Factor B, respectively). In Table 12.4(b), the marginal means represent the two main effects of Message Type and Authoritarianism. Also, the bottom right-hand corner of Table 12.4(a) contains the mean of the dependent variable for the total sample ($\overline{X}_T$). $\overline{X}_T$ is the mean of all of the data, combined across all of the cells; it can be determined by calculating the mean of the marginal means for either of the independent variables:

$$\overline{X}_T = \frac{\Sigma X_A}{a} = \frac{4.31 + 4.56}{2} = \frac{8.87}{2} = 4.44 \quad \text{or} \quad \overline{X}_T = \frac{\Sigma \overline{X}_B}{b} = \frac{4.37 + 4.50}{2} = \frac{8.87}{2} = 4.44$$

The notation for means in the two-factor research design is provided in Table 12.5.

Examining the Descriptive Statistics

Figure 12.8 provides a bar graph of the means of the combinations of the independent variables. In this graph, Authoritarianism has been placed along the X-axis, with the light- and dark-colored bars representing the different levels of Message Type (Reward and Threat). Authoritarianism has been placed along the X-axis to highlight how participants low and high in authoritarianism differ in their reaction to the two types of messages, which is the study's primary research hypothesis. In displaying the data, a bar graph has been used rather than a line graph because the independent variable Authoritarianism consists of distinct groups (Low, High) that are not on a numeric continuum.

Looking at Figure 12.8, we can begin to determine whether there is an interaction effect. It appears from the graph that the effect of message type on ratings of message quality may depend on the participant's level of authoritarianism. Low-authoritarian participants reacted more positively to the Reward message ($M = 4.62$) than the Threat message ($M = 4.13$); however, the opposite was true for the high-authoritarian participants, where the Threat mean ($M = 5.00$) was higher than the Reward mean ($M = 4.00$). The descriptive statistics provide initial support for the study's research hypothesis; however, we must still determine whether the interaction effect is statistically significant.

TABLE 12.3 ● RATING OF MESSAGE QUALITY FOR PARTICIPANTS IN THE MESSAGE VOTING STUDY

(a) Raw Data

Message Type:	Reward	Reward	Threat	Threat
Authoritarianism:	Low	High	Low	High
	4	5	3	5
	4	3	5	4
	5	4	5	6
	6	4	4	5
	4	5	3	5
	5	3	5	6
	5	3	4	4
	4	5	4	5

(b) Descriptive Statistics

Message Type:	Reward	Reward	Threat	Threat
Authoritarianism:	Low	High	Low	High
N_{AB}	8	8	8	8
Mean ($\bar{X}_{AB}$)	4.62	4.00	4.13	5.00
Standard deviation (s_{AB})	.74	.93	.83	.76

TABLE 12.4 ● TABLE OF MEANS FOR THE TWO-FACTOR RESEARCH DESIGN, VOTING MESSAGE STUDY

(a) Notation

Factor A

		a_1	a_2	
Factor B	b_1	$\bar{X}_{AB}$	$\bar{X}_{AB}$	$\bar{X}_B$
	b_2	$\bar{X}_{AB}$	$\bar{X}_{AB}$	$\bar{X}_B$
		$\bar{X}_A$	$\bar{X}_A$	$\bar{X}_T$

(b) Voting Message Study

Message type

		Reward	Threat	
Authoritarianism	Low	4.62	4.13	4.37
	High	4.00	5.00	4.50
		4.31	4.56	4.44

TABLE 12.5 ● NOTATION FOR MEANS IN THE TWO-FACTOR RESEARCH DESIGN

Aspect of Design	Notation	Voting Message Study Example
Level within a factor	$\bar{X}_A$	$\bar{X}_A = 4.31$, $\bar{X}_A = 4.56$
	$\bar{X}_B$	$\bar{X}_B = 4.37$, $\bar{X}_B = 4.50$
Combination of factors	$\bar{X}_{AB}$	$\bar{X}_{AB} = 4.62$, $\bar{X}_{AB} = 4.00$
		$\bar{X}_{AB} = 4.13$, $\bar{X}_{AB} = 5.00$
Total sample	$\bar{X}_T$	$\bar{X}_T = 4.44$

FIGURE 12.8 ● BAR GRAPH OF MESSAGE QUALITY BY MESSAGE TYPE AND AUTHORITARIANISM IN THE VOTING MESSAGE STUDY

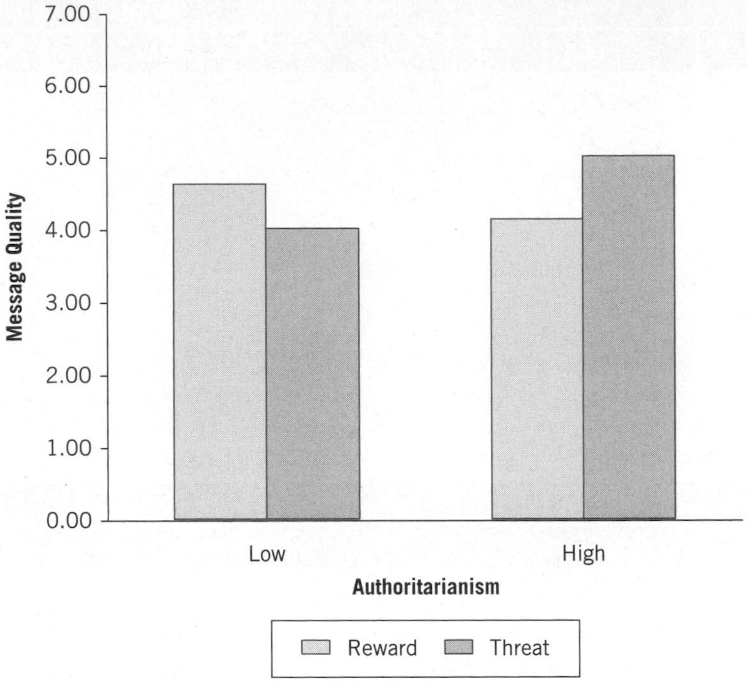

In addition to the interaction effect, it's useful to examine the descriptive statistics to understand the main effects. Looking back at Table 12.4(b), the mean ratings for the two message types (4.31 and 4.56) are somewhat similar to one another. This suggests that, when combined over the two levels of Authoritarianism, the two message types do not appear to greatly differ in their ratings of message quality. A small difference also exists for the main effect of Authoritarianism (4.37 vs. 4.50). This indicates that, when message type is ignored, participants with low and high levels of authoritarianism rate the quality of the messages in a similar manner. As we mentioned earlier, however, it is important to remember that the interpretation of main effects depends on whether an interaction effect is present.

12.4 INTRODUCTION TO ANALYSIS OF VARIANCE (ANOVA) FOR THE TWO-FACTOR RESEARCH DESIGN

Once the descriptive statistics have been calculated and examined, the next step is to calculate inferential statistics designed to test the study's research hypotheses. To introduce the analysis of the two-factor research design, let's return to the one-way ANOVA covered in Chapter 11, which tested a single-factor research design consisting of one independent variable. Conducting the one-way ANOVA involves calculating a statistic known as the *F*-ratio, which is the ratio of two types of variance: between-group variance and within-group variance.

The numerator of the *F*-ratio, between-group variance (MS_{BG}), refers to differences among the means of the groups comprising an independent variable. This variance is attributable to two factors: the hypothesized effect of the independent variable as well as factors not taken into account by the researcher, referred to as "error." The denominator of the F-ratio, within-group variance (MS_{WG}), is the variability of the dependent variable within the groups comprising an independent variable. It represents variance in the dependent variable not explained or accounted for by the independent variable and therefore represents error. To summarize:

$$F = \frac{\text{between} - \text{group variance}}{\text{within} - \text{group variance}}$$
$$= \frac{\text{effect} + \text{error}}{\text{error}}$$
$$= \frac{MS_{BG}}{MS_{WG}}$$

F-Ratios in the Two-Factor Research Design

As with the single-factor design, analyzing the two-factor research design involves calculating the F-ratio statistic, a statistic that contains between-group and within-group variance. Before analyzing the two-factor research design, however, we must redefine the term *group*. Rather than being a level of one independent variable, in the two-factor research design, a group represents a *combination* of two independent variables. For example, in the voting message study, one group is the Low/Reward combination, which consists of the low-authoritarian participants who receive the Reward message.

In the voting message study, between-group variance refers to differences in the ratings of message quality among the four combinations of Message Type and Authoritarianism. Why might these four combinations differ from each other? As it turns out, differences between the combinations may be due to three factors: the effect of Message Type (Reward vs. Threat), the effect of Authoritarianism (Low vs. High), and the Message Type by Authoritarianism interaction effect. In other words, between-group variance in the two-factor research design consists of three effects: the main effect of Factor A, the main effect of Factor B, and the A × B interaction effect. This may be expressed using the following formula:

$$\begin{aligned}\text{Between-group variance} &= \text{main effect of Factor A}\\ &+ \text{main effect of Factor B}\\ &+ \text{A} \times \text{B interaction effect}\end{aligned}$$

To fully understand between-group variance in the two-factor research design, we must analyze each of these three effects.

What is *within*-group variance in the two-factor research design? In Chapter 11, within-group variance was based on the variability of scores within the groups that comprise a single independent variable. However, because a group in the two-factor research design is a combination of two independent variables, within-group variance in this design is based on the variability of scores within the different combinations of the two factors.

To summarize, analyzing the two-factor research design involves determining the statistical significance of three sources of between-group variance: the main effect of Factor A, the main effect of Factor B, and the A × B interaction effect. This is accomplished by calculating three F-ratios, each of which divides its source of between-group variance by within-group variance, which is the variability of scores within the different combinations of the two factors. The next section discusses the steps needed to calculate and interpret these F-ratios.

12.5 INFERENTIAL STATISTICS: TWO-WAY ANALYSIS OF VARIANCE (ANOVA)

Although the descriptive statistics in Table 12.3(b) appear to indicate support for the voting message study's research hypothesis, we must still conduct inferential statistical analyses to determine whether any observed effects are statistically significant. These analyses are conducted using essentially the same steps as in earlier chapters:

- state the null and alternative hypotheses (H_0 and H_1),
- make decisions about the null hypotheses,
- draw conclusions from the analyses, and
- relate the results of the analyses to the research hypothesis.

The main difference between these steps and those presented in earlier chapters is the use of plural nouns rather than singular. For example, rather than "make a *decision* about the null *hypothesis,*" we must now "make *decisions* about the null *hypotheses*" for the two main effects and the interaction effect.

State the Null and Alternative Hypotheses (H_0 and H_1)

In the two-way ANOVA, three effects are tested: the main effect of Factor A, the main effect of Factor B, and the A × B interaction effect. Each of these three effects has its own set of null and alternative hypotheses.

Main Effect of Factor A

For the main effect of Message Type in the voting message study, we are testing whether ratings of message quality are different for those exposed to the Reward message versus the Threat message, ignoring the person's level of authoritarianism. In other words, Factor A is treated as if it were a single-factor design. For that reason, the statistical hypotheses for Factor A look exactly the same as in the one-way ANOVA discussed in Chapter 11:

$$H_0\text{: all } \mu\text{s are equal}$$

$$H_1\text{: not all } \mu\text{s are equal}$$

Main Effect of Factor B

Similar to the main effect of Factor A, testing the main effect for Factor B involves ignoring the other factor. In this example, testing the main effect of Authoritarianism determines whether ratings of message quality are different for people low versus high in authoritarianism, ignoring the type of message the person receives. Because Factor B is also treated as if it were a single-factor design, the null and alternative hypotheses are the same as for the main effect of Factor A:

$$H_0\text{: all } \mu\text{s are equal}$$

$$H_1\text{: not all } \mu\text{s are equal}$$

A × B Interaction Effect

The A × B interaction effect, unlike the main effects, takes into consideration the joint influence of the two factors. Therefore, the statistical hypotheses for the A × B interaction effect are different from those for the main effects:

$$H_0\text{: an interaction effect does not exist}$$

$$H_1\text{: an interaction effect exists}$$

The null hypothesis for the interaction effect states that no interaction effect exists between the two independent variables. In the voting message study, this implies that the effect of message type on ratings of message quality does not depend on the person's level of authoritarianism. The alternative hypothesis, mutually exclusive from the null hypothesis, states that some type of interaction effect does in fact exist. You may have noticed that the alternative hypothesis does not specify the precise nature of the interaction effect. As we illustrated in Figure 12.2, a significant interaction effect can take on a wide variety of different shapes. For this reason, it would be inappropriate to state any particular form of the interaction in the alternative hypothesis.

Make Decisions About the Null Hypotheses

Once the three sets of null and alternative hypotheses have been stated, the next step is to make decisions regarding whether to reject any of the null hypotheses. These decisions are made using the following steps:

- calculate the degrees of freedom (*df*);
- set alpha (α), identify the critical values, and state decision rules;
- calculate statistics: *F*-ratios for the two-way ANOVA;
- make decisions whether to reject the null hypotheses;
- determine the levels of significance; and
- calculate measures of effect size.

Each of these steps is discussed below; within each step, the main effects and the interaction effect are discussed separately.

Calculate the Degrees of Freedom (df)

The first step in calculating and evaluating the *F*-ratios in a two-way ANOVA is to calculate the appropriate degrees of freedom (*df*). In the two-factor research design, there are four degrees of freedom that correspond to the three effects comprising between-group variance (main effect of Factor A, main effect of Factor B, and the A × B interaction effect) and within-group variance.

Main effect of Factor A (df_A). The number of degrees of freedom for the main effect of Factor A is essentially the same as the degrees of freedom for the single-factor design. The degrees of freedom for Factor A (df_A) is the following:

$$df_A = a - 1 \tag{12-1}$$

where a is the number of levels of Factor A. In this example, Factor A (Message Type) consists of two levels (Reward and Threat). Therefore, df_A is equal to

$$\begin{aligned} df_A &= a - 1 \\ &= 2 - 1 \\ &= 1 \end{aligned}$$

Main effect of Factor B (df_B). The formula for the degrees of freedom associated with Factor B (df_B) is a simple modification of that for df_A:

$$df_B = b - 1 \tag{12-2}$$

where b is the number of levels of Factor B. Authoritarianism, which has been labeled Factor B, consists of two levels: Low and High. We may then conclude that

$$\begin{aligned} df_{B} &= b - 1 \\ &= 2 - 1 \\ &= 1 \end{aligned}$$

A × B interaction effect ($df_{A \times B}$). The number of degrees of freedom for the A × B interaction effect is the product of the degrees of freedom for each of the two factors. More specifically, we may state the following:

$$df_{A \times B} = (a - 1)\,(b - 1) \tag{12-3}$$

where a is the number of levels of Factor A and b is the number of levels of Factor B. Explaining the precise logic behind the formula for $df_{A \times B}$ is beyond the framework of this book and will not be described here. (See Keppel & Wickens [2004] for a detailed explanation.) Because both Message Type and Authoritarianism in the voting message study consist of two groups ($a = 2$ and $b = 2$), $df_{A \times B}$ may be formulated as

$$\begin{aligned} df_{A \times B} &= (a - 1)\,(b - 1) \\ &= (2 - 1)\,(2 - 1) = (1)\,(1) \\ &= 1 \end{aligned}$$

Within-group variance (df_{WG}). Finally, we must determine the degrees of freedom associated with within-group variance. In the single-factor design, the degrees of freedom (df_{WG}) was the number of scores in each level of the independent variable minus 1 ($N_i - 1$) multiplied by the number of levels of the independent variable. In the two-factor research design, the degrees of freedom for within-group variance is the number of scores in each *combination* of the two independent variables minus 1 ($N_{AB} - 1$) multiplied by the number of *combinations* of the two independent variables. Therefore, the formula for df_{WG} is the following:

$$df_{WG} = (a)\,(b)\,(N_{AB} - 1) \tag{12-4}$$

where a is the number of levels of Factor A, b is the number of levels of Factor B, and N_{AB} is the number of scores in each A × B combination.

In the voting message study, Message Type and Authoritarianism both consist of two levels ($a = 2$, $b = 2$), with eight scores in each combination of the two variables ($N_{AB} = 8$). Therefore, df_{WG} is equal to

$$\begin{aligned} df_{WG} &= (a)\,(b)\,(N_{AB} - 1) \\ &= (2)\,(2)\,(8 - 1) = (4)\,(7) \\ &= 28 \end{aligned}$$

Set Alpha (α), Identify the Critical Values, and State Decision Rules

As in earlier chapters, in the two-way ANOVA alpha is traditionally set at .05 ($\alpha = .05$). Once alpha has been determined, the critical values of the F-ratio for the two main effects and the interaction effect may be identified.

Main effect of Factor A. For the main effect of Factor A, the two degrees of freedom associated with the between-group and within-group variance of its F-ratio are df_{A} and df_{WG}. For the voting message study, these degrees of freedom have been calculated as $df_{A} = 1$ and $df_{WG} = 28$. Using the table of critical values for the

F-ratio, we move to the 1 column under "Degrees of freedom for Numerator" and then down the rows of the table until we reach the row corresponding to 28 for the "Degrees of freedom for Denominator." For $\alpha = .05$, we find a critical value of 4.20. Consequently, the critical value for the main effect of Message Type may be stated as follows:

For $\alpha = .05$ and $df = 1, 28$, critical value = 4.20

Using the identified critical value of 4.20, the following decision rule may be stated for the main effect of Message Type:

If $F_A > 4.20$, reject H_0; otherwise, do not reject H_0

The critical value and regions of rejection and nonrejection for this main effect are illustrated in Figure 12.9.

Main effect of Factor B. The two degrees of freedom associated with the main effect of Factor B are df_B and df_{WG}. For the voting message study, these have been calculated as $df_B = 1$ and $df_{WG} = 28$. Therefore, the critical value for the main effect of Authoritarianism may be stated as

For $\alpha = .05$ and $df = 1, 28$, critical value = 4.20

For the main effect of Authoritarianism, the decision rule regarding the null hypothesis may be stated as

If $F_B > 4.20$, reject H_0; otherwise, do not reject H_0

A × B interaction effect. For the interaction effect between Factor A and Factor B, the two degrees of freedom are $df_{A \times B}$ and df_{WG}. For the voting message study, these degrees of freedom are $df_{A \times B} = 1$ and $df_{WG} = 28$. Consequently, the critical value for the Message Type × Authoritarianism interaction effect may be stated as

For $\alpha = .05$ and $df = 1, 28$, critical value = 4.20

FIGURE 12.9 ● CRITICAL VALUE AND REGIONS OF REJECTION AND NONREJECTION FOR THE MAIN EFFECT OF MESSAGE TYPE IN THE VOTING MESSAGE STUDY

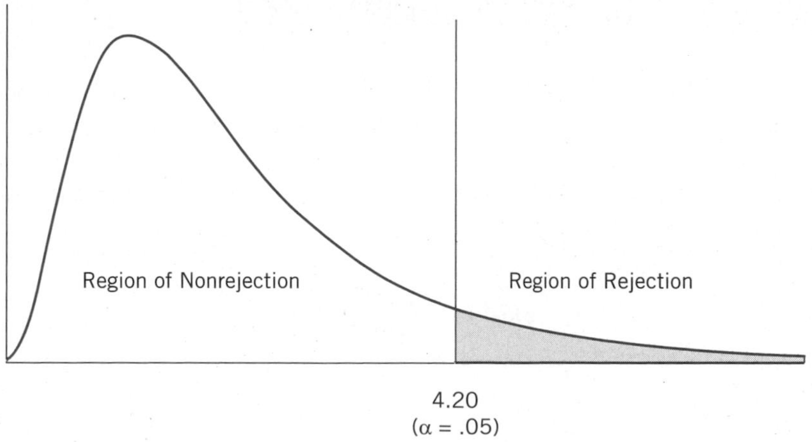

Using the critical value of 4.20 for the Message Type × Authoritarianism interaction effect, the decision rule regarding the null hypothesis is

$$\text{If } F_{\text{A} \times \text{B}} > 4.20, \text{ reject H}_0\text{; otherwise, do not reject H}_0$$

In this example, which is a 2 × 2 design, the degrees of freedom for the numerator of the two main effects (df_{A} and df_{B}) and the A × B interaction effect ($df_{\text{A} \times \text{B}}$) all happen to be equal to 1. This makes the identification of the critical values a relatively straightforward matter. However, what if we were using a 3 × 4 research design such as the one illustrated in Table 12.6? Assuming there are three scores in each A × B combination ($N_{\text{AB}} = 3$), Table 12.6 calculates the degrees of freedom and identifies the critical value for the two main effects and the interaction effect. Note that the three critical values in this table are all different from each other. For this reason, it is essential to correctly calculate the appropriate degrees of freedom for each effect.

Calculate Statistics: F-Ratios for the Two-Way ANOVA

The next step in making decisions about the null hypotheses is to calculate values of an inferential statistic. Conducting a two-way ANOVA requires calculating the *F*-ratio statistic, in which a source of between-group variance is divided by within-group variance:

$$F = \frac{\text{between-group variance}}{\text{within-group variance}}$$

Three *F*-ratios are calculated as part of the two-way ANOVA, one for each source of between-group variance: main effect of Factor A (F_{A}), main effect of Factor B (F_{B}), and the A × B interaction effect ($F_{\text{A} \times \text{B}}$). In this section, we first calculate four types of variance (three related to between-group variance and one representing within-group variance) before calculating values for the three *F*-ratios.

Calculate the mean squares (MS). The first step in conducting a two-way ANOVA is to calculate the mean squares (*MS*) that comprise the three *F*-ratios. This section discusses how to calculate mean squares for the main effect of Factor A (MS_{A}), main effect of Factor B (MS_{B}), A × B interaction effect ($MS_{\text{A} \times \text{B}}$), and within-group variance (MS_{WG}).

Main effect of Factor A (MS_A). One source of between-group variance in the two-factor research design is associated solely with Factor A, without considering Factor B and the A × B interaction effect. The between-group variance associated with Factor A (MS_{A}) is calculated as follows:

$$MS_{\text{A}} = \frac{SS_{\text{A}}}{df_{\text{A}}} = \frac{(b)(N_{\text{AB}})[\Sigma(\bar{X}_{\text{A}} - \bar{X}_{\text{T}})^2]}{a - 1} \tag{12-5}$$

where b is the number of levels of Factor B, N_{AB} is the number of scores within each A × B combination, $\bar{X}_{\text{A}}$ is the mean for each level of Factor A, $\bar{X}_{\text{T}}$ is the mean for the total sample, and a is the number of levels of Factor A.

Although it may not be apparent, the formula for MS_{A} is actually very similar to the formula for MS_{BG} in the one-way ANOVA (Formula 11-3):

$$MS_{\text{BG}} = \frac{N_i \Sigma(\bar{X}_i - \bar{X}_T)^2}{\#\,\text{groups} - 1}$$

TABLE 12.6 DEGREES OF FREEDOM AND CRITICAL VALUES FOR 3 × 4 DESIGN (N_{AB} = 3 AND α = .05)

		Factor A		
		a_1	a_2	a_3
	b_1			
Factor B	b_2			
	b_3			
	b_4			

(a) Calculate the degrees of freedom (*df*)

(1) Main effect of Factor A

$df_A = a - 1 = 3 - 1 = \mathbf{2}$

(2) Main effect of Factor B

$df_B = b - 1 = 4 - 1 = \mathbf{3}$

(3) A × B interaction effect

$df_{A \times B} = (a - 1)(b - 1) = (3 - 1)(4 - 1) = (2)(3) = \mathbf{6}$

(4) Within-group variance

$df_{WG} = (a)(b)(N_{AB} - 1) = (3)(4)(3 - 1) = (12)(2) = \mathbf{24}$

(b) Set alpha (α), identify the critical values, and state decision rules

(1) Main effect of Factor A

For α = .05 and *df* = 2, 24, critical value = **3.40**

(2) Main effect of Factor B

For α = .05 and *df* = 3, 24, critical value = **3.01**

(3) A × B interaction effect

For α = .05 and *df* = 6, 24, critical value = **2.51**

LEARNING CHECK 2:

Reviewing What You've Learned So Far

1. Review questions
   a. What do the letters A and B represent in an "A × B factorial research design"?
   b. In the two-factor research design, what is between-group variance composed of?
   c. In the two-factor research design, what is within-group variance?
   d. Why are there three sets of statistical hypotheses in a two-factor research design?
   e. Why don't the null and alternative hypotheses for the interaction effect specify the nature of the interaction effect?
   f. Why are the degrees of freedom (*df*), critical values, and decision rules for main effects similar to those in the single-factor research design?
2. For each of the research situations below, identify the two independent variables and indicate the type of A × B design (e.g., 2 × 2, 3 × 3).
   a. An instructor teaches two courses: one taught in the typical classroom setting and one taught online. She believes students taking the classroom course are more likely to keep up with assigned course readings than students in the online course; furthermore, she believes this difference is greater for students living off-campus than on-campus.
   b. When a child misbehaves, who is responsible for the child's bad behavior? A family therapist believes that parents are more likely to attribute responsibility to the child than to themselves, particularly when the child is a teenager rather than an adolescent.
   c. Before taking an exam, do you engage in rituals like kissing a luck charm or wearing a particular piece of clothing? Two researchers (Rudski & Edwards, 2007) hypothesized that students engage in rituals more when an exam is important (vs. unimportant) to students' grade in the course, as well as when an exam was seen as "very difficult" (as opposed to "easy" or "somewhat difficult").
3. For each of the following tables of cell means, calculate marginal means ($\bar{X}_A$ and $\bar{X}_B$) and the total mean ($\bar{X}_T$).

a.

Factor B	Factor A: a_1	a_2
b_1	3.00	4.00
b_2	2.00	2.00

b.

Factor B	Factor A: a_1	a_2
b_1	8.50	6.50
b_2	7.00	5.00
b_3	11.00	9.00

c.

Factor B	Factor A: a_1	a_2	a_3
b_1	36.50	41.25	38.00
b_2	39.00	46.75	33.75
b_3	43.25	52.00	55.50

d.

Factor B	Factor A: a_1	a_2	a_3	a_4
b_1	1.53	2.10	2.49	2.36
b_2	2.27	2.19	1.93	1.87

4. For each of the following, calculate four degrees of freedom (df_A, df_B, $df_{A \times B}$, df_{WG}) and identify the three critical values of the three *F*-ratios (F_A, F_B, $F_{A \times B}$) (assume $\alpha = .05$).
   a. $a = 2, b = 2, N_{AB} = 5$
   b. $a = 3, b = 2, N_{AB} = 6$
   c. $a = 3, b = 3, N_{AB} = 10$
   d. $a = 4, b = 3, N_{AB} = 7$

The numerator in both formulas is based on the difference between the mean of a level of an independent variable and the total mean: $\bar{X}_i - \bar{X}_T$ for the one-way ANOVA and $\bar{X}_A - \bar{X}_T$ for Factor A in the two-way ANOVA. The critical change in analyzing the main effect of Factor A is the addition of the symbol *b* in the numerator, which takes into account the fact that this design includes a second independent variable (Factor B).

To calculate MS_A for the main effect of Message Type, *b* (the number of levels of Authoritarianism) is equal to 2 and N_{AB} is equal to 8; the marginal means for the two levels of Message Type $(\bar{X}_A)$, as well as the total mean $(\bar{X}_T)$, are calculated in Table 12.4(b), and *a* (the number of levels of Message Type) is equal to 2. Using this information, the calculation of MS_A is provided below:

$$\begin{aligned} MS_A &= \frac{(b)(N_{AB})[\Sigma(\bar{X}_A - \bar{X}_T)^2]}{a-1} \\ &= \frac{(2)(8)[(4.31-4.44)^2 + (4.56-4.44)^2]}{2-1} \\ &= \frac{16[(-.13)^2 + (.12)^2]}{1} \\ &= \frac{16(.02+.01)}{1} = \frac{16(.03)}{1} = \frac{.48}{1} \\ &= .48 \end{aligned}$$

Main effect of Factor B (MS_B). The second effect tested in the two-way ANOVA is the main effect of Factor B. The formula for the mean squares for Factor B (MS_B) is a simple modification of the formula for MS_A, substituting the letter A for the letter B, and vice versa:

$$MS_B = \frac{SS_B}{df_B}\frac{(a)(N_{AB})[\Sigma(\bar{X}_B - \bar{X}_T)^2]}{b-1} \quad (12\text{-}6)$$

where *a* is the number of levels of Factor A, N_{AB} is the number of scores within each A × B combination, $\bar{X}_B$ is the mean for each level of Factor B, $\bar{X}_T$ is the mean for the total sample, and *b* is the number of levels of Factor B.

For the voting message study, Factor B represents Authoritarianism. Based on the information used to calculate MS_A, as well as the marginal means for low and high Authoritarianism $(\bar{X}_B)$ from Table 12.4(b), MS_B is calculated as follows:

$$\begin{aligned} MS_B &= \frac{(a)(N_{AB})[\Sigma(\bar{X}_B - \bar{X}_T)^2]}{b-1} \\ &= \frac{(2)(8)[(4.37-4.44)^2 + (4.50-4.44)^2}{2-1} \\ &= \frac{16[(-.07)^2 + (.06)^2]}{1} \\ &= \frac{16(.005+.004)}{1} = \frac{16(.009)}{1} = \frac{.14}{1} \\ &= .14 \end{aligned}$$

A × B interaction effect ($MS_{A \times B}$). How do you calculate the between-group variance associated with the A × B interaction effect? As we've learned, between-group variance in the two-factor research design consists of three sources: the main effect of Factor A, the main effect of Factor B, and the A × B interaction effect. In terms of the Sum of Squares (*SS*), the composition of between-group variance may be expressed using the following formula:

$$SS_{BG} = SS_A + SS_B + SS_{A \times B}$$

With a little algebraic manipulation, the sums of squares for the A × B interaction ($SS_{A \times B}$) can be represented by the following:

$$SS_{A \times B} = SS_{BG} - SS_A - SS_B$$

Therefore, the sums of squares for the A × B interaction ($SS_{A \times B}$) is equal to the total amount of between-group variance (SS_{BG}) minus that associated with the two main effects (SS_A and SS_B).

Based on the above formula, calculating $SS_{A \times B}$ requires first calculating SS_{BG}. Formula 12-7 provides the computational formula for SS_{BG}:

$$SS_{BG} = N_{AB}\Sigma(\overline{X}_{AB} - \overline{X}_T)^2 \quad (12\text{-}7)$$

where N_{AB} is the number of scores within each A × B combination, $\overline{X}_{AB}$ is the mean for each combination, and $\overline{X}_T$ is the mean for the total sample. Formula 12-7 is actually very similar to the formula for between-group variance in the single-factor design discussed in Chapter 11 ($SS_{BG} = N_i\Sigma(\overline{X}_i - \overline{X}_T)^2$). In both situations, between-group variance is based on the difference between the mean of a group and the total mean.

The above formula for SS_{BG} can be used to help create the following formula for $MS_{A \times B}$:

$$MS_{A \times B} = \frac{SS_{A \times B}}{df_{A \times B}} \frac{[N_{AB}\Sigma(\overline{X}_{AB} - \overline{X}_T)^2] - SS_A - SS_B}{(a-1)(b-1)} \quad (12\text{-}8)$$

where N_{AB} is the number of scores within each A × B combination, $\overline{X}_{AB}$ is the mean for each combination of the two factors, $\overline{X}_T$ is the mean for the total sample, SS_A is the sums of squares for Factor A, SS_B is the sums of squares for Factor B, a is the number of levels of Factor A, and b is the number of levels of Factor B. Although it may not be obvious, $SS_{A \times B}$, which is the numerator of $MS_{A \times B}$, is equal to $SS_{BG} - SS_A - SS_B$, meaning that the variance associated with the interaction effect is what remains after the variance for the two main effects is subtracted from the total amount of between-group variance.

Much of the information needed to calculate $MS_{A \times B}$ can be obtained from the earlier calculation of MS_A and MS_B, and the values for $\overline{X}_{AB}$ are found in the descriptive statistics calculated in Table 12.3(b). In the voting message study, $MS_{A \times B}$ for the Message Type × Authoritarianism interaction effect is calculated as follows:

$$\begin{aligned}
MS_{A \times B} &= \frac{[N_{AB}\Sigma(\overline{X}_{AB} - \overline{X}_T)^2] - SS_A - SS_B}{(a-1)(b-1)} \\
&= \frac{8[(4.62-4.44)^2 + (4.00-4.44)^2 + (4.13-4.44)^2 + (5.00-4.44)^2] - .48 - .14}{(2-1)(2-1)} \\
&= \frac{8[(.18)^2 + (-.44)^2 + (-.31)^2 + (.56)^2] - .48 - .14}{1} \\
&= \frac{8[(.03 + .19 + .10 + .31] - .48 - .14}{1} \\
&= \frac{8(.63) - .48 - .14}{1} = \frac{5.04 - .48 - .14}{1} = \frac{4.42}{1} \\
&= 4.42
\end{aligned}$$

Within-group variance (MS_{WG}). Within-group variance is an estimate of error variance and is based on the variability of scores within the combinations of Factors A and B. The formula for within-group variance for the two-factor research design (MS_{WG}) is provided in Formula 12-9:

$$MS_{WG} = \frac{SS_{WG}}{df_{WG}} = \frac{(N_{AB}-1)\Sigma s^2{}_{AB}}{(a)(b)(N_{AB}-1)} \quad (12\text{-}9)$$

where N_{AB} is the number of scores in each combination, s_{AB} is the standard deviation for each combination, a is the number of levels of Factor A, and b is the number of levels of Factor B.

Consistent with between-group variance, the formula for MS_{WG} in the two-factor research design is very similar to the formula for MS_{WG} in the one-way ANOVA (Formula 11-5):

$$MS_{WG} = \frac{(N_i - 1)\Sigma s_i^2}{\Sigma(N_i - 1)}$$

Within-group variance in both types of research designs is based on the variability of scores within each group, represented by the standard deviation (s).

For the voting message study, the values for N_{AB}, a, and b can be found in earlier calculations; the standard deviations for the combinations are found in the descriptive statistics (Table 12.3(b)). Using this information, MS_{WG} for the voting message study can be stated in the following way:

$$\begin{aligned} MS_{WG} &= \frac{(N_{AB} - 1)\Sigma s_{AB}^2}{(a)(b)(N_{AB} - 1)} \\ &= \frac{(8-1)[(.74)^2 + (.93)^2 + (.83)^2 + (.76)^2]}{(2)(2)(8-1)} \\ &= \frac{7[.55 + .86 + .69 + .58]}{28} = \frac{7(2.68)}{28} = \frac{18.76}{28} \\ &= .67 \end{aligned}$$

Calculate the F-ratios (F). Once the four variances have been calculated, the next step is to calculate three F-ratios, one for each of the two main effects and one for the interaction effect. These F-ratios are calculated by dividing a particular source of between-group variance by the within-group variance.

Main effect of Factor A (F_A). To calculate the F-ratio for the main effect of Factor A (F_A), we divide the between-group variance associated with this effect (MS_A) by within-group variance (MS_{WG}):

$$F_A = \frac{MS_A}{MS_{WG}} \qquad (12\text{-}10)$$

For the voting message study, the F-ratio for the main effect of Message Type is equal to

$$\begin{aligned} F_A &= \frac{MS_A}{MS_{WG}} \\ &= \frac{.48}{.67} \\ &= .72 \end{aligned}$$

Main effect of Factor B (F_B). The F-ratio for Factor B (F_B) is calculated in a similar manner as the main effect of Factor A:

$$F_B = \frac{MS_B}{MS_{WG}} \qquad (12\text{-}11)$$

The F-ratio for the main effect of Authoritarianism in the voting message study is equal to

$$\begin{aligned} F_B &= \frac{MS_B}{MS_{WG}} \\ &= \frac{.14}{.67} \\ &= .21 \end{aligned}$$

A × B interaction effect $(F_{A \times B})$. Finally, the F-ratio for the A × B interaction effect $(F_{A \times B})$ is calculated using the following formula:

$$F_{A \times B} = \frac{MS_{A \times B}}{MS_{WG}} \tag{12-12}$$

For the Message Type × Authoritarianism interaction effect, the F-ratio is equal to

$$\begin{aligned} F_{A \times B} &= \frac{MS_{A \times B}}{MS_{WG}} \\ &= \frac{4.42}{.67} \\ &= 6.60 \end{aligned}$$

The symbols and computational formulas for a two-way ANOVA are provided in the ANOVA summary tables in Table 12.7. The ANOVA summary table in Table 12.7(c) is the one you would use to report the results of the analysis for the voting message study. As in our discussion of the one-way ANOVA in Chapter 11, the word *Error* is used to represent within-group variance as this is variance in the dependent variable not explained or accounted for by the independent variables and is therefore attributed to random, chance factors.

Make Decisions Whether to Reject the Null Hypotheses

Once values for the three F-ratios (F_A, F_B, $F_{A \times B}$) have been calculated, the next step is to make the decision whether to reject the null hypothesis for each of the three effects. Each of these decisions for the voting message study is described below.

Main effect of Factor A. The decision of whether to reject the null hypothesis regarding the main effect of Factor A is made by comparing the calculated value of F_A with its critical value. For the main effect of Message Type, we may state that

$$F_A = .72 < 4.20 \therefore \text{ do not reject } H_0 \ (p > .05)$$

Because the calculated value of .72 for the main effect of Message Type is less than the critical value of 4.20, it falls in the region of nonrejection (see Figure 12.10) and the null hypothesis is not rejected. Therefore, the main effect of Message Type is not significant, meaning that the message quality ratings of those receiving the Reward ($M = 4.31$) and Threat ($M = 4.56$) messages (not taking into account the level of authoritarianism) do not significantly differ.

Main effect of Factor B. The decision regarding the statistical significance of the main effect of Factor B is made by comparing the value of F_B with its critical value. In the voting message study, the decision regarding the main effect of Authoritarianism is represented as follows:

$$F_B = .21 < 4.20 \therefore \text{ do not reject } H_0 \ (p > .05)$$

Because the null hypothesis for Factor B was not rejected, we may conclude that, ignoring the type of message, the message quality ratings for the High ($M = 4.37$) and Low ($M = 4.50$) authoritarianism participants are not significantly different.

A × B interaction effect. In this step, we make the decision regarding whether there is a significant interaction effect between Factor A and Factor B. In the voting message study, we may state the following:

$$F_{A \times B} = 6.60 > 4.20 \therefore \text{ reject } H_0 \ (p < .05)$$

TABLE 12.7 ● SUMMARY TABLES FOR THE TWO-WAY ANOVA

a. Notation and symbols

Source	*SS*	*df*	*MS*	*F*
Factor A (A)	SS_A	df_A	MS_A	F_A
Factor B (B)	SS_B	df_B	MS_B	F_B
A × B interaction (A × B)	$SS_{A \times B}$	$df_{A \times B}$	$MS_{A \times B}$	$F_{A \times B}$
Within-group (WG)	SS_{WG}	df_{WG}	MS_{WG}	
Total (T)	SS_T	df_T		

b. Formulas

Source	*SS*	*df*	*MS*	*F*
Factor A	$(b)(N_{AB})[\Sigma\bar{X}_A - \bar{X}_T)^2]$	$a - 1$		
Factor B	$(a)(N_{AB})[\Sigma\bar{X}_B - \bar{X}_T)^2]$	$b - 1$		
A × B interaction	$(N_{AB})\Sigma(\bar{X}_i -)^2 - SS_A - SS_B$	$(a - 1)(b - 1)$		
Within-group	$(N_{AB} - 1)\ \Sigma\ s^2{}_{AB}$	$(a)(b)(N_{AB} - 1)$		
Total	$SS_A + SS_B + SS_{A \times B} + SS_{WG}$	$df_A + df_B + df_{A \times B} + df_{WG}$		

c. Voting Message Study Example

Source	*SS*	*df*	*MS*	*F*
Message Type	.48	1	.48	.72
Authoritarianism	.12	1	.12	.21
Message Type × Authoritarianism	4.42	1	4.42	6.60
Error	18.76	28	.67	
Total	23.78	31		

In this example, because the value of 6.60 for $F_{A \times B}$ is greater than the critical value of 4.20, we make the decision to reject the null hypothesis and conclude that a significant interaction effect exists between Message Type and Authoritarianism.

In the voting message example, although neither of the two main effects was statistically significant, a significant interaction effect was nevertheless found. This reinforces our earlier observation that the absence of main effects provides no indication of whether an interaction effect is either present or absent.

Determine the Levels of Significance

If the decision is made to reject the null hypothesis, it is useful for the sake of precision to determine whether the probability of a calculated value of a statistic is not only less than .05 ($p < .05$) but also less than .01 ($p < .01$). In this step, we determine the appropriate level of significance for the two main effects and the interaction effect.

FIGURE 12.10 ● MAKING THE DECISION REGARDING THE NULL HYPOTHESIS FOR THE MAIN EFFECT OF MESSAGE TYPE IN THE VOTING MESSAGE STUDY

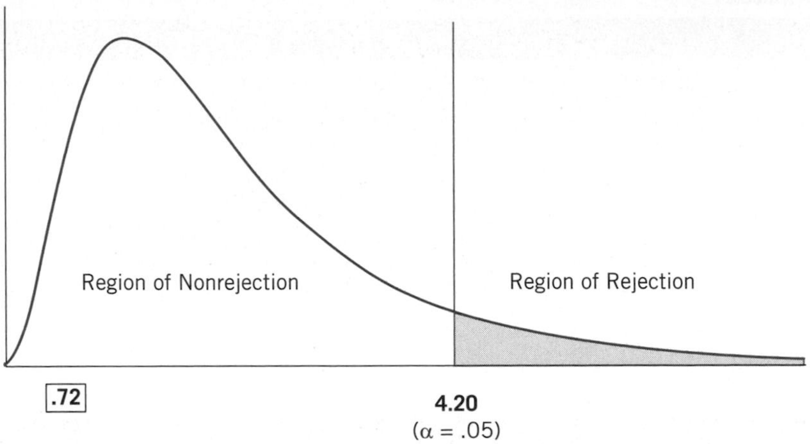

Main effect of Factor A. For the main effect of Factor A in the voting message study, the null hypothesis was not rejected because the probability of the calculated value of $F_A = .72$ was greater than the alpha value of .05 ($p > .05$). Consequently, because the probability of F_A is greater than .05, it is not necessary to determine whether its probability is less than .01.

Main effect of Factor B. Because the probability of F_B for the main effect of Authoritarianism ($F_B = .21$) was not less than .05, the null hypothesis was not rejected. As a result, there is no reason to determine whether the probability of F_B is less than .01.

A × B interaction effect. Unlike the two main effects, the Message Type × Authoritarianism interaction effect was statistically significant. As a result, it is appropriate to make a precise assessment of the probability of the calculated *F*-ratio by comparing it with the .01 critical value. Looking at the Appendix, we observe that for $df_{A \times B} = 1$ and $df_{WG} = 28$, the .01 critical value for the *F*-ratio is equal to 7.64. The level of significance for the Message Type × Authoritarianism interaction effect may be stated as follows:

$$F_{A \times B} = 6.60 < 7.64 \therefore p < .05 \text{ (but not } < .01)$$

Because the *F*-ratio value of 6.60 is between the $\alpha = .05$ (4.20) and .01 (7.64) critical values, the probability of the *F*-ratio is less than .05 but not less than .01. Therefore, the appropriate level of significance for the *F*-ratio for the Message Type × Authoritarianism interaction effect is $p < .05$.

Calculate Measures of Effect Size (R^2)

The purpose of calculating inferential statistics such as the *F*-ratio is to make one of two decisions: reject or not reject the null hypothesis. However, in addition to determining whether or not an effect is statistically significant, it's useful to calculate a measure of effect size that estimates the size or magnitude of a particular effect. As we discussed in Chapter 11, there are different measures of effect size. Although debate continues regarding the appropriate measure of effect size for factorial research designs, because of its relative simplicity, this book will rely on R^2, which is the percentage of the total variance in the dependent variable accounted for by a particular effect.

Main effect of Factor A. The main effect of Factor A is treated as though it were a single-factor design. Consequently, the formula for R^2 for Factor A is essentially the same as that presented in Chapter 11:

$$R_A^2 = \frac{SS_A}{SS_T} \tag{12-13}$$

where SS_A is the sum of squares for Factor A and SS_T is the sum of squares for the total amount of variance. Using the values for SS_A and SS_T provided in the ANOVA summary table in Table 12.7(c), we may calculate the R^2 for Message Type in the voting message study in the following way:

$$\begin{aligned} R_A^2 &= \frac{SS_A}{SS_T} \\ &= \frac{.48}{23.78} \\ &= .02 \end{aligned}$$

The R^2 for Message Type implies that 2% of the variance in ratings of message quality is explained by the type of message, reward versus threat, presented to the participant. Using Cohen's (1988) rules of thumb, this is considered a relatively small effect.

Main effect of Factor B. As you might expect, the formula for R^2 for the main effect of Factor B is a simple modification of the formula used for Factor A:

$$R_B^2 = \frac{SS_B}{SS_T} \tag{12-14}$$

where SS_B is the sum of squares for Factor B and SS_T is the sum of squares for the total amount of variance. Again relying on the summary table in Table 12.7(c), the R^2 for the main effect of Authoritarianism may be stated as

$$\begin{aligned} R_B^2 &= \frac{SS_B}{SS_T} \\ &= \frac{.12}{23.78} \\ &= .01 \end{aligned}$$

Therefore, 1% of the variance in ratings of message quality is explained by the participant's level of authoritarianism, which is also a relatively small effect.

A × B interaction effect. The R^2 for the A × B interaction effect is calculated in a manner similar to that for the two main effects:

$$R_{A \times B}^2 = \frac{SS_{A \times B}}{SS_T} \tag{12-15}$$

where $SS_{A \times B}$ is the sum of squares for the A × B interaction effect and SS_T is the sum of squares for the total amount of variance. In the voting message study, the R^2 for the Message Type × Authoritarianism interaction effect may be stated as

$$R^2_{A \times B} = \frac{SS_{A \times B}}{SS_T}$$
$$= \frac{4.42}{23.78}$$
$$= .19$$

An R^2 value of .19 is an effect of relatively large magnitude.

LEARNING CHECK 3:

Reviewing What You've Learned So Far

1. Review questions
   a. Which descriptive statistic is used to calculate between-group variance in the two-way ANOVA? Which descriptive statistic is used to calculate within-group variance?
   b. Why is the word *Error* included in an ANOVA summary table?
2. For each of the following, calculate the *F*-ratios (*F*) for the two-way ANOVA, create an ANOVA summary table, and calculate measures of effect size (R^2).
   a. $SS_A = 20.00$, $df_A = 1$, $SS_B = 10.00$, $df_B = 1$, $SS_{A \times B} = 15.00$, $df_{A \times B} = 1$, $SS_{WG} = 80.00$, $df_{WG} = 20$
   b. $SS_A = 54.00$, $df_A = 1$, $SS_B = 25.00$, $df_B = 1$, $SS_{A \times B} = 16.00$, $df_{A \times B} = 1$, $SS_{WG} = 400.00$, $df_{WG} = 44$
   c. $SS_A = 4.60$, $df_A = 2$, $SS_B = 9.79$, $df_B = 1$, $SS_{A \times B} = 10.31$, $df_{A \times B} = 2$, $SS_{WG} = 43.87$, $df_{WG} = 42$
3. Does the type of car you own influence how attractive you are to someone of the opposite sex? Two researchers (Dunn & Searle, 2010) showed men and women a picture of a person of the opposite sex driving one of two cars: a mid-priced sedan or a luxury car. Each participant then rated the attractiveness of the driver on a 1 to 10 scale (the higher the rating, the higher the attractiveness). As research suggested that men focus on a woman's physical attractiveness but women are concerned with a man's wealth or status, it was hypothesized that men would give similar ratings of attractiveness to the drivers of the two types of cars, whereas women would give lower ratings to the driver of the mid-priced car than the luxury car. The ratings of attractiveness given by men and women participants to the drivers of the two types of cars are summarized below:
   a. Create a figure to represent the cell means (place Sex on the X-axis).
   b. Conduct the two-way ANOVA and create an ANOVA summary table.
   c. Report the decisions regarding the null hypotheses and the level of significance for the main effects and the interaction effect.
   d. Calculate measures of effect size (R^2) for the main effects and the interaction effect.

Sex (Factor A): Type of Car (Factor B):	Man Mid-priced	Man Luxury	Woman Mid-priced	Woman Luxury
N_{AB}	7	7	7	7
Mean ($\bar{X}_{AB}$)	6.43	6.86	4.29	7.29
Standard deviation (s_{AB})	1.51	1.57	1.11	1.38

Draw Conclusions From the Analyses

Given the complexity of the analysis of the two-factor research design, it's important to describe the results of the analysis as accurately and completely as possible. Below is one way of describing the two-way ANOVA for the voting message study:

> Participants' ratings of message quality were analyzed using a 2 (Message Type: threat vs. reward) × 2 (Authoritarianism: low vs. high) ANOVA. The main effect of Message Type was not significant, $F(1, 28) = .72, p > .05, R^2 = .02$, implying that the message quality ratings of the Reward ($M = 4.31$) and Threat ($M = 4.56$) conditions were not significantly different. In terms of the main effect of Authoritarianism, a nonsignificant difference was found in the ratings of participants low ($M = 4.37$) or high ($M = 4.50$) in authoritarianism, $F(1, 28) = .21, p > .05, R^2 = .01$. However, a significant Message Type × Authoritarianism interaction effect was found, $F(1, 28) = 6.60, p < .05, R^2 = .19$, implying that the effect of receiving a message emphasizing threat versus reward on the perceived quality of the message depends on the participant's level of authoritarianism.

The above paragraph starts by informing the reader of the dependent variable that was analyzed ("ratings of message quality"), the statistical procedure used to analyze the dependent variable, the independent variables, and the groups that comprise each independent variable ("a 2 (Message Type: threat vs. reward) × 2 (Authoritarianism: low vs. high) ANOVA"). The next sentences summarize the tests of significance for both the main effects and the interaction effect. For each effect tested, information is provided regarding descriptive statistics, the nature and direction of the differences between groups, and the inferential statistic. Reporting the inferential statistic involves providing the statistic that has been calculated, the degrees of freedom, the calculated value of the statistic, the level of significance of the statistic, and the measure of effect size.

Relate the Results of the Analyses to the Research Hypothesis

Box 12.1 summarizes the steps in the two-way ANOVA, using the voting message study to illustrate these steps. The last step in this process is to relate the results of the analyses to the study's research hypothesis. In the voting message study, the researchers hypothesized an interaction effect between message type and authoritarianism such that the reward message was hypothesized to be more persuasive than the threat message for people low on authoritarianism, whereas the threat message was hypothesized to be more persuasive than the reward message for people high on authoritarianism. Given that a significant Message Type × Authoritarianism interaction was found in conducting the two-way ANOVA, do we know whether the research hypothesis has been supported? As you may have guessed, the answer to this question is no.

SUMMARY BOX 12.1 CONDUCTING THE TWO-WAY ANOVA, VOTING MESSAGE STUDY EXAMPLE

State the null and alternative hypotheses (H_0 and H_1)

Main effect of Factor A

H_0: all μs are equal H_1: not all μs are equal

Main effect of Factor B

H_0: all μs are equal H_1: not all μs are equal

A × B interaction effect

H_0: an interaction effect does not exist H_1: an interaction effect exists

(Continued)

(Continued)

Make decisions about the null hypotheses

Calculate the degrees of freedom (df)

Main effect of Factor A

$$df_A = a - 1 = 2 - 1 = 1$$

Main effect of Factor B

$$df_B = b - 1 = 2 - 1 = 1$$

A × B interaction effect

$$df_{A \times B} = (a - 1)(b - 1) = (2 - 1)(2 - 1) = 1$$

Within-group variance

$$df_{WG} = (a)(b)(N_{AB} - 1) = (2)(2)(8 - 1) = 28$$

Set alpha (α), identify the critical values, and state decision rules

Main effect of Factor A

If $F_A > 4.20$, reject H_0; otherwise, do not reject H_0

Main effect of Factor B

If $F_B > 4.20$, reject H_0; otherwise, do not reject H_0

A × B interaction effect

If $F_{A \times B} > 4.20$, reject H_0; otherwise, do not reject H_0

Calculate statistics: F-ratios for the two-way ANOVA

Calculate the mean squares (MS)

Main effect of Factor A

$$\begin{aligned} MS_A &= \frac{SS_A}{df_A} = \frac{(b)(N_{AB})[\Sigma(\bar{X}_A - \bar{X}_T)^2]}{a-1} \\ &= \frac{(2)(8)[(4.31-4.44)^2 + (4.56-4.44)^2}{2-1} \\ &= \frac{16[(-.13)^2 + (.12)^2]}{1} = \frac{16(.03)}{1} = \frac{.48}{1} \\ &= .48 \end{aligned}$$

Main effect of Factor B

$$\begin{aligned} MS_B &= \frac{SS_B}{df_B} = \frac{(a)(N_{AB})[\Sigma(\bar{X}_B - \bar{X}_T)^2]}{b-1} \\ &= \frac{(2)(8)[(4.37-4.44)^2 + (4.50-4.44)^2}{2-1} \\ &= \frac{16[(-.07)^2 + (.06)^2]}{1} = \frac{16(.009)}{1} = \frac{.14}{1} \\ &= .14 \end{aligned}$$

A × B interaction effect

$$MS_{A \times B} = \frac{SS_{A \times B}}{df_{A \times B}} = \frac{[N_{AB}\Sigma(\overline{X}_{AB} - \overline{X}_T)^2] - SS_A - SS_B}{(a-1)(b-1)}$$

$$= \frac{8[(4.62-4.44)^2 + (4.00-4.44)^2 + (4.13-4.44)^2 + (5.00-4.44)^2 - .48 - .14}{(2-1)(2-1)}$$

$$= \frac{8(.03+.19+.10+.31) - .48 - .14}{1}$$

$$= \frac{5.04 - .48 - .14}{1} = \frac{4.42}{1}$$

$$= 4.42$$

Within-group variance

$$MS_{WG} = \frac{SS_{WG}}{df_{WG}} = \frac{(N_{AB}-1)\Sigma s_{AB}^2}{(a)(b)(N_{AB}-1)}$$

$$= \frac{(8-1)[(.74)^2 + (.93)^2 + (.83)^2 + (.76)^2]}{(2)(2)(8-1)}$$

$$= \frac{7(.55+.86+.69+.58)}{28} = \frac{7(2.68)}{28} = \frac{18.76}{28} = .67$$

Calculate the F-ratios (F)

Main effect of Factor A

$$F_A = \frac{MS_A}{MS_{WG}} = \frac{.48}{.67} = .72$$

Main effect of Factor B

$$F_B = \frac{MS_B}{MS_{WG}} = \frac{.14}{.67} = .21$$

A × B interaction effect

$$F_{A \times B} = \frac{MS_{A \times B}}{MS_{WG}} = \frac{4.42}{.67} = 6.60$$

Make decisions about the null hypotheses

Main effect of Factor A

$$F_A = .72 < 4.20 \therefore \text{ do not reject } H_0 \ (p > .05)$$

Main effect of Factor B

$$F_B = .21 < 4.20 \therefore \text{ do not reject } H_0 \ (p > .05)$$

A × B interaction effect

$$F_{A \times B} = 6.60 > 4.20 \therefore \text{ reject } H_0 \ (p < .05)$$

(Continued)

(Continued)

Determine the levels of significance

Main effect of Factor A

Not applicable (H_0 not rejected)

Main effect of Factor B

Not applicable (H_0 not rejected)

A × B interaction effect

$$F_{A \times B} = 6.60 < 7.64 \therefore p < .05 \text{ (but not } < .01)$$

Calculate measures of effect size (R^2)

Main effect of Factor A

$$R_A^2 = \frac{SS_A}{SS_T} = \frac{.48}{23.78} = .02$$

Main effect of Factor B

$$R_B^2 = \frac{SS_B}{SS_T} = \frac{.12}{23.78} = .01$$

A × B interaction effect

$$R_{A \times B}^2 = \frac{SS_{A \times B}}{SS_T} = \frac{4.42}{23.78} = .19$$

Draw conclusions from the analyses

Participants' ratings of message quality were analyzed using a 2 (Message Type: threat vs. reward) × 2 (Authoritarianism: low vs. high) ANOVA. The main effect of Message Type was not significant, $F(1, 28) = .72$, $p > .05$, $R^2 = .02$, implying that the ratings of the Reward ($M = 4.31$) and Threat ($M = 4.56$) conditions were not significantly different. In terms of the main effect of Authoritarianism, a nonsignificant difference was found in the ratings of participants low ($M = 4.37$) or high ($M = 4.50$) in authoritarianism, $F(1, 28) = .21$, $p > .05$, $R^2 = .01$. However, a significant Message Type × Authoritarianism interaction effect was found, $F(1, 28) = 6.60$, $p < .05$, $R^2 = .19$, implying that the effect of receiving a message emphasizing threat versus reward on the perceived quality of the message depends on the participant's level of authoritarianism.

Relate the results of the analyses to the research hypothesis

At this point in the analysis, we are unable to determine whether the study's research hypothesis has been supported until additional analyses are conducted.

Rejecting the null hypothesis for an A × B interaction effect does not provide either support or a lack of support for a research hypothesis. Because of how the alternative hypothesis is stated (H_1: an interaction effect exists), the most specific conclusion we can draw is simply that an interaction effect is present. Rejecting the null hypothesis does not provide insight into the precise nature or source of the interaction effect. Therefore, at this point in the analysis, we are unable to determine whether the study's research hypothesis has been supported until additional statistical analyses are conducted.

In Chapter 11, the most specific conclusion that could be drawn from rejecting the null hypothesis in a one-way ANOVA is that the means for the groups are not all equal to each other. To determine the precise nature and source of the significant effect, additional analyses were conducted. These analyses, referred to as analytical comparisons, involved making comparisons between groups that are part of a larger research design. A similar situation exists in the two-factor research design when a significant A × B interaction effect is found. The next section introduces analyses designed to determine the source and nature of a significant A × B interaction effect.

12.6 INVESTIGATING A SIGNIFICANT A × B INTERACTION EFFECT: ANALYSIS OF SIMPLE EFFECTS

A significant A × B interaction implies that the effect of one independent variable changes at the different levels of the other independent variable. However, to more precisely determine the source of a significant interaction effect, additional analyses must be conducted that break down a larger factorial research design into a number of smaller designs.

As a way of introducing the logic that underlies these analyses, let's return to the research hypothesis in the voting message study:

> High authoritarian recipients would perceive the threat message as more valid and persuasive than the reward message. . . . A threat-related persuasive message would lead to less persuasion among low authoritarian recipients, relative to the reward message. (Lavine et al., 1999, p. 340)

What specific comparisons are needed to test this research hypothesis? It appears that two specific comparisons are necessary: the difference between the Reward and Threat conditions for those high in authoritarianism and the difference between the Reward and Threat conditions for those low in authoritarianism. This analysis plan, illustrated in Figure 12.11(a), shows we are testing the effect of one independent variable (Message Type) separately for each level of the other independent variable (Authoritarianism).

The analyses in Figure 12.11(a) illustrate what is known as the analysis of simple effects. A **simple effect** is the effect of one independent variable at one of the levels of another independent variable. For the voting message study, we would be testing the simple effect of Message Type at each level of Authoritarianism.

The researchers in the voting message study did in fact compare the perceived quality of the message for the two message types (reward vs. threat) separately for the low- and high-authoritarian participants. Below is their conclusion regarding their analysis of the simple effects:

> Follow-up contrasts revealed that high authoritarians did indeed perceive the threat message as more persuasive than the reward message . . . low authoritarians perceived the reward message as containing more persuasive arguments than the threat message. (Lavine et al., 1999, p. 343)

On the basis of the results of their simple effects analyses, the researchers concluded that their research hypothesis had been supported.

The above example tested the simple effect of Factor A (Message Type) at each level of Factor B (Authoritarianism). As Figure 12.11(b) illustrates, we might instead have chosen to test the simple effect of Authoritarianism at each level of Message Type. Because it's possible to test the simple effects in either of these two ways, which should you choose? The choice of which simple effects to test does not alter the statistical significance of the interaction. Rather, the different simple effects provide different perspectives on the same set of data.

The simple effect of Message Type at each level of Authoritarianism (Figure 12.11(a)) focuses on the effectiveness of different types of message (reward vs. threat) for people who have a particular level of authoritarianism. That is, the simple effects of Message Type emphasize the effectiveness of different types of messages. On the

FIGURE 12.11 ● ILLUSTRATION OF SIMPLE EFFECTS ANALYSES, VOTING MESSAGE STUDY.

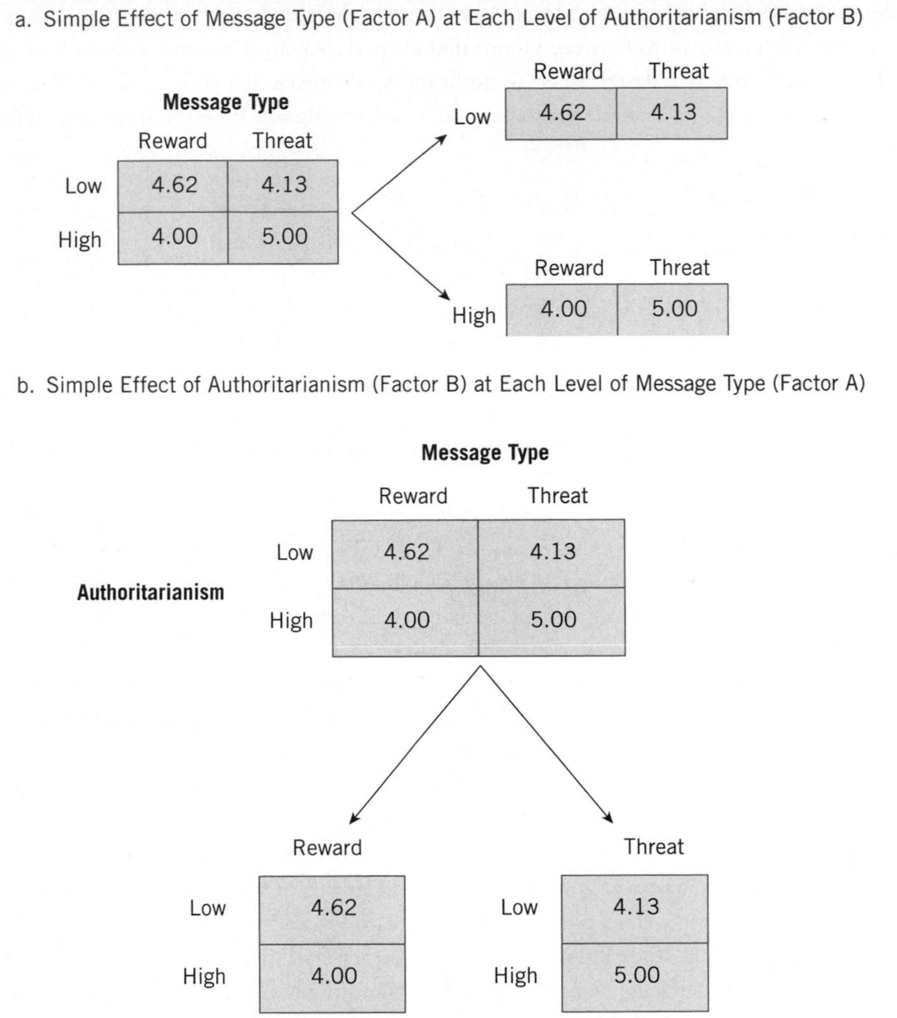

other hand, the simple effect of Authoritarianism at each level of Message Type (Figure 12.11(b)) focuses on how different types of people (people low vs. high in authoritarianism) respond to a particular type of message. The simple effects of Authoritarianism emphasize the responses of different types of people. Ultimately, the choice of which simple effects to test depends on the purpose of the study and how a study's research hypotheses are stated.

In summary, rejecting the null hypothesis for the A × B interaction effect does not reveal the precise source and nature of the interaction. To directly test one's research hypotheses, additional analyses are necessary. A common way to conduct these analyses is to examine the simple effects of one variable at the different levels of the other variable. Due to space considerations, this book will not discuss the calculations necessary to conduct these analyses. Those wishing to explore this topic in greater detail are referred to Keppel and Wickens (2004) and Keppel, Saufley, and Tokunaga (1992).

The two-factor research design and two-way ANOVA are perhaps the most challenging research methodology and statistical procedures covered in this book. The inclusion of more than one independent variable not only increases the number of calculations but also requires a greater understanding of different effects that can exist in the same study. Although this may be challenging, hopefully you've gained an appreciation of the ability of factorial designs to capture the complexity of phenomena researchers choose to study.

12.7 LOOKING AHEAD

Much of the presentation and discussion of the two-way ANOVA builds off of the information provided in earlier chapters. Chapter 9 included a discussion of the *t*-test, which is used to compare the means of two groups. In Chapter 11, the one-way ANOVA was presented to compare the means of three or more groups that comprise one independent variable. The *t*-test, one-way ANOVA, and two-way ANOVA are used when the independent variable or variables are categorical in nature, consisting of groups. The goal of these three statistical procedures is to compare the means of groups on a continuous dependent variable, a variable that is measured at the interval or ratio level of measurement. The next chapter introduces a statistical procedure used when both the independent and dependent variables are continuous in nature.

LEARNING CHECK 4:

Reviewing What You've Learned So Far

1. Review questions
   a. What conclusion can you draw from a significant A × B interaction? What conclusion can you *not* draw?
   b. What is the difference between a main effect and a simple effect?
   c. What is the purpose of analyzing simple effects in the A × B research design?
2. For each of the research situations below, draw a figure such as those in Figure 12.11 that illustrate simple effects analyses appropriate to test the study's research hypothesis.
   a. An instructor teaches two courses: one taught in the typical classroom setting and one taught online. She believes students taking the classroom course are more likely to keep up with assigned course readings than students in the online course; furthermore, she believes this difference is greater for students living off-campus than on-campus.
   b. When a child misbehaves, who is responsible for the child's bad behavior? A family therapist believes that parents are more likely to attribute responsibility to the child than to themselves, particularly when the child is a teenager rather than an adolescent.
   c. Before taking an exam, do you engage in rituals like kissing a luck charm or wearing a particular piece of clothing? Two researchers (Rudski & Edwards, 2007) hypothesized that students engage in rituals more when an exam is important (vs. unimportant) to students' grade in the course, as well as when an exam was seen as "very difficult" (as opposed to "easy" or "somewhat difficult").

12.8 Summary

Unlike ***single-factor research designs,*** which consist of a single independent variable, ***factorial research designs*** consist of all possible combinations of the levels of two or more independent variables. A factorial research design consisting of two independent variables is referred to as a ***two-factor (A × B) research design.***

Conceptually, factorial research designs allow researchers to determine whether an ***interaction effect*** exists between independent variables. An interaction effect is present when the effect of one of the independent variables on the dependent variable changes (is not the same) at the different levels of another independent variable.

In addition to testing for interaction effects, a factorial research design also involves testing for ***main effects;*** a main effect is the effect of a particular independent variable on the dependent variable of a particular factor. The two-factor research design contains two main effects, each of which is treated as a separate single-factor design.

There are two important aspects of the relationship between main effects and interaction effects. First, the presence or absence of main effects provides no indication of whether an interaction effect is present or absent and vice versa. Second, whether and how one interprets main effects depends on the presence or absence of interaction effects. If an interaction effect is present, the effects of both independent variables must be considered simultaneously rather than in isolation of each other. If an interaction effect is not present, the two-factor design may be treated as if it were two separate single-factor designs, one for each main effect.

A significant A × B interaction effect implies that the effect of one independent variable changes at the different levels of the other independent variable. However, to more precisely determine the source of a significant interaction effect, additional analyses must be conducted that break down a larger factorial research design into a number of smaller designs. For example, ***simple effects*** analyses may be conducted to test the effect of one independent variable at each of the different levels of another independent variable.

12.9 Important Terms

single-factor research design (p. 413)
factorial research design (p. 413)
interaction effect (p. 413)
cell (p. 414)
cell mean (p. 414)
main effect (p. 415)
marginal mean (p. 417)
two-factor (A × B) research design (p. 425)
simple effect (p. 449)

12.10 Formulas Introduced in This Chapter

Degrees of Freedom for Main Effect of Factor A, Two-Way ANOVA (df_A)

$$df_A = a - 1 \tag{12-1}$$

Degrees of Freedom for Main Effect of Factor B, Two-Way ANOVA (df_B)

$$df_B = b - 1 \tag{12-2}$$

Degrees of Freedom for A × B Interaction Effect, Two-Way ANOVA ($df_{A \times B}$)

$$df_{A \times B} = (a - 1)\,(b - 1) \tag{12-3}$$

Degrees of Freedom for Within-Group Variance, Two-Way ANOVA (df_{WG})

$$df_{WG} = (a)\,(b)\,(N_{AB} - 1) \tag{12-4}$$

Between-Group Variance for Main Effect of Factor A, Two-Way ANOVA (MS_A)

$$MS_A = \frac{SS_A}{df_A} = \frac{(b)(N_{AB})[\Sigma(\bar{X}_A - \bar{X}_T)^2]}{a - 1} \tag{12-5}$$

Between-Group Variance for Main Effect of Factor B, Two-Way ANOVA (MS_B)

$$MS_B = \frac{SS_B}{df_B}\frac{(a)(N_{AB})[\Sigma(\bar{X}_B - \bar{X}_T)^2]}{b-1} \tag{12-6}$$

Sums of Squares for Between-Group Variance, Two-Way ANOVA (SS_{BG})

$$SS_{BG} = N_{AB}\Sigma(\bar{X}_{AB} - \bar{X}_T)^2 \tag{12-7}$$

Between-Group Variance for A × B Interaction, Two-Way ANOVA ($MS_{A \times B}$)

$$MS_{A\times B} = \frac{SS_{A\times B}}{df_{A\times B}}\frac{[N_{AB}\Sigma(\bar{X}_{AB} - \bar{X}_T)^2] - SS_A - SS_B}{(a-1)(b-1)} \tag{12-8}$$

Within-Group Variance, Two-Way ANOVA (MS_{WG})

$$MS_{WG} = \frac{SS_{WG}}{df_{WG}} = \frac{(N_{AB} - 1)\Sigma s^2{}_{AB}}{(a)(b)(N_{AB} - 1)} \tag{12-9}$$

***F*-Ratio for Main Effect of Factor A, Two-Way ANOVA (F_A)**

$$F_A = \frac{MS_A}{MS_{WG}} \tag{12-10}$$

***F*-Ratio for Main Effect of Factor B, Two-Way ANOVA (F_B)**

$$F_B = \frac{MS_B}{MS_{WG}} \tag{12-11}$$

***F*-Ratio for A × B Interaction, Two-Way ANOVA ($F_{A \times B}$)**

$$F_{A\times B} = \frac{MS_{A\times B}}{MS_{WG}} \tag{12-12}$$

Measure of Effect Size for Main Effect of Factor A, Two-Way ANOVA (R^2_A)

$$R^2_A = \frac{SS_A}{SS_T} \tag{12-13}$$

Measure of Effect Size for Main Effect of Factor B, Two-Way ANOVA (R^2_B)

$$R^2_B = \frac{SS_B}{SS_T} \tag{12-14}$$

Measure of Effect Size for A × B Interaction, Two-Way ANOVA ($R^2_{A \times B}$)

$$R^2_{\text{A} \times \text{B}} = \frac{SS_{\text{A} \times \text{B}}}{SS_{\text{T}}} \tag{12-15}$$

12.11 Using SPSS

Two-Way Analysis of Variance (ANOVA): The Voting Message Study (12.1)

1. Define independent and dependent variables (name, # decimals, labels for the variables, labels for values of the independent variable) and enter data for the variables.

   NOTE: Numerically code values of the independent variables (i.e., Message Type [1 = Reward, 2 = Threat], Authoritarianism [1 = Low, 2 = High]) and provide labels for these values in **Values** box within **Variable View**.

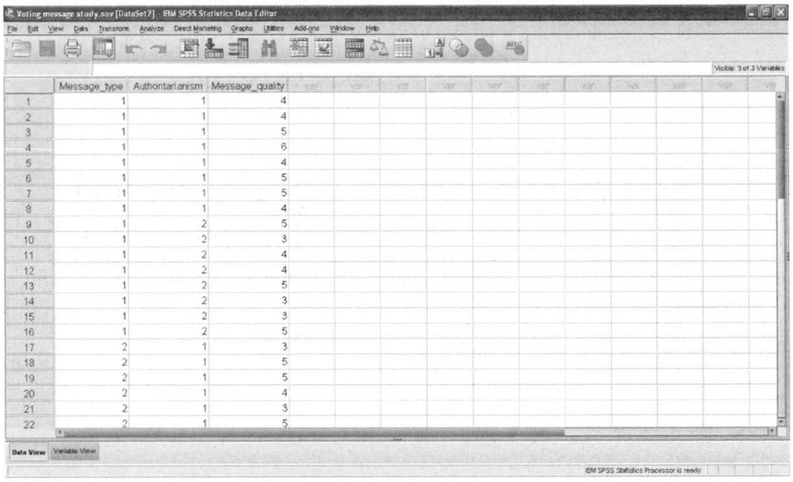

2. Select the two-way ANOVA procedure within SPSS.

   How? (1) Click **Analyze** menu, (2) click **General Linear Model**, and (3) click **Univariate**.

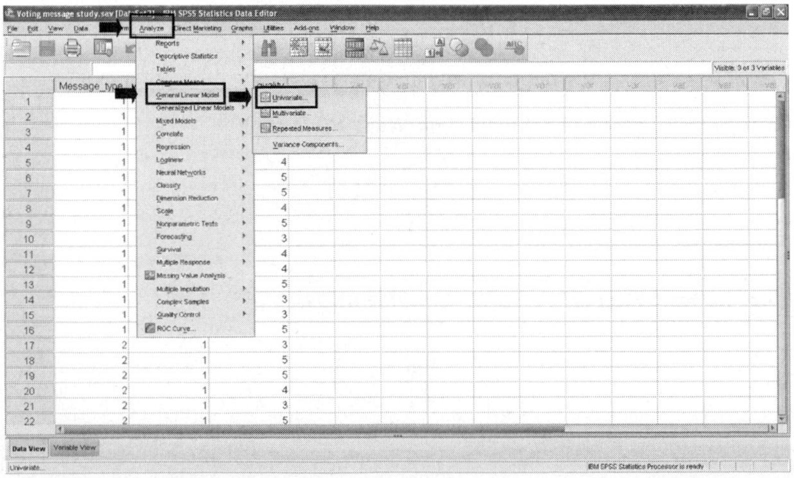

3. Identify the dependent variable and the independent variables, and ask for descriptive statistics.

   How? (1) Click dependent variable and → **Dependent Variable**, (2) click independent variable and → **Fixed Factor(s)**, (3) click **Options** and click **Descriptive Statistics**, (4) click Continue, and (5) click OK.

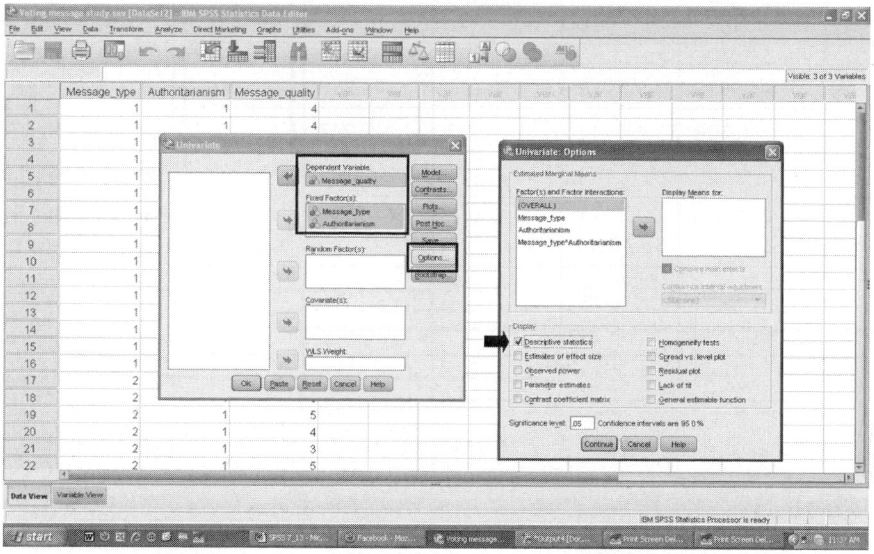

4. Examine output.

Univariate Analysis of Variance

Descriptive Statistics

Dependent Variable: Message_quality

Message type	Authoritarianism	Mean	Std. Deviation	N
Reward	Low	4.62	.744	8
	High	4.00	.926	8
	Total	4.31	.873	16
Threat	Low	4.13	.835	8
	High	5.00	.756	8
	Total	4.56	.892	16
Total	Low	4.38	.806	16
	High	4.50	.966	16
	Total	4.44	.878	32

Cell means

Marginal means (Factor A)

Marginal means (Factor B)

Total mean

Tests of Between-Subjects Effects

Dependent Variable: Message_quality

Source	Type III Sum of Squares	*df*	Mean Square	F	Sig.	
						Level of significance (Note: 'Sig.'= probability (p)).
~~Corrected Model~~	~~5.125^a~~	~~3~~	~~1.708~~	~~2.551~~	~~.070~~	
~~Intercept~~	~~630.125~~	~~1~~	~~630.125~~	~~940.987~~	~~.000~~	
Message_type	.500	1	.500	.747	.395	Main effect, Factor A
Authoritarianism	.125	1	.125	.187	.669	Main effect, Factor B
Message_type* Authoritarianism	4.500	1	4.500	6.720	.015	A x B interaction
Error	18.750	28	.670			Within group (error)
~~Total~~	~~634.000~~	~~32~~				
Corrected Total	23.875	31				Total variance

12.12 Exercises

1. Create a line graph for each of the following tables of cell means (assume the scores have a possible range of 0 to 10 and put Factor A along the X (horizontal) axis).

a.

Factor A

Factor B	a_1	a_2
b_1	4	10
b_2	4	4

b.

Factor A

Factor B	a_1	a_2
b_1	6	10
b_2	2	6

c.

Factor A

Factor B	a_1	a_2	a_3
b_1	8	8	8
b_2	4	4	4

d.

Factor A

Factor B	a_1	a_2	a_3
b_1	6	8	10
b_2	6	4	2

2. Create a line graph for each of the following tables of cell means (assume the scores have a possible range of 0 to 10 and put Factor A along the X (horizontal) axis).

a.

Factor A

Factor B	a_1	a_2
b_1	4	4
b_2	4	8

b.

Factor A

Factor B	a_1	a_2
b_1	6	10
b_2	6	2

c.

		Factor A		
		a_1	a_2	a_3
Factor B	b_1	4	6	8
	b_2	4	6	8

d.

		Factor A		
		a_1	a_2	a_3
Factor B	b_1	2	6	10
	b_2	6	6	6

3. Create a table of cell means for each of the following graphs.

a.

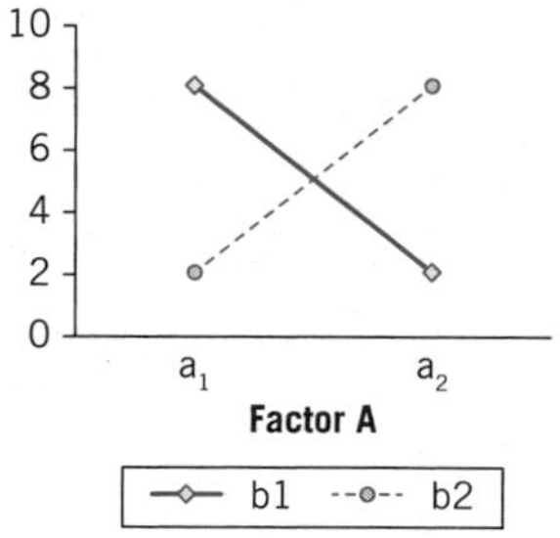

b.

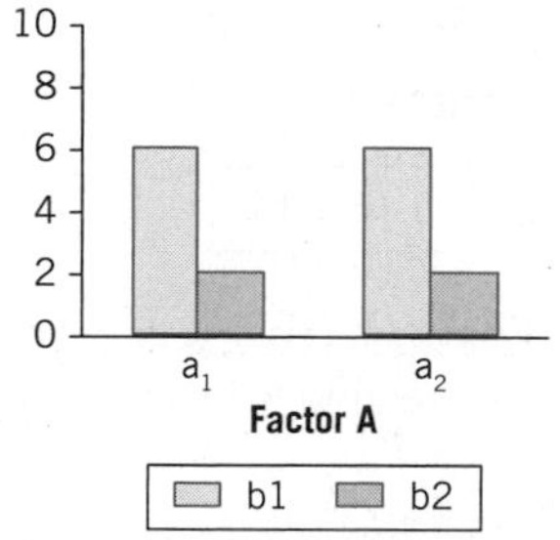

c.

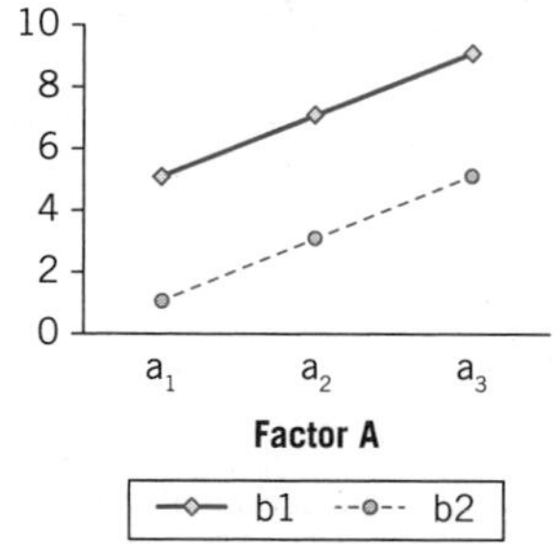

d.

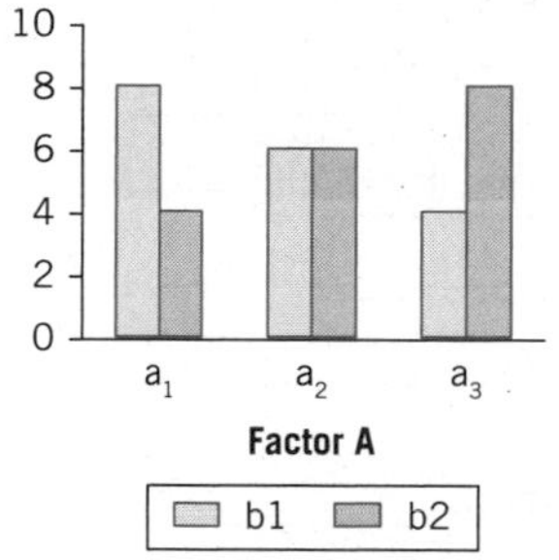

4. Create a table of cell means for each of the following graphs.

a.

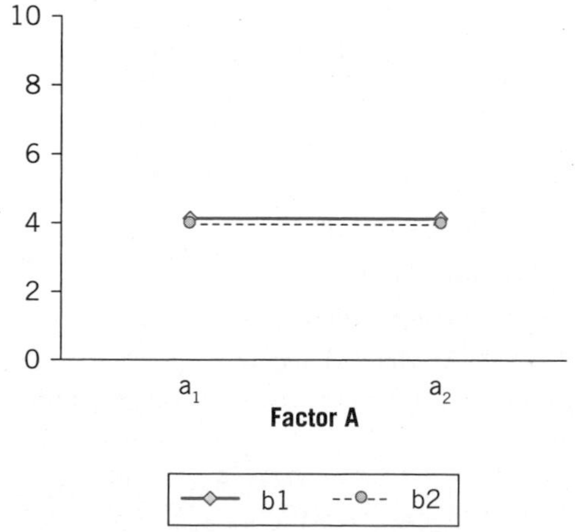

b.

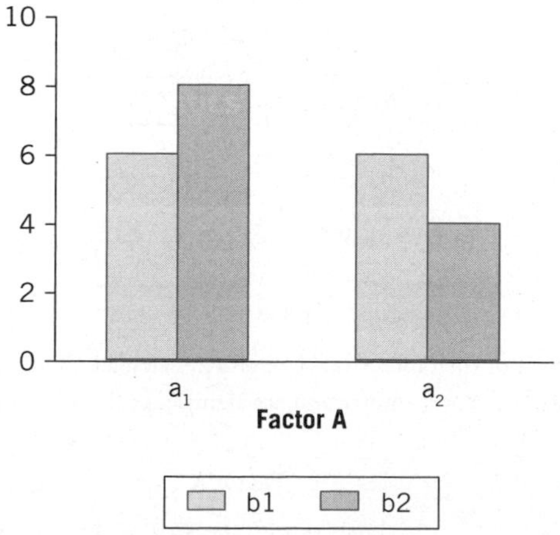

c.

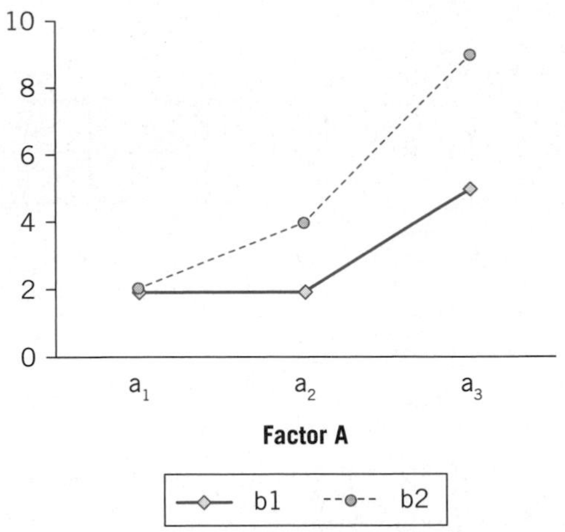

d.

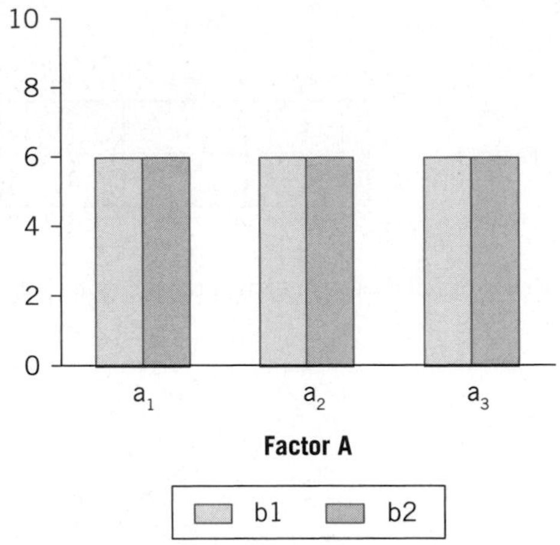

5. For each of the tables from Exercise 1, calculate the marginal means ($\overline{X}_A$ and $\overline{X}_B$) and determine whether the main effects and A × B interaction are significant (in this exercise, assume any difference between means is significant).

a.

Factor A

Factor B		a_1	a_2	$\overline{X}_B$
	b_1	4	10	____
	b_2	4	4	____
	$\overline{X}_A$	____	____	

Main effect, Factor A? __________

Main effect, Factor B? __________

A × B interaction effect? __________

b.

Factor A

Factor B		a_1	a_2	$\overline{X}_B$
	b_1	6	10	____
	b_2	2	6	____
	$\overline{X}_A$	____	____	

Main effect, Factor A? __________

Main effect, Factor B? __________

A × B interaction effect? __________

c.

Factor A

Factor B		a_1	a_2	a_3	$\overline{X}_B$
	b_1	8	8	8	____
	b_2	4	4	4	____
	$\overline{X}_A$	____	____	____	

Main effect, Factor A? __________

Main effect, Factor B? __________

A × B interaction effect? __________

d.

Factor A

Factor B		a_1	a_2	a_3	$\overline{X}_B$
	b_1	6	8	10	____
	b_2	6	4	2	____
	$\overline{X}_A$	____	____	____	

Main effect, Factor A? __________

Main effect, Factor B? __________

A × B interaction effect? __________

6. For each of the tables from Exercise 2, calculate the marginal means ($\overline{X}_A$ and $\overline{X}_B$) and determine whether the main effects and A × B interaction are significant (in this exercise, assume any difference between means is significant).

a.

Factor A

Factor B		a_1	a_2	$\overline{X}_B$
	b_1	4	4	____
	b_2	8	8	____
	$\overline{X}_A$	____	____	

b.

Factor A

Factor B		a_1	a_2	$\overline{X}_B$
	b_1	6	10	____
	b_2	6	2	____
	$\overline{X}_A$	____	____	

Main effect, Factor A? ________

Main effect, Factor B? ________

A × B interaction effect? ________

Main effect, Factor A? ________

Main effect, Factor B? ________

A × B interaction effect? ________

c.

Factor A

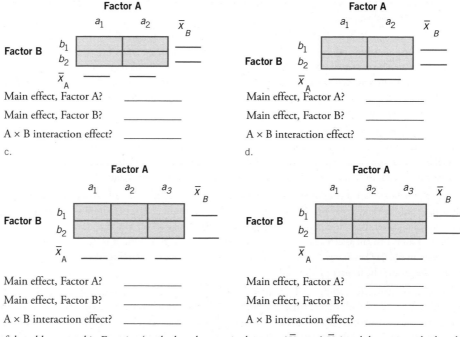

Factor B	a_1	a_2	a_3	$\bar{X}_B$
b_1	4	6	8	____
b_2	4	6	8	____
$\bar{X}_A$	____	____	____	

Main effect, Factor A? ________

Main effect, Factor B? ________

A × B interaction effect? ________

d.

Factor A

Factor B	a_1	a_2	a_3	$\bar{X}_B$
b_1	6	8	10	____
b_2	6	4	2	____
$\bar{X}_A$	____	____	____	

Main effect, Factor A? ________

Main effect, Factor B? ________

A × B interaction effect? ________

7. For each of the tables created in Exercise 3, calculate the marginal means ($\bar{X}_A$ and $\bar{X}_B$) and determine whether the main effects and the A × B interaction are significant (in this exercise, assume any difference between means is significant).

a.

Factor A

Factor B	a_1	a_2	$\bar{X}_B$
b_1			____
b_2			____
$\bar{X}_A$	____	____	

Main effect, Factor A? ________

Main effect, Factor B? ________

A × B interaction effect? ________

b.

Factor A

Factor B	a_1	a_2	$\bar{X}_B$
b_1			____
b_2			____
$\bar{X}_A$	____	____	

Main effect, Factor A? ________

Main effect, Factor B? ________

A × B interaction effect? ________

c.

Factor A

Factor B	a_1	a_2	a_3	$\bar{X}_B$
b_1				____
b_2				____
$\bar{X}_A$	____	____	____	

Main effect, Factor A? ________

Main effect, Factor B? ________

A × B interaction effect? ________

d.

Factor A

Factor B	a_1	a_2	a_3	$\bar{X}_B$
b_1				____
b_2				____
$\bar{X}_A$	____	____	____	

Main effect, Factor A? ________

Main effect, Factor B? ________

A × B interaction effect? ________

8. For each of the tables created in Exercise 4, calculate the marginal means ($\bar{X}_A$ and $\bar{X}_B$) and determine whether the main effects and the A × B interaction are significant (in this exercise, assume any difference between means is significant).

a.

Factor A

Factor B	a_1	a_2	$\bar{X}_B$
b_1			____
b_2			____
$\bar{X}_A$	____	____	

Main effect, Factor A? ________

Main effect, Factor B? ________

A × B interaction effect? ________

b.

Factor A

Factor B	a_1	a_2	$\bar{X}_B$
b_1			____
b_2			____
$\bar{X}_A$	____	____	

Main effect, Factor A? ________

Main effect, Factor B? ________

A × B interaction effect? ________

c.

Factor A

		a_1	a_2	a_3	$\bar{X}_B$
Factor B	b_1				_____
	b_2				_____
	$\bar{X}_A$	_____	_____	_____	

Main effect, Factor A? __________

Main effect, Factor B? __________

A × B interaction effect? __________

d.

Factor A

		a_1	a_2	a_3	$\bar{X}_B$
Factor B	b_1				_____
	b_2				_____
	$\bar{X}_A$	_____	_____	_____	

Main effect, Factor A? __________

Main effect, Factor B? __________

A × B interaction effect? __________

9. For each of the research situations below, identify the two independent variables and indicate the type of A × B design (e.g., 2 × 2, 3 × 3).

 a. An instructor gives a final exam in which she randomly assigns students to one of two conditions: handwritten or typed. She also assesses and classifies students' typing ability: low or high typing ability. She hypothesizes that students with low typing ability will do better on the handwritten exam than the typed, but the opposite will be true for those with high typing ability.

 b. A researcher examines how reading to young children affects their interest in books. The researcher believes that the age of the child being read to and the type of book being read might influence their interest. She conducts a study in which groups of 3-year-old, 5-year-old, and 7-year-old children read a fiction book or a nonfiction book. She hypothesizes that for the youngest children, reading fiction books rather than nonfiction books will lead to a greater interest in reading. However, as children get older, this difference between the two types of books will grow smaller such that for the oldest children, the type of book will not make a difference.

 c. A researcher hypothesizes that the effect of alcohol consumption on one's motor skills depends on the person's weight (underweight, normal, or overweight) but that this differs for men versus women. More specifically, for underweight people, men and women are equally affected. However, the more a male weighs, the greater the effect of alcohol, whereas for females, the effect stays the same no matter how much females weigh.

 d. A researcher believes that the influence of drinking coffee on the ability to fall asleep depends on the amount of coffee one drinks and the time of day one drinks the coffee. She has groups of students drink one, two, or three cups of coffee in the morning, afternoon, or evening. She hypothesizes that drinking coffee in the morning only has a slight effect on the later ability to fall asleep, and this is the same regardless of the amount of coffee one drinks. However, drinking coffee later in the day has a greater effect on the ability to fall asleep; furthermore, the more coffee one drinks later in the day, the greater the impact on this ability.

 e. A researcher is interested in studying the relationship between students' motivation and their performance on different types of tests. She hypothesizes that students who have low motivation to achieve their goals will perform at a level similar to students with high achievement motivation when a test is easy. However, on a hard test, students with low achievement motivation will do worse than they did on the easy test, whereas students high on achievement motivation will do better than they did on the easy test.

10. For each of the research situations below, identify the two independent variables and indicate the type of A × B design (e.g., 2 × 2, 3 × 3).

 a. A gardener reads an article that states that lawns grow better when they are watered once a week for 30 minutes as opposed to three times a week for 10 minutes. From his experience, the gardener hypothesizes that which watering schedule is better depends on the size of the lawn (small vs. large). For small lawns, the type of watering schedule does not make a difference. However, for large lawns, it is better to water once a week for 30 minutes than three times a week for 10 minutes.

 b. A teacher is interested in seeing how attending class and reading the assigned chapters is related to her students' performance on tests. She assesses each student in terms of how often they attend class (rarely, sometimes, regularly) and how often they read the assigned chapters (rarely, sometimes, regularly). She hypothesizes that students who attend class rarely will do poorly on tests, and they will do poorly regardless of how often they read the assigned chapters. She also believes students who attend class regularly will do well on tests, and they will do well regardless of how often they do the readings. However, for students who attend occasionally, she hypothesizes that the more they do the required readings, the better they will do on tests.

c. A fire chief reads research about how automobile accidents may be related to the color of the vehicles involved. Consequently, he wonders whether he should change the color of fire engines from red to yellow. Although he thinks both colors will be equally visible during the day, at night the yellow trucks will be easier to see than red trucks.

d. Let's say you wish to test the effects of consuming alcohol; more specifically, there is a psychological effect such that the expectation that one is drinking alcohol can influence aggression. To test this hypothesis, you have a group of people drink either a nonalcoholic beer or a regular beer. Regardless of what they are actually drinking, some of these people are told they are drinking nonalcoholic beer and some are told they are drinking regular beer. You later measure them on their level of aggression. You hypothesize that, regardless of which type of beer they are drinking, those who are told they are drinking regular beer will be more aggressive than those told they are drinking nonalcoholic beer.

a.

		Factor A	
		a_1	a_2
Factor B	b_1	5.00	9.00
	b_2	3.00	7.00

b.

		Factor A	
		a_1	a_2
Factor B	b_1	3.50	1.25
	b_2	2.25	3.50
	b_3	5.75	4.25

c.

		Factor A		
		a_1	a_2	a_3
Factor B	b_1	4.28	4.61	4.95
	b_2	2.30	6.87	3.14
	b_3	7.06	5.52	2.26

11. For each of the following tables of cell means, calculate marginal means ($\overline{X}_A$ and $\overline{X}_B$) and the total mean ($\overline{X}_T$).

a.

		Factor A	
		a_1	a_2
Factor B	b_1	11.83	16.30
	b_2	9.04	21.47

b.

		Factor A		
		a_1	a_2	a_3
Factor B	b_1	86.63	79.47	85.29
	b_2	71.90	80.36	88.11

c.

		Factor A			
		a_1	a_2	a_3	a_4
Factor B	b_1	107.32	113.84	96.59	89.01
	b_2	101.66	104.27	106.30	108.53
	b_3	112.51	98.92	100.46	111.14

12. For each of the following tables of cell means, calculate marginal means ($\overline{X}_A$ and $\overline{X}_B$) and the total mean ($\overline{X}_T$).

13. For each of the following, calculate four degrees of freedom (df_A, df_B, $df_{A \times B}$, df_{WG}) and identify the three critical values of the three F-ratios (F_A, F_B, $F_{A \times B}$) (assume $\alpha = .05$).
    a. $a = 2$, $b = 2$, $N_{AB} = 4$
    b. $a = 2$, $b = 2$, $N_{AB} = 18$
    c. $a = 3$, $b = 2$, $N_{AB} = 9$
    d. $a = 2$, $b = 4$, $N_{AB} = 6$

14. For each of the following, calculate four degrees of freedom (df_A, df_B, $df_{A \times B}$, df_{WG}) and identify the three critical values of the three F-ratios (F_A, F_B, $F_{A \times B}$) (assume $\alpha = .05$).
    a. $a = 2$, $b = 2$, $N_{AB} = 10$
    b. $a = 2$, $b = 3$, $N_{AB} = 5$
    c. $a = 3$, $b = 3$, $N_{AB} = 12$
    d. $a = 3$, $b = 4$, $N_{AB} = 5$

15. For each of the following, calculate the F-ratios (F) for the two-way ANOVA, create an ANOVA summary table, and calculate measures of effect size (R^2).
    a. $SS_A = 12.00$, $df_A = 1$, $SS_B = 24.00$, $df_B = 1$, $SS_{A \times B} = 8.00$, $df_{A \times B} = 1$, $SS_{WG} = 64.00$, $df_{WG} = 16$
    b. $SS_A = 3.91$, $df_A = 1$, $SS_B = 5.64$, $df_B = 1$, $SS_{A \times B} = 3.12$, $df_{A \times B} = 1$, $SS_{WG} = 67.04$, $df_{WG} = 92$
    c. $SS_A = 4.23$, $df_A = 1$, $SS_B = 3.72$, $df_B = 2$, $SS_{A \times B} = 1.81$, $df_{A \times B} = 2$, $SS_{WG} = 8.62$, $df_{WG} = 18$
    d. $SS_A = 52.78$, $df_A = 3$, $SS_B = 18.58$, $df_B = 1$, $SS_{A \times B} = 78.06$, $df_{A \times B} = 3$, $SS_{WG} = 297.21$, $df_{WG} = 32$
    e. $SS_A = 78.43$, $df_A = 1$, $SS_B = 61.90$, $df_B = 3$, $SS_{A \times B} = 27.39$, $df_{A \times B} = 3$, $SS_{WG} = 695.32$, $df_{WG} = 56$

16. For each of the following, calculate the F-ratios (F) for the two-way ANOVA, create an ANOVA summary table, and calculate measures of effect size (R^2).
    a. $SS_A = 16.50$, $df_A = 1$, $SS_B = 9.75$, $df_B = 1$, $SS_{A \times B} = 31.26$, $df_{A \times B} = 1$, $SS_{WG} = 337.82$, $df_{WG} = 56$
    b. $SS_A = 174.37$, $df_A = 2$, $SS_B = 387.02$, $df_B = 1$, $SS_{A \times B} = 250.87$, $df_{A \times B} = 2$, $SS_{WG} = 2011.65$, $df_{WG} = 54$
    c. $SS_A = 21.90$, $df_A = 2$, $SS_B = 13.29$, $df_B = 2$, $SS_{A \times B} = 36.09$, $df_{A \times B} = 4$, $SS_{WG} = 143.12$, $df_{WG} = 63$
    d. $SS_A = 125.87$, $df_A = 3$, $SS_B = 176.31$, $df_B = 3$, $SS_{A \times B} = 359.00$, $df_{A \times B} = 9$, $SS_{WG} = 812.59$, $df_{WG} = 48$

17. For the examples in Exercise 15, indicate the type of A × B design (e.g., 2 × 2, 3 × 3) and the number of scores in each combination of the two factors (N_{AB}).

18. For the examples in Exercise 16, indicate the type of A × B design (e.g., 2 × 2, 3 × 3) and the number of scores in each combination of the two factors (N_{AB}).

19. One study examined parents' use of discipline with their children (McKee et al., 2007). They asked, "Do rates of harsh verbal and physical discipline differ by gender of parent and child?" (p. 188). They wrote that "with regard to frequency of use of harsh discipline, we propose that boys will receive more harsh discipline than girls, particularly from fathers" (p. 188). To examine this hypothesis, they asked a sample of boys and girls to indicate how often their mother or father used harsh discipline with them. The frequency of harsh discipline by mothers and fathers reported by the boys and girls is summarized below:
    a. Create a figure to represent the cell means.
    b. Conduct the two-way ANOVA and create an ANOVA summary table.
    c. Report the decisions regarding the null hypotheses and the level of significance for the main effects and the interaction effect.

d. Calculate measures of effect size (R^2) for the main effects and the interaction effect.

Child (Factor A):	Boy	Boy	Girl	Girl
Parent (Factor B):	Mother	Father	Mother	Father
N_{AB}	6	6	6	6
Mean ($\bar{X}_{AB}$)	2.83	4.67	2.17	2.00
Standard deviation (s_{AB})	1.17	.52	1.17	1.10

20. Does playing violent video games lead to greater aggressive behavior? A team of researchers believed greater amount of time playing these games leads to higher feelings of hostility (Barlett, Harris, & Baldassaro, 2007). Furthermore, they hypothesized this increase in hostility is greater for people playing games with realistic-looking weapons rather than standard game controllers. They had groups of people play a violent video game for either 15 minutes of 30 minutes; some of these people used a standard game controller and others used a gun-shaped controller; this is a 2 (Controller: standard vs. gun) × 2 (Time: 15 min vs. 30 min) design. After playing the game, their level of hostility was measured.

Time (Factor A):	15 min	15 min	30 min	30 min
Controller (Factor B):	Standard	Gun	Standard	Gun
N_{AB}	5	5	5	5
Mean ($\bar{X}_{AB}$)	10.60	13.60	16.40	25.60
Standard deviation (s_{AB})	3.51	3.05	2.70	3.36

a. Create a figure to represent the cell means.

b. Conduct the two-way ANOVA and create an ANOVA summary table.

c. Report the decisions regarding the null hypotheses and the level of significance for the main effects and the interaction effect.

d. Calculate measures of effect size (R^2) for the main effects and the interaction effect.

21. A researcher is interested in examining how teachers' expectations of their students' scholastic abilities might affect students' self-perceptions of their own ability and whether this effect varies as a function of a student's age. She has a sample of teachers rate their students as having either low or high expectations of success; these teachers are from the first, third, and fifth grades. She then has students rate their own abilities on a scale of 1 (low) to 15 (high). She hypothesizes that students with low expectations by the teacher would display lower ratings of their own ability than students with high teacher expectations. Furthermore, she believes this difference will increase as students progress through the educational system, growing larger from first to third to fifth grade. Below are descriptive statistics of students' ratings of their ability.

Grade (Factor A):	First	First	Third	Third	Fifth	Fifth
Expectation (Factor B):	Low	High	Low	High	Low	High
N_{AB}	3	3	3	3	3	3
Mean ($\bar{X}_{AB}$)	5.00	6.33	3.67	8.67	2.33	9.67
Standard deviation (s_{AB})	1.00	1.53	1.53	1.53	1.15	1.53

a. Create a figure to represent the cell means.

b. Conduct the two-way ANOVA and create an ANOVA summary table.

c. Report the decisions regarding the null hypotheses and the level of significance for the main effects and the interaction effect.

d. Calculate measures of effect size (R^2) for the main effects and the interaction effect.

22. Concerns regarding the possibility of getting skin cancer have influenced beliefs regarding the positive benefits of getting a suntan. One study examined whether men and women have similar beliefs regarding whether a woman's suntan influences perceptions of her physical attractiveness (Banerjee, Campo, & Greene, 2008). They showed men and women one of three photographs: a woman with no tan, a medium tan, or a dark tan; next, they rated the woman in terms of her physical attractiveness. The researchers hypothesized that, for men (but not for women), the darker a woman's tan, the more she is seen as physically attractive.

Gender (Factor A):	Male	Male	Male	Female	Female	Female
Tan (Factor B):	No Tan	Medium	Dark	No Tan	Medium	Dark
N_{AB}	4	4	4	4	4	4
Mean ($\bar{X}_{AB}$)	4.75	5.00	7.50	7.25	6.75	6.50
Standard deviation (s_{AB})	.96	.82	1.29	1.26	.96	1.29

a. Create a figure to represent the cell means.

b. Conduct the two-way ANOVA and create an ANOVA summary table.

c. Report the decisions regarding the null hypotheses and the level of significance for the main effects and the interaction effect.

d. Calculate measures of effect size (R^2) for the main effects and the interaction effect.

23. Characteristics of students may influence how favorably they're perceived by their teachers. One study examined whether teachers' perceptions of students may be influenced by the student's gender (male, female) and socioeconomic status (SES) (lower class, upper class) (Auwarter & Aruguete, 2008). They had a sample of teachers read a passage about a troubled student; in these passages, the researchers varied the student's gender and SES. After reading the passage, the teacher rated the student on his or her personal characteristics on a 1 to 5 scale; the higher the rating, the more positively the teacher viewed the student. The researchers hypothesized that male students would be rated more highly if they were of upper rather than lower SES, but the reverse was true for female students.

SES (Factor A):	Lower	Lower	Upper	Upper
Gender (Factor B):	Male	Female	Male	Female
	3	4	4	4
	4	2	3	2
	3	4	4	2
	2	5	3	4
	5	3	3	3
	3	4	4	2
	2	5	5	3

a. Calculate descriptive statistics ($N_{AB}, \bar{X}_{AB}, s_{AB}$) for each combination of the two factors.

b. Create a figure to represent the cell means.

c. Conduct the two-way ANOVA and create an ANOVA summary table.

d. Report the decisions regarding the null hypotheses and the level of significance for the main effects and the interaction effect.

e. Calculate measures of effect size (R^2) for the main effects and the interaction effect.

24. When children see adults arguing, what do they think or expect will happen? Two researchers had elementary school children watch a videotape of a man and a woman arguing (El-Sheikh & Elmore-Staton, 2007). In some of the videotapes, the adults were intoxicated, but in some, the adults were sober. After watching the videotape, each child was asked how likely either the man or the woman would be verbally or physically aggressive (the higher the score, the greater the likelihood of aggression). These researchers hypothesized that children expect higher levels of aggression

when they thought the adults were intoxicated versus if they were sober, particularly if the intoxicated adult was a man as opposed to a woman.

Condition (Factor A):	Intoxicated	Intoxicated	Sober	Sober
Sex of Adult (Factor B):	Man	Woman	Man	Woman
	2	5	2	3
	4	4	1	5
	3	5	3	3
	6	6	2	2
	3	3	3	3
	4	5	3	5

a. Calculate descriptive statistics ($N_{AB}, \bar{X}_{AB}, s_{AB}$) for each combination of the two factors.

b. Create a figure to represent the cell means.

c. Conduct the two-way ANOVA and create an ANOVA summary table.

d. Report the decisions regarding the null hypotheses and the level of significance for the main effects and the interaction effect.

e. Calculate measures of effect size (R^2) for the main effects and the interaction effect.

25. One way of improving job performance in organizations is through the use of "self-managed work teams": teams of employees given a certain amount of control (autonomy) over how they conduct their work activities. Previous research has suggested too little autonomy leads to bored, unmotivated employees; however, too much autonomy may result in a lack of structure and supervision. A researcher hypothesizes that the effect of autonomy on team performance may in fact depend on the size of the work team. She designs an experiment whereby teams are given a simple task to perform (the construction of a toy boat). Participants are randomly assigned to one of three sized work teams: Small (4 members), Medium (7 members), or Large (10 members). Each group is given either low or high autonomy in determining how the boats are to be constructed. She counts the number of toy boats built by each team in the designated time.

Team Size (Factor A):	Small	Small	Medium	Medium	Large	Large
Autonomy (Factor B):	Low	High	Low	High	Low	High
	2	4	4	6	7	1
	3	5	2	2	5	2
	1	5	4	5	4	2
	0	3	3	3	3	3

a. Calculate descriptive statistics ($N_{AB}, \bar{X}_{AB}, s_{AB}$) for each combination of the two factors.

b. Create a figure to represent the cell means.

c. Conduct the two-way ANOVA and create an ANOVA summary table.

d. Report the decisions regarding the null hypotheses and the level of significance for the main effects and the interaction effect.

e. Calculate measures of effect size (R^2) for the main effects and the interaction effect.

26. Earlier in this chapter, we mentioned a study looking at students' use of rituals related to taking exams (Rudski & Edwards, 2007). The researchers believed that two factors related to students' use of rituals are the importance of the exam to the student's grade in the course and the perceived difficulty of the exam. More specifically, they hypothesized that when an exam is low in importance, students will only engage in rituals if the exam is very difficult; however, when an exam is high in importance, students will engage in a high level of rituals regardless of the difficulty

of the exam. Listed below are the number of rituals performed by students preparing to take exams either low or high in importance to their course grades and believed to be easy, somewhat difficult, or very difficult:

Importance (Factor A):	Low	Low	Low	High	High	High
Difficulty (Factor B):	Easy	Somewhat	Very	Easy	Somewhat	Very
	0	2	4	3	4	3
	2	3	3	5	3	4
	1	0	5	3	5	4
	0	2	4	2	4	5
	3	2	3	5	4	4
	2	1	4	4	5	6

a. Calculate descriptive statistics $(N_{AB}, \overline{X}_{AB}, s_{AB})$ for each combination of the two factors.

b. Create a figure to represent the cell means.

c. Conduct the two-way ANOVA and create an ANOVA summary table.

d. Report the decisions regarding the null hypotheses and the level of significance for the main effects and the interaction effect.

e. Calculate measures of effect size (R^2) for the main effects and the interaction effect.

27. For each of the research situations described in Exercises 19 to 22, draw a figure that illustrates simple effects analyses appropriate to test the study's research hypothesis.

28. For each of the research situations described in Exercises 23 to 25, draw a figure that illustrates simple effects analyses appropriate to test the study's research hypothesis.

Answers to Learning Checks

Learning Check 1

2.

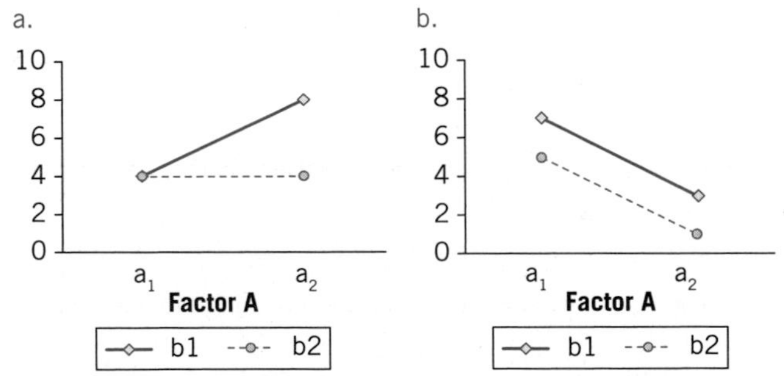

c.

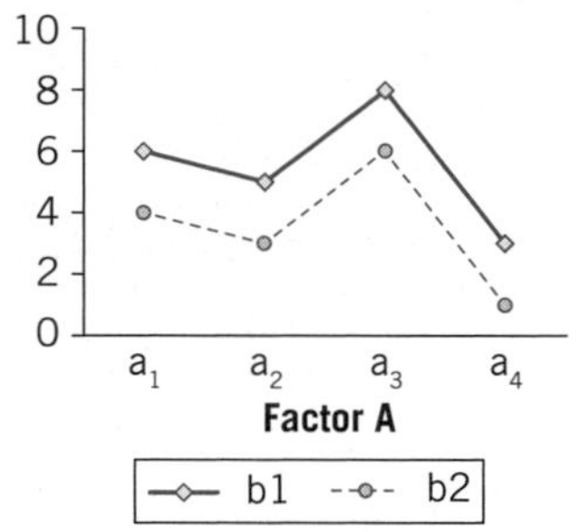

d.

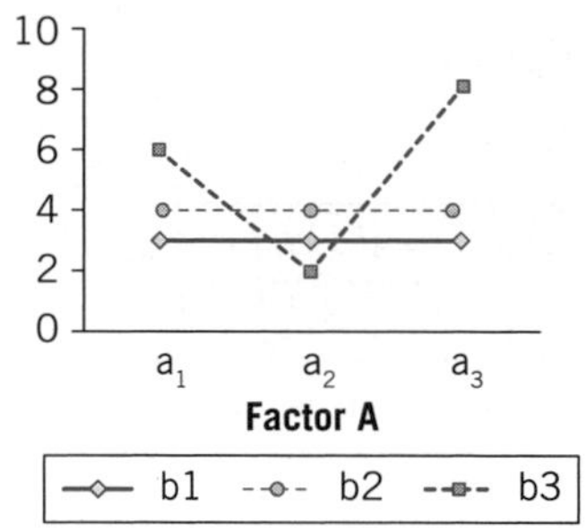

3. a. Present
   b. Absent
   c. Absent
   d. Present

4. a.

		Day			
		Monday	*Wednesday*	*Friday*	$\bar{X}_A$
Gender	Male	4	4	2	**3.33**
	Female	2	2	4	**2.67**
	$\bar{X}_B$	**3.00**	**3.00**	**3.00**	

b.

		Day			
		Monday	*Wednesday*	*Friday*	$\bar{X}_A$
Gender	Male	4	4	2	**4.00**
	Female	3.50	3.50	3.50	**3.50**
	$\bar{X}_B$	**3.75**	**3.75**	**3.75**	

c.

		Day			
		Monday	*Wednesday*	*Friday*	$\bar{X}_A$
Gender	Male	4	3	3	**3.33**
	Female	3	2	2	**2.33**
	$\bar{X}_B$	**3.50**	**2.50**	**2.50**	

d.

		Day			
		Monday	*Wednesday*	*Friday*	$\bar{X}_A$
Gender	Male	4	4	2	**3.00**
	Female	2	2	2	**2.00**
	$\bar{X}_B$	**3.00**	**2.50**	**2.00**	

Learning Check 2

2. a. Type of Course (Classroom, Online) and Living Arrangement (Off-campus, On-campus) (2 × 2)
   b. Attribution of Responsibility (Child, Parent) and Age of Child (Adolescent, Teenager) (2 × 2)
   c. Perceived Difficulty of Exam (Easy, Somewhat Difficult, Very Difficult) and Importance of Exam (Important, Unimportant) (3 × 2)

3. a.

		Factor A		
		a_1	a_2	$\bar{X}_B$
Factor B	b_1	3.00	4.00	3.50
	b_2	2.00	2.00	2.00
	$\bar{X}_A$	2.50	3.00	$\bar{X}_T = 2.75$

b.

		Factor A		
		a_1	a_2	$\bar{X}_B$
Factor B	b_1	8.50	6.50	7.50
	b_2	7.00	5.00	6.00
	b_3	11.00	9.00	10.00
	$\bar{X}_A$	8.83	6.83	$\bar{X}_T = 7.83$

c.

Factor A

Factor B	a_1	a_2	a_2	a_3	$\overline{X}_B$
b_1	1.53	2.10	2.49	2.36	2.12
b_2	2.27	2.19	1.93	1.87	2.07
$\overline{X}_A$	1.90	2.15	2.21	2.12	$\overline{X}_T = 2.09$

d.

Factor A

Factor B	a_1	a_2	a_3	$\overline{X}_B$
b_1	36.50	41.25	38.00	38.58
b_2	39.00	46.75	33.75	39.83
b_3	43.25	52.00	55.50	50.25
$\overline{X}_A$	39.58	46.67	42.42	$\overline{X}_T = 42.89$

4. a. Degrees of freedom: $df_A = 1$, $df_B = 1$, $df_{A \times B} = 1$, $df_{WG} = 16$
   Critical values: $F_A = 4.49$, $F_B = 4.49$, $F_{A \times B} = 4.49$
   b. Degrees of freedom: $df_A = 2$, $df_B = 1$, $df_{A \times B} = 2$, $df_{WG} = 30$
   Critical values: $F_A = 3.32$, $F_B = 4.17$, $F_{A \times B} = 3.32$
   c. Degrees of freedom: $df_A = 2$, $df_B = 2$, $df_{A \times B} = 4$, $df_{WG} = 81$
   Critical values: $F_A = 3.11$, $F_B = 3.11$, $F_{A \times B} = 2.48$
   d. Degrees of freedom: $df_A = 3$, $df_B = 2$, $df_{A \times B} = 6$, $df_{WG} = 72$
   Critical values: $F_A = 2.76$, $F_B = 3.15$, $F_{A \times B} = 2.25$

Learning Check 3

2. a.

Source	*SS*	*df*	*MS*	*F*	R^2
Factor A	20.00	1	20.00	5.00	.16
Factor B	10.00	1	10.00	2.50	.08
A × B interaction	15.00	1	15.00	3.75	.12
Error	80.00	20	4.00		
Total	125.00	23			

b.

Source	*SS*	*df*	*MS*	*F*	R^2
Factor A	54.00	1	54.00	5.94	.11
Factor B	25.00	1	25.00	2.75	.05
A × B interaction	16.00	1	16.00	1.76	.03
Error	400.00	44	9.09		
Total	495.00	47			

c.

Source	*SS*	*df*	*MS*	*F*	R^2
Factor A	4.60	2	2.30	2.20	.07
Factor B	9.79	1	9.79	9.37	.14
A × B interaction	10.31	2	5.16	4.94	.15
Error	43.87	42	1.04		
Total	68.57	47			

3. a.

b.

Source	*SS*	*df*	*MS*	*F*
Sex	5.12	1	5.12	2.60
Type of car	20.59	1	20.59	10.45
Sex x Type of car	11.55	1	11.55	5.86
Error	47.29	24	1.97	
Total	84.55	27		

c. Main effect of Sex: $F_A = 2.60 < 4.26$ ∴ do not reject H_0 ($p > .05$)

Main effect of Type of Car: $F_B = 10.45 > 7.82$ ∴ reject H_0 ($p < .01$)

Sex × Type of Car interaction: $F_{A \times B} = 5.86 > 4.26$ ∴ reject H_0 ($p < .05$)

d. Main effect of Sex: = .06

Main effect of Type of Car: = .24

Sex × Type of Car interaction: = .14

Learning Check 4

2.

a.

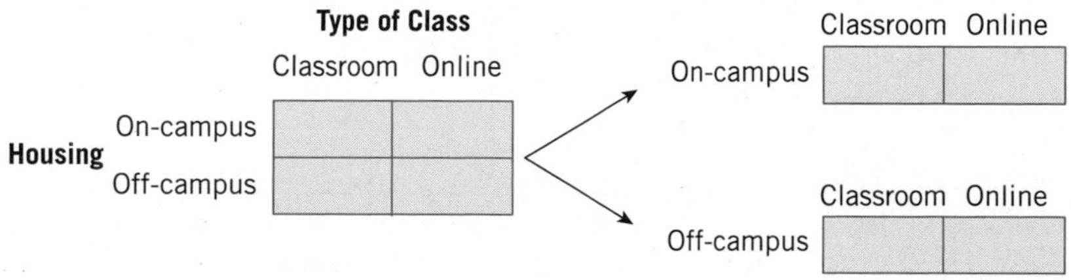

b.

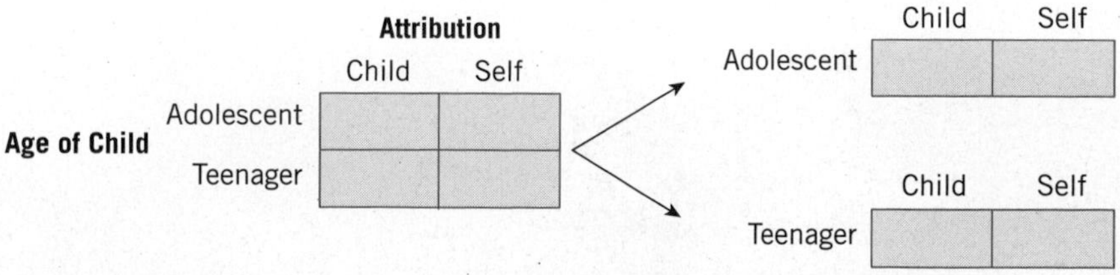

c.

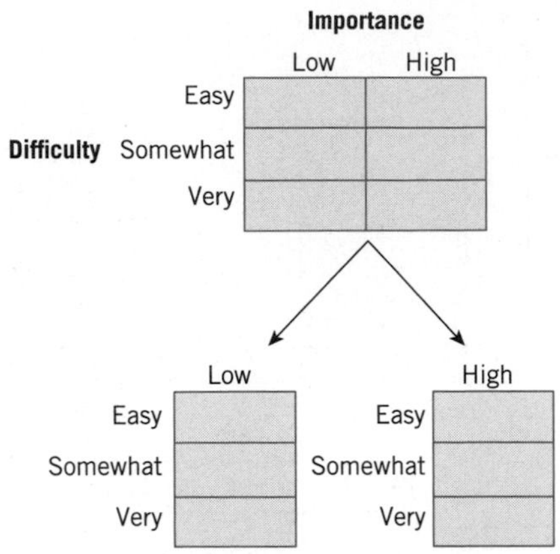

Answers to Odd-Numbered Exercises

1.

a.

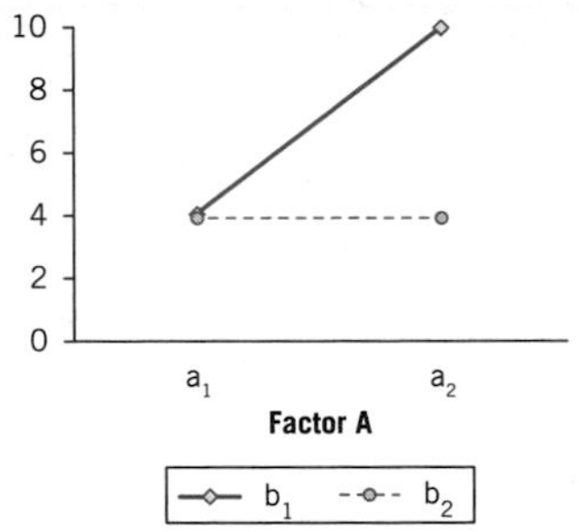

b.

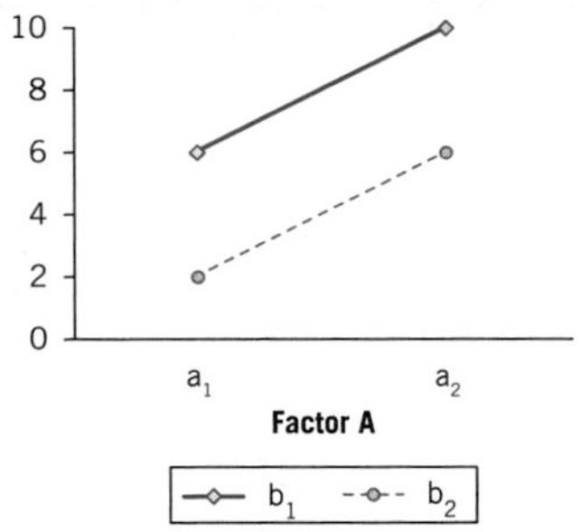

c.

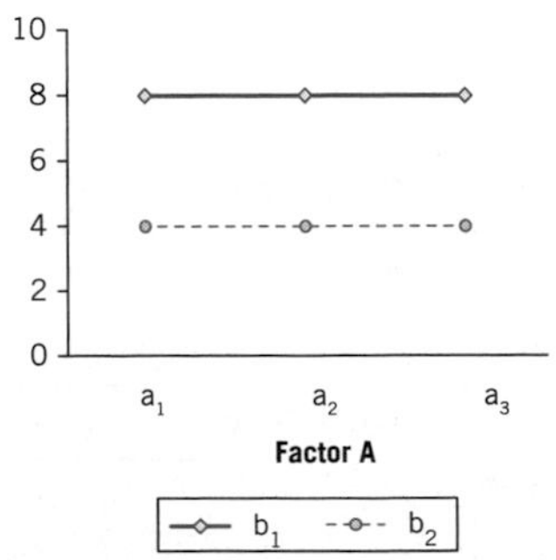

d.

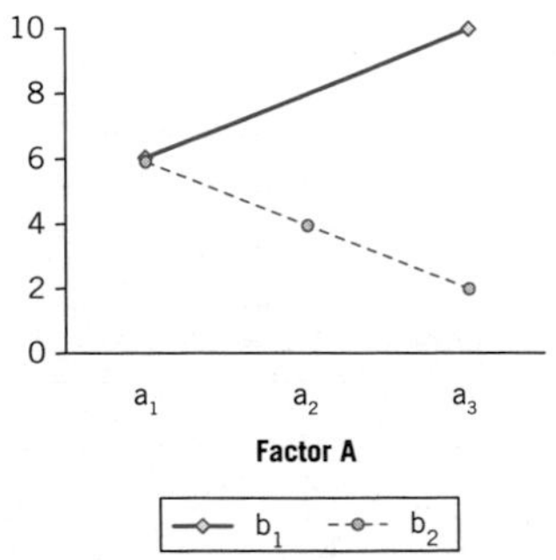

3. a.

	Factor A: a_1	a_2
Factor B: b_1	8	2
b_2	2	8

b.

	Factor A: a_1	a_2
Factor B: b_1	6	6
b_2	2	2

c.

	Factor A: a_1	a_2	a_3
Factor B: b_1	5	7	9
b_2	1	3	5

d.

	Factor A: a_1	a_2	a_3
Factor B: b_1	8	6	4
b_2	4	6	8

5. a.

	Factor A: a_1	a_2	$\overline{X}_B$
Factor B: b_1	4	10	**7**
b_2	4	4	**4**
$\overline{X}_A$	**4**	**7**	

Main effect, Factor A? **No**
Main effect, Factor B? **No**
A x B interaction effect? **Yes**

b.

	Factor A: a_1	a_2	$\overline{X}_B$
Factor B: b_1	6	10	**8**
b_2	2	6	**4**
$\overline{X}_A$	**4**	**8**	

Main effect, Factor A? **No**
Main effect, Factor B? **No**
A x B interaction effect? **Yes**

c.

	Factor A: a_1	a_2	a_3	$\overline{X}_B$
Factor B: b_1	8	8	8	**8**
b_2	4	4	4	**4**
$\overline{X}_A$	**6**	**6**	**6**	

Main effect, Factor A? **No**
Main effect, Factor B? **No**
A × B interaction effect? **Yes**

d.

	Factor A: a_1	a_2	a_3	$\overline{X}_B$
Factor B: b_1	6	8	10	**8**
b_2	6	4	2	**4**
$\overline{X}_A$	**6**	**6**	**6**	

Main effect, Factor A? **No**
Main effect, Factor B? **No**
A × B interaction effect? **Yes**

7\. a.

		Factor A a_1	Factor A a_2	$\bar{X}_B$
Factor B	b_1	8	2	**5**
	b_2	2	8	**5**
	$\bar{X}_A$	**5**	**5**	

Main effect, Factor A? **No**
Main effect, Factor B? **No**
A × B interaction effect? **Yes**

b.

		Factor A a_1	Factor A a_2	$\bar{X}_B$
Factor B	b_1	6	6	**6**
	b_2	2	2	**2**
	$\bar{X}_A$	**5**	**5**	

Main effect, Factor A? **No**
Main effect, Factor B? **Yes**
A × B interaction effect? **No**

c.

		Factor A a_1	Factor A a_2	Factor A a_3	$\bar{X}_B$
Factor B	b_1	5	7	9	**7**
	b_2	1	3	5	**3**
	$\bar{X}_A$	**3**	**5**	**7**	

Main effect, Factor A? **No**
Main effect, Factor B? **No**
A × B interaction effect? **Yes**

d.

		Factor A a_1	Factor A a_2	Factor A a_3	$\bar{X}_B$
Factor B	b_1	8	6	4	**6**
	b_2	4	6	8	**6**
	$\bar{X}_A$	**6**	**6**	**6**	

Main effect, Factor A? **No**
Main effect, Factor B? **No**
A × B interaction effect? **Yes**

9\. a. Exam (Handwritten, Typed) and Typing Ability (Low, High) (2 × 2)
b. Age (3, 5, 7) and Type of Book (Fiction, Nonfiction) (3 × 2)
c. Weight (Underweight, Normal, Overweight) and Gender (Male, Female) (3 × 2)
d. Amount of Coffee (One, Two, or Three Cups) and Time of Day (Morning, Afternoon, Evening) (3 × 3)
e. Achievement Motivation (Low, High) and Type of Test (Easy, Hard) (2 × 2)

11\. a.

		Factor A a_1	Factor A a_2	$\bar{X}_B$
Factor B	b_1	5.00	9.00	**7.00**
	b_2	3.00	7.00	**5.00**
	$\bar{X}_A$	**4.00**	**8.00**	$\bar{X}_T$ = **6.00**

b.

		Factor A a_1	Factor A a_2	$\bar{X}_B$
Factor B	b_1	3.50	1.25	**2.38**
	b_2	2.25	3.50	**2.88**
	b_3	5.75	4.25	**5.00**
	$\bar{X}_A$	**3.83**	**3.00**	$\bar{X}_T$ = **3.42**

c.

		Factor A a_1	Factor A a_2	Factor A a_3	$\bar{X}_B$
Factor B	b_1	4.28	4.61	4.95	**4.61**
	b_2	2.30	6.87	3.14	**4.10**
	b_3	7.06	5.52	2.26	**4.95**
	$\bar{X}_A$	**4.55**	**5.67**	**3.45**	$\bar{X}_T$ = **4.55**

13\. a. Degrees of freedom: $df_A = 1$, $df_B = 1$, $df_{A \times B} = 1$, $df_{WG} = 12$
Critical values: $F_A = 4.75$, $F_B = 4.75$, $F_{A \times B} = 4.75$
b. Degrees of freedom: $df_A = 1$, $df_B = 1$, $df_{A \times B} = 1$, $df_{WG} = 68$
Critical values: $F_A = 4.00$, $F_B = 4.00$, $F_{A \times B} = 4.00$

c. Degrees of freedom: $df_A = 2$, $df_B = 1$, $df_{A \times B} = 2$, $df_{WG} = 48$

Critical values: $F_A = 3.23$, $F_B = 4.08$, $F_{A \times B} = 3.23$

d. Degrees of freedom: $df_A = 1$, $df_B = 3$, $df_{A \times B} = 3$, $df_{WG} = 40$

Critical values: $F_A = 4.08$, $F_B = 2.84$, $F_{A \times B} = 2.84$

15. a.

Source	*SS*	*df*	*MS*	*F*	R^2
Factor A	12.00	1	12.00	3.00	.11
Factor B	24.00	1	24.00	6.00	.22
A × B interaction	8.00	1	8.00	2.00	.07
Error	64.00	16	4.00		
Total	108.00	19			

b.

Source	*SS*	*df*	*MS*	*F*	R^2
Factor A	3.91	1	3.91	5.37	.05
Factor B	5.64	1	5.64	7.74	.07
A × B interaction	3.12	1	3.12	4.28	.04
Error	67.04	92	.73		
Total	79.71	95			

c.

Source	*SS*	*df*	*MS*	*F*	R^2
Factor A	4.23	1	4.23	8.83	.23
Factor B	3.72	2	1.86	3.88	.20
A × B interaction	1.81	2	.91	1.89	.10
Error	8.62	18	.48		
Total	18.38	23			

d.

Source	*SS*	*df*	*MS*	*F*	R^2
Factor A	52.78	3	17.59	1.89	.12
Factor B	18.58	1	18.58	2.00	.04
A × B interaction	78.06	3	26.02	2.80	.17
Error	297.21	32	9.29		
Total	446.63	39			

e.

Source	*SS*	*df*	*MS*	*F*	R^2
Factor A	78.43	1	78.43	6.32	.09
Factor B	61.90	3	20.63	1.66	.07
A × B interaction	27.39	3	9.13	.74	.03
Error	695.32	56	12.42		
Total	863.04	63			

17. a. 2×2 design, $N_{AB} = 5$
 b. 2×2 design, $N_{AB} = 24$
 c. 2×3 design, $N_{AB} = 4$
 d. 4×2 design, $N_{AB} = 5$
 e. 2×4 design, $N_{AB} = 8$

19. a.

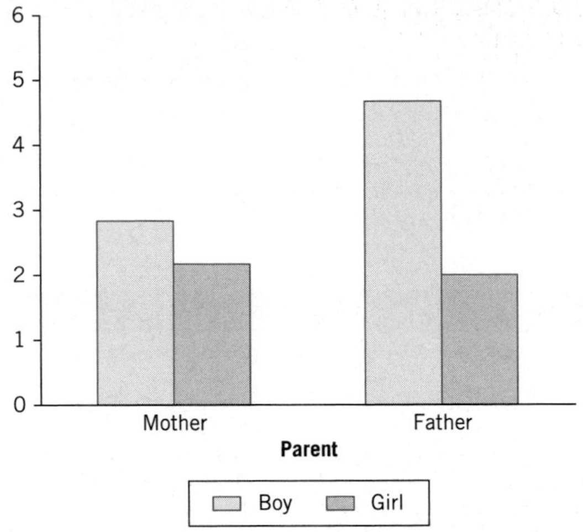

b.

Source	*SS*	*df*	*MS*	*F*
Factor A	16.67	1	16.67	15.88
Factor B	4.17	1	4.17	3.97
A × B interaction	6.00	1	6.00	5.17
Error	12.00	20	1.05	
Total	47.83	23		

c. Main effect of Child: $F_A = 15.88 > 8.10$ ∴ reject H_0 ($p < .01$)
 Main effect of Parent: $F_B = 3.97 < 4.35$ ∴ do not reject H_0 ($p > .05$)
 Child × Parent interaction: $F_{A \times B} = 5.71 > 4.35$ ∴ reject H_0 ($p < .05$)

d. Main effect of Child: $R^2_A = .35$
 Main effect of Parent: $R^2_B = .09$
 Child × Parent interaction: $R^2_{A \times B} = .13$

21. a.

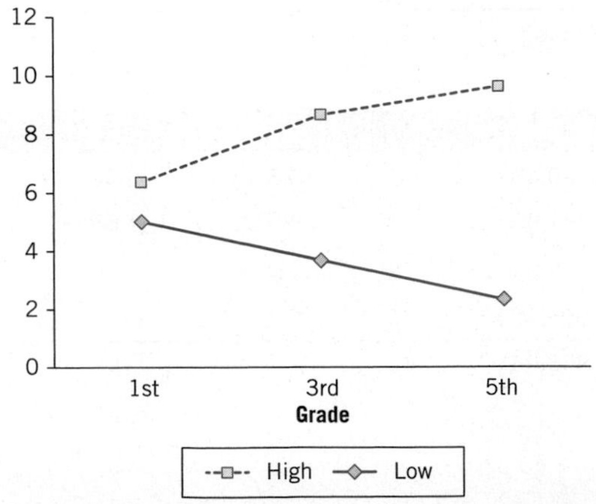

b.

Source	*SS*	*df*	*MS*	*F*
Grade	.78	2	.39	.20
Expectation	93.39	1	93.39	48.04
Grade × Expectation	27.44	2	13.72	7.06
Error	23.33	12	1.94	
Total	144.94	17		

c. Main effect of Grade: $F_A = .20 < 3.89 \therefore$ do not reject H_0 ($p > .05$)

Main effect of Expectation: $F_B = 48.04 > 9.33 \therefore$ reject H_0 ($p < .01$)

Grade × Expectation interaction: $F_{A \times B} = 7.06 > 6.93 \therefore$ reject H_0 ($p < .01$)

d. Main effect of Grade: $R_A^2 = .01$

Main effect of Expectation: $R_B^2 = .64$

Grade × Expectation interaction: $R_{A \times B}^2 = .19$

23. a.

SES (Factor A):	Lower	Lower	Upper	Upper
Gender (Factor B):	Male	Female	Male	Female
N_{AB}	7	7	7	7
Mean ($\bar{X}_{AB}$)	3.14	3.86	3.71	2.86
Standard deviation (s_{AB})	1.07	1.07	.76	.90

b.

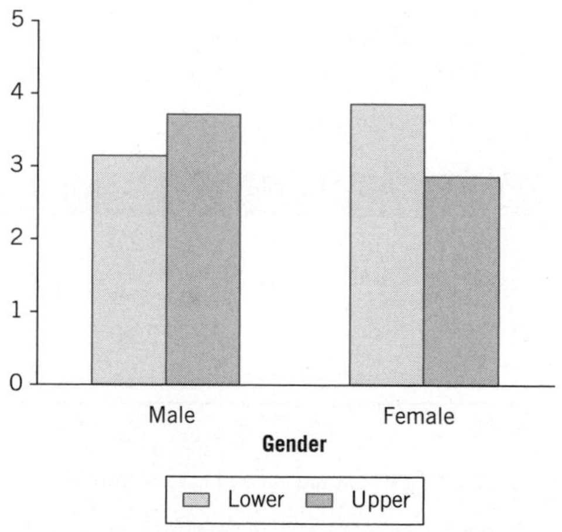

c.

Source	*SS*	*df*	*MS*	*F*
SES	.32	1	.32	.35
Gender	.04	1	.04	.04
SES × Gender	4.32	1	4.32	4.71
Error	22.00	24	.92	
Total	26.68	27		

d. Main effect of SES: $F_A = .35 < 4.23$ ∴ do not reject H_0 ($p > .05$)

Main effect of Gender: $F_B = .04 < 4.23$ ∴ do not reject H_0 ($p > .05$)

SES × Gender interaction: $F_{A \times B} = 4.71 > 4.23$ ∴ reject H_0 ($p < .05$)

e. Main effect of SES: $R_A^2 = .01$

Main effect of Gender: $R_B^2 = .001$

SES × Gender interaction: $R_{A\times B}^2 = .16$

25. a.

Team size (Factor A):	Small	Small	Medium	Medium	Large	Large
Autonomy (Factor B):	Low	High	Low	High	Low	High
N_{AB}	4	4	4	4	4	4
Mean ($\bar{X}_{AB}$)	1.50	4.25	3.25	4.00	2.00	3.33
Standard deviation (S_{AB})	1.29	.96	.96	1.83	1.71	1.82

b.

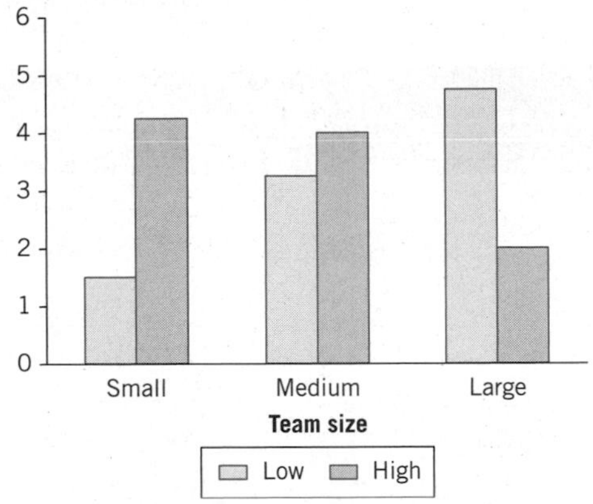

c.

Source	*SS*	*df*	*MS*	*F*
Team size	2.33	2	1.17	.67
Autonomy	.38	1	.38	.22
Team size × Autonomy	31.00	2	15.50	8.93
Error	13.25	18	1.74	
Total	64.96	23		

d. Main effect of Team Size: $F_A = .67 < 3.55$ ∴ do not reject H_0 ($p > .05$)

Main effect of Autonomy: $F_B = .22 < 4.41$ ∴ do not reject H_0 ($p > .05$)

Team Size × Autonomy interaction: $F_{A \times B} = 8.93 > 6.01$ ∴ reject H_0 ($p < .01$)

e. Main effect of Team Size: $R_A^2 = .04$

Main effect of Autonomy: $R_B^2 = .01$

Team Size × Autonomy interaction: $R_{A \times B}^2 = .48$

27. a. Simple effect of Child (Factor A) at each level of Parent (Factor B).

Parent	Child: Boy	Child: Girl
Mother	2.83	2.17
Father	4.67	2.00

	Boy	Girl
Mother	2.83	2.17

	Boy	Girl
Father	4.67	2.00

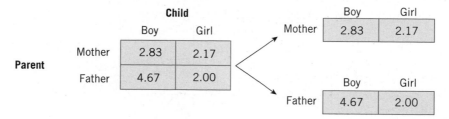

b. Simple effect of Time (Factor A) at each level of Controller (Factor B).

Controller	Time: 15 mins	Time: 30 mins
Standard	10.60	16.40
Gun	13.60	25.60

	15 mins	30 mins
Standard	10.60	16.40

	15 mins	30 mins
Gun	13.60	25.60

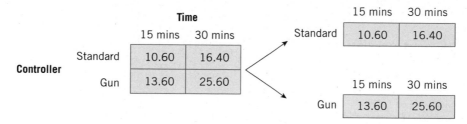

c. Simple effect of Expectation (Factor B) at each level of Grade (Factor A).

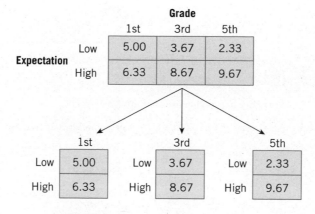

Expectation	Grade: 1st	Grade: 3rd	Grade: 5th
Low	5.00	3.67	2.33
High	6.33	8.67	9.67

	1st
Low	5.00
High	6.33

	3rd
Low	3.67
High	8.67

	5th
Low	2.33
High	9.67

d. Simple effect of Tan (Factor B) at each level of Grade (Factor A).

Gender

		Male	Female
Tan	No tan	4.75	7.25
	Medium	5.00	6.75
	Dark	7.50	6.50

	Male
No tan	4.75
Medium	5.00
Dark	7.50

	Female
No tan	7.25
Medium	6.75
Dark	6.50

SAGE edge™

Sharpen your skills with **SAGE edge at edge.sagepub.com/tokunaga2e**

SAGE edge for students provides a personalized approach to help you accomplish your coursework goals in an easy-to-use learning environment. Log on to access:

- eFlashcards
- Web Quizzes
- SPSS Data Files
- Video and Audio Resources

13

CORRELATION

CHAPTER OUTLINE

The research examples in earlier chapters of this book share one feature: They all involved comparing the means of groups. With independent variables such as Group (Intruder, No Intruder), Condition (Observation, Prediction, Explanation), and Message Type (Reward, Threat), the research and statistical hypotheses in these studies addressed whether differences in group means on the dependent variable were statistically significant. The data in these studies were analyzed with different versions of the analysis of variance (ANOVA), which is used when the independent variable or variables are categorical in nature. This chapter introduces statistical procedures used when the independent variable is continuous rather than categorical.

13.1 AN EXAMPLE FROM THE RESEARCH: IT'S GOOD FOR YOU!

Would you describe yourself as "high maintenance" or "low maintenance"? More specifically, when you were growing up, did you eat what your parents put on your plate or were you a picky eater? If you were picky, was there anything your parents should (or should not) have done to influence your eating behavior?

A team of researchers led by Lisa Fries was interested in seeing whether a picky eater's refusal to eat a new food was related to parents' feeding practices, which they defined as behaviors parents use to influence what a child eats (Fries et al., 2017). As you can imagine, parents employ a wide variety of feeding practices when trying to get their children to eat; the example in this chapter will focus on coercive prompts, in which parents offer the child rewards and/or pressure the child to eat. In their study, examples of coercive prompts included parents saying such things as, "If you eat your broccoli, you can have ice cream," "You only have a few bites left, then you can go," and "If you won't eat, I'm taking away the iPad." In terms of their research hypothesis, the researchers stated the following: "Based on the existing literature on feeding practices, we hypothesized that more controlling feeding practices would be associated with more observed food refusals" (Fries et al., 2017, p. 94).

For their study, which we will refer to as the picky eater study, the researchers recruited families who "had at least one toddler between the ages of 12–36 months, had one legal guardian over the age of 21 who was a stay-at-home caregiver for the child, had an annual household income over $30,000, and spoke only English in the home" (Fries et al., 2017, p. 94). They collected data from families with children the researchers categorized as either picky eaters or nonpicky eaters; in this example, we'll re-create their findings for the picky eaters using a sample of 18 families.

In deciding how to collect their data, the researchers stated that much of the research on picky eaters relied on parents' description of their children's behavior. But rather than relying on parents' recollections, they noted "there are very few studies that have explored picky eating through observing child behaviors" (Fries et al., 2017, p. 93). Furthermore, rather than observing the interaction between parents and children in a lab setting, the researchers wanted the behavior to occur in the child's home. Consequently, the families were given video cameras and asked to videotape a meal in which they fed their child a fruit or vegetable the child had never tasted before.

The researchers viewed the video recordings and recorded two pieces of information for each family. The first piece of information was the percentage of the total number of feeding practices that were coercive prompts. For example, if the parents performed 12 feeding practices, 4 of which were coercive prompts, the score for coercive prompts was 33 (4/12 = 33%). The second piece of information was the number of times the child refused to eat by displaying behaviors such as turning his or her head, pushing the food away, crying, and spitting out the food.

Therefore, the data in this example consist of two pieces of information for each of 18 families: the percentage of coercive prompts and the number of food refusals. Both of these variables are continuous in nature, with the possible values falling along a numeric continuum. For example, the coercive prompts had a possible range of 0 (0%) to 100 (100%); higher scores reflect a greater use of rewards and pressure in getting children to eat new foods. Note that this example differs from those in previous chapters, in which the independent variables were categorical and comprised different groups.

The coercive prompts and food refusals for the 18 families are listed in Table 13.1(a). Descriptive statistics for the two variables are provided in Table 13.1(b). In earlier chapters, descriptive statistics were calculated on a dependent variable for the groups that make up an independent variable. In Chapter 9, for example, the average time taken to leave a parking space was calculated for the Intruder and No Intruder conditions. In the picky eater study, however, statistics must be calculated on two variables rather than on one: coercive prompts and food refusals. As such, the two variables must be distinguished from each other. In Table 13.1(b), the coercive prompts variable has been labeled Variable X and the food refusals variable has been labeled Variable Y. The means for the two variables are represented by the symbols $\overline{X}$ and $\overline{Y}$, and the two standard deviations (s_X and s_Y) have the subscripts X and Y.

TABLE 13.1 ● COERCIVE PROMPTS AND FOOD REFUSALS IN THE PICKY EATER STUDY

(a) Raw Data

Family	Coercive Prompts	Food Refusals	Family	Coercive Prompts	Food Refusals
1	10	7	10	54	25
2	37	23	11	20	5
3	71	16	12	81	32
4	30	8	13	24	26
5	62	35	14	32	15
6	11	12	15	15	19
7	47	19	16	42	28
8	17	9	17	19	16
9	66	22	18	45	11

(b) Descriptive Statistics

	Coercive Prompts (X)	Food Refusals (Y)
N	18	18
Mean	$\overline{X} = 37.94$	$\overline{Y} = 18.22$
Standard deviation	$s_X = 21.92$	$s_Y = 8.79$

It's useful to create a figure to gain an initial understanding of data that have been collected. In previous chapters, a bar graph was used to compare the means of groups. However, the goal of the picky eater study is not to compare groups on a dependent variable but rather to see if scores on one variable are associated with scores on another variable. We might ask, for example, if a lower percentage of coercive prompts is associated with a relatively smaller number of food refusals.

When both variables being analyzed are continuous in nature, meaning they are measured along a numeric continuum, the relationship between scores on the variables may be illustrated with a figure known as a **scatterplot**: a graphical display of paired scores on two variables. Figure 13.1 presents a scatterplot of the data for the coercive prompts and food refusals variables. Coercive prompts (the *X* variable) is located along the horizontal axis; food refusals (the *Y* variable) has been placed along the vertical axis. The location of each family within the scatterplot is based on its two respective pieces of information, with each dot in the scatterplot representing a particular family. For example, the dot in the lower left of the scatterplot represents the pair of scores for the first family in Table 13.1(a), which had a score of 10 for coercive prompts and a score of 7 for food refusals.

Examining a figure such as a scatterplot provides an initial indication of how scores on two variables are associated with each other. Looking at the scatterplot in Figure 13.1, we see that a low (or high) use of coercive prompts is associated with a low (or high) number of food refusals. The next section discusses how the relationship between two variables may be described and characterized.

FIGURE 13.1 ● SCATTERPLOT OF DATA FOR THE PICKY EATER STUDY

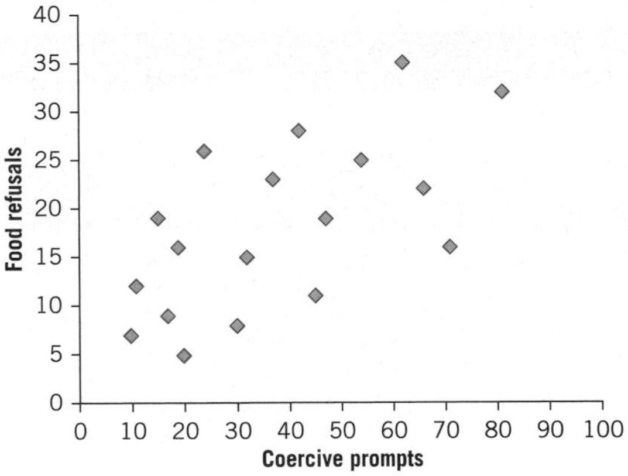

13.2 INTRODUCTION TO THE CONCEPT OF CORRELATION

A scatterplot provides an initial indication of whether scores on one variable are related to scores on another variable. The purpose of this section is to introduce the concept of **correlation**, which may be defined as a mutual or reciprocal relationship between variables such that systematic changes in the values of one variable are accompanied by systematic changes in the values of another variable. This section will discuss how the relationship between variables may be described.

Describing the Relationship Between Variables

In this book, the relationship between variables will be described along three aspects: the nature of the relationship, the direction of the relationship, and the strength of the relationship. Each of these three aspects is described and illustrated below.

Nature of the Relationship

The nature of the relationship between variables pertains to the manner in which changes in scores on one variable correspond to changes in scores on another variable. The nature of the relationship between two variables may take on different forms. For example, a **linear relationship** is a relationship between variables appropriately represented by a straight line, such that increases or decreases in one variable are associated with corresponding increases or decreases in another variable. The scatterplot in Figure 13.1 for the picky eater study illustrates a linear relationship, in that increases in coercive prompts are associated with increases in food refusals.

Figure 13.2(a) also illustrates a linear relationship between two variables. It's considered a linear relationship because the data in the scatterplot are located close to a line that's been generated to represent the relationship. Note that the line is tilted at an angle, meaning it's not parallel to the X-axis. The angle of the line indicates that changes in one variable are associated with changes in the other variable, which is part of the definition of correlation.

What would be the nature of the relationship if scores on one of the variables do not change? As an extreme example, what if every child, regardless of the number of coercive prompts, had the same number of food refusals? In this situation, knowing the number of coercive prompts would be unnecessary—you would, for example, predict the same number of food refusals for a child who receives 4 prompts, 19 prompts, or 35 prompts. This scenario, which would result in a perfectly horizontal line parallel to the X-axis, illustrates an important principle in assessing the relationship between two variables: To relate one variable to another, there must be differences in the scores of the variables.

It's also possible for the nature of the relationship between two variables to not be linear. A **nonlinear relationship** is a relationship between variables that is not appropriately represented by a straight line. For example, how many hours of sleep do you need each night to be at your best? One hour? Fourteen hours? For most of us, the answer lies somewhere between these two extremes, such that too little sleep makes one inattentive but too much sleep makes one lethargic. A proposed nonlinear relationship between "number of hours of sleep" and "alertness" is portrayed in Figure 13.2(b).

Although nonlinear relationships certainly do exist, linear relationships are more common. For this reason, the correlational statistics discussed in this chapter assume the relationship between two variables is linear in nature. Statistics have been developed to measure nonlinear relationships; discussion of these statistics may be found in advanced statistics books (see, e.g., Cohen & Cohen, 1983, chap. 6; Hays, 1988, pp. 698–701; Keppel & Zedeck, 1989, pp. 50–54; McNemar, 1969, pp. 315–316).

FIGURE 13.2 ● NATURE OF THE RELATIONSHIP BETWEEN TWO VARIABLES

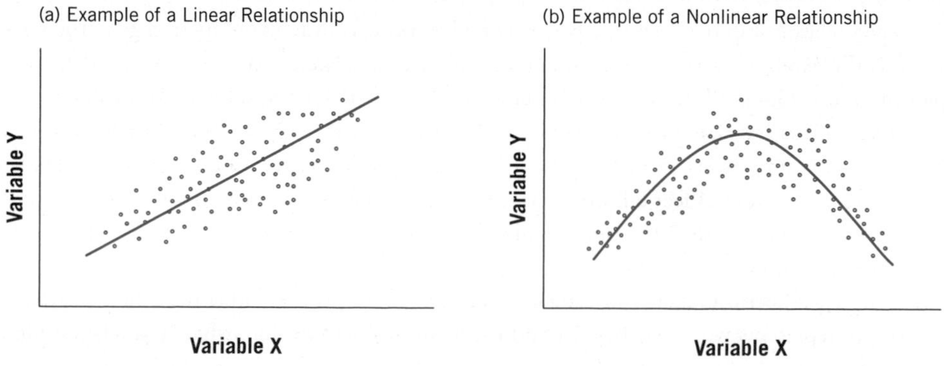

FIGURE 13.3 ● DIRECTION OF THE RELATIONSHIP BETWEEN TWO VARIABLES

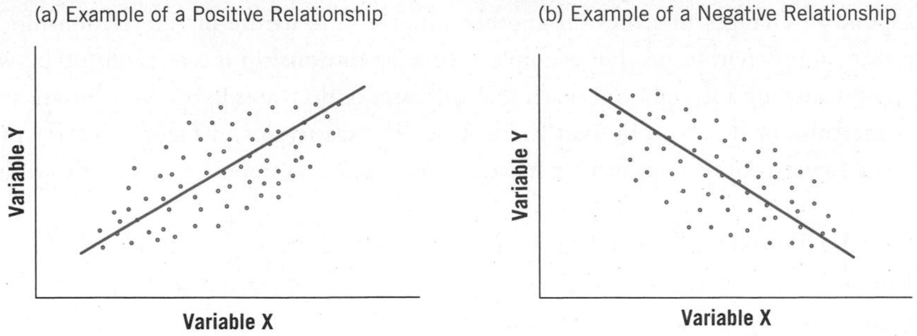

Direction of the Relationship

The second aspect of the relationship between variables refers to the direction in which changes in one variable are associated with changes in another. A **positive relationship** is a relationship in which increases in scores for one variable are associated with increases in scores of another variable ("as *X* goes up, *Y* goes up"). The scatterplot in Figure 13.3(a) is an example of a positive relationship. Note that the word *positive* indicates the direction of the relationship rather than its "value" or "goodness."

One study found a positive relationship between rumination (thinking about anger-provoking events after they've occurred) and dangerous driving behavior, such that the *higher* a person's tendency to ruminate, the *more* the person reported doing dangerous driving behaviors like speeding and tailgating (Suhr & Dula, 2017). The combined use of the words *higher* and *more* describes a positive relationship, one in which scores on variables move in the same direction.

A **negative relationship**, on the other hand, is a relationship in which increases in scores for one variable are associated with *decreases* in scores of another variable ("as *X* goes up, *Y* goes *down*"). The word *negative* does not imply the relationship is "bad" or is "worse" than a positive relationship. One example of a negative relationship was found in a study of families of college freshmen, where the *more* their parents engaged in "helicopter parenting," defined as the "hovering of parents over their children, ready to take responsibility for their decisions and their problems" (Darlow, Norvilitis, & Schuetze, 2017, p. 2), the *less* these freshmen adjusted to the social and academic demands of college. In a negative relationship, such as the one depicted in Figure 13.3(b), the scores on two variables move in opposite directions.

Strength of the Relationship

The third way a relationship between variables may be described is in terms of its strength. The strength of a relationship is the extent to which scores on one variable are associated with scores on another variable. The scatterplot in Figure 13.4(a) illustrates what is known as a "perfect" relationship: a relationship in which each score for one variable is associated with only one score for the other variable. In a perfect relationship, all of the data fall exactly on a line. What if, for example, there was a perfect relationship between people's height and weight? If so, everyone who is 4 feet tall would have the exact same weight (perhaps 90 pounds), everyone who is 6 feet tall would weigh exactly 205 pounds, and so on. In such a case, knowing a person's height enables one to perfectly predict that person's weight.

As you may suspect, perfect relationships are the incredibly rare exception rather than the rule. The scatterplot in Figure 13.4(b) represents what may be referred to as a "strong" relationship, which is a relationship in which a score on one variable is associated with a relatively small range of scores on another variable. If we draw a line around all of the points in this scatterplot, it has a narrow elliptical shape like a football. When there is a strong relationship between two variables, knowing the score on one variable allows us to predict the score on the other variable within a small range. The frequently observed correlation between height and weight (e.g., the expectation that a man who is 5′ 10″ will typically weigh between 170 and 190 pounds) is an example of a strong relationship.

Figures 13.4(c) demonstrates a "moderate" relationship; note that as a relationship becomes weaker, a figure drawn around the data becomes less elliptical and more rounded. Finally, the last scatterplot (Figure 13.4(d)) illustrates a zero relationship between two variables. A "zero" relationship exists when each score on one variable is associated with a wide range of scores on another variable. If we were to encapsulate all of the points in this scatterplot, it would look like a circle. For example, what if we related a person's height with his or her IQ? If we assume there is no relationship between these two variables, any particular height will be associated with a wide range of IQs. When there is no relationship between two variables, knowing the score on one variable doesn't allow us to predict the score on the other variable with any degree of precision.

FIGURE 13.4 ● STRENGTH OF THE RELATIONSHIP BETWEEN TWO VARIABLES

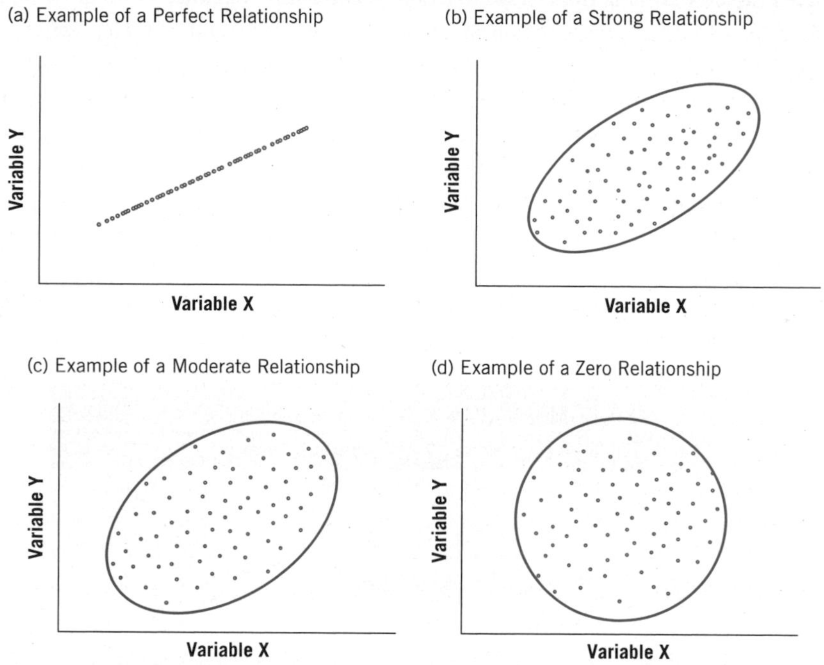

Measuring the Relationship Between Variables: Correlational Statistics

Scatterplots are very useful in gaining an initial sense of the relationship between variables. To measure and test these relationships, however, researchers calculate what are known as **correlational statistics**, which are statistics designed to measure relationships between variables. There are a number of different correlational statistics, with the choice of statistic depending on how variables are measured. The most commonly used correlational statistic was developed in 1895 by Karl Pearson, an influential mathematician who, among his many achievements, founded the world's first statistics department at London's University College in 1911. This statistic is introduced and defined in the next section.

Introduction to the Pearson Correlation Coefficient (r)

The **Pearson correlation coefficient**, represented by the symbol r, is a statistic that measures the linear relationship between two continuous variables measured at the interval or ratio level of measurement. The Pearson r is designed to measure the three aspects of a relationship mentioned earlier: nature, direction, and strength.

First, in terms of the nature of the relationship, the Pearson r assesses the degree of *linear* relationship between two variables, meaning that it assumes the relationship may be represented by a straight line, with increases or decreases in one variable corresponding to increases or decreases in the other variable (see Figure 13.2(a)). If the relationship between the variables is not linear, the Pearson correlation coefficient would not be the appropriate statistic to measure the relationship. This highlights the importance of examining scatterplots of data before calculating any statistics on the data.

Second, the sign (+ or –) of the Pearson correlation coefficient indicates the direction of the relationship, with the possible values of r being either positive or negative. For example, a positive value of r (e.g., .14 or .63) indicates a positive relationship like the one depicted in Figure 13.3(a). Negative values of r (e.g., –.09 or –.28) represent negative relationships like the one in Figure 13.3(b). Similar to the z-score and t-statistic, the sign of the correlation is typically only reported if it is a negative (–) value.

Finally, the strength of the relationship is represented by the numeric value of the correlation, with the possible values of the Pearson r ranging from –1.00 to +1.00. A correlation of either +1.00 or –1.00 represents the strongest relationship possible: a perfect relationship. Note that both +1.00 and –1.00 represent perfect relationships, in that the sign (+ or –) indicates the direction of the relationship rather than its strength. Figure 13.4(a) presents a perfect positive relationship in which $r = 1.00$. Between the two maximum values of –1.00 and 1.00, the specific value of the Pearson r represents the strength of the relationship. For example, a correlation with the value of .35 represents a stronger relationship than a correlation of .29, a weaker relationship than a correlation of –.46, and the same strength as a correlation of –.35. The midpoint of the range of possible values of the Pearson r is zero (.00), which implies the complete absence of a relationship between the two variables (see Figure 13.4(d)).

Jacob Cohen (1988), who also pioneered the measures of effect size introduced in Chapter 10, provided the following guidelines for interpreting values of the Pearson r in terms of its strength:

	Pearson r	
Strength of Relationship	**Negative**	**Positive**
Weak	–.10 to –.30	.10 to .30
Moderate	–.30 to –.50	.30 to .50
Strong	–.50 to –1.00	.50 to 1.00

According to these guidelines, a relationship between two variables is considered "weak" if the calculated value of the Pearson r is between .10 and .30 (ignoring the ± sign). A relationship is considered "moderate" if r is between .30 and .50, and "strong" if r exceeds ±.50. Keep in mind that although these guidelines aid in the interpretation of a calculated value of r, they do not take into account critical factors such as the size of the sample or whether the relationship is statistically significant.

Calculating Correlational Statistics: The Role of Variance and Covariance

The Pearson correlation coefficient is a statistic measuring the linear relationship between two variables. Before we present and illustrate formulas used to calculate the Pearson r, this section discusses the conceptual basis of these formulas to help explain what's involved in measuring the relationship between variables.

Correlational statistics are based on variance—a concept you've been introduced to in earlier chapters of this book. In relating two variables with each other, we begin with the knowledge that both variables contain variance, meaning there are differences among scores for the two variables. In the picky eater study, for example, there were differences in parents' use of coercive prompts and differences in children's food refusals. However, in addition to variance, the relationship between two variables is also based on the extent to which differences in the scores for the two variables are systematically connected to each other. In other words, measuring the relationship between variables involves both the extent to which two variables vary on their own and the extent to which they vary together.

Covariance refers to the extent to which variables vary together, which means the variables have shared variance (i.e., variance in common with each other). Height and weight provide a useful example of variables that have both variance and covariance. There is variance in each of these two variables in that people differ in how tall they are and in how much they weigh. In addition, height and weight have covariance: Generally speaking, someone who is taller (or shorter) than the average person generally weighs more (or less) than the average person. For this reason, we can say that height and weight vary together, or covary.

The following conceptual formula for the Pearson correlation coefficient takes into account both the amount of covariance between Variables X and Y (the extent to which X and Y vary together) and the variance of Variables X and Y (the extent to which X and Y vary on their own):

$$r = \frac{\text{covariance}(X,Y)}{\sqrt{\text{variance}(X)\,\text{variance}\,(Y)}}$$

$$= \frac{\text{extent to which } X \text{ and } Y \text{ vary together}}{\text{extent to which } X \text{ and } Y \text{ vary on their own}}$$

We see from the above formula that, like all of the statistics we've calculated in this book, the Pearson correlation coefficient is calculated by dividing one quantity by another. More specifically, the Pearson r is a ratio of the covariance between two variables (the extent to which variables vary together) to the variance of the two variables (the extent to which the variables vary on their own). When the amount of covariance is high relative to the variance of the two variables, a strong relationship exists between the two variables and the value for r will be high (closer to ± 1.00). However, if the amount of covariance is relatively low, there will be a weak relationship between the two variables, resulting in a smaller value of r. If there is absolutely no covariance between two variables, such as the relationship between height and IQ, the Pearson r will be equal to zero (.00).

LEARNING CHECK 1:

Reviewing What You've Learned So Far

1. Review questions
   a. For what reasons would you create a scatterplot for a set of data?
   b. What three aspects can be used to describe the relationship between variables?
   c. What is the difference between a linear relationship and a nonlinear relationship?
   d. What is the difference between a positive relationship and a negative relationship? Is a positive relationship better than a negative relationship?
   e. What is the difference between a perfect relationship, a moderate relationship, and a zero relationship?
   f. What does the Pearson correlation coefficient (r) measure?
   g. What are the main features of the Pearson r?
   h. What is covariance? What roles do covariance and variance play in calculating the Pearson r?

(Continued)

(Continued)

2. For each of the following situations, create a scatterplot of the data and describe the nature, direction, and strength of the relationship.
   a. A teacher hypothesizes that the more days of school a student misses, the worse the student will do on the final exam (possible scores on the exam range from 0 to 20).

Student	# Days Missed	Final Exam Score
1	3	16
2	2	18
3	5	13
4	8	7

Student	# Days Missed	Final Exam Score
5	4	12
6	7	11
7	6	14

   b. A researcher predicts that the more often a student raises his or her hand in class, the more favorably the teacher will view the student. She counts the number of times students raise their hands during a week and then has the teacher rate each student on a 1 (low) to 10 (high) scale.

Student	# Times Raise Hands	Teacher Rating
1	9	4
2	4	2
3	8	9
4	5	8
5	6	1

Student	# Times Raise Hands	Teacher Rating
6	7	8
7	4	7
8	3	5
9	6	6
10	7	3

13.3 INFERENTIAL STATISTICS: PEARSON CORRELATION COEFFICIENT

Returning to the picky eater study introduced at the beginning of the chapter, the scatterplot in Figure 13.1 indicates there is a relationship between coercive prompts and food refusals, which suggests that higher parents' use of coercion is associated with more children's resistance to eating new foods. To determine whether the relationship is statistically significant, an inferential statistic must be calculated. This requires following the steps used in earlier chapters:

- state the null and alternative hypotheses (H_0 and H_1),
- make a decision about the null hypothesis,
- draw a conclusion from the analysis, and
- relate the result of the analysis to the research hypothesis.

The difference between this chapter and previous ones is that the inferential statistic calculated is not a version of ANOVA but rather a correlation coefficient, which measures the relationship between variables rather than differences between groups.

State the Null and Alternative Hypotheses (H_0 and H_1)

In previous chapters, we've examined research studies that tested differences between the means of groups. In these studies, the null hypothesis was that the means of the groups in the larger population were equal to each other, implying no (zero) difference between the groups. Alternatively, the alternative hypothesis stated that the population means were not equal to each other, implying that differences between groups were not equal to zero. A similar logic is the basis for the null and alternative hypotheses for the correlation between two variables.

In testing a correlation between two variables, the null hypothesis states that the correlation in the population (represented by the Greek letter ρ [rho]) is equal to zero, implying that the two variables are not related to each other:

$$H_0: \rho = 0$$

What is a mutually exclusive alternative to the belief that the population correlation ρ is zero? One alternative hypothesis is that the correlation in the population (ρ) is *not* equal to zero:

$$H_1: \rho \neq 0$$

Because the above alternative hypothesis $\rho \neq 0$ is nondirectional (two-tailed), the null hypothesis will be rejected if the sample's value for the correlation is of sufficient size in either a positive or negative direction. As in the case of the *t*-test, it's possible to state a directional (one-tailed) alternative hypothesis in which the null hypothesis will only be rejected if the calculated correlation is in the hypothesized direction. For example, if we believed that the relationship between coercive prompts and food refusals scores could *only* be positive, we could state the alternative hypothesis as "H_1: $\rho > 0$." However, it's more customary to use a two-tailed test to allow for the possibility of a relationship in the unexpected, opposite direction.

Make a Decision About the Null Hypothesis

Once the null and alternative hypotheses have been stated, the next step is to make a decision whether to reject the null hypothesis that the population correlation is equal to zero. This decision is made using the following steps:

- calculate the degrees of freedom (*df*);
- set alpha (α), identify the critical values, and state a decision rule;
- calculate a statistic: Pearson correlation coefficient (*r*);
- make a decision whether to reject the null hypothesis;
- determine the level of significance; and
- calculate a measure of effect size (r^2).

Each of these steps is discussed below, using the picky eater study to illustrate how each step may be completed.

Calculate the Degrees of Freedom (*df*)

The first step in making the decision about the null hypothesis is to calculate the degrees of freedom (*df*) associated with the correlation coefficient. In previous examples, the degrees of freedom was equal to the number of scores for a variable minus 1 ($N - 1$). However, in a study such as the picky eater study, data are collected on *two* variables rather than one. Therefore, the formula for degrees of freedom for the correlation coefficient is as follows:

$$df = N - 2 \tag{13-1}$$

where N equals the number of pairs of scores involved in the computation of the correlation.

In the picky eater study, data were collected from 18 families. Therefore, the degrees of freedom is equal to

$$\begin{aligned} df &= N - 2 \\ &= 18 - 2 \\ &= 16 \end{aligned}$$

Set Alpha (α), Identify the Critical Values, and State a Decision Rule

The next step is to set alpha (α), which is the probability of the correlation needed to reject the null hypothesis, and identify the critical values of the correlation. As in earlier chapters, alpha is traditionally set at .05, meaning the null hypothesis is rejected when the probability of obtaining the value of the correlation when the null hypothesis is true is less than .05. Because the alternative hypothesis in the picky eater study is nondirectional (H_1: $\rho \neq 0$), alpha may be stated as "$\alpha = .05$ (two-tailed)."

Table 6 in the back of this book provides a table of critical values for *r*. This table contains a series of rows corresponding to different degrees of freedom (*df*). For each *df*, the critical values for *r* have been provided for both a directional (one-tailed) and a nondirectional (two-tailed) alternative hypothesis. For the picky eater study, we identify the critical values by moving down the *df* column of Table 6 until we reach the row for $df = 16$. Moving to the right until we reach the $\alpha = .05$ column for a two-tailed test, we find a critical value of .468. Therefore, the critical values may be stated as follows:

$$\text{For } \alpha = .05 \text{ (two-tailed) and } df = 16, \text{ critical values} = \pm.468$$

Figure 13.5 illustrates the critical values and the regions of rejection and nonrejection for the picky eater study. Note that the theoretical distribution for the Pearson *r* resembles that of the *t*-statistic in that both are symmetrical with a hypothetical mean of zero (0).

FIGURE 13.5 ● CRITICAL VALUES AND REGIONS OF REJECTION AND NONREJECTION FOR THE PICKY EATER STUDY

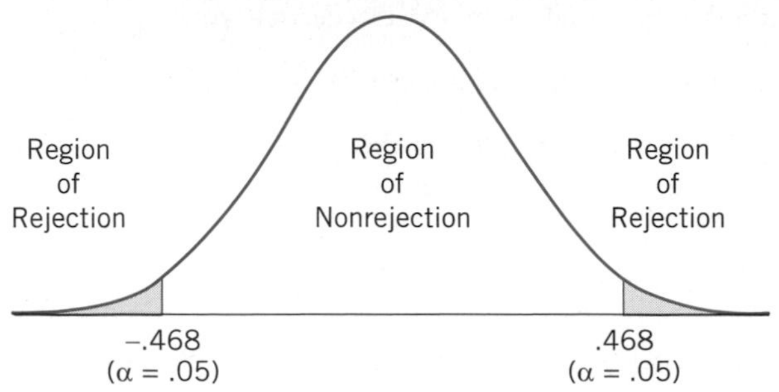

Once the critical values of r have been identified, we may state a decision rule specifying when the null hypothesis will be rejected. For the picky eater study, the following decision rule is stated:

$$\text{If } r < -.468 \text{ or } > .468, \text{ reject } H_0; \text{ otherwise, do not reject } H_0$$

This decision rule implies that we'll reject the null hypothesis if the value of the Pearson r calculated from the sample is either less than −.468 or greater than +.468.

Calculate a Statistic: Pearson Correlation Coefficient (r)

The next step in making the decision about the null hypothesis is to calculate a value of an inferential statistic. To introduce the formula for the Pearson r, let's examine the conceptual formula presented earlier:

$$r = \frac{\text{covariance}(X,Y)}{\sqrt{\text{variance}(X)\,\text{variance}\,(Y)}}$$

Looking at the above formula, we find that three quantities must be represented to calculate the Pearson r: the covariance between Variables X and Y, the variance of Variable X, and the variance of Variable Y. Each of these is discussed below.

Represent the covariance between X *and* Y: SP_{XY}. Covariance refers to the extent to which the variance in the two variables is related to each other. Recall from Chapter 4 that variance is the average squared deviation of a score from the mean. Therefore, the variance of the X variable is based on the deviation of each X score from the mean of the X scores $(X - \bar{X})$, and the variance of the Y variable is based on the deviation of each Y score from its mean $(Y - \bar{Y})$. Because covariance measures the degree to which two variables vary together, covariance involves relating these two deviations to each other.

Measuring the covariance in a set of data consists of connecting the two deviations $(X - \bar{X})$ and $(Y - \bar{Y})$ for each participant. One way to do this is to multiply the two deviations with each other $[(X - \bar{X})(Y - \bar{Y})]$. As the number created by multiplication is called a "product," the covariance between two variables may be represented by the sum of these products, represented by the symbol SP_{XY} (sum of products between variables X and Y). The definitional formula for SP_{XY} is provided in Formula 13-2:

$$SP_{XY} = \Sigma[(X - \bar{X})(Y - \bar{Y})] \qquad (13\text{-}2)$$

where X is a score on the X variable, $\bar{X}$ is the mean of the X variable, Y is a score on the Y variable, and $\bar{Y}$ is the mean of the Y variable.

The value for SP_{XY} for the picky eater study is calculated in Table 13.2. The first step in calculating SP_{XY} is to calculate the deviations of each participant's scores on the X variable $(X - \bar{X})$ and the Y variable $(Y - \bar{Y})$. For example, for the first family, $X - \bar{X} = 10 - 37.94 = -27.94$, and $Y - \bar{Y} = 7 - 18.22 = -11.22$. Once the two deviations have been determined, we multiply them together $[(X - \bar{X})(Y - \bar{Y})]$; for the first family, this product is equal to (−27.94) (−11.22) = 313.60. The products for the 18 families in the picky eater study are provided in the last column of Table 13.2.

Once each of the products have been calculated, the value for SP_{XY} is the sum of the products. For the picky eater study, SP_{XY} is calculated as follows:

$$\begin{aligned} SP_{XY} &= \Sigma[(X - \bar{X})(Y - \bar{Y})] \\ &= 313.60 + -4.51 + \ldots + 42.10 + -50.96 \\ &= 2057.22 \end{aligned}$$

TABLE 13.2 ● CALCULATING THE SUM OF PRODUCTS (SP_{XY}) IN THE PICKY EATER STUDY

Family	Coercive Prompts (X)	Food Refusals (Y)	$X-\bar{X}$	$Y-\bar{Y}$	$[(X-\bar{X})(Y-\bar{Y})]$
1	10	7	−27.94	−11.22	313.60
2	37	23	−.94	4.78	−4.51
3	71	16	33.06	−2.22	−73.46
4	30	8	−7.94	−10.22	81.21
5	62	35	24.06	16.78	403.60
6	11	12	−26.94	−6.22	167.65
7	47	19	9.06	.78	7.04
8	17	9	−20.94	−9.22	193.15
9	66	22	28.06	3.78	105.99
10	54	25	16.06	6.78	108.82
11	20	5	−17.94	−13.22	237.27
12	81	32	43.06	13.78	593.21
13	24	26	−13.94	7.78	−108.46
14	32	15	−5.94	−3.22	19.15
15	15	19	−22.94	.78	−17.85
16	42	28	4.06	9.78	39.65
17	19	16	−18.94	−2.22	42.10
18	45	11	7.06	−7.22	−50.96
	$\bar{X} = 37.94$	$\bar{Y} = 18.22$			$SP_{XY} = 2{,}057.22$

Represent the variance of Variable X: SS_X. The second component of the Pearson r involves the variance of Variable X. Because variance is based on the squared deviation of scores from the mean, the variance of the X variable may be represented by the sum of the squared deviations for the X variable $(\Sigma(X-\bar{X})^2)$. Formula 13-3 shows the definitional formula for this sum of squared deviations (represented by SS_X):

$$SS_X = \Sigma(X-\bar{X})^2 \tag{13-3}$$

where X is a score on the X variable and $\bar{X}$ is the mean of the X variable.

For the picky eater study, SS_X is calculated in the middle columns of Table 13.3. Calculating SS_X involves squaring the deviations that were calculated for the X variable in Table 13.2. For the first family, for example, $(X-\bar{X})^2 = (-27.94)^2 = 780.89$. Once the deviations have been squared, they are summed to calculate SS_X:

$$\begin{aligned} SS_X &= \Sigma(X-\bar{X})^2 \\ &= 780.89 + .89 + \ldots + 358.89 + 49.78 \\ &= 8{,}164.93 \end{aligned}$$

Represent the variance of Variable Y: SS_Y. In a manner similar to Variable X, the variance for the Y variable is represented by the sum of the squared deviations for the Y variable ($\Sigma(Y-\overline{Y})^2$). The definitional formula for the sum of squares for the Y variable (SS_Y) is presented in Formula 13-4:

$$SS_Y = \Sigma(Y-\overline{Y})^2 \quad (13\text{-}4)$$

where Y is a score on the Y variable, and $\overline{Y}$ is the mean of the Y variable.

SS_Y is calculated for the picky eater study in the final columns of Table 13.3:

$$\begin{aligned} SS_Y &= \Sigma(Y-\overline{Y})^2 \\ &= 125.94 + 22.83 + \ldots + 4.94 + 52.16 \\ &= 1{,}313.10 \end{aligned}$$

Before moving on, it's useful to point out something that may prevent errors in calculating the Pearson r. The first component of the Pearson r, SP_{XY}, is calculated by multiplying together two deviations: $(X-\overline{X})$ and $(Y-\overline{Y})$. Because either deviation can either be positive or negative, the product of these deviations, and ultimately the value for SP_{XY}, can be positive or negative. However, because SS_X and SS_Y are based on *squared* deviations, which can only be positive numbers, the values for SS_X and SS_Y must always be positive.

TABLE 13.3 ● CALCULATING DEVIATIONS TO REPRESENT VARIANCES IN THE PICKY EATER STUDY

Family	Coercive Prompts (X)	Food Refusals (Y)	$X-\overline{X}$	$(X-\overline{X})^2$	$Y-\overline{Y}$	$(Y-\overline{Y})^2$
1	10	7	−27.94	780.89	−11.22	125.94
2	37	23	−.94	.89	4.78	22.83
3	71	16	33.06	1,092.67	−2.22	4.94
4	30	8	−7.94	63.11	−10.22	104.49
5	62	35	24.06	578.67	16.78	281.49
6	11	12	−26.94	726.00	−6.22	38.72
7	47	19	9.06	82.00	.78	.60
8	17	9	−20.94	438.67	−9.22	85.05
9	66	22	28.06	787.11	3.78	14.27
10	54	25	16.06	257.78	6.78	45.94
11	20	5	−17.94	322.00	−13.22	174.83
12	81	32	43.06	1,853.78	13.78	189.83
13	24	26	−13.94	194.45	7.78	60.49
14	32	15	−5.94	35.34	−3.22	10.38
15	15	19	−22.94	526.45	.78	.60
16	42	28	4.06	16.45	9.78	95.60
17	19	16	−18.94	358.89	−2.22	4.94
18	45	11	7.06	49.78	−7.22	52.16
	$\overline{X} = 37.94$	$\overline{Y} = 18.22$		$\Sigma(X-\overline{X})^2 = \mathbf{8164.93}$		$\Sigma(Y-\overline{Y})^2 = \mathbf{1{,}313.10}$

Calculate the Pearson correlation coefficient (r). Now that formulas for the covariance and variances have been provided, they may be inserted into the conceptual formula for the Pearson r provided earlier:

$$r = \frac{\text{covariance}(X,Y)}{\sqrt{\text{variance}(X)\text{variance}(Y)}} = \frac{SP_{XY}}{\sqrt{(SS_X)(SS_Y)}}$$

The symbols SP_{XY}, SS_X, and SS_Y may be replaced by the mathematical operations in Formulas 13-2 to 13-4 to create the definitional formula for the Pearson correlation coefficient:

$$r = \frac{SP_{XY}}{\sqrt{(SS_X)(SS_Y)}} \frac{\Sigma[(X - \bar{X})(Y - \bar{Y})]}{\sqrt{(\Sigma(X - \bar{X})^2)(\Sigma(Y - \bar{Y})^2)}} \tag{13-5}$$

where X is a score on the X variable, $\bar{X}$ is the mean of the X variable, Y is a score on the Y variable, and $\bar{Y}$ is the mean of the Y variable.

Using the values of SP_{XY}, SS_X, and SS_Y calculated earlier, we can determine the Pearson r for the picky eater study in the following manner:

$$\begin{aligned} r &= \frac{SP_{XY}}{\sqrt{(SS_X)(SS_Y)}} \\ &= \frac{2{,}057.22}{\sqrt{(8{,}164.93)(1{,}313.10)}} = \frac{2{,}057.22}{\sqrt{10{,}721{,}369.58}} = \frac{2{,}057.22}{3{,}274.35} \\ &= .63 \end{aligned}$$

One part of the above calculations that may result in computational errors is the need to calculate the square root of the denominator ($\sqrt{(SS_X)(SS_Y)}$). The square root is calculated because, unlike SP_{XY} in the numerator, the denominator is based on *squared* deviations rather than deviations. The purpose of the square root is to base both the numerator and denominator on deviations.

In calculating the Pearson r, it's also important to remember that the calculated value of r must be between −1.00 and 1.00. One way to check for errors in calculating r is to compare the calculated value of r with the scatterplot of the data. A value of $r = .63$ implies a strong positive linear relationship between two variables, which appears to match the scatterplot in Figure 13.1. If the square root of the denominator had not been calculated, the value of r would have been equal to 2,057.22 ÷ 10,721,369.58, or .0002; a value of $r = .0002$ implies a zero relationship, which is certainly not suggested by the scatterplot.

LEARNING CHECK 2:

Reviewing What You've Learned So Far

1. Review questions
   a. In testing the Pearson r, what is implied by the null hypothesis (H_0)?
   b. What three things are represented in calculating the Pearson r?
   c. Which of these determines whether the value for the Pearson r is a positive or negative number: SP_{XY}, SS_X, or SS_Y? Why?
   d. Why is there a square root symbol in the denominator of the formula for the Pearson r?

2. For each of the following, calculate the degrees of freedom (*df*) and identify the critical values of *r* (assume $\alpha = .05$ [two-tailed]).
   a. $N_i = 6$
   b. $N_i = 9$
   c. $N_i = 18$
3. For each of the following, calculate the Pearson correlation (*r*).
   a. $SP_{XY} = 7.00$, $SS_X = 11.00$, $SS_Y = 14.00$
   b. $SP_{XY} = -1.25$, $SS_X = 2.00$, $SS_Y = 4.00$
   c. $SP_{XY} = 12.57$, $SS_X = 135.31$, $SS_Y = 109.68$

Make a Decision Whether to Reject the Null Hypothesis

Once the value for the Pearson *r* has been calculated, the decision is made regarding the null hypothesis by comparing the sample's value of *r* with the identified critical value. For the picky eater study, this decision is stated as follows:

$$r = .63 > .468 \therefore \text{ reject } H_0 \ (p < .05)$$

Because the calculated value of .63 for the Pearson *r* is greater than the critical value .468, the decision is made to reject the null hypothesis that there is no relationship between the two variables. For the picky eater study, this decision implies that there is a statistically significant linear relationship between coercive prompts and food refusals.

Determine the Level of Significance

If the decision has been made to reject the null hypothesis, the next step is to determine whether the probability of the correlation is not only less than .05 but also less than .01. Turning to Table 6, for $df = 16$ and a two-tailed probability of .01, we find a critical value of .590. For the picky eater example, we may state

$$r = .63 > .590 \therefore p < .01$$

Because the calculated value of *r* exceeds the .01 critical value of .590, the probability of the correlation is not only less than .05 but also less than .01 ($p < .01$). The level of significance for the correlation in the picky eater study is illustrated in Figure 13.6.

Calculate a Measure of Effect Size (r^2)

Along with the inferential statistic, we learned in Chapter 10 that researchers often report a measure of effect size as an index or estimate of the size or magnitude of the effect. More specifically, effect size may be measured as the percentage of variance in one variable that is accounted for by another variable, ranging from .00 (0%) to 1.00 (100%).

For the Pearson correlation coefficient, a measure of effect size is r^2, which is the square of the correlation coefficient *r*:

$$\text{Estimate of effect} = r^2 \qquad (13\text{-}6)$$

If r^2 looks familiar, that's because it is simply another version of the R^2 statistic presented in earlier chapters. For the picky eater study, the measure of effect size is calculated as follows:

$$r^2 = (.63)^2$$
$$= .40$$

Therefore, we can conclude that 40% of the variability in children's food refusals is explained by differences in parents' coercive prompts.

To help interpret or characterize this measure of effect size, Cohen (1988, pp. 79–81) has suggested the following classification for correlational research:

$$\text{Small effect: } r^2 = .01\ (r = \pm.10)$$

$$\text{Medium effect: } r^2 = .09\ (r = \pm.30)$$

$$\text{Large effect: } r^2 = .25\ (r = \pm.50)$$

In the picky eater example, the r^2 value of .40 implies the relationship between coercive prompts and food refusals represents a large effect.

Draw a Conclusion From the Analysis

For the picky eater study, we've calculated the correlation coefficient and made the decision to reject the null hypothesis. The results of this analysis may be reported in the following way:

> In the sample of 18 families, parents' coercive prompts and children's food refusals had a statistically significant strong positive linear relationship, $r(16) = .63$, $p < .01$, $r^2 = .40$.

This sentence includes the following relevant information about the analysis:

- the sample ("18 families"),
- the two variables being related to each other ("coercive prompts" and "food refusals"),
- the nature and direction of the relationship ("a statistically significant strong positive linear relationship"), and
- information about the inferential statistic ("$r(16) = .63$, $p < .01$, $r^2 = .40$"), which indicates the inferential statistic calculated (r), the degrees of freedom (16), the value of the statistic (.63), the level of significance ($p < .01$), and the measure of effect size ($r^2 = .40$).

FIGURE 13.6 ● DETERMINING THE LEVEL OF SIGNIFICANCE FOR THE PICKY EATER STUDY

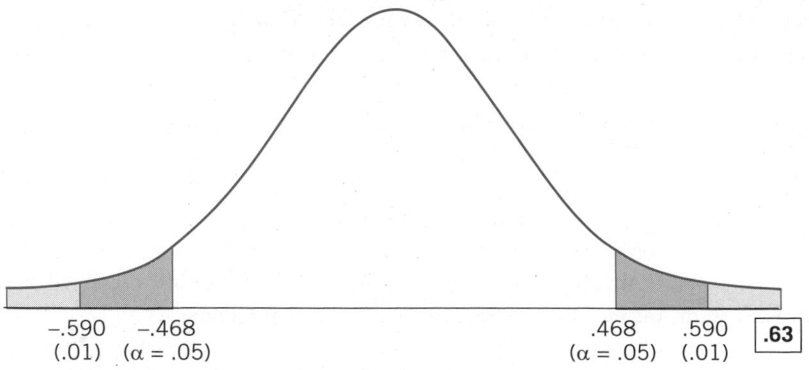

Relate the Result of the Analysis to the Research Hypothesis

The picky eater study revealed a significant positive relationship between coercive prompts and food refusals. The implication of this finding for the study's research hypothesis may be stated the following way:

> The result of this analysis supports the research hypothesis that parents' use of rewards and pressure to get their children to eat is positively related to food refusals by their children.

Box 13.1 lists the steps involved in calculating the Pearson correlation coefficient, using the picky eater study to illustrate each step.

Assumptions of the Pearson Correlation

Similar to the statistics discussed in earlier chapters, the Pearson correlation coefficient is based on assumptions about the distribution of scores for the variables included in the analysis. To appropriately use the Pearson *r*, researchers must determine whether the data in their samples meet these assumptions. This section describes the following assumptions: the assumption of continuous variables, the assumption of normality, the assumption of linearity, and the assumption of homoscedasticity.

The first assumption of the Pearson correlation, the assumption of continuous variables, is that the variables being related to each other are measured at the interval or ratio level of measurement, meaning that the values of the variables are equally spaced along a numeric continuum. One way in which this assumption may not be met is if the variables are measured at the ordinal level of measurement, meaning the values may be placed along a relative order but are not equally spaced from each other. Later in this chapter we discuss a correlational statistic that may be used with ordinal variables.

The second assumption is the assumption of normality, which is the assumption that scores for the variables are approximately normally distributed (bell shaped) in the population. The implication of this assumption is that distributions of scores for samples drawn from the populations should also be approximately normal. This assumption is checked by examining the distribution of scores for the two variables.

The third assumption is the assumption of linearity, which is that the relationship between variables is linear in nature and may be represented with a straight line. One way to determine whether this assumption has been met is to examine a scatterplot of the variables (see Figure 13.2(a) for an example).

The fourth assumption of the Pearson correlation is the **assumption of homoscedasticity**. This assumption relates to the shape of the distribution of a variable at each value of another variable. More specifically, the assumption of homoscedasticity is the assumption that a variable is normally distributed at each value of another variable. For example, if you were relating age and income, homoscedasticity would exist if income is normally distributed at each value of age (18, 19, 20, etc.). One way to determine whether this assumption has been met is to encapsulate the data in a scatterplot—the shape of this encapsulation should be elliptical (football shaped), such as in Figure 13.2(a). This assumption may not be met if the encapsulation resembles a cone or a triangle (Δ), which may occur if one or both variables are skewed rather than normally distributed.

A Computational Formula for the Pearson *r*

The formulas for the Pearson *r* presented in the previous section were referred to as definitional formulas. They were called definitional formulas because they literally reflect the definitions of variance and covariance as being based on deviations from the mean. Definitional formulas have been emphasized throughout this book because they clearly represent the concepts being measured. However, because calculations involving deviations such as $(X-\bar{X})$ and $(Y-\bar{Y})$ can be tedious, this section provides a computational formula for the Pearson correlation coefficient. A computational formula is an algebraic manipulation of a definitional formula that minimizes the complexity of mathematical calculations.

SUMMARY BOX 13.1 CALCULATING THE PEARSON CORRELATION COEFFICIENT USING DEFINITIONAL FORMULAS (PICKY EATER STUDY EXAMPLE)

State the null and alternative hypotheses (H_0 and H_1)

$$H_0: \rho = 0 \quad H_1: \rho \neq 0$$

Make a decision about the null hypothesis

Calculate the degrees of freedom (*df*)

$$df = N - 2 = 18 - 2 = 16$$

Set alpha (α), identify the critical values, and state a decision rule

If $r < -.468$ or $> .468$, reject H_0; otherwise, do not reject H_0

Calculate a statistic: Pearson correlation coefficient (*r*)

Represent the covariance between *X* and *Y* (SP_{XY})

$$\begin{aligned} SP_{XY} &= \Sigma[(X - \overline{X})(Y - \overline{Y})] \\ &= [(10 - 37.94)(7 - 18.22)] + \ldots + [(45 - 37.94)(11 - 18.22)] \\ &= [(-27.94)(-11.22)] + \ldots + [(7.06)(-7.22)] \\ &= 313.60 + \ldots + -50.96 \\ &= 2057.22 \end{aligned}$$

Represent the variance of Variable *X* (SS_X)

$$\begin{aligned} SS_X &= \Sigma(X - \overline{X})^2 \\ &= (10 - 37.94)^2 + \ldots + (45 - 37.94)^2 \\ &= (-27.94)^2 + \ldots + (7.06)^2 \\ &= 780.89 + \ldots + 49.78 \\ &= 8{,}164.93 \end{aligned}$$

Represent the variance of Variable *Y* (SS_Y)

$$\begin{aligned} SS_Y &= \Sigma(Y - \overline{Y})^2 \\ &= (7 - 18.22)^2 + \ldots + (11 - 18.22)^2 \\ &= (-11.22)^2 + \ldots + (-7.22)^2 \\ &= 125.94 + \ldots + 52.16 \\ &= 1{,}313.10 \end{aligned}$$

Calculate the Pearson correlation coefficient (*r*)

$$\begin{aligned} r &= \frac{SP_{XY}}{\sqrt{(SS_X)(SS_Y)}} = \frac{\Sigma(X - \overline{X})(Y - \overline{Y})}{\sqrt{\Sigma(X - \overline{X})^2(Y - \overline{Y})^2}} \\ &= \frac{2{,}057.22}{\sqrt{(8{,}164.93)1{,}313.10}} = \frac{2{,}057.22}{\sqrt{10{,}721{,}369.58}} = \frac{2{,}057.22}{3{,}274.35} = .63 \end{aligned}$$

Make a decision whether to reject the null hypothesis

$$r = .63 > .468 \therefore \text{reject } H_0 \ (p < .05)$$

Determine the level of significance

$$r = .63 > .590 \therefore p < .01$$

Calculate a measure of effect size (r^2)

$$r^2 = (.63)^2 = .40$$

Draw a conclusion from the analysis

In the sample of 18 families, parents' coercive prompts and children's food refusals had a statistically significant strong positive linear relationship, $r(16) = .63$, $p < .01$, $r^2 = .40$.

Relate the result of the analysis to the research hypothesis

The result of this analysis supports the research hypothesis that parents' use of rewards and pressure to get their children to eat is positively related to food refusals by their children.

LEARNING CHECK 3:

Reviewing What You've Learned So Far

1. Review questions
   a. What does r^2 for the Pearson correlation represent?
2. Students applying to graduate school may be asked to take an admissions test. To study how well these tests predict performance in grad school, one study hypothesized a positive relationship between scores on the Graduate Management Admissions Test (GMAT) and performance in an MBA (Masters of Business Administration) program such that higher GMAT scores are associated with higher levels of academic performance (Ahmadi, Raiszadeh, & Helms, 1997). The researchers obtained the GMAT score (range: 200 to 800) and grade point average (GPA) (range: 0.00 to 4.00) from students in an MBA program. The findings from the study are reproduced using the following sample of 26 students:

Student	GMAT	GPA	Student	GMAT	GPA
1	710	3.65	7	330	2.80
2	570	3.40	8	590	3.60
3	460	3.25	9	460	2.50
4	450	3.70	10	430	2.70
5	320	3.15	11	680	3.00
6	530	3.55	12	530	3.05
13	360	2.50	20	600	3.15
14	590	3.80	21	470	3.50

(Continued)

(Continued)

Student	GMAT	GPA	Student	GMAT	GPA
15	470	2.95	22	420	3.60
16	430	3.35	23	630	3.85
17	660	3.50	24	480	2.75
18	490	2.60	25	400	3.05
19	370	3.55	26	620	3.55

a. State the null and alternative hypotheses (H_0 and H_1).
b. Make a decision about the null hypothesis.
   (1) Calculate the degrees of freedom (*df*).
   (2) Set alpha (α), identify the critical values, and state a decision rule.
   (3) Calculate the Pearson correlation (*r*).
   (4) Make a decision whether to reject the null hypothesis.
   (5) Determine the level of significance.
   (6) Calculate a measure of effect size (r^2).
c. Draw a conclusion from the analysis.
d. Relate the result of the analysis to the research hypothesis.

Calculating the Pearson *r* using computational formulas involves the same three steps as in our earlier calculations: representing the covariance between Variables *X* and *Y*, representing the variance of Variable *X*, and representing the variance of Variable *Y*. Each of these steps is described below, using the picky eater study as an example.

Represent the Covariance Between *X* and *Y* (SP_{XY})

The first step in calculating the Pearson *r* is to represent the covariance between the *X* and *Y* variables by calculating the sum of products (SP_{XY}). A computational formula for SP_{XY} is provided below:

$$SP_{XY} = \Sigma XY - \frac{(\Sigma X)(\Sigma Y)}{N} \tag{13-7}$$

where *X* is a score on Variable *X*, *Y* is a score on Variable *Y*, and *N* is the number of pairs of scores.

Three quantities are needed to compute SP_{XY} using the computational formula: ΣXY (the sum of the products of scores on Variables *X* and *Y*), ΣX (the sum of the scores for Variable *X*), and ΣY (the sum of the scores for Variable *Y*). Note that Formula 13-7 differs from the definitional formula for SP_{XY} (Formula 13-2) in that it does not require calculating the deviations $X - \overline{X}$ and $Y - \overline{Y}$.

The first part of SP_{XY}, ΣXY, requires multiplying each score on the *X* variable by its corresponding score on the *Y* variable. The products for the 18 families in the picky eater study are listed in the last column of Table 13.4. For example, multiplying the first family's coercive prompts of 10 and food refusals of 7 results in a product of 70 ($10 \times 7 = 70$). Connecting the two variables with each other by calculating products is similar to the definitional formula; in this case, however, the product is created by multiplying scores rather than deviations. At the bottom of this column, the products have been summed such that ΣXY is equal to 14,503.

In addition to calculating ΣXY, Table 13.4 also sums the scores for the X ($\Sigma X = 683$) and Y ($\Sigma Y = 328$) variables as they are also part of Formula 13-7. Inserting the values of ΣXY, ΣX, and ΣY into the formula, we calculate SP_{XY} for the picky eater study data in the following manner:

$$\begin{aligned} SP_{XY} &= \Sigma XY - \frac{(\Sigma X)(\Sigma Y)}{N} \\ &= 14{,}503 - \frac{(683)(328)}{18} \\ &= 14{,}503 - \frac{224{,}024}{18} = 14{,}503 - 12{,}445.78 \\ &= 2{,}057.22 \end{aligned}$$

Comparing the above value of 2,057.22 with the calculations in the previous section, we see the same value was calculated for SP_{XY} using the computational formula and the definitional formula.

Represent the Variance of Variable X (SS_X)

The sums of squares representing the variance of the X variable (SS_X) may be calculated using the following computational formula:

$$SS_X = \Sigma X^2 - \frac{(\Sigma X)^2}{N} \qquad (13\text{-}8)$$

where X is a score on Variable X, and N is the number of pairs of scores.

TABLE 13.4 ● CALCULATING THE SUM OF PRODUCTS (SP_{XY}) IN THE PICKY EATER STUDY USING THE COMPUTATIONAL FORMULA

Family	Coercive Prompts (X)	Food Refusals (Y)	(XY)
1	10	7	70
2	37	23	851
3	71	16	1,136
4	30	8	240
5	62	35	2,170
6	11	12	132
7	47	19	893
8	17	9	153
9	66	22	1,452
10	54	25	1,350
11	20	5	100
12	81	32	2,592
13	24	26	624
14	32	15	480
15	15	19	285
16	42	28	1,176
17	19	16	304
18	45	11	495
	$\Sigma X = 683$	$\Sigma Y = 328$	$\Sigma XY =$ **14,503**

Two quantities are needed to compute SS_X: ΣX^2 (the sum of squared scores) and $(\Sigma X)^2$ (the squared sum of scores). The first quantity, ΣX^2, is created by squaring each of the X variable scores and then summing the squared scores (first squaring, then summing). The second quantity, $(\Sigma X)^2$, is calculated by summing the scores for the X variable and then squaring this sum (first summing, then squaring).

ΣX^2 and $(\Sigma X)^2$ for the coercive prompts variable in the picky eater study are calculated in Table 13.6. The fourth column of this table provides the squared value of each score (X^2); the bottom of this column shows that the sum of the squared scores (ΣX^2) is 34,081. Once the sum of the scores ($\Sigma X = 683$) is also calculated, SS_X for the coercive prompts variable may be calculated as follows:

$$\begin{aligned} SS_X &= \Sigma X^2 - \frac{(\Sigma X)^2}{N} \\ &= 34{,}081 - \frac{(683)^2}{18} \\ &= 34{,}081 - \frac{466489}{18} = 34{,}081 - 25916.07 \\ &= 8{,}164.93 \end{aligned}$$

Represent the Variance of Variable Y (SS_Y)

The formula for the sum of squares for the Y variable (SS_Y) is a simple modification of the formula for SS_X:

$$SS_Y = \Sigma Y^2 - \frac{(\Sigma Y)^2}{N} \qquad (13\text{-}9)$$

where Y is a score on Variable Y and N is the number of pairs of scores. The last column of Table 13.5 lists the squared values for the food refusals variable; looking at the bottom of this column, we see $\Sigma Y^2 = 7290$. Using this quantity, as well as the sum of the scores ($\Sigma Y = 328$), SS_Y for the food refusals variable is calculated below:

$$\begin{aligned} SS_Y &= \Sigma Y^2 - \frac{(\Sigma Y)^2}{N} \\ &= 7{,}290 - \frac{(328)^2}{18} \\ &= 7{,}290 - \frac{107584}{18} = 7{,}290 - 5{,}976.90 \\ &= 1{,}313.10 \end{aligned}$$

Calculate the Pearson Correlation Coefficient (r)

Once SP_{XY}, SS_X, and SS_Y have been calculated, the value for the Pearson r may be calculated. Formula 13-10 presents the computational formula for the Pearson r:

$$r = \frac{SP_{XY}}{\sqrt{(SS_X)(SS_Y)}} = \frac{\Sigma XY - \frac{(\Sigma X)(\Sigma Y)}{N}}{\sqrt{(\Sigma X^2 - \frac{(\Sigma X)^2}{N})(\Sigma Y)^2 - \frac{(\Sigma Y)^2}{N})}} \qquad (13\text{-}10)$$

Using Formula 13-10, we calculate the correlation between coercive prompts and food refusals in the picky eater study in the following way:

$$\begin{aligned} r &= \frac{SP_{XY}}{\sqrt{(SS_X)(SS_Y)}} \\ &= \frac{2{,}057.22}{\sqrt{(8{,}164.93)(1{,}313.10)}} = \frac{2{,}057.22}{\sqrt{10{,}721{,}369.58}} = \frac{2{,}057.22}{3{,}274.35} \\ &= .63 \end{aligned}$$

TABLE 13.5 ● CALCULATING SQUARED SCORES TO REPRESENT VARIANCES IN THE PICKY EATER STUDY

Family	Coercive Prompts (X)	Food Refusals (Y)	(X^2)	(Y^2)
1	10	7	100	49
2	37	23	1,369	529
3	71	16	5,041	256
4	30	8	900	64
5	62	35	3,844	1,225
6	11	12	121	144
7	47	19	2,209	361
8	17	9	289	81
9	66	22	4,356	484
10	54	25	2,916	625
11	20	5	400	25
12	81	32	6,561	1,024
13	24	26	576	676
14	32	15	1,024	225
15	15	19	225	361
16	42	28	1,764	784
17	19	16	361	256
18	45	11	2,025	121
	$\Sigma X = 683$	$\Sigma Y = 328$	$\Sigma X^2 =$ **34,081**	$\Sigma Y^2 =$ **7,290**

Note that the computational formula in Formula 13-10 results in the same value for the Pearson r as the definitional formula (Formula 13-5).

The steps needed to calculate the Pearson r using the computational formula are summarized in Box 13.2. We discussed the definitional formula to illustrate the conceptual basis of the Pearson correlation; the computational formula has been provided to help you apply the Pearson correlation to your own work and study.

SUMMARY BOX 13.2 CALCULATING THE PEARSON CORRELATION COEFFICIENT USING COMPUTATIONAL FORMULAS (PICKY EATER STUDY EXAMPLE)

Represent the covariance between X and Y (SP_{XY})

$$SP_{XY} = \sum XY - \frac{\sum(X)\sum(Y)}{N}$$
$$= 14{,}503 - \frac{(683)(328)}{18} = 14{,}503 - \frac{224{,}024}{18}$$
$$= 14{,}503 - 12{,}445.78 = 2{,}057.22$$

(Continued)

(Continued)

Represent the variance of Variable X (SS_X)

$$SS_X = \Sigma X^2 - \frac{(\Sigma X)^2}{N}$$
$$= 34{,}081 - \frac{(683)^2}{18} = 34{,}081 - \frac{466{,}489}{18}$$
$$= 34{,}081 - 25{,}916.07 = 8{,}164.93$$

Represent the variance of Variable Y (SS_Y)

$$SS_Y = \Sigma Y^2 - \frac{(\Sigma Y)^2}{N}$$
$$= 7{,}290 - \frac{(328)^2}{18} = 7{,}290 - \frac{107{,}584}{18}$$
$$= 7{,}290 - 5{,}976.90 = 1{,}313.10$$

Calculate the Pearson correlation coefficient (r)

$$r = \frac{\Sigma XY - \frac{(\Sigma X)(\Sigma Y)}{N}}{\sqrt{(\Sigma X^2 - \frac{(\Sigma X)^2}{N})(\Sigma Y)^2 - \frac{(\Sigma Y)^2}{N})}}$$
$$= \frac{2{,}057.22}{\sqrt{(8{,}164.93)(1{,}313.10)}} = \frac{2{,}057.22}{\sqrt{10{,}721{,}369.58}} = \frac{2{,}057.22}{3{,}274.35} = .63$$

When the concept of correlational statistics was introduced earlier in this chapter, we noted that a number of different correlational statistics exist and that the choice of statistic depends on how variables in a study are measured. The next section describes a correlational statistic that is used when both variables are measured at the ordinal level of measurement.

LEARNING CHECK 4:

Reviewing What You've Learned So Far

1. Review questions
   a. What is the purpose of a computational formula?
2. Calculate the Pearson correlation coefficient for each set of data below using the computational formulas:
   a.

Participant	X	Y
1	8	7
2	3	1
3	7	6

Participant	X	Y
4	6	2
5	8	3
6	4	5

b.

Participant	X	Y
1	17	11
2	23	7
3	16	5
4	22	9

Participant	X	Y
5	29	4
6	14	10
7	20	3
8	27	7

c.

Participant	X	Y
1	4.09	51
2	3.13	28
3	2.58	43
4	4.26	86
5	1.19	75
6	4.82	88

Participant	X	Y
7	1.63	93
8	3.67	31
9	4.70	74
10	1.52	82
11	4.47	77
12	2.07	61

13.4 CORRELATING TWO SETS OF RANKS: THE SPEARMAN RANK-ORDER CORRELATION

In Chapter 1, we learned that variables may be measured at one of four levels of measurement: nominal, ordinal, interval, or ratio. In the examples included thus far in this chapter, the variables have been measured at the interval or ratio level of measurement. The values of interval or ratio variables are equally spaced along a numeric continuum. In the picky eater study, the food refusals variable was measured at the ratio level of measurement, with its values (0, 1, 2, etc.) equally spaced from each other.

This section correlates two variables measured at the ordinal level of measurement. The values of ordinal variables can be placed in an order relative to the other values, but the values are not assumed to be equally spaced. Some examples of ordinal variables are "size" (extra small, small, medium, large, extra large), "height" (short, average, tall), and "difficulty" (very easy, somewhat easy, somewhat difficult, very difficult). In ordinal variables, one value represents more or less of a variable than do other values, but it's not possible to specify the precise difference between values. For example, the difference between a "medium" and "large" soda may not be the same as between "large" and "extra large."

Introduction to Ranked Variables

A common form of ordinal data is ranks. A **rank** is defined as a relative position in a graded or evaluated group. The numeric value for a ranked variable reflects a relative location within an order. You may have been assigned a number as part of a ranked data set, for example, if your high school transcripts listed your class rank as "32nd out of 145" or if you finished "first," "second," or "third" in a race. Note that ranks are ordinal variables in that although you can say that a runner who finished "fourth" ran faster than the runner who finished "fifth," you cannot specify the exact difference between the times of the two runners.

There are perhaps two primary situations in which ranks may be used in a study. The first situation is one in which the collected data consist of ranks. For example, a newspaper editor may ask two movie critics to rank eight movies from 1 (most favorite) to 8 (least favorite). A second situation is when data originally collected at the interval or ratio level of measurement are converted to ranks. For example, a teacher may give a 40-item quiz to 15 students such that each student's score ranges from 0 to 40 but then, on the basis of their quiz scores, rank orders the students from 1 (highest) to 15 (lowest).

This section below describes a research study whose goal was to correlate two sets of ranks to assess the degree of similarity between two groups of research participants. As you will see, calculating and interpreting a correlational statistic used with ordinal data has both similarities and differences from the Pearson correlation.

An Example From the Research: Do Students and Teachers Think Alike?

Imagine you're a college senior about to graduate and enter the workforce. What would be most important to you in selecting your first job: a high salary, challenge and responsibility, or friendly coworkers? A study conducted by J. Stuart Devlin and Robin Peterson at New Mexico State University examined the extent to which college business students and their professors agreed on the desirability of different characteristics of students' first jobs after graduation (Devlin & Peterson, 1994).

The 127 senior business majors and 28 business school professors in the study, which will be referred to as the business student study, rated the desirability of 13 characteristics of an entry-level position (see Table 13.6). Some of the characteristics reflected aspects of jobs that provide motivation for good job performance, such as the extent to which a job offers opportunities for growth and development, challenge, and freedom; other characteristics pertained to rewards and benefits organizations give to employees, such as salary, job title, and job security.

The students and professors rated the desirability of the 13 characteristics on a 5-point scale (1 = low, 5 = high). For each characteristic, the average desirability rating was calculated separately for the students and professors. Based on their average ratings, the 13 characteristics were ranked from 1 = most important to 13 = least important for the two groups; Table 13.6 provides the rankings for the business students and professors.

Looking at the two sets of ranks in Table 13.6, we see that the students and professors agreed on the relative desirability of some of the characteristics (freedom on the job, opportunity for advancement, type of work). However, the students ranked characteristics of the workplace and the job itself (e.g., opportunity for self-development, challenge and responsibility) more highly than did faculty, who reported a higher preference for employee benefits and other personal concerns (e.g., salary, location of work, and job security).

Looking at the professors' ranks in the right column of Table 13.6, two of the characteristics (job title and training) have the same rank of 12.5. These characteristics were assigned the same rank because they had the same average desirability. In a situation such as this, each of the tied scores is assigned the average of the tied rankings. In this case, the average of the 12th and 13th ranked characteristics is 12.5. As another example of tied ranks, if two characteristics had been tied for the 2nd and 3rd ranks, both characteristics would have been assigned a rank of 2.5, with the next highest ranked characteristic being assigned a rank of 4.

Inferential Statistics: The Spearman Rank-Order Correlation Coefficient

The degree of similarity between the beliefs of the students and professors regarding the desirability of different job characteristics was tested using the **Spearman rank-order correlation (r_s)**, named after the British psychologist Charles Spearman in 1904. The Spearman rank-order correlation is a statistic measuring the relationship between two variables measured at the ordinal level of measurement.

The Spearman rank-order correlation possesses many of the same qualities as the Pearson correlation coefficient. First, the possible values of the Spearman r_s range from −1.00 to +1.00. A positive relationship indicates that the two sets of ranks are similar to one another; a negative correlation implies the two sets of rankings are dissimilar to one another. For the business student study, a negative value for r_s would mean that characteristics

TABLE 13.6 ● RANKING OF IMPORTANCE OF ENTRY-LEVEL JOB CHARACTERISTICS, STUDENTS AND PROFESSORS

Job Characteristic	Student Ranking	Professor Ranking
Challenge and responsibility	3	6
Company reputation	12	9
Freedom on the job	10	10
Job security	8	5
Job title	13	12.5
Location of work	11	3
Opportunity for advancement	1	2
Opportunity for self-development	2	8
Salary	7	1
Training	6	12.5
Type of work	5	4
Working conditions	4	7
Working with people	9	11

ranked *higher* by students were ranked *lower* by professors. Second, the numeric value of the Spearman r_s reflects the strength of the relationship. The closer the value r_s is to +1.00 or −1.00, the greater is the similarity or difference between the two sets of ranks. A value of zero for r_s indicates no relationship whatsoever between the two sets of ranks.

The primary distinction between the Spearman r_s and the Pearson *r* is in how they are evaluated. Statistical procedures such as Pearson correlation, *t*-test, and ANOVA that use interval or ratio variables are based on the assumption that scores for the variables are normally distributed (bell shaped) in the larger population. This assumption influences the shape of the theoretical distributions of the Pearson *r*, *t*-statistic, and *F*-ratio, which in turn determines the critical values used to test the statistical significance of these statistics. However, statistics such as the Spearman r_s that analyze ordinal variables do not make assumptions regarding the shape of the distribution. More specifically, the Spearman r_s does not require variables to be normally distributed. For example, the ranks in the business student study (the numbers 1 through 13) are not normally distributed. Consequently, the critical values of the Spearman r_s differ from those for the Pearson *r*.

The steps involved in calculating, evaluating, and interpreting the Spearman rank-order correlation are as follows:

- state the null and alternative hypotheses (H_0 and H_1),
- make a decision about the null hypothesis,
- draw a conclusion from the analysis, and
- relate the result of the analysis to the research hypothesis.

Each of these steps is illustrated below using the business student study as the example. In moving through these steps, the similarities and differences between the Spearman r_s and the Pearson *r* will be noted.

State the Null and Alternative Hypotheses (H_0 and H_1)

The null and alternative hypotheses for r_s are stated in a manner similar to the Pearson correlation coefficient:

$$H_0: \rho = 0$$

$$H_1: \rho \neq 0$$

where ρ (rho) is the correlation between the two sets of ranks in the population. The value of zero for ρ represents the absence of a relationship between the two sets of ranks. In the business student study, a nondirectional alternative hypothesis ($H_1: \rho \neq 0$) was used to allow for the possibility that the ranks of the students and professors may either be similar (i.e., a positive relationship) or dissimilar (i.e., a negative relationship).

Make a Decision About the Null Hypothesis

Once the null and alternative hypotheses have been stated, the next step is to make a decision whether to reject the null hypothesis that the population correlation between the two sets of ranks is equal to zero. For the business student study, this involves determining whether a significant relationship exists between the rankings of students and professors. This question is answered by following these steps:

- calculate the degrees of freedom (*df*);
- set alpha (α), identify the critical values, and state a decision rule;
- calculate a statistic: Spearman rank-order correlation (r_s);
- make a decision whether to reject the null hypothesis;
- determine the level of significance; and
- calculate a measure of effect size.

Calculate the degrees of freedom (df). The degrees of freedom for the Spearman rank-order correlation are the same as for the Pearson correlation:

$$df = N - 2$$

where N is the number of pairs of scores being ranked. In the business student study, the rankings of 13 characteristics were paired to assess the similarity of students and professors. Therefore, the degrees of freedom is equal to

$$\begin{aligned} df &= N - 2 \\ &= 13 - 2 \\ &= 11 \end{aligned}$$

Set alpha, identify the critical values, and state a decision rule. Alpha (α), the probability of a statistic needed to reject the null hypothesis, is set to the traditional .05 level. The critical values for the Spearman rank-order correlation are provided in Table 7 at the back of the book. For the business student study, move down the *df* column until you reach the value *11*. For α = .05 (two-tailed), we find the critical value is .560. This can be expressed in the following manner:

For $\alpha = .05$ (two-tailed) and $df = 11$, critical values $= \pm.560$

The null hypothesis is rejected when the calculated value of r_s exceeds an identified critical value. For the business student study, we may state the following decision rule:

If $r_s < -.560$ or $> .560$, reject H_0; otherwise, do not reject H_0

Calculate a statistic: Spearman rank-order correlation (r_S). The relationship between the two sets of ranks is measured by calculating a value for the Spearman rank-order correlation. The formula for r_s is provided in Formula 13-11:

$$r_s = 1 - \frac{6\Sigma D^2}{N(N^2-1)} \tag{13-11}$$

where D is the difference between the ranks of each variable and N is the number of pairs of variables. Note that this formula contains a constant (the number 6) used to calculate r_s regardless of the number of pairs of variables involved in the analysis. Without going into detail, the number 6 in the numerator and N and N^2 in the denominator are needed to represent all of the possible differences that can occur between the two rankings for each variable in a set of variables.

You may also wonder why calculating the Spearman r_s involves subtracting a quantity from the number 1. The calculated value of the Spearman r_s is based on the difference (D) between the two rankings on each variable. If there happened to be perfect agreement between two sets of rankings, what would be the value of D for each variable? The answer is zero (i.e., zero difference). The greater the differences between the rankings, the greater the values for D and consequently the lower the value for r_s. In essence, calculating the Spearman correlation involves starting with a perfect relationship ($r_s = 1.00$) and then subtracting any differences between the rankings.

The first step in calculating r_s is to calculate ΣD^2, which is the sum of the squared differences between the rankings for each variable. The difference between the rankings for each variable (D) is calculated by subtracting one ranking from the other. The fourth column in Table 13.7 lists the values for D for each of the 13 job characteristics in the business student study. For example, for the characteristic "Challenge and responsibility," D is equal to the students' ranking (3) minus the professors' ranking (6): $3 - 6 = -3$. Each value of D is then squared; for "Challenge and responsibility," $D^2 = (-3)^2 = 9$. At the bottom of the last column, of Table 13.7, the squared differences are summed. For the business student study, the sum of the squared differences (ΣD^2) is equal to 220.50.

Using the sum of the squared difference scores (ΣD^2) from Table 13.7, we calculate r_s for the 13 characteristics ranked by the students and professors as follows:

$$\begin{aligned} r_s &= 1 - \frac{6\Sigma D^2}{N(N^2-1)} \\ &= 1 - \frac{6(220.50)}{13(13^2-1)} \\ &= 1 - \frac{1323.00}{13(168)} = 1 - \frac{1323.00}{2184} = 1 - .61 \\ &= .39 \end{aligned}$$

Therefore, the correlation between the student's and professors' rankings of the desirability of the 13 job characteristics was .39. Using Cohen's guidelines presented earlier in this chapter, we determine that there is a moderate relationship, or degree of similarity, between the rankings of the students and professors.

TABLE 13.7 ● DIFFERENCE (D) OF RANKINGS OF DESIRABILITY OF ENTRY-LEVEL JOB CHARACTERISTICS, STUDENTS AND PROFESSORS

Job Characteristic	Student Ranking	Professor Ranking	Difference (D)	Difference² (D²)
Challenge and responsibility	3	6	−3	9
Company reputation	12	9	3	9
Freedom on the job	10	10	0	0
Job security	8	5	3	9
Job title	13	12.5	.5	.25
Location of work	11	3	8	64
Opportunity for advancement	1	2	−1	1
Opportunity for self-development	2	8	−6	36
Salary	7	1	6	36
Training	6	12.5	−6.5	42.25
Type of work	5	4	1	1
Working conditions	4	7	−3	9
Working with people	9	11	−2	4
				$\Sigma D^2 = 220.50$

Make a decision whether to reject the null hypothesis. Do business students and their professors agree on the relative desirability of different entry-level job characteristics? Comparing the value of $r_s = .39$ with the critical values, we reach the following conclusion:

$$r_s = .39 \text{ is not} < -.560 \text{ or} > .560 \therefore \text{ do not reject } H_0\ (p > .05)$$

That is, the decision is made to not reject the null hypothesis, and we subsequently conclude that the relationship between students' and professors' rankings of the 13 characteristics is not statistically significant. This implies that the students and professors did not significantly agree on the relative desirability of different entry-level job characteristics.

Determine the level of significance. In this example, the null hypothesis was not rejected. Therefore, it is not necessary to compare the value of the statistic with the .01 critical value of .703.

Calculate measure of effect size (r_s^2). As with the Pearson r, a measure of effect size may be calculated by squaring the Spearman rank order correlation:

$$\text{Estimate of effect} = r_s^2 \tag{13-12}$$

For the business student study, we may state that

$$\begin{aligned} r_s^2 &= (.39)^2 \\ &= .15 \end{aligned}$$

One way to interpret the value of r_s^2 is to conclude that 15% of the variability in the students' rankings may be explained by the professors' rankings. Using Cohen's guidelines described earlier, an r_s^2 value of .15 implies this is a medium-sized effect.

Draw a Conclusion From the Analysis

What conclusion may be drawn about the nonsignificant relationship between the rankings of the students and professors? Here is one way that the results of the analysis may be reported:

> The rankings of the 13 job characteristics for the 127 business students and the 28 business professors were not significantly related to each other, indicating that students and professors did not agree in their perceptions of the relative desirability of the characteristics, $r_s(11) = .39$, $p > .05$, $r_s^2 = .15$.

Relate the Result of the Analysis to the Research Hypothesis

The nonsignificance of the Spearman rank-order correlation suggests that business students and professors do not agree on the relative desirability of various entry-level job characteristics. Assuming that one part of being a professor is to understand students' work-related values, what are possible implications of this finding? The authors of this study stated the following implications of their research:

> Apparently, professors were not very accurate in gauging student preferences. . . . Professors would be well advised to become more informed on student values so that they can be more effective in counseling students about careers. (Devlin & Peterson, 1994, pp. 156–157)

Box 13.3 lists the steps followed in correlating two sets of ranks using the Spearman rank-order correlation, applying these steps to the business student study.

SUMMARY BOX 13.3 CALCULATING THE SPEARMAN RANK-ORDER CORRELATION (BUSINESS STUDENT STUDY EXAMPLE)

State the null and alternative hypotheses (H_0 and H_1)

$$H_0: \rho = 0 \quad H_1: \rho \neq 0$$

Make a decision about the null hypothesis

Calculate the degrees of freedom (*df*)

$$df = N - 2 = 13 - 2 = 11$$

Set alpha (α), identify the critical values, and state a decision rule

If $r_s < -.560$ or $> .560$, reject H_0; otherwise, do not reject H_0

Calculate a statistic: Spearman rank-order correlation (r_s)

$$\begin{aligned} r_s &= 1 - \frac{6\Sigma D^2}{N(N^2-1)} \\ &= 1 - \frac{6(220.50)}{13(13^2-1)} = 1 - \frac{1323.00}{13(168)} = 1 - \frac{1323.00}{2184} \\ &= 1 - .61 = .39 \end{aligned}$$

(Continued)

(Continued)

Make a decision whether to reject the null hypothesis

$$r_s = .39 \text{ is not} < -.560 \text{ or} > .560 \therefore \text{ do not reject } H_0 \ (p > .05)$$

Determine the level of significance

Not applicable (H_0 not rejected)

Calculate a measure of effect size (r_s^2)

$$r_s^2 = (.39)^2 = .15$$

Draw a conclusion from the analysis

The rankings of the 13 job characteristics for the 127 business students and the 28 business professors were not significantly related to each other, indicating that students and professors did not agree in their perceptions of the relative desirability of the characteristics, $r_s(11) = .39, p > .05, r_s^2 = .15$.

Relate the result of the analysis to the research hypothesis

"Apparently, professors were not very accurate in gauging student preferences. . . . Professors would be well advised to become more informed on student values so that they can be more effective in counseling students about careers" (Devlin & Peterson, 1994, pp. 156–157).

13.5 CORRELATIONAL STATISTICS VS. CORRELATIONAL RESEARCH

The main purpose of this chapter was to introduce what are known as correlational statistics. Correlational statistics, such as the Pearson *r*, are statistics designed to measure relationships between variables. However, in Chapter 1, we introduced another concept that included the word *correlation:* correlational research. The final section of this chapter discusses the difference between correlational research and correlational statistics.

LEARNING CHECK 5:

Reviewing What You've Learned So Far

1. Review questions
   a. How do variables measured at the ordinal level of measurement differ from variables measured at the interval or ratio level of measurement?
   b. What research situations may involve the use of ranked variables?
   c. How are ranks assigned when there is a tie between two or more ranks?
   d. What are the main similarities and differences between the Spearman rank-order correlation (r_s) and the Pearson correlation (r)?
2. Calculate the Spearman rank-order correlation (r_s) for each of the following pairs of ranks.
   a.

Rank Order 1	Rank Order 2
4	4
1	2
3	1
2	3

b.

Rank Order 1	Rank Order 2
2	3
4	6
6	5
3	1
1	2
5	4

c.

Rank Order 1	Rank Order 2
9	9
7	6.5
2	5
6	8
3.5	2
8	6.5
3.5	3
1	1
5	4

3. One study examined the preferences of girls and boys for different musical instruments (Abeles, 2009). As part of the study, the number of sixth-grade girls and boys playing eight different instruments was recorded. Based on these numbers, the instruments can be ranked from 1 (*most preferred*) to 8 (*least preferred*). Imagine the study hypothesized that boys and girls would display similar preferences for the different instruments.

Instrument	Sixth-Grade Girls	Sixth-Grade Boys
Cello	4	6
Clarinet	1.5	7
Drums	8	5
Flute	3	8
Saxophone	6	4
Trombone	7	2
Trumpet	5	1
Violin	1.5	3

a. State the null and alternative hypotheses (H_0 and H_1).
b. Make a decision about the null hypothesis.
   (1) Calculate the degrees of freedom (*df*).
   (2) Set alpha (α), identify the critical values, and state a decision rule.
   (3) Calculate the Spearman rank-order correlation (r_s).
   (4) Make a decision whether to reject the null hypothesis.
   (5) Determine the level of significance.
   (6) Calculate a measure of effect size (r_s^2).
c. Draw a conclusion from the analysis.
d. Relate the result of the analysis to the research hypothesis.

Correlational research refers to methods of conducting research that examine the relationship between variables without the ability to infer cause-effect relationships. In Chapter 1, we provided examples of correlational research methods, which include quasi-experiments, survey research, observational research, and archival research. The types of variables included in correlational research may be biographical (e.g., age, gender, education), physiological (e.g., height, weight, reaction time), or psychological (e.g., intelligence, personality, attitudes and opinions). As diverse as these variables may be, they have in common the fact that they are measured by the researcher rather than controlled or manipulated. For example, a researcher cannot randomly assign someone to be "male" or "female."

Correlational research methods differ from another type of research method, experimental research, which is used to infer cause-effect relationships between variables. The ability to infer causality is accomplished by isolating the effect of an independent variable through experimental control and manipulation, as well as random assignment to conditions. Chapter 11's chess study is an example of experimental research, in which participants were randomly assigned to one of three methods of learning chess (Observation, Prediction, or Explanation). Because correlational research methods do not involve experimental control or random assignment, they are unable to test causal relationships between variables. For example, although researchers may find a relationship between high school GPA and college GPA, they may not conclude that having a low GPA in high school "causes" a student to have a low college GPA.

Does ANOVA = Causality?

An important distinction must be made between correlational research and correlational statistics:

> Correlational research refers to methods used to *collect* data, whereas correlational statistics refers to methods used to *analyze* data.

Correlational research refers to a type of research methodology that examines the relationship between variables without the ability to make causal inferences. On the other hand, correlational statistics refers to mathematical procedures that measure the relationship between variables. Although both correlational research and correlational statistics are interested in the relationship between variables, these two concepts are very different from each other.

The distinction between correlational research and correlational statistics can be a source of confusion to students. Correlational research frequently, although not necessarily, involves the use of continuous variables such as Time and Exam score. As a result, data collected using correlational research are often analyzed using a correlational statistic. On the other hand, because experimental research often involves assigning participants to groups or conditions, the independent variables are typically categorical in nature. Consequently, researchers analyze differences between these groups using a version of ANOVA, such as a t-test or one-way ANOVA.

Students sometimes make the leap in inference that because ANOVA is typically used to analyze experiments and correlational statistics are used to analyze correlational research, using ANOVA enables one to make causal inferences but using correlational statistics do not. This is simply not true—the ability to draw causal inferences is a function of how data are *collected,* not how they are *analyzed.*

The use of ANOVA to analyze data does not necessarily lead to the ability to make causal inferences. For example, in the voting message study discussed in Chapter 12 (two-way ANOVA), one of the independent variables was authoritarianism, defined as the extent to which people "perceive a great deal of threat from their immediate and broader environment" (Lavine et al., 1999, p. 338). Because authoritarianism is a personality characteristic, it could not be manipulated by the researcher. Consequently, part of the voting message study was an example of correlational research.

Just as ANOVA may be used to analyze correlational research, an experiment may be analyzed using correlational statistics. For example, suppose a researcher wants to examine whether the temperature of a swimming pool affects swimmers' speed. Swimmers of equal ability are randomly assigned to swim in one of 10 pools that have been manipulated to differ in temperature by 2 degrees each. In this situation, the researcher has used the critical features of experimental research: random assignment and experimental control. However, because the independent variable in this study, pool temperature, is measured at the interval level of measurement, a correlational statistical procedure could be used to analyze the data.

It may surprise you to learn that ANOVA and correlational statistics are much more similar than they are different. Entire textbooks have been written to demonstrate that any research situation that can be analyzed with ANOVA may also be analyzed with correlational statistics with exactly the same result (Keppel & Zedeck, 1989).

"Correlation Is Not Causality"—What Does That Mean?

Students are often taught that that "correlation is not causality." This is true, in that a correlational research methodology does not allow one to make causal inferences. However, this assertion has absolutely *nothing* to do with correlational statistics. The ability to infer causal relationships is *not* a function of how data are analyzed; it's a function of how data are collected. How data are analyzed is primarily a function of how variables are measured. ANOVA is used when the independent variables are categorical; correlation is used when the independent variables are continuous. ANOVA is not "better" than correlation—it is simply different.

One issue that may confuse students in learning about correlational research is the common practice of labeling one of the variables the "independent variable" and the other the "dependent variable." For example, researchers relating age and income may call age the "independent variable" and income the "dependent variable." Although these labels are appropriate in experimental research because changes in the dependent variable are perceived to be "dependent on" changes in the independent variable, in correlational research these labels infer a direction of influence that may be inappropriate. To avoid confusion, researchers sometimes refer to the X and Y variables in a correlational research study as the **predictor variable** and **criterion variable,** respectively, particularly when the goal of the study is to predict one variable from another.

13.6 LOOKING AHEAD

The main purpose of this chapter was to introduce correlational statistical procedures. The research examples in this chapter have involved the calculation and interpretation of a correlation coefficient that measures the relationship between two variables. However, in addition to *measurement,* another goal of science is *prediction.* Researchers frequently use identified relationships to predict one variable from another. Using the findings of the picky eater study, for example, how might you use knowledge of how often parents use coercive prompts to predict the extent to which a child will refuse to eat a new food? The next chapter discusses how researchers use the relationship between two variables to predict scores on one variable from another.

13.7 Summary

One way to visually examine the relationship between scores on two continuous variables measured at the interval or ratio level of measurement is to create what is known as a ***scatterplot***, a graphical display of paired scores on two variables.

Correlation may be defined as a mutual or reciprocal relationship between variables such that systematic changes in the values of one variable are accompanied by systematic changes in the values of another variable.

The relationship between variables may be described along three aspects: the nature of the relationship, the direction of the relationship, and the strength of the relationship.

The nature of the relationship between variables pertains to the manner in which differences in scores on one variable correspond to differences in scores on another variable. A ***linear relationship*** is a relationship between variables appropriately represented by a straight line, such that increases or decreases in one variable are associated with corresponding increases or decreases in another variable. A ***nonlinear relationship*** is a relationship between variables that is not appropriately represented by a straight line.

In terms of the direction of a relationship, when scores on variables move in the same direction, with increases in one variable associated with increases in another variable, this is referred to as a ***positive relationship***. When scores on variables move in the opposite direction, with increases in one variable associated with decreases in another variable, this is called a ***negative relationship***.

The strength of a relationship is the extent to which scores on one variable are associated with scores on another variable. A perfect relationship is a relationship in which each score for one variable is associated with only one score for the other variable, a strong relationship is a relationship in which a score on one variable is associated with a relatively small range of scores on another variable, and a zero relationship exists when all of the scores on one variable are associated with a wide range of scores on another variable.

Correlational statistics are statistics designed to measure the relationship between variables. There are a number of different correlational statistics; the choice of statistic depends on how variables in a study have been measured.

The most commonly used correlational statistic is the ***Pearson correlation coefficient (r)***, which is a statistic that measures the linear relationship between two continuous variables measured at the interval and/or ratio level of measurement. The possible values for the Pearson r range from −1.00 to +1.00. The sign (+ or −) of the correlation indicates the direction of the relationship, and the numeric value of the correlation represents the strength of the relationship.

Correlational statistics are based on the concept of variance, which represents differences among the scores for variables, and ***covariance***, which is the extent to which two variables have shared variance (i.e., variance in common with each other). The Pearson correlation coefficient r is a ratio of the covariance between two variables to the variance of the two variables.

Unlike the Pearson r, which measures the relationship between two variables measured at the interval or ratio level of measurement, the ***Spearman rank-order correlation*** (**r_s**) measures the relationship between two variables measured at the ordinal level of measurement.

Correlational statistics such as the Pearson r and the Spearman r_s are distinct from correlational research methods, which are methods of conducting research designed to examine the relationship between variables without the ability to infer cause-effect relationships. Correlational research refers to methods used to collect data, whereas correlational statistics refers to methods used to analyze data. Researchers sometimes refer to the X and Y variables in a correlational research study as the ***predictor variable*** and ***criterion variable***, respectively, particularly when the goal of the study is to predict one variable from another.

13.8 Important Terms

scatterplot (p. 482)
correlation (p. 482)
linear relationship (p. 483)
nonlinear relationship (p. 483)
positive relationship (p. 484)
negative relationship (p. 485)
correlational statistics (p. 485)
Pearson correlation coefficient (r) (p. 486)
covariance (p. 487)
rank (p. 505)
Spearman rank-order correlation (r_s) (p. 506)
predictor variable (p. 515)
criterion variable (p. 515)

13.9 Formulas Introduced in This Chapter

Degrees of Freedom for Pearson Correlation (*df*)

$$df = N - 2 \tag{13-1}$$

Sum of Products (Definitional Formula) (SP_{XY})

$$SP_{XY} = \Sigma[(X - \bar{X})(Y - \bar{Y})] \quad (13\text{-}2)$$

Sum of Squares for the *X* Variable (Definitional Formula) (SS_X)

$$SS_X = \Sigma(X - \bar{X})^2 \quad (13\text{-}3)$$

Sum of Squares for the *Y* Variable (Definitional Formula) (SS_Y

$$SS_Y = \Sigma(Y - \bar{Y})^2 \quad (13\text{-}4)$$

Pearson Correlation (Definitional Formula) (r)

$$r = \frac{\Sigma[(X - \bar{X})(Y - \bar{Y})]}{\sqrt{(\Sigma(X - \bar{X})^2)(\Sigma(Y - \bar{Y})^2)}} \quad (13\text{-}5)$$

Measure of Effect Size for the Pearson Correlation Coefficient (r^2)

$$\text{Estimate of effect} = r^2 \quad (13\text{-}6)$$

Sum of Products (Computational Formula) (SP_{XY})

$$SP_{XY} = \Sigma XY - \frac{(\Sigma X)(\Sigma Y)}{N} \quad (13\text{-}7)$$

Sum of Squares for the *X* Variable (Computational Formula) (SS_X)

$$SS_X = \Sigma X^2 - \frac{(\Sigma X)^2}{N} \quad (13\text{-}8)$$

Sum of Squares for the *Y* Variable (Computational Formula) (SS_Y)

$$SS_Y = \Sigma Y^2 - \frac{(\Sigma Y)^2}{N} \quad (13\text{-}9)$$

Pearson Correlation (Computational Formula) (r)

$$r = \frac{\Sigma XY - \frac{(\Sigma X)(\Sigma Y)}{N}}{\sqrt{(\Sigma X^2 - \frac{(\Sigma X)^2}{N})(\Sigma Y)^2 - \frac{(\Sigma Y)^2}{N})}} \quad (13\text{-}10)$$

Spearman Rank-Order Correlation (r_s)

$$r_s = 1 - \frac{6\Sigma D^2}{N(N^2 - 1)} \quad (13\text{-}11)$$

Measure of Effect Size for the Spearman Rank-Order Correlation (r_s^2)

$$\text{Estimate of effect} = r_s^2 \quad (13\text{-}12)$$

13.10 Using SPSS

Pearson Correlation: The Picky Eater Study (13.1)

1. Define independent and dependent variables (name, # decimals, labels for the variables) and enter data for the variables.

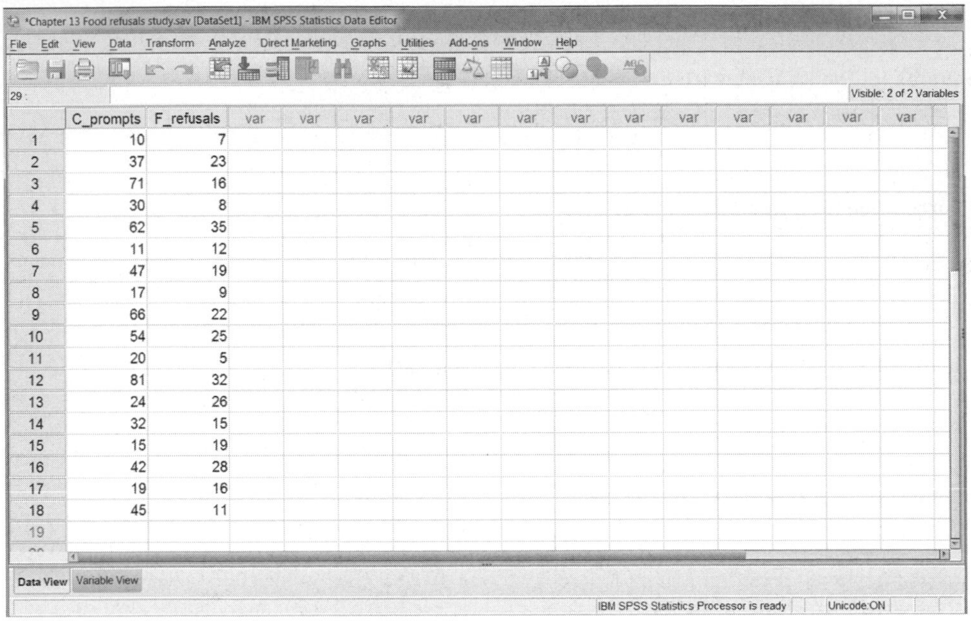

2. Select the Pearson correlation procedure within SPSS.

   How? (1) Click **Analyze** menu, (2) click **Correlate,** (3) click **Bivariate**.

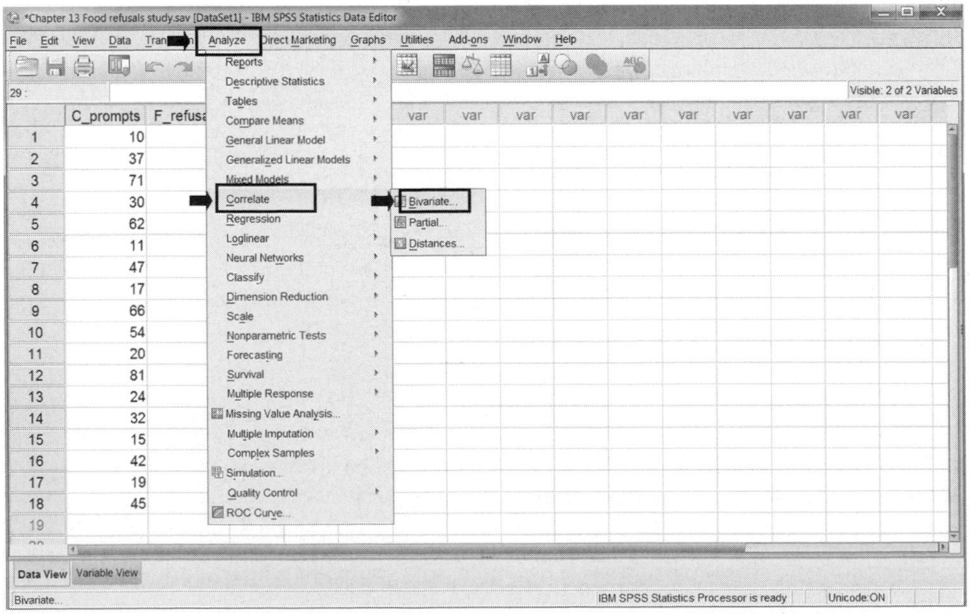

3. Select the variables to be correlated, and ask for descriptive statistics.

   How? (1) Click variables and → **Variables**, (2) click **Options** and click **Means and standard deviations**, (3) click, **Continue**, and (4) click **OK.**

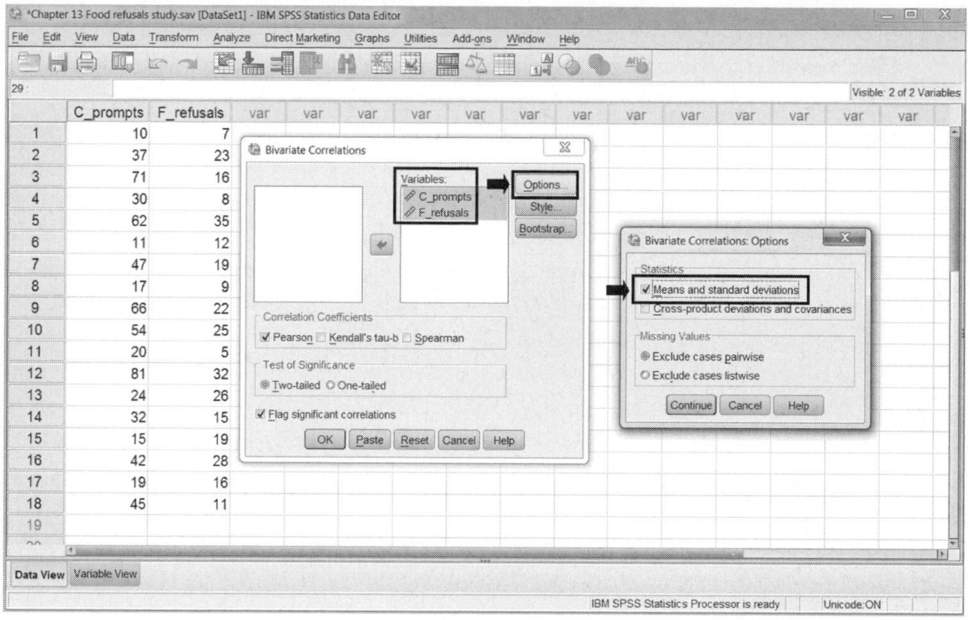

4. Examine output.

Correlations

Descriptive Statistics

	Mean	Std. Deviation	N
C_prompts	37.94	21.916	18
F_refusals	18.22	8.789	18

→ X variable (C_prompts)
→ Y variable (F_refusals)

Correlations

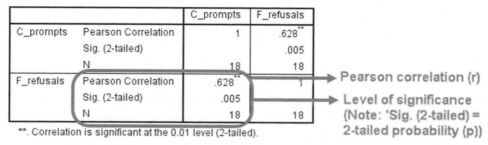

		C_prompts	F_refusals
C_prompts	Pearson Correlation	1	.628**
	Sig. (2-tailed)		.005
	N	18	18
F_refusals	Pearson Correlation	.628**	1
	Sig. (2-tailed)	.005	
	N	18	18

**. Correlation is significant at the 0.01 level (2-tailed).

→ Pearson correlation (r)
→ Level of significance (Note: 'Sig. (2-tailed) = 2-tailed probability (p))

Spearman Rank-Order Correlation: The Business Student Study (13.4)

1. Define the two sets of ranks (name, # decimals, labels for the variables) and enter the two sets of ranks.

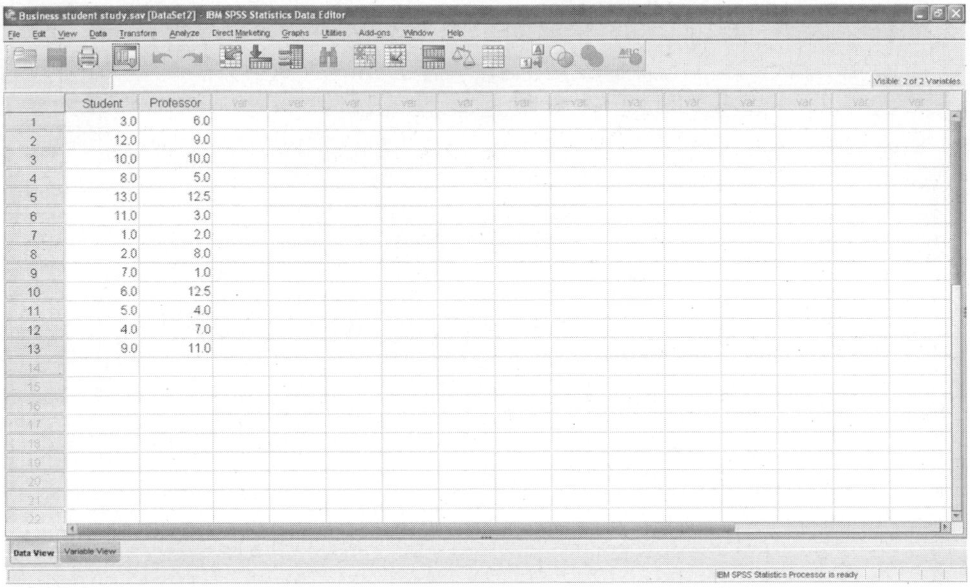

2. Begin the Spearman rank-order correlation procedure within SPSS.

   How? (1) Click **Analyze** menu, (2) click **Correlate,** and (3) click **Bivariate.**

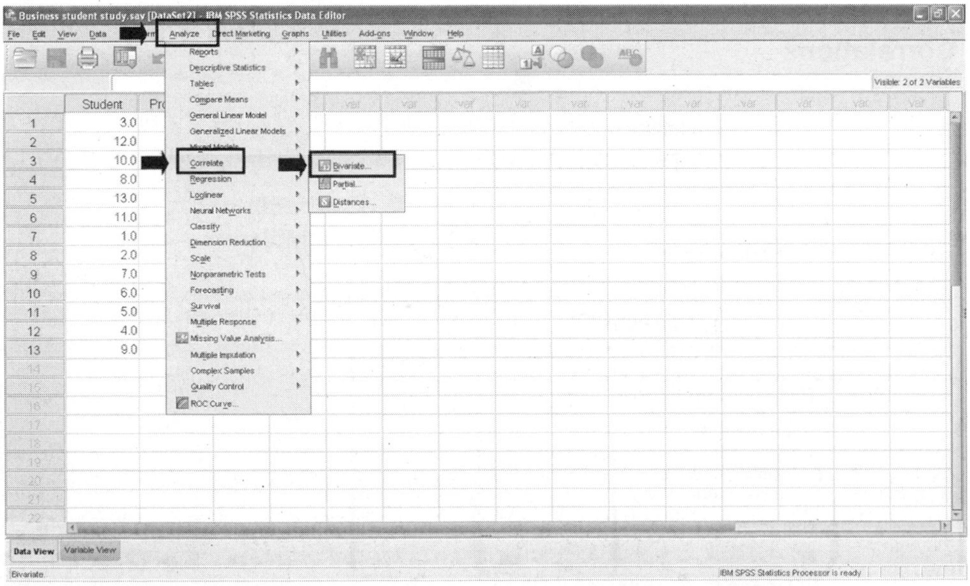

3. Select the variables to be correlated, and ask for the Spearman rank-order correlation.

   How? (1) Click variables and → **Variables**, (2) click **Spearman**, (3) click **OK**.

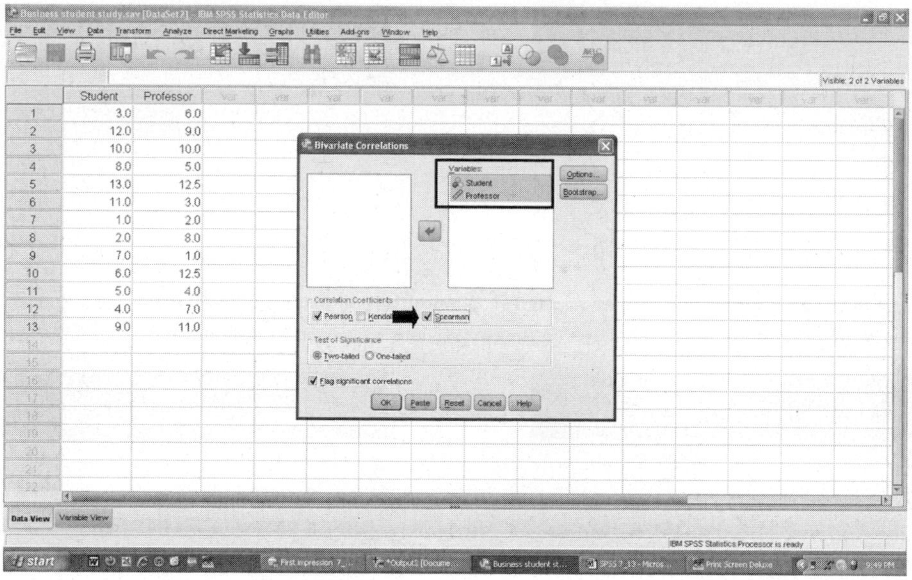

4. Examine output.

Nonparametric Correlations

Correlations

			Student	Professor
Spearman's rho	Student	Correlation Coefficient	1.000	.393
		Sig. (2-tailed)	.	.184
		N	13	13
	Professor	Correlation Coefficient	.393	1.000
		Sig. (2-tailed)	.184	.
		N	13	13

→ Spearman correlaton

→ Level of significance (Note: 'Sig. (2-tailed) = 2-tailed probability (p))

13.11 Exercises

1. For each of the following situations, create a scatterplot of the data and describe the nature, direction, and strength of the relationship.

   a. A researcher hypothesizes a positive relationship between the number of children in a family and the number of televisions a family owns.

Family	# Children	# Televisions
1	3	5
2	1	3
3	2	5

Family	# Children	# Televisions
4	4	6
5	3	3
6	5	8

b. Is there a relationship between the amount of time spent watching sports on television and the amount of time actually playing sports? The following data are from a small sample of children who reported their average number of daily hours spent watching sports on television and playing sports.

Child	# Hours Watching	# Hours Playing
1	.5	.2
2	6	.3
3	1	1.0
4	3	3.0
5	1	.4

Child	# Hours Watching	# Hours Playing
6	5	.5
7	2	2.0
8	2	1.8
9	4	1.6

c. A test developer gives people two versions of a survey (a longer version that contains 40 items and a shorter version that contains 20 items) and wants to see if scores on the two versions are related to each other.

Participant	Longer Version	Shorter Version
1	37	14
2	17	11
3	26	17
4	15	9
5	27	13
6	33	16

Participant	Longer Version	Shorter Version
7	20	5
8	14	15
9	24	18
10	22	13
11	29	12

2. a. A set of parents hypothesize there is a negative relationship between the average number of hours their children spend on the Internet and their scores on a recent school quiz (range = 0–100).

Student	# Hours Internet	Quiz Score
1	7	65
2	2	96
3	4	92
4	8	49
5	5	73
6	3	90
7	6	57

Student	# Hours Internet	Quiz Score
8	3	78
9	1	81
10	4	59
11	6	76
12	1	95
13	2	66

b. A college counselor hypothesizes that the farther students live from campus (in miles), the less they feel they are part of the school community (1 = low, 10 = high).

Student	Miles From Campus	Part of Community
1	2	9
2	19	7
3	32	3
4	23	9
5	40	5
6	5	2
7	20	1
8	16	2

Student	Miles From Campus	Part of Community
9	10	8
10	30	5
11	3	4
12	36	8
13	17	4
14	7	6
15	25	4
16	12	4

c. A publishing company develops a new college admissions test and hypothesizes that scores on the test are positively related to academic achievement in college. The company administers the test to a sample of college students (possible scores range from 0–100) and collects college grade point averages (GPAs) of the students.

Student	Test Score	College GPA
1	68	2.3
2	84	3.7
3	69	3.2
4	78	2.7
5	91	3.4
6	26	1.5
7	57	2.9

Student	Test Score	College GPA
8	62	3.6
9	39	1.8
10	74	3.9
11	43	2.6
12	52	2.0
13	40	3.3
14	35	2.5

3. For each of the following, calculate the degrees of freedom (*df*) and identify the critical value of *r*.
   a. $N = 8$, $\alpha = .05$ (two-tailed)
   b. $N = 19$, $\alpha = .05$ (two-tailed)
   c. $N = 36$, $\alpha = .05$ (two-tailed)
   d. $N = 50$, $\alpha = .05$ (one-tailed)

4. For each of the following, calculate the degrees of freedom (*df*) and identify the critical value of *r*.
   a. $N = 21$, $\alpha = .05$ (two-tailed)
   b. $N = 40$, $\alpha = .05$ (two-tailed)
   c. $N = 12$, $\alpha = .05$ (one-tailed)
   d. $N = 120$, $\alpha = .05$ (one-tailed)

5. For each of the following, calculate the Pearson correlation (*r*).
   a. $SP_{XY} = -3.00$, $SS_X = 8.00$, $SS_Y = 10.00$
   b. $SP_{XY} = 4.00$, $SS_X = 9.00$, $SS_Y = 5.00$
   c. $SP_{XY} = -1.50$, $SS_X = 7.50$, $SS_Y = 10.50$
   d. $SP_{XY} = 13.26$, $SS_X = 5.73$, $SS_Y = 142.08$
   e. $SP_{XY} = -86.74$, $SS_X = 110.43$, $SS_Y = 132.79$

6. For each of the following, calculate the Pearson correlation (*r*).
   a. $SP_{XY} = .82$, $SS_X = 6.91$, $SS_Y = 9.24$
   b. $SP_{XY} = -43.29$, $SS_X = 411.89$, $SS_Y = 17.50$
   c. $SP_{XY} = 205.96$, $SS_X = 485.73$, $SS_Y = 612.52$
   d. $SP_{XY} = -16.80$, $SS_X = 47.02$, $SS_Y = 89.65$
   e. $SP_{XY} = 145.97$, $SS_X = 108.24$, $SS_Y = 378.33$

7. Calculate the Pearson correlation coefficient for each of the following sets of data.

a.

Variable *X*	Variable *Y*
15	24
10	8
8	15
23	21

b.

Variable *X*	Variable *Y*
3	1
6	8
7	1
1	8
5	7
8	2
5	1

c.

Variable X	Variable Y
17	41
23	11
9	17
15	32
12	23
20	46
19	20
10	37
14	13
8	39

d.

Variable X	Variable Y
19	4
17	3
21	6
20	4
23	2
26	7
22	5
24	4
26	5
18	1
21	3

Variable X	Variable Y
28	6
23	4
24	2

e.

Variable X	Variable Y
3.8	3.4
2.1	2.9
3.6	3.1
2.8	3.5
2.9	2.4
4.0	3.6
1.9	2.1
3.3	2.8
3.7	3.9
2.5	3.7
3.9	2.6
2.4	1.9
1.7	2.8
2.2	2.7
3.3	3.0
3.5	2.9
3.1	2.4

8. Calculate the Pearson correlation coefficient for each of the following sets of data.

a.

Variable X	Variable Y
6	3
4	4
1	2
5	4
3	1

b.

Variable X	Variable Y
12	10
7	14
3	10
9	13
5	21
10	17
8	18
6	22

c.

Variable X	Variable Y
75	83
68	63
91	88
72	74
84	90
61	73
80	67
95	88
77	72
89	93
82	70

d.

Variable *X*	Variable *Y*
21	3.3
28	1.9
19	4.2
23	2.7
20	3.0
26	2.1
18	3.9
22	2.9
19	3.6
21	4.5
25	3.3
27	2.6
22	2.3

e.

Variable *X*	Variable *Y*
132	125
178	118
124	110
140	92
186	116
153	90
139	104
185	106
171	96
190	105
157	124
162	91
151	102
180	92
145	116
131	87

9. Calculate the Pearson correlation coefficient for each situation in Exercise 1.

10. Calculate the Pearson correlation coefficient for each situation in Exercise 2.

11. A teacher hypothesizes there is a positive relationship between her students' scores on a midterm examination and on the final examination.

Student	Midterm Exam	Final Exam
1	69	73
2	87	91
3	65	84
4	78	81
5	52	67
6	61	69

a. Draw a scatterplot of the data.

b. Calculate descriptive statistics (the mean and standard deviation) for the two variables.

c. State the null and alternative hypotheses (H_0 and H_1) (use a nondirectional H_1).

d. Make a decision about the null hypothesis.

 (1) Calculate the degrees of freedom (*df*).

 (2) Set alpha (α), identify the critical values, and state a decision rule.

 (3) Calculate a statistic: Pearson correlation (*r*).

 (4) Make a decision whether to reject the null hypothesis.

 (5) Determine the level of significance.

 (6) Calculate a measure of effect size (r^2).

e. Draw a conclusion from the analysis.

f. Relate the result of the analysis to the research hypothesis.

12. What makes a person a "leader"? A social psychologist hypothesizes that the extent to which people are perceived to be leaders is positively related to the amount of talking they do, regardless of what they are actually saying. She conducts an experiment in which groups of participants solve a problem together. One of the participants in each group is a confederate who speaks a different number of sentences for different groups, half of which are designed to help the group solve the problem and half of which are not helpful. Following the group work, the confederate is rated by the other participants on his leadership ability (1 = low, 10 = high). The following data represent the findings of this fictional researcher.

Group	# Sentences Spoken	Rating of Leadership
1	4	5
2	10	6
3	29	7
4	22	9
5	6	2
6	7	3
7	18	8
8	15	4

a. Draw a scatterplot of the data.
b. Calculate descriptive statistics (the mean and standard deviation) for the two variables.
c. State the null and alternative hypotheses (H_0 and H_1) (use a nondirectional H_1).
d. Make a decision about the null hypothesis.
   (1) Calculate the degrees of freedom (*df*).
   (2) Set alpha (α), identify the critical values, and state a decision rule.
   (3) Calculate a statistic: Pearson correlation (*r*).
   (4) Make a decision whether to reject the null hypothesis.
   (5) Determine the level of significance.
   (6) Calculate a measure of effect size (r^2).
e. Draw a conclusion from the analysis.
f. Relate the result of the analysis to the research hypothesis.

13. What makes a joke funny? Sigmund Freud's research led him to theorize that people find jokes funnier when the jokes contain higher levels of sexuality or hostility. Kirsh and Olczak (2001) questioned whether gender affects this relationship between humor and hostility by measuring men's and women's reactions to comic books of differing levels of violence. The following data are representative of the women in their study. The researchers hypothesized that, because women find violence less humorous than do men, there is a negative relationship between the level of violence in a comic book (1 = low to 7 = high) and how humorous women rate the comic book (1 = not at all humorous to 6 = extremely humorous). In other words, the higher the level of violence in the comic book, the less humorous the comic is to women.

Woman	Level of Violence	Humorous Rating
1	2	4
2	7	3
3	6	3
4	3	4
5	4	2
6	7	1
7	5	3
8	2	6
9	1	5
10	6	2

a. Draw a scatterplot of the data.

b. Calculate descriptive statistics (the mean and standard deviation) for the two variables.

c. State the null and alternative hypotheses (H_0 and H_1) (use a nondirectional H_1).

d. Make a decision about the null hypothesis.

(1) Calculate the degrees of freedom (*df*).

(2) Set alpha (α), identify the critical values, and state a decision rule.

(3) Calculate a statistic: Pearson correlation (*r*).

(4) Make a decision whether to reject the null hypothesis.

(5) Determine the level of significance.

(6) Calculate a measure of effect size (r^2).

e. Draw a conclusion from the analysis.

f. Relate the result of the analysis to the research hypothesis.

14. Earlier in this chapter, we discussed a study that examined the relationship between the amount of time students take to complete a test and their scores on the test (Herman, 1997). The two variables in this study were Time (the number of minutes taken by each student to finish the examination) and Exam score (the number of correct answers for each student, ranging from 0 to 100). The researcher in this study wished to test the hypothesis that the more time students take to finish an exam, the lower their score on the exam. In other words, there is a negative relationship between Time and Exam score. The data for the 32 students in this study are listed below (note that the scatterplot and descriptive statistics for this data were provided earlier in this chapter):

Student	Time	Exam Score	Student	Time	Exam Score
1	34	36	17	62	74
2	39	72	18	62	73
3	42	92	19	64	70
4	43	74	20	68	85
5	44	89	21	73	77
6	47	63	22	77	73
7	48	88	23	77	80
8	51	86	24	82	73
9	52	71	25	92	61
10	53	79	26	98	75
11	56	76	27	99	62
12	56	74	28	102	76
13	57	83	29	103	58
14	57	63	30	108	58
15	59	77	31	112	86
16	59	82	32	121	90

a. State the null and alternative hypotheses (H_0 and H_1) (use a nondirectional H_1).

b. Make a decision about the null hypothesis.

(1) Calculate the degrees of freedom (*df*).

(2) Set alpha (α), identify the critical values, and state a decision rule.

(3) Calculate a statistic: Pearson correlation (*r*).

(4) Make a decision whether to reject the null hypothesis.

(5) Determine the level of significance.

(6) Calculate a measure of effect size (r^2).

c. Draw a conclusion from the analysis.

d. Relate the result of the analysis to the research hypothesis.

15. Calculate SP_{XY}, SS_X, SS_Y, and r using the computational formulas for the data in Exercise 1.

16. Calculate SP_{XY}, SS_X, SS_Y, and r using the computational formulas for the data in Exercise 2.

17. Calculate the Spearman rank-order correlation (r_s) for each of the following pairs of ranks.

a.

Rank Order 1	Rank Order 2
5	5
3	1
4	4
1	2
2	3
6	5
7	7
8	9
9	8
10	10

b.

Rank Order 1	Rank Order 2
1	1
2	3
3	4
4	6
5	2

c.

Rank Order 1	Rank Order 2
2.5	1
5	3
4	5.5
1	2
2.5	4
6	5.5

18. Calculate the Spearman rank-order correlation (r_s) for each of the following pairs of ranks.

a.

Rank Order 1	Rank Order 2
1	5
2	3
3	1
4	8
5	2
6	7
7	4
8	6
9	7.5
10	10
11	7.5

b.

Rank Order 1	Rank Order 2
1	3
2	6
3.5	2
3.5	4
5	1
6	9
7.5	5
7.5	11

c.

Rank Order 1	Rank Order 2
6	8
2	4
10	10
3	1
8	12
4	5
12	13
5	3
13	14
9	6
1	2
11	7
7	9
14	11

19. Do students' needs change as they progress through their education? One study asked third-year and fourth-year college students preparing to be teachers what knowledges they would like to gain in order to effectively work with misbehaving students (Psunder, 2009). The seven knowledges below were rank ordered based on how often they were mentioned (1 = most often mentioned to 7 = least often mentioned). As the fourth-year students had gained more actual teaching experience than the third-year students, it was believed their desired knowledges would be different from the third-year students.

Knowledge	Third-Year Students	Fourth-Year Students
Working with problematic students	1	2
How to solve problems and conflicts	2	4
Cooperation with parents of problematic children	3	1
Discipline (practical suggestions)	4	5
How to make classes interesting	5	6
How to encourage students' activity	6	7
Taking measures in accordance with school legislation	7	3

a. State the null and alternative hypotheses (H_0 and H_1).

b. Make a decision about the null hypothesis.

(1) Calculate the degrees of freedom (*df*).

(2) Set alpha (α), identify the critical values, and state a decision rule.

(3) Calculate a statistic: Spearman rank-order correlation (r_s).

(4) Make a decision whether to reject the null hypothesis.

(5) Determine the level of significance.

(6) Calculate a measure of effect size (r_s^2).

c. Draw a conclusion from the analysis.

d. Relate the result of the analysis to the research hypothesis.

20. As people do not agree on who are the most important artists of all time, one study examined whether textbooks written in English and French agree on who are the most important artists who were French born or spent a substantial amount of time in France (O'Hagan & Kelly, 2005). To measure the importance of each artist, the researchers counted the number of illustrations of each artist's work that appeared in a sample of English and French textbooks; the artists were then rank ordered from highest to lowest based on the number of illustrations. The rankings of the top 16 artists are provided below. Imagine that the researcher hypothesized that the rankings of artists are similar for English and French textbooks.

Artist	English Textbooks	French Textbooks	Artist	English Textbooks	French Textbooks
Picasso	1	1	Degas	9	10
Matisse	2	2	Renoir	10	9
Cezanne	3	3	Duchamp	11	16
Manet	4	6	Courbet	12	11
Monet	5	4	Miro	13	15
Braque	6	7.5	Seurat	14	14
van Gogh	7	5	Leger	15	12
Gauguin	8	7.5	Toulouse-Lautrec	16	13

a. State the null and alternative hypotheses (H_0 and H_1).
b. Make a decision about the null hypothesis.
   (1) Calculate the degrees of freedom (*df*).
   (2) Set alpha (α), identify the critical values, and state a decision rule.
   (3) Calculate a statistic: Spearman rank-order correlation (r_s).
   (4) Make a decision whether to reject the null hypothesis.
   (5) Determine the level of significance.
   (6) Calculate a measure of effect size (r_s^2).
c. Draw a conclusion from the analysis.
d. Relate the result of the analysis to the research hypothesis.

Answers to Learning Checks

Learning Check 1

2. a. There is a strong, negative linear relationship such that the more days a student misses during a semester, the lower the student's score on the final exam.

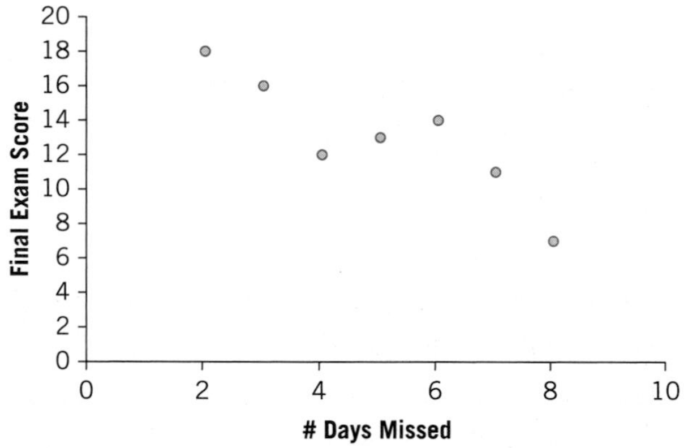

b. There is a very weak relationship between the number of times a student raises his or her hand and how favorably the student is viewed by the teacher.

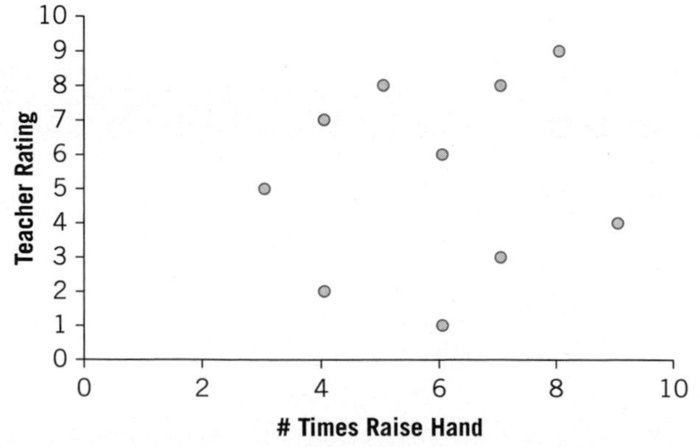

Learning Check 2

2. a. $df = 4$, critical value = ±.811
   b. $df = 7$, critical value = ±.666
   c. $df = 16$, critical value = ±.468
3. a. $r = .56$
   b. $r = -.44$
   c. $r = .10$

Learning Check 3

2. a. $H_0: \rho = 0$, $H_1: \rho \neq 0$
   b. (1) $df = 24$
      (2) If $r < -.388$ or $> .388$, reject H_0; otherwise, do not reject H_0
      (3) $SP_{XY} = 487.87$, $SS_X = 298{,}203.85$, $SS_Y = 4.24$, $r = .43$
      (4) $r = .43 > .388 \therefore$ reject H_0 ($p < .05$)
      (5) $r = .43 < .496 \therefore p < .05$ (but not $< .01$)
      (6) $r^2 = .18$ (a medium effect)
   c. There was a statistically significant positive relationship between students' GMAT scores and their grade point average (GPA), $r(24) = .43$, $p < .05$, $r^2 = .18$.
   d. This analysis supports the hypothesis that the higher the score on the GMAT, the greater the likelihood the student will achieve a high level of academic performance.

Learning Check 4

2. a.
$$SP_{XY} = 157 - \frac{(36)(24)}{6} = 13.00 \qquad SS_X = 238 - \frac{(36)^2}{6} = 22.00$$
$$SS_Y = 124 - \frac{(24)^2}{6} = 28.00 \qquad r = \frac{13.00}{\sqrt{(22.00)(28.00)}} = .52$$

   b.
$$SP_{XY} = 1131 - \frac{(168)(56)}{8} = -45.00 \qquad SS_X = 3724 - \frac{(168)^2}{8} = 196.00$$
$$SS_Y = 450 - \frac{(24)^2}{8} = 58.00 \qquad r = \frac{-45.00}{\sqrt{(196.00)(58.00)}} = -.42$$

   c.
$$SP_{XY} = 2495 - \frac{(38.13)(789)}{12} = -11.85 \qquad SS_X = 140.77 - \frac{(38.13)^2}{12} = 19.61$$
$$SS_Y = 57459 - \frac{(789)^2}{12} = 5582.25 \qquad r = \frac{-11.85}{\sqrt{(19.61)(5582.25)}} = -.04$$

Learning Check 5

2. a. $r_s = 1 - \frac{6(6)}{4(4^2-1)} = .40$

   b. $r_s = 1 - \frac{6(12)}{6(6^2-1)} = .66$

   c. $r_s = 1 - \frac{6(19)}{9(9^2-1)} = .84$

3. a. $H_0: \rho = 0$, $H_1: \rho \neq 0$

   b. (1) $df = 6$

   (2) If $r_s < -.738$ or $> .738$, reject H_0; otherwise, do not reject H_0

   (3) $N = 8$, $\Sigma D^2 = 115.50$, $r_s = -.38$

   (4) $r_s = -.38$ is not $< -.738$ or $> .738$ $\therefore$ do not reject H_0 ($p > .05$)

   (5) Not applicable (H_0 not rejected)

   (6) $r_s^2 = .14$

   c. There was not a significant relationship between the rank ordering of boys and girls in their preferences for the eight musical instruments, $r_s(6) = -.38$, $p > .05$, $r^2 = .14$.

   d. This analysis does not support the hypothesis that boys and girls would display similar preferences for the different types of musical instruments.

Answers to Odd-Numbered Exercises

1. a. There is a strong, positive linear relationship between the number of children in a family and the number of televisions a family owns such that the larger the number of children, the more televisions a family owns.

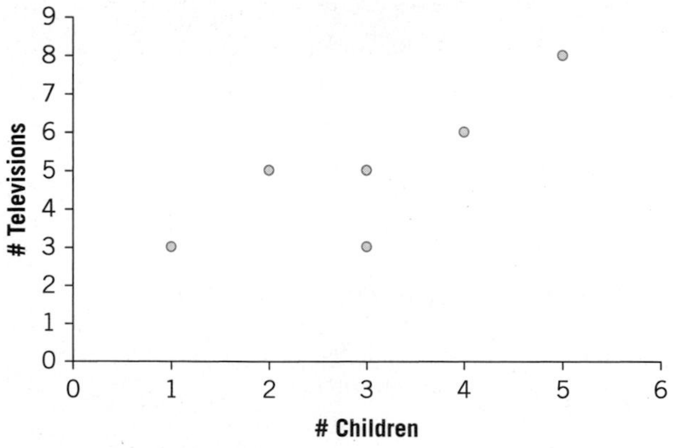

b. There is a nonlinear relationship between the average number of daily hours spent watching sports on television and playing sports.

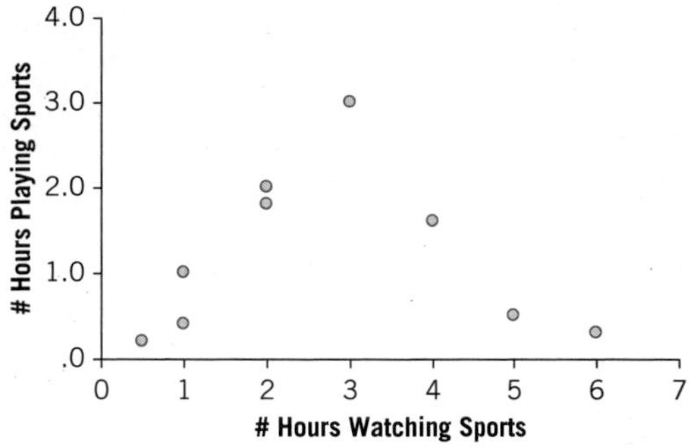

c. There is a moderate positive relationship between scores on the longer and shorter versions of the survey.

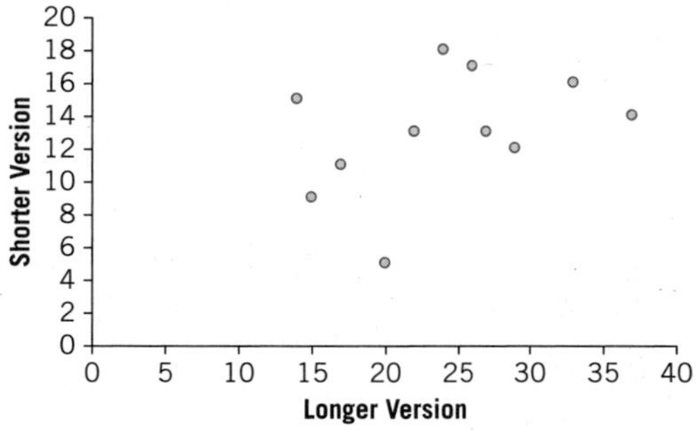

3. a. $df = 6$, critical value = ±.707
   b. $df = 17$, critical value = ±.456
   c. $df = 34$, critical value = ±.349
   d. $df = 48$, critical value = .257 or −.257

5. a. $r = -.34$
   b. $r = .60$
   c. $r = -.17$
   d. $r = .46$
   e. $r = -.72$

7. a. $SP_{XY} = 91.00$, $SS_X = 134.00$, $SS_Y = 150.00$, $r = .64$
   b. $SP_{XY} = -18.00$, $SS_X = 34.00$, $SS_Y = 72.00$, $r = -.36$
   c. $SP_{XY} = -78.30$, $SS_X = 228.10$, $SS_Y = 1434.90$, $r = -.14$
   d. $SP_{XY} = 38.00$, $SS_X = 132.86$, $SS_Y = 38.00$, $r = .53$
   e. $SP_{XY} = 2.91$, $SS_X = 8.74$, $SS_Y = 5.07$, $r = .44$

9. a. $SP_{XY} = 11.00$, $SS_X = 10.00$, $SS_Y = 18.00$, $r = .82$
   b. $SP_{XY} = -.60$, $SS_X = 29.56$, $SS_Y = 7.38$, $r = -.04$
   c. $SP_{XY} = 105.00$, $SS_X = 538.00$, $SS_Y = 140.00$, $r = .38$

11. a.

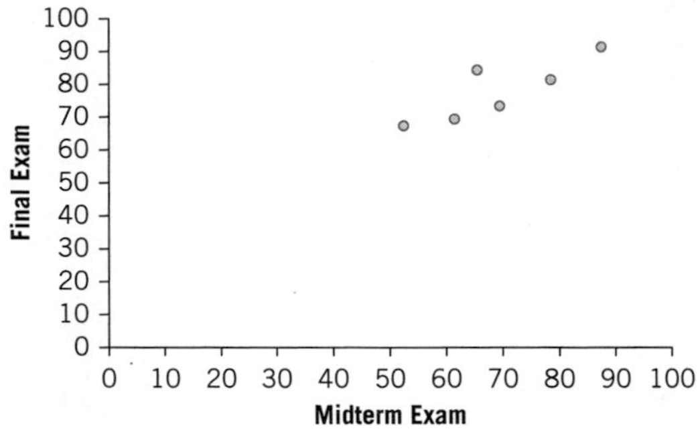

   b. Midterm exam (X): $N = 6, \overline{X} = 68.67, s_X = 12.44$
      Final exam (Y): $N = 6, \overline{Y} = 77.50, s_Y = 9.38$
   c. H_0: $\rho = 0$, H_1: $\rho \neq 0$
   d. (1) $df = 4$
      (2) If $r < -.811$ or $> .811$, reject H_0; otherwise, do not reject H_0
      (3) $SP_{XY} = 495.00$, $SS_X = 773.33$, $SS_Y = 439.50$, $r = .85$
      (4) $r = .85 > .811$ $\therefore$ reject H_0 ($p < .05$)
      (5) $r = .85 < .917$ $\therefore$ $p < .05$ but not $< .01$
      (6) $r^2 = .72$ (a large effect)
   e. There was a statistically significant positive relationship between students' midterm exam and final exam scores, $r(4) = .85, p < .05, r^2 = .72$.
   f. This analysis supports the teacher's hypothesis that there is a positive relationship between her students' scores on a midterm examination and on the final examination.

13. a.

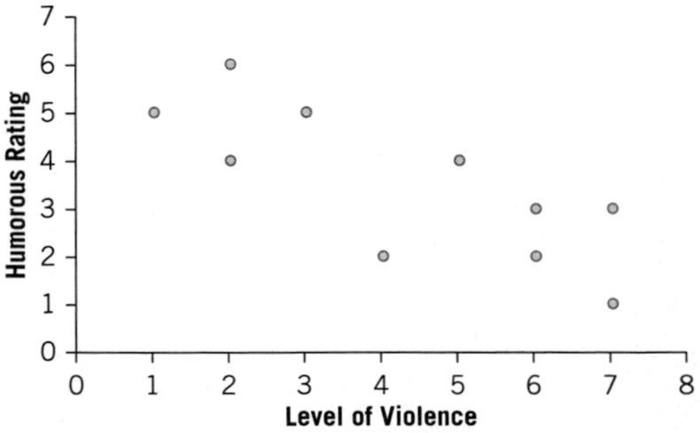

b. Level of violence (*X*): $N = 10, \bar{X} = 4.30, s_X = 2.21$

Humorous rating (*Y*): $N = 10, \bar{Y} = 3.50, s_Y = 1.58$

c. $H_0: \rho = 0, H_1: \rho \neq 0$

d. (1) $df = 8$

(2) If $r < -.632$ or $> .632$, reject H_0; otherwise, do not reject H_0

(3) $SP_{XY} = -24.50, SS_X = 44.10, SS_Y = 22.50, r = -.78$

(4) $r = -.78 < -.632 \therefore$ reject H_0 ($p < .05$)

(5) $r = -.78 < -.765 \therefore p < .01$

(6) $r^2 = .61$ (a large effect)

e. There was a statistically significant negative relationship between the level of violence in comic books and women's ratings of humor, $r(8) = -.78, p < .01, r^2 = .61$.

f. This analysis supports the hypothesis that there is a negative relationship between the level of violence in a comic book and how humorous women rate the comic book.

15. a. $SP_{XY} = 101 - \frac{(18)(30)}{6} = 11.00$ $\quad SS_X = 64 - \frac{(18)^2}{6} = 10.00$

$SS_Y = 168 - \frac{(30)^2}{6} = 18.00$ $\quad r = \frac{11.00}{\sqrt{(10.00)(18.00)}} = .82$

b. $SP_{XY} = 28.80 - \frac{(24.50)(10.80)}{9} = -.60$ $\quad SS_X = 96.25 - \frac{(24.50)^2}{9} = 29.56$

$SS_Y = 20.34 - \frac{(10.80)^2}{9} = 7.38$ $\quad r = \frac{6.22}{\sqrt{(8.71)(7.41)}} = -.04$

c. $SP_{XY} = 3537 - \frac{(264)(143)}{11} = 105.00$ $\quad SS_X = 6874 - \frac{(264)^2}{11} = 538.00$

$SS_Y = 1999 - \frac{(143)^2}{11} = 140.00$ $\quad r = \frac{6.22}{\sqrt{(8.71)(7.41)}} = .38$

17. a. $r_s = 1 - \frac{6(6)}{5(5^2 - 1)} = .70$

b. $r_s = 1 - \frac{6(18)}{10(10^2 - 1)} = .89$

c. $r_s = 1 - \frac{6(12)}{6(6^2 - 1)} = .66$

19. a. $H_0: \rho = 0, H_1: \rho \neq 0$

b. (1) $df = 5$

(2) If $r_s < -.786$ or $> .786$, reject H_0; otherwise, do not reject H_0

(3) $N = 7, \Sigma D^2 = 28, r_s = .50$

(4) $r_s = .50$ is not $< -.786$ or $> .786 \therefore$ do not reject H_0 ($p > .05$)

(5) Not applicable (H_0 not rejected)

(6) $r_s^2 = .25$

c. There was a nonsignificant relationship between the rank ordering of desired knowledges for the third-year and fourth-year students, $r_s(5) = .50$, $p > .05$, $r^2 = .25$.

d. This analysis supports the hypothesis that the desired knowledges of the third-year and fourth-year students are different from each other.

SAGE edge™

Sharpen your skills with **SAGE edge at edge.sagepub.com/tokunaga2e**

SAGE edge for students provides a personalized approach to help you accomplish your coursework goals in an easy-to-use learning environment. Log on to access:

- eFlashcards
- Web Quizzes
- SPSS Data Files
- Video and Audio Resources

14

LINEAR REGRESSION AND MULTIPLE CORRELATION

CHAPTER OUTLINE

Chapter 13 introduced the concept of correlation, which was defined as a mutual or reciprocal relationship between two variables. The research examples in that chapter involved calculating a statistic known as a correlation coefficient. More specifically, Chapter 13 discussed the Pearson correlation coefficient (r), a statistic that measures the linear relationship between two continuous variables measured at the interval or ratio level of measurement. One main purpose of the Pearson r is to describe the relationship between two variables. However, in addition to description, another goal of science is *prediction*. Researchers frequently use identified relationships between variables to predict one variable from another. For example, in the picky eater study, a significant relationship was found between parents' coercive prompts and children's food refusals. Consequently, how might you use knowledge of how often parents use coercive prompts to predict the extent to which a child will refuse to eat a new food? This chapter discusses how researchers use the relationship between variables to predict scores on one variable from another.

14.1 PREDICTING ONE VARIABLE FROM ANOTHER: LINEAR REGRESSION

When a relationship is found to exist between variables, this provides an opportunity to predict one variable from another. For example, if college admissions officers determine there's a relationship between students' high school grade point average (GPA) and their college GPA, this information can be used to predict an applicant's college GPA from his or her high school GPA. If political scientists find a relationship between income and attitudes about a presidential candidate, they may wish to use a voter's income to predict the likelihood of voting for that candidate.

For the purpose of our discussion, moving from "relationship" to "prediction" involves moving from "correlation" to "regression." **Regression** is defined as the use of a relationship between two or more correlated variables to predict values of one variable from values of other variables. Whereas the main purpose of correlation is to measure the relationship between variables, the goal of regression is to predict one variable from other variables.

Linear regression refers to a statistical procedure in which a straight line is fitted to a set of data to best represent the relationship between two variables. When we introduced the concept of correlation (Section 13.2 of Chapter 13), we noted that the nature of the relationship between two variables may be represented using a straight line. This notion of a straight line is the basis of how one variable can be used to predict another.

To introduce linear regression, let's examine the formula for a straight line. You may have learned the following formula in a high school geometry class:

$$Y = mX + b$$

In this equation, Y is a value for a Y variable, m is the slope of the line, X is a value for an X variable, and b is the Y-intercept. The slope (m) is the angle or tilt of the line relative to the X-axis (the horizontal axis). The slope is sometimes referred to as the "rise over run" or "change in Y over change in X"—the rate of change in the Y variable associated with the rate of change in the X variable. A value of zero (0) for the slope represents a horizontal line that is parallel to the X-axis. The Y-intercept (b) is the point at which the line crosses the Y-axis (the vertical axis) when X is equal to 0.

The Linear Regression Equation

Applying the concept of a straight line to linear regression involves creating a line that best represents, or fits, the relationship between two variables in a set of data. This "best-fit" line is known as the linear regression equation. The **linear regression equation** is a mathematical equation that predicts a score on one variable from a score on another variable based on the relationship between two variables.

The formula for the linear regression equation is a modification of the formula for a straight line:

$$Y' = a + bX \qquad (14\text{-}1)$$

where Y' represents a predicted score for the Y variable, a is the Y-intercept, b is the slope, and X is a score for the X variable. The main difference between Formula 14-1 and the formula for a line involves the use of Y' rather than Y. Y' represents a *predicted* value for the Y variable rather than an *actual* value. In other words, the purpose of a linear regression equation is to predict scores for a variable based on its relationship with another variable found in a sample of data. For example, a political scientist who finds income is related to attitudes about a presidential candidate in one sample might use this relationship to predict attitudes about the candidate from the income of participants in another sample.

Calculating the Linear Regression Equation

Developing a linear regression equation requires calculating the slope (*b*) and the Y-intercept (*a*). This section calculates the linear regression equation using the data from the picky eater study. As a reminder, in this study, we calculated a Pearson correlation of $r = .63$ between parents' coercive prompts (the *X* variable) and children's food refusals (the *Y* variable). In order to calculate the linear regression equation, we'll also need descriptive statistics for the two variables:

	Coercive Prompts (*X*)	Food Refusals (*Y*)
N	18	18
Mean	$\overline{X} = 37.94$	$\overline{Y} = 18.22$
Standard deviation	$s_X = 21.92$	$s_Y = 8.79$

Calculate the Slope of the Equation (*b*)

The first step in calculating the linear regression equation is to calculate the slope (*b*) using the following formula:

$$b = r\frac{s_Y}{s_X} \tag{14-2}$$

where *r* is the correlation between Variables *X* and *Y*, s_Y is the standard deviation of Variable *Y*, and s_X is the standard deviation of Variable *X*.

There are several aspects of Formula 14-2 worth noting. First, the stronger the relationship between the two variables (represented by values of the Pearson *r* closer to ±1.00), the greater is the angle or slope of the equation (the value of *b*). Conversely, the weaker the relationship, the more the slope approaches zero (.00), which is represented by a horizontal line parallel to the X-axis. Second, because the Pearson *r* can be positive or negative, *b* can also be positive or negative. A positive value for the slope means the equation angles upward from left to right, whereas a negative slope indicates the equation angles downward from left to right. Third, the standard deviation of Variable *Y* (s_Y) is divided by the standard deviation of Variable *X* (s_X) in order to represent the "rise over run" (change in Variable *Y* divided by change in Variable *X*) aspect of the slope described earlier.

To begin the calculation of the slope for the picky eater study, recall that the correlation between coercive prompts and food refusals was $r = .63$. Using the standard deviations for the two variables provided in the above table, the slope may be calculated as follows:

$$\begin{aligned} b &= r\frac{s_Y}{s_X} \\ &= .63\,\frac{8.79}{21.92} = .63(.40) \\ &= .25 \end{aligned}$$

Calculate the Y-Intercept of the Equation (*a*)

In linear regression, the Y-intercept *a* is the predicted value for the *Y* variable when *X* is equal to zero. Formula 14-3 provides a computational formula for the Y-intercept:

$$a = \overline{Y} - b\overline{X} \tag{14-3}$$

where $\overline{Y}$ is the mean of the Y variable, b is the slope of the equation, and $\overline{X}$ is the mean of the X variable.

Using the value of b = .25 calculated earlier and the means for the two variables from the above table, we calculate the Y-intercept for the picky eater study in the following manner:

$$\begin{aligned} a &= \overline{Y} - b\overline{X} \\ &= 18.22 - .25(37.94) = 18.22 - 9.56 \\ &= 8.66 \end{aligned}$$

Report the Linear Regression Equation

Once the values for the slope and the Y-intercept have been calculated, the linear regression equation may be reported. For the picky eater study, the following equation predicts the number of children's food refusals from the number of parents' coercive prompts:

$$Y' = a + bX$$

$$\text{Food refusals}' = 8.66 + .25\ (\text{Coercive prompts})$$

Interpreting our regression equation, we obtain a predicted number of food refusals by multiplying the number of coercive prompts by .25 and then adding 8.66.

Drawing the Linear Regression Equation

It is possible to draw a linear regression equation in a scatterplot of the data used to create the equation. One way to draw this equation is to determine any two points in a line produced by the equation and connect these two points. Determining these two points involves calculating the predicted score on the Y variable for any two possible scores on the X variable. For the picky eater study, let's arbitrarily choose values of 8 and 75 for the coercive prompts variable. Inserting these two scores into the regression equation, we may determine two predicted values of food refusals:

Coercive Prompts = 8	Coercive Prompts = 75
Food refusals′ = 8.66 + .25 (*8*)	Food refusals′ = 8.66 + .25 (*75*)
= 8.66 + 2.00	= 8.66 + 18.75
= 10.66	= 27.41

Once the two predicted Y values have been calculated, the next step is to locate them in a scatterplot of the two variables. In Figure 14.1, the coordinates of these two points ((8, 10.66) and (75, 27.41)) have been marked in the scatterplot of the picky eater study (Figure 13.1 of Chapter 13), and a line that connects them has been drawn. There are several things to note about this line. First, the line has a positive or upward slope because the relationship between the two variables is strong and positive (r = .63). Second, the line passes through the center of the scatterplot of scores, indicating that the equation "fits" the data as closely as possible. Because the relationship between the two variables is less than perfect, some data points fall above the line and some fall below. The equation minimizes as much as possible the differences between the data points and the line. In other words, the equation is the "best-fitting line" for the data in this particular sample.

Box 14.1 summarizes the steps in calculating the linear regression equation, using the picky eater study as the example. In the next section, we discuss a second research example to accomplish two goals. The first goal is to further illustrate the process of calculating and interpreting linear regression equations. The second goal is to answer the question, how do you predict one variable from another when they're not related to each other?

FIGURE 14.1 ● LINEAR REGRESSION EQUATION FOR THE PICKY EATER STUDY

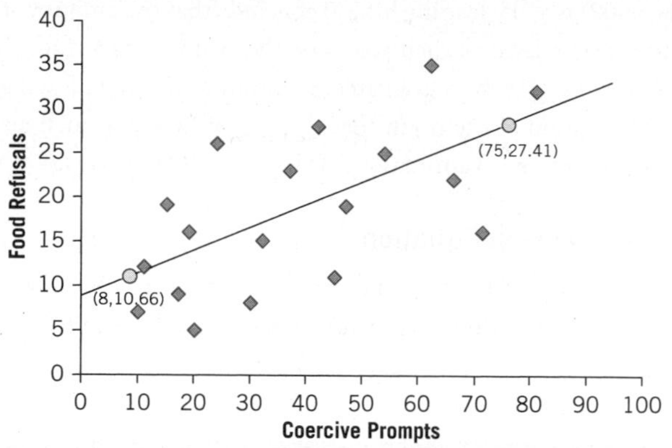

SUMMARY BOX 14.1 CALCULATING THE LINEAR REGRESSION EQUATION (PICKY EATER STUDY EXAMPLE)

Calculate the slope of the equation (b)

$$b = r\frac{S_Y}{S_X}$$
$$= .63\frac{8.79}{21.92} = .63(.40) = .25$$

Calculate the Y-intercept of the equation (a)

$$a = \overline{Y} - b\overline{X}$$
$$= 18.22 - .25(37.94) = 18.22 - 9.56 = 8.66$$

Report the linear regression equation

$$Y' = a + bX$$
$$\text{Food refusals}' = 8.66 + .25\ (\text{Coercive prompts})$$

A Second Example: Put Down Your Pencils

Imagine you're in a classroom taking an exam (probably not a very difficult thing to imagine). Working quickly, you glance up and see students turning in their exams. The following thoughts may enter your mind: "Am I going too slowly?" "What if I can't finish in time?" "Will I do worse than the students who've already finished?"

Is there a relationship between the amount of time students take to complete a test and their scores on the test? Although teachers and students have their own beliefs about this relationship, William Herman at the State University of New York examined the relationship empirically (Herman, 1997). The participants in this study were 32 students in one of Dr. Herman's courses. The data from these students are included in the exercises at the end of the chapter.

The students in Dr. Herman's course were given a 100-item multiple-choice test. For each student, Dr. Herman recorded information on two variables. The first variable, Time, was the number of minutes taken to finish the examination. The second variable, Exam score, was the number of correct answers, ranging from 0 to 100.

To save space, the calculation and testing of the Pearson correlation between Time and Exam score will not be shown here (a scatterplot of the data is provided in Figure 14.2). Instead, we will simply report what Dr. Herman found: $r(30) = -.02, p > .05$, leading him to conclude that the amount of time students took to finish the exam was not significantly related to their scores on the exam. In fact, there was virtually *zero* relationship between the two variables! Dr. Herman evaluated his findings in the following way: "This finding lends support for the conscious destruction of the myth that the rate of finishing such an examination is related to how well students will perform on these examinations" (Herman, 1997, p. 117).

Calculating the Linear Regression Equation

Even though no relationship was found between Time and Exam score, a linear regression equation can still be created between the two variables. To develop this equation, descriptive statistics of the two variables are needed:

	Time (X)	Exam Score (Y)
N	32	32
Mean	$\overline{X} = 68.66$	$\overline{Y} = 74.25$
Standard deviation	$s_X = 23.96$	$s_Y = 11.62$

The first step in calculating the equation is to determine the slope (b) using Formula 14-2. Using the correlation between the two variables ($r = -.02$), along with the standard deviations for the two variables, the slope is calculated as follows:

$$\begin{aligned} b &= r\frac{S_Y}{S_X} \\ &= -.02\frac{11.62}{23.96} = -.02(.49) \\ &= -.01 \end{aligned}$$

In this example, the value for the slope ($b = -.01$) is a negative number because the relationship between the two variables is negative ($r = -.02$).

Once the slope has been calculated, the next step is to calculate the Y-intercept (a) using Formula 14-3. Using the value of $b = -.01$ and the means for the two variables, the Y-intercept is calculated as follows:

$$\begin{aligned} a &= \overline{Y} - b\overline{X} \\ &= 74.25 - (-.01)(68.66) = 74.25 - (-.69) \\ &= 74.94 \end{aligned}$$

Using the calculated values for b and a, we may report the following linear regression equation predicting a student's exam score from the number of minutes taken to complete the exam:

$$Y' = a + bX$$
$$\text{Exam score}' = 74.94 + (-.01)(\text{Time})$$

Drawing the Linear Regression Equation

To draw the linear regression equation into the scatterplot of the two variables, we need to identify two points along the line. Arbitrarily picking values of Time of 20 minutes and 120 minutes, two predicted exam scores are calculated as follows:

Time = 20	Time = 120
Exam score′ = 74.94 + (−.01) (*20*)	Exam score′ = 74.94 + (−.01) (*120*)
= 74.94 − .20	= 74.94 − 1.20
= 74.74	= 73.74

Figure 14.2 uses these two points to draw the regression line into the scatterplot of the data from the 32 students. Looking at the figure, we see that the regression equation is nearly a horizontal line—the line has almost no slope because there is virtually no relationship between the two variables. The implication of a horizontal line is that, regardless of the amount of time it takes a student to complete the test, virtually the same exam score is predicted. For example, there's almost no difference between the predicted score of a student who takes 20 minutes and a student who takes 120 minutes (74.74 vs. 73.74).

From the regression equation, we see that the Y-intercept ($a = 74.94$) and the predicted exam scores for 20 and 120 minutes (74.74 and 73.74) are all almost identical to the mean of the Exam scores ($\overline{Y} = 74.25$). This illustrates an important point: When there is no relationship between two variables, the most accurate prediction we can make, regardless of the score on one variable, is the mean of the other variable. To further illustrate this, imagine you're asked the question, "I'm thinking of a friend of mine who is 5′ 11″. What's his IQ?" Because there's no relationship between height and IQ, the best prediction you could make would be the mean IQ in the population (for example, 100). Furthermore, you would predict this mean of 100 for every person regardless of his or her height. Ultimately, there must be a relationship between variables to improve accuracy of prediction beyond that provided by the mean.

FIGURE 14.2 ● LINEAR REGRESSION EQUATION PREDICTING EXAM SCORE FROM TIME

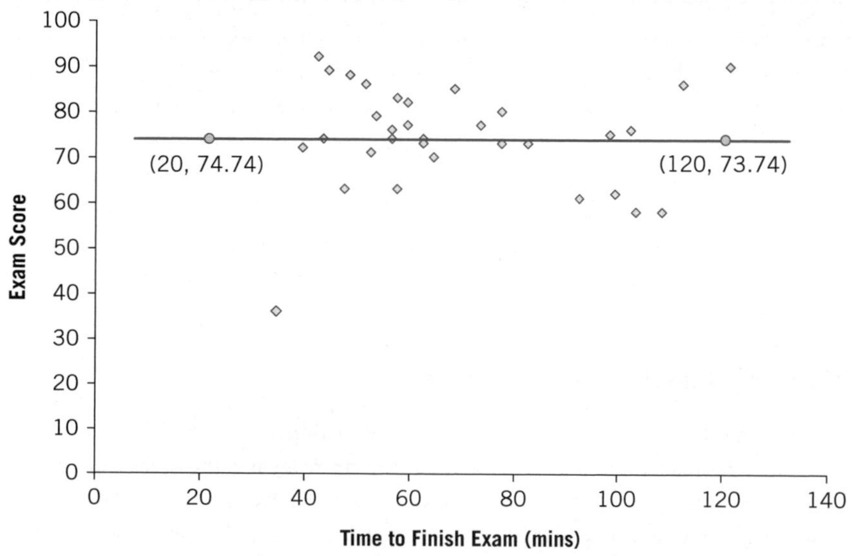

LEARNING CHECK 1:

Reviewing What You've Learned So Far

1. Review questions
   a. What is the difference between correlation and regression?
   b. What is the purpose of a linear regression equation?
   c. What are the main components of a linear regression equation?
   d. What determines whether the slope (*b*) of a linear regression equation is positive or negative?
   e. What do you need to do to draw a linear regression equation into a scatterplot?
   f. What does the linear regression equation look like when there is no relationship between two variables?
   g. When there is no relationship between two variables, what is the most accurate prediction you can make for the *Y* variable?
2. For each of the following, calculate the linear regression equation and draw it into a scatterplot.
   a. Variable *X*: range = 1–10, $\bar{X}_i = 6.00$, $s_X = 2.00$; Variable *Y*: range = 2–12, $\bar{Y} = 9.00$, $s_Y = 3.00$; Pearson correlation: $r = .30$
   b. Variable *X*: range = 1–5, $\bar{X}_i = 2.59$, $s_X = .88$; Variable *Y*: range = 1–5, $\bar{Y} = 3.78$, $s_Y = .93$; Pearson correlation: $r = .47$
   c. Variable *X*: range = 1–50, $\bar{X}_i = 32.76$, $s_X = 4.60$; Variable *Y*: range = 0–5, $\bar{Y} = 1.84$, $s_Y = .42$; Pearson correlation: $r = -.63$
3. Calculate predicted scores (Y') for each of the research situations in 2(a–c).
   a. For the research situation in 2a, what is the predicted score (Y') for someone with $X = 6$?
   b. For the research situation in 2b, what is the predicted score (Y') for someone with $X = 3$?
   c. For the research situation in 2c, what is the predicted score (Y') for someone with $X = 41$?

14.2 CORRELATION WITH TWO OR MORE PREDICTORS: INTRODUCTION TO MULTIPLE CORRELATION AND REGRESSION

This chapter discusses two related statistical concepts: correlation and regression. Correlation involves measuring the relationship between variables; an example of a correlational statistic is the Pearson correlation (*r*), which measures the linear relationship between two variables, one of which may be called the predictor and the other the criterion. Regression, on the other hand, involves the prediction of a criterion variable. One example of regression is linear regression, which predicts a criterion from one predictor variable. It is possible, however, to include more than one predictor in the same analysis. In this section, we provide a brief introduction of correlation and regression with two or more predictors; for a much more detailed discussion, we encourage you to read Tabachnick and Fidell (2013) and Cohen, Cohen, West, and Aiken (2003).

The Limitations of Using Only One Predictor

In the picky eater study, based on a calculated Pearson correlation of $r = .63$, we concluded that coercive prompts were significantly related to food refusals. Squaring the Pearson *r*, the r^2 measure of effect size of .40 implied that 40% of the variance in food refusals is explained by coercive prompts. By subtraction, an r^2 of .40 indicates that 60% of the variance in food refusals is *not* explained. Assuming one goal of science is to explain as much of the variance in variables as possible, we ask ourselves two questions: Why were we unable to explain more of the variance in the criterion, and how can we explain more of the variance?

In terms of the first question, there are both conceptual and statistical reasons why one predictor may explain only a small amount of the variance in a criterion. Conceptually, the criterion variables studied by researchers are often complex, implying that they can't be explained by one predictor. For example, why do children refuse to eat new foods? As suggested by the picky eater study, it could be because of how parents try to get them to eat. However, it could also be because of many other factors: The age of the child, the gender of the child, the child's personality, and the child's relationship with his or her parents are only a few of the many variables that may be related to children's food refusals. Consequently, it's unrealistic to expect any one predictor to explain a large portion of the variance in a criterion. Statistically, would you expect any one predictor to explain all (100%) of the variance in a criterion? Probably not, as this would require the Pearson r to be 1.00 (a perfect relationship); the impact of random, chance factors that are part of every study works against this possibility. Furthermore, even a strong Pearson r such as .70 explains only half (49%) of the variance in a criterion.

In terms of the second question, how can we explain a greater amount of the variance in a criterion? One strategy would be to collect data on more than one predictor and relate the combined predictors to the criterion in a single analysis. Assuming each predictor is related to the criterion, including multiple predictors in one analysis should increase the amount of explained variance in the criterion. The next section introduces statistics designed to measure the relationship between multiple predictors and a criterion.

Introduction to Multiple Correlation

Multiple correlation may be defined as a family of statistical procedures assessing the relationship between two or more predictor variables and a criterion variable. Compared with Pearson correlation, which involves a single predictor, multiple correlation has advantages and disadvantages of both a conceptual and statistical nature. Conceptually, as mentioned earlier, multiple correlation provides a greater ability to predict a complex criterion than does a single predictor.

One statistical advantage of multiple correlation is that it provides a single statistic that represents the combined relationship of all of the predictors with the criterion. More important, multiple correlation measures this combined relationship while taking into account the relationships among the predictors. For example, imagine you have two predictors (X_1 and X_2) that are both correlated with the criterion (Y): the correlation between X_1 and Y is .50 ($r^2 = .25$), and the correlation between X_2 and Y is .30 ($r^2 = .09$). If you wanted to measure the combined ability of X_1 and X_2 to explain variance in Y, can you add the two r^2 values together (.25 + .09) and say the two predictors explain 34% of the variance in Y? The answer is "It depends." If the two predictors are related to each other, which means they overlap with each other, part of the variance in Y explained by X_1 is redundant with part of the variance in Y explained by X_2, and vice versa. As a result, we can't just add together the two r^2 values because this would "double-count" some of the explained variance. Fortunately, as you will see below, the calculation of a multiple correlation statistic takes into account the relationships among predictors.

Compared with Pearson correlation, what are disadvantages of multiple correlation? As you may have guessed, conducting a multiple correlation analysis is more complicated than Pearson correlation. Particularly when more than two predictors are included in an analysis, statistical software such as SPSS is typically used to conduct the calculations. However, in order to illustrate the basic concepts of multiple correlation, the next section provides an example of the simplest version of multiple correlation in which two predictors are related to a criterion.

Multiple Correlation With Two Predictors: An Example

Imagine you work at a college admissions office and are responsible for evaluating college applications and deciding who to admit for the upcoming school year. As one of your goals is to admit students who are likely to be successful in college, you want to base your evaluation not on your own personal beliefs but rather on information correlated to success in college. After reviewing relevant research, one variable you hypothesize is related to success in college is success in high school, which is traditionally measured using high school GPA. However,

TABLE 14.1 ● DESCRIPTIVE STATISTICS AND CORRELATION MATRIX FOR COLLEGE ADMISSIONS EXAMPLE

(a) Descriptive Statistics

	High School GPA (X_1)	# Activities (X_2)	College GPA (Y)
N	76	76	76
Mean	3.37	5.92	3.07
Standard deviation	.51	.93	.38

(b) Correlation Matrix

	High School GPA	# Activities	College GPA
High school GPA	—		
# Activities	.24*	—	
College GPA	.58**	.41**	—

in addition to grades in classes, you hypothesize that participation in activities such as clubs, sports, and music is also related to success in college. Consequently, you collect the following information from a sample of 76 college students currently attending your university: the two predictor variables of high school GPA and the number of activities participated in during high school and the criterion variable of college GPA.

Table 14.1(a) presents the descriptive statistics for the three variables; Table 14.1(b) provides a **correlation matrix**, which is a table of Pearson correlations between variables. In a correlation matrix, asterisks (*) are often used to indicate the level of significance of the correlations. Looking at the correlations, we see that both high school GPA ($r(74) = .58, p < .01$) and the number of activities ($r(74) = .41, p < .01$) are significantly and positively related to college GPA, which suggests that higher levels of academic achievement in high school and higher participation in activities are related to greater success in college. However, it's important to note that the two predictors are also significantly related to each other ($r(74) = .24, p < .05$), which suggests that students with higher high school GPAs are involved in a greater number of activities. This last correlation implies that the two predictors are somewhat overlapping and redundant with each other, which affects the total amount of variance in college GPA they explain.

A visual representation of the relationships between the three variables is provided in the Venn diagram in Figure 14.3; the three circles in this figure represent variance in each of the variables. The diagonal lines slanting right and left within the circles signify the relationship between the two predictors (high school GPA and activities) and the criterion variable (college GPA). Given the goal of the predictors is to explain variance in the criterion, the area shaded by the diagonal lines should be as large as possible. However, the horizontal lines indicate the relationship between the two predictors, with the darker area where the three circles overlap representing variance in the criterion that's explained by either of the predictors. Given that overlap among predictors represents redundancy, you would want the area shaded by the horizontal lines to be as small as possible.

Measuring the Relationship Between Two Predictors and a Criterion: The Multiple Correlation (R)

To measure the relationship between two or more predictor variables and a criterion variable, a correlational statistic known as the **multiple correlation coefficient**, represented by R, may be calculated. The formula for the multiple correlation coefficient varies depending on the number of predictors in the analysis; Formula 14-4 presents the formula for the multiple correlation coefficient for two predictors:

FIGURE 14.3 ● VENN DIAGRAM ILLUSTRATION OF CORRELATIONS AMONG VARIABLES

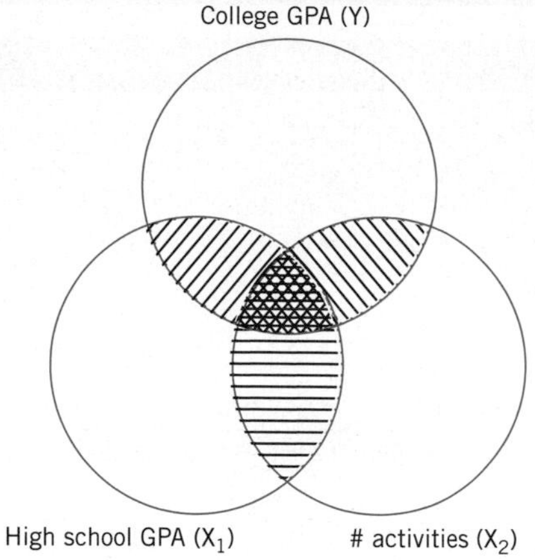

$$R = \sqrt{\frac{[(r_{X_1Y})^2 + (r_{X_2Y})^2] - (2r_{X_1Y}r_{X_2Y}r_{X_1X_2})}{1-(r_{X_1X_2})^2}} \qquad (14\text{-}4)$$

where r_{X_1Y} is the correlation between predictor X_1 and the criterion Y, r_{X_2Y} is the correlation between predictor X_2 and the criterion Y, and $r_{X_1X_2}$ is the correlation between predictors X_1 and X_2.

Several aspects of Formula 14-4 are noteworthy. First, the calculation of the multiple correlation begins by determining the amount of variance in the criterion explained by the combined predictors $[(r_{X_1Y})^2 + (r_{X_2Y})^2]$, but then it takes into account any overlap between the two predictors ($r_{X_1X_2}$). Second, this formula involves computing a square root, the result of which must be a positive number. Consequently, unlike the Pearson r, the multiple correlation R can *only* be a positive number ranging from .00 to 1.00. Third, although any of the three Pearson correlations used to calculate R may be negative, they are squared before calculating R because you can't calculate the square root of a negative number.

Using the three Pearson correlations in the correlation matrix in Table 14.1(b), the multiple correlation between the two predictors (high school GPA and activities) and college GPA is calculated below:

$$R = \sqrt{\frac{[(r_{X_1Y})^2 + (r_{X_2Y})^2] - (2r_{X_1Y}r_{X_2Y}r_{X_1X_2})}{1-(r_{X_1X_2})^2}}$$

$$= \sqrt{\frac{[(.58)^2 + (.41)^2] - (2 * .58 * .41 * .24)}{1-(.24)^2}}$$

$$= \sqrt{\frac{[.34 + 17] - .12}{1 - .06}} = \sqrt{\frac{.39}{.94}} = \sqrt{.42} = .65$$

Therefore, the multiple correlation between the combination of high school GPA and number of activities with college GPA is .65 ($R = .65$). As mentioned earlier, because the two predictors are correlated with each other ($r = .24$), the squared multiple correlation ($R^2 = .42$) is less than the sum of the two squared Pearson correlations ($(.58)^2 + (.41)^2 = .51$). Box 14.2 summarizes the calculations of the multiple correlation coefficient.

SUMMARY BOX 14.2 CALCULATING THE MULTIPLE CORRELATION COEFFICIENT, COLLEGE ADMISSIONS EXAMPLE

	High School GPA	# Activities	College GPA
High school GPA (X_1)	—		
# Activities (X_2)	.24*	—	
College GPA (Y)	.58**	.41**	—

$$\begin{aligned} R &= \sqrt{\frac{[(r_{X_1Y})^2 + (r_{X_2Y})^2] - (2r_{X_1Y}r_{X_2Y}r_{X_1X_2})}{1-(r_{X_1X_2})^2}} \\ &= \sqrt{\frac{[(.58)^2 + (.41)^2] - (2*.58*.41*.24)}{1-(.24)^2}} \\ &= \sqrt{\frac{[.34+17]-.12}{1-.06}} = \sqrt{\frac{.39}{.94}} = \sqrt{.42} = .65 \end{aligned}$$

Testing the Multiple Correlation for Statistical Significance

It is possible to determine whether a multiple correlation R is statistically significant, meaning it is significantly different from zero (.00). Because correlation is calculated by analyzing variances (variance in the criterion that is explained or not explained by the predictors), the multiple correlation R is tested for significance by transforming R into an F-ratio, the same statistic used in Chapters 11 and 12 for the analysis of variance (ANOVA):

$$F = \frac{R^2 / \text{k}}{(1-R^2)/(N-k-1)} \tag{14-5}$$

where R is the multiple correlation coefficient, k is the number of predictor variables, and N is the total sample size. Just like the F-ratio in ANOVA, the numerator of this formula represents an "effect" (the amount of variance in the criterion explained by the predictors [R^2]), and the denominator represents "error" (variance in the criterion *not* explained by the predictors [$1 - R^2$]). The degrees of freedom (df) for this F-ratio are k (the number of predictors) for the numerator and $N - k - 1$ for the denominator.

In this example, data on two predictors ($k = 2$) were collected from a sample of 76 college students ($N = 76$). Inserting these values, along with our value of $R = .65$, the F-ratio for this example is calculated as follows:

$$\begin{aligned} F &= \frac{R^2 / \text{k}}{(1-R^2)/(N-k-1)} \\ &= \frac{(.65)^2 / 2}{(1-(.65)^2)/(76-2-1)} \\ &= \frac{.41/2}{(1-.41)/73} = \frac{.21}{.59/73} = \frac{.21}{.008} \\ &= 26.25 \end{aligned}$$

In this example, the degrees of freedom (df) are 2 ($k = 2$) for the numerator of the F-ratio and 73 ($N - k - 1 = 76 - 2 - 1$) for the denominator of the F-ratio. The critical value for the F-ratio may be identified using the table at the back of the book; for $\alpha = .05$ and $df = 2, 60$, the critical values are 3.15 ($\alpha = .05$) and 4.98 ($\alpha = .01$). Because our F-ratio of 26.25 exceeds both of these critical values, we make the decision to reject the null hypothesis and conclude the multiple correlation of .65 is statistically significant with a probability less than .01. A conclusion regarding this analysis may be stated as follows:

> In a sample of 76 college students, the combination of high school GPA and the number of activities participated in during high school was significantly related to college GPA, $R = .65$, $F(2, 73) = 26.25$, $p < .01$, $R^2 = .42$.

Predicting a Criterion From Two Predictors: The Multiple Regression Equation

Earlier in this chapter, we discussed that when a relationship exists between variables, the relationship can be used to predict one variable from the other variables. When there is more than one predictor, the Pearson correlations for the variables can be used to develop a **multiple regression equation**, defined as a mathematical equation that predicts a score on a criterion variable from scores on two or more predictor variables.

The formula for the multiple regression equation with two predictors is an extension of the formula for the linear regression equation ($Y' = a + bX$) and is presented in Formula 14-6:

$$Y' = a + b_1X_1 + b_2X_2 \qquad (14\text{-}6)$$

where Y' represents a predicted score for the criterion Y, a is the estimate of the Y-intercept, b_1 is the regression coefficient for predictor X_1, X_1 is a score for predictor X_1, b_2 is the regression coefficient for predictor X_2, and X_2 is a score for predictor X_2. A **regression coefficient** such as b_1 represents the rate of change in the criterion variable Y associated with changes in a predictor variable.

The first step in calculating the multiple regression equation is to calculate the two regression coefficients (b_1 and b_2) using Formula 14-7:

$$b_1 = \frac{r_{X_1Y} - (r_{X_2Y}r_{X_1X_2})}{1-(r_{X_1X_2})^2}\left(\frac{s_Y}{s_{X_1}}\right) \qquad b_2 = \frac{r_{X_2Y} - (r_{X_1Y}r_{X_1X_2})}{1-(r_{X_1X_2})^2}\left(\frac{s_Y}{s_{X_2}}\right) \qquad (14\text{-}7)$$

where r_{X_1Y} is the correlation between predictor X_1 and the criterion Y, r_{X_2Y} is the correlation between predictor X_2 and the criterion Y, $r_{X_1X_2}$ is the correlation between predictors X_1 and X_2, s_Y is the standard deviation for the criterion Y, s_{X_1} is the standard deviation for predictor X_1, and s_{X_2} is the standard deviation for predictor X_2.

Using the descriptive statistics in Table 14.1(a) and the Pearson correlations in Table 14.2(b), the regression coefficients b_1 and b_2 are calculated for high school GPA and number of activities:

High School GPA (X_1)

$$\begin{aligned} b_1 &= \frac{r_{X_1Y} - (r_{X_2Y}r_{X_1X_2})}{1-(r_{X_1X_2})^2}\left(\frac{s_Y}{s_{X_1}}\right) \\ &= \frac{.58-((.41)(.24))}{1-(.24)^2}\left(\frac{.38}{.51}\right) \\ &= \frac{.58-.10}{1-.06}(.75) = \frac{.48}{.94}(.75) \\ &= .38 \end{aligned}$$

 # Activities (X_2)

$$\begin{aligned} b_2 &= \frac{r_{X_2Y} - (r_{X_1Y}r_{X_1X_2})}{1-(r_{X_1X_2})^2}\left(\frac{s_Y}{s_{X_2}}\right) \\ &= \frac{.41-((.58)(.24))}{1-(.24)^2}\left(\frac{.38}{.93}\right) \\ &= \frac{.41-.14}{1-.06}(.41) = \frac{.27}{.94}(.41) \\ &= .12 \end{aligned}$$

Therefore, the regression coefficient for high school GPA (X_1) is .38, and the regression coefficient for the number of activities (X_2) is .12.

Once the regression coefficients (b_1 and b_2) have been calculated, the next step is to calculate the estimate of the Y-intercept (a) using the formula below:

$$a = \bar{Y} - b_1\bar{X}_1 - b_2\bar{X}_2 \quad (14\text{-}8)$$

where $\bar{Y}$ is the mean of the criterion Y, b_1 is the regression coefficient for predictor X_1, $\bar{X}_1$ is the mean of predictor X_1, b_2 is the regression coefficient for predictor X_2, and $\bar{X}_2$ is the mean of predictor X_2.

Inserting the regression coefficients we've just calculated and the means for the variables provided in Table 14.1(a), the estimate of the Y-intercept a for this example is calculated below:

$$\begin{aligned} a &= \bar{Y} - b_1\bar{X}_1 - b_2\bar{X}_2 \\ &= 3.07 - (.38)(3.37) - (.12)(5.92) \\ &= 3.07 - 1.28 - .71 \\ &= 1.08 \end{aligned}$$

Once the regression coefficients and Y-intercept have been calculated, the multiple regression equation may be reported. For this example, the following equation predicts a student's college GPA from high school GPA and the number of activities participated in during high school:

$$Y' = a + b_1\bar{X}_1 + b_2\bar{X}_2$$
$$\text{College GPA}' = 1.08 + .38\ (\text{High school GPA}) + .12\ (\#\text{activities})$$

Interpreting our regression equation, a predicted college GPA is determined by multiplying a student's high school GPA by .38 and the student's number of activities by .12 and then adding these two products to 1.08. To illustrate this, below we make predictions for two students with different combinations of high school GPA and activities:

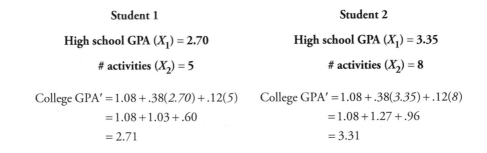

Student 1

High school GPA (X_1) = 2.70

 # activities (X_2) = 5

$$\begin{aligned} \text{College GPA}' &= 1.08 + .38(2.70) + .12(5) \\ &= 1.08 + 1.03 + .60 \\ &= 2.71 \end{aligned}$$

Student 2

High school GPA (X_1) = 3.35

 # activities (X_2) = 8

$$\begin{aligned} \text{College GPA}' &= 1.08 + .38(3.35) + .12(8) \\ &= 1.08 + 1.27 + .96 \\ &= 3.31 \end{aligned}$$

Box 14.3 summarizes the calculation of the multiple regression equation.

Conclusion

Hopefully, the relative complexity of the above calculations has not discouraged you from using multiple correlation in your own research as it is a powerful statistical tool. In addition to measuring the combined relationship between predictors and a criterion variable, multiple correlation provides researchers the ability to test a variety of effects. For example, a researcher can determine whether a predictor has an incremental effect above and beyond another predictor. In this example, we can address a question such as, "Once high school GPA has been taken into account, is the number of activities participated in during high school related to college GPA?" Researchers can also determine whether interaction effects exist

SUMMARY BOX 14.3 CALCULATING THE MULTIPLE REGRESSION EQUATION, COLLEGE ADMISSIONS EXAMPLE

Calculate the regression coefficients (b_1 and b_2)

$$b_1 = \frac{r_{X_1Y} - (r_{X_2Y} r_{X_1X_2})}{1-(r_{X_1X_2})^2}\left(\frac{s_Y}{s_{X_1}}\right) \qquad b_2 = \frac{r_{X_2Y} - (r_{X_1Y} r_{X_1X_2})}{1-(r_{X_1X_2})^2}\left(\frac{s_Y}{s_{X_2}}\right)$$

$$= \frac{.58-((.41)(.24))}{1-(.24)^2}\left(\frac{.38}{.51}\right) \qquad = \frac{.41-((.58)(.24))}{1-(.24)^2}\left(\frac{.38}{.93}\right)$$

$$= \frac{.58-.10}{1-.06}(.75) = \frac{.48}{.94}(.75) \qquad = \frac{.41-.14}{1-.06}(.41) = \frac{.27}{.94}(.41)$$

$$= .38 \qquad = .12$$

Calculate the Y-intercept (a)

$$a = \overline{Y} - b_1\overline{X}_1 - b_2\overline{X}_2$$
$$= 3.07 - (.38)(3.37) - (.12)(5.92)$$
$$= 3.07 - 1.28 - .71 = 1.08$$

Report the multiple regression equation

$$Y' = a + b_1\overline{X}_1 + b_2\overline{X}_2$$
College GPA′ = 1.08 + .38 (High school GPA) + .12 (#activities)

LEARNING CHECK 2:

Reviewing What You've Learned So Far

1. Review questions
   a. What is the difference between Pearson correlation and multiple correlation?
   b. What are conceptual and statistical limitations of using only one variable to predict another variable?
   c. What are some advantages and disadvantages of multiple correlation versus Pearson correlation?
2. Calculate the multiple correlation coefficient (R) and F-ratio for each of these research situations; identify the critical values and make a decision about the null hypothesis:

a.

	X_1	X_2	Y		X_1	X_2	Y
N	24	24	24	X_1	—		
Mean	19.54	18.46	36.57	X_2	.36	—	
SD	10.19	11.80	3.84	Y	.53	.33	—

b.

	X_1	X_2	Y		X_1	X_2	Y
N	32	32	32	X_1	—		
Mean	71.22	21.09	20.16	X_2	.16	—	
SD	9.41	10.55	2.42	Y	.27	.57	—

3. Calculate the multiple regression equation for the research situations in Question 2 above.
   a. For the research situation in 2a, what is the predicted score (Y') for someone with $X_1 = 13$ and $X_2 = 15$?
   b. For the research situation in 2b, what is the predicted score (Y') for someone with $X_1 = 39$ and $X_2 = 18$?

between predictors. In Chapter 12, we stated that an interaction effect exists when the effect of one independent variable on the dependent variable changes at the different levels of another independent variable. In this example, an interaction effect would be, "Does the number of activities a student participates in during high school influence the relationship between high school GPA and college GPA?" To learn more about multiple correlation, we encourage you to read further (i.e., Cohen et al., 2003; Tabachnick & Fidell, 2013).

14.3 LOOKING AHEAD

The main purpose of this chapter was to use our discussion of correlation to introduce regression and multiple correlation. In Chapter 13, we noted that the Pearson correlation coefficient measures the linear relationship between two variables measured at the interval and/or ratio level of measurement, whereas the Spearman rank-order correlation measures the relationship between two ordinal variables. As it is crucial to remember that the choice of statistical procedure is based on the nature of the variables included in the analysis, the next and final chapter of the book discusses statistical procedures used when both the independent and dependent variables are categorical in nature, measured at the nominal level of measurement.

14.4 Summary

Whereas the main purpose of correlation is to measure and test the relationship between variables, the main goal of ***regression*** is to use a relationship between two or more correlated variables to predict values of one variable from values of other variables.

Linear regression refers to a statistical procedure in which a straight line is fitted to a set of data to best represent the relationship between the two variables. The ***linear regression equation*** ($Y' = a + bX$) is a mathematical equation based on the relationship between two variables that predicts a score on one variable using a score on the other variable. In this equation, Y' is a predicted score for the Y variable, a is an estimate of the Y-intercept, $\boldsymbol{b}$ is an estimate of the slope, and X is a score for the X variable.

It is possible to relate more than one predictor variable to a criterion variable. ***Multiple correlation*** is a family of statistical procedures assessing the relationship between two or more predictor variables and a criterion variable. The ***multiple correlation coefficient (R)*** is a correlational statistic that measures the relationship between two or more predictor variables and a criterion variable. A ***multiple regression equation*** is a mathematical equation that predicts a score on a criterion variable from scores on two or more predictor variables.

14.5 Important Terms

regression (p. 538)
linear regression (p. 538)
linear regression equation (p. 538)
multiple correlation (p. 545)
correlation matrix (p. 546)
multiple correlation coefficient (R) (p. 546)
multiple regression equation (p. 549)
regression coefficient (p. 549)

14.6 Formulas Introduced in This Chapter

Linear Regression Equation

$$Y' = a + bX \tag{14-1}$$

Estimate of Slope (*b*)

$$b = r\frac{S_Y}{S_X} \tag{14-2}$$

Estimate of Y-Intercept (*a*)

$$a = \overline{Y} - b\overline{X} \tag{14-3}$$

Multiple Correlation Coefficient (*R*)

$$R = \sqrt{\frac{[(r_{X_1Y})^2 + (r_{X_2Y})^2 - (2r_{X_1Y}r_{X_2Y}r_{X_1X_2})}{1-(r_{X_1X_2})^2}} \tag{14-4}$$

***F*-Ratio for the Multiple Correlation Coefficient (*F*)**

$$F = \frac{R^2 / \text{k}}{(1-R^2)/(N-k-1)} \tag{14-5}$$

Multiple Regression Equation

$$Y' = a + b_1X_1 + b_2X_2 \tag{14-6}$$

Regression Coefficient

$$b_1 = \frac{r_{X_1Y} - (r_{X_2Y}r_{X_1X_2})}{1-(r_{X_1X_2})^2}\left(\frac{s_Y}{s_{X_1}}\right) \qquad b_2 = \frac{r_{X_2Y} - (r_{X_1Y}r_{X_1X_2})}{1-(r_{X_1X_2})^2}\left(\frac{s_Y}{s_{X_2}}\right) \tag{14-7}$$

Estimate of Y-Intercept (*a*) (Multiple Regression)

$$a = \overline{Y} - b_1\overline{X}_1 - b_2\overline{X}_2 \tag{14-8}$$

14.7 Using SPSS

Linear Regression: The Picky Eater Study (14.1)

1. Begin the Linear regression procedure within SPSS.

   How? (1) Click **Analyze** menu, (2) click **Regression**, and (3) click **Linear**.

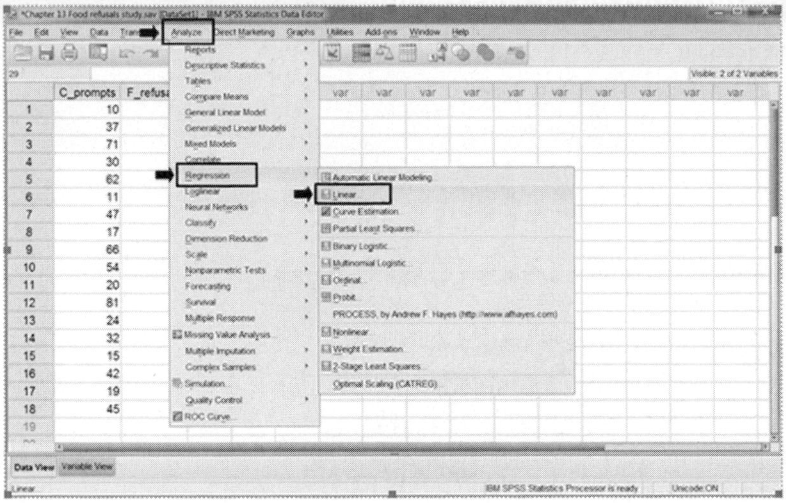

2. Identify the dependent variable and the independent variable.

   How? (1) Click dependent variable and → **Dependent**, (2) Click independent variable and → **Independent(s)**, and (3) click **OK**.

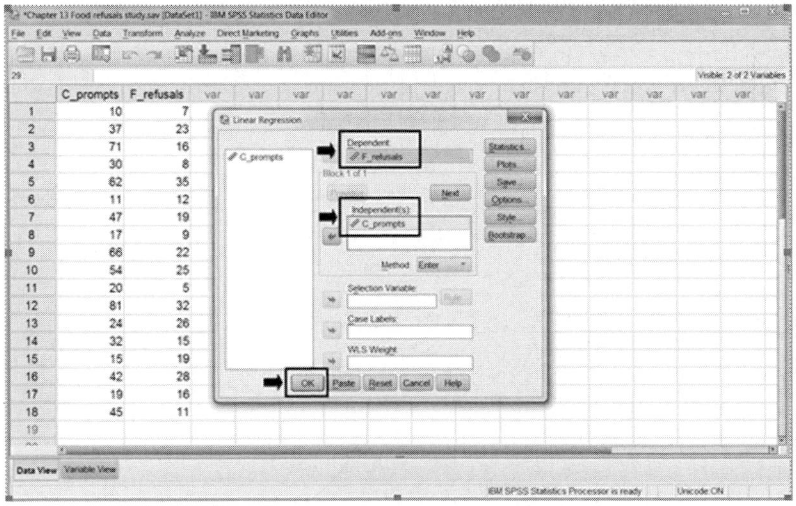

3. Examine output.

Coefficients[a]   Y-intercept (a)   Slope (b)

Model		Unstandardized Coefficients B	Unstandardized Coefficients Std. Error	Standardized Coefficients Beta	t	Sig.
1	(Constant)	8.662	3.394		2.552	.021
	C_prompts	.252	.078	.628	3.230	.005

a. Dependent Variable: F_refusals

Multiple Regression: The College Admission Example (14.2)

1. Begin the Linear regression procedure within SPSS.

   How? (1) Click **Analyze** menu, (2) click **Regression**, and (3) click **Linear**.

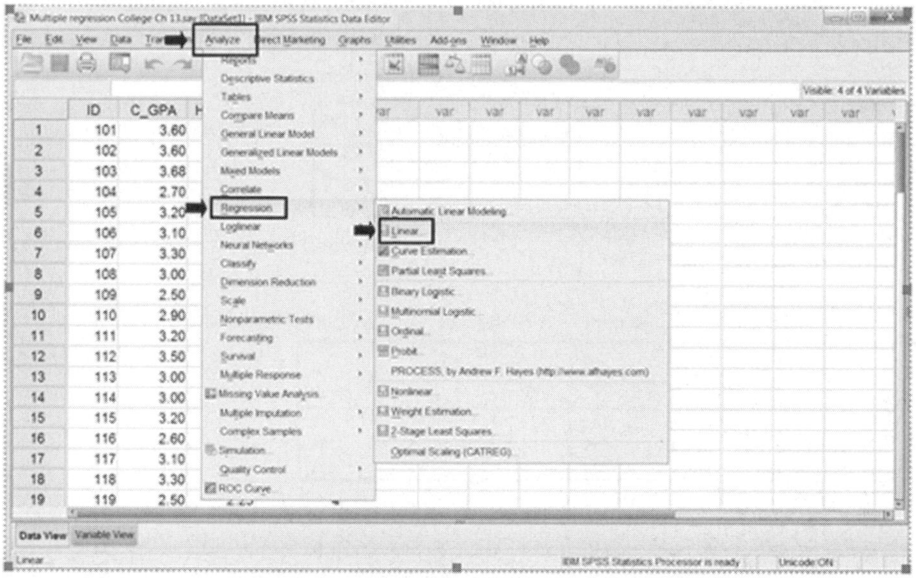

2. Identify the dependent variable and the independent variable.

   How? (1) Click dependent variable and → **Dependent**, (2) Click independent variable and → **Independent(s)**, and (3) click **OK**.

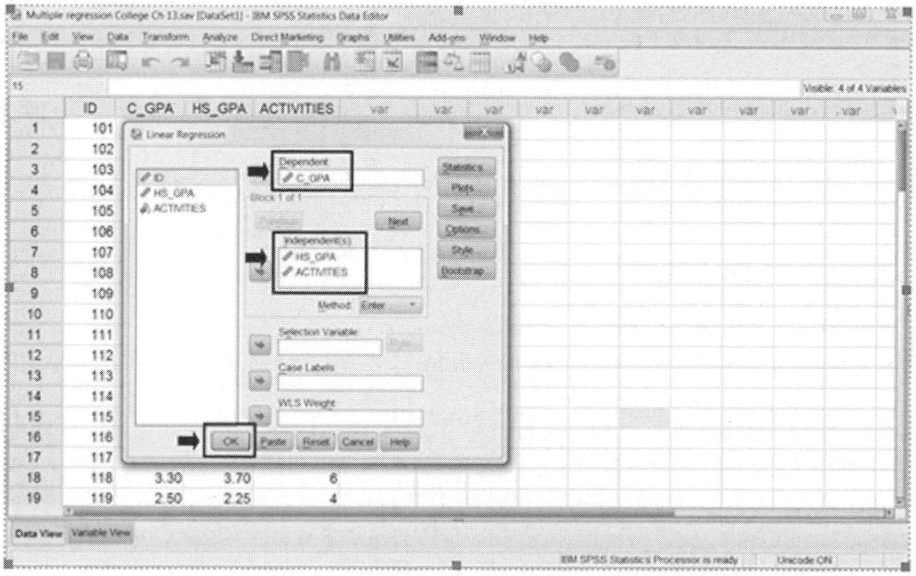

3. Examine output.

Multiple correlation (R)
Measure of effect size (R2)

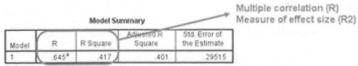

Model Summary

Model	R	R Square	Adjusted R Square	Std. Error of the Estimate
1	.645[a]	.417	.401	.29515

a. Predictors: (Constant), ACTIVITIES, HS_GPA

degrees of freedom (df)

ANOVA[a]

Model		Sum of Squares	df	Mean Square	F	Sig.
1	Regression	4.541	2	2.271	26.065	.000[b]
	Residual	6.359	73	.087		
	Total	10.900	75			

a. Dependent Variable: C_GPA

b. Predictors: (Constant), ACTIVITIES, HS_GPA

F-ratio

Level of significance
(Note: "Sig." = probability (p))

Y-intercept (a)

Regression coefficients (b)

Coefficients[a]

Model		Unstandardized Coefficients B	Std. Error	Standardized Coefficients Beta	t	Sig.
1	(Constant)	1.082	.283		3.825	.000
	HS_GPA	.384	.069	.512	5.559	.000
	ACTIVITIES	.118	.038	.289	3.138	.002

a. Dependent Variable: C_GPA

14.8 Exercises

1. Calculate the linear regression equation for each of these situations:
   a. Variable *X*: Range $= 1-6$, $\overline{X}_i = 5.00$, $s_X = 2.00$; Variable *Y*: Range $= 1-6$, $\overline{Y} = 4.00$, $s_Y = 1.00$, Correlation: $r = .40$
   b. Variable *X*: Range $= 0-6$, $\overline{X}_i = 3.59$, $s_X = 1.32$; Variable *Y*: Range $= 1-10$, $\overline{Y} = 7.52$, $s_Y = 2.46$, Correlation: $r = -.28$
   c. Variable *X*: Range $= 0-100$, $\overline{X}_i = 73.87$, $s_X = 12.65$; Variable *Y*: Range $= 0-125$, $\overline{Y} = 86.32$, $s_Y = 16.80$, Correlation: $r = .07$
   d. Variable *X*: Range $= 1-20$, $\overline{X}_i = 14.41$, $s_X = 2.09$; Variable *Y*: Range $= 1-50$, $\overline{Y} = 37.88$, $s_Y = 6.23$, Correlation: $r = -.51$
   e. Variable *X*: Range $= 0-250$, $\overline{X}_i = 134.91$, $s_X = 25.96$; Variable *Y*: Range $= 0-250$, $\overline{Y} = 179.33$, $s_Y = 21.63$, Correlation: $r = .34$

2. Calculate the linear regression equation for each of these situations:
   a. Variable *X*: Range $= 0-60$, $\overline{X}_i = 45.00$, $s_X = 5.00$; Variable *Y*: Range $= 0-25$, $\overline{Y} = 11.00$, $s_Y = 3.00$, Correlation: $r = .30$
   b. Variable *X*: Range $= 0-100$, $\overline{X}_i = 37.22$, $s_X = 11.76$; Variable *Y*: Range $= 0-150$, $\overline{Y} = 104.32$, $s_Y = 15.92$ Correlation: $r = .62$
   c. Variable *X*: Range $= 1-6$, $\overline{X}_i = 4.32$, $s_X = .97$; Variable *Y*: Range $= 1-5$, $\overline{Y} = 2.51$, $s_Y = .46$, Correlation: $r = -.09$
   d. Variable *X*: Range $= 1-5$, $\overline{X}_i = 3.91$, $s_X = 1.04$; Variable *Y*: Range $= 1-110$, $\overline{Y} = 76.40$, $s_Y = 9.87$, Correlation: $r = .25$
   e. Variable *X*: Range $= 1-12$, $\overline{X}_i = 8.75$, $s_X = 1.78$; Variable *Y*: Range $= 0-35$, $\overline{Y} = 24.65$, $s_Y = 4.82$, Correlation: $r = -.54$

3. Calculate the following predicted scores (Y') for the regression equations calculated in 1(a–e).
   a. For the regression equation in 1a, what is the predicted score (Y') for $X = 3$?
   b. For the regression equation in 1b, what is the predicted score (Y') for $X = 3.25$?
   c. For the regression equation in 1c, what is the predicted score (Y') for $X = 50$?
   d. For the regression equation in 1d, what is the predicted score (Y') for $X = 15$?
   e. For the regression equation in 1e, what is the predicted score (Y') for $X = 130$?
4. Calculate the following predicted scores (Y') for the regression equations calculated in 2(a–e).
   a. For the regression equation in 2a, what is the predicted score (Y') for $X = 40$?
   b. For the regression equation in 2b, what is the predicted score (Y') for $X = 30$?
   c. For the regression equation in 2c, what is the predicted score (Y') for $X = 4$?
   d. For the regression equation in 2d, what is the predicted score (Y') for $X = 3$?
   e. For the regression equation in 2e, what is the predicted score (Y') for $X = 7$?
5. Calculate predicted scores (Y') for each of the following regression equations.
   a. Using the equation $Y' = 1.74 + .11(X)$, what is the predicted score (Y') for $X = 5$?
   b. Using the equation $Y' = 26.23 + 1.06(X)$, what is the predicted score (Y') for $X = 14$?
   c. Using the equation $Y' = 89.66 + -.54(X)$, what is the predicted score (Y') for $X = 47$?
   d. Using the equation $Y' = 2.23 + -.15(X)$, what is the predicted score (Y') for $X = 2$?
   e. Using the equation $Y' = 208.52 + .26(X)$, what is the predicted score (Y') for $X = 305$?
6. Calculate predicted scores (Y') for each of the following regression equations.
   a. Using the equation $Y' = 1.68 + .36(X)$, what is the predicted score (Y') for $X = 11$?
   b. Using the equation $Y' = 35.21 + -.05(X)$, what is the predicted score (Y') for $X = 125$?
   c. Using the equation $Y' = 22.09 + .14(X)$, what is the predicted score (Y') for $X = 14$?
   d. Using the equation $Y' = 2.62 + .42(X)$, what is the predicted score (Y') for $X = 9$?
   e. Using the equation $Y' = 112.25 + -.34(X)$, what is the predicted score (Y') for $X = 90$?
7. Create a figure that illustrates the regression equations calculated in 1(a–c) (use the provided ranges for the variables).
8. Create a figure that illustrates the regression equations calculated in 2(a–c) (use the provided ranges for the variables).
9. In Chapter 13, Exercise 11 related midterm and final exam scores for a sample of students:

   Midterm (X): Range $= 0 - 100$, $\bar{X}_i = 68.67$, $s_X = 12.44$; Final exam (Y): Range $= 0 - 100$, $\bar{Y} = 77.50$, $s_Y = 9.38$, Correlation: $r = .85$

   a. Calculate the regression equation predicting final exam scores from midterm scores and draw it in into a scatterplot.
   b. What final exam score would you predict for a student with a midterm exam score of 75?
10. In Chapter 13, Exercise 12 related the number of sentences spoken by a person in a group to the extent the person is rated to be a leader:

    # sentences spoken (X): Range $= 0 - 35$, $\bar{X}_i = 13.88$, $s_X = 8.74$

    Rating of leadership (Y): Range $= 1 - 10$, $\bar{Y} = 5.50$, $s_Y = 2.45$, Correlation: $r = .72$

    a. Calculate the regression equation predicting rating of leadership from sentences spoken and draw it in into a scatterplot.
    b. What rating of leadership would you predict for a person who speaks 25 sentences?
11. In Chapter 13, Exercise 13 studied the relationship between the level of violence in a comic book and how humorous the comic book is rated by a sample of women:

    Level of violence (X): Range $= 1 - 7$, $\bar{X}_i = 4.30$, $s_X = 2.21$

    Humorous rating (Y): Range $= 1 - 6$, $\bar{Y} = 3.50$, $s_Y = 1.58$ Correlation: $r = -.78$

a. Calculate the linear regression equation predicting humorous ratings from level of violence and draw it into a scatterplot.

b. What humorous rating would you predict for a comic book with a level of violence of 4?

12. In Chapter 13, Exercise 1(b) proposed a relationship between the amount of time spent watching sports on television and the amount of time actually playing sports:

hours watching sports (X): Range $= 0 - 7$, $\bar{X}_i = 2.72$, $s_X = 1.92$

hours playing sports (Y): Range $= 0 - 4$, $\bar{Y} = 1.20$, $s_Y = .96$ Correlation: $r = -.04$

a. Calculate the linear regression equation predicting number of hours playing sports from number of hours watching sports and draw it into a scatterplot.

b. How many hours playing sports would you predict for someone who watches sports for 5 hours?

13. Calculate the multiple correlation coefficient (R) for each of the following situations (the descriptive statistics are on the left and the correlation matrix is on the right):

a.

	X_1	X_2	Y		X_1	X_2	Y
N	25	25	25	X_1	—		
Mean	3.53	2.76	3.30	X_2	.17	—	
SD	1.04	.97	1.25	Y	.31	.44	—

b.

	X_1	X_2	Y		X_1	X_2	Y
N	43	43	43	X_1	—		
Mean	6.91	1.35	14.22	X_2	.12	—	
SD	1.78	.46	2.90	Y	.46	.08	—

c.

	X_1	X_2	Y		X_1	X_2	Y
N	19	19	19	X_1	—		
Mean	7.42	22.80	3.16	X_2	.39	—	
SD	2.09	3.27	1.17	Y	.51	.48	—

d.

	X_1	X_2	Y		X_1	X_2	Y
N	64	64	64	X_1	—		
Mean	63.60	78.27	84.02	X_2	.42	—	
SD	9.54	11.93	10.71	Y	.17	.33	—

e.

	X_1	X_2	Y		X_1	X_2	Y
N	33	33	33	X_1	—		
Mean	14.54	4.56	23.64	X_2	−.12	—	
SD	2.07	1.93	4.65	Y	−.41	.34	—

14. Calculate the multiple correlation coefficient (R) for each of the following situations (the descriptive statistics are on the left and the correlation matrix is on the right):

a.

	X_1	X_2	Y		X_1	X_2	Y
N	22	22	22	X_1	—		
Mean	5.76	6.12	4.86	X_2	.16	—	
SD	1.14	2.03	1.66	Y	.28	.47	—

b.

	X_1	X_2	Y		X_1	X_2	Y
N	35	35	35	X_1	—		
Mean	1.80	2.23	1.98	X_2	.09	—	
SD	.54	.71	.39	Y	.55	.12	—

c.

	X_1	X_2	Y		X_1	X_2	Y
N	16	16	16	X_1	—		
Mean	21.87	13.50	7.13	X_2	.27	—	
SD	3.85	2.25	1.87	Y	.31	.22	—

d.

	X_1	X_2	Y		X_1	X_2	Y
N	54	54	54	X_1	—		
Mean	121.87	135.08	52.18	X_2	−.42	—	
SD	12.81	14.21	8.92	Y	.19	−.37	—

e.

	X_1	X_2	Y		X_1	X_2	Y
N	30	30	30	X_1	—		
Mean	16.84	.86	6.72	X_2	−.15	—	
SD	3.12	.15	1.34	Y	−.55	.36	—

15. Calculate the F-ratio for each situation below, all of which assume there are two predictor variables ($k = 2$); identify the critical values and make a decision about the null hypothesis.

a. $R = .63$; $N = 18$

b. $R = .47$; $N = 20$

c. $R = .39$; $N = 36$

d. $R = .45$; $N = 47$

e. $R = .29$; $N = 77$

16. Calculate the F-ratio for each situation below, all of which assume there are two predictor variables ($k = 2$); identify the critical values and make a decision about the null hypothesis.

a. $R = .58$; $N = 19$

b. $R = .45$; $N = 28$

c. $R = .41; N = 40$

d. $R = .32; N = 65$

e. $R = .35; N = 84$

17. Calculate the *F*-ratio for each of the research situations in Exercise 13(a–e); identify the critical values and make a decision about the null hypothesis.

18. Calculate the *F*-ratio for each of the research situations in Exercise 14(a–e); identify the critical values and make a decision about the null hypothesis.

19. A political consultant is interested in seeing if a voter's age and number of years of education can predict the likelihood of voting for a particular presidential candidate (from 0–100):

	Age	Education	Likelihood		Age	Education	Likelihood
	(X_1)	(X_2)	(Y)		(X_1)	(X_2)	(Y)
N	36	36	36	Age	—		
Mean	38.47	10.96	59.20	Education	.38	—	
SD	9.23	2.37	11.46	Likelihood	.46	.33	—

a. Calculate the multiple correlation coefficient (*R*).

b. Calculate the *F*-ratio, identify the critical values, and make a decision about the null hypothesis.

c. Draw a conclusion based on the results of the analysis.

20. An instructor hypothesizes that the number of times a student is absent from her class is related to the number of friends who are also taking the class and the distance (in miles) the student lives from campus:

	Friends	Distance	Absences		Friends	Distance	Absences
	(X_1)	(X_2)	(Y)		(X_1)	(X_2)	(Y)
N	29	29	29	Friends	—		
Mean	2.31	9.67	3.82	Distance	.09	—	
SD	.69	3.13	2.29	Absences	.26	.32	—

a. Calculate the multiple correlation coefficient (*R*).

b. Calculate the *F*-ratio, identify the critical values, and make a decision about the null hypothesis.

c. Draw a conclusion based on the results of the analysis.

21. Calculate the regression coefficients (*b*s) for each of the research situations in Exercise 13(a–c).

22. Calculate the regression coefficients (*b*s) for each of the research situations in Exercise 14(a–c).

23. Calculate the multiple regression equation for each of the research situations in Exercise 13(a–e).

24. Calculate the multiple regression equation for each of the research situations in Exercise 14(a–e).

25. For each of the regression equations below, calculate the predicted score (Y') for the given scores for X_1 and X_2.

a. $Y' = 2.50 + .39(X_1) + .22(X_2); X_1 = 5, X_2 = 16$

b. $Y' = 22.59 + 1.14(X_1) + .18(X_2); X_1 = 17, X_2 = 26$

c. $Y' = 17.20 + .16(X_1) + 1.23(X_2); X_1 = 2, X_2 = 8$

d. $Y' = 106.54 + -.92(X_1) + .68(X_2); X_1 = 70, X_2 = 42$

e. $Y' = 14.81 + 1.09(X_1) + .40(X_2); X_1 = 15, X_2 = 21$

26. For each of the regression equations below, calculate the predicted score (Y') for the given scores for X_1 and X_2.
   a. $Y' = 1.25 + .54(X_1) + .44(X_2)$; $X_1 = 2$, $X_2 = 4$
   b. $Y' = 13.82 + 1.27(X_1) + .13(X_2)$; $X_1 = 10$, $X_2 = 17$
   c. $Y' = 88.80 + .11(X_1) + .73(X_2)$; $X_1 = 145$, $X_2 = 82$
   d. $Y' = 37.09 + -.30(X_1) + .25(X_2)$; $X_1 = 25$, $X_2 = 43$
   e. $Y' = 4.64 + .63(X_1) + -.19(X_2)$; $X_1 = 4$, $X_2 = 18$
27. Calculate the multiple regression equation for the research situation in Exercise 19.
   a. What is the predicted likelihood of voting (Y') for someone who is 41 years old and has 11 years of education?
   b. What is the predicted likelihood of voting (Y') for someone who is 64 years old and has 12 years of education?
   c. What is the predicted likelihood of voting (Y') for someone who is 23 years old and has 15 years of education?
28. Calculate the multiple regression equation for the research situation in Exercise 20.
   a. What are the predicted absences (Y') for someone with four friends and lives 7 miles from campus?
   b. What are the predicted absences (Y') for someone with two friends and lives 12 miles from campus?
   c. What are the predicted absences (Y') for someone with one friend and lives 8 miles from campus?

Answers to Learning Checks

Learning Check 1

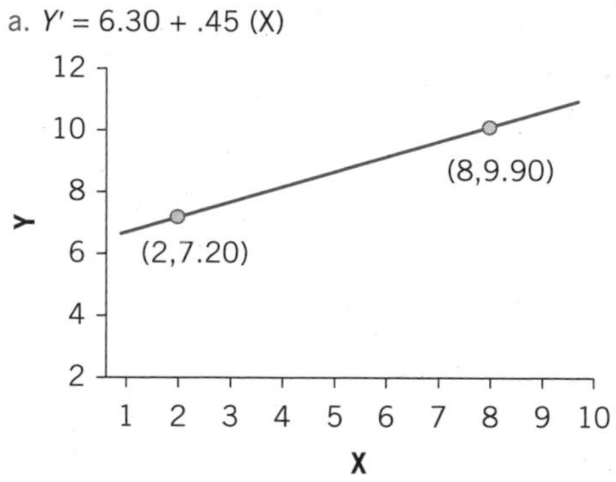

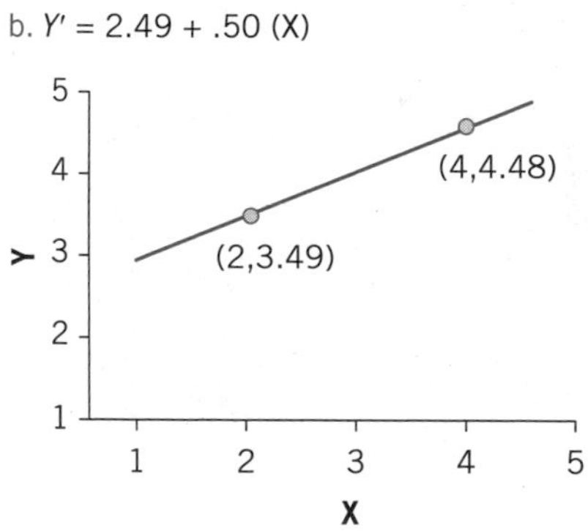

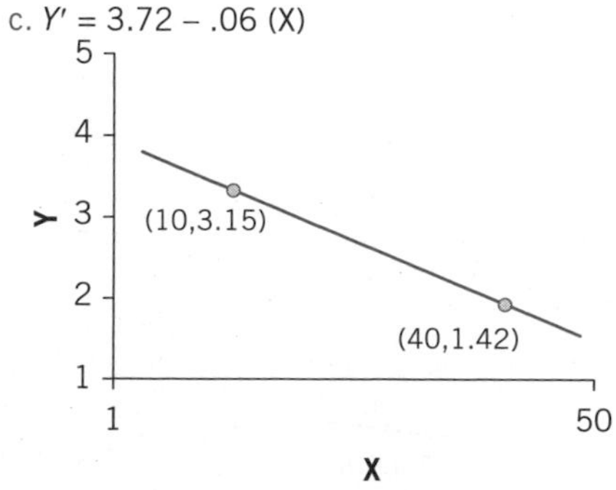

2. a. $Y' = 6.30 + .45(X)$
   b. $Y' = 2.49 + .50(X)$
   c. $Y' = 3.72 - .06(X)$

3. a. For $X = 8$, $Y' = 9.90$
   b. For $X = 4$, $Y' = 4.48$
   c. For $X = 41$, $Y' = 1.37$

Learning Check 2

2. a. $R = \sqrt{\dfrac{[(.53)^2 + (.33)^2] - (2 * .53 * .33 * .36)}{1 - (.36)^2}} = .56$

   $F = \dfrac{(.56)^2 / 2}{(1 - (.56)^2)(24 - 2 - 1)} = 4.85$; critical value (2, 21) = 3.47 (.05), 5.78 (.01); Reject H_0 ($p < .05$)

   b. $R = \sqrt{\dfrac{[(.27)^2 + (.57)^2] - (2 * .27 * .57 * .16)}{1 - (.16)^2}} = .60$

   $F = \dfrac{(.60)^2 / 2}{(1 - (.60)^2)(32 - 2 - 1)} = 8.18$, critical value (2, 29) = 3.33 (.05), 5.42 (.01); Reject H_0 ($p < .01$)

3. a. $Y' = 32.13 + .18(X_1) + .05(X_2)$; for $X_1 = 13$ and $X_2 = 15$, $Y' = 35.22$
   b. $Y' = 13.86 + .05(X_1) + .13(X_2)$; for $X_1 = 39$ and $X_2 = 18$, $Y' = 40.05$

Answers to Odd-Numbered Exercises

1. a. $Y' = 3.00 + .20(X)$
   b. $Y' = 9.39 + -.52(X)$
   c. $Y' = 79.45 + .09(X)$
   d. $Y' = 59.79 + -1.52(X)$
   e. $Y' = 141.11 + .28(X)$

3. a. $Y' = 3.60$
   b. $Y' = 7.70$
   c. $Y' = 84.10$
   d. $Y' = 36.98$
   e. $Y' = 177.94$

5. a. $Y' = 2.72$
   b. $Y' = 41.13$
   c. $Y' = 53.78$
   d. $Y' = 1.92$
   e. $Y' = 287.83$

7. a. $Y' = 3.00 + .20(X)$

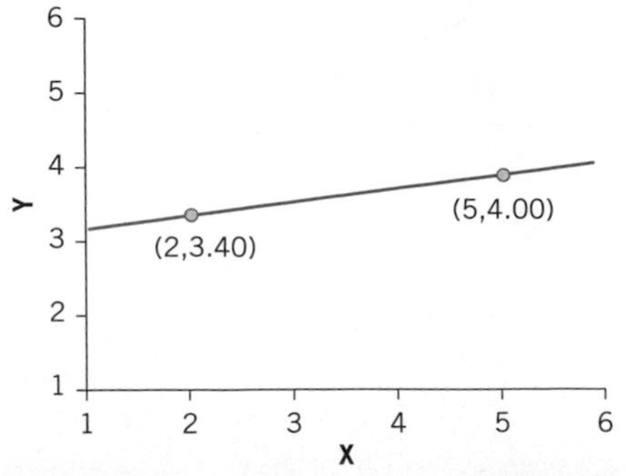

b. $Y' = 9.39 + -.52(X)$

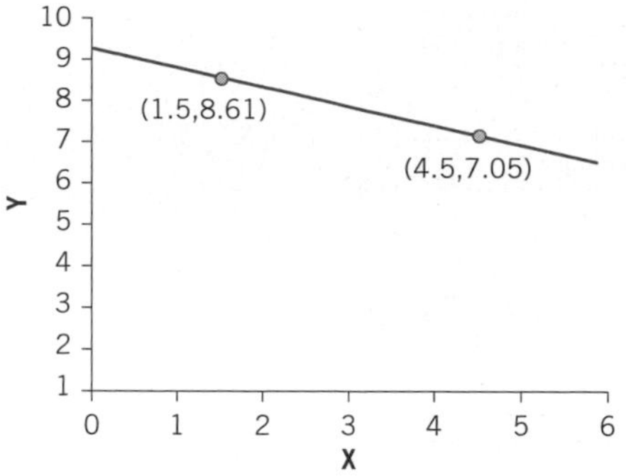

c. $Y' = 79.45 + .09(X)$

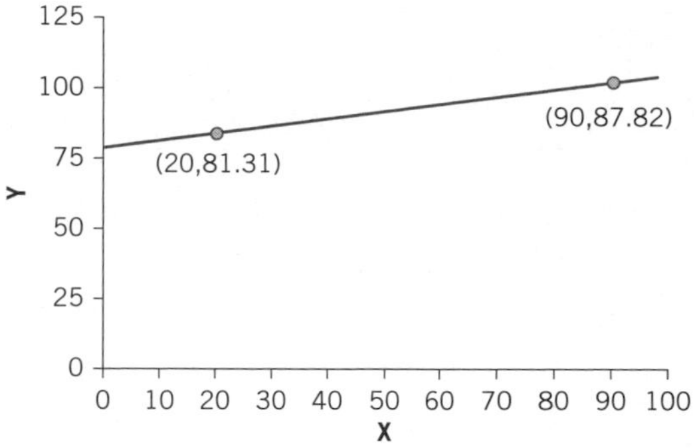

9. a. Final′ = 33.49 + .64 (Midterm)

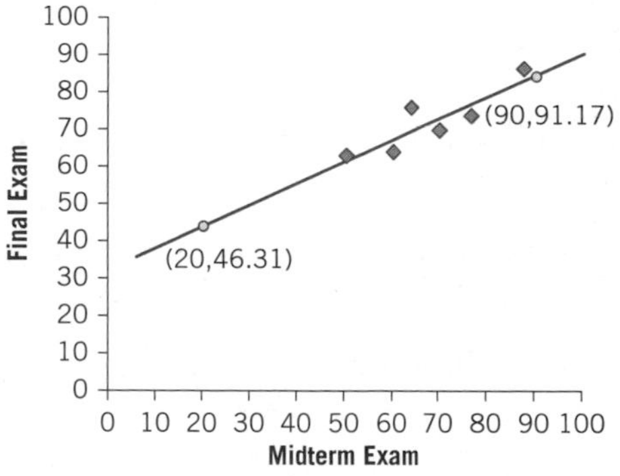

b. For Midterm = 75, Final′ = 81.56

11. a. Humorous rating′ = 5.90 + –.56 (Level of violence)

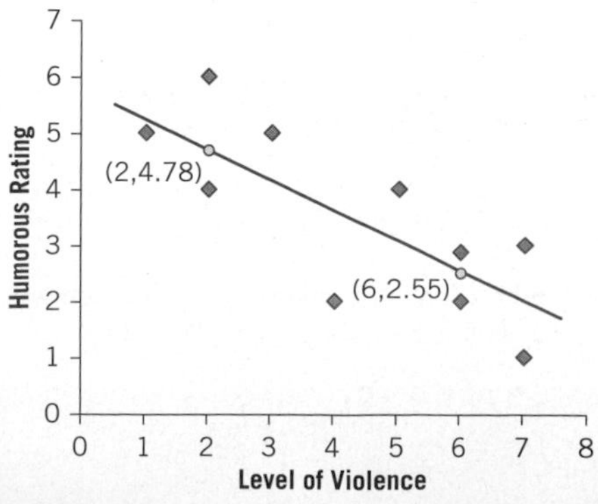

b. For Level of violence = 4, Humorous rating′ = 3.67

13. a. $R=\sqrt{\frac{[(.31)^2+(.44)^2]-(2*.31*.44*.17)}{1-(.17)^2}}=.51$

b. $R=\sqrt{\frac{[(.46)^2+(.08)^2]-(2*.46*.08*.12)}{1-(.12)^2}}=.47$

c. $R=\sqrt{\frac{[(.51)^2+(.48)^2]-(2*.51*.48*.39)}{1-(.39)^2}}=.58$

d. $R=\sqrt{\frac{[(.17)^2+(.33)^2]-(2*.17*.33*.42)}{1-(.42)^2}}=.35$

e. $R=\sqrt{\frac{[(-.41)^2+(.34)^2]-(2*-.41*.34*-.12)}{1-(.12)^2}}=.50$

15. a. $F==4.94$; critical value (2, 15) = 3.68 (.05), 6.36 (.01); Reject H_0 ($p<.05$)

b. $F=\frac{(.52)^2/2}{(1-(.52)^2)(20-2-1)}=3.15$; critical value (2, 17) = 3.59 (.05), 6.11 (.01); Do not reject H_0 ($p>.05$)

c. $F=\frac{(.39)^2/2}{(1-(.39)^2)(36-2-1)}=2.96$; critical value (2, 30) = 3.32 (.05), 5.39 (.01); Do not reject H_0 ($p>.05$)

d. $F=\frac{(.45)^2/2}{(1-(.45)^2)(47-2-1)}=5.59$; critical value (2, 40) = 3.23 (.05), 5.18 (.01); Reject H_0 ($p<.01$)

e. $F=\frac{(.29)^2/2}{(1-(.29)^2)(77-2-1)}=3.40$; critical value (2, 60) = 3.15 (.05), 4.98 (.01); Reject H_0 ($p<.05$)

17. a. $F=\frac{(.51)^2/2}{(1-(.51)^2)(25-2-1)}=3.82$; critical value (2, 22) = 3.44 (.05), 5.72 (.01); Reject H_0 ($p<.05$)

b. $F=\frac{(.47)^2/2}{(1-(.47)^2)(43-2-1)}=5.50$; critical value (2, 40) = 3.23 (.05), 5.18 (.01); Reject H_0 ($p<.01$)

c. $F=\frac{(.58)^2/2}{(1-(.58)^2)(19-2-1)}=4.15$; critical value (2, 16) = 3.63 (.05), 6.23 (.01); Reject H_0 ($p<.05$)

d. $F=\frac{(.35)^2/2}{(1-(.35)^2)(64-2-1)}=4.29$; critical value (2, 60) = 3.15 (.05), 4.98 (.01); Reject H_0 ($p<.05$)

e. $F=\frac{(.50)^2/2}{(1-(.50)^2)(33-2-1)}=5.20$; critical value (2, 30) = 3.32 (.05), 5.39 (.01); Reject H_0 ($p<.05$)

19. a. $R=\sqrt{\frac{[(-.46)^2+(.33)^2]-(2*.46*.33*.38)}{1-(.38)^2}}=.48$

b. $F=\frac{(.48)^2/2}{(1-(.48)^2)(36-2-1)}=5.22$; critical value (2, 30) = 3.32 (.05), 5.39 (.01); Reject H_0 ($p<.05$)

c. In a sample of 36 voters, the combination of age and years of education was significantly related to the likelihood of voting for the candidate, $R=.65$, $F(2, 33)=5.22$, $p<.05$, $R^2=.23$.

21. a. $b_1 = \frac{.31-((.44)(.17))}{1-(.17)^2}\left(\frac{1.25}{1.04}\right) = .30$ $b_2 = \frac{.44-((.31)(.17))}{1-(.17)^2}\left(\frac{1.25}{.97}\right) = .52$

b. $b_1 = \frac{.46-((.08)(.12))}{1-(.12)^2}\left(\frac{2.90}{1.78}\right) = .73$ $b_2 = \frac{.08-((.46)(.12))}{1-(.12)^2}\left(\frac{2.90}{.46}\right) = .13$

c. $b_1 = \frac{.51-((.48)(.39))}{1-(.39)^2}\left(\frac{1.17}{2.09}\right) = .21$ $b_2 = \frac{.48-((.51)(.39))}{1-(.39)^2}\left(\frac{1.17}{3.27}\right) = .12$

23. a. $Y' = .80 + .30(X_1) + .52(X_2)$
b. $Y' = 9.00 + .73(X_1) + .13(X_2)$
c. $Y' = -1.14 + .21(X_1) + .12(X_2)$
d. $Y' = 58.78 + .04(X_1) + .29(X_2)$
e. $Y' = 32.52 + -.83(X_1) + .70(X_2)$

25. a. $Y' = 7.97$
b. $Y' = 46.65$
c. $Y' = 27.36$
d. $Y' = 70.70$
e. $Y' = 39.56$

27. a. Likelihood′ = 31.04 + .47(Age) + .92(Education); Likelihood′ = 60.43
b. Likelihood′ = 72.16
c. Likelihood′ = 55.65

15

CHI-SQUARE

CHAPTER OUTLINE

Chapter 13 discussed correlational statistics, which are statistics designed to measure the relationship between variables. One correlational statistic, the Pearson correlation coefficient, measures the relationship between variables measured at the interval or ratio level of measurement. Another correlational statistic, the Spearman rank-order correlation, relates ordinal variables with each other. The last chapter of this book introduces a statistic used when the variables in a study are categorical, measured at the nominal level of measurement. Two research situations will be examined: one that consists of a single categorical variable and one that contains two categorical variables. As you have seen throughout this book, how variables in a study are measured determines which statistical procedure may be appropriate to analyze the data. However, the steps involved in analyzing the data remain the same.

15.1 AN EXAMPLE FROM THE RESEARCH (ONE CATEGORICAL VARIABLE): ARE YOU MY TYPE?

The study of personality has received a great deal of attention for many years. People are very interested in understanding how they are similar to and different from others. One group of people for whom the topic of personality has been given particular emphasis is college students. You may have read that students' personalities are related to the colleges they attend, the majors they select, and the careers they pursue after graduation. Research has investigated the relationship between college students' personality characteristics and behaviors that are both academic (e.g., exam scores, cheating on tests, the likelihood of graduating) and nonacademic (e.g., involvement in extracurricular activities, drug and alcohol use, sexual activity).

One line of research has examined whether the personalities of college students are different from other groups of people as well as from earlier generations of students. One such study, conducted by Kenneth Stewart and Paul Bernhardt at Frostburg State University, measured college students on two dimensions of personality (Stewart & Bernhardt, 2010). One dimension was the degree to which a person is extraverted or introverted. People who are extraverted tend to be assertive, talkative, and self-confident, whereas introverts are reserved, quiet, and reluctant to express their emotions. The second dimension was the degree to which one accepts or questions social norms, which are rules widely accepted within a group. People defined as "norm favoring" conform to social norms and are described as efficient and self-disciplined; people who are "norm questioning" challenge traditional values and tend more toward unconventionality and skepticism.

On the basis of their combination of extraversion/introversion and norm favoring/norm questioning, the college students in the study, which will be referred to as the personality study, were classified as one of four types: Alphas, Betas, Gammas, or Deltas. These four personality types were developed by psychologist Harrison Gough and his colleagues in their measure of personality known as the California Psychological Inventory (CPI) (Gough & Bradley, 2005). *Alphas* (extraverted and norm favoring) are task oriented, assertive, and talkative. They have the potential to be charismatic leaders, but they may also be authoritarian and manipulative. *Betas* (introverted and norm favoring) have a high degree of self-control, place others' needs before their own, but are also somewhat detached. *Gammas* (extraverted and norm questioning) openly question traditional beliefs and values. They may be creative and visionary, but they may also be reckless and rebellious. Finally, *Deltas* (introverted and norm questioning) are somewhat detached from others and absorbed in their own thoughts. Deltas can be imaginative and artistic, but they may also be withdrawn and aloof.

The developers of the CPI believed that the four personality types are evenly distributed in the population, with approximately 25% of individuals classified as Alphas, Betas, Gammas, or Deltas. However, based on their review of research, the researchers in the personality study hypothesized that the four types are not equally distributed among college students.

To test their hypothesis, the researchers collected CPI responses from 588 college students and classified each student as an Alpha, Beta, Gamma, or Delta. This variable, which will be referred to as Type, is a categorical variable measured at the nominal level of measurement. As it would require a great deal of space to list the data for all 588 students, Table 15.1 summarizes the data using a frequency distribution table. You may observe from the table, for example, that 251 of the 588 students were classified as Alphas, representing 42.7% (251/588) of the total sample.

Figure 15.1 presents a bar chart comparing the percentages of the four types; a bar chart displays the frequencies (f) or percentages (%) of groups that comprise a categorical variable. The bars in a bar chart do not touch, indicating that the groups are distinct from each other and cannot be placed along a numeric continuum. The bar chart in Figure 15.1 displays percentages rather than frequencies because reporting frequencies can be somewhat uninformative. For example, the 251 Alphas would have been evaluated differently if they represented 251 of a sample of 5,880 students rather than 588.

Looking at Figure 15.1, Alphas (extraverted and norm favoring) and Gammas (extraverted and norm questioning) were the most frequently occurring personality types among the sample of college students—combining these two types, three fourths of the students (75.7%) were extraverted rather than introverted. Among the introverts, students were almost equally likely to be a norm-favoring Beta (11.0%) or a norm-questioning Delta (13.3%).

In earlier chapters, after creating tables and figures to examine data, descriptive statistics such as the mean and standard deviation were calculated. Measures of central tendency and variability are appropriate for continuous variables measured at the interval or ratio level of measurement; however, they do not describe categorical variables. For example, if you asked a group of people where they live, it wouldn't make sense to calculate the "average" city because cities don't fall on a numeric continuum. Instead, you'd summarize their responses by reporting the number and/or percentage of the sample living in each city. Consequently, analyzing categorical variables involves comparing frequencies and percentages rather than means and standard deviations.

For the sample of 588 students in the personality study, Table 15.1 and Figure 15.1 suggest that Alphas and Gammas are much more prevalent than Betas and Deltas. This provides initial support for the researchers' hypothesis that the distribution of personality types among college students differs from the distribution proposed by the developers of the CPI. The next section introduces a statistic used to determine whether the difference between the two distributions is statistically significant.

15.2 INTRODUCTION TO THE CHI-SQUARE STATISTIC

The primary issue of concern illustrated by the personality study is whether the distribution of frequencies of groups in a sample differs from a proposed distribution. Are the frequencies of the four types in the sample of 588 college students different from frequencies expected to occur based on the beliefs of the developers of the CPI? Like many of the inferential statistics discussed in this book, answering this question requires calculating a statistic that evaluates the difference between a sample and a hypothesized population.

Although the research situation in this chapter may seem new to you, it's actually similar to the research situations examined in Chapter 7 (the test of one mean). In that chapter, the goal was to evaluate the difference

TABLE 15.1 ● FREQUENCY DISTRIBUTION TABLE FOR FOUR TYPES IN THE PERSONALITY STUDY

Type	f	%
Alpha	251	42.7%
Beta	65	11.0%
Gamma	194	33.0%
Delta	78	13.3%
Total	588	100.0%

FIGURE 15.1 ● BAR CHART OF FOUR TYPES FOR THE PERSONALITY STUDY

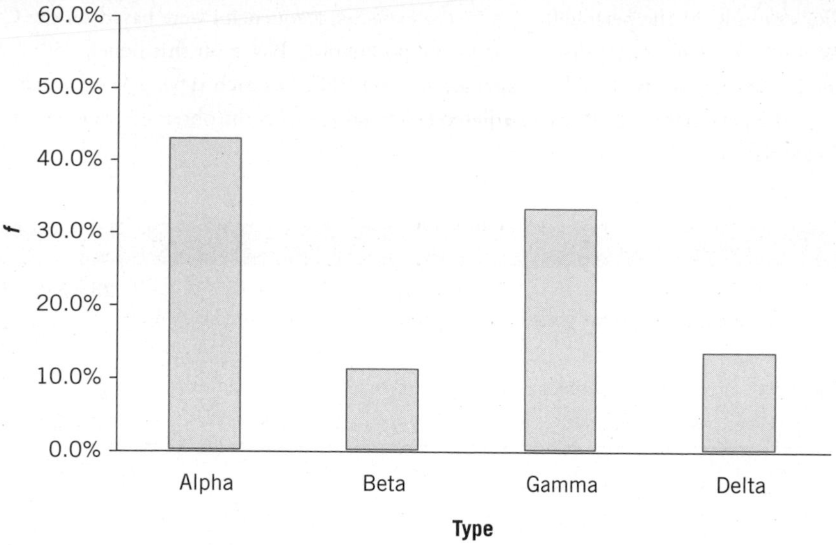

between a sample mean ($\bar{X}$) and a hypothesized population mean (μ). To evaluate this difference, a sampling distribution was created of all of the possible sample means for samples of size N randomly selected from a population that has a mean equal to μ. Using this distribution of sample means, the difference between $\bar{X}$ and μ was transformed into a statistic known as the t-statistic. A distribution of t-statistics was then used to determine the probability of the t-statistic calculated from the sample of data. If this probability was low ($p < .05$), the null hypothesis was rejected, meaning that the difference between $\bar{X}$ and μ was statistically significant.

The situation illustrated by the personality study is very similar to those presented in Chapter 7, with one critical difference: The goal in this study was to evaluate differences between frequencies rather than means. The personality study was conducted to evaluate the difference between the frequencies of the four types in the sample of 588 students and the frequencies expected to occur according to the developers of the CPI. Evaluating this difference requires a distribution of all of the possible combinations of four frequencies that can occur for samples of $N = 588$ drawn randomly from the population. To illustrate how this distribution is created, imagine that from a population known to contain 25% of each of the four types, we randomly draw a sample of 588 students and count the number of Alphas, Betas, Gammas, and Deltas. If we were to do this an infinite number of times, we could develop a distribution of all of the possible combinations of the four frequencies for $N = 588$. This distribution can be used to transform the difference between a sample's frequencies and the expected frequencies into a statistic. The probability of this statistic can then be determined to decide whether the difference between the frequencies in the sample and the proposed frequencies is statistically significant.

Observed and Expected Frequencies: The Basis of the Chi-Square Statistic

To calculate the statistic used to analyze frequencies, two pieces of information must be obtained for each group in the study. The first piece of information is the **observed frequency (f_o)**, which is the frequency of a group in the sample of data. The observed frequencies for the personality study may be found in the frequency distribution table in Table 15.1. For example, the observed frequency of the Beta type is 65 ($f_o = 65$).

The second piece of information needed for each group is the **expected frequency** (**f_e**), defined as the frequency of a group expected to occur in a sample of data. Expected frequencies are sometimes based on theory or research. For example, in the personality study, the expected frequencies were based on the CPI developers' belief that the four types are evenly distributed in the population. Based on this belief, 25% (.25) of the 588 students in the personality study would be expected to be classified as each type. The expected frequencies for the four types in the personality study are calculated below (along with the observed frequencies, which will be used for later calculations):

Type	Observed Frequency (f_o)	Expected Frequency (f_e)
Alpha	251	**.25 * 588 = 147**
Beta	65	**.25 * 588 = 147**
Gamma	194	**.25 * 588 = 147**
Delta	78	**.25 * 588 = 147**
Total	588	**588**

Based on the above calculations, according to the developers of the CPI, in a sample of 588 students, it is expected that 147 students belong to each of the four types. As a way of checking the accuracy of calculations, the sum of the observed and expected frequencies is the same ($N = 588$).

Once the observed and expected frequencies have been determined, the next step is to calculate the difference between the two frequencies for each group:

Type	Observed Frequency (f_o)	Expected Frequency (f_e)	Observed – Expected ($f_o - f_e$)
Alpha	251	147	**251 – 147 = 104**
Beta	65	147	**65 – 147 = –82**
Gamma	194	147	**194 – 147 = 47**
Delta	78	147	**78 – 147 = –69**
Total	588	588	**0**

To evaluate the data for the entire sample, we must combine the differences between the observed and expected frequencies across all of the groups. However, looking at the above table, we encounter a problem: The differences sum to 0 ($104 + -82 + 47 + -69 = 0$).

Similar to the differences between group means and the mean of the total sample in analysis of variance (ANOVA), the differences between observed and expected frequencies in any set of data will *always* sum to 0. Because it would be incorrect to say there are zero differences between the observed and expected frequencies, these differences must be squared to eliminate any negative differences (see table below).

These squared differences between observed and expected frequencies are the basis of the **chi-square statistic** (χ^2) (pronounced "kye," which rhymes with "eye"). The chi-square statistic tests the difference between the distribution of observed frequencies in a sample and a proposed distribution of expected frequencies. The statistic calculated in this chapter is sometimes called the *Pearson's chi-square statistic.* It is named after Karl Pearson, who also developed the Pearson correlation coefficient discussed in Chapter 13. Later in this chapter, we will provide formulas needed to calculate different chi-square statistics.

Type	Observed Frequency (f_o)	Expected Frequency (f_e)	(Observed – Expected)^2 $(f_o - f_e)^2$
Alpha	251	147	**$(104)^2$ = 10,816**
Beta	65	147	**$(-82)^2$ = 6,724**
Gamma	194	147	**$(47)^2$ = 2,209**
Delta	78	147	**$(-69)^2$ = 4,761**
Total	588	588	

Characteristics of the Distribution of Chi-Square Statistics

In order for the probability of any particular value of the chi-square statistic to be determined, it must be located within a distribution of chi-square statistics; an example of a chi-square distribution is provided in Figure 15.2. Several aspects of this distribution are important to understand. First, like the *t*-statistic, *F*-ratio, and Pearson correlation, there is not just one chi-square statistic distribution; instead, there are different distributions depending on the number of groups that comprise the variable of interest. Second, because calculating the chi-square statistic involves squared differences between observed and expected frequencies ($(f_o - f_e)^2$), the chi-square statistic is always a positive number. Consequently, like the *F*-ratio, the distribution of chi-square statistics is positively skewed rather than symmetrical; as the number of groups comprising the variable of interest increases, the degree of skew decreases such that the chi-square distribution approaches a normal distribution. It's important for you to understand that the precise shape of this distribution changes somewhat dramatically as a function of the number of groups; however, for the sake of simplicity, we will use the distribution in Figure 15.2 throughout this chapter.

15.3 INFERENTIAL STATISTIC: CHI-SQUARE GOODNESS-OF-FIT TEST

To test the differences between the observed and expected frequencies in a set of data, a chi-square statistic is calculated. When the data comprise a single variable, such as the Type variable in the personality study, the appropriate statistical procedure is known as the **chi-square goodness-of-fit test**. It has been given this name because it tests the "fit" between observed frequencies in a set of data and expected frequencies derived from theory or research.

FIGURE 15.2 ● EXAMPLE DISTRIBUTION OF A CHI-SQUARE STATISTIC

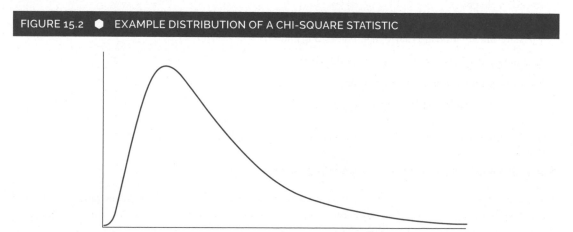

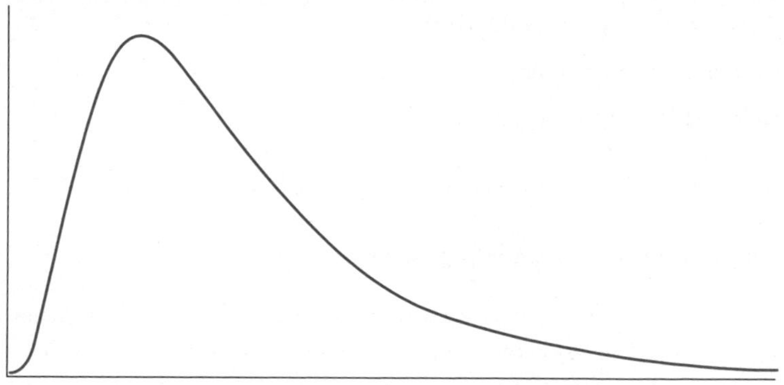

LEARNING CHECK 1:

Reviewing What You've Learned So Far

1. Review questions
   a. In creating a bar chart for a categorical variable, why would you choose to have each bar represent a percentage (%) rather than a frequency (*f*)?
   b. Why is it inappropriate to calculate the mean and standard deviation for categorical variables?
   c. What is the difference between an observed frequency and an expected frequency?
   d. What is the main purpose of the chi-square statistic (χ^2)?
   e. What are the main characteristics of the distribution of chi-square statistics?
   f. What are some assumptions underlying the chi-square statistic?
2. For each of the following situations, create a frequency distribution table and bar chart.
   a. A smartphone manufacturer wants to learn whether customers prefer a keyboard or a touch screen:

Type of Phone	Type of Phone	Type of Phone	Type of Phone
Touch screen	Keyboard	Touch screen	Touch screen
Keyboard	Touch screen	Touch screen	Keyboard
Touch screen	Touch screen	Keyboard	Touch screen
Touch screen	Touch screen	Touch screen	Touch screen

   b. A café owner is interested in seeing the type of milk her customers add to their drinks: whole, low fat, or nonfat:

Type of Milk	Type of Milk	Type of Milk	Type of Milk	Type of Milk
Nonfat	Low fat	Whole	Low fat	Low fat
Low fat	Low fat	Low fat	Whole	Nonfat
Whole	Low fat	Whole	Whole	Low fat
Low fat	Whole	Nonfat	Low fat	Whole

The steps involved in calculating the chi-square statistic in the chi-square goodness-of-fit test are the same as in earlier chapters:

- state the null and alternative hypotheses (H_0 and H_1),
- make a decision about the null hypothesis,
- draw a conclusion from the analysis, and
- relate the result of the analysis to the research hypothesis.

Each of these steps is presented below, using the personality study as an example.

State the Null and Alternative Hypotheses (H_0 and H_1)

What are the statistical hypotheses when the data being analyzed consist of frequencies rather than group means? Rather than stating differences between means, the null and alternative hypotheses for the chi-square goodness-of-fit test relate to the difference between the distributions of observed and expected frequencies. For example, the null hypothesis is stated as follows:

H_0: The distribution of observed frequencies fits the distribution of expected frequencies.

For the personality study, this null hypothesis implies that the frequencies of the four types in the sample of 588 students are consistent with expected frequencies based on the developers of the CPI.

If the null hypothesis is that the distribution of observed frequencies fits the distribution of expected frequencies, what is the mutually exclusive alternative to this hypothesis? One way to state the alternative hypothesis is as follows:

H_1: The distribution of observed frequencies does not fit the distribution of expected frequencies.

Rejecting the null hypothesis in the personality study implies that the distribution of frequencies of the four types in the sample of data does not fit the distribution of expected frequencies.

Make a Decision About the Null Hypothesis

The next step is to make the decision regarding whether to reject the null hypothesis. This consists of steps with which you are quite familiar:

- calculate the degrees of freedom (*df*);
- set alpha (α), identify the critical value, and state a decision rule;
- calculate a statistic: chi-square (χ^2);
- make a decision whether to reject the null hypothesis;
- determine the level of significance; and
- calculate a measure of effect size (Cramér's ϕ).

These steps are discussed below, using the personality study to demonstrate how each step may be completed.

Calculate the Degrees of Freedom (*df*)

The chi-square goodness-of-fit test involves differences between frequencies for the groups that comprise a variable. Similar to the one-way ANOVA discussed in Chapter 11, the degrees of freedom for this test is the number of groups that comprise the variable, minus 1:

$$df = \#\text{ groups} - 1 \tag{15-1}$$

In the personality study, there are four types (Alpha, Beta, Gamma, and Delta). Therefore, the degrees of freedom is equal to

$$\begin{aligned} df &= \#\text{group} - 1 \\ &= 4 - 1 \\ &= 3 \end{aligned}$$

Set Alpha (α), Identify the Critical Value, and State a Decision Rule

Alpha (α), the probability of the statistic needed to reject the null hypothesis, is typically set to .05. Because a chi-square statistic can only be a positive number, the theoretical distribution is positively skewed, with the region of rejection located only at the right end of the distribution. As such, it is not necessary to indicate whether alpha is one-tailed or two-tailed.

Table 8 in the back of this book provides a table of critical values for the chi-square statistic. This table consists of a series of rows corresponding to different degrees of freedom (*df*); the critical value for different values of alpha (α) are provided for each *df*. For the personality study, the critical value is identified by moving down the *df* column until you reach the row associated with $df = 3$; moving to the right, for $\alpha = .05$, a critical value of 7.81 is found. Therefore, the following critical value may be obtained for the personality study:

$$\text{For } \alpha = .05 \text{ and } df = 3, \text{ critical value} = 7.81$$

Figure 15.3 illustrates the critical value and the regions of rejection and nonrejection for the personality study.

Once the critical value for the chi-square statistic has been identified, a decision rule may be stated regarding the values of the statistic that lead to the rejection of the null hypothesis. For the personality study,

$$\text{If } \chi^2 > 7.81, \text{ reject } H_0; \text{ otherwise, do not reject } H_0$$

Calculate a Statistic: Chi-Square (χ^2)

The next step in conducting the chi-square goodness-of-fit test is to calculate a value of the chi-square statistic. Two steps are involved in calculating this statistic: (1) calculate expected frequencies and (2) calculate the chi-square statistic. Although some of the calculations presented in this section will be the same as those presented earlier in introducing the chi-square statistic, they will be repeated here to formally introduce relevant formulas.

Calculate expected frequencies (f_e). The first step in calculating the chi-square statistic is to calculate the expected frequency (f_e) for each of the groups comprising the variable of interest. The formula for an expected frequency in the chi-square goodness-of-fit test is provided in Formula 15-2:

$$f_e = (\text{hypothesized proportion})\ (N) \qquad (15\text{-}2)$$

where "hypothesized proportion" is the proportion of the population hypothesized to be in a particular group, and N is the total sample size.

In the personality study, the developers of the CPI believed the four types are equally distributed in the population. Consequently, the hypothesized proportion for each group is .25 (25%). Applying these hypothesized

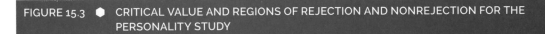

FIGURE 15.3 ● CRITICAL VALUE AND REGIONS OF REJECTION AND NONREJECTION FOR THE PERSONALITY STUDY

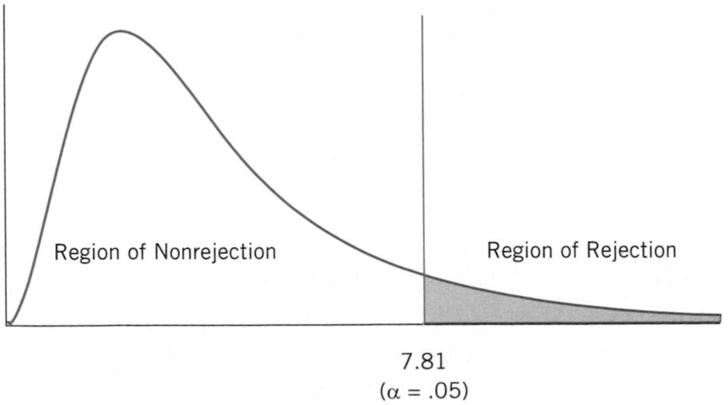

proportions to the 588 students in the personality study ($N = 588$), the expected frequencies for the four types are calculated below:

Type	f_o	f_e
Alpha	251	**.25 (588) = 147.00**
Beta	65	**.25 (588) = 147.00**
Gamma	194	**.25 (588) = 147.00**
Delta	78	**.25 (588) = 147.00**
Total	588	**588.00**

If the developers of the CPI are correct, in a sample of 588 students, each of the four types should contain 147.00 students. As a way of checking your calculations, the sum of the expected frequencies (Σf_e) should be equal to the total sample size (N):

$$\Sigma f_e = N$$
$$147.00 + 147.00 + 147.00 + 147.00 = 588$$
$$588 = 588$$

Calculate the chi-square statistic (χ^2). Once the expected frequencies have been determined, the chi-square statistic (χ^2) that tests the differences between observed and expected frequencies can be calculated using the following formula:

$$\chi^2 = \Sigma \frac{(f_o - f_e)^2}{f_e} \tag{15-3}$$

where f_o is the observed frequency for each group and f_e is the expected frequency for each group. As we discussed earlier in this chapter, the main component of the chi-square statistic is the squared difference between the observed and expected frequency for each group ($(f_o - f_e)^2$). To calculate the chi-square statistic, this squared difference is divided by each group's expected frequency (f_e) before combining these calculations across the groups.

For the personality study, using the observed and expected frequencies calculated earlier, the chi-square statistic is calculated as follows:

$$\begin{aligned}
\chi^2 &= \Sigma \frac{(f_o - f_e)^2}{f_e} \\
&= \frac{(251 - 147.00)^2}{147.00} + \frac{(65 - 147.00)^2}{147.00} + \frac{(194 - 147.00)^2}{147.00} + \frac{(78 - 147.00)^2}{147.00} \\
&= \frac{(104.00)^2}{147.00} + \frac{(-82.00)^2}{147.00} + \frac{(47.00)^2}{147.00} + \frac{(-69.00)^2}{147.00} \\
&= \frac{10816.00}{147.00} + \frac{6724.00}{147.00} + \frac{2209.00}{147.00} + \frac{4761.00}{147.00} \\
&= 73.58 + 45.74 + 15.03 + 32.39 \\
&= 166.74
\end{aligned}$$

Looking at the above calculations, you may wonder why the squared difference between each observed and expected frequency ($(f_o - f_e)^2$) is divided by its expected frequency (f_e). The purpose of this division is to take into account the number of observations in each group when evaluating the difference between an observed and expected frequency. For example, a difference of 5 between f_o and f_e should be given more weight if the expected frequency for a group is 10 than if it is 100. As a simple analogy, imagine you've decided to host a party. Your hosting would probably be more affected if you expected 5 people but 10 people showed up than if you expected 100 people but 105 showed up.

LEARNING CHECK 2:

Reviewing What You've Learned So Far

1. Review questions
   a. In what type of research situation would the chi-square goodness-of-fit test be conducted?
   b. What is implied by the null and alternative hypotheses in the goodness-of-fit test?
   c. What are the two steps involved in calculating the chi-square statistic?
   d. In the goodness-of-fit test, what determines the values for the "hypothesized proportions"?
   e. In calculating the chi-square statistic, why is each squared difference divided by its expected frequency?
2. For each of the following variables, calculate the degrees of freedom (*df*) and identify the critical value ($\alpha = .05$).
   a. Answer (Correct, Incorrect)
   b. Location (Top, Center, Bottom)
   c. Sport (Baseball, Football, Basketball, Soccer, Lacrosse)
3. One study compared two techniques designed to help children teach themselves how to do math problems (Grafman & Cates, 2010). The first method, Cover, Copy, and Compare (CCC), required students to read the problem, cover it with their hand, write their solution, and then compare their solution with one provided to them. The second method (MCCC) was identical to CCC except that students first copied down the problem before starting the CCC procedure. The researchers asked students which of the two methods they preferred: 34 chose CCC and 10 chose MCCC.
   a. Calculate expected frequencies (f_e) (assume the two hypothesized proportions are equal).
   b. Calculate the chi-square statistic (χ^2) for the goodness-of-fit test.

Make a Decision Whether to Reject the Null Hypothesis

Once the value of the chi-square statistic has been calculated, a decision is made regarding whether to reject the null hypothesis. This decision is made by comparing the chi-square statistic with the critical value. For the personality study,

$$\chi^2 = 166.74 > 7.81 \therefore \text{ reject } H_0 \ (p < .05)$$

Because the obtained value of the chi-square statistic is greater than the $\alpha = .05$ critical value of 7.81, it lies in the region of rejection. Consequently, the decision is made to reject the null hypothesis and conclude that the distribution of observed frequencies does not fit the distribution of expected frequencies.

Determine the Level of Significance

In the personality study, because the null hypothesis was rejected, the next step is to determine whether the probability of the chi-square statistic is less than .01. To do so, let's return to the table of critical values in Table 8. For $df = 3$, the .01 critical value is 11.34. Comparing the chi-square value of 166.74 with the .01 critical value,

$$\chi^2 = 166.74 > 11.34 \therefore p < .01$$

Consequently, it can be concluded that the probability of the chi-square statistic in the personality study is not only less than .05 but also less than .01. The level of significance for the personality study is illustrated in Figure 15.4.

FIGURE 15.4 ● DETERMINING THE LEVEL OF SIGNIFICANCE FOR THE PERSONALITY STUDY

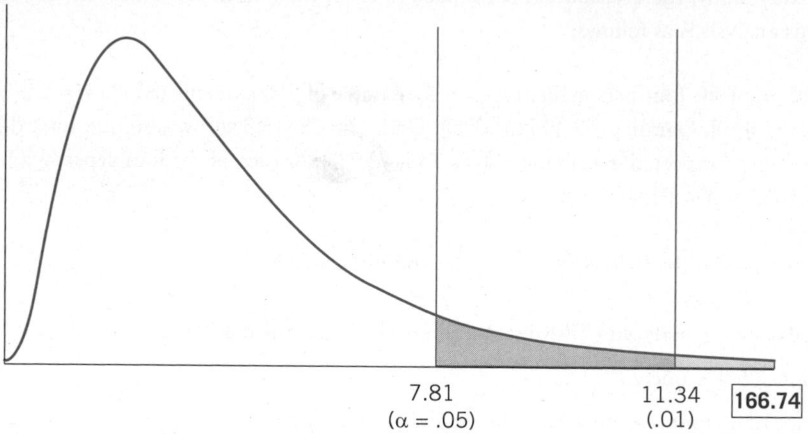

Calculate a Measure of Effect Size (Cramér's ϕ)

As was discussed in earlier chapters, it's useful to supplement an inferential statistic with a statistic that provides an estimate of the size or magnitude of the effect. There are several different measures of effect size for the chi-square statistic; in this book, we will use **Cramér's ϕ** (phi), sometimes referred to as *Cramér's V*. The formula for Cramér's ϕ for the chi-square goodness-of-fit test is provided as follows:

$$\phi = \sqrt{\frac{\chi^2}{N(\#\text{ groups} - 1)}} \tag{15-4}$$

where χ^2 is the calculated value of the chi-square statistic and N is the sample size.

For the personality study, N is equal to 588, and the number of groups is equal to 4. Using the calculated value of 166.74 for the chi-square statistic, Cramér's ϕ is calculated as follows:

$$\begin{aligned} \Phi &= \sqrt{\frac{\chi^2}{N(\#\text{ groups} - 1)}} \\ &= \sqrt{\frac{166.74}{588(4-1)}} = \sqrt{\frac{166.74}{1764}} = \sqrt{.09} \\ &= .30 \end{aligned}$$

As with other measures of effect size discussed in this book, the possible values for Cramér's ϕ range from .00 to 1.00. To assist in the interpretation of ϕ, Cohen (1988, pp. 224–226) defines small, medium, and large values of ϕ as the following:

Small effect: $\phi = .10$
Medium effect: $\phi = .30$
Large effect: $\phi = .50$

Accordingly, the value of .30 for ϕ indicates that the difference between the observed and expected frequencies for the four personality types represents a medium-sized effect.

Draw a Conclusion From the Analysis

For the personality study, the decision has been made to reject the null hypothesis. One way to communicate the result of this analysis is as follows:

> The distribution of the four personality types in the sample of 588 students (Alpha [f= 251 (42.7%)], Beta [f= 65 (11.0%)], Gamma [f= 194 (33.0%)], Delta [f= 78 (13.3%)]) was significantly different from the CPI developers' expected distribution of f= 251 (42.7%) for each of the four types, $\chi^2(3, N = 588) = 166.74, p < .01, \phi = .30$.

This sentence provides the following information about the analysis:

- the variable being analyzed ("The distribution of the four personality types"),
- the sample ("588 students"),
- the groups that comprise the variable being analyzed ("Alpha . . . Beta . . . Gamma . . . Delta"),
- descriptive statistics (i.e., "f= 251 (42.7%)"),
- the nature and direction of the relationship ("was significantly different from the CPI developers' expected distribution"), and
- information about the inferential statistic ("$\chi^2(3, N = 588) = 166.74, p < .01, \phi = .30$"), which indicates the inferential statistic calculated (χ^2), the degrees of freedom (3), the value of the statistic (166.74), the level of significance ($p < .01$), and the measure of effect size ($\phi = .30$).

In reporting the chi-square statistic, the sample size (N= 588) is typically included. This is because, unlike other inferential statistics, the degrees of freedom for the chi-square statistic is based on the number of groups rather than the number of participants.

Relate the Result of the Analysis to the Research Hypothesis

What is the implication of rejecting the null hypothesis in the personality study? In terms of the purpose of this study, the following could be stated:

> The result of this analysis supports the research hypothesis that the distribution of personality types among college students does not match or fit the distribution in the population proposed by the developers of the CPI.

Box 15.1 lists the steps involved in conducting the chi-square goodness-of-fit test. In the personality study, the hypothesized proportions of the four types were presumed to be the same (.25). The next section discusses how to conduct the goodness-of-fit test with unequal hypothesized proportions.

SUMMARY BOX 15.1 CONDUCTING THE CHI-SQUARE GOODNESS-OF-FIT TEST (PERSONALITY STUDY EXAMPLE)

State the null and alternative hypotheses (H_0 and H_1)

H_0: The distribution of observed frequencies fits the distribution of expected frequencies

H_1: The distribution of observed frequencies does not fit the distribution of expected frequencies

Make a decision about the null hypothesis

Calculate the degrees of freedom (df)

$$df = \#\text{ groups} - 1 = 4 - 1 = 3$$

Set alpha (α), identify the critical value, and state a decision rule

$$\text{If } \chi^2 > 7.81, \text{ reject } H_0; \text{ otherwise, do not reject } H_0$$

Calculate a statistic: chi-square (χ^2)

Calculate expected frequencies (f_e)

f_e = (hypothesized proportion) (N)

= Alpha: .25 (588) = 147.00 Beta: .25 (588) = 147.00

Gamma: .25 (588) = 147.00 Delta: .25 (588) = 147.00

Calculate the chi-square statistic (χ^2)

$$\chi^2 = \Sigma\frac{(f_o - f_e)^2}{f_e}$$
$$= \frac{(251-147.00)^2}{147.00} + \frac{(65-147.00)^2}{147.00} + \frac{(194-147.00)^2}{147.00} + \frac{(78-147.00)^2}{147.00}$$
$$= \frac{10{,}816.00}{147.00} + \frac{6{,}724.00}{147.00} + \frac{2{,}209.00}{147.00} + \frac{4{,}761.00}{147.00}$$
$$= 73.58 + 45.74 + 15.03 + 32.39 = 166.74$$

Make a decision whether to reject the null hypothesis

$$\chi^2 = 166.74 > 7.81 \therefore \text{ reject } H_0 \ (p < .05)$$

Determine the level of significance

$$\chi^2 = 166.74 > 11.34 \therefore p < .01$$

Calculate a measure of effect size (Cramér's ϕ)

$$\phi = \sqrt{\frac{\chi^2}{N(\#\text{groups} - 1)}}$$
$$= \sqrt{\frac{166.74}{588(4-1)}} = \sqrt{\frac{166.74}{1764}} = \sqrt{.09} = .30$$

Draw a conclusion from the analysis

The distribution of the four personality types in the sample of 588 students (Alpha [f = 251 (42.7%)], Beta [f = 65 (11.0%)], Gamma [f = 194 (33.0%)], Delta [f = 78 (13.3%)]) was significantly different from the CPI developers' expected distribution of f = 147 (25%) for each of the four types, $\chi^2(3, N = 588) = 166.74$, $p < .01$, $\phi = .30$.

Relate the result of the analysis to the research hypothesis

The result of this analysis supports the research hypothesis that the distribution of personality types among college students does not match or fit the distribution in the population proposed by the developers of the CPI.

LEARNING CHECK 3:

Reviewing What You've Learned So Far

1. Review questions
   a. What conclusion can be drawn when the null hypothesis for the goodness-of-fit test is rejected?
2. The company that makes M&M's candy (www.mms.com) once conducted a survey asking people which new color they would prefer: purple, aqua, or pink. Below are the votes of a hypothetical sample that represents their actual findings:

Color	Color	Color	Color	Color	Color
Pink	Pink	Aqua	Purple	Purple	Aqua
Purple	Purple	Pink	Pink	Pink	Purple
Pink	Purple	Purple	Purple	Aqua	Aqua
Aqua	Pink	Pink	Purple	Pink	Purple
Pink	Purple	Purple	Aqua	Pink	Purple
Purple	Pink	Aqua	Purple	Purple	

Conduct the chi-square goodness-of-fit test to determine whether there were any differences between the votes for three colors (assume equal hypothesized proportions).

a. Construct a frequency distribution table and bar chart for these data.
b. State the null and alternative hypotheses (H_0 and H_1).
c. Make a decision about the null hypothesis.
   (1) Calculate the degrees of freedom (*df*).
   (2) For $\alpha = .05$, identify the critical value and state a decision rule.
   (3) Calculate a value of the chi-square statistic (χ^2).
   (4) Make a decision whether to reject the null hypothesis.
   (5) Determine the level of significance.
   (6) Calculate a measure of effect size (Cramér's ϕ).
d. Draw a conclusion from the analysis.

Assumptions Underlying the Chi-Square Statistic

Like the other statistics discussed in this book, the appropriate use of the chi-square statistic is based on meeting certain assumptions. This section discusses two assumptions: the independence of observations and minimum expected frequencies.

In using the chi-square statistic, the **assumption of independence of observations** is the assumption that each observation in a set of data is independent of all other observations. In other words, an observation cannot be related to any other observation. This assumption is typically met by assigning each research participant to only one group. For example, in the personality study, the assumption of independence of observations is met by assigning each student to only one of the four types; a violation of this assumption would occur if a student could somehow be classified to more than one type at the same time.

As with all of the inferential statistics discussed in this book, the size of the sample plays a critical role: The smaller the sample size, the less confidence we have that the sample is representative of the population from which the sample was drawn. In using the chi-square statistic, the **assumption of minimum expected frequencies** are two rules of thumb pertaining to small sample sizes. The first rule of thumb is that none of the expected frequencies should be equal to zero (0). The second rule of thumb is that most if not all of the expected frequencies should be greater than 5. Failure to meet either or both of these rules of thumb may limit the ability of the chi-square statistic to approximate the population from which the sample was drawn.

Different statistics have been developed to use with samples with small expected frequencies. For example, when any of the expected frequencies are less than 5, it has been recommended that the **Fisher's exact test** be used rather than the chi-square statistic (Hays, 1988). On the other hand, if any of the expected frequencies are between 5 and 10, **Yates's correction for continuity** (Hays, 1988) may be used to test the differences between the observed and expected frequencies.

Conducting the Chi-Square Goodness-of-Fit Test With Unequal Hypothesized Proportions

One purpose of the personality study was to compare the observed frequencies in the sample of 588 college students with expected frequencies derived from the developers of the CPI. In addition to comparing their sample of college students with the general population of adults, the researchers wished to see whether the distribution of the four personality types among college students may have changed over time. Consequently, they compared their sample with the distribution of personality types in a sample of 7,361 college students from the 1980s.

In the 1980s sample, the four lifestyles were distributed the following way:

Type	%
Alpha	35.2%
Beta	22.3%
Gamma	24.0%
Delta	18.5%
Total	100.0%

Looking at these percentages, it appears that Alpha (extraverted and norm favoring) was the most common personality type (35.2%), with similar percentages across the other three types.

The goal of this analysis was to test the difference between the observed frequencies of the sample of 588 college students with expected frequencies based on the proportions found in the 1980s sample. As it turns out, the chi-square goodness-of-fit test can accommodate the situation where the hypothesized proportions are not all the same. The purpose of this example is to demonstrate how to conduct the chi-square goodness-of-fit test with unequal hypothesized proportions.

Many of the steps in conducting the chi-square goodness-of-fit test have the same result regardless of whether the hypothesized proportions are equal or unequal. Using the personality study as the example, the null and alternative hypotheses, degrees of freedom, alpha, critical value, and the decision rule are identical in both situations. Consequently, we will not report these steps below. What may change, however, is the calculated value of the chi-square statistic. Furthermore, if the calculated value of the statistic changes, what may also change is the decision made about the null hypothesis, the conclusions drawn from the analysis, and whether the analysis supports the research hypothesis.

To compare the distribution of the four personality types for the sample of 588 students with the 1980s sample, a value of the chi-square statistic must be calculated. As before, there are two steps in calculating the chi-square statistic for the goodness-of-fit test with unequal hypothesized proportions: calculate expected frequencies (f_e) and calculate the chi-square statistic (χ^2).

Calculate Expected Frequencies (f_e)

The expected frequency (f_e) for each group is again calculated by multiplying the group's hypothesized proportion by the total sample (*N*). Using the proportions for the 1980s sample reported above, the expected frequencies for the personality study data are calculated as follows:

Type	f_o	f_e
Alpha	251	**.352 (588) = 206.98**
Beta	65	**.223 (588) = 131.12**
Gamma	194	**.240 (588) = 141.12**
Delta	78	**.185 (588) = 108.78**
Total	588	**588.00**

Note that these expected frequencies are different from those calculated when the hypothesized proportions were the same (.25) for all four types. However, the sum of the expected frequencies is again equal to the total sample size (206.98 + 131.12 + 141.12 + 108.78 = 588).

Calculate the Chi-Square Statistic (χ^2)

Once the expected frequencies have been determined, the next step is to calculate the chi-square statistic that tests the differences between the observed and expected frequencies. Regardless of whether or not the hypothesized proportions are equal, the chi-square statistic is calculated using the same formula (Formula 15-3):

$$
\begin{aligned}
\chi^2 &= \Sigma \frac{(f_o - f_e)^2}{f_e} \\
&= \frac{(251-206.98)^2}{206.98} + \frac{(65-131.12)^2}{131.12} + \frac{(194-141.12)^2}{141.12} + \frac{(78-108.78)^2}{108.78} \\
&= \frac{(44.02)^2}{206.98} + \frac{(-66.12)^2}{131.12} + \frac{(52.88)^2}{141.12} + \frac{(-30.78)^2}{108.78} \\
&= \frac{1{,}937.76}{206.98} + \frac{4{,}371.85}{131.12} + \frac{2{,}796.29}{141.12} + \frac{947.41}{108.78} \\
&= 9.36 + 33.34 + 19.82 + 8.71 \\
&= 71.23
\end{aligned}
$$

In calculating the chi-square statistic, care must be taken to use the appropriate expected frequency (f_e) for each group.

LEARNING CHECK 4:

Reviewing What You've Learned So Far

1. Review questions
   a. What steps in conducting the chi-square goodness-of-fit test change when the hypothesized proportions are unequal rather than equal? What steps do not change?
2. In Learning Check 3, preferences for three M&M colors (purple, aqua, and pink) were compared. As it turns out, the final worldwide vote conducted by www.mms.com found that 42% chose purple, 38% chose aqua, and 20% chose pink. Conduct the chi-square goodness-of-fit test comparing this sample's proportions with those reported by the M&Ms company.

a. State the null and alternative hypotheses (H_0 and H_1).
b. Make a decision about the null hypothesis.
   (1) Calculate the degrees of freedom (*df*).
   (2) For $\alpha = .05$, identify the critical value and state a decision rule.
   (3) Calculate a value of the chi-square statistic (χ^2).
   (4) Make a decision whether to reject the null hypothesis.
   (5) Determine the level of significance.
   (6) Calculate a measure of effect size (Cramér's ϕ).
c. Draw a conclusion from the analysis.

Comparing the result of this analysis with the earlier one, the calculated value of χ^2 changed from 166.74 to 71.23. Changing the expected frequencies may alter the decisions and conclusions resulting from the remaining steps in hypothesis testing (making the decision whether to reject the null hypothesis, drawing a conclusion from the analysis, etc.). For this reason, it is important that the hypothesized proportions are based on sound theoretical or empirical reasoning.

This section has demonstrated how to analyze a single categorical variable using the chi-square goodness-of-fit test. The next section introduces a somewhat more complicated research situation, one that consists of two categorical variables rather than one.

15.4 AN EXAMPLE FROM THE RESEARCH (TWO CATEGORICAL VARIABLES): SEEING RED

You may have a favorite color you like to wear—perhaps it's blue or green or yellow. As such, you may wonder just how and when you developed this preference. Research regarding people's color preferences has found that young children apparently prefer the color red (Maier, Barchfeld, Elliot, & Peckrun, 2009). For example, 3-month-old infants have been found to spend more time looking at a red stimulus than one that was yellow, blue, or green (R. J. Adams, 1987), and children in a daycare center preferred classrooms with red walls to walls that were purple, blue, green, yellow, orange, or gray (Read & Upington, 2009). To try to explain these findings, one study hypothesized that children prefer the color red because it's associated with positive emotional states (Zentner, 2001). In that study, a sample of 3- and 4-year-olds not only exhibited a preference for the color red but also were more likely to assign it a happy face than a face that was sad or angry.

Believing that children may or may not like a color because it's associated with a particular experience or feeling, a team of researchers led by Markus Maier at Stony Brook University in New York wished to see whether these preferences change if these associations change (Maier et al., 2009). More specifically, they hypothesized that infants' preferences for the color red may be altered by associating the color with a face that is angry versus one that is happy:

> Specifically, we predicted that in the presence of a happy (hospitable) face, red would be preferred because it signals a potentially desirable outcome . . . but that in the presence of an angry (hostile) face, red would not be preferred because it signals a potentially undesirable outcome. (Maier et al., 2009, p. 735)

To test their hypothesis, the researchers conducted an experiment with "a total of 40 infants (25 girls). . . . The mean age of participants was 17.75 months (SD = 2.49), with a range of 15 to 26 months. All infants were members of a day care center" (Maier et al., 2009, pp. 736–737). In the study, infants were brought into a room and seated on the lap of a daycare worker in front of a table. On the table, covered by a white cloth, were three toys that were identical except for their color: red, green, or gray. Before the infant was shown the toys, he or she was shown one of two pictures of the face of a man. For half of the infants, the man had a happy expression; for the other half, the man was angry. Once the experimenters had determined that the infant had looked at the picture, they removed the cloth and showed the infant the toys. When the infant picked up one of the toys, the color of the toy was recorded.

The study, which will be referred to as the color preference study, features two variables of interest. The first variable, Face, is the expression of the face shown to the infant (Happy or Angry). The second variable, Toy Color, is the color of the toy preferred by the infant (Red, Green, or Gray). Both of these variables are categorical, consisting of distinct groups. The colors of the toys selected by the 20 infants in each of the Happy and Angry groups are listed in Table 15.2.

In order to organize the data, two tables have been constructed in Table 15.3. Table 15.3(a) is a **contingency table**, which is a table in which the rows and columns represent the values of categorical variables and the cells of the table contain observed frequencies for combinations of the variables. It is called a contingency table because the goal of the analysis is to determine whether the distribution of frequencies for one variable is contingent on (depends on) another variable. For the color preference study, we are examining whether the distribution of the colors of the toy is contingent on whether the child is shown a happy or angry face.

A contingency table may be described in terms of the number of its rows and columns. For example, the contingency table for the color preference study is a "3 × 2" table because it consists of three rows (Toy Color: Red, Green, Gray) and two columns (Face: Happy, Angry). Each cell within the body of a contingency table contains the observed frequency for a particular combination of variables. For example, the number 15 in the upper-left cell of Table 15.3(a) implies that 15 infants who were exposed to the happy face selected the red toy.

The main purpose of creating a contingency table is to prepare for the calculation of a chi-square statistic. However, in reporting the data for a study, it is useful to calculate and report percentages related to each observed frequency. Table 15.3(b) provides the percentage of infants selecting the three toy colors for each of the two faces. For example, the percentage 75.0% in the upper-left cell of this table represents the percentage of the 20 infants shown the happy face who selected the red toy (15/20 = 75.0%). It should be noted that, rather than

TABLE 15.2 ● THE COLOR OF THE SELECTED TOY FOR INFANTS IN THE HAPPY AND ANGRY GROUPS

Happy				Angry			
Infant	Toy Color	Infant	Toy Color	Infant	Toy Color	Infant	Toy Color
1	Red	11	Red	1	Green	11	Gray
2	Red	12	Green	2	Gray	12	Red
3	Green	13	Red	3	Green	13	Green
4	Red	14	Red	4	Red	14	Green
5	Gray	15	Red	5	Gray	15	Red
6	Red	16	Red	6	Gray	16	Gray
7	Red	17	Red	7	Red	17	Green
8	Red	18	Gray	8	Green	18	Gray
9	Gray	19	Red	9	Red	19	Gray
10	Red	20	Red	10	Green	20	Red

TABLE 15.3 ● TABLES OF FREQUENCIES AND PERCENTAGES FOR THE COLOR PREFERENCE STUDY

(a) Contingency Table of Frequencies

		Face: Happy	Face: Angry	Total
Toy color	Red	15	6	21
	Green	2	7	9
	Gray	3	7	10
	Total	20	20	40

(b) Frequencies and Percentages of Toy Colors for Each Type of Face

		Face: Happy	Face: Angry
Toy color	Red	15 75.0%	6 30.0%
	Green	2 10.0%	7 35.0%
	Gray	3 15.0%	7 35.0%
		20 100.0%	20 100.0%

FIGURE 15.5 ● BAR CHART OF THE PERCENTAGE OF TOY COLOR (RED, GREEN, GRAY) FOR EACH TYPE OF FACE (HAPPY, ANGRY)

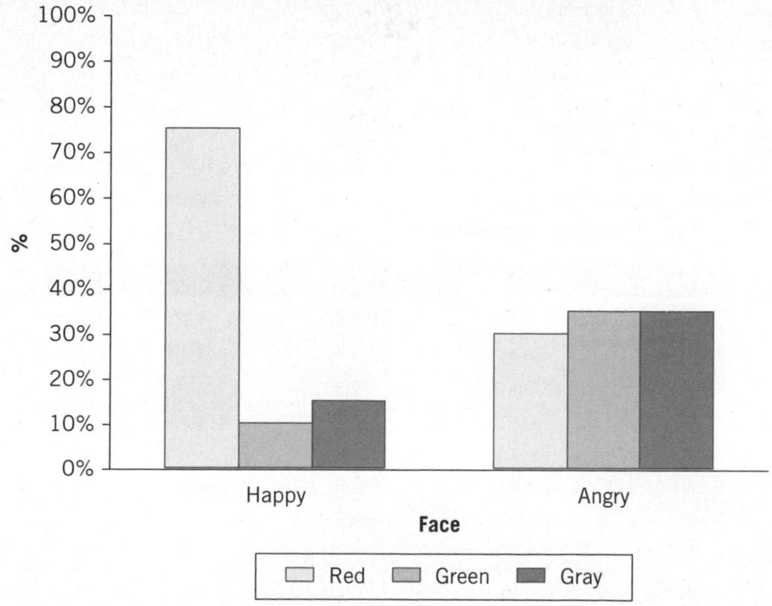

having the two columns of percentages add up to 100.0%, percentages across the three rows instead could have been calculated to represent the two faces within each of the three toy colors. For example, for the 21 infants who selected the red toy, we could have calculated the percentage of infants who were shown the happy face (15/21 or 71.4%) versus the angry face (6/21 or 29.6%). However, column percentages were calculated because the research hypothesis of the color preference study addressed differences between color preferences rather than differences between faces.

To illustrate the percentages in Table 15.3(b), a bar chart is provided in Figure 15.5. The bars in this bar chart represent the three different toy colors for the two types of faces. The toy colors have been grouped together to highlight differences in the distribution of color preferences for the two faces. Looking at the bar chart, a wide majority (75%) of the infants shown the happy face chose the red toy. However, for the infants shown the angry face, there were very little differences between the three colors. This appears to provide initial support for the research hypothesis that red would be preferred when associated with a happy face but not preferred when associated with an angry face. However, because we need to calculate an inferential statistic to determine whether these differences are statistically significant, the next section introduces a chi-square statistic used to analyze two categorical variables.

LEARNING CHECK 5:

Reviewing What You've Learned So Far

1. Review questions
   a. What is the purpose of a contingency table?
2. For each of the following situations, create a contingency table and a bar chart.
   a. A teacher wants to see whether preferences for chocolate versus vanilla ice cream are different for boys versus girls.

Boy		Girl	
Ice Cream	Ice Cream	Ice Cream	Ice Cream
Chocolate	Chocolate	Chocolate	Vanilla
Chocolate	Vanilla	Vanilla	Chocolate
Chocolate	Chocolate	Chocolate	Chocolate
Vanilla	Chocolate	Chocolate	Vanilla
Chocolate	Chocolate	Vanilla	Chocolate

   b. A college instructor examines whether freshmen, sophomores, juniors, and seniors differ in whether they tend to sit in the front or the back of the classroom:

Freshman	Sophomore	Junior	Senior
Front	Front	Back	Back
Front	Back	Back	Back
Back	Front	Front	Front
Front	Front	Front	Back
Front	Front	Back	Back
Back	Back	Front	Back
Front	Back	Back	Back

c. A political pollster examines whether opinions about a proposed law (in favor, against, unsure) are different for members of three political parties (Parties A, B, and C).

Party A		Party B		Party C	
Opinion	**Opinion**	**Opinion**	**Opinion**	**Opinion**	**Opinion**
Against	Against	In favor	Against	In favor	In favor
In favor	Unsure	Against	Unsure	In favor	In favor
Against	Against	Unsure	In favor	Unsure	In favor
Against	Unsure	In favor	Unsure	In favor	Against
Unsure	Against	Unsure		Against	In favor
Against	Against	Unsure		In favor	

15.5 INFERENTIAL STATISTIC: CHI-SQUARE TEST OF INDEPENDENCE

To test differences between observed and expected frequencies in a set of data, a chi-square statistic is calculated. When the data consist of two categorical variables, the appropriate statistical procedure is the **chi-square test of independence**. As its name suggests, the chi-square test of independence determines whether the distribution of frequencies for one variable is independent of another variable. Variables are not independent when the distribution of frequencies for one variable changes as a function of a second variable, which implies the two variables are related. For the color preference study, the question is whether the distribution of toy colors is the same for infants shown the happy face versus the angry face. If the two distributions are not the same, it can be concluded that Toy Color and Face are related to each other, such that the distribution of frequencies for toy color changes as a function of whether the child is shown a happy or angry face.

The steps involved in conducting the chi-square test of independence are the same as the other inferential statistical procedures discussed in this book:

- state the null and alternative hypotheses (H_0 and H_1),
- make a decision about the null hypothesis,
- draw a conclusion from the analysis, and
- relate the result of the analysis to the research hypothesis.

As you will see, adding a second categorical variable makes completing the steps for the chi-square test of independence slightly more complicated than the goodness-of-fit test.

State the Null and Alternative Hypotheses (H_0 and H_1)

The purpose of the chi-square test of independence is to determine whether two categorical variables are independent of each other, which is the same as determining whether the two variables are related. Therefore, the null and alternative hypotheses may be stated as the following:

H_0: The two variables are independent (i.e., there is no relationship between the two variables).

H_1: The two variables are not independent (i.e., there is a relationship between the two variables).

For the color preference example, the null hypothesis implies that an infant's color preference is independent of whether he or she is shown a happy or angry face. On the other hand, the alternative hypothesis implies that that an infant's color preference depends on the type of face he or she is shown.

Make a Decision About the Null Hypothesis

Once the null and alternative hypotheses have been stated, you move on to making a decision whether to reject the null hypothesis and conclude that the two variables are related. The steps used to make this decision for the chi-square test of independence are the same as for the goodness-of-fit test:

- calculate the degrees of freedom (*df*);
- set alpha (α), identify the critical value, and state a decision rule;
- calculate a statistic: chi-square (χ^2);
- make a decision whether to reject the null hypothesis;
- determine the level of significance; and
- calculate a measure of effect size (Cramér's ϕ).

Calculate the Degrees of Freedom (*df*)

Calculating the degrees of freedom (*df*) for two categorical variables is very similar to the situation presented in conducting the two-way ANOVA (Chapter 12). In that chapter, the degrees of freedom for the A × B interaction effect ($df_{A \times B}$) was equal to $(a - 1)(b - 1)$, where a and b were the number of groups that comprised two independent variables. Using similar logic, the degrees of freedom for the chi-square test of independence may be stated in the following way:

$$df = (\#\text{ rows} - 1)\ (\#\text{ columns} - 1) \tag{15-5}$$

where "# rows" and "# columns" relate to the contingency table created for the two variables. More specifically, the numbers of rows and columns are equal to the number of groups that comprise the two variables.

For the color preference study, the contingency table in Table 15.4(a) has three rows (Toy Color: Red, Green, Gray) and two columns (Face: Happy, Angry). Using Formula 15-5, the degrees of freedom is equal to

$$\begin{aligned} df &= (\#\text{rows} - 1)(\#\text{columns} - 1) \\ &= (3 - 1)(2 - 1) = (2)(1) \\ &= 2 \end{aligned}$$

We may state, therefore, that there are two degrees of freedom in the color preference example.

Set Alpha (α), Identify the Critical Value, and State a Decision Rule

Alpha, the probability of the statistic needed to reject the null hypothesis, is again set at the traditional value of .05. Using the table of critical values for the chi-square statistic in Table 8, the critical value for the color preference example is identified by moving down the *df* column until we reach the $df = 2$ row. Moving to the right, for $\alpha = .05$, we find a critical value of 5.99. Therefore,

$$\text{For } \alpha = .05 \text{ and } df = 2, \text{ critical value} = 5.99$$

The decision rule explicitly describes the values of the statistic leading to the rejection of the null hypothesis. For the color preference example,

$$\text{If } \chi^2 > 5.99, \text{ reject } H_0; \text{ otherwise, do not reject } H_0$$

Calculate a Statistic: Chi-square (χ^2)

Calculating the chi-square statistic for the chi-square test of independence involves the same two steps as for the goodness-of-fit test: calculating expected frequencies (f_e) and calculating the chi-square statistic (χ^2). However, the addition of a second variable changes how the expected frequencies are calculated.

Calculate expected frequencies (f_e). When there is one categorical variable, the expected frequency for a group is based on a hypothesized proportion based on theory or research. However, when there are two categorical variables, each expected frequency is based on the number of participants in the sample. More specifically, because each observed frequency is a combination of the two variables in a sample of data, the formula for an expected frequency (f_e) in the chi-square test of independence is based on these combinations:

$$f_e = \frac{(\text{row total})(\text{column total})}{N} \quad (15\text{-}6)$$

where "row total" and "column total" refer to the different rows and columns in a contingency table, and N is equal to the total sample size. Looking at Table 15.4(a), for the color preference example, the "row total" refers to the number of infants selecting a particular toy color (Red, Green, or Gray), the "column total" refers to the number of infants exposed to a particular face (Happy or Angry), and N refers to the total number of infants in the sample ($N = 40$).

To illustrate how expected frequencies are calculated for the chi-square test of independence, let's compute the expected frequency for the Happy/Red combination. Looking at the contingency table in Table 15.3(a), this combination is located in the top row (which has a total of 21 infants) and in the left column (which has a total of 20 infants). Using these two totals as well as the total sample size ($N = 40$), how many infants would be expected to be in the Happy/Red combination? The expected frequency for the Happy/Red combination is

$$\begin{aligned} f_e &= \frac{(\text{row total})(\text{column total})}{N} \\ &= \frac{(21)(20)}{40} = \frac{420}{40} \\ &= 10.50 \end{aligned}$$

Therefore, given that 21 of the 40 infants chose the red toy, and 20 of the 40 infants were exposed to the happy face, 10.50 of the 40 infants in this sample are expected to be in the Happy/Red combination.

Table 15.4 provides the expected frequencies for the six combinations in the color preference study. In this example, the calculation of the expected frequencies is made easier by the fact that the two column totals (20 and 20) are the same. The exercises at the end of this chapter contain examples where this is not the case in order to provide practice in using the correct row and column totals in your calculations.

To check that the expected frequencies have been calculated correctly, the sum of the expected frequencies for each row or column should match the corresponding row or column total. For example, for the Red toy, the sum of the expected frequencies for the Happy and Angry faces is equal to the number of infants who chose the red toy ($10.50 + 10.50 = 21$). Another check of calculations is to ensure that the sum of the expected frequencies equals the total sample size ($10.50 + 4.50 + 5.00 + 10.50 + 4.50 + 5.00 = 40$).

Calculate the chi-square statistic (χ^2). Once the expected frequencies have been determined, the chi-square statistic is calculated using the same formula (Formula 15-3) as for the goodness-of-fit test. Using the observed and expected frequencies in Table 15.5, the chi-square statistic for the color preference study is calculated as follows:

$$\begin{aligned}
\chi^2 &= \Sigma\frac{(f_o - f_e)^2}{f_e} \\
&= \frac{(15-10.50)^2}{10.50} + \frac{(2-4.50)^2}{4.50} + \frac{(3-5.00)^2}{5.00} + \frac{(6-10.50)^2}{10.50} + \frac{(7-4.50)^2}{4.50} + \frac{(7-5.00)^2}{5.00} \\
&= \frac{(4.50)^2}{10.50} + \frac{(-2.50)^2}{4.50} + \frac{(-2.00)^2}{5.00} + \frac{(-4.50)^2}{10.50} + \frac{(2.50)}{4.50} + \frac{(2.00)}{5.00} \\
&= \frac{20.25}{10.50} + \frac{6.25}{4.50} + \frac{4.00}{5.00} + \frac{20.25}{10.50} + \frac{6.25}{4.50} + \frac{4.00}{5.00} \\
&= 1.93 + 1.39 + .80 + 1.93 + 1.39 + .80 \\
&= 8.24
\end{aligned}$$

In calculating the chi-square statistic, it's important to be sure to divide each squared difference between an observed and expected frequency ($(f_o - f_e)^2$) by the correct expected frequency (f_e).

Make a Decision Whether to Reject the Null Hypothesis

For the color preference study, do you reject the null hypothesis that the two variables are independent of each other? Comparing the calculated value of the chi-square statistic with the identified critical value,

$$\chi^2 = 8.24 > 5.99 \therefore \text{ reject } H_0 \ (p < .05)$$

In this example, the decision to reject the null hypothesis implies that Toy Color and Face are related to each other such that an infant's preference for the red, green, or gray toy depends on the type of face (happy or angry) the infant is shown.

Determine the Level of Significance

Because the null hypothesis has been rejected in this example, it's appropriate to determine whether the probability of the chi-square statistic is less than .01. Turning to Table 8, for $df = 2$ and $\alpha = .01$, the critical value for

TABLE 15.4 ● TABLE OF OBSERVED FREQUENCIES (f_o) AND EXPECTED FREQUENCIES (f_e) FOR THE COLOR PREFERENCE STUDY

		Face		
		Happy	Angry	Row Total
Toy Color	Red	$f_o = 15$ $f_e = \frac{(21)(20)}{40} = 10.50$	$f_o = 6$ $f_e = \frac{(21)(20)}{40} = 10.50$	21
	Green	$f_o = 2$ $f_e = \frac{(9)(20)}{40} = 4.50$	$f_o = 7$ $f_e = \frac{(9)(20)}{40} = 4.50$	9
	Gray	$f_o = 3$ $f_e = \frac{(10)(20)}{40} = 5.00$	$f_o = 7$ $f_e = \frac{(10)(20)}{40} = 5.00$	10
	Column Total	20	20	$N = 40$

the chi-square statistic is 9.21. Comparing the color preference study's chi-square value with this critical value, the following conclusion is drawn:

$$\chi^2 = 8.24 < 9.21 \therefore p < .05 \text{ (but not} < .01)$$

The level of significance for the color preference study is illustrated in Figure 15.6. Because the chi-square statistic falls between the .05 and .01 critical values, its probability is less than .05 but *not* less than .01. Consequently, the level of significance would be reported as "$p < .05$."

Calculate a Measure of Effect Size (Cramér's ϕ)

The Cramér's ϕ measure of effect size was calculated earlier for the goodness-of-fit test. However, the formula for Cramér's ϕ for the chi-square test of independence is slightly different due to the existence of two variables rather than one variable and is presented as follows:

$$\Phi = \sqrt{\frac{\chi^2}{N(k-1)}} \quad (15\text{-}7)$$

where χ^2 is the calculated value of the chi-square statistic, N is the total sample size, and k is the smaller of the number of rows or number of columns in the contingency table. For the color preference example, the values for N and k are obtained by looking at the contingency table in Table 15.3(a). In this table, you find that N is equal to 40. Also, because the contingency table consists of three rows but only two columns, k is equal to 2. Using the value for the chi-square statistic calculated from these data, Cramér's ϕ is calculated as follows:

$$\begin{aligned}\Phi &= \sqrt{\frac{\chi^2}{N(k-1)}} \\ &= \sqrt{\frac{8.24}{40(2-1)}} = \sqrt{\frac{8.24}{40}} = \sqrt{.21} \\ &= .46\end{aligned}$$

Using Cohen's guidelines presented earlier in this chapter, the value of $\phi = .46$ for the color preference study indicates that the relationship between Toy Color and Face represents a medium effect.

Draw a Conclusion From the Analysis

The results of a chi-square test of independence such as the one in the color preference study may be reported in the following way:

> The frequencies representing the color preferences (red, green, or gray) of 40 infants shown either a happy or angry face were analyzed using the chi-square test of independence. This analysis found a significant relationship between Toy Color and Face such that the distribution of the three color preferences is different for infants shown the happy face versus the angry face, $\chi^2(2, N = 40) = 8.24$, $p < .05$, $\phi = .46$.

The above sentence provides the following information:

- the variable being analyzed ("The frequencies representing the color preferences"),
- the sample ("40 infants"),
- the variables and groups involved in the analysis ("Toy Color . . . red, green, or gray" and "Face . . . happy or angry face"),

FIGURE 15.6 ● DETERMINING THE LEVEL OF SIGNIFICANCE FOR THE COLOR PREFERENCE STUDY

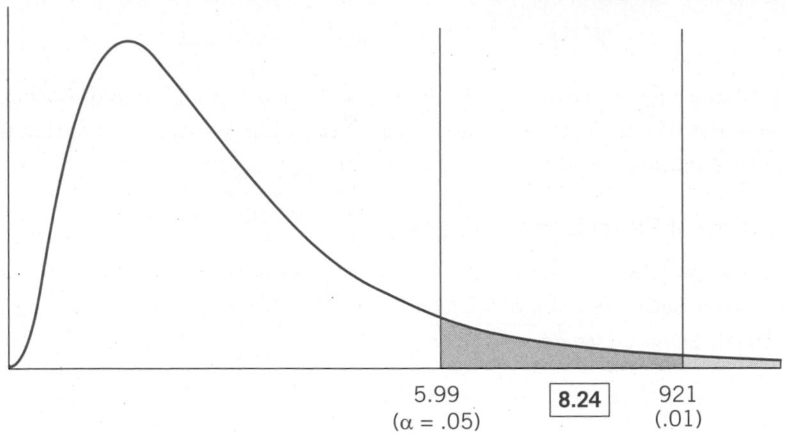

- the statistical procedure used to analyze the variable ("chi-square test of independence"),
- the nature and direction of the relationship ("a significant relationship . . . such that the distribution of the three color preferences is different for infants shown the happy face versus the angry face"), and
- information about the inferential statistic ("$\chi^2(2, N = 40) = 8.24$, $p < .05$, $\phi = .46$"), which indicates the inferential statistic calculated (χ^2), the degrees of freedom (2), the sample size ($N = 40$), the value of the statistic (8.24), the level of significance ($p < .05$), and the measure of effect size ($\phi = .46$).

Relate the Result of the Analysis to the Research Hypothesis

The results of the chi-square test of independence in the color preference study found that infants' preferences depended on which type of face they were shown. Looking at the bar chart in Figure 15.5, the red toy was preferred by the infants shown the happy face; however, there were very little differences between the three colors for the infants shown the angry face. As a result, the researchers concluded that their research hypothesis had been supported, summarizing their findings in the following way:

> Our findings indicate that infants' preference for red changes with the context in which it is presented. Specifically, in a hospitable context, red is preferred, whereas in a hostile context, red is not preferred. (Maier et al., 2009, p. 737)

LEARNING CHECK 6:

Reviewing What You've Learned So Far

1. Review questions
   a. What are the main differences between the chi-square goodness-of-fit test and the chi-square test of independence?
   b. What is implied by the null and alternative hypotheses in the chi-square test of independence?
   c. What determines the expected frequencies in the chi-square test of independence?

2. Each of the following situations consists of two variables; for each situation, calculate the degrees of freedom (*df*) and determine the critical value of χ^2 (assume $\alpha = .05$).
   a. Price (Regular, Sale) and Buying Decision (Buy, Not Buy)
   b. Size (Small, Medium, Large, Extra Large) and Perceived Value (Low, High)
   c. Location (Orchestra, Mezzanine, Balcony) and Rating (Excellent, Good, Fair, Poor)
3. Another study looking at children's color preferences examined whether color preferences may be related to one's sex (Walsh, Toma, Tuveson, & Sondhi, 1990). In this study, a group of 5- and 9-year-old children (60 boys and 60 girls) chose their favorite color of a Skittles candy (green, red, orange, or yellow). The following is a contingency table showing the distribution of preferred colors for the boys and girls in this sample:

		Color				
		Green	Red	Orange	Yellow	Total
Sex	Boy	18 30.0%	20 33.3%	13 21.7%	9 15.0%	60 100.0%
	Girl	14 23.3%	29 48.3%	10 16.7%	7 11.7%	60 100.0%
	Total	32	49	23	16	120

Conduct the chi-square test of independence to address whether there is a difference in the distribution of color preferences for boys versus girls.

   a. Construct a bar chart of these data.
   b. State the null and alternative hypotheses (H_0 and H_1).
   c. Make a decision about the null hypothesis.
      (1) Calculate the degrees of freedom (*df*).
      (2) Set alpha, identify the critical value, and state a decision rule.
      (3) Calculate a value of the chi-square statistic (χ^2).
      (4) Make a decision whether to reject the null hypothesis.
      (5) Determine the level of significance.
      (6) Calculate a measure of effect size (Cramér's ϕ).
   d. Draw a conclusion from the analysis.

Box 15.2 summarizes the steps involved in analyzing two categorical variables using the chi-square test of independence, using the color preference study as an example. The next section ends this chapter (and this book) by placing the chi-square statistic within a larger body of statistical procedures known as nonparametric statistics and illustrating the relationship between nonparametric statistics and statistics presented earlier in this book.

15.6 PARAMETRIC AND NONPARAMETRIC STATISTICAL TESTS

Throughout this book, you've seen that variables may be measured in different ways and at different levels of measurement. How a variable has been measured influences how it can be statistically analyzed. As discussed in the current chapter, when a study's variables are categorical (measured at the nominal level of measurement), the chi-square statistic is used to analyze frequencies of groups that comprise the study's variables. On the other hand, Chapters 9, 11, and 12 discussed different versions of the ANOVA, which is used when the independent variables are categorical but the dependent variable is continuous, measured at the interval or ratio level of measurement. The goal of ANOVA is to test differences between the means of groups. Furthermore, the Pearson correlation coefficient (Chapter 13), used when both variables are continuous, measures the linear relationship between variables. Based on their similarities and differences, these statistical procedures may be grouped into two broad categories known as parametric and nonparametric statistical tests. The following sections compare these two categories of tests and discuss situations in which nonparametric statistical tests might be used.

Parametric vs. Nonparametric Statistical Tests

Statistical procedures such as the *t*-test, ANOVA, and Pearson correlation are used to analyze continuous dependent variables measured at the interval or ratio level of measurement. These statistical procedures are examples of **parametric statistical tests**, which are statistical tests based on several assumptions about the populations from which samples are taken.

SUMMARY BOX 15.2 CONDUCTING THE CHI-SQUARE TEST OF INDEPENDENCE (COLOR PREFERENCE STUDY EXAMPLE)

State the null and alternative hypotheses (H_0 and H_1)

H_0: The two variables are independent (i.e., there is no relationship between the two variables).

H_1: The two variables are not independent (i.e., there is a relationship between the two variables).

Make a decision about the null hypothesis

Calculate the degrees of freedom (*df*)

$$df = (\text{\# rows} - 1)(\text{\# columns} - 1) = (3 - 1)(2 - 1) = 2$$

Set alpha (α), identify the critical value, and state a decision rule

If $\chi^2 > 5.99$, reject H_0; otherwise, do not reject H_0

Calculate a statistic: chi-square (χ^2)

Calculate expected frequencies (f_e)

$$f_e = \frac{(\text{row total})(\text{column total})}{N}$$

$$\text{Happy/Red:} \quad f_e = \frac{(21)(20)}{40} = 10.50$$

$$\text{Happy/Green:} \quad f_e = \frac{(9)(20)}{40} = 4.50$$

$$\text{Happy/Gray:} \quad f_e = \frac{(10)(20)}{40} = 5.00$$

$$\text{Angry/Red:} \quad f_e = \frac{(21)(20)}{40} = 10.50$$

$$\text{Angry/Green:} \quad f_e = \frac{(9)(20)}{40} = 4.50$$

$$\text{Angry/Gray:} \quad f_e = \frac{(10)(20)}{40} = 5.00$$

Calculate the chi-square statistic (χ^2)

$$\chi^2 = \Sigma\frac{(f_o - f_e)^2}{f_e}$$
$$= \frac{(15-10.50)^2}{10.50} + \frac{(2-4.50)^2}{4.50} + \frac{(3-5.00)^2}{5.00} + \frac{(6-10.50)^2}{10.50} + \frac{(7-4.50)^2}{4.50} + \frac{(7-5.00)^2}{5.00}$$
$$= \frac{20.25}{10.50} + \frac{6.25}{4.50} + \frac{4.00}{5.00} + \frac{20.25}{10.50} + \frac{6.25}{4.50} + \frac{4.00}{5.00}$$
$$= 1.93 + 1.39 + .80 + 1.93 + 1.39 + .80 = 8.24$$

Make a decision whether to reject the null hypothesis

$$\chi^2 = 8.24 > 5.99 \therefore \text{ reject } H_0 \ (p < .05)$$

Determine the level of significance

$$\chi^2 = 8.24 < 9.21 \therefore p < .05 \text{ (but not } < .01)$$

Calculate a measure of effect size (Cramér's ϕ)

$$\Phi = \sqrt{\frac{\chi^2}{N(k-1)}}$$
$$= \sqrt{\frac{8.24}{40(2-1)}} = \sqrt{\frac{8.24}{40}} = \sqrt{.21} = .46$$

Draw a conclusion from the analysis

The frequencies representing the color preferences (red, green, or gray) of 40 infants shown either a happy or angry face were analyzed using the chi-square test of independence. This analysis found a significant relationship between Toy Color and Face such that the distribution of the three color preferences is different for infants shown the happy face versus the angry face, $\chi^2(2, N = 40) = 8.24$, $p < .05$, $\phi = .46$.

Relate the result of the analysis to the research hypothesis

"Our findings indicate that infants' preference for red changes with the context in which it is presented. Specifically, in a hospitable context, red is preferred, whereas in a hostile context, red is not preferred" (Maier et al., 2009, p. 737).

The first assumption of parametric statistical tests is that data from a sample are being used to estimate a population parameter. For example, the null hypothesis in the *t*-test includes the population parameter μ, which is the mean of a variable in the population. The second assumption of parametric statistical tests pertains to the shape of the distribution of variables that are the basis of parameters. More specifically, parametric statistics assume that the distribution of scores for a variable is normally distributed in the larger population. In the case of ANOVA, this assumption is taken one step further in that it's assumed that the dependent variable is normally distributed for each group involved in the analysis. The third assumption of parametric tests relates to the shape of the distribution of data in samples drawn from populations. Because parametric tests assume variables are normally distributed in the population, it's also assumed that data collected from samples of the population are also normally distributed.

In contrast to parametric tests, **nonparametric statistical tests** are statistical tests *not* based on the assumptions that underlie parametric statistical tests. One important feature of nonparametric statistical tests is that they do not involve the estimation of population parameters. For example, the null hypothesis for the chi-square goodness-of-fit test (H_0: the distribution of observed frequencies fits the distribution of expected frequencies) pertains to a distribution of frequencies rather than a parameter such as μ. A second key feature of nonparametric statistical tests is that they don't make assumptions regarding the shape of the distribution of variables in populations or samples. More specifically, nonparametric tests neither assume nor require variables to be normally distributed. In the chi-square goodness-of-fit test, for example, the distribution of expected frequencies can take on any shape because it's based on hypothesized proportions derived from theory or research. In the chi-square test of independence, the expected frequencies are based on the row and column frequencies for a sample rather than on assumptions regarding how frequencies for the two variables are distributed in the population. For these reasons, the chi-square statistic is an example of a nonparametric statistical test. Because nonparametric tests do not make assumptions regarding the shape of the distribution, they are sometimes referred to as "distribution-free" tests.

Reasons for Using Nonparametric Statistical Tests

There are two primary reasons why nonparametric statistical tests may be used in a research study—these reasons relate to how variables have been measured as well as concerns regarding the shape of the distribution of scores in the sample. First, nonparametric statistical tests are used in research situations in which the dependent variable is not measured at the interval or ratio level of measurement. In this book we've introduced two nonparametric statistical tests: the chi-square statistic discussed in this chapter and the Spearman rank-order correlation in Chapter 13. The chi-square statistic is used for categorical variables measured at the nominal level of measurement. The Spearman correlation tests the relationship between two variables measured at the ordinal level of measurement.

A second situation where nonparametric statistical tests are used is when data have been collected on interval or ratio variables, but there is concern that the sample data are not normally distributed. Parametric statistical tests such as the *t*-test, *F*-ratio, and Pearson *r* assume data collected on variables is normally distributed. If this assumption is not met, incorrect conclusions may be drawn from statistical analyses conducted on the data. For example, if one or both of the variables used to calculate a Pearson correlation are severely asymmetrical (skewed) rather than symmetrical, the relationship between the two variables may not be linear and the Pearson correlation does not provide an accurate indication of the relationship between the variables. In a situation such as this, a nonparametric statistical test rather than a parametric test may be used to analyze the data.

Examples of Nonparametric Statistical Tests

Several parametric statistical tests have a nonparametric alternative. Although these nonparametric tests are designed for use with variables measured at the ordinal or nominal levels of measurement, they can also analyze variables measured at the interval or ratio level of measurement. Researchers may choose to use a nonparametric test if they are concerned about the shape of the distribution of data in their samples, particularly when their sample is small or the sample sizes of groups differ greatly from one another. In this situation, nonparametric tests are used to lower the possibility of drawing inappropriate conclusions from statistical analyses.

The left-hand column of Table 15.5 lists several parametric tests covered in this book; the right-hand column indicates the nonparametric alternative. Information regarding these nonparametric tests may be found in advanced statistical textbooks; these tests are mentioned at the conclusion of this book to give you a sense of the variety of statistical procedures available to researchers based on their particular needs.

Given the assumptions that must be met to use parametric statistical tests, you may wonder why researchers don't always use nonparametric tests, regardless of how the variables in a research study have been measured. As it turns out, when a ratio or interval variable is in fact normally distributed, the nonparametric test has less

TABLE 15.5 ● PARAMETRIC AND CORRESPONDING NONPARAMETRIC STATISTICAL TESTS

Parametric Test	Corresponding Nonparametric Test
t-test for two independent means (Chapter 9)	Mann-Whitney test
t-test for dependent means (Chapter 9)	Wilcoxon matched-pairs signed-rank test
One-way ANOVA (Chapter 11)	Kruskal-Wallis test
Pearson correlation (Chapter 13)	Spearman rank-order correlation

statistical power than its parametric counterpart, which means the nonparametric test is more likely to make a Type II error, defined in Chapter 10 as failing to detect an effect that does in fact exist. Therefore, using nonparametric tests may be an overly conservative strategy that makes it difficult to detect effects.

15.7 LOOKING AHEAD

This chapter introduced the analysis of research situations in which all of the variables in the research studies were categorical in nature. Although analyzing categorical variables differs in some ways from the analyses discussed in previous chapters, it's important to remember that the basic process of analyzing data does not change. Regardless of how variables are measured or how a research hypothesis is stated, the same steps are used to conduct and interpret statistical analyses. We've ended this book by noting a few of the many other statistical procedures used by researchers to answer the questions they have. Looking ahead, we hope this book has given you a sense of how statistics can help answer questions you may have. More important, we wish you the best of luck in identifying and exploring these questions.

15.8 Summary

The ***chi-square statistic*** (χ^2) tests the difference between ***observed frequencies***, which is the frequency of a group in the sample of data, and ***expected frequencies***, which is the frequency of a group expected to occur in a sample of data.

When the data comprise a single categorical variable, the appropriate statistical procedure is the ***chi-square goodness-of-fit*** test, which tests the "fit" between observed frequencies in a set of data and expected frequencies derived from theory or research.

There are two important assumptions of the chi-square statistic: independence of observations and minimum expected frequencies. Independence of observations implies that each observation is independent of all other observations in the set of data. Minimum expected frequencies means that none of the expected frequencies should be equal to zero, and most if not all of the expected frequencies should be greater than 5. When one or more of the expected frequencies are smaller than 5, ***Fisher's exact test*** may be used to analyze the data. If any of the expected frequencies fall between 5 and 10, ***Yates's correction for continuity*** may be used.

When there are two categorical variables, a ***contingency table*** is created, which is a table whose rows and columns represent the values of categorical variables and whose cells contain observed frequencies for combinations of the variables. In this situation, the appropriate statistical test is the ***chi-square test of independence***, which tests whether the distribution of frequencies for one variable is either independent of or dependent on another variable.

Statistical procedures such as ANOVA and Pearson correlation that use continuous dependent variables measured at the interval or ratio level of measurement are examples of ***parametric statistical tests***. Parametric statistical tests involve using data from a sample to estimate population parameters such as μ, and they assume that the distribution of scores for a variable is normally distributed in the larger population.

The chi-square statistic is an example of a ***nonparametric statistical test***, which is a test that does not involve the estimation of population parameters or assume or require variables to be normally distributed. Two primary uses of nonparametric tests are when the dependent variable is measured at a level of measurement other than interval or ratio, as well as when data have been collected from interval or ratio variables but the distribution in the sample may not be normally distributed. Several parametric statistical tests such as the *t*-test or *F*-ratio for the one-way ANOVA have a nonparametric alternative.

15.9 Important Terms

observed frequency (f_o) (p. 569)
expected frequency (f_e) (p. 570)
chi-square statistic (Pearson's chi-square statistic) (χ^2) (p. 570)
chi-square goodness-of-fit test (p. 571)
Cramér's ϕ (phi) (p. 577)
assumption of independence of observations (p. 580)
Fisher's exact test (p. 581)
Yates's correction for continuity (p. 581)
contingency table (p. 584)
chi-square test of independence (p. 587)
parametric statistical tests (p. 594)
nonparametric statistical tests (p. 596)

15.10 Formulas Introduced in This Chapter

Degrees of Freedom for the Chi-Square Goodness-of-Fit Test

$$df = \#\text{ groups} - 1 \tag{15-1}$$

Expected Frequency for the Chi-Square Goodness-of-Fit Test

$$f_e = (\text{hypothesized proportion})\ (N) \tag{15-2}$$

Chi-Square Statistic (χ^2)

$$\chi^2 = \Sigma \frac{(f_o - f_e)^2}{f_e} \tag{15-3}$$

Cramér's ϕ for the Chi-Square Goodness-of-Fit Test

$$\Phi = \sqrt{\frac{\chi^2}{N(\#\text{groups} - 1)}} \tag{15-4}$$

Degrees of Freedom for the Chi-Square Test of Independence

$$df = (\#\text{ rows} - 1)\ (\#\text{ columns} - 1) \tag{15-5}$$

Expected Frequency for the Chi-Square Test of Independence

$$f_e = \frac{(\text{row total})(\text{column total})}{N} \tag{15-6}$$

Cramér's ϕ for the Chi-Square Test of Independence

$$\Phi = \sqrt{\frac{\chi^2}{N(k-1)}} \tag{15-7}$$

15.11 Using SPSS

Chi-Square Goodness-of-Fit Test—Equal Hypothesized Proportions: The Personality Study (15.1)

1. Define variable (name, # decimals, label for the variable, labels for values of the variable) and enter data for the variable.

   NOTE: Numerically code values of the variable (i.e., 1 = Alpha, 2 = Beta, 3 = Gamma, 4 = Delta) and provide labels for these values in **Values** box within **Variable View**.

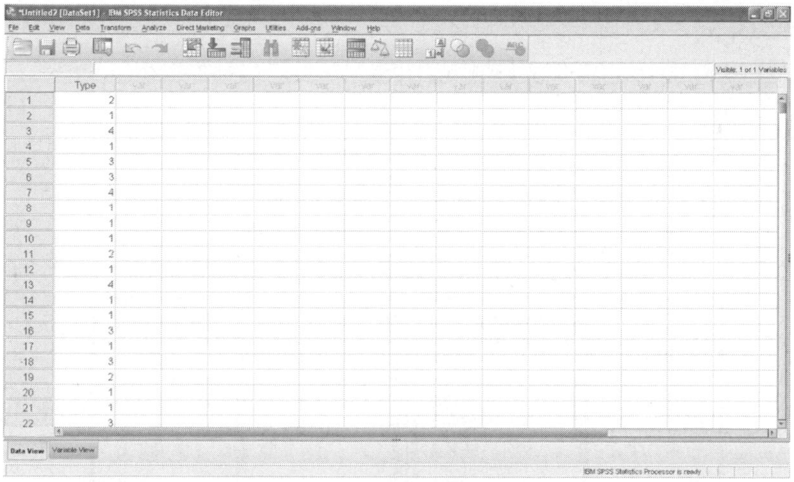

2. Select the Chi-square procedure within SPSS.

   How? (1) Click **Analyze** menu, (2) click **Nonparametric tests**, (3) click **Legacy Dialogs**, and (4) click **Chi-square**.

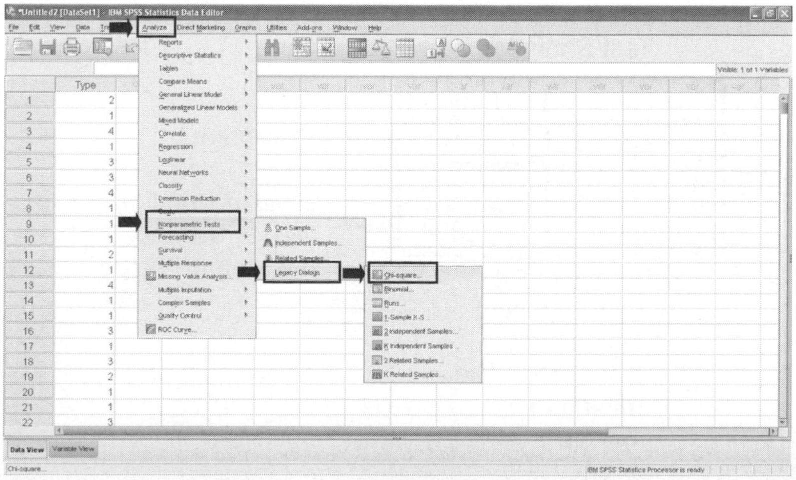

3. Select the variable to be analyzed.

   How? (1) Click variable and →, and (2) click **OK**.

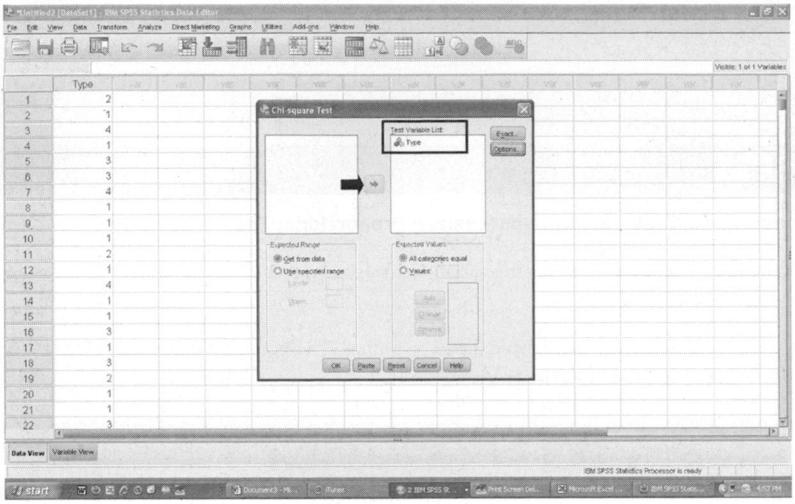

4. Examine output.

Observed frequencies → Observed N; Expected frequencies → Expected N

Type

	Observed N	Expected N	Residual
Alpha	251	147.0	104.0
Beta	65	147.0	-82.0
Gamma	194	147.0	47.0
Delta	78	147.0	-69.0
Total	588		

Test Statistics

	Type
Chi-Square	166.735[a]
df	3
Asymp. Sig.	.000

a. 0 cells (0.0%) have expected frequencies less than 5. The minimum expected cell frequency is 147.0.

Chi-Square → Chi-square

Asymp. Sig. → Level of significance (NOTE: 'Asymp. Sig.' = probability (p))

Chi-Square Goodness-of-Fit Test—Unequal Hypothesized Proportions: The Personality Study (15.3)

1. Select the Chi-square procedure within SPSS.

   How? (1) Click **Analyze** menu, (2) click **Nonparametric tests**, (3) click **Legacy Dialogs**, and (4) click **Chi-square**.

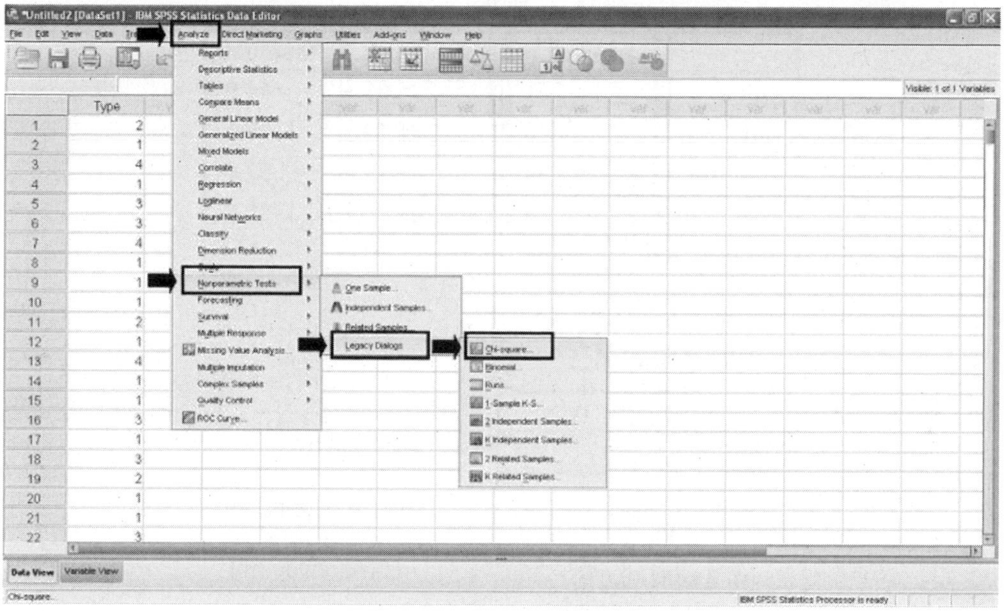

2. Select the variable to be analyzed and set hypothesized frequency for each group.

   How? (1) Click variable and →, (2) click **Values**, (3) enter hypothesized frequencies, and (4) click **OK**.

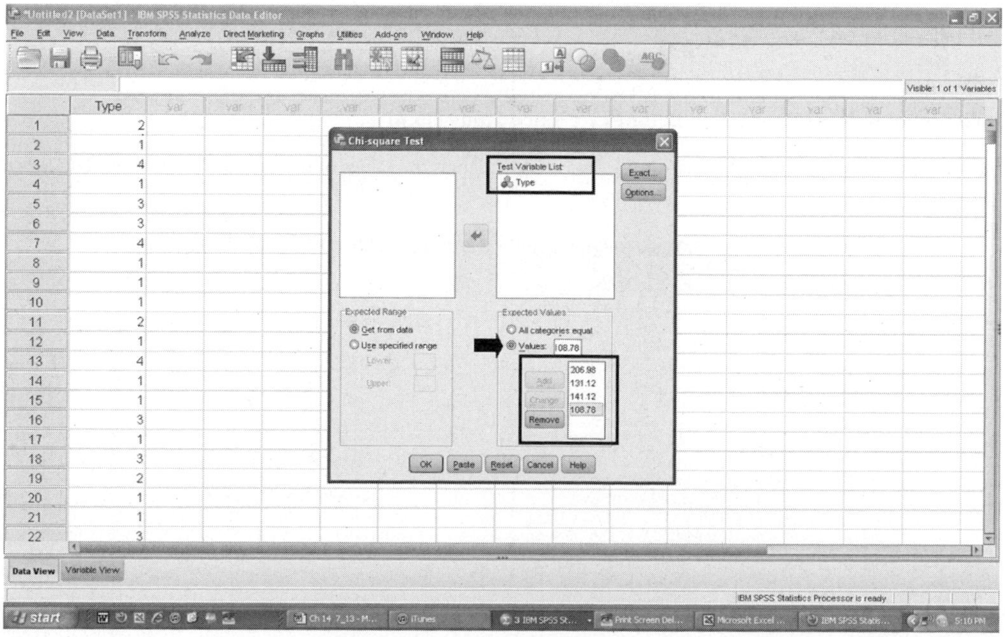

3. Examine output.

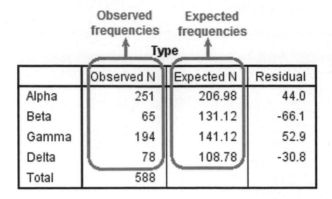

Observed frequencies → Observed N; Expected frequencies → Expected N

Type	Observed N	Expected N	Residual
Alpha	251	206.98	44.0
Beta	65	131.12	-66.1
Gamma	194	141.12	52.9
Delta	78	108.78	-30.8
Total	588		

Test Statistics

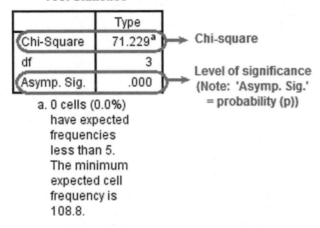

	Type
Chi-Square	71.229[a]
df	3
Asymp. Sig.	.000

a. 0 cells (0.0%) have expected frequencies less than 5. The minimum expected cell frequency is 108.8.

Chi-Square → Chi-square

Asymp. Sig. → Level of significance (Note: 'Asymp. Sig.' = probability (p))

Chi-Square Test of Independence: The Color Preference Study (15.4)

1. Define variables (names, # decimals, labels for the variables, labels for values of the variables) and enter data for the variable.

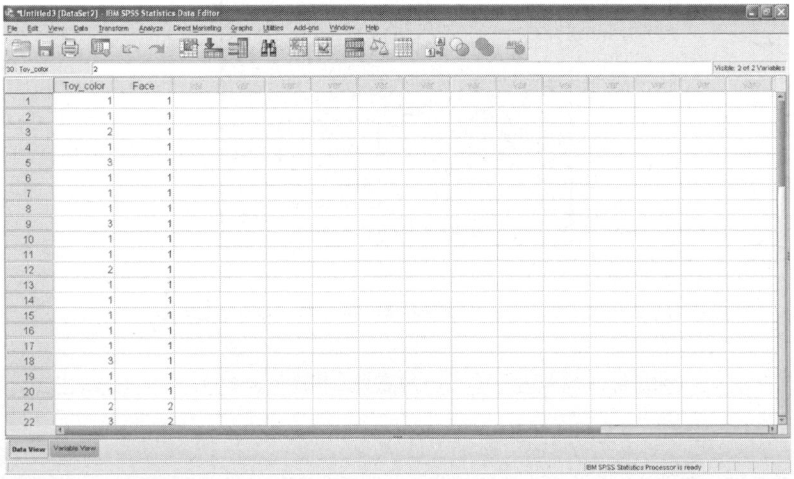

NOTE: Numerically code values of the variables (i.e., Toy Color [1 = Red, 2 = Green, 3 = Gray], Face [1 = Happy, 2 = Angry]) and provide labels for these values in **Values** box within **Variable View**.

2. Select the Chi-square procedure within SPSS.

   How? (1) Click **Analyze** menu, (2) click **Descriptive Statistics,** and (3) click **Crosstabs**.

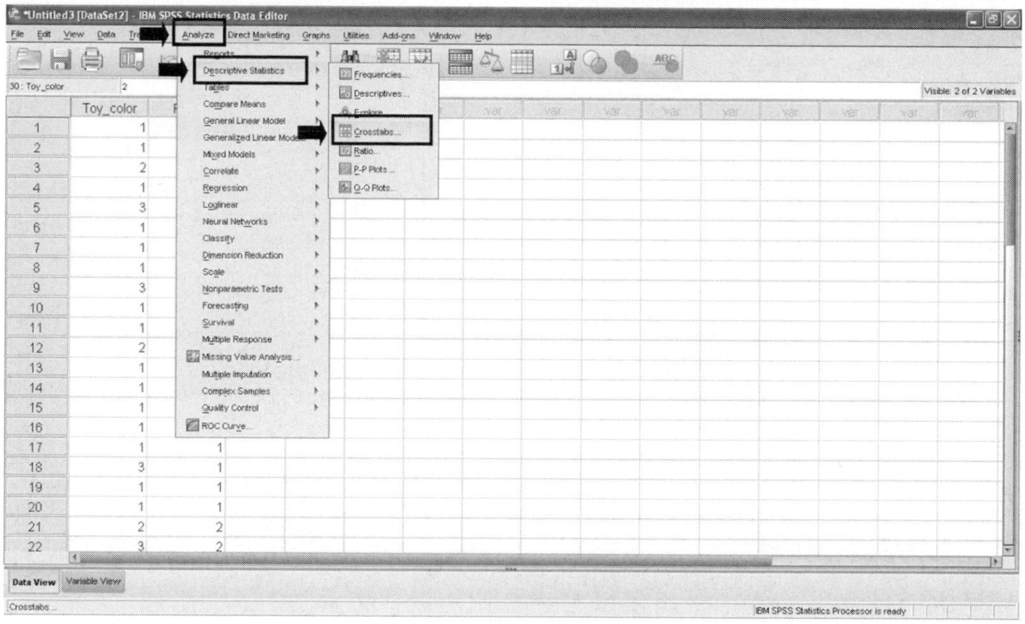

3. Assign the variables to the rows and columns of the contingency table and ask for row or column percentages.

   How? (1) Click row variable and → **Row(s)**, (2) click column variable and → **Column(s)**, (3) click **Cells**, (4) click desired Percentages, and (5) click **Continue**.

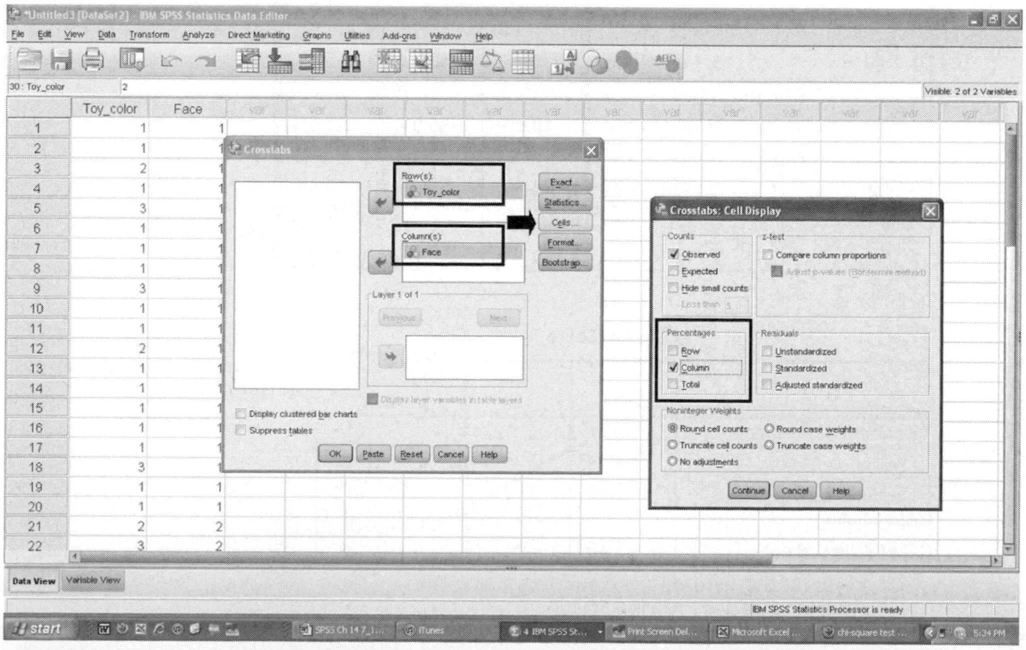

4. Ask for the chi-square statistic.

   How? (1) Click **Statistics**, (2) click **Chi-Square**, (3) click **Continue**, and (4) click **OK**.

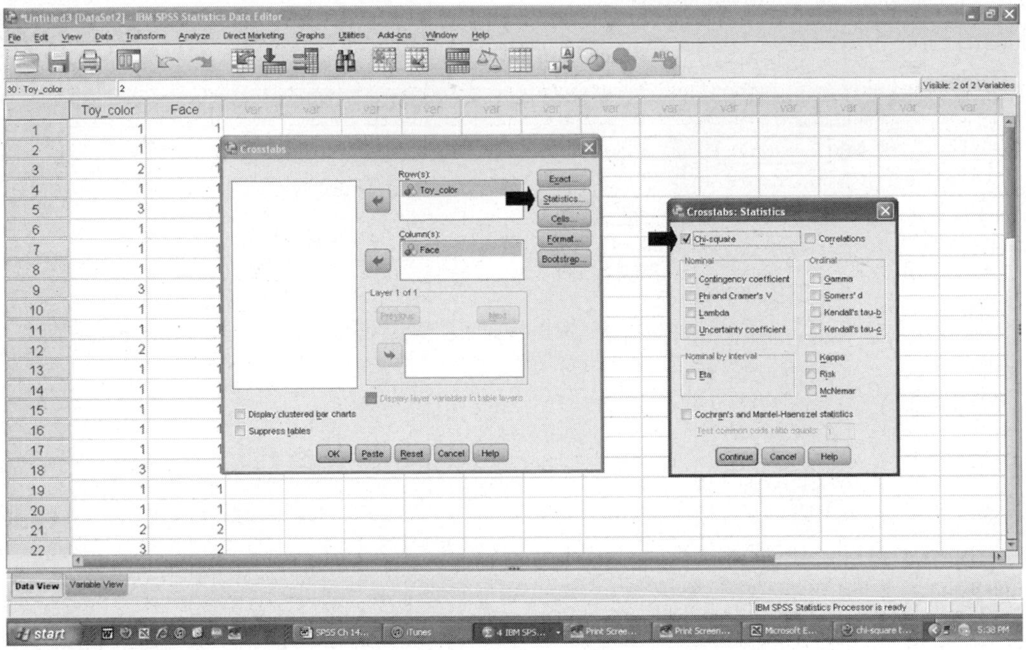

5. Examine output.

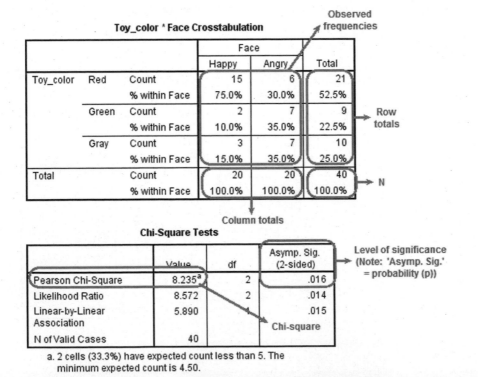

Toy_color * Face Crosstabulation

			Face: Happy	Face: Angry	Total
Toy_color	Red	Count	15	6	21
		% within Face	75.0%	30.0%	52.5%
	Green	Count	2	7	9
		% within Face	10.0%	35.0%	22.5%
	Gray	Count	3	7	10
		% within Face	15.0%	35.0%	25.0%
Total		Count	20	20	40
		% within Face	100.0%	100.0%	100.0%

Observed frequencies

Row totals

N

Column totals

Chi-Square Tests

	Value	df	Asymp. Sig. (2-sided)
Pearson Chi-Square	8.235[a]	2	.016
Likelihood Ratio	8.572	2	.014
Linear-by-Linear Association	5.890	1	.015
N of Valid Cases	40		

a. 2 cells (33.3%) have expected count less than 5. The minimum expected count is 4.50.

Chi-square

Level of significance (Note: 'Asymp. Sig.' = probability (p))

15.12 Exercises

1. The owner of a car dealership interested in assessing young drivers' preferences asks 26 college students to indicate which type of automobile they are more likely to purchase: traditional gasoline powered or hybrid. Their responses are listed below:

Type of Automobile	Type of Automobile	Type of Automobile	Type of Automobile	Type of Automobile
Gasoline	Gasoline	Hybrid	Hybrid	Hybrid
Hybrid	Hybrid	Gasoline	Hybrid	Hybrid
Hybrid	Gasoline	Hybrid	Gasoline	
Gasoline	Hybrid	Hybrid	Hybrid	
Gasoline	Hybrid	Gasoline	Hybrid	
Hybrid	Gasoline	Hybrid	Gasoline	

a. Construct a frequency distribution table for these data and a bar chart of these data (put the percentage [%] of each type of automobile on the Y-axis).

2. A student interested in consumer behavior believes women are more concerned about their personal appearance than are men; consequently, she hypothesizes that women are less likely to go to a chain hairstyling store than are men. To test her hypothesis, she stands outside of a hairstyling store for 2 hours and counts the number of men and women who purchase haircuts. Her results are listed below:

Sex	Sex	Sex	Sex	Sex	Sex
Man	Woman	Man	Man	Man	Woman
Man	Woman	Man	Woman	Man	Man
Woman	Man	Man	Man	Woman	Man
Man	Man	Woman	Man	Man	Man
Woman	Man	Man	Man	Woman	
Man	Man	Man	Woman	Man	

a. Construct a frequency distribution table for these data and a bar chart of these data (put the percentage [%] of men and women on the Y-axis).

3. A pollster asks a sample of voters to indicate whether they are in favor, undecided, or against a proposed law. Here are their answers:

Opinion	Opinion	Opinion	Opinion	Opinion
Undecided	In favor	In favor	Against	Undecided
In favor	Against	In favor	In favor	In favor
Against	In favor	Against	In favor	In favor
Undecided	Against	Undecided	Undecided	Against
Against	Undecided	In favor	Against	Against
Against	In favor	In favor	Against	In favor
In favor	In favor	Against	In favor	Undecided
In favor	Against	In favor	In favor	
Against	In favor	Against	Against	

a. Construct a frequency distribution table for these data and a bar chart of these data (put the percentage [%] of the three opinions on the Y-axis).

4. A Vermont university (www.middlebury.edu) surveyed its graduating seniors regarding their plans immediately following graduation. The majority of their responses fell into one of four categories: work full-time, go to graduate school, travel, or no plans. Below are the choices of a hypothetical sample of graduating seniors (which mirrors the actual findings):

Plans After Graduation	Plans After Graduation	Plans After Graduation	Plans After Graduation
Graduate school	Work full-time	Work full-time	Work full-time
Work full-time	Work full-time	Work full-time	Graduate school
Work full-time	No plans	Travel	Work full-time
Travel	Work full-time	Work full-time	Work full-time
Work full-time	Graduate school	Work full-time	Graduate school
No plans	Work full-time	Work full-time	Work full-time
Work full-time	Work full-time	No plans	Work full-time
Work full-time	Graduate school	Work full-time	Travel

a. Construct a frequency distribution table for these data and a bar chart of these data (put the percentage [%] of each group of seniors on the Y-axis).

5. An instructor wants to examine the distribution of grades students receive in her classes:

Grade	Grade	Grade	Grade	Grade	Grade	Grade
C	F	D	D	B	C	A
B	D	B	A	C	B	B
F	A	C	C	D	A	C
B	C	F	B	A	F	B
C	B	A	C	B	B	

a. Construct a frequency distribution table for these data and a bar chart of these data (put the percentage [%] of each grade on the Y-axis).

6. Given the importance of advertising and products related to the Super Bowl, one company (www.itsasurvey.com) commissioned a survey asking people to indicate their favorite snack to eat during the game. You're interested in seeing whether your friends expressed the same preferences, so you visit several parties and see what people are choosing to eat. Here are your findings:

Favorite Snack	Favorite Snack	Favorite Snack	Favorite Snack
Chicken fingers	Pizza	Chicken fingers	Chicken fingers
Pizza	Nachos	Pizza	Veggies/dip
Nachos	Veggies/dip	Veggies/dip	Pizza
Potato chips	Pizza	Potato chips	Veggies/dip
Pizza	Chicken fingers	Pizza	Pizza
Nachos	Potato chips	Potato chips	Potato chips
Veggies/dip	Veggies/dip	Veggies/dip	Pizza
Nachos	Chicken fingers	Pizza	Potato chips
Chicken fingers	Pizza	Veggies/dip	Veggies/dip
Pizza	Chicken fingers	Nachos	Nachos
Potato chips	Veggies/dip	Potato chips	Chicken fingers
Chicken fingers	Pizza	Pizza	Pizza

a. Construct a frequency distribution table for these data (put the percentage [%] of each snack on the Y-axis).

7. For each of the following variables, calculate the degrees of freedom (*df*) and identify the critical value ($\alpha = .05$).
   a. Drink (Coffee, Tea)
   b. Season of Year (Winter, Spring, Summer, Fall)
   c. Movie (Comedy, Drama, Romance, Adventure, Horror)
   d. Instrument (Guitar, Drums, Trumpet, Saxophone, Flute, Violin, Cello)

8. For each of the following variables, calculate the degrees of freedom (*df*) and identify the critical value ($\alpha = .05$).
   a. Direction (Left, Right, Up, Down)
   b. Shape (Circle, Triangle, Square, Rectangle, Oval, Pentagon, Star)
   c. Location (Front, Middle, Back)
   d. Pie Flavor (Apple, Peach, Strawberry, Pumpkin, Pecan, Cherry)

9. Calculate the expected frequencies (f_e) and chi-square statistic (χ^2) for the goodness-of-fit test for the data in Exercises 1 to 3 (assume equal hypothesized proportions).
   a. Exercise 1 (type of automobile)
   b. Exercise 2 (hairstyling)
   c. Exercise 3 (proposed law)

10. Calculate the expected frequencies (f_e) and chi-square statistic (χ^2) for the goodness-of-fit test for the data in Exercises 4 to 6 (assume equal hypothesized proportions).
   a. Exercise 4 (plans after graduation)
   b. Exercise 5 (class grades)
   c. Exercise 6 (Super Bowl snack)

11. "Sex in advertising prompts much discussion and media attention, typically how advertisers blend sex with their brands to gain attention" (Reichert & Lambiase, 2003, p. 120). As part of their study, they asked the following question: "Are sexual ads more likely to appear in women's or men's magazines?" (p. 125). To address this question, they examined the full-page advertisements in a sample of women's and men's magazines and found 107 sexual ads that were divided into women's and men's magazines as follows:

Type of Magazine	*f*	%
Women's	59	55.1%
Men's	48	44.9%
Total	107	100.0%

To determine whether sexual ads were more likely to be found in women's or men's magazines, conduct the chi-square goodness-of-fit test (assume equal proportions):

   a. State the null and alternative hypotheses (H_0 and H_1).
   b. Make a decision about the null hypothesis.
      (1) Calculate the degrees of freedom (*df*).
      (2) Set alpha (α), identify the critical value, and state a decision rule.
      (3) Calculate a statistic: chi-square (χ^2).
      (4) Make a decision whether to reject the null hypothesis.
      (5) Determine the level of significance.
      (6) Calculate a measure of effect size (Cramér's ϕ).
   c. Draw a conclusion from the analysis.
   d. Relate the result of the analysis to the research hypothesis.

12. An important principle in Sthapatya Veda, a holistic science of architecture, is that buildings should be designed in alignment with the earth's magnetic field and the movement of the sun. Consequently, "a southern orientation of the entrance or sleeping with the head to the north is held to create negative influences" (Travis et al., 2005, p. 555). These researchers "tested the prediction that homes with south entrances would have higher incidents of burglaries" (p. 555) by recording the addresses of 95 burglaries in the crime section of a local newspaper. Next, they went to each address and recorded whether the main entrance of the house was oriented to the east, north, west, or south. The number (f) and percentage of homes in each of the four orientations are provided below:

Orientation	*f*	%
East	20	21.1%
North	20	21.1%
West	18	18.9%
South	37	38.9%
Total	95	100.0%

Conduct the chi-square goodness-of-fit test for these data (assume an equal proportion of burglaries for the four orientations):

a. State the null and alternative hypotheses (H_0 and H_1).
b. Make a decision about the null hypothesis.
   (1) Calculate the degrees of freedom (df).
   (2) Set alpha (α), identify the critical value, and state a decision rule.
   (3) Calculate a statistic: chi-square (χ^2).
   (4) Make a decision whether to reject the null hypothesis.
   (5) Determine the level of significance.
   (6) Calculate a measure of effect size (Cramér's ϕ).
c. Draw a conclusion from the analysis.
d. Relate the result of the analysis to the research hypothesis.

13. Historians are interested in the accuracy of eyewitness accounts of traumatic events. One study examined survivors' recall of the sinking of the ship *Titanic* (Riniolo, Koledin, Drakulic, & Payne, 2003). The researchers reviewed the transcripts of survivors' testimony at governmental hearings (www.titanicinquiry.org) to see whether they testified that the ship was intact or breaking apart during the ship's final plunge (it was in fact breaking apart). Listed below is the testimony of 20 survivors:

Testimony	Testimony	Testimony	Testimony
Breaking apart	Intact	Breaking apart	Breaking apart
Intact	Breaking apart	Intact	Breaking apart
Breaking apart	Breaking apart	Breaking apart	Intact
Breaking apart	Intact	Breaking apart	Breaking apart
Breaking apart	Breaking apart	Breaking apart	Breaking apart

To test the hypothesis that survivors were able to accurately recall the state of the ship at the time of the final plunge, conduct the chi-square goodness-of-fit test for these data (assume there was an equal likelihood of saying the ship was intact or breaking apart).

a. State the null and alternative hypotheses (H_0 and H_1).
b. Make a decision about the null hypothesis.
   (1) Calculate the degrees of freedom (df).
   (2) Set alpha (α), identify the critical value, and state a decision rule.

(3) Calculate a statistic: chi-square (χ^2).

(4) Make a decision whether to reject the null hypothesis.

(5) Determine the level of significance.

(6) Calculate a measure of effect size (Cramér's ϕ).

c. Draw a conclusion from the analysis.

d. Relate the result of the analysis to the research hypothesis.

14. One study investigated what children think about people who share their money (McCrink, Bloom, & Santos, 2010). In this study, sixteen 5-year-olds were shown two puppets: One had a large number of coins, and the other had a small number of coins. Next, the two puppets gave the child some of their coins: One puppet gave the child 3 of its 12 coins, and the other puppet gave 1 of its 4 coins; although the number of coins given by the two puppets differed, the percentage was the same (i.e., 3/12 = 1/4 = 25%). The child then selected the puppet he or she thought was nicer: The "rich" puppet that started with more and gave more or the "poor" puppet that started with less and ended with less. Each child was categorized into one of three groups based on which puppet was thought to be nicer: the rich puppet, the poor puppet, or unsure. The researchers hypothesized that, even though the percentages were the same, children will think a puppet who gives a greater number of coins is nicer than a puppet who gives fewer coins. The selections for the sixteen 5-year-olds in this sample are presented below; conduct the chi-square goodness-of-fit test (assume equal hypothesized proportions):

Puppet	Puppet	Puppet	Puppet	Puppet	Puppet
Rich	Unsure	Rich	Rich	Rich	Rich
Rich	Rich	Rich	Unsure	Unsure	
Unsure	Rich	Unsure	Rich	Rich	

a. Create a frequency distribution table and bar chart of these data.

b. State the null and alternative hypotheses (H_0 and H_1).

c. Make a decision about the null hypothesis.

(1) Calculate the degrees of freedom (*df*).

(2) Set alpha (α), identify the critical value, and state a decision rule.

(3) Calculate a statistic: chi-square (χ^2).

(4) Make a decision whether to reject the null hypothesis.

(5) Determine the level of significance.

(6) Calculate a measure of effect size (Cramér's ϕ).

d. Draw a conclusion from the analysis.

e. Relate the result of the analysis to the research hypothesis.

15. Exercise 2 tested the hypothesis that women are less likely to go to a chain hairstyling store than are men. However, rather than assume men and women are equally likely to go to chain hairstyle shops, you learn that the Supercuts chain of hairstyling shops reported that 65% of their customers are men while only 35% are women (www.answers.com). Calculate the chi-square goodness-of-fit using hypothesized unequal proportions of .65 for men and .35 for women.

a. State the null and alternative hypotheses (H_0 and H_1).

b. Make a decision about the null hypothesis.

(1) Calculate the degrees of freedom (*df*).

(2) Set alpha (α), identify the critical value, and state a decision rule.

(3) Calculate a statistic: chi-square (χ^2).

(4) Make a decision whether to reject the null hypothesis.

(5) Determine the level of significance.

(6) Calculate a measure of effect size (Cramér's ϕ).

c. Draw a conclusion from the analysis.

d. Relate the result of the analysis to the research hypothesis.

16. Exercise 6 looked at people's preferences for snacks to eat during the Super Bowl. A survey commissioned by www.itsasurvey.com found that 10% preferred chicken fingers, 27% nachos, 37% pizza, 13% potato chips, and 13% veggies and dip. Calculate the chi-square goodness-of-fit test comparing your sample's choices with those from itsasurvey.

a. State the null and alternative hypotheses (H_0 and H_1).

b. Make a decision about the null hypothesis.

(1) Calculate the degrees of freedom (*df*).

(2) Set alpha (α), identify the critical value, and state a decision rule.

(3) Calculate a statistic: chi-square (χ^2).

(4) Make a decision whether to reject the null hypothesis.

(5) Determine the level of significance.

(6) Calculate a measure of effect size (Cramér's ϕ).

c. Draw a conclusion from the analysis.

17. Do people arrested for driving while intoxicated (DWI) believe they're problem drinkers? Researchers in one study asked 199 people arrested for DWI, "Have you ever thought you might have a drinking problem?" (Adams & Dennis, 2011). Nineteen of the offenders (9.5%) answered "yes" and 180 (91.5%) answered "no." To test the hypothesis that people arrested for DWI do not believe they have a drinking problem, conduct a goodness-of-fit test using hypothesized proportions of 71.8% "yes" and 28.2% "no," which are based on an expert's opinion as to the percentage of offenders in DWI programs who do in fact have drinking problems.

a. State the null and alternative hypotheses (H_0 and H_1).

b. Make a decision about the null hypothesis.

(1) Calculate the degrees of freedom (*df*).

(2) Set alpha (α), identify the critical value, and state a decision rule.

(3) Calculate a statistic: chi-square (χ^2).

(4) Make a decision whether to reject the null hypothesis.

(5) Determine the level of significance.

(6) Calculate a measure of effect size (Cramér's ϕ).

c. Draw a conclusion from the analysis.

d. Relate the result of the analysis to the research hypothesis.

18. Each of the following situations consists of two variables; for each situation, calculate the degrees of freedom (*df*) and determine the critical value of χ^2 (assume $\alpha = .05$).

a. Condition (Drug, Placebo) and Age (Young, Old)

b. Grade (Third, Sixth) and Score (Below Average, Average, Above Average)

c. Time of Day (Morning, Afternoon, Evening) and Activity (Running, Biking, Swimming)

d. Type of Car (Luxury, Sports, Sedan, Station Wagon) and Major (Business, Engineering, Social Sciences, Humanities)

19. a. Flavor (Chocolate, Vanilla, Strawberry) and Attitude (Like, Dislike)

b. Beverage (Water, Tea, Soda, Beer) and Food (American, Asian, Italian, Mexican)

c. Gender (Male, Female) and Education (Middle School, High School, College, Grad School)

d. Grade (Freshman, Sophomore, Junior, Senior) and Residence (On-campus, Off-campus, Home)

20. According to one study, deception is an important strategy in dating relationships. "If one does not honestly possess the desired characteristics to attract a mate, one must portray the image that he/she . . . has these qualities. This . . . may require some skill in deception" (Benz, Anderson, & Miller, 2005, p. 306). Do people believe men use deception more in presenting their financial situation (i.e., how much money they make) or their physical appearance (i.e., type of clothing worn)? This study hypothesized that women are more likely than men to believe that men use deception more often regarding their financial situation than their physical appearance. The contingency table of the two types of deception for the male and female respondents is presented below:

Type of Deception Used by Men

		Financial	Physical	Total
Sex	Man	20	9	29
		69.0%	31.0%	100%
	Woman	54	7	61
		88.5%	11.5%	100%
	Total	74	16	90

Use these data to address the question of whether men and women differ in their beliefs regarding men's deception.

a. Draw a bar chart of the data.

b. State the null and alternative hypotheses (H_0 and H_1).

c. Make a decision about the null hypothesis.

  (1) Calculate the degrees of freedom (*df*).

  (2) Set alpha (α), identify the critical value, and state a decision rule.

  (3) Calculate a statistic: chi-square (χ^2).

  (4) Make a decision whether to reject the null hypothesis.

  (5) Determine the level of significance.

  (6) Calculate a measure of effect size (Cramér's ϕ).

d. Draw a conclusion from the analysis.

e. Relate the result of the analysis to the research hypothesis.

21. In terms of beliefs regarding women's deception in dating situations, the Benz et al. (2005) study hypothesized that men, more so than women, believe women use deception more often regarding their physical appearance than their financial situation. Below is the contingency table for this analysis:

Type of Deception Used by Women

		Financial	Physical	Total
Sex	Man	0	29	29
		0.0%	100.0%	100%
	Woman	9	52	61
		14.8%	85.2%	100%
	Total	9	81	90

Use these data to address the question of whether men and women differ in their beliefs regarding women's deception.

a. Draw a bar chart of the data.

b. State the null and alternative hypotheses (H_0 and H_1).

c. Make a decision about the null hypothesis.

(1) Calculate the degrees of freedom (*df*).

(2) Set alpha (α), identify the critical value, and state a decision rule.

(3) Calculate a statistic: chi-square (χ^2).

(4) Make a decision whether to reject the null hypothesis.

(5) Determine the level of significance.

(6) Calculate a measure of effect size (Cramér's ϕ).

d. Draw a conclusion from the analysis.

e. Relate the result of the analysis to the research hypothesis.

22. Exercise 11 discussed a study looking at sexual ads in women's and men's magazines. Although the researchers found sexual ads were equally likely to appear in women's and men's magazines, they wondered whether different types of sexual ads appeared in the two magazines. To study this question, they coded sexual ads into one of three categories: sexual attractiveness (using the product will make the person appear more attractive), sexual behavior (using the product will increase the likelihood of sexual activity), or sex esteem (using the product will make one feel sexier). The contingency table of the three types of ads in their sample of women's and men's magazines is presented below:

		Type of Sexual Advertisement			
		Sexual Attractiveness	Sexual Behavior	Sex Esteem	Total
Type of Magazine	Women's	11	21	15	47
		23.4%	44.7%	31.9%	100%
	Men's	2	24	5	31
		6.5%	77.4%	16.1%	100%
	Total	13	45	20	78

Use these data to address the question of whether the distribution of the three types of sexual ads is different for women's and men's magazines.

a. Draw a bar chart of the data.

b. State the null and alternative hypotheses (H_0 and H_1).

c. Make a decision about the null hypothesis.

(1) Calculate the degrees of freedom (*df*).

(2) Set alpha (α), identify the critical value, and state a decision rule.

(3) Calculate a statistic: chi-square (χ^2).

(4) Make a decision whether to reject the null hypothesis.

(5) Determine the level of significance.

(6) Calculate a measure of effect size (Cramér's ϕ).

d. Draw a conclusion from the analysis.

23. The personality study discussed earlier in this chapter not only compared the distribution of four types (Alpha, Beta, Gamma, Delta) in a sample of 588 college students with a sample of college students classified during the 1980s but also compared it with a sample of 147 graduate students, saying that "we reasoned that graduate students would differ from typical undergraduates on personality characteristics that support academic achievement" (Stewart & Bernhardt,

2010, p. 582). The following contingency table shows the distribution of the types for the college students and the graduate students:

		Type				
		Alpha	Beta	Gamma	Delta	Total
Type of Student	College	251 42.7%	65 11.0%	194 33.0%	78 13.3%	588 100.0%
	Graduate	88 59.9%	16 10.9%	35 23.8%	8 5.4%	147 100.0%
	Total	339	81	229	86	735

Conduct the chi-square test of independence to address whether the distribution of the four types depends on the type of student (college student vs. graduate student).

a. Draw a bar chart of the data.

b. State the null and alternative hypotheses (H_0 and H_1).

c. Make a decision about the null hypothesis.

   (1) Calculate the degrees of freedom (*df*).

   (2) Set alpha (α), identify the critical value, and state a decision rule.

   (3) Calculate a statistic: chi-square (χ^2).

   (4) Make a decision whether to reject the null hypothesis.

   (5) Determine the level of significance.

   (6) Calculate a measure of effect size (Cramér's ϕ).

d. Draw a conclusion from the analysis.

24. Students and teachers sometimes disagree regarding what factors and information teachers should take into account in assigning grades to their students. "Although faculty and students agree that grades should reflect *achievement* performance, they do not agree on the relative impact *effort* should have on grades" (J. B. Adams, 2005, pp. 21–22). If a student's performance in a general elective course is failing but at the same time he or she puts in a lot of effort, what grade should that student receive? This study asked this question to a sample of faculty and students; their responses are organized below:

		Assigned Grade				
		B	C	D	F	Total
Group	Faculty	0 0.0%	8 13.6%	28 47.5%	23 39.0%	59 100.0%
	Student	16 10.1%	102 64.2%	38 23.9%	3 1.9%	159 100.0%
	Total	16	110	66	26	218

Conduct the chi-square test of independence to address whether there is a difference in the distribution of grades assigned by faculty versus students.

a. Draw a bar chart of the data.

b. State the null and alternative hypotheses (H_0 and H_1).

c. Make a decision about the null hypothesis.

(1) Calculate the degrees of freedom (*df*).

(2) Set alpha (α), identify the critical value, and state a decision rule.

(3) Calculate a statistic: chi-square (χ^2).

(4) Make a decision whether to reject the null hypothesis.

(5) Determine the level of significance.

(6) Calculate a measure of effect size (Cramér's ϕ).

d. Draw a conclusion from the analysis.

e. Relate the result of the analysis to the research hypothesis.

25. Do children think bedtime stories are real? One study examined whether children's beliefs regarding the reality of people or events in stories depended on the type of book (Woolley & Cox, 2007). In the study, 3-year-old children were read four books that were one of three types: realistic (people interacting with family and friends), fantastical (people interacting with monsters), or religious (people interacting with religious figures). After listening to each book, the children indicated whether they believed the people or events could exist or happen in real life. The children can be grouped into one of two categories based on the number of books with people or events they believed could exist in real life: none of the books versus one or more books. Imagine the researchers hypothesized that children were more likely to believe realistic books were real than fantastical or religious books. Listed below are the data from this study:

Realistic		Fantastical		Religious	
# Books	# Books	# Books	# Books	# Books	# Books
None	1 or more	None	None	None	None
1 or more	None	None	1 or more	None	
None	1 or more	None	None	1 or more	
None	1 or more	1 or more	None	None	
None	None	None	None	1 or more	
1 or more	1 or more	None	1 or more	None	
1 or more		None	None	None	
None		1 or more	None	1 or more	
1 or more		None	None	None	
None		1 or more	None	None	
None		None		None	

Conduct the chi-square test of independence to test whether children's beliefs regarding the reality of people or events in books depend on the type of book they are read.

a. Construct a contingency table and bar chart for these data.

b. State the null and alternative hypotheses (H_0 and H_1).

c. Make a decision about the null hypothesis.

(1) Calculate the degrees of freedom (*df*).

(2) Set alpha (α), identify the critical value, and state a decision rule.

(3) Calculate a statistic: chi-square (χ^2).

(4) Make a decision whether to reject the null hypothesis.

(5) Determine the level of significance.

(6) Calculate a measure of effect size (Cramér's ϕ).

d. Draw a conclusion from the analysis.

e. Relate the result of the analysis to the research hypothesis.

26. Another aspect of the study in Exercise 25 examined age-related changes in children's beliefs; more specifically, the study hypothesized that older children were more likely to believe that the people or events in books could really exist or happen than younger children. To test this hypothesis, children who were 3, 4, or 5 years old were read four books in which people interacted with religious figures such as God. Listed below are the number of books (none, one, two or more) that each child indicated he or she believed the people or events in the book could exist or happen in real life:

3 Years Old		4 Years Old		5 Years Old	
# Books	# Books	# Books	# Books	# Books	# Books
None	None	1	None	2 or more	2 or more
2 or more	None	None	2 or more	2 or more	None
None		2 or more	1	1	1
2 or more		None	None	2 or more	2 or more
None		None	1	None	2 or more
None		1	None	2 or more	1
None		None		2 or more	2 or more
2 or more		2 or more		1	None
None		None		2 or more	2 or more
None		None		2 or more	

Conduct the chi-square test of independence to address whether the beliefs regarding the reality of people or events in religious books depend on the age of the child.

a. Construct a contingency table and bar chart for these data.
b. State the null and alternative hypotheses (H_0 and H_1).
c. Make a decision about the null hypothesis.
   (1) Calculate the degrees of freedom (*df*).
   (2) Set alpha (α), identify the critical value, and state a decision rule.
   (3) Calculate a statistic: chi-square (χ^2).
   (4) Make a decision whether to reject the null hypothesis.
   (5) Determine the level of significance.
   (6) Calculate a measure of effect size (Cramér's ϕ).
d. Draw a conclusion from the analysis.
e. Relate the result of the analysis to the research hypothesis.

Answers to Learning Checks

Learning Check 1

2. a.

Type of Phone	*f*	%
Keyboard	4	25.0%
Touch screen	12	75.0%
Total	16	100.0%

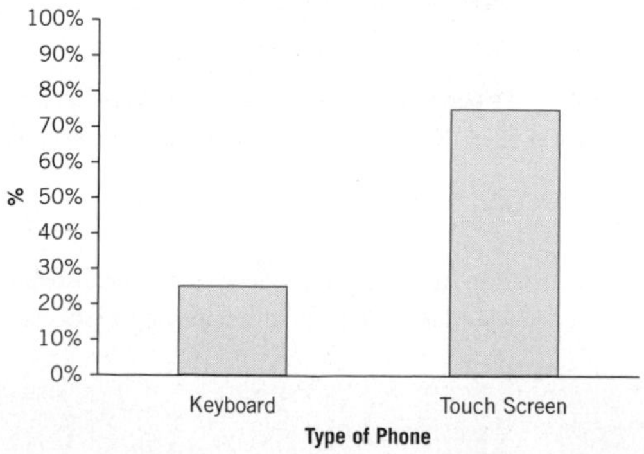

b.

Type of Milk	f	%
Whole	7	35.0%
Low fat	10	50.0%
Non-fat	3	15.0%
Total	20	100.0%

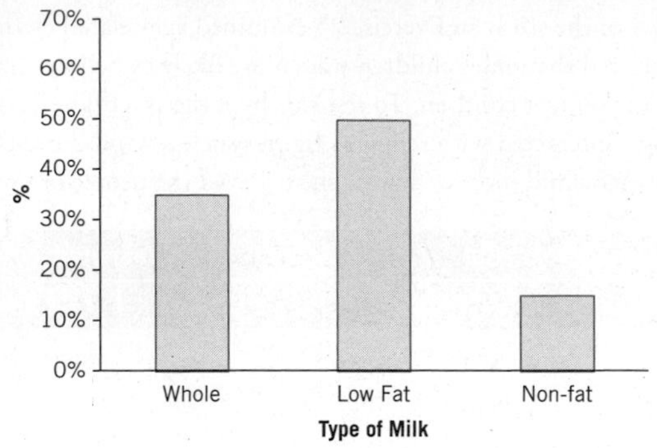

Learning Check 2

2. a. $df = 1$, critical value = 3.84
   b. $df = 2$, critical value = 5.99
   c. $df = 4$, critical value = 9.49

3. a. CCC: $f_e = 22.00$; MCCC: $f_e = 22.00$
   b. $\chi^2 = 13.10$

Learning Check 3

2. a.

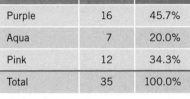

Color	f	%
Purple	16	45.7%
Aqua	7	20.0%
Pink	12	34.3%
Total	35	100.0%

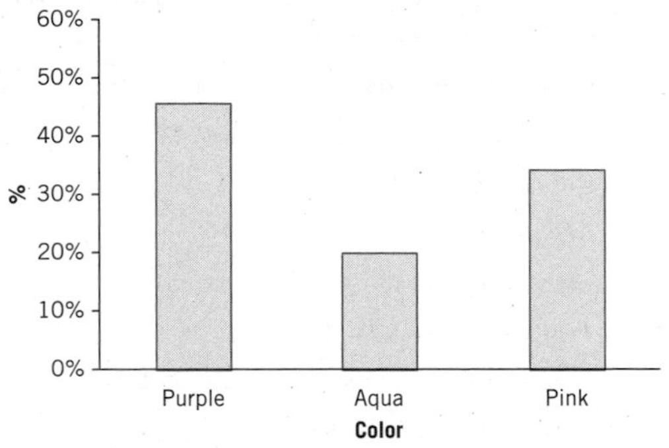

   b. H_0: The distribution of observed frequencies fits the distribution of expected frequencies. H_1: The distribution of observed frequencies does not fit the distribution of expected frequencies.
   c. (1) $df = 2$
      (2) If $\chi^2 > 5.99$, reject H_0; otherwise, do not reject H_0
      (3) Purple: $f_e = 11.67$; Aqua: $f_e = 11.67$; Pink: $f_e = 11.67$; $\chi^2 = 3.49$
      (4) $\chi^2 = 3.49 < 5.99$ $\therefore$ do not reject H_0 ($p > .05$)
      (5) Not necessary (H_0 not rejected)
      (6) $\phi = .22$
   d. In the sample of 35 people, votes for three proposed M&M colors (Purple [$f = 16$ (45.7%)], Aqua [$f = 7$ (20.0%)], and Pink [$f = 12$ (34.3%)]) were not significantly different from each other, $\chi^2(2, N = 35) = 3.49$, $p > .05$, $\phi = .22$.

Learning Check 4

2. a. H_0: The distribution of observed frequencies fits the distribution of expected frequencies. H_1: The distribution of observed frequencies does not fit the distribution of expected frequencies.

b. (1) $df = 2$

(2) If $\chi^2 > 5.99$, reject H_0; otherwise, do not reject H_0

(3) Purple: $f_e = 14.30$; Aqua: $f_e = 13.30$; Pink: $f_e = 7.00$; $\chi^2 = 6.66$

(4) $\chi^2 = 6.66 > 5.99 \therefore$ reject H_0 ($p < .05$)

(5) $\chi^2 = 6.66 < 9.21 \therefore p < .05$ (but not < .01)

(6) $\phi = .31$

c. In the sample of 35 people, votes for three proposed M&M colors (Purple [$f = 16$ (45.7%)], Aqua [$f = 7$ (20.0%)], and Pink [$f = 12$ (34.3%)]) were significantly different from frequencies based on unequal hypothesized proportions (Purple [42%], Aqua [38%], Pink [20%]), $\chi^2(2, N = 35) = 6.66, p > .05, \phi = .31$.

Learning Check 5

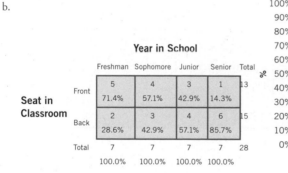

2. a.

Sex		Flavor: Chocolate	Vanilla	Total
	Boy	8	2	10
		80.0%	20.0%	100.0%
	Girl	6	4	10
		60.0%	40.0%	100.0%
	Total	14	6	20

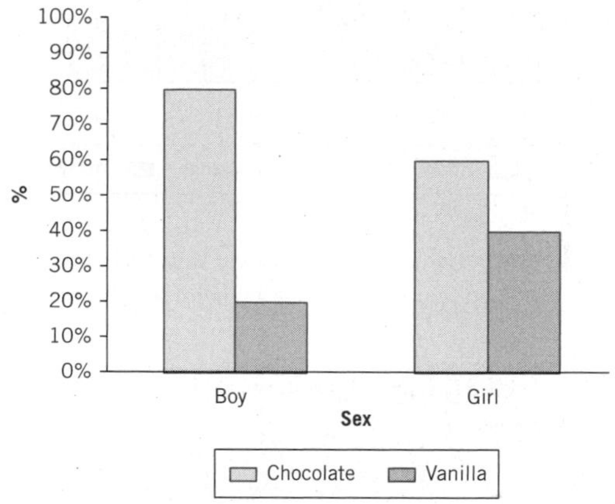

b.

Seat in Classroom		Year in School: Freshman	Sophomore	Junior	Senior	Total
	Front	5	4	3	1	13
		71.4%	57.1%	42.9%	14.3%	
	Back	2	3	4	6	15
		28.6%	42.9%	57.1%	85.7%	
	Total	7	7	7	7	28
		100.0%	100.0%	100.0%	100.0%	

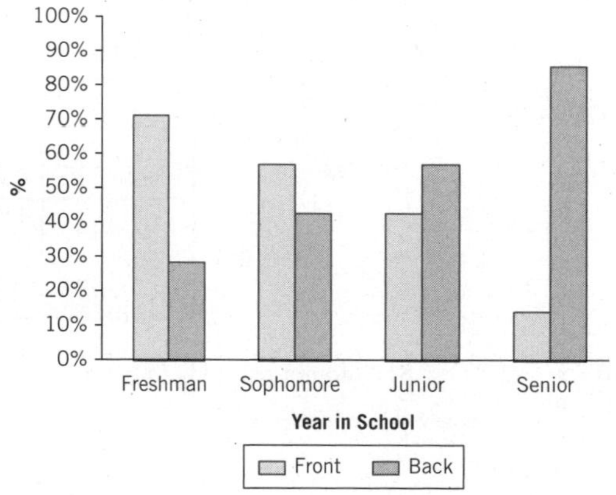

c.

Political Party		Opinion: In Favor	Against	Unsure	Total
	Party A	1	8	3	12
		8.3%	66.7%	25.0%	100.0%
	Party B	3	2	5	10
		30.0%	20.0%	50.0%	100.0%
	Party C	8	2	1	11
		72.7%	18.2%	9.1%	100.0%
	Total	12	12	9	33

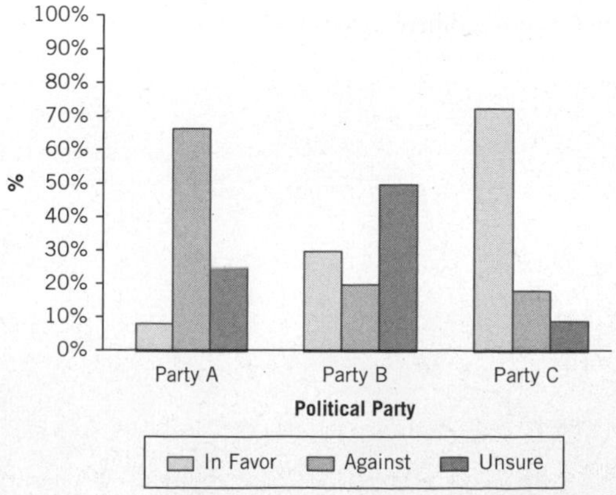

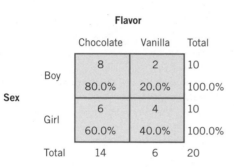

Learning Check 6

2. a. $df = 1$, critical value = 3.84
   b. $df = 3$, critical value = 7.81
   c. $df = 6$, critical value = 12.59
3. a.

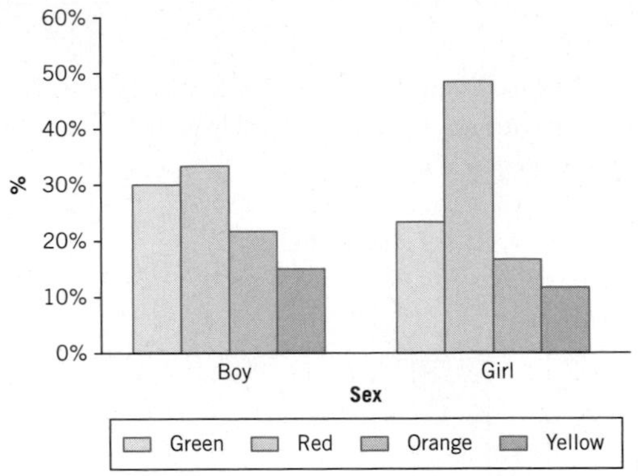

   b. H_0: The two variables are independent (i.e., there is no relationship between the two variables). H_1: The two variables are not independent (i.e., there is a relationship between the two variables).
   c. (1) $df = 3$
   (2) If $\chi^2 > 7.81$, reject H_0; otherwise, do not reject H_0
   (3) $\chi^2 = 2.79$

Color

Sex		Green	Red	Orange	Yellow
	Boy	$f_o = 18$ $f_e = 16.00$	$f_o = 20$ $f_e = 24.50$	$f_o = 13$ $f_e = 11.50$	$f_o = 9$ $f_e = 8.00$
	Girl	$f_o = 14$ $f_e = 16.00$	$f_o = 29$ $f_e = 24.50$	$f_o = 10$ $f_e = 11.50$	$f_o = 7$ $f_e = 8.00$

   (4) $\chi^2 = 2.79 < 7.81 \therefore$ do not reject H_0 ($p > .05$)
   (5) Not necessary (H_0 not rejected)
   (6) $\phi = .15$
   d. In a sample of 120 five- and nine-year-olds, the distribution of preferred colors for the 60 boys and 60 girls in this sample was not different, $\chi^2(3) = 2.79$, $p > .05$, $\phi = .15$.

Answers to Odd-Numbered Exercises

1. a.

Type of Automobile	*f*	%
Gasoline	10	38.5%
Hybrid	16	61.5%
Total	26	100.0%

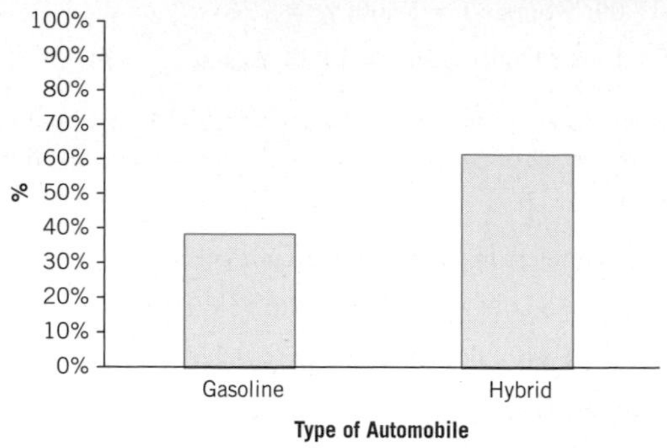

3. a.

Opinion	*f*	%
In Favor	20	46.5%
Undecided	7	16.3%
Against	16	37.2%
Total	43	100.0%

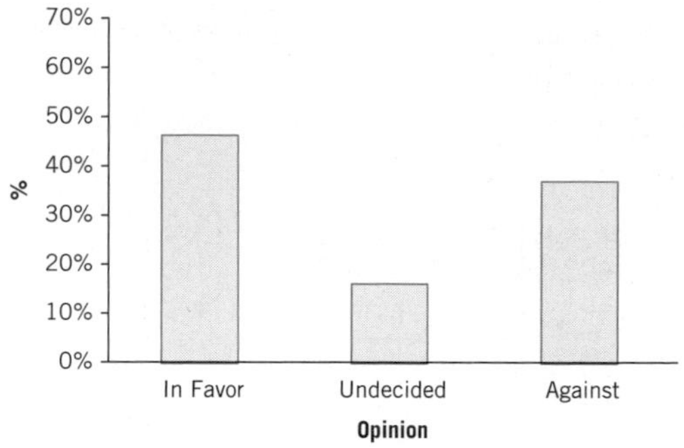

5. a.

Grade	*f*	%
A	6	17.6%
B	11	32.4%
C	9	26.5%
D	4	11.8%
F	4	11.8%
Total	34	100.0%

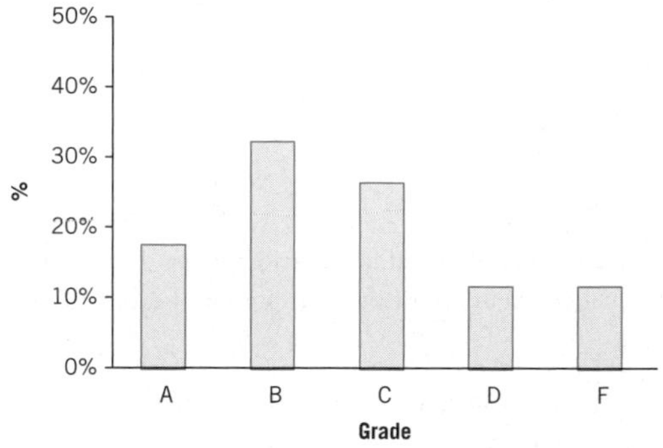

7. a. $df = 1$, critical value = 3.84
   b. $df = 3$, critical value = 7.81
   c. $df = 4$, critical value = 9.49
   d. $df = 6$, critical value = 12.59

9. a. Gasoline: $f_e = 13.00$; Hybrid: $f_e = 13.00$; $\chi^2 = 1.38$
   b. Man: $f_e = 17.00$; Woman: $f_e = 17.00$; $\chi^2 = 5.76$
   c. In favor: $f_e = 14.33$; Undecided: $f_e = 14.33$; Against: $f_e = 14.33$; $\chi^2 = 6.18$

11. a. H_0: The distribution of observed frequencies fits the distribution of expected frequencies. H_1: The distribution of observed frequencies does not fit the distribution of expected frequencies.
    b. (1) $df = 1$
       (2) If $\chi^2 > 3.84$, reject H_0; otherwise, do not reject H_0
       (3) Women's: $f_e = 53.50$; Men's: $f_e = 53.50$; $\chi^2 = 1.13$
       (4) $\chi^2 = 1.13 < 3.84$ ∴ do not reject H_0 ($p > .05$)
       (5) Not necessary (H_0 not rejected)
       (6) $\phi = .10$
    c. "Although more sexual ads appeared in women's (55%, N = 59) than in men's (45%, N = 48) magazines, the overall difference was not significant, χ^2 (1, $N = 107$) = 1.13, $p > .05$" (Reichert & Lambiase, 2003, p. 128).
    d. This analysis does not support the research hypothesis that sexual ads are more likely to appear in women's versus men's magazines.

13. a.

Testimony	f	%
Intact	5	25.0%
Breaking Up	15	75.0%
Total	20	100.0%

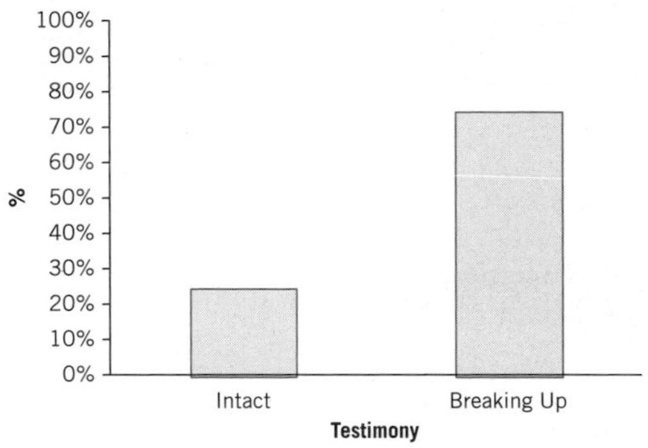

    b. H_0: The distribution of observed frequencies fits the distribution of expected frequencies. H_1: The distribution of observed frequencies does not fit the distribution of expected frequencies.
    c. (1) $df = 1$
       (2) If $\chi^2 > 3.84$, reject H_0; otherwise, do not reject H_0
       (3) Intact: $f_e = 10.00$; Breaking up: $f_e = 10.00$; $\chi^2 = 5.00$
       (4) $\chi^2 = 5.00 > 3.84$ ∴ reject H_0 ($p < .05$)
       (5) $\chi^2 = 5.00 < 6.63$ ∴ $p < .05$ (but not < .01)
       (6) $\phi = .50$
    d. In the sample of 20 *Titanic* survivors, a significantly greater number of survivors testified that the ship was breaking apart ($f = 15$ [75.0%]) than intact ($f = 5$ [25.0%]) during the ship's final plunge, $\chi^2(1, N = 20) = 5.00$, $p < .05$, $\phi = .50$.
    e. This analysis supports the research hypothesis that hypothesis that *Titanic* survivors were able to accurately recall the state of the ship at the time of the final plunge.

15. a. H_0: The distribution of observed frequencies fits the distribution of expected frequencies. H_1: The distribution of observed frequencies does not fit the distribution of expected frequencies.
    b. (1) $df = 1$
       (2) If $\chi^2 > 3.84$, reject H_0; otherwise, do not reject H_0
       (3) Men: $f_e = 22.10$; Women: $f_e = 11.90$; $\chi^2 = .46$
       (4) $\chi^2 = .46 < 3.84$ ∴ do not reject H_0 ($p > .05$)

(5) Not necessary (H_0 not rejected)

(6) $\phi = .12$

c. The number of men (24 [70.6%]) and women (10 [29.4%]) shopping at the hairstyling chain store in this sample did not differ significantly from frequencies based on hypothesized proportions of 65% for men and 35% for women, $\chi^2(1, N = 34) = .46, p > .05, \phi = .12$.

d. This analysis does not support the research hypothesis that women are less likely to go to a chain hairstyling store than are men.

17. a. H_0: The distribution of observed frequencies fits the distribution of expected frequencies. H_1: The distribution of observed frequencies does not fit the distribution of expected frequencies.

b. (1) $df = 1$

(2) If $\chi^2 > 3.84$, reject H_0; otherwise, do not reject H_0

(3) Yes: $f_e = 142.88$; No: $f_e = 56.12$; $\chi^2 = 380.88$

(4) $\chi^2 = 380.88 > 3.84 \therefore$ reject H_0 ($p < .05$)

(5) $\chi^2 = 380.88 > 6.63 \therefore p < .01$

(6) $\phi = 1.38$

c. In the sample of 199 people arrested for DWI, the percentage of people who answered "yes" ($f = 19$ [9.5%]) and "no" ($f = 180$ [91.5%]) to the question, "Have you ever thought you might have a drinking problem?" were significantly different from the expert's opinion of 71.8% "yes" and 28.2% "no," $\chi^2(1, N = 199) = 380.88, p < .01, \phi = 1.38$.

d. This analysis supports the research hypothesis that people arrested for DWI do not believe they have a drinking problem.

19. a. $df = 2$, critical value = 5.99

b. $df = 9$, critical value = 16.92

c. $df = 3$, critical value = 7.81

d. $df = 6$, critical value = 12.59

21. a.

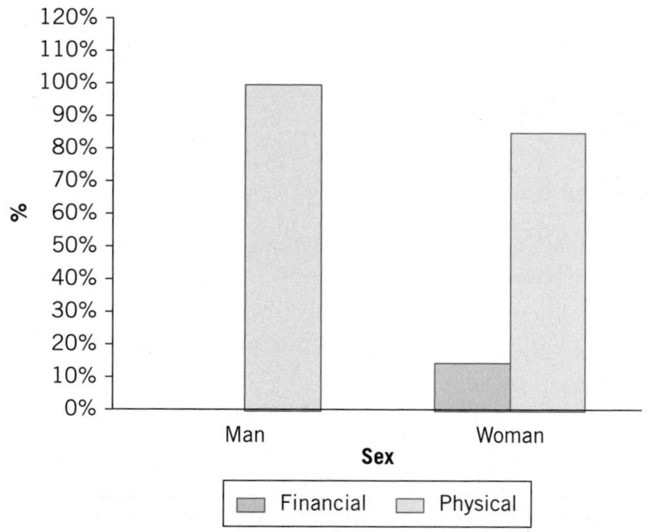

b. H_0: The two variables are independent (i.e., there is no relationship between the two variables). H_1: The two variables are not independent (i.e., there is a relationship between the two variables).

c. (1) $df = 1$

(2) If $\chi^2 > 3.84$, reject H_0; otherwise, do not reject H_0

(3) $\chi^2 = 4.75$

Type of Deception Used by Women

Sex		Financial	Physical
	Man	$f_o = 0$ $f_e = 2.90$	$f_o = 29$ $f_e = 26.10$
	Woman	$f_o = 9$ $f_e = 6.10$	$f_o = 52$ $f_e = 54.90$

(4) $\chi^2 = 4.75 > 3.84$ ∴ reject H_0 ($p < .05$)

(5) $\chi^2 = 4.75 < 6.63$ ∴ $p < .05$ (but not < .01)

(6) $\phi = .23$

d. This analysis found a significant relationship between Sex and Type of Deception used by women such that a higher percentage of female respondents (100.0%) than male respondents (85.2%) believed women were more likely to use physical deception than financial deception, χ^2 (1, N = 90) = 4.75, $p < .05$, $\phi = .23$.

e. The result of this analysis supports the hypothesis that men, more so than women, believe women use deception more often regarding their physical appearance than their financial situation.

23. a.

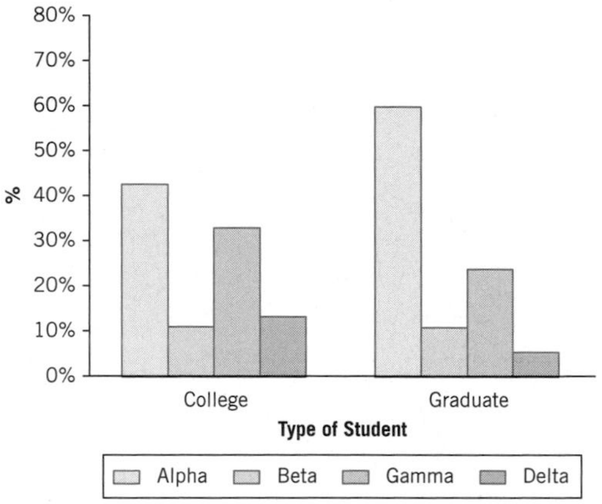

b. H_0: The two variables are independent (i.e., there is no relationship between the two variables). H_1: The two variables are not independent (i.e., there is a relationship between the two variables).

c. (1) $df = 3$

(2) If $\chi^2 > 7.81$, reject H_0; otherwise, do not reject H_0

(3) $\chi^2 = 16.86$

Type

Type of Student		Alpha	Beta	Gamma	Delta
	Women's	$f_o = 251$ $f_e = 271.20$	$f_o = 65$ $f_e = 64.80$	$f_o = 194$ $f_e = 183.20$	$f_o = 78$ $f_e = 68.80$
	Men's	$f_o = 88$ $f_e = 67.80$	$f_o = 16$ $f_e = 16.20$	$f_o = 35$ $f_e = 45.80$	$f_o = 8$ $f_e = 17.20$

(4) $\chi^2 = 16.86 > 7.81$ ∴ reject H_0 ($p < .05$)

(5) $\chi^2 = 16.86 > 11.34$ ∴ $p < .01$

(6) $\phi = .15$

d. "A Chi-square test of independence was conducted to determine if the distribution of the four Lifestyle types differed depending on graduate status (comparing 2004–08 undergraduates to 2004–08 graduates). The test was significant, showing the distribution did depend on graduate status, χ^2 (3, N=735) = 16.86, p = .001" (Stewart & Bernhardt, 2010, p. 591).

e. The result of this analysis supports the research hypothesis that the distribution of the four types depends on the type of student (college student vs. graduate student).

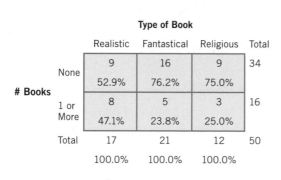

25. a.

		Type of Book			
		Realistic	Fantastical	Religious	Total
# Books	None	9 52.9%	16 76.2%	9 75.0%	34
	1 or More	8 47.1%	5 23.8%	3 25.0%	16
	Total	17 100.0%	21 100.0%	12 100.0%	50

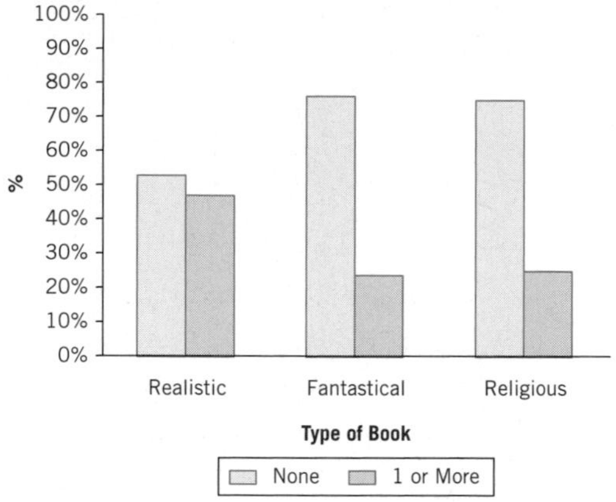

b. H_0: The two variables are independent (i.e., there is no relationship between the two variables). H_1: The two variables are not independent (i.e., there is a relationship between the two variables).

c. (1) $df = 2$

(2) If $\chi^2 > 5.99$, reject H_0; otherwise, do not reject H_0

(3) $\chi^2 = 2.69$

		Type of Book		
		Realistic	Fantastical	Religious
# Books	None	$f_o = 9$ $f_e = 11.56$	$f_o = 16$ $f_e = 14.28$	$f_o = 9$ $f_e = 8.16$
	1 or More	$f_o = 8$ $f_e = 5.44$	$f_o = 5$ $f_e = 6.72$	$f_o = 3$ $f_e = 3.84$

(4) $\chi^2 = 2.69 < 5.99$ ∴ do not reject H_0 ($p > .05$)

(5) Not necessary (H_0 not rejected)

(6) $\phi = .23$

d. In a sample of fifty 3-year-old children, a chi-square test of independence indicated a nonsignificant relationship between the type of book read to them (realistic, fantastical, or religious) and the number of books (none, one or more) they believed contained people or events that could exist or happen in real life, χ^2 (2, N = 50) = 2.69, $p > .05$, $\phi = .23$.

e. The result of this analysis does not support the research hypothesis that children are more likely to believe that realistic books are real than that fantastical or religious books are real.

TABLES

Table 1. Proportions of Area Under the Standard Normal Distribution

Table 2. Binomial Probabilities
Adapted from Burington, R. S., & May, D. C. (1970). *Handbook of probability and statistics with tables* (2nd ed.). New York, NY: McGraw-Hill.

Table 3. Critical Values of *t*
Adapted from Pearson, E. S., & Hartley, H. O. (Eds.). (1966). *Biometrika tables for statisticians* (3rd ed., Vol. 1). New York, NY: Cambridge University Press.

Table 4. Critical Values of *F*
Adapted from Pearson, E. S., & Hartley, H. O. (Eds.). (1966). *Biometrika tables for statisticians* (3rd ed., Vol. 1). New York, NY: Cambridge University Press.

Table 5. The Studentized Range Statistic (q_T)
Adapted from Pearson, E. S., & Hartley, H. O. (Eds.). (1966). *Biometrika tables for statisticians* (3rd ed., Vol. 1). New York, NY: Cambridge University Press.

Table 6. Critical Values of the Pearson Correlation (*r*)

Table 7. Critical Values of the Spearman Rank-Order Correlation (r_S)
Adapted from Ramsey, P. H. (1989). Critical values for Spearman's rank order correlation. *Journal of Educational Statistics, 14,* 245–253.

Table 8. Critical Values of Chi-Square (χ^2)
Adapted from Pearson, E. S., & Hartley, H. O. (Eds.). (1966). *Biometrika tables for statisticians* (3rd ed., Vol. 1). New York, NY: Cambridge University Press.

TABLE 1 ● PROPORTIONS OF AREA UNDER THE STANDARD NORMAL DISTRIBUTION

Instructions: The "*z*" column lists *z*-score values. The "Area between mean and *z*" column lists the proportion of the area between the mean and the *z*-score value. The "Area beyond *z*" column lists the proportion of the area beyond the *z*-score value in the tail of the distribution. Note: Because the distribution is symmetrical, areas for negative *z*-scores are the same as for positive *z*-scores.

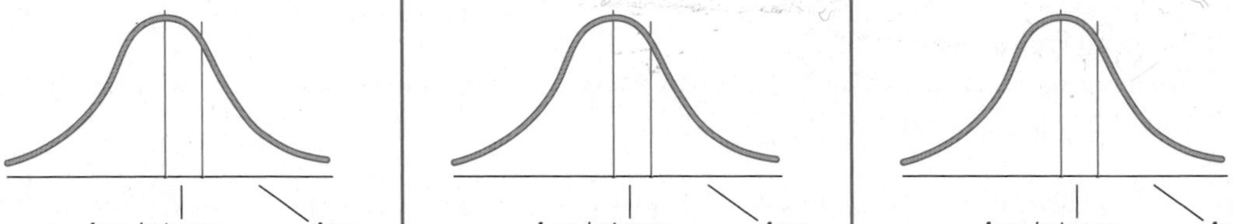

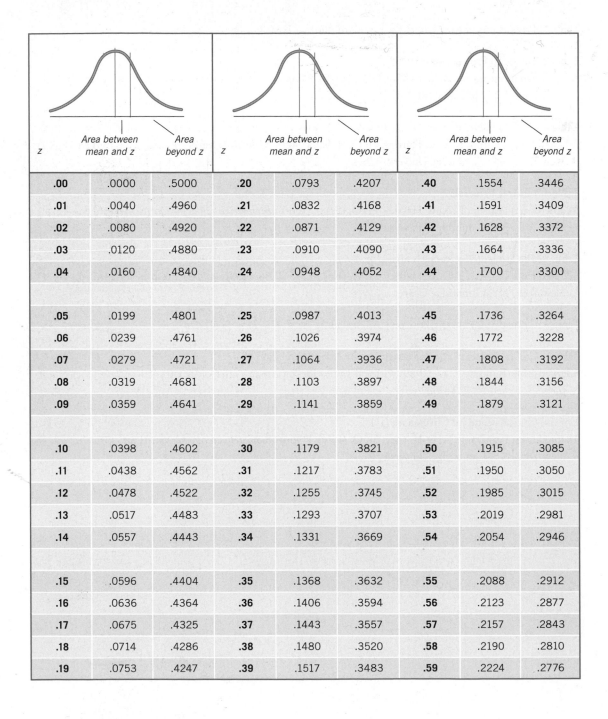

z	*Area between mean and z*	*Area beyond z*	*z*	*Area between mean and z*	*Area beyond z*	*z*	*Area between mean and z*	*Area beyond z*
.00	.0000	.5000	**.20**	.0793	.4207	**.40**	.1554	.3446
.01	.0040	.4960	**.21**	.0832	.4168	**.41**	.1591	.3409
.02	.0080	.4920	**.22**	.0871	.4129	**.42**	.1628	.3372
.03	.0120	.4880	**.23**	.0910	.4090	**.43**	.1664	.3336
.04	.0160	.4840	**.24**	.0948	.4052	**.44**	.1700	.3300
.05	.0199	.4801	**.25**	.0987	.4013	**.45**	.1736	.3264
.06	.0239	.4761	**.26**	.1026	.3974	**.46**	.1772	.3228
.07	.0279	.4721	**.27**	.1064	.3936	**.47**	.1808	.3192
.08	.0319	.4681	**.28**	.1103	.3897	**.48**	.1844	.3156
.09	.0359	.4641	**.29**	.1141	.3859	**.49**	.1879	.3121
.10	.0398	.4602	**.30**	.1179	.3821	**.50**	.1915	.3085
.11	.0438	.4562	**.31**	.1217	.3783	**.51**	.1950	.3050
.12	.0478	.4522	**.32**	.1255	.3745	**.52**	.1985	.3015
.13	.0517	.4483	**.33**	.1293	.3707	**.53**	.2019	.2981
.14	.0557	.4443	**.34**	.1331	.3669	**.54**	.2054	.2946
.15	.0596	.4404	**.35**	.1368	.3632	**.55**	.2088	.2912
.16	.0636	.4364	**.36**	.1406	.3594	**.56**	.2123	.2877
.17	.0675	.4325	**.37**	.1443	.3557	**.57**	.2157	.2843
.18	.0714	.4286	**.38**	.1480	.3520	**.58**	.2190	.2810
.19	.0753	.4247	**.39**	.1517	.3483	**.59**	.2224	.2776

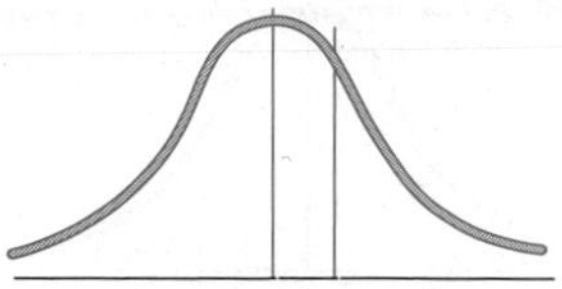
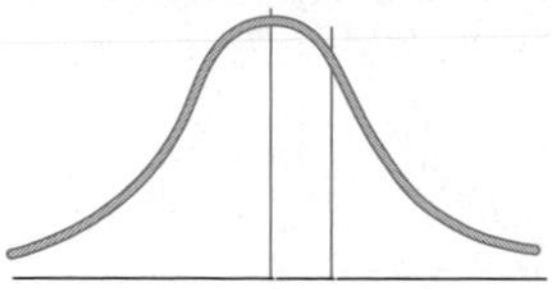
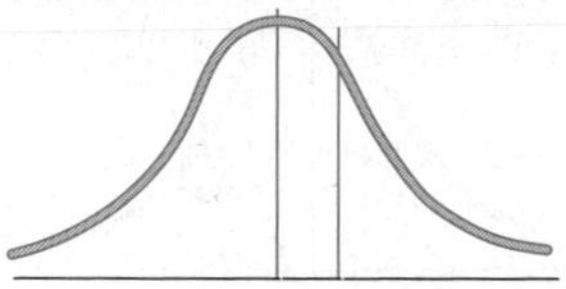

z	Area between mean and *z*	Area beyond *z*	*z*	Area between mean and *z*	Area beyond *z*	*z*	Area between mean and *z*	Area beyond *z*
.60	.2257	.2743	**.90**	.3159	.1841	**1.20**	.3849	.1151
.61	.2291	.2709	**.91**	.3186	.1814	**1.21**	.3869	.1131
.62	.2324	.2676	**.92**	.3212	.1788	**1.22**	.3888	.1112
.63	.2357	.2643	**.93**	.3238	.1762	**1.23**	.3907	.1093
.64	.2389	.2611	**.94**	.3264	.1736	**1.24**	.3925	.1075
.65	.2422	.2578	**.95**	.3289	.1711	**1.25**	.3944	.1056
.66	.2454	.2546	**.96**	.3315	.1685	**1.26**	.3962	.1038
.67	.2486	.2514	**.97**	.3340	.1660	**1.27**	.3980	.1020
.68	.2517	.2483	**.98**	.3365	.1635	**1.28**	.3997	.1003
.69	.2549	.2451	**.99**	.3389	.1611	**1.29**	.4015	.0985
.70	.2580	.2420	**1.00**	.3413	.1587	**1.30**	.4032	.0968
.71	.2611	.2389	**1.01**	.3438	.1562	**1.31**	.4049	.0951
.72	.2642	.2358	**1.02**	.3461	.1539	**1.32**	.4066	.0934
.73	.2673	.2327	**1.03**	.3485	.1515	**1.33**	.4082	.0918
.74	.2704	.2296	**1.04**	.3508	.1492	**1.34**	.4099	.0901
.75	.2734	.2266	**1.05**	.3531	.1469	**1.35**	.4115	.0885
.76	.2764	.2236	**1.06**	.3554	.1446	**1.36**	.4131	.0869
.77	.2794	.2206	**1.07**	.3577	.1423	**1.37**	.4147	.0853
.78	.2823	.2177	**1.08**	.3599	.1401	**1.38**	.4162	.0838
.79	.2852	.2148	**1.09**	.3621	.1379	**1.39**	.4177	.0823
.80	.2881	.2119	**1.10**	.3643	.1357	**1.40**	.4192	.0808
.81	.2910	.2090	**1.11**	.3665	.1335	**1.41**	.4207	.0793
.82	.2939	.2061	**1.12**	.3686	.1314	**1.42**	.4222	.0778
.83	.2967	.2033	**1.13**	.3708	.1292	**1.43**	.4236	.0764
.84	.2995	.2005	**1.14**	.3729	.1271	**1.44**	.4251	.0749
.85	.3023	.1977	**1.15**	.3749	.1251	**1.45**	.4265	.0735
.86	.3051	.1949	**1.16**	.3770	.1230	**1.46**	.4279	.0721
.87	.3078	.1922	**1.17**	.3790	.1210	**1.47**	.4292	.0708
.88	.3106	.1894	**1.18**	.3810	.1190	**1.48**	.4306	.0694
.89	.3133	.1867	**1.19**	.3830	.1170	**1.49**	.4319	.0681

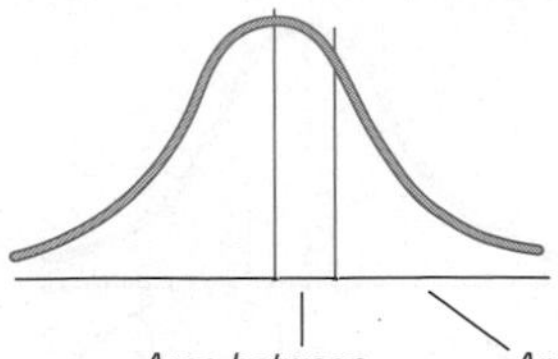
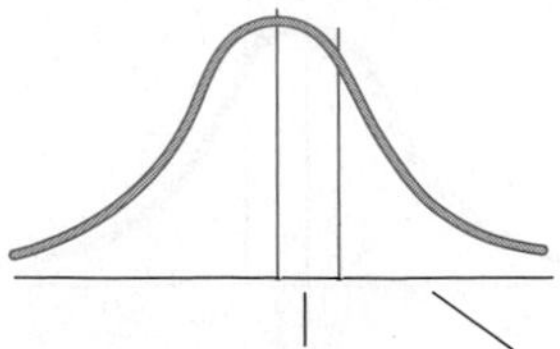
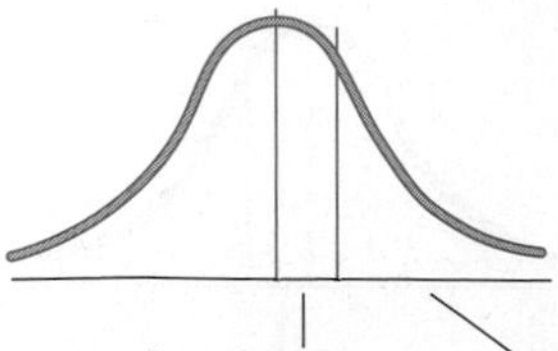

z	*Area between mean and z*	*Area beyond z*	*z*	*Area between mean and z*	*Area beyond z*	*z*	*Area between mean and z*	*Area beyond z*
1.50	.4332	.0668	**1.80**	.4641	.0359	**2.10**	.4821	.0179
1.51	.4345	.0655	**1.81**	.4649	.0351	**2.11**	.4826	.0174
1.52	.4357	.0643	**1.82**	.4656	.0344	**2.12**	.4830	.0170
1.53	.4370	.0630	**1.83**	.4664	.0336	**2.13**	.4834	.0166
1.54	.4382	.0618	**1.84**	.4671	.0329	**2.14**	.4838	.0162
1.55	.4394	.0606	**1.85**	.4678	.0322	**2.15**	.4842	.0158
1.56	.4406	.0594	**1.86**	.4686	.0314	**2.16**	.4846	.0154
1.57	.4418	.0582	**1.87**	.4693	.0307	**2.17**	.4850	.0150
1.58	.4429	.0571	**1.88**	.4699	.0301	**2.18**	.4854	.0146
1.59	.4441	.0559	**1.89**	.4706	.0294	**2.19**	.4857	.0143
1.60	.4452	.0548	**1.90**	.4713	.0287	**2.20**	.4861	.0139
1.61	.4463	.0537	**1.91**	.4719	.0281	**2.21**	.4864	.0136
1.62	.4474	.0526	**1.92**	.4726	.0274	**2.22**	.4868	.0132
1.63	.4484	.0516	**1.93**	.4732	.0268	**2.23**	.4871	.0129
1.64	.4495	.0505	**1.94**	.4738	.0262	**2.24**	.4875	.0125
1.65	.4505	.0495	**1.95**	.4744	.0256	**2.25**	.4878	.0122
1.66	.4515	.0485	**1.96**	.4750	.0250	**2.26**	.4881	.0119
1.67	.4525	.0475	**1.97**	.4756	.0244	**2.27**	.4884	.0116
1.68	.4535	.0465	**1.98**	.4761	.0239	**2.28**	.4887	.0113
1.69	.4545	.0455	**1.99**	.4767	.0233	**2.29**	.4890	.0110
1.70	.4554	.0446	**2.00**	.4772	.0228	**2.30**	.4893	.0107
1.71	.4564	.0436	**2.01**	.4778	.0222	**2.31**	.4896	.0104
1.72	.4573	.0427	**2.02**	.4783	.0217	**2.32**	.4898	.0102
1.73	.4582	.0418	**2.03**	.4788	.0212	**2.33**	.4901	.0099
1.74	.4591	.0409	**2.04**	.4793	.0207	**2.34**	.4904	.0096
1.75	.4599	.0401	**2.05**	.4798	.0202	**2.35**	.4906	.0094
1.76	.4608	.0392	**2.06**	.4803	.0197	**2.36**	.4909	.0091
1.77	.4616	.0384	**2.07**	.4808	.0192	**2.37**	.4911	.0089
1.78	.4625	.0375	**2.08**	.4812	.0188	**2.38**	.4913	.0087
1.79	.4633	.0367	**2.09**	.4817	.0183	**2.39**	.4916	.0084

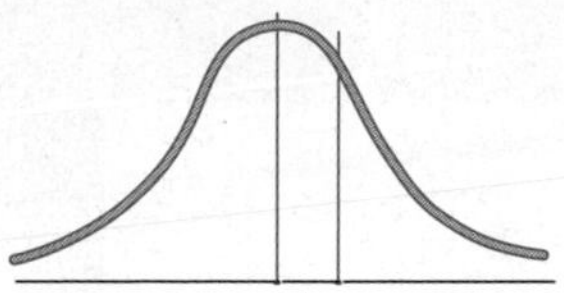

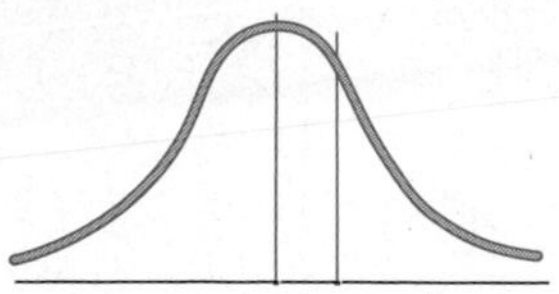

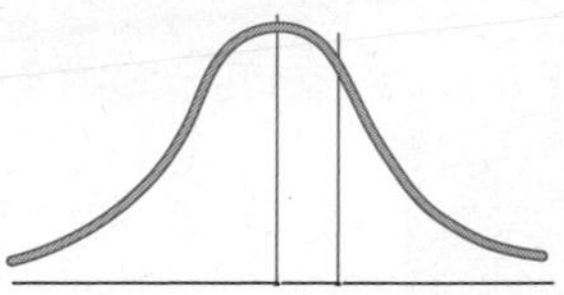

z	Area between mean and z	Area beyond z	z	Area between mean and z	Area beyond z	z	Area between mean and z	Area beyond z
2.40	.4918	.0082	**2.70**	.4965	.0035	**3.00**	.4987	.0013
2.41	.4920	.0080	**2.71**	.4966	.0034	**3.01**	.4987	.0013
2.42	.4922	.0078	**2.72**	.4967	.0033	**3.02**	.4987	.0013
2.43	.4925	.0075	**2.73**	.4968	.0032	**3.03**	.4988	.0012
2.44	.4927	.0073	**2.74**	.4969	.0031	**3.04**	.4988	.0012
2.45	.4929	.0071	**2.75**	.4970	.0030	**3.05**	.4989	.0011
2.46	.4931	.0069	**2.76**	.4971	.0029	**3.06**	.4989	.0011
2.47	.4932	.0068	**2.77**	.4972	.0028	**3.07**	.4989	.0011
2.48	.4934	.0066	**2.78**	.4973	.0027	**3.08**	.4990	.0010
2.49	.4936	.0064	**2.79**	.4974	.0026	**3.09**	.4990	.0010
2.50	.4938	.0062	**2.80**	.4974	.0026	**3.10**	.4990	.0010
2.51	.4940	.0060	**2.81**	.4975	.0025	**3.11**	.4991	.0009
2.52	.4941	.0059	**2.82**	.4976	.0024	**3.12**	.4991	.0009
2.53	.4943	.0057	**2.83**	.4977	.0023	**3.13**	.4991	.0009
2.54	.4945	.0055	**2.84**	.4977	.0023	**3.14**	.4992	.0008
2.55	.4946	.0054	**2.85**	.4978	.0022	**3.15**	.4992	.0008
2.56	.4948	.0052	**2.86**	.4979	.0021	**3.20**	.4993	.0007
2.57	.4949	.0051	**2.87**	.4979	.0021	**3.25**	.4994	.0006
2.58	.4951	.0049	**2.88**	.4980	.0020	**3.30**	.4995	.0005
2.59	.4952	.0048	**2.89**	.4981	.0019	**3.35**	.4996	.0004
2.60	.4953	.0047	**2.90**	.4981	.0019	**3.40**	.4997	.0003
2.61	.4955	.0045	**2.91**	.4982	.0018	**3.45**	.4997	.0003
2.62	.4956	.0044	**2.92**	.4982	.0018	**3.50**	.4998	.0002
2.63	.4957	.0043	**2.93**	.4983	.0017	**3.60**	.4998	.0002
2.64	.4959	.0041	**2.94**	.4984	.0016	**3.70**	.4999	.0001
2.65	.4960	.0040	**2.95**	.4984	.0016	**3.80**	.4999	.0001
2.66	.4961	.0039	**2.96**	.4985	.0015	**3.90**	.49995	.00005
2.67	.4962	.0038	**2.97**	.4985	.0015	**4.00**	.49997	.00003
2.68	.4963	.0037	**2.98**	.4986	.0014			
2.69	.4964	.0036	**2.99**	.4986	.0014			

TABLE 2 ● BINOMIAL DISTRIBUTION

		P											
	x	.01	.05	.10	.15	.20	.25	.30	.35	.40	.45	.50	
n=1	0	.9900	.9500	.9000	.8500	.8000	.7500	.7000	.6500	.6000	.5500	.5000	1
	1	.0100	.0500	.1000	.1500	.2000	.2500	.3000	.3500	.4000	.4500	.5000	0
n=2	0	.9801	.9025	.8100	.7225	.6400	.5625	.4900	.4225	.3600	.3025	.2500	2
	1	.0198	.0950	.1800	.2550	.3200	.3750	.4200	.4550	.4800	.4950	.5000	1
	2	.0001	.0025	.0100	.0225	.0400	.0625	.0900	.1225	.1600	.2025	.2500	0
n=3	0	.9703	.8574	.7290	.6141	.5120	.4219	.3430	.2746	.2160	.1664	.1250	3
	1	.0294	.1354	.2430	.3251	.3840	.4219	.4410	.4436	.4320	.4084	.3750	2
	2	.0003	.0071	.0270	.0574	.0960	.1406	.1890	.2389	.2880	.3341	.3750	1
	3		.0001	.0010	.0034	.0080	.0156	.0270	.0429	.0640	.0911	.1250	0
n=4	0	.9606	.8145	.6561	.5220	.4096	.3164	.2401	.1785	.1296	.0915	.0625	4
	1	.0388	.1715	.2916	.3685	.4096	.4219	.4116	.3845	.3456	.2995	.2500	3
	2	.0006	.0135	.0486	.0975	.1536	.2109	.2646	.3105	.3456	.3675	.3750	2
	3		.0005	.0036	.0115	.0256	.0469	.0756	.1115	.1536	.2005	.2500	1
	4			.0001	.0005	.0016	.0039	.0081	.0150	.0256	.0410	.0625	0
n=5	0	.9510	.7738	.5905	.4437	.3277	.2373	.1681	.1160	.0778	.0503	.0313	5
	1	.0480	.2036	.3281	.3915	.4096	.3955	.3602	.3124	.2592	.2059	.1563	4
	2	.0010	.0214	.0729	.1382	.2048	.2637	.3087	.3364	.3456	.3369	.3125	3
	3		.0011	.0081	.0244	.0512	.0879	.1323	.1811	.2304	.2757	.3125	2
	4			.0005	.0022	.0064	.0146	.0284	.0488	.0768	.1128	.1563	1
	5				.0001	.0003	.0010	.0024	.0053	.0102	.0185	.0313	0
		.99	.95	.90	.85	.80	.75	.70	.65	.60	.55	.50	*x*
		P											

		P											
	x	.01	.05	.10	.15	.20	.25	.30	.35	.40	.45	.50	
n=6	0	.9415	.7351	.5314	.3771	.2621	.1780	.1176	.0754	.0467	.0277	.0156	6
	1	.0571	.2321	.3543	.3993	.3932	.3560	.3025	.2437	.1866	.1359	.0938	5
	2	.0014	.0305	.0984	.1762	.2458	.2966	.3241	.3280	.3110	.2780	.2344	4
	3		.0021	.0146	.0415	.0819	.1318	.1852	.2355	.2765	.3032	.3125	3
	4		.0001	.0012	.0055	.0154	.0330	.0595	.0951	.1382	.1861	.2344	2
	5			.0001	.0004	.0015	.0044	.0102	.0205	.0369	.0609	.0938	1
	6					.0001	.0002	.0007	.0018	.0041	.0083	.0156	0
n=7	0	.9321	.6983	.4783	.3206	.2097	.1335	.0824	.0490	.0280	.0152	.0078	7
	1	.0659	.2573	.3720	.3960	.3670	.3115	.2471	.1848	.1306	.0872	.0547	6
	2	.0020	.0406	.1240	.2097	.2753	.3115	.3177	.2985	.2613	.2140	.1641	5
	3		.0036	.0230	.0617	.1147	.1730	.2269	.2679	.2903	.2918	.2734	4
	4		.0002	.0026	.0109	.0287	.0577	.0972	.1442	.1935	.2388	.2734	3
	5			.0002	.0012	.0043	.0115	.0250	.0466	.0774	.1172	.1641	2
	6				.0001	.0004	.0013	.0036	.0084	.0172	.0320	.0547	1
	7						.0001	.0002	.0006	.0016	.0037	.0078	0
n=8	0	.9227	.6634	.4305	.2725	.1678	.1001	.0576	.0319	.0168	.0084	.0039	8
	1	.0746	.2793	.3826	.3847	.3355	.2670	.1977	.1373	.0896	.0548	.0313	7
	2	.0026	.0515	.1488	.2376	.2936	.3115	.2965	.2587	.2090	.1569	.1094	6
	3	.0001	.0054	.0331	.0839	.1468	.2076	.2541	.2786	.2787	.2568	.2188	5
	4		.0004	.0046	.0185	.0459	.0865	.1361	.1875	.2322	.2627	.2734	4
	5			.0004	.0026	.0092	.0231	.0467	.0808	.1239	.1719	.2188	3
	6				.0002	.0011	.0038	.0100	.0217	.0413	.0703	.1094	2
	7					.0001	.0004	.0012	.0033	.0079	.0164	.0313	1
	8							.0001	.0002	.0007	.0017	.0039	0
		.99	.95	.90	.85	.80	.75	.70	.65	.60	.55	.50	x
		P											

		P											
	x	.01	.05	.10	.15	.20	.25	.30	.35	.40	.45	.50	
n=9	0	.9135	.6302	.3874	.2316	.1342	.0751	.0404	.0207	.0101	.0046	.0020	9
	1	.0830	.2985	.3874	.3679	.3020	.2253	.1556	.1004	.0605	.0339	.0176	8
	2	.0034	.0629	.1722	.2597	.3020	.3003	.2668	.2162	.1612	.1110	.0703	7
	3	.0001	.0077	.0446	.1069	.1762	.2336	.2668	.2716	.2508	.2119	.1641	6
	4		.0006	.0074	.0283	.0661	.1168	.1715	.2194	.2508	.2600	.2461	5
	5			.0008	.0050	.0165	.0389	.0735	.1181	.1672	.2128	.2461	4
	6			.0001	.0006	.0028	.0087	.0210	.0424	.0743	.1160	.1641	3
	7					.0003	.0012	.0039	.0098	.0212	.0407	.0703	2
	8						.0001	.0004	.0013	.0035	.0083	.0176	1
	9								.0001	.0003	.0008	.0020	0
n=10	0	.9044	.5987	.3487	.1969	.1074	.0563	.0282	.0135	.0060	.0025	.0010	10
	1	.0914	.3151	.3874	.3474	.2684	.1877	.1211	.0725	.0403	.0207	.0098	9
	2	.0042	.0746	.1937	.2759	.3020	.2816	.2335	.1757	.1209	.0763	.0439	8
	3	.0001	.0105	.0574	.1298	.2013	.2503	.2668	.2522	.2150	.1665	.1172	7
	4		.0010	.0112	.0401	.0881	.1460	.2001	.2377	.2508	.2384	.2051	6
	5		.0001	.0015	.0085	.0264	.0584	.1029	.1536	.2007	.2340	.2461	5
	6			.0001	.0012	.0055	.0162	.0368	.0689	.1115	.1596	.2051	4
	7				.0001	.0008	.0031	.0090	.0212	.0425	.0746	.1172	3
	8					.0001	.0004	.0014	.0043	.0106	.0229	.0439	2
	9							.0001	.0005	.0016	.0042	.0098	1
	10									.0001	.0003	.0010	0
		.99	.95	.90	.85	.80	.75	.70	.65	.60	.55	.50	x
		P											

		P											
	x	.01	.05	.10	.15	.20	.25	.30	.35	.40	.45	.50	
n=11	0	.8953	.5688	.3138	.1673	.0859	.0422	.0198	.0088	.0036	.0014	.0005	11
	1	.0995	.3293	.3835	.3248	.2362	.1549	.0932	.0518	.0266	.0125	.0054	10
	2	.0050	.0867	.2131	.2866	.2953	.2581	.1998	.1395	.0887	.0513	.0269	9
	3	.0002	.0137	.0710	.1517	.2215	.2581	.2568	.2254	.1774	.1259	.0806	8
	4		.0014	.0158	.0536	.1107	.1721	.2201	.2428	.2365	.2060	.1611	7
	5		.0001	.0025	.0132	.0388	.0803	.1321	.1830	.2207	.2360	.2256	6
	6			.0003	.0023	.0097	.0268	.0566	.0985	.1471	.1931	.2256	5
	7				.0003	.0017	.0064	.0173	.0379	.0701	.1128	.1611	4
	8					.0002	.0011	.0037	.0102	.0234	.0462	.0806	3
	9						.0001	.0005	.0018	.0052	.0126	.0269	2
	10								.0002	.0007	.0021	.0054	1
	11										.0002	.0005	0
n=12	0	.8864	5404	.2824	.1422	.0687	.0317	.0138	.0057	.0022	.0008	.0002	12
	1	.1074	.3413	.3766	.3012	.2062	.1267	.0712	.0368	.0174	.0075	.0029	11
	2	.0060	.0988	.2301	.2924	.2835	.2323	.1678	.1088	.0639	.0339	.0161	10
	3	.0002	.0173	.0852	.1720	.2362	.2581	.2397	.1954	.1419	.0923	.0537	9
	4		.0021	.0213	.0683	.1329	.1936	.2311	.2367	.2128	.1700	.1208	8
	5		.0002	.0038	.0193	.0532	.1032	.1585	.2039	.2270	.2225	.1934	7
	6			.0005	.0040	.0155	.0401	.0792	.1281	.1766	.2124	.2256	6
	7				.0006	.0033	.0115	.0291	.0591	.1009	.1489	.1934	5
	8				.0001	.0005	.0024	.0078	.0199	.0420	.0762	.1208	4
	9					.0001	.0004	.0015	.0048	.0125	.0277	.0537	3
	10							.0002	.0008	.0025	.0068	.0161	2
	11								.0001	.0003	.0010	.0029	1
	12										.0001	.0002	0
		.99	.95	.90	.85	.80	.75	.70	.65	.60	.55	.50	x
		P											

		P											
	x	.01	.05	.10	.15	.20	.25	.30	.35	.40	.45	.50	
n=13	0	.8775	.5133	.2542	.1209	.0550	.0238	.0097	.0037	.0013	.0004	.0001	13
	1	.1152	.3512	.3672	.2774	.1787	.1029	.0540	.0259	.0113	.0045	.0016	12
	2	.0070	.1109	.2448	.2937	.2680	.2059	.1388	.0836	.0453	.0220	.0095	11
	3	.0003	.0214	.0997	.1900	.2457	.2517	.2181	.1651	.1107	.0660	.0349	10
	4		.0028	.0277	.0838	.1535	.2097	.2337	.2222	.1845	.1350	.0873	9
	5		.0003	.0055	.0266	.0691	.1258	.1803	.2154	.2214	.1989	.1571	8
	6			.0008	.0063	.0230	.0559	.1030	.1546	.1968	.2169	.2095	7
	7			.0001	.0011	.0058	.0186	.0442	.0833	.1312	.1775	.2095	6
	8				.0001	.0011	.0047	.0142	.0336	.0656	.1089	.1571	5
	9					.0001	.0009	.0034	.0101	.0243	.0495	.0873	4
	10						.0001	.0006	.0022	.0065	.0162	.0349	3
	11							.0001	.0003	.0012	.0036	.0095	2
	12									.0001	.0005	.0016	1
	13											.0001	0
		.99	.95	.90	.85	.80	.75	.70	.65	.60	.55	.50	x
		P											

		P											
	x	.01	.05	.10	.15	.20	.25	.30	.35	.40	.45	.50	
n=14	0	.8687	.4877	.2288	.1028	.0440	.0178	.0068	.0024	.0008	.0002	.0001	14
	1	.1229	.3593	.3559	.2539	.1539	.0832	.0407	.0181	.0073	.0027	.0009	13
	2	.0081	.1229	.2570	.2912	.2501	.1802	.1134	.0634	.0317	.0141	.0056	12
	3	.0003	.0259	.1142	.2056	.2501	.2402	.1943	.1366	.0845	.0462	.0222	11
	4		.0037	.0349	.0998	.1720	.2202	.2290	.2022	.1549	.1040	.0611	10
	5		.0004	.0078	.0352	.0860	.1468	.1963	.2178	.2066	.1701	.1222	9
	6			.0013	.0093	.0322	.0734	.1262	.1759	.2066	.2088	.1833	8
	7			.0002	.0019	.0092	.0280	.0618	.1082	.1574	.1952	.2095	7
	8				.0003	.0020	.0082	.0232	.0510	.0918	.1398	.1833	6
	9					.0003	.0018	.0066	.0183	.0408	.0762	.1222	5
	10						.0003	.0014	.0049	.0136	.0312	.0611	4
	11							.0002	.0010	.0033	.0093	.0222	3
	12								.0001	.0005	.0019	.0056	2
	13									.0001	.0002	.0009	1
	14											.0001	0
		.99	.95	.90	.85	.80	.75	.70	.65	.60	.55	.50	x
		P											

		P											
	x	.01	.05	.10	.15	.20	.25	.30	.35	.40	.45	.50	
n=15	0	.8601	.4633	.2059	.0874	.0352	.0134	.0047	.0016	.0005	.0001		15
	1	.1303	.3658	.3432	.2312	.1319	.0668	.0305	.0126	.0047	.0016	.0005	14
	2	.0092	.1348	.2669	.2856	.2309	.1559	.0916	.0476	.0219	.0090	.0032	13
	3	.0004	.0307	.1285	.2184	.2501	.2252	.1700	.1110	.0634	.0318	.0139	12
	4		.0049	.0428	.1156	.1876	.2252	.2186	.1792	.1268	.0780	.0417	11
	5		.0006	.0105	.0449	.1032	.1651	.2061	.2123	.1859	.1404	.0916	10
	6			.0019	.0132	.0430	.0917	.1472	.1906	.2066	.1914	.1527	9
	7			.0003	.0030	.0138	.0393	.0811	.1319	.1771	.2013	.1964	8
	8				.0005	.0035	.0131	.0348	.0710	.1181	.1647	.1964	7
	9				.0001	.0007	.0034	.0116	.0298	.0612	.1048	.1527	6
	10					.0001	.0007	.0030	.0096	.0245	.0515	.0916	5
	11						.0001	.0006	.0024	.0074	.0191	.0417	4
	12							.0001	.0004	.0016	.0052	.0139	3
	13								.0001	.0003	.0010	.0032	2
	14										.0001	.0005	1
	15												0
		.99	.95	.90	.85	.80	.75	.70	.65	.60	.55	.50	x
		P											

		P											
	x	.01	.05	.10	.15	.20	.25	.30	.35	.40	.45	.50	
n=16	0	.8515	.4401	.1853	.0743	.0281	.0100	.0033	.0010	.0003	.0001		16
	1	.1376	.3706	.3294	.2097	.1126	.0535	.0228	.0087	.0030	.0009	.0002	15
	2	.0104	.1463	.2745	.2775	.2111	.1336	.0732	.0353	.0150	.0056	.0018	14
	3	.0005	.0359	.1423	.2285	.2463	.2079	.1465	.0888	.0468	.0215	.0085	13
	4		.0061	.0514	.1311	.2001	.2252	.2040	.1553	.1014	.0572	.0278	12
	5		.0008	.0137	.0555	.1201	.1802	.2099	.2008	.1623	.1123	.0667	11
	6		.0001	.0028	.0180	.0550	.1101	.1649	.1982	.1983	.1684	.1222	10
	7			.0004	.0045	.0197	.0524	.1010	.1524	.1889	.1969	.1746	9
	8			.0001	.0009	.0055	.0197	.0487	.0923	.1417	.1812	.1964	8
	9				.0001	.0012	.0058	.0185	.0442	.0840	.1318	.1746	7
	10					.0002	.0014	.0056	.0167	.0392	.0755	.1222	6
	11						.0002	.0013	.0049	.0142	.0337	.0667	5
	12							.0002	.0011	.0040	.0115	.0278	4
	13							.0000	.0002	.0008	.0029	.0085	3
	14									.0001	.0005	.0018	2
	15										.0001	.0002	1
	16												0
		.99	.95	.90	.85	.80	.75	.70	.65	.60	.55	.50	x
		P											

		P											
	x	.01	.05	.10	.15	.20	.25	.30	.35	.40	.45	.50	
n=17	0	.8429	.4181	.1668	.0631	.0225	.0075	.0023	.0007	.0002			17
	1	.1447	.3741	.3150	.1893	.0957	.0426	.0169	.0060	.0019	.0005	.0001	16
	2	.0117	.1575	.2800	.2673	.1914	.1136	.0581	.0260	.0102	.0035	.0010	15
	3	.0006	.0415	.1556	.2359	.2393	.1893	.1245	.0701	.0341	.0144	.0052	14
	4		.0076	.0605	.1457	.2093	.2209	.1868	.1320	.0796	.0411	.0182	13
	5		.0010	.0175	.0668	.1361	.1914	.2081	.1849	.1379	.0875	.0472	12
	6		.0001	.0039	.0236	.0680	.1276	.1784	.1991	.1839	.1432	.0944	11
	7			.0007	.0065	.0267	.0668	.1201	.1685	.1927	.1841	.1484	10
	8			.0001	.0014	.0084	.0279	.0644	.1134	.1606	.1883	.1855	9
	9				.0003	.0021	.0093	.0276	.0611	.1070	.1540	.1855	8
	10					.0004	.0025	.0095	.0263	.0571	.1008	.1484	7
	11					.0001	.0005	.0026	.0090	.0242	.0525	.0944	6
	12						.0001	.0006	.0024	.0081	.0215	.0472	5
	13							.0001	.0005	.0021	.0068	.0182	4
	14								.0001	.0004	.0016	.0052	3
	15									.0001	.0003	.0010	2
	16											.0001	1
	17												0
		.99	.95	.90	.85	.80	.75	.70	.65	.60	.55	.50	x
		P											

		P											
	x	.01	.05	.10	.15	.20	.25	.30	.35	.40	.45	.50	
n=18	0	.8345	.3972	.1501	.0536	.0180	.0056	.0016	.0004	.0001			18
	1	.1517	.3763	.3002	.1704	.0811	.0338	.0126	.0042	.0012	.0003	.0001	17
	2	.0130	.1683	.2835	.2556	.1723	.0958	.0458	.0190	.0069	.0022	.0006	16
	3	.0007	.0473	.1680	.2406	.2297	.1704	.1046	.0547	.0246	.0095	.0031	15
	4		.0093	.0700	.1592	.2153	.2130	.1681	.1104	.0614	.0291	.0117	14
	5		.0014	.0218	.0787	.1507	.1988	.2017	.1664	.1146	.0666	.0327	13
	6		.0002	.0052	.0301	.0816	.1436	.1873	.1941	.1655	.1181	.0708	12
	7			.0010	.0091	.0350	.0820	.1376	.1792	.1892	.1657	.1214	11
	8			.0002	.0022	.0120	.0376	.0811	.1327	.1734	.1864	.1669	10
	9				.0004	.0033	.0139	.0386	.0794	.1284	.1694	.1855	9
	10				.0001	.0008	.0042	.0149	.0385	.0771	.1248	.1669	8
	11					.0001	.0010	.0046	.0151	.0374	.0742	.1214	7
	12						.0002	.0012	.0047	.0145	.0354	.0708	6
	13							.0002	.0012	.0045	.0134	.0327	5
	14								.0002	.0011	.0039	.0117	4
	15									.0002	.0009	.0031	3
	16										.0001	.0006	2
	17											.0001	1
	18												0
		.99	.95	.90	.85	.80	.75	.70	.65	.60	.55	.50	x
		P											

		P											
	x	.01	.05	.10	.15	.20	.25	.30	.35	.40	.45	.50	
n=19	0	.8262	.3774	.1351	.0456	.0144	.0042	.0011	.0003	.0001			19
	1	.1586	.3774	.2852	.1529	.0685	.0268	.0093	.0029	.0008	.0002		18
	2	.0144	.1787	.2852	.2428	.1540	.0803	.0358	.0138	.0046	.0013	.0003	17
	3	.0008	.0533	.1796	.2428	.2182	.1517	.0869	.0422	.0175	.0062	.0018	16
	4		.0112	.0798	.1714	.2182	.2023	.1491	.0909	.0467	.0203	.0074	15
	5		.0018	.0266	.0907	.1636	.2023	.1916	.1468	.0933	.0497	.0222	14
	6		.0002	.0069	.0374	.0955	.1574	.1916	.1844	.1451	.0949	.0518	13
	7			.0014	.0122	.0443	.0974	.1525	.1844	.1797	.1443	.0961	12
	8			.0002	.0032	.0166	.0487	.0981	.1489	.1797	.1771	.1442	11
	9				.0007	.0051	.0198	.0514	.0980	.1464	.1771	.1762	10
	10				.0001	.0013	.0066	.0220	.0528	.0976	.1449	.1762	9
	11					.0003	.0018	.0077	.0233	.0532	.0970	.1442	8
	12						.0004	.0022	.0083	.0237	.0529	.0961	7
	13						.0001	.0005	.0024	.0085	.0233	.0518	6
	14							.0001	.0006	.0024	.0082	.0222	5
	15								.0001	.0005	.0022	.0074	4
	16									.0001	.0005	.0018	3
	17										.0001	.0003	2
	18												1
	19												0
		.99	.95	.90	.85	.80	.75	.70	.65	.60	.55	.50	x
		P											

		P											
	x	.01	.05	.10	.15	.20	.25	.30	.35	.40	.45	.50	
n=20	0	.8179	.3585	.1216	.0388	.0115	.0032	.0008	.0002				20
	1	.1652	.3774	.2702	.1368	.0576	.0211	.0068	.0020	.0005	.0001		19
	2	.0159	.1887	.2852	.2293	.1369	.0669	.0278	.0100	.0031	.0008	.0002	18
	3	.0010	.0596	.1901	.2428	.2054	.1339	.0716	.0323	.0123	.0040	.0011	17
	4		.0133	.0898	.1821	.2182	.1897	.1304	.0738	.0350	.0139	.0046	16
	5		.0022	.0319	.1028	.1746	.2023	.1789	.1272	.0746	.0365	.0148	15
	6		.0003	.0089	.0454	.1091	.1686	.1916	.1712	.1244	.0746	.0370	14
	7			.0020	.0160	.0545	.1124	.1643	.1844	.1659	.1221	.0739	13
	8			.0004	.0046	.0222	.0609	.1144	.1614	.1797	.1623	.1201	12
	9			.0001	.0011	.0074	.0271	.0654	.1158	.1597	.1771	.1602	11
	10				.0002	.0020	.0099	.0308	.0686	.1171	.1593	.1762	10
	11					.0005	.0030	.0120	.0336	.0710	.1185	.1602	9
	12					.0001	.0008	.0039	.0136	.0355	.0727	.1201	8
	13						.0002	.0010	.0045	.0146	.0366	.0739	7
	14							.0002	.0012	.0049	.0150	.0370	6
	15								.0003	.0013	.0049	.0148	5
	16									.0003	.0013	.0046	4
	17										.0002	.0011	3
	18											.0002	2
	19												1
	20												0
		.99	.95	.90	.85	.80	.75	.70	.65	.60	.55	.50	x
		P											

TABLE 3 ● CRITICAL VALUES OF *t*

Instructions: The columns of this table are associated with different values of α for a one-tailed and two-tailed test; the rows are associated with the degrees of freedom for the *t*-statistic.

	Level of Significance for One-Tailed Test				*Level of Significance for Two-Tailed Test*			
df	*.10*	*.05*	*.01*	*.001*	*.10*	*.05*	*.01*	*.001*
1	3.078	6.314	31.821	318.310	6.314	12.706	63.657	636.619
2	1.886	2.920	6.965	22.326	2.920	4.303	9.925	31.598
3	1.638	2.353	4.541	1.213	2.353	3.182	5.841	12.941
4	1.533	2.132	3.747	7.173	2.132	2.776	4.604	8.610
5	1.476	2.015	3.365	5.893	2.015	2.571	4.032	6.859
6	1.440	1.943	3.143	5.208	1.943	2.447	3.707	5.959
7	1.415	1.895	2.998	4.785	1.895	2.365	3.499	5.405
8	1.397	1.860	2.896	4.501	1.860	2.306	3.355	5.041
9	1.383	1.833	2.821	4.297	1.833	2.262	3.250	4.781
10	1.372	1.812	2.764	4.144	1.812	2.228	3.169	4.587
11	1.363	1.796	2.718	4.025	1.796	2.201	3.106	4.437
12	1.356	1.782	2.681	3.930	1.782	2.179	3.055	4.318
13	1.350	1.771	2.650	3.852	1.771	2.160	3.012	4.221
14	1.345	1.761	2.624	3.787	1.761	2.145	2.977	4.140
15	1.341	1.753	2.602	3.733	1.753	2.131	2.947	4.073
16	1.337	1.746	2.583	3.686	1.746	2.120	2.921	4.015
17	1.333	1.740	2.567	3.646	1.740	2.110	2.898	3.965
18	1.330	1.734	2.552	3.610	1.734	2.101	2.878	3.922
19	1.328	1.729	2.639	3.579	1.729	2.093	2.861	3.883
20	1.325	1.725	2.528	3.552	1.725	2.086	2.845	3.850
21	1.323	1.721	2.518	3.527	1.721	2.080	2.831	3.819
22	1.321	1.717	2.508	3.505	1.717	2.074	2.819	3.792
23	1.319	1.714	2.500	3.485	1.714	2.069	2.807	3.767
24	1.318	1.711	2.492	3.467	1.711	2.064	2.797	3.745
25	1.316	1.708	2.485	3.450	1.708	2.060	2.787	3.725
26	1.315	1.706	2.479	3.435	1.706	2.056	2.779	3.707
27	1.314	1.703	2.473	3.421	1.703	2.052	2.771	3.690
28	1.313	1.701	2.467	3.408	1.701	2.048	2.763	3.674
29	1.311	1.699	2.462	3.396	1.699	2.045	2.756	3.659
30	1.310	1.697	2.457	3.385	1.697	2.042	2.750	3.646
40	1.303	1.684	2.423	3.307	1.684	2.021	2.704	3.551
50	1.299	1.676	2.403	3.261	1.676	2.009	2.678	3.496
60	1.296	1.671	2.390	3.232	1.671	2.000	2.660	3.460
90	1.291	1.662	2.368	3.183	1.662	1.987	2.632	3.403
120	1.289	1.658	2.358	3.232	1.658	1.980	2.617	3.373
∞	1.282	1.645	2.326	3.183	1.645	1.960	2.576	3.291

TABLE 4 ● CRITICAL VALUES OF *F*

Instructions: The columns of this table are associated with the degrees of freedom for the numerator of the *F* ratio; the rows are associated with the degrees of freedom for the denominator of the *F* ratio. The numbers listed in **boldface** type are critical values for α = .05; the numbers listed in lightface roman type are critical values for α = .01.

Degrees of Freedom for Denominator	Degrees of Freedom for Numerator: 1	2	3	4	5	6	7	8	9	10	20	∞
1	**161.4**	**199.5**	**215.7**	**224.6**	**230.2**	**234.0**	**236.8**	**238.9**	**240.5**	**241.9**	**248.0**	**254.3**
	4052	4999	5403	5625	5764	5159	5928	5981	6021	6056	6208	6366
2	**18.51**	**19.00**	**19.16**	**19.25**	**19.30**	**19.33**	**19.36**	**19.37**	**19.38**	**19.39**	**19.44**	**19.50**
	98.49	99.00	99.17	99.25	99.30	99.33	99.36	99.37	99.39	99.40	99.45	99.50
3	**10.13**	**9.55**	**9.28**	**9.12**	**9.01**	**8.94**	**8.88**	**8.84**	**8.81**	**8.78**	**8.66**	**8.53**
	34.12	30.82	29.46	28.71	28.24	27.91	27.67	27.49	27.34	27.23	26.69	26.12
4	**7.71**	**6.94**	**6.59**	**6.39**	**6.26**	**6.16**	**6.09**	**6.04**	**6.00**	**5.96**	**5.80**	**5.63**
	21.20	18.00	16.69	15.98	15.52	15.21	14.98	14.80	14.66	14.54	14.02	13.46
5	**6.61**	**5.79**	**5.41**	**5.19**	**5.05**	**4.95**	**4.88**	**4.82**	**4.78**	**4.74**	**4.56**	**4.36**
	16.26	13.27	12.06	11.39	10.97	10.67	10.45	10.29	10.15	10.05	9.55	9.02
6	**5.99**	**5.14**	**4.76**	**4.53**	**4.39**	**4.28**	**4.21**	**4.15**	**4.10**	**4.06**	**3.87**	**3.67**
	13.74	10.92	9.78	9.15	8.75	8.47	8.26	8.10	7.98	7.87	7.39	6.88
7	**5.59**	**4.74**	**4.35**	**4.12**	**3.97**	**3.87**	**3.79**	**3.73**	**3.68**	**3.63**	**3.44**	**3.23**
	12.25	9.55	8.45	7.85	7.46	7.19	7.00	6.84	6.71	6.62	6.15	5.65
8	**5.32**	**4.46**	**4.07**	**3.84**	**3.69**	**3.58**	**3.50**	**3.44**	**3.39**	**3.34**	**3.15**	**2.93**
	11.26	8.65	7.59	7.01	6.63	6.37	6.19	6.03	5.91	5.82	5.36	4.86
9	**5.12**	**4.26**	**3.86**	**3.63**	**3.48**	**3.37**	**3.29**	**3.23**	**3.18**	**3.13**	**2.93**	**2.71**
	10.56	8.02	6.99	6.42	6.06	5.80	5.62	5.47	5.35	5.26	4.80	4.31
10	**4.96**	**4.10**	**3.71**	**3.48**	**3.33**	**3.22**	**3.14**	**3.07**	**3.02**	**2.97**	**2.77**	**2.54**
	10.04	7.56	6.55	5.99	5.64	5.39	5.21	5.06	4.95	4.85	4.41	3.91
11	**4.84**	**3.98**	**3.59**	**3.36**	**3.20**	**3.09**	**3.01**	**2.95**	**2.90**	**2.86**	**2.65**	**2.40**
	9.65	7.20	6.22	5.67	5.32	5.07	4.88	4.74	4.63	4.54	4.10	3.60
12	**4.75**	**3.89**	**3.49**	**3.26**	**3.11**	**3.00**	**2.92**	**2.85**	**2.80**	**2.76**	**2.54**	**2.30**
	9.33	6.93	5.95	5.41	5.06	4.82	4.65	4.50	4.39	4.30	3.86	3.36
13	**4.67**	**3.80**	**3.41**	**3.18**	**3.02**	**2.92**	**2.84**	**2.77**	**2.72**	**2.67**	**2.46**	**2.21**
	9.07	6.70	5.74	5.20	4.86	4.62	4.44	4.30	4.19	4.10	3.67	3.16
14	**4.60**	**3.74**	**3.34**	**3.11**	**2.96**	**2.85**	**2.77**	**2.70**	**2.65**	**2.60**	**2.39**	**2.13**
	8.86	6.51	5.56	5.03	4.69	4.46	4.28	4.14	4.03	3.94	3.51	3.00
15	**4.54**	**3.68**	**3.29**	**3.06**	**2.90**	**2.79**	**2.70**	**2.64**	**2.59**	**2.55**	**2.33**	**2.07**
	8.68	6.36	5.42	4.89	4.56	4.32	4.14	4.00	3.89	3.80	3.36	2.87
16	**4.49**	**3.63**	**3.24**	**3.01**	**2.85**	**2.74**	**2.66**	**2.59**	**2.54**	**2.49**	**2.28**	**2.01**
	8.53	6.23	5.29	4.77	4.44	4.2	4.03	3.89	3.78	3.69	3.25	2.75

Degrees of Freedom for Denominator	Degrees of Freedom for Numerator: 1	2	3	4	5	6	7	8	9	10	20	∞
17	**4.45**	**3.59**	**3.20**	**2.96**	**2.81**	**2.70**	**2.62**	**2.55**	**2.50**	**2.45**	**2.23**	**1.96**
	8.40	6.11	5.18	4.67	4.34	4.10	3.93	3.79	3.68	3.59	3.16	2.65
18	**4.41**	**3.55**	**3.16**	**2.93**	**2.77**	**2.66**	**2.58**	**2.51**	**2.46**	**2.41**	**2.19**	**1.92**
	8.28	6.01	5.09	4.58	4.25	4.01	3.85	3.71	3.60	3.51	3.07	2.57
19	**4.38**	**3.52**	**3.13**	**2.90**	**2.74**	**2.63**	**2.55**	**2.48**	**2.43**	**2.38**	**2.15**	**1.88**
	8.18	5.93	5.01	4.50	4.17	3.94	3.77	3.63	3.52	3.43	3.00	2.49
20	**4.35**	**3.49**	**3.10**	**2.87**	**2.71**	**2.60**	**2.52**	**2.45**	**2.40**	**2.35**	**2.12**	**1.84**
	8.10	5.85	4.94	4.43	4.10	3.87	3.71	3.56	3.45	3.37	2.94	2.42
21	**4.32**	**3.47**	**3.07**	**2.84**	**2.68**	**2.57**	**2.49**	**2.42**	**2.37**	**2.32**	**2.09**	**1.81**
	8.02	5.78	4.87	4.37	4.04	3.81	3.65	3.51	3.40	3.31	2.88	2.36
22	**4.30**	**3.44**	**3.05**	**2.82**	**2.66**	**2.55**	**2.47**	**2.40**	**2.35**	**2.30**	**2.07**	**1.78**
	7.94	5.72	4.82	4.31	3.99	3.76	3.59	3.45	3.35	3.26	2.83	2.31
23	**4.28**	**3.42**	**3.03**	**2.80**	**2.64**	**2.53**	**2.45**	**2.38**	**2.32**	**2.28**	**2.04**	**1.76**
	7.88	5.66	4.76	4.26	3.94	3.71	3.54	3.41	3.30	3.21	2.78	2.26
24	**4.26**	**3.40**	**3.01**	**2.78**	**2.62**	**2.51**	**2.43**	**2.36**	**2.30**	**2.26**	**2.02**	**1.73**
	7.82	5.61	4.72	4.22	3.90	3.67	3.50	3.36	3.25	3.17	2.74	2.21
25	**4.24**	**3.38**	**2.99**	**2.76**	**2.60**	**2.49**	**2.41**	**2.34**	**2.28**	**2.24**	**2.00**	**1.71**
	7.77	5.57	4.68	4.18	3.86	3.63	3.46	3.32	3.21	3.13	2.70	2.17
26	**4.22**	**3.37**	**2.98**	**2.74**	**2.59**	**2.47**	**2.39**	**2.32**	**2.27**	**2.22**	**1.99**	**1.69**
	7.72	5.53	4.64	4.14	3.82	3.59	3.42	3.29	3.17	3.09	2.66	2.13
27	**4.21**	**3.35**	**2.96**	**2.73**	**2.57**	**2.46**	**2.37**	**2.30**	**2.25**	**2.20**	**1.97**	**1.67**
	7.68	5.49	4.60	4.11	3.79	3.56	3.39	3.26	3.14	3.06	2.63	2.10
28	**4.20**	**3.34**	**2.95**	**2.71**	**2.56**	**2.44**	**2.36**	**2.29**	**2.24**	**2.19**	**1.96**	**1.65**
	7.64	5.45	4.57	4.07	3.76	3.53	3.36	3.23	3.11	3.03	2.60	2.06
29	**4.18**	**3.33**	**2.93**	**2.70**	**2.54**	**2.43**	**2.35**	**2.28**	**2.22**	**2.18**	**1.94**	**1.64**
	7.60	5.42	4.54	4.04	3.73	3.50	3.33	3.20	3.08	3.00	2.57	2.03
30	**4.17**	**3.32**	**2.92**	**2.69**	**2.53**	**2.42**	**2.34**	**2.27**	**2.21**	**2.16**	**1.93**	**1.62**
	7.56	5.39	4.51	4.02	3.70	3.47	3.30	3.17	3.06	2.98	2.55	2.01
40	**4.08**	**3.23**	**2.84**	**2.61**	**2.45**	**2.34**	**2.25**	**2.18**	**2.12**	**2.07**	**1.84**	**1.51**
	7.31	5.18	4.31	3.83	3.51	3.29	3.12	2.99	2.88	2.80	2.37	1.81
50	**4.03**	**3.18**	**2.79**	**2.56**	**2.40**	**2.29**	**2.20**	**2.13**	**2.07**	**2.02**	**1.78**	**1.44**
	7.17	5.06	4.20	3.72	3.41	3.18	3.02	2.88	2.78	2.70	2.26	1.68
60	**4.00**	**3.15**	**2.76**	**2.52**	**2.37**	**2.25**	**2.17**	**2.10**	**2.04**	**1.99**	**1.75**	**1.39**
	7.08	4.98	4.13	3.65	3.34	3.12	2.95	2.82	2.72	2.63	2.20	1.60
80	**3.96**	**3.11**	**2.72**	**2.48**	**2.33**	**2.21**	**2.12**	**2.05**	**1.99**	**1.95**	**1.70**	**1.32**
	6.96	4.88	4.04	3.56	3.25	3.04	2.87	2.74	2.64	2.55	2.11	1.49
120	**3.92**	**3.07**	**2.68**	**2.45**	**2.29**	**2.17**	**2.09**	**2.02**	**1.96**	**1.91**	**1.66**	**1.25**
	6.85	4.79	3.95	3.48	3.17	2.96	2.79	2.66	2.56	2.47	2.03	1.38
∞	**3.84**	**2.99**	**2.60**	**2.37**	**2.21**	**2.09**	**2.01**	**1.94**	**1.88**	**1.83**	**1.57**	**1.00**
	6.64	4.60	3.78	3.32	3.02	2.80	2.64	2.51	2.41	2.32	1.87	1.00

TABLE 5 ⬢ THE STUDENTIZED RANGE STATISTIC (q_T)

Instructions: The columns of this table are associated with number of groups in the analysis (*k*); the rows are associated with the degrees of freedom for the error term (denominator) of the *F* ratio (df_{error}). The numbers listed in **boldface** type are critical values for α = .05; the numbers listed in lightface roman type are critical values for α = .01.

	k = *number of groups*										
df_{error}	2	3	4	5	6	7	8	9	10	11	12
1	**17.97**	**26.98**	**32.82**	**37.08**	**40.41**	**43.12**	**45.40**	**47.36**	**49.07**	**50.59**	**51.96**
	90.03	135.00	164.30	185.60	202.20	215.80	227.20	237.00	245.60	253.20	260.00
2	**6.08**	**8.33**	**9.80**	**10.88**	**11.74**	**12.44**	**13.03**	**13.54**	**13.99**	**14.39**	**14.75**
	14.04	19.02	22.29	24.72	26.63	28.20	29.53	30.68	31.69	32.59	33.40
3	**4.50**	**5.91**	**6.82**	**7.50**	**8.04**	**8.48**	**8.85**	**9.18**	**9.46**	**9.72**	**9.95**
	8.26	10.62	12.17	13.33	14.24	15.00	15.64	16.20	16.69	17.13	17.53
4	**3.93**	**5.04**	**5.76**	**6.29**	**6.71**	**7.05**	**7.35**	**7.60**	**7.83**	**8.03**	**8.21**
	6.51	8.12	9.17	9.96	10.58	11.10	11.55	11.93	12.27	12.57	12.84
5	**3.64**	**4.60**	**5.22**	**5.67**	**6.03**	**6.33**	**6.58**	**6.80**	**6.99**	**7.17**	**7.32**
	5.70	6.98	7.80	8.42	8.91	9.32	9.67	9.97	10.24	10.48	10.70
6	**3.46**	**4.34**	**4.90**	**5.30**	**5.63**	**5.90**	**6.12**	**6.32**	**6.49**	**6.65**	**6.79**
	5.24	6.33	7.03	7.56	7.97	8.32	8.61	8.87	9.10	9.30	9.48
7	**3.34**	**4.16**	**4.68**	**5.06**	**5.36**	**5.61**	**5.82**	**6.00**	**6.16**	**6.30**	**6.43**
	4.95	5.92	6.54	7.01	7.37	7.68	7.94	8.17	8.37	8.55	8.71
8	**3.26**	**4.04**	**4.53**	**4.89**	**5.17**	**5.40**	**5.60**	**5.77**	**5.92**	**6.05**	**6.18**
	4.75	5.64	6.20	6.62	6.96	7.24	7.47	7.68	7.86	8.03	8.18
9	**3.20**	**3.95**	**4.41**	**4.76**	**5.02**	**5.24**	**5.43**	**5.59**	**5.74**	**5.87**	**5.98**
	4.60	5.43	5.96	6.35	6.66	6.91	7.13	7.33	7.49	7.65	7.78
10	**3.15**	**3.88**	**4.33**	**4.65**	**4.91**	**5.12**	**5.30**	**5.46**	**5.60**	**5.72**	**5.83**
	4.48	5.27	5.77	6.14	6.43	6.67	6.87	7.05	7.21	7.36	7.49
11	**3.11**	**3.82**	**4.26**	**4.57**	**4.82**	**5.03**	**5.20**	**5.35**	**5.49**	**5.61**	**5.71**
	4.39	5.15	5.62	5.97	6.25	6.48	6.67	6.84	6.99	7.13	7.25
12	**3.08**	**3.77**	**4.20**	**4.51**	**4.75**	**4.95**	**5.12**	**5.27**	**5.39**	**5.51**	**5.61**
	4.32	5.05	5.50	5.84	6.10	6.32	6.51	6.67	6.81	6.94	7.06
13	**3.06**	**3.73**	**4.15**	**4.45**	**4.69**	**4.88**	**5.05**	**5.19**	**5.32**	**5.43**	**5.53**
	4.26	4.96	5.40	5.73	5.98	6.19	6.37	6.53	6.67	6.79	6.90
14	**3.03**	**3.70**	**4.11**	**4.41**	**4.64**	**4.83**	**4.99**	**5.13**	**5.25**	**5.36**	**5.46**
	4.21	4.89	5.32	5.63	5.88	6.08	6.26	6.41	6.54	6.66	6.77

	k = number of groups										
df_{error}	2	3	4	5	6	7	8	9	10	11	12
15	**3.01**	**3.67**	**4.08**	**4.37**	**4.59**	**4.78**	**4.94**	**5.08**	**5.20**	**5.31**	**5.40**
	4.17	4.84	5.25	5.56	5.80	5.99	6.16	6.31	6.44	6.55	6.66
16	**3.00**	**3.65**	**4.05**	**4.33**	**4.56**	**4.74**	**4.90**	**5.03**	**5.15**	**5.26**	**5.35**
	4.13	4.79	5.19	5.49	5.72	5.92	6.08	6.22	6.35	6.46	6.56
17	**2.98**	**3.63**	**4.02**	**4.30**	**4.52**	**4.70**	**4.86**	**4.99**	**5.11**	**5.21**	**5.31**
	4.10	4.74	5.14	5.43	5.66	5.85	6.01	6.15	6.27	6.38	6.48
18	**2.97**	**3.61**	**4.00**	**4.28**	**4.49**	**4.67**	**4.82**	**4.96**	**5.07**	**5.17**	**5.27**
	4.07	4.70	5.09	5.38	5.60	5.79	5.94	6.08	6.20	6.31	6.41
19	**2.96**	**3.59**	**3.98**	**4.25**	**4.47**	**4.65**	**4.79**	**4.92**	**5.04**	**5.14**	**5.23**
	4.05	4.67	5.05	5.33	5.55	5.73	5.89	6.02	6.14	6.25	6.34
20	**2.95**	**3.58**	**3.96**	**4.23**	**4.45**	**4.62**	**4.77**	**4.90**	**5.01**	**5.11**	**5.20**
	4.02	4.64	5.02	5.29	5.51	5.69	5.84	5.97	6.09	6.19	6.28
22	**2.93**	**3.55**	**3.93**	**4.20**	**4.41**	**4.58**	**4.72**	**4.85**	**4.96**	**5.06**	**5.14**
	3.99	4.59	4.96	5.22	5.43	5.61	5.75	5.88	5.99	6.09	6.19
24	**2.92**	**3.53**	**3.90**	**4.17**	**4.37**	**4.54**	**4.68**	**4.81**	**4.92**	**5.01**	**5.10**
	3.96	4.55	4.91	5.17	5.37	5.54	5.69	5.81	5.92	6.02	6.11
26	**2.91**	**3.51**	**3.88**	**4.14**	**4.35**	**4.51**	**4.65**	**4.77**	**4.88**	**4.98**	**5.06**
	3.93	4.51	4.87	5.12	5.32	5.49	5.63	5.75	5.86	5.95	6.04
28	**2.90**	**3.50**	**3.86**	**4.12**	**4.32**	**4.49**	**4.62**	**4.74**	**4.85**	**4.94**	**5.03**
	3.91	4.48	4.83	5.08	5.28	5.44	5.58	5.70	5.80	5.90	5.98
30	**2.89**	**3.49**	**3.85**	**4.10**	**4.30**	**4.46**	**4.60**	**4.72**	**4.82**	**4.92**	**5.00**
	3.89	4.45	4.80	5.05	5.24	5.40	5.54	5.65	5.76	5.85	5.93
40	**2.86**	**3.44**	**3.79**	**4.04**	**4.23**	**4.39**	**4.52**	**4.63**	**4.73**	**4.82**	**4.90**
	3.82	4.37	4.70	4.93	5.11	5.26	5.39	5.50	5.60	5.69	5.76
50	**2.84**	**3.42**	**3.76**	**4.00**	**4.19**	**4.34**	**4.47**	**4.58**	**4.68**	**4.77**	**4.85**
	3.79	4.32	4.63	4.86	5.04	5.19	5.31	5.41	5.51	5.59	5.67
60	**2.83**	**3.40**	**3.74**	**3.98**	**4.16**	**4.31**	**4.44**	**4.55**	**4.65**	**4.73**	**4.81**
	3.76	4.28	4.59	4.82	4.99	5.13	5.25	5.36	5.45	5.53	5.60
90	**2.81**	**3.37**	**3.70**	**3.94**	**4.12**	**4.27**	**4.39**	**4.50**	**4.59**	**4.67**	**4.75**
	3.72	4.23	4.53	4.74	4.91	5.05	5.16	5.26	5.35	5.43	5.49
120	**2.80**	**3.36**	**3.68**	**3.92**	**4.10**	**4.24**	**4.36**	**4.47**	**4.56**	**4.64**	**4.71**
	3.70	4.20	4.50	4.71	4.87	5.01	5.12	5.21	5.30	5.37	5.44
∞	**2.77**	**3.31**	**3.63**	**3.86**	**4.03**	**4.17**	**4.28**	**4.39**	**4.47**	**4.55**	**4.62**
	3.64	4.12	4.40	4.60	4.76	4.88	4.99	5.08	5.16	5.23	5.29

TABLE 6 ● CRITICAL VALUES OF THE PEARSON CORRELATION COEFFICIENT (*r*)

Instructions: The columns of this table are associated with different values of α for a one-tailed and two-tailed test; the rows are associated with the degrees of freedom for the Pearson correlation coefficient (*r*)

df	*Level of Significance for One-Tailed Test*				*Level of Significance for Two-Tailed Test*			
(N − 2)	*.10*	*.05*	*.01*	*.001*	*.10*	*.05*	*.01*	*.001*
1	.951	.988	.9995	.9999	.988	.997	.9999	.99999
2	.800	.900	.980	.998	.900	.950	.990	.999
3	.687	.805	.934	.986	.805	.878	.959	.991
4	.608	.729	.882	.963	.729	.811	.917	.974
5	.551	.669	.833	.935	.669	.754	.875	.951
6	.507	.622	.789	.905	.622	.707	.834	.925
7	.472	.582	.750	.875	.582	.666	.798	.898
8	.443	.549	.716	.847	.549	.632	.765	.872
9	.419	.521	.685	.820	.521	.602	.735	.847
10	.398	.497	.658	.795	.497	.576	.708	.823
11	.380	.476	.634	.772	.476	.553	.684	.801
12	.365	.458	.612	.750	.458	.532	.661	.780
13	.351	.441	.592	.730	.441	.514	.641	.760
14	.338	.426	.574	.711	.426	.497	.623	.742
15	.327	.412	.558	.694	.412	.482	.606	.725
16	.317	.400	.542	.678	.400	.468	.590	.708
17	.308	.389	.528	.662	.389	.456	.575	.693
18	.299	.378	.516	.648	.378	.444	.561	.679
19	.291	.369	.503	.635	.369	.433	.549	.665
20	.284	.360	.492	.622	.360	.423	.537	.652
21	.277	.352	.482	.610	.352	.413	.526	.640
22	.271	.344	.472	.599	.344	.404	.515	.629
23	.265	.337	.462	.588	.337	.396	.505	.618
24	.260	.330	.453	.578	.330	.388	.496	.607
25	.255	.323	.445	.568	.323	.381	.487	.597
26	.250	.317	.437	.559	.317	.374	.479	.588
27	.245	.311	.430	.550	.311	.367	.471	.579
28	.241	.306	.423	.541	.306	.361	.463	.570
29	.237	.301	.416	.533	.301	.355	.456	.562
30	.233	.296	.409	.526	.296	.349	.449	.554
40	.202	.257	.358	.463	.257	.304	.393	.490
50	.181	.231	.322	.419	.231	.273	.354	.443
60	.165	.211	.295	.385	.211	.250	.325	.408
90	.135	.173	.242	.318	.173	.205	.267	.338
120	.117	.150	.210	.283	.150	.178	.232	.294

TABLE 7 ● CRITICAL VALUES OF THE SPEARMAN RANK-ORDER CORRELATION (r_S)

Instructions: The columns of this table are associated with different values of α for a one-tailed and two-tailed test; the rows are associated with the degrees of freedom for the Spearman rank-order correlation (r_S).

	Level of Significance for One-Tailed Test				Level of Significance for Two-Tailed Test			
df (N – 2)	.10	.05	.01	.001	.10	.05	.01	.001
4	.657	.829	.943		.829	.886	.9999	
5	.571	.714	.893	.9999	.714	.786	.929	.9999
6	.524	.643	.833	.952	.643	.738	.881	.976
7	.483	.600	.783	.917	.600	.700	.833	.933
8	.455	.564	.745	.879	.564	.648	.794	.903
9	.427	.536	.709	.845	.536	.618	.755	.873
10	.406	.503	.678	.818	.503	.587	.727	.846
11	.385	.484	.648	.791	.484	.560	.703	.824
12	.367	.464	.626	.771	.464	.538	.679	.802
13	.354	.446	.604	.750	.446	.521	.654	.779
14	.341	.429	.582	.729	.429	.503	.635	.762
15	.328	.414	.566	.711	.414	.488	.618	.743
16	.317	.401	.550	.692	.401	.472	.600	.725
17	.309	.391	.535	.675	.391	.460	.584	.709
18	.299	.380	.522	.662	.380	.447	.570	.693
19	.292	.370	.509	.647	.370	.436	.556	.678
20	.284	.361	.497	.633	.361	.425	.544	.665
21	.278	.353	.486	.621	.353	.416	.532	.652
22	.271	.344	.476	.609	.344	.407	.521	.640
23	.265	.337	.466	.597	.337	.398	.511	.628
24	.259	.331	.457	.586	.331	.390	.501	.618
25	.255	.324	.449	.576	.324	.383	.492	.607
26	.250	.318	.441	.567	.318	.375	.483	.597
27	.245	.312	.433	.558	.312	.368	.475	.588
28	.240	.306	.425	.549	.306	.362	.467	.579
29	.236	.301	.419	.540	.301	.356	.459	.570
30	.232	.296	.412	.532	.296	.350	.452	.562
40	.202	.257	.359	.468	.257	.305	.396	.495
50	.180	.231	.323	.422	.231	.274	.356	.447
60	.165	.211	.296	.387	.211	.250	.326	.411
90	.135	.173	.243	.319	.173	.205	.268	.339

TABLE 8 ● CRITICAL VALUES OF CHI-SQUARE (χ^2)

Instructions: The columns of this table are associated with different values of α; the rows are associated with the degrees of freedom for the chi-square statistic (χ^2).

	Level of significance			
df	*.10*	*.05*	*.01*	*.001*
1	2.71	3.84	6.63	10.83
2	4.61	5.99	9.21	13.82
3	6.25	7.81	11.34	16.27
4	7.78	9.49	13.28	18.46
5	9.24	11.07	15.09	20.52
6	10.64	12.59	16.81	22.46
7	12.02	14.07	18.48	24.32
8	13.36	15.51	20.09	26.12
9	14.68	16.92	21.67	27.88
10	15.99	18.31	23.21	29.59
11	17.28	19.68	24.72	31.26
12	18.55	21.03	26.22	32.91
13	19.81	22.36	27.69	34.53
14	21.06	23.68	29.14	36.12
15	22.31	25.00	30.58	37.70
16	23.54	26.30	32.00	39.29
17	24.77	27.59	33.41	40.75
18	25.99	28.87	34.81	42.31
19	27.20	30.14	36.19	43.82
20	28.41	31.41	37.57	45.32
21	29.62	32.67	38.93	46.80
22	30.81	33.92	40.29	48.27
23	32.01	35.17	41.64	49.73
24	33.20	36.42	42.98	51.18
25	34.38	37.65	44.31	52.62
26	35.56	38.88	45.64	54.05
27	36.74	40.11	46.96	55.48
28	37.92	41.34	48.28	56.89
29	39.09	42.69	49.59	58.30
30	40.26	43.77	50.89	59.70
40	51.81	55.76	63.69	73.40
50	63.17	67.50	76.15	86.66
60	74.40	79.08	88.38	99.61
80	96.58	101.88	112.33	124.84
100	118.50	124.34	135.81	149.45

APPENDIX: REVIEW OF BASIC MATHEMATICS

The purpose of this appendix is to provide a review of some of the symbols, terms, and basic math skills needed to perform calculations presented in this book. Many students will know some or all of the material covered in this appendix; however, others may need a more extensive review. It's important for you to know that none of the statistical procedures conducted in this book require higher (i.e., geometry, trigonometry, calculus) math skills.

This appendix covers seven aspects of a basic math review: (1) symbols; (2) terms; (3) addition, subtraction, multiplication, and division; (4) order of operations; (5) fractions, decimals, and percentages; (6) solving simple equations with one unknown; and (7) solving more complex equations with one unknown. Rules, steps, and examples are provided within each of these sections, as well as practice problems designed to help you assess your understanding; the answers to the problems are provided at the end of this appendix. If you have difficulty with this review, we encourage you to refer to your previous math books and get assistance from your instructor or perhaps a tutor.

I. SYMBOLS

Symbol	Meaning	Example	Read
$+$	Addition	$4 + 7$	4 plus 7
$-$	Subtraction	$14 - 12$	14 minus 12
$\pm$	Addition and subtraction	5 ± 2	5 plus or minus 2
$\times$	Multiplication	3×9	3 times 9
( ) ( )		(2) (16)	2 times 16
		7(8)	7 times 8
//	Division	30/5	30 divided by 5
$\div$		$121 \div 12$	121 divided by 12
$\frac{X}{Y}$		$\frac{476}{147}$	476 divided by 147
$=$	Equals	$X = 9$	*X* is equal to 9
$\neq$	Not equals	$X \neq 14$	*X* is not equal to 14
$<$	Less than	$X < 310$	*X* is less than 310
$>$	Greater than	$X > 6$	*X* is greater than 6
$\leq$	Less than or equal to	$X \leq 3$	*X* is less than or equal to 3
$\geq$	Greater than or equal to	$X \geq 5$	*X* is greater than or equal to 5
$()^2$	Square	$(16)^2$ or 16^2	16 squared (16 times 16)
$\sqrt{}$	Square root	$\sqrt{175}$	The square root of 175
$\|\|$	Absolute value	$\|13\|$	The absolute value of 13
. . .	Continuation of pattern	1, 2, 3 … 7, 8, 9	1, 2, 3, 4, 5, 6, 7, 8, 9

II. TERMS

Term	Meaning	Example	Read
Sum	Result of addition	$4 + 7 = 11$	The sum of 4 and 7 equals 11
Difference	Result of subtraction	$14 - 12 = 2$	The difference between 14 and 12 equals 2
Product	Result of multiplication	$3 \times 9 = 27$	The product of 3 and 9 equals 27
Quotient	Result of division	$\frac{24}{12} = 2$	The quotient of 24 divided by 12 equals 2
Numerator	Top expression in division	$\frac{84}{7}$	The numerator of 84/7 is 84
Denominator	Bottom expression in division	$\frac{84}{7}$	The denominator of 84/7 is 7
Fraction	Number represented with numerator and denominator	$\frac{2}{5}, \frac{1}{4}$	2 divided by 5, 1 divided by 4
Percent	Part per 100	33%	33 percent
Square	Result of multiplying a number by itself	$4^2 = 4 \times 4$	4 squared equals 4 times 4
Square root	Square root of a value is equal to a number that, when multiplied by itself, results in the original value	$\sqrt{25} = 5$	The square root of 25 equals 5
Equation	Mathematical statement consisting of two equal quantities	$11 = 8 + 3$	11 equals 8 plus 3

III. ADDITION, SUBTRACTION, MULTIPLICATION, DIVISION

A. All Positive Numbers

RULE: Adding, multiplying, or dividing numbers that are all positive results in a positive number.

Examples: $3 + 6 = 9$

$6 \times 4 = 24$

$\frac{30}{6} = 5$

RULE: Subtracting a smaller positive number from a larger positive number results in a positive number.

Examples: $6 - 4 = 2$

$24 - 9 = 13$

RULE: Subtracting a larger positive number from a smaller positive number results in a negative number.

Examples: $7 - 10 = -3$

$14 - 22 = -8$

B. Adding All Negative Numbers

STEPS: (1) Add the numbers as though they were positive.

(2) Place a negative sign (–) before the sum.

Examples: $-1 + -4 + -7 = -(1 + 4 + 7) = -12$

$-5 + -16 + -9 + -3 = -(5 + 16 + 9 + 3) = -33$

C. Adding a Combination of Positive and Negative Numbers

STEPS: (1) Add the positive and negative numbers separately.

(2) Subtract the smaller sum from the larger sum.

(3) If the larger sum is negative, assign the minus sign to the difference.

Examples: $3 + -2 + -5 + 8 = (3 + 8)$ and $-(2 + 5) = 11 - 7 = 4$

$-10 + -1 + 9 + 7 + -12 = (9 + 7)$ and $-(10 + 1 + 12) = 23 - 16 = -7$

$16 + -11 + -6 = 16$ and $-(11 + 6) = 17 - 16 = -1$

D. Subtracting Negative Numbers

STEPS: (1) Change the sign of the negative number to positive.

(2) Add the numbers together.

Examples: $2 - (-5) = 2 + 5 = 7$

$-43 - (-16) - (-7) = -43 -16 + 7 = -20$

$12 - (-8) - (-4) - (-17) = 12 + 8 + 4 + 17 = 41$

E. Multiplying Two Negative Numbers

RULE: The product of two negative numbers is a positive number.

Examples: $-4 \times -9 = 36$

$(-7)(-3) = 21$

$-2(-5) = 10$

F. Multiplying a Combination of Positive and Negative Numbers

RULE: A product involving an *odd* number of negative numbers is a negative number.

Examples: $5 \times -3 = -15$

$8(-9) = -72$

$(4)(-5)(-3)(-2) = -120$

RULE: A product involving an *even* number of negative numbers is a positive number.

Examples: $4 \times -5 \times -2 = 40$

$(-3)(6)(4)(-1) = 72$

$6(3)(-7) = 126$

G. Dividing Two Negative Numbers

RULE: The quotient of two negative numbers is a positive number.

Examples: $\frac{-6}{-2} = 3$

$-48 \div -12 = 4$

$\frac{-36}{-4} = 9$

H. Dividing One Positive Number and One Negative Number

RULE: The quotient of one positive number and one negative number is a negative number.

Examples: $\frac{30}{-5} = -6$

$\frac{-8}{2} = -4$

$-80 \div 10 = -8$

Practice Problems

Addition, Subtraction, Multiplication, and Division

1. All positive numbers
   a. $5 + 8$
   b. $13 + 9$
   c. 6×7
   d. $(3)(11)$
   e. $3(24)$
   f. $\frac{27}{3}$
   g. $72 \div 9$
   h. $\frac{87}{3}$
   i. $9 - 2$
   j. $16 - 6$
   k. $6 - 7$
   l. $15 - 19$
2. Adding all negative numbers
   a. $-8 + -3$
   b. $-17 + -5 + -14$
   c. $-23 + -56 + -3$
   d. $-11 + -5 + -2 + -39$
3. Adding a combination of positive and negative numbers
   a. $6 + 3 + -4 + -7$
   b. $11 + -5 + 10 + -1$
   c. $-9 + 4 + -5 + -17 + 25$
   d. $8 + 32 + -13 + 19 + 7 + -14$
4. Subtracting negative numbers
   a. $21 - (-7)$
   b. $43 - (-2) - (-19)$
   c. $-15 - (-7) - (-5)$
   d. $-24 - (-13) - (-16) - (-6)$

5. Multiplying two negative numbers
   a. -2×-6
   b. $(-5)(-10)$
   c. $(-9)(-7)$
   d. $-6(-15)$
6. Multiplying combination of positive and negative numbers
   a. -6×9
   b. $(-5)(-19)$
   c. 13×-11
   d. -7×-8
   e. $2(-4)(-12)$
   f. $-5 \times -3 \times -7$
   g. $(-4)(3)(-12)(-2)$
   h. $(2)(2)(-4)(-9)$
7. Dividing two negative numbers
   a. $\frac{-32}{-8}$
   b. $\frac{-63}{-3}$
   c. $\frac{-57}{-3}$
   d. $\frac{-112}{-4}$
8. Dividing one positive number and one negative number
   a. $\frac{49}{-7}$
   b. $\frac{-90}{-10}$
   c. $\frac{-85}{5}$
   d. $\frac{154}{-14}$

IV. ORDER OF OPERATIONS

The order of operations is a rule used within algebra to specify the order in which operations should be performed in a given mathematical expression.

First: Parentheses ( ) and Brackets [ ]

Simplify the inside of parentheses and brackets and then remove the parentheses before continuing.

Second: Squares and/or Square Roots

Simplify the square or square root of a number before multiplying, dividing, adding, or subtracting it.

Third: Multiplication and/or Division

Simplify multiplication and division in the order they appear from left to right.

Fourth: Addition and/or Subtraction

Simplify addition and subtraction in the order they appear from left to right.

a. Parentheses ( ) and brackets [ ]

Examples: $(1 + 3)^2 = (4)^2 = 16$

$$[(4 - 1)^2 - (3 - 1)^2] = [(3)^2 - (2)^2] = [9 - 4] = 5$$

$$\sqrt{(7 \times 3) + (6 \times 10)} + \sqrt{21 + 60} + \sqrt{81} = 9$$

$(3 + 2) \times 10 = 5 \times 10 = 50$

$3 + (2 \times 10) = 3 + 20 = 23$

$12/(6 - 2) = 12/4 = 3$

$(12/6) - 2 = 2 - 2 = 0$

$(3 + 9 + 12 + 8)/8 = 32/8 = 4$

$[(12 \times 4)/(6 \times 2)]/2 = [48/12]/2 = 4/2 = 2$

$[(6 - 4)(9 - 4)] + 11 = [(2)(5)] + 11 = 10 + 11 = 21$

$$\sqrt{(4-1)^2 + (5-1)^2} = \sqrt{(3)^2 + (4)^2} = \sqrt{9+16} = \sqrt{25} = 5$$

$$\frac{[(5-3)+(6-1)]^2}{7} = \frac{[2+5]^2}{7} = \frac{(7)^2}{7} = \frac{49}{7} = 7$$

b. Squares and/or square roots

Examples: $2^2 + 5^2 = 4 + 25 = 29$

$8 + 3^2 = 8 + 9 = 17$

$8 + 3^2 = 8 + 9 = 17$

$\sqrt{26+38} = \sqrt{64} = 8$

$7(\sqrt{36}) = 7(6) = 42$

$\sqrt{4} + \sqrt{49} = 2 + 7 = 9$

$\sqrt{\frac{774}{86}} = \sqrt{9} = 3$

$31 - \sqrt{88+12} = 31 - \sqrt{100} = 31 - 10 = 21$

$(9)\sqrt{(3)^2 + (4)^2} = (9)\sqrt{9+16} = (9)\sqrt{25} = (9)5 = 45$

c. Multiplication and/or division

Examples: $\frac{12}{6} \times 2 = 2 \times 2 = 4$

$\frac{3 \times 6}{3} = \frac{18}{3} = 6$

$\frac{16}{8} + 7 = 2 + 7 = 9$

$2 \times 8 - 10 = 16 - 10 = 6$

$19 + 21 \div 3 = 19 + 7 = 26$

$14 - 3 \times 3 = 14 - 9 = 5$

$5 + 11 - 8 + \frac{18}{6} = 5 + 11 - 8 + 3 = 11$

Practice Problems

Order of Operations

1. Parentheses ( ) and brackets [ ]

a. $(9-2)^2$
b. $(24 \div 3)^2$
c. $(4 \times 3)^2$
d. $[(2+4)(3+2)]$
e. $\sqrt{(8-3)(6-1)}$
f. $(6-2) \times 2$
g. $6-(2 \times 2)$
h. $(16-4) \div 2$
i. $16-(4 \div 2)$
j. $(8+4+7+5)/6$
k. $4(9+2+6+1)$
l. $3[(2 \times 6)+(4 \times 2)]]$
m. $\dfrac{[(3 \times 6)+(4 \times 8)]}{10}$
n. $\sqrt{(11 \times 2)-(3 \times 2)}$
o. $\dfrac{\sqrt{64}}{\sqrt{16}}$
p. $(3-1)[(4-1)^2+(5-1)^2]$
q. $(7-2)\dfrac{[(5-3)+(6-1)]^2}{7}$

2. Squares and/or square roots

a. 4^2+6^2
b. 7^2-13
c. $(3)5^2$
d. $\dfrac{9^2}{3}$
e. $\dfrac{8}{\sqrt{16}}$
f. $\sqrt{36}+\sqrt{121}$
g. $\sqrt{81}-7$
h. $4^2-\sqrt{81}$
i. $\sqrt{(6-2)^2+(5-2)^2}$
j. $(6-1)\sqrt{\dfrac{[(5+1)-(3+1)]^2}{4}}$

3. Multiplication and/or division

a. $3 \times 7 \times 4$
b. $3 \times 6 \div 9$
c. $7 \times 12 \div 3$
d. $33 \div 11 \times 8$
e. $6 \times 12 + 3$
f. $9 \times 7 - 14$
g. $\dfrac{36}{9}-10$
h. $13 + 3 \times 4$
i. $17 - 39 \div 13$
j. $3 + 2 + 4 + 6 \div 2$
k. $6 - 2 + 4 \times 7$
l. $16 \div 2 + 7 \times 3 - 11$

V. FRACTIONS, DECIMALS, PERCENTAGES

A. Fraction

DEFINITION: A fraction is a numerical quantity expressed with a numerator and denominator indicating that one number is being divided by another.

$$\text{Examples: } \frac{1}{4}, \frac{2}{3}, \frac{-11}{14}, \frac{4}{2}, \frac{15}{4}, \frac{20}{8}$$

B. Decimal

DEFINITION: A decimal is the result of dividing a numerator by a denominator represented numerically with decimal places.

$$\text{Examples: } \frac{1}{4} = .25, \frac{2}{3} = .67, \frac{-11}{14} = -.79, \frac{4}{2} = 2.00, \frac{15}{4} = 3.75, \frac{20}{8} = 2.50$$

C. Percentage

RULE: A percentage is calculated by multiplying a decimal by 100 and placing a percent sign (%) after the result.

Examples: $.25 \times 100 = 25\%$, $.67 \times 100 = 67\%$, $-.79 \times 100 = -79\%$, $2.00 \times 100 = 200\%$

RULE: A percentage is converted to a decimal by removing the percent sign and dividing the percentage by 100.

$$\text{Examples: } 50\% = \frac{50}{100} = .50, 72\% = \frac{72}{100} = .72, 150\% = \frac{150}{100} = 1.50$$

Practice Problems

Fractions, Decimals, and Percentages

1. Convert the following fractions to decimals.
   a. $\frac{1}{5}$ d. $\frac{-9}{27}$ g. $\frac{14}{8}$
   b. $\frac{3}{4}$ e. $\frac{18}{6}$ h. $\frac{36}{40}$
   c. $\frac{12}{20}$ f. $\frac{9}{2}$ i. $\frac{75}{12}$
2. Convert the following decimals to percentages.
   a. .05 c. .16 e. 1.36
   b. .68 d. −.52 f. 2.33
3. Convert the following percentages to decimals.
   a. 1% c. 47%
   b. 95% d. 99%

VI. SOLVING SIMPLE EQUATIONS WITH ONE UNKNOWN (X)

RULE: Solve for the unknown (*X*) by isolating *X* on one side of the equation and then simplifying the steps on the other side of the equation.

A. *X* Has a Value Subtracted From It (Solve for *X* Using Addition)

Steps: Isolate *X* by adding a number to both sides of the equation.

Example: $X - 6 = 4$ Check: $X - 6 = 4$

$X - 6 + 6 = 4 + 6$ $10 - 6 = 4$

$X = 10$ $4 = 4$

B. *X* Has a Value Added to It (Solve for *X* Using Subtraction)

Steps: Isolate *X* by subtracting a number from both sides of the equation.

Example: $X + 12 = 19$	Check: $X + 12 = 19$
$X + 12 - 12 = 19 - 12$	$7 + 12 = 19$
$X = 7$	$19 = 19$

C. *X* Is Multiplied by a Value (Solve for *X* Using Division)

Steps: Isolate *X* by dividing both sides of the equation by a number.

Examples: $4(X) = 36$	Check: $4(X) = 36$
$\frac{4(X)}{4} = \frac{36}{4}$	$4(9) = 36$
$X = 9$	$36 = 36$

D. *X* Is Divided by a Value (Solve for *X* Using Multiplication)

Steps: Isolate *X* by multiplying both sides of the equation by a number.

Examples: $\frac{X}{7} = 4$	Check: $\frac{X}{7} = 4$
$(7)\frac{X}{7} = 4(7)$	$\frac{28}{7} = 4$
$X = 28$	$4 = 4$

Practice Problems

Simple Equations With One Unknown

1. *X* has a value subtracted from it (solve for *X* using addition).
   a. $X - 5 = 3$
   b. $X - 13 = 8$
   c. $X - 37 = 79$
   d. $X - 231 = 590$
   e. $X - 1.54 = 3.21$
   f. $X - 29.64 = 105.31$
2. *X* has a value added to it (solve for *X* using subtraction).
   a. $X + 5 = 9$
   b. $X + 14 = 23$
   c. $X + 53 = 104$
   d. $X + 167 = 331$
   e. $X +.83 = 6.72$
   f. $X + 13.06 = 73.29$
3. *X* is multiplied by a value (solve for *X* using division).
   a. $3(X) = 15$
   b. $6(X) = 42$
   c. $1.50(X) = 10.50$
   d. $2.25(X) = 8.46$
   e. $-4(X) = -36$
   f. $-5(X) = 25$
4. *X* is divided by a value (solve for *X* using multiplication).
   a. $\frac{X}{3} = 6$
   b. $\frac{X}{6} = 10$
   c. $\frac{X}{11} = 9$
   d. $\frac{X}{13} = 12$
   e. $\frac{X}{2.50} = 6.00$
   f. $\frac{X}{3.88} = 9.50$

VII. SOLVING MORE COMPLEX EQUATIONS WITH ONE UNKNOWN (*X*)

RULE: Solve for the unknown (*X*) by using a combination of simple operations. Below are some possible complex equations.

A. *X* Is Multiplied by a Value and Then Has a Value Either Added to It or Subtracted From It (Solve for *X* Using Either Subtraction or Addition and Then Division)

Example: $2(X) + 7 = 19$

$2(X) + 7 - 7 = 19 - 7$ (subtract 7 from both sides of the equation)

$2(X) = 12$

$\frac{2(X)}{2} = \frac{12}{2}$ (divide both sides of the equation by 2)

$X = 6$

Check: $2(X) + 7 = 19$

$2(6) + 7 = 19$

$12 + 7 = 19$

$19 = 19$

B. *X* Is Divided by a Value and Then Has a Value Either Added to It or Subtracted From It (Solve for *X* Using Either Addition or Subtraction and Then Multiplication)

Example: $\frac{X}{6} - 5 = 2$

$\frac{X}{6} - 5 + 5 = 2 + 5$ (add 5 to both sides of the equation)

$\frac{X}{6} = 7$

$\frac{6(X)}{6} = 7(6)$ (multiply both sides of the equation by 6)

$X = 42$

Check: $\frac{42}{6} - 5 = 2$

$7 - 5 = 2$

$2 = 2$

C. *X* Has a Value Added to It or Subtracted From It and Is Then Multiplied by a Value (Solve for *X* Using Division and Then Either Subtraction or Addition)

Example: $4(X + 3) = 48$

$\frac{4(X+3)}{4} = \frac{48}{4}$ (divide both sides of the equation by 4)

$X + 3 = 12$

$X + 3 - 3 = 12 - 3$ (subtract 3 from both sides of the equation)

$X = 9$

Check: $4(9 + 3) = 48$

$4(12) = 48$

$48 = 48$

D. *X* Has a Value Either Added to It or Subtracted From It and Is Then Divided by a Value (Solve for *X* Using Multiplication and Then Either Subtraction or Addition)

Example: $\frac{X-11}{9} = 3$

$(9)\frac{X-11}{9} = 3(9)$ (multiply both sides of the equation by 4)

$X - 11 = 27$

$X - 11 + 11 = 27 + 11$ (add 11 to both sides of the equation)

$X = 38$

Check: $\frac{38-11}{9} = 3$

$\frac{27}{9} = 3$

$3 = 3$

PRACTICE PROBLEMS

More Complex Equations With One Unknown

1. *X* is multiplied by a value and then has a value either added to it or subtracted from it.
   a. $3(X) + 4 = 25$
   b. $5(X) - 3 = 17$
   c. $4(X) + 2.50 = 16.50$
   d. $6 + 2(X) = 36$
   e. $29 - 4(X) = 5$
   f. $7.25 - 1.50(X) = 3.47$

(Continued)

(Continued)

2. X is divided by a value and then has a value either added to it or subtracted from it.

a. $\frac{X}{2} - 4 = 1$

b. $\frac{X}{4} + 5 = 11$

c. $\frac{X}{3} + 1.75 = 8.25$

d. $9 + \frac{X}{8} = 13$

e. $27 - \frac{X}{5} = 16$

f. $2.53 + \frac{X}{2} = 3.60$

3. X has a value added to it or subtracted from it and is then multiplied by a value.

a. $3(X + 2) = 21$

b. $10(X - 7) = 40$

c. $6(X - 6) = 78$

d. $.50(X + 2) = 14.50$

e. $1.34(X - 2.75) = 21.24$

f. $4.00(X + 7.44) = 56.52$

4. X has a value added to it or subtracted from it and is then divided by a value.

a. $\frac{X - 5}{2} = 4$

b. $\frac{X - 7}{3} = 9$

c. $\frac{X - 11}{14} = 3$

d. $\frac{X - 1.75}{4} = 11.00$

e. $\frac{X - 5.85}{3.70} = 12.50$

f. $\frac{X - 25.88}{7.20} = 34.90$

ANSWERS TO PROBLEMS

Addition, Subtraction, Multiplication, Division

1. All positive numbers

a. 13
b. 22
c. 42
d. 33
e. 72
f. 9
g. 8
h. 27
i. 7
j. 10
k. −1
l. −4

2. Adding all negative numbers

a. −11
b. −36
c. −82
d. −67

3. Adding a combination of positive and negative numbers

a. −2
b. 15
c. −2
d. 39

4. Subtracting negative numbers

a. 28
b. 64
c. −3
d. 11

5. Multiplying two negative numbers

a. 12
b. 50
c. 63
d. 90

6. Multiplying combination of positive and negative numbers

a. −54
b. 95
c. −143
d. −56
e. 96
f. 75
g. −288
h. 144

7. Dividing two negative numbers
   a. 4
   b. 21
   c. 19
   d. 28
8. Dividing one positive number and one negative number
   a. −7
   b. −9
   c. −17
   d. −11

Order of Operations

1. Parentheses ( ) and brackets [ ]
   a. $(7)^2 = 49$
   b. $(8)^2 = 64$
   c. $(12)^2 = 144$
   d. $(6)(5) = 30$
   e. $\sqrt{(5)(5)} = \sqrt{25} = 5$
   f. $4 \times 2 = 8$
   g. $6 - 4 = 2$
   h. $12 \div 2 = 6$
   i. $16 - 2 = 14$
   j. $24/6 = 4$
   k. $4(18) = 72$
   l. $3[(12)+(8)] = 3(20) = 60$
   m. $\frac{[18+32]}{10} = \frac{50}{10} = 5$
   n. $\sqrt{22-6} = \sqrt{16} = 4$
   o. $\frac{16}{4} = 4$
   p. $(2)[9 + 16] = (2)[25] = 50$
   q. $(5)\frac{[2+5]^2}{7} = (5)\frac{49}{7} = (5)(7) = 35$
2. Squares and/or square roots
   a. $16 + 36 = 52$
   b. $49 - 13 = 36$
   c. $(3)25 = 75$
   d. $81/3 = 27$
   e. $\frac{8}{4} = 2$
   f. $6 + 11 = 17$
   g. $9 - 7 = 2$
   h. $16 - 9 = 7$
   i. $\sqrt{16+9} = \sqrt{25} = 5$
   j. $(5)\sqrt{\frac{16}{4}} = (5)(2) = 10$
3. Multiplication and/or division
   a. 84
   b. 28
   c. 84
   d. 24
   e. 75
   f. 49
   g. −6
   h. 25
   i. 14
   j. 12
   k. 32
   l. 18

Fractions, Decimals, and Percentages

1. Convert the following fractions to decimals.
   a. .20
   b. .75
   c. .60
   d. −.33
   e. 3.00
   f. 4.50
   g. 1.75
   h. .90
   i. 6.25
2. Convert the following decimals to percentages.
   a. 5%
   b. 68%
   c. 16%
   d. −52%
   e. 136%
   f. 233%
3. Convert the following percentages to decimals.
   a. .01
   b. .95
   c. .47
   d. .99

Simple Equations With One Unknown

1. *X* has a value subtracted from it (solve for *X* using addition).

   a. 8 c. 116 e. 4.75
   b. 21 d. 821 f. 134.95

2. *X* has a value added to it (solve for *X* using subtraction).

   a. 4 c. 51 e. 5.89
   b. 9 d. 164 f. 60.23

3. *X* is multiplied by a value (solve for *X* using division).

   a. 5 c. 7.00 e. 9
   b. 7 d. 3.76 f. −5

4. *X* is divided by a value (solve for *X* using multiplication).

   a. 18 c. 99 e. 15.00
   b. 60 d. 156 f. 36.86

More Complex Equations With One Unknown

1. *X* is multiplied by a value and then has a value either added to it or subtracted from it.

   a. 7 c. 3.50 e. 6
   b. 4 d. 15 f. 2.52

2. *X* is divided by a value and then has a value either added to it or subtracted from it.

   a. 10 c. 19.50 e. 55
   b. 24 d. 32 f. 12.26

3. *X* has a value added to it or subtracted from it and is then multiplied by a value.

   a. 5 c. 19 e. 18.60
   b. 11 d. 27 f. 6.69

4. *X* has a value added to it or subtracted from it and is then divided by a value.

   a. 13 c. 31 e. 52.10
   b. 20 d. 45.75 f. 277.16

GLOSSARY

Addition rule: combined probability of mutually exclusive outcomes is the sum of the individual probabilities.

Alpha: probability of a statistic used to make the decision whether to reject the null hypothesis.

Alternative hypothesis: statistical hypothesis that a hypothesized change, difference, or relationship among groups or variables does exist in the population.

Analysis of variance (ANOVA): family of statistical procedures used to test differences between two or more group means.

Analytical comparisons: comparisons between groups that are part of a larger research design.

ANOVA summary table: table summarizing the calculations of an analysis of variance.

Archival research: research methods involving the use of records or documents of the activities of individuals, groups, or organizations.

Assumption of homoscedasticity: assumption that a variable is normally distributed at each value of another variable.

Assumption of independence of observations: assumption that each observation in a set of data is independent of all other observations.

Assumption of interval or ratio scale of measurement: assumption that variables being analyzed are measured at the interval or ratio scale of measurement.

Assumption of normality: assumption that scores for a variable are approximately normally distributed in the population.

Assumption of random sampling: assumption that a sample is created by random selection from the population.

Asymmetric distribution: distribution in which the frequencies change in a different manner moving away in both directions from the most frequently occurring values.

Bar chart: figure that uses bars to represent the frequency or percentage of a sample corresponding to each value of a variable.

Bar graph: figure that uses bars to represent the mean of the dependent variable for each level of the independent variable.

Between-group variability: differences among the means of groups that comprise an independent variable.

Between-subjects research design: research design in which each research participant appears in only one level or category of the independent variable.

Biased estimate: statistic based on a sample that systematically underestimates or overestimates the population from which the sample was drawn.

Bimodal distribution: distribution in which two values clearly distinct from each other have the greatest frequency.

Binomial distribution: distribution of probabilities for a binomial variable.

Binomial variable: variable consisting of exactly two categories.

Cell: combination of independent variables within a factorial research design.

Cell mean: mean of the dependent variable for a combination of independent variables.

Central limit theorem: theorem stating that sample means are approximately normally distributed for an infinite number of random samples drawn from a population.

Chi-square goodness-of-fit test: statistic that tests the difference between the distribution of observed frequencies and expected frequencies for a single variable.

Chi-square statistic: statistic that tests the difference between the distribution of observed frequencies and expected frequencies.

Chi-square test of independence: statistic that tests the difference between the distribution of observed frequencies and expected frequencies for two categorical variables.

Cohen's *d*: estimate of the magnitude of the difference between the means of two groups measured in standard deviation units.

Complex comparisons: analytical comparisons involving more than two groups.

Computational formula: algebraic manipulation of a definitional formula designed to minimize the complexity of mathematical calculations.

Confidence interval: range or interval of values with a stated probability of containing an unknown population parameter.

Confidence interval for the mean: interval of values for a variable with a stated probability of containing an unknown population mean.

Confidence limits: lower and upper boundaries of a confidence interval.

Confounding variable: variable related to an independent variable that provides an alternative explanation for the relationship between the independent and dependent variables.

Contingency table: table in which the rows and columns represent the values of categorical variables and the cells contain observed frequencies for combinations of the variables.

Control group: group of participants in an experiment not exposed to the independent variable.

Correlation: mutual or reciprocal relationship between two variables such that systematic changes in the values of one variable are accompanied by systematic changes in the values of another variable.

Correlation matrix: table of Pearson correlations between variables.

Correlational statistic: statistic designed to measure relationships between variables.

Covariance: extent to which two variables vary together such that they have shared variance.

Cramér's ϕ(phi): measure of effect size for the chi-square statistic.

Criterion variable: label for the dependent variable in a nonexperimental (correlational) research study; variable predicted by a predictor variable.

Critical value: value of a statistic that separates the regions of rejection and nonrejection.

Cumulative percentage: percentage of a sample at or below a particular value of a variable.

Decision rule: rule specifying the values of the statistic that result in the decision to reject or not reject the null hypothesis.

Definitional formula: formula based on the actual or literal definition of a concept.

Degrees of freedom: number of values or quantities free to vary when a statistic is used to estimate a parameter.

Dependent variable: variable measured by the researcher expected to change or vary as a function of the independent variable.

Descriptive statistic: statistic used to summarize and describe a set of data for a variable.

Directional alternative hypothesis: alternative hypothesis that indicates the direction of the change, difference, or relationship.

Dunnett test: statistical procedure used to control familywise error when comparing each group with a single reference group.

Error variance: variability that cannot be explained or accounted for.

Expected frequency: frequency of a group expected to occur in a sample of data under the assumption the null hypothesis is true.

Experimental research methods: research methods designed to test causal relationships between variables.

***F* ratio:** statistic used to test differences between group means.

Factorial research design: research design consisting of all possible combinations of two or more independent variables.

Familywise error: probability of making at least one Type I error across a set of comparisons.

Fisher's exact test: chi-square statistic used if any of the expected frequencies in a sample of data are less than 5.

Flat distribution: distribution where the data are spread evenly across the values of a variable.

Frequency: number of participants in a sample corresponding to a value of a variable.

Frequency distribution table: table summarizing the number and percentage of participants for the different values of a variable.

Frequency polygon: line graph that uses connected data points to represent the frequencies of values of a variable.

Grouped frequency distribution table: table that groups values of a variable into intervals and provides the frequency and percentage within each interval.

Heteroscedasticity: see "Assumptions" terms.

Histogram: figure that uses connected bars to represent the frequencies of values of a variable.

Homogeneity of variance: assumption that the variance of scores for a variable is the same for different populations.

Independent variable: variable manipulated or controlled by the researcher.

Index of qualitative variation: measure of variability for nominal variables.

Inferential statistic: statistical procedure used to test hypotheses and draw conclusions from data collected during research studies.

Interaction effect: effect of one independent variable on the dependent variable changes at the different levels of another independent variable.

Interquartile range: range of the middle 50% of the scores in a set of data.

Interval estimation: estimation of a population parameter with a range, or interval, of values within which one has a certain degree of confidence the population parameter falls.

Interval level of measurement: values of variables equally spaced along a numeric continuum.

Linear regression: statistical procedure in which a straight line is fitted to a set of data to best represent the relationship between two variables.

Linear regression equation: mathematical equation that predicts one variable from another variable.

Linear relationship: relationship between variables appropriately represented by a straight line.

Linear transformation: mathematical transformation of a variable comprising addition, subtraction, multiplication, or division.

Longitudinal research: research design in which the same information is collected from a research participant over two or more administrations.

Main effect: effect of an independent variable on the dependent variable within a factorial research design.

Marginal mean: mean representing a main effect within a factorial research design.

Mean: the arithmetic average of a set of scores.

Measure of central tendency: descriptive statistic that is the most typical, common, or frequently occurring value for a variable.

Measure of effect size: statistic that measures the magnitude of the relationship between variables.

Measure of variability: descriptive statistic of the amount of differences in a set of data for a variable.

Measurement: assignment of numbers or categories to objects or events according to rules.

Median: value of a variable that splits a distribution of scores in half, with the same number of scores above the median as below.

Modality: value or values of a variable that have the highest frequency in a set of data.

Mode: most frequently occurring score or value of a variable in a set of data.

Multimodal distribution: distribution where more than two values have the greatest frequency.

Multiple correlation: family of statistical procedures assessing the relationship between two or more predictor variables and a criterion variable.

Multiple correlation coefficient: statistic measuring the relationship between two or more predictor variables and a criterion variable.

Multiple regression equation: mathematical equation predicting a criterion variable from two or more predictor variables.

Negative relationship: relationship in which increases in scores for one variable are associated with decreases in scores of another variable.

Negatively skewed distribution: distribution in which the higher frequencies are at the upper end of the distribution, with the tail on the lower (left) end of the distribution.

Nominal level of measurement: values of variables differing in category or type.

Nondirectional alternative hypothesis: alternative hypothesis that does not indicate the direction of the change, difference, or relationship between groups or variables.

Nonexperimental research methods: research methods designed to measure naturally occurring relationships between variables.

Nonlinear relationship: relationship between variables not appropriately represented by a straight line.

Nonparametric statistical tests: statistical tests that do not involve the estimation of population parameters or the assumption that scores for a variable are normally distributed in the larger population.

Normal curve table: table containing the percentage of the standard normal distribution associated with different *z*-scores.

Normal distribution: distribution based on a population of an infinite number of scores calculated from a mathematical formula.

Normally distributed variable: distribution for a variable considered to be unimodal, symmetric, and neither peaked nor flat.

Null hypothesis: statistical hypothesis that a hypothesized change, difference, or relationship among groups or variables does not exist in the population.

Observational research: research methods involving the systematic and objective observation of naturally occurring behavior or events.

Observed frequency: frequency of a group in the sample of data.

One-way ANOVA: statistical procedure used to test differences between the means of three or more groups comprising a single independent variable.

Ordinal level of measurement: values of variables can be placed in an order relative to the other values.

Outlier: rare, extreme score that lies outside of the range of the majority of scores in a set of data.

Parameter: numeric characteristic of a population.

Parametric statistical tests: statistical tests based on assumption that data from a sample are being used to estimate population parameters and that the distribution of scores for a variable is normally distributed in the larger population.

Peaked distribution: distribution where much of the data are in a small number of values of a variable.

Pearson correlation coefficient: statistic measuring the linear relationship between two variables measured at the interval or ratio level of measurement.

Pie chart: figure that uses a circle divided into proportions to represent the percentage of the sample corresponding to each value of a variable.

Planned comparisons: comparisons built into a research design prior to data collection.

Point estimate: single value used to estimate an unknown population parameter.

Population: total number of possible units or elements that could potentially be included in a study.

Population mean: mean associated with an entire population.

Population standard deviation: square root of the population variance that represents the average deviation of a score from the population mean.

Population standard error of the mean: standard deviation of the sampling distribution of the mean when the population standard deviation is known.

Population variance: average squared deviation of a score from the population mean.

Positive relationship: relationship in which increases in scores for one variable are associated with increases in scores of another variable.

Positively skewed distribution: distribution in which the higher frequencies are at the lower end of the distribution, with the tail on the upper (right) end of the distribution.

Predictor variable: label for the independent variable in a nonexperimental (correlational) research study; variable used to predict a criterion variable.

Pretest-posttest research design: research design in which data are collected from participants before and after the introduction of an intervention or experimental manipulation.

Probability: likelihood of occurrence of a particular outcome of an event given all possible outcomes.

Quasi-experimental research: research methods comparing naturally formed or preexisting groups.

r squared (r^2): percentage of variance in one variable accounted for by another variable.

R^2: percentage of variance in the dependent variable associated with differences between groups comprising an independent variable.

Random assignment: assignment of participants to categories of an independent variable such that each participant has an equal chance of being assigned to each category.

Range: mathematical difference between the lowest and highest scores in a set of data.

Rank: relative position in a graded or evaluated group.

Ratio level of measurement: values of variables equally spaced along a numeric continuum with a true zero point.

Real limits: values of a variable located halfway between the top of one interval and the bottom of the next interval.

Real lower limit: smallest value of a variable in a particular interval of values.

Real upper limit: largest value of a variable in a particular interval of values.

Region of nonrejection: values of a statistic that result in the decision to not reject the null hypothesis.

Region of rejection: values of a statistic that result in the decision to reject the null hypothesis.

Regression: use of a relationship between two or more correlated variables to predict values of one variable from values of other variables.

Regression coefficient: constant representing the rate of change in a criterion Variable *Y* associated with changes in a predictor variable.

Research hypothesis: statement regarding an expected or predicted relationship between variables.

Robust: ability of a statistical procedure to withstand moderate violations of the assumption of normality.

Sample: subset or portion of a population.

Sample mean: mean calculated from data for a sample.

Sampling distribution: distribution of statistics for samples randomly drawn from populations.

Sampling distribution of the difference: distribution of all possible values of the difference between two sample means when an infinite number of pairs of samples of size *N* are randomly selected from two populations.

Sampling distribution of the mean: distribution of sample means for an infinite number of samples of size *N* randomly selected from the population.

Sampling error: differences between statistics calculated from a sample and statistics pertaining to the population from which the sample is drawn due to random, chance factors.

Scatterplot: graphical display of paired scores on two variables measured at the interval or ratio level of measurement.

Scheffé test: statistical procedure used to control familywise error when conducting all possible simple and complex comparisons in a set of data.

Scientific method: method of investigation that uses the objective and systematic collection and analysis of empirical data to test theories and hypotheses.

Simple comparisons: analytical comparisons between two groups.

Simple effect: effect of one independent variable at one of the levels of another independent variable.

Single-factor research design: research design consisting of one independent variable.

Spearman rank-order correlation: statistic measuring the relationship between two variables measured at the ordinal level of measurement.

Standard deviation: square root of the variance that represents the average deviation of a score from the mean.

Standard error of the difference: standard deviation of the sampling distribution of the difference.

Standard error of the mean: standard deviation of the sampling distribution of the mean; average deviation of a sample mean from the population mean.

Standard normal distribution: normal distribution measured in standard deviation units with a mean equal to 0 and a standard deviation equal to 1.

Standardized distribution: distribution of *z*-scores for scores in a frequency distribution.

Standardized score: *z*-score within a standardized distribution.

Statistic: numeric characteristic of a sample.

Statistical hypothesis: statement about an expected outcome or relationship involving population parameters.

Statistical power: probability of rejecting the null hypothesis when the alternative hypothesis is true.

Statistics: branch of mathematics dealing with the collection, analysis, interpretation, and presentation of masses of numerical data.

Student *t*-distribution: distribution of the *t* statistic based on an infinite number of samples of size *N* randomly drawn from the population.

Survey research: research methods obtaining information directly from a group of people regarding their opinions, beliefs, or behavior.

Symmetric distribution: distribution in which the frequencies change in a similar manner moving away in both directions from the most frequently occurring values.

Symmetry: how the frequencies of values of a variable change in relation to the most common or frequently occurring values.

Theory: set of propositions used to describe or explain a phenomenon.

***t*-test for dependent means:** inferential statistic testing the difference between two means that are based on the same participant or paired participants.

***t*-test for independent means:** inferential statistic that tests the difference between the means of two samples drawn from two populations.

***t*-test for one mean:** statistical procedure testing the difference between a sample mean and a hypothesized population mean when the population standard deviation is not known.

Tukey test: statistical procedure used to control familywise error when conducting all possible simple comparisons between groups.

Two-factor (A × B) research design: factorial research design consisting of two independent variables.

Type I error: error that occurs when the null hypothesis is true but the decision is made to reject the null hypothesis.

Type II error: error that occurs when the alternative hypothesis is true but the decision is made to not reject the null hypothesis.

Unbiased estimate: statistic based on a sample that is equally likely to underestimate or overestimate the population from which the sample was drawn.

Unimodal distribution: distribution where one value occurs with the greatest frequency.

Unplanned comparisons: comparisons not having been built into a research design prior to data collection.

Variability: the amount of differences in a distribution of data for a variable.

Variable: property or characteristic of an object, event, or person that can take on different values.

Variance: average squared deviation from the mean.

Within-group variability: variability of scores within groups that comprise an independent variable.

Within-subjects research design: research design in which each participant appears in all levels or categories of the independent variable.

Yates's correction for continuity: chi-square statistic used when any of the expected frequencies in a sample of data are between 5 and 10.

***z*-score:** scores for the standard normal distribution measured in standard deviation units.

***z*-test for one mean:** statistical procedure testing the difference between a sample mean and a population mean when population standard deviation is known.

REFERENCES

Abar, C. C. (2012). Examining the relationship between parenting types and patterns of student alcohol-related behavior during the transition to college. *Psychology of Addictive Behaviors, 26*, 20–29.

Abeles, H. (2009). Are musical instrument gender associations changing? *Journal of Research in Music Education, 57*, 127–139.

Adams, A. E., & Dennis, M. (2011). A comparison of actual and perceived problem drinking among driving while intoxicated (DWI) offenders. *Journal of Alcohol & Drug Education, 55*, 53–69.

Adams, J. B. (2005). What makes the grade? Faculty and student perceptions. *Teaching of Psychology, 32*, 21–24.

Adams, R. J. (1987). An evaluation of color preferences in early infancy. *Infant Behavior and Development, 10*, 143–150.

Ahmadi, M., Raiszadeh, F., & Helms, M. (1997). An examination of the admission criteria for the MBA programs: A case study. *Education*, 117, 540–546.

Alterovitz, S. S., & Mendelsohn, G. A. (2009). Partner preferences across the life span: Online dating by older adults. *Psychology and Aging, 24*, 513–517.

American Psychological Association. (2010). *Publication manual of the American Psychological Association* (6th ed.). Washington, DC: Author.

Anderson, C. A., & Dill, K. E. (2000). Video games and aggressive thoughts, feelings, and behavior in the laboratory and in life. *Journal of Personality and Social Psychology, 78*, 772–790.

Arano, K. G. (2006). Credit card usage among students: Evidence from a survey of Fort Hays State University Students. *Kansas Policy Review, 28*, 31–38.

Auwarter, A. E., & Aruguete, M. S. (2008). Effects of student gender and socioeconomic status on teacher perceptions. *Journal of Educational Research, 101*, 243–246.

Banerjee, S. C., Campo, S., & Green, K. (2008). Fact or wishful thinking? Biased expectations in "I think I look better when I'm tanned." *American Journal of Health Behavior, 32*, 243–252.

Barlett, C. P., Harris, R. J., & Baldassaro, R. (2007). Longer you play, the more hostile you feel: Examination of first person shooter video games and aggression during video game play. *Aggressive Behavior, 33*, 458–466.

Bear, J. B., & Collier, B. (2016). Where are the women in Wikipedia? Understanding the different psychological experiences of men and women in Wikipedia. *Sex Roles, 74*, 254–265.

Benz, J. J., Anderson, M. K., & Miller, R. L. (2005). Attributions of deception in dating situations. *The Psychological Record, 55*, 305–314.

Berg, E. M., & Lippman, L. G. (2001). Does humor in radio advertising affect recognition of novel product brand names? *Journal of General Psychology, 128*, 194–205.

Beyerstein, B. L. (1999). Whence cometh the myth that we only use 10% of our brains? In S. Della Sala (Ed.), *Mind-myths: Exploring popular assumptions about the mind and brain* (pp. 3–24). New York, NY: John Wiley.

Bhat, R. A., & Kushtagi, P. (2006). A re-look at the duration of human pregnancy. *Singapore Medical Journal, 47*, 1044–1048.

Brenner, V., & Fox, R. A. (1998). Parental discipline and behavior problems in young children. *Journal of Genetic Psychology: Research and Theory on Human Development, 159*, 251–256.

Busato, V. V., Prins, F. J., Elshout, J. J., & Hamaker, C. (2000). Intellectual ability, learning style, personality, achievement motivation and academic success of psychology students in higher education. *Personality and Individual Differences, 29*, 1057–1068.

Carlton, P., & Pyle, R. (2007). A program for parents of teens with anorexia nervosa and eating disorder not otherwise specified. *International Journal of Psychiatry in Clinical Practice, 11*, 9–15.

Cohen, J. (1988). *Statistical power analysis for the behavioral sciences* (2nd ed.). Hillsdale, NJ: Lawrence Erlbaum.

Cohen, J., & Cohen, P. (1983). *Applied multiple regression/correlation analysis for the behavioral sciences* (2nd ed.). Hillsdale, NJ: Lawrence Erlbaum.

Cohen, J., Cohen, P., West, S. G., & Aiken, L. S. (2003). *Applied multiple regression/correlation analysis for the behavioral sciences* (3rd ed.). Mahwah, NJ: Lawrence Erlbaum.

Connolly J., Baird K., Bravo V., Lovald B., Pepler D., & Craig W. (2015). Adolescents' use of affiliative and aggressive strategies during conflict with romantic partners and best-friends. *European Journal of Developmental Psychology, 12*, 549–564.

Cumming, G. (2014). The new statistics: Why and how. *Psychological Science, 25*, 7–29.

Darlow, V., Norvilitis, J. M., & Schuetze, P. (2017). The relationship between helicopter parenting and adjustment to college. *Journal of Child and Family Studies*. Advance online publication. doi:10.1007/s10826-017-0751-3

de Bruin, A. B., Rikers, R. M., & Schmidt, H. G. (2007). The effect of self-explanation and prediction on the development of principled understanding of chess in novices. *Contemporary Educational Psychology, 32*, 188–205.

Deitz, D. K., Cook, R. F., Billings, D. W., & Hendrickson, A. (2009). A web-based mental health program: Reaching parents at work. *Journal of Pediatric Psychology, 34*, 488–494.

Detweiler, J. B., Bedell, B. T., Salovey, P., Pronin, E., & Rothman, A. J. (1999). Message framing and sunscreen use: Gain-framed messages motivate beach-goers. *Health Psychology, 18*, 189–196.

Devlin, J., & Peterson, R. (1994). Student perceptions of entry-level employment goals: An international comparison. *Journal of Education for Business, 69*, 154–158.

Diener, E., & Seligman, M. E. P. (2002). Very happy people. *Psychological Science, 13*, 81–84.

Dietz, T. L. (1998). An examination of violence and gender role portrayals in video games: Implications for gender socialization and aggressive behavior. *Sex Roles, 38*, 425–442.

Doll, G. A. (2009). An exploratory study of resistance training and functional ability in older adults. *Activities, Adaptation & Aging, 33*, 179–190.

Dunn, M. J., & Searle, J. (2010). Effect of manipulated prestige-car ownership on both sex attractiveness ratings. *British Journal of Psychology, 101*, 69–80.

Dunnett, C. W. (1955). A multiple comparisons procedure for comparing several treatments with a control. *Journal of the American Statistical Association, 50*, 1096–1121.

DuRant, R. H., Treiber, F., Getts, A., McCloud, K., Linder, C. W., & Woods, E. R. (1996). Comparison of two violence prevention curricula for middle school adolescents. *Journal of Adolescent Health, 19*, 111–117.

DuVernet, A., Poteet, M. L., Parker, B. N., Conley, K. M., & Herman, A. E. (2017). Overview of results from the 2016 SIOP Income and Employment Survey. *TIP (The Industrial-Organizational Psychologist)*, 54(3).

Edmonds, E., O'Donoghue, C., Spano, S., & Algozzine, R. F. (2009). Learning when school is out. *Journal of Educational Research, 102*, 213–221.

Eisenman, D. P., Glik, D., Ong, M., Zhou, Q., Tseng, C., Long, A., . . . Asch, S. (2009). Terrorism-related fear and avoidance behavior in a multiethnic urban population. *American Journal of Public Health, 99*, 168–174.

Elaad, E., Ginton, A., & Jungman, N. (1992). Detection measures in real-life criminal Guilty Knowledge Test. *Journal of Applied Psychology, 77*, 757–767.

El-Sheikh, M., & Elmore-Staton, L. (2007). The alcohol–aggression link: Children's aggression expectancies in marital arguments as a function of the sobriety or intoxication of the arguing couple. *Aggressive Behavior, 33*, 458–466.

Erdodi, L. A. (2012). What makes a test difficult? Exploring the effect of item content on student's performance. *Journal of Instructional Psychology, 39*, 171–176.

Eysenck, H. J. (1982). The biological basis of cross-cultural differences in personality: Blood group antigens. *Psychological Reports, 51*, 531–540.

Fontaine, K. R., Cheskin, L. J., & Barofsky, I. (1996). Health-related quality of life in obese persons seeking treatment. *Journal of Family Practice, 43*, 265–270.

Fries, L. R., Martin, N., & van der Horst, K. (2017). Parent-child mealtime interactions associated with toddlers' refusals of novel and familiar foods. *Physiology & Behavior, 176*, 93-100.

Frohlich, C. (1994). Baseball: Pitching no-hitters. *Chance, 7(3)*, 24–30.

Gallucci, N. T. (1997). An evaluation of the characteristics of undergraduate psychology majors. *Psychological Reports, 81*, 879–889.

Gardner, D. G., Van Dyne, L., & Pierce, J. L. (2004). The effects of pay level on organization-based self-esteem and performance: A field study. *Journal of Occupational Psychology, 77*, 307–322.

Gough, H. G., & Bradley, P. (2005). *CPI 260™ manual*. Mountain View, CA: CPP, Inc.

Grafman, J. M., & Cates, G. L. (2010). The differential effects of two self-managed math instruction procedures: Cover, copy, and compare versus copy, cover, and compare. *Psychology in the Schools, 47*, 153–165.

Gupta, S. (1992). Season of birth in relation to personality and blood groups. *Personality and Individual Differences, 13*, 631–633.

Gurm, H., & Litaker, D. G. (2000). Framing procedural risks to patients: Is 99% safe the same as a risk of 1 in 100? *Academic Medicine, 75*, 840–842.

Hardoon, K. K., Baboushkin, H. R., Derevensky, J. L., & Gupta, R. (2001). Underlying cognitions in the selection of lottery tickets. *Journal of Clinical Psychology, 57*, 749–763.

Hays, W. L. (1988). *Statistics* (4th ed.). New York, NY: Holt, Rinehart & Winston.

Herman, W. E. (1997). The relationship between time to completion and achievement on multiple choice exams. *Journal of Research and Development in Education, 30*, 113–117.

Herrick, R. (2001). The effects of political ambition on legislative behavior: A replication. *The Social Science Journal, 38*, 469–474.

Hetsroni, A. (2000). Choosing a mate in television dating games: The influence of setting, culture, and gender. *Sex Roles, 42*, 83–106.

Higbee, K. L., & Clay, S. L. (1998). College students' beliefs in the ten-percent myth. *Journal of Psychology, 132*, 469–476.

Ho, J., & Lindquist, M. (2001). Time saved with the use of emergency warning lights and siren while responding to requests for emergency medical aid in a rural environment. *Prehospital Emergency Care, 5*, 159–162.

Hong, J., & Sun, Y. (2012). Warm it up with love: The effect of physical coldness on liking of romance movies. *Journal of Consumer Research, 39*, 293–306.

Jaccard, J., Hand, D., Ku, L., Richardson, K., & Abella, R. (1981). Attitudes toward male oral contraceptives: Implications for models of the relationship between beliefs and attitudes. *Journal of Applied Social Psychology, 11*, 181–191.

James, W. (1911). *The energies of men*. New York, NY: Henry Holt.

John, D., Bassett, D., Thompson, D., Fairbrother, J., & Baldwin, D. (2009). Effect of using a treadmill workstation on performance of simulated office work tasks. *Journal of Physical Activity and Health, 6*, 617–624.

Katkowski, D. A., & Medsker, G. J. (2001). Income and employment of SIOP members in 2000. *The Industrial-Organizational Psychologist, 39(1)*, 21–36.

Kebbell, M. R., & Milne, R. (1998). Police officers' perceptions of eyewitness performance in forensic investigations. *Journal of Social Psychology, 138*, 323–330.

Keppel, G., Saufley, W. H., & Tokunaga, H. (1992). *Introduction to design and analysis: A student's handbook* (2nd ed.). New York, NY: W. H. Freeman.

Keppel, G., & Wickens, T. D. (2004). *Design and analysis: A researcher's handbook* (4th ed.). Upper Saddle River, NJ: Pearson Prentice Hall.

Keppel, G., & Zedeck, S. (1989). *Data analysis for research designs: Analysis of variance and multiple regression/correlation approaches*. New York, NY: W. H. Freeman.

Kirsh, S. J., & Olczak, P. V. (2001). Rating comic book violence: Contributions of gender and trait hostility. *Social Behavior and Personality, 29*, 833–836.

Larimer, M. E., Turner, A. P., Anderson, B. K., Fader, J. S., Kilmer, J. R., Palmer, R. S., & Cronce, J. M. (2001). Evaluating a brief alcohol intervention with fraternities. *Journal of Studies on Alcohol, 62*, 370–380.

Latané, B., & Rodin, J. (1969). A lady in distress: Inhibiting effects of friends and strangers on bystander intervention.

Journal of Experimental Social Psychology, 5, 189–202.

Lavine, H., Burgess, D., Snyder, M., Transue, J., Sullivan, J. L., Haney, B., & Wagner, S. H. (1999). Threat, authoritarianism, and voting: An investigation of personality and persuasion. *Personality & Social Psychology Bulletin, 25*, 337–347.

Lewis, B. A., & O'Neill, H. K. (2000). Alcohol expectancies and social deficits relating to problem drinking among college students. *Addictive Behaviors, 25*, 295–299.

Loftus, G. R. (1996). Psychology will be a much better science when we change the way we analyze data. *Current Directions in Psychological Science, 5*, 161–171.

Lohse, G. L., & Rosen, D. L. (2001). Signaling quality and credibility in Yellow Pages advertising: The influence of color and graphics on choice. *Journal of Advertising, 30*, 73–85.

Loney, D. M., & Cutler, B. L. (2016). Coercive interrogation of eyewitnesses can produce false accusations. *Journal of Police and Criminal Psychology, 31*, 29–36.

Ma, S. (2008). Paternal race/ethnicity and birth outcomes. *American Journal of Public Health, 98*, 2285–2292.

Maier, M. A., Barchfeld, P., Elliot, A. J., & Peckrun, R. (2009). Context specificity of implicit preferences: The case of human preference for red. *Emotion, 9*, 734–738.

Matvienko, O., & Ahrabi-Fard, I. (2010). The effects of a 4-week after-school program on motor skills and fitness of kindergarten and first-grade students. *American Journal of Health Promotion, 24*, 299–303.

McCrink, K., Bloom, P., & Santos, L. R. (2010). Children's and adults' judgments of equitable resource distributions. *Developmental Science, 13*, 37–45.

McKee, L., Roland, E., Coffelt, N., Olson, A. L., Forehand, R., Massari, C., & Zens, M. S. (2007). Harsh discipline and child problem behaviors: The roles of positive parenting and gender. *Journal of Family Violence, 22*, 187–196.

McLeish, A. C., & Del Ben, K. S. (2008). Symptoms of depression and posttraumatic stress disorder in an outpatient population before and after Hurricane Katrina. *Depression and Anxiety, 25*, 416–421.

McNemar, Q. (1969). *Psychological statistics* (4th ed.). New York, NY: John Wiley.

Michalski, D., Kohout, J., Wicherski, M., & Hart, B. (2011). *2009 Doctorate employment survey*. Washington, DC: American Psychological Association.

Miranda, A., Villaescusa, M. I., & Vidal-Abarca, E. (1997). Is attribution retraining necessary? Use of self-regulation procedures for enhancing the reading comprehension strategies of children with learning disabilities. *Journal of Learning Disabilities, 30*, 503–512.

Miyazaki, A. D., Langenderfer, J., & Sprott, D. E. (1999). Government-sponsored lotteries: Exploring purchase and nonpurchase motivations. *Psychology & Marketing, 16*, 1–20.

Mondin, G. W., Morgan, W. P., Piering, P. N., Stegner, A. J., Stotesbery, C. L., Trine, M. R., & Wu, M. (1996). Psychological consequences of exercise deprivation in habitual exercisers. *Medicine & Science in Sports & Exercise, 28*, 1199–1203.

Nickerson, R. S. (2000). Null hypothesis significance testing: A review of an old and continuing controversy. *Psychological Methods, 5*, 241–301.

O'Hagan, J., & Kelly, E. (2005). Identifying the most important artists in a historical context: Methods used and initial results. *Historical Methods, 38*, 118–125.

Osberg, J. S., & Di Scala, C. (1992). Morbidity among pediatric motor vehicle crash victims: The effectiveness of seat belts. *American Journal of Public Health, 82*, 422–425.

Pellegrini, R. J. (1973). The astrological 'theory' of personality: An unbiased test by a biased observer. *Journal of Psychology: Interdisciplinary and Applied, 85*, 21–28.

Proudfoot, D., Kay, A. C., & Koval, C. Z. (2015). A gender bias in the attribution of creativity: Archival and experimental evidence for the perceived association between masculinity and creative thinking. *Psychological Science, 26*, 1751–1761.

Pryor, T., & Wiederman, M. W. (1998). Personality features and expressed concerns of adolescents with eating disorders. *Adolescence, 33*, 291–300.

Psunder, M. (2009). Future teachers' knowledge and awareness of their role in student misbehaviour. *The New Educational Review, 19*, 247–262.

Read, M. A., & Upington, D. (2009). Young children's color preferences in the interior environment. *Early Childhood Education Journal, 36*, 491–496.

Reed, J. M., Marchand-Martella, N. E., Martella, R. C., & Kolts, R. L. (2007). Assessing the effects of the Reading Success Level A Program with fourth-grade students at a Title I elementary school. *Education and Treatment of Children, 30*, 45–68.

Reichert, T., & Lambiase, J. (2003). How to get "kissably close": Examining how advertisers appeal to consumers' sexual needs and desires. *Sexuality and Culture, 7*, 120–136.

Ridge, R. D., & Reber, J. S. (2002). 'I think she's attracted to me': The effect of men's beliefs on women's behavior in a job interview scenario. *Basic and Applied Social Psychology, 24*, 1–14.

Riniolo, T. C., Koledin, M., Drakulic, G. M., & Payne, R. A. (2003). An archival study of eyewitness memory of the *Titanic*'s final plunge. *Journal of General Psychology, 130*, 89–95.

Rist, C. (2001, May). The physics of . . . baseball: Unraveling the mystery of why it's so easy to hit a home run. *Discover Magazine*, pp. 33–34.

Rosenthal, R., & Rosnow, R. L. (1991). *Essentials of behavioral research: Methods and data analysis*. New York, NY: McGraw-Hill.

Rossi, J. S. (1997). A case study in the failure of psychology as a cumulative science: The spontaneous recovery of verbal learning. In L. L. Harlow, S. A. Mulaik, & J. H. Steiger (Eds.), *What if there were no significance tests?* (pp. 175–197). Mahwah, NJ: Lawrence Erlbaum.

Ruback, R. B., & Juieng, D. (1997). Territorial defense in parking lots: Retaliation against waiting drivers. *Journal of Applied Social Psychology, 27*, 821–834.

Rudski, J. M., & Edwards, A. (2007). Malinowski goes to college: Factors influencing students' use of ritual and superstition. *Journal of General Psychology, 134*, 389–403.

Scheffé, H. (1953). A method for judging all contrasts in the analysis of variance. *Biometrika, 40*, 87–104.

Schmidt, F. L. (1996). Statistical significance testing and cumulative knowledge in psychology: Implications for training of researchers. *Psychological Methods, 1*, 115–129.

Schuetze, P., Lewis, A., & DiMartino, D. (1999). Relation between time spent in daycare and exploratory behaviors in

9-month-old infants. *Infant Behavior & Development*, *22*, 267–276.

Sebastian, R. J., & Bristow, D. (2008). Formal or informal? The impact of style of dress and forms of address on business students' perceptions of professors. *Journal of Education for Business*, *83*, 196–201.

Shaw, T., & Duys, D. K. (2005). Work values of mortuary science students. *Career Development Quarterly*, *53*, 348–352.

Skevington, S. M., & Tucker, C. (1999). Designing response scales for cross-cultural use in health care: Data from the development of the UK WHOQOL. *British Journal of Medical Psychology*, *72*, 51–61.

Snyder, C. R. (1997). Unique invulnerability: A classroom demonstration in estimating personal mortality. *Teaching of Psychology*, *24*, 197–199.

Stewart, K. D., & Bernhardt, P. C. (2010). Comparing millennials to pre-1987 students and with one another. *North American Journal of Psychology*, *12*, 579–602.

Suhr, K. A., & Dula, C. S. (2017). The dangers of rumination on the road: Predictors of risky driving. *Accident Analysis and Prevention*, *99*, 153–160.

Tabachnick, B. G., & Fidell, L. S. (2013). *Using multivariate statistics* (6th ed.). Boston, MA: Pearson.

Tokunaga, H. T. (1985). The effect of bereavement upon death-related attitudes and fears. *Omega*, *16*, 267–280.

Tom, G., & Ruiz, S. (1997). Everyday low price or sale price. *Journal of Psychology: Interdisciplinary and Applied*, *131*, 401–406.

Travis, F., Bonshek, A., Butler, V., Rainforth, M., Alexander, C. N., Khare, R., & Lipman, J. (2005). Can a building's orientation affect the quality of life of the people within? Testing principles of Maharishi Sthapatya Veda. *Journal of Social Behavior and Personality*, *17*, 553–564.

Tuckman, B. W. (1996). The relative effectiveness of incentive motivation and prescribed learning strategy in improving college students' course performance. *Journal of Experimental Education*, *64*, 197–210.

Tukey, J. W. (1953). *The problem of multiple comparisons*. Unpublished manuscript, Princeton University.

Walsh, L. M., Toma, R. B., Tuveson, R. V., & Sondhi, L. (1990). Color preference and food choice among children. *Journal of Psychology*, *124*, 645–653.

Warner, B., & Rutledge, J. (1999). Checking the *Chips Ahoy*! guarantee. *Chance*, *12(1)*, 10–14.

Watson, J. C., & Bedard, D. L. (2006). Clients' emotional processing in psychotherapy: A comparison between cognitive-behavioral and process-experiential therapies. *Journal of Consulting and Clinical Psychology*, *74*, 152–159.

Welch, B. L. (1938). The significance of the difference between two means when the population variances are unequal. *Biometrika*, *29*, 350–362.

Williams, V. A., Johnson, C. E., & Danhauer, J. L. (2009). Hearing aid outcomes: Effects of gender and experience on patients' use and satisfaction. *Journal of the American Academy of Audiology*, *20*, 422–432.

Woolley, J. D., & Cox, V. (2007). Development of beliefs about storybook reality. *Developmental Science*, *10*, 681–693.

Zentner, M. R. (2001). Preferences for colors and color-emotion combinations in early childhood. *Developmental Science*, *4*, 389–398.

Zimmerman, D.W. (2004). A note on preliminary test of equality of variances. *British Journal of Mathematical and Statistical Psychology*, *57*, 173–181.

INDEX